Handbook of Experimental Pharmacology

Volume 155

Editorial Board

G.V.R. Born, London
M. Eichelbaum, Stuttgart
D. Ganten, Berlin
H. Herken, Berlin
F. Hofmann, München
L. Limbird, Nashville, TN
W. Rosenthal, Berlin
G. Rubanyi, Richmond, CA
K. Starke, Freiburg i. Br.

Springer

*Berlin
Heidelberg
New York
Barcelona
Hong Kong
London
Milan
Paris
Tokyo*

CNS Neuroprotection

Contributors

M.F. Beal, K.J. Becker, A. Blesch, D.W. Choi, T.M. Dawson,
V.L. Dawson, H.J. Federoff, G.Z. Feuerstein, A.C. Foster,
S. Gandy, M.P. Goldberg, J.M. Hallenbeck, M.W. Halterman,
M. Jackson, G.A. Kerchner, A.H. Kim, W. Koroshetz,
M.P. Mattson, L.P. Miller, J.W. Phillis, R.M. Poole, H.D. Rosas,
J.D. Rothstein, M. Sasaki, C.P. Taylor, M.H. Tuszynski,
K.K.W. Wang, X. Wang, J.B. Wiesner

Editors

Frank W. Marcoux and Dennis W. Choi

Springer

Dr. FRANK W. MARCOUX
Vice President, Biology,
Pfizer Global Research and Development
Ann Arbor Laboratories
2800 Plymouth Road
Ann Arbor, MI 48105
USA
e-mail: frank.marcoux@pfizer.com

Dr. DENNIS W. CHOI
Washington University School of Medicine
Department of Neurology Box 8111
660 South Euclid Avenue
St. Louis, MO 63110
USA
e-mail: choid@neuro.wustl.edu

With 49 Figures and 22 Tables

ISBN 3-540-42412-1 Springer-Verlag Berlin Heidelberg New York

Library of Congress Cataloging-in-Publication Data
CNS neuroprotection / editors, Frank W. Marcoux and Dennis W. Choi ; contributors, M.F. Beal . . . [et al.]. – 1st ed.
  p. cm. – (Handbook of experimental pharmacology ; v. 155)
  Includes bibliographical references and index.
  ISBN 3540424121 (alk. paper)
  1. Central nervous system – Degeneration – Chemoprevention. 2. Central nervous system – Degeneration – Molecular aspects. I. Marcoux, Frank W., 1952– II. Choi, Dennis W. III. Beal, M. Flint. IV. Series.
QP905 .H3 vol. 155
[RC365]
615'.5 s – dc21
[616.8'0461]   2002021066

This work is subject to copyright. All rights are reserved, whether the whole or part of the material is concerned, specifically the rights of translation, reprinting, re-use of illustrations, recitation, broadcasting, reproduction on microfilms or in any other way, and storage in data banks. Duplication of this publication or parts thereof is permitted only under the provisions of the German Copyright Law of September 9, 1965, in its current version, and permission for use must always be obtained from Springer-Verlag. Violations are liable for Prosecution under the German Copyright Law.

Springer-Verlag Berlin Heidelberg New York
a member of BertelsmannSpringer Science+Business Media GmbH

© Springer-Verlag Berlin Heidelberg 2002
Printed in Germany

The use of general descriptive names, registered names, etc. in this publication does not imply, even in the absence of a specific statement, that such names are exempt from the relevant protective laws and regulations and free for general use.

Product liability: The publishers cannot guarantee the accuracy of any information about dosage and application contained in this book. In every individual case the user must check such information by consulting the relevant literature.

Cover design: design & production GmbH, Heidelberg
Typesetting: SNP Best-set Typesetter Ltd., Hong Kong

SPIN: 10693431   27/3020xv – 5 4 3 2 1 0 – printed on acid-free paper

Preface

"CNS neuroprotection" was a common subject of papers, symposia, and reviews during the previous "decade of the brain." Indeed, in recent years, experimental study of putative neuroprotective agents prompted clinical trials of numerous drug candidates in acute and chronic human neurodegenerative conditions. While the outcomes of these trials have not been as successful as initially hoped, these were early explorations, and the pipeline of relevant ideas continues to grow in strength and depth. We predict that early in this new millennium, crippling disorders such as stroke and Alzheimer's disease will be treated effectively by therapeutic neuroprotective strategies.

This volume of the *Handbook of Experimental Pharmacology* titled *CNS Neuroprotection* provides a pharmacological perspective on currently promising neuroprotective approaches, and a clinical perspective on the challenges involved in establishing the efficacy of these approaches through appropriate clinical trials.

Section I, "Mechanistic Approaches to CNS Neuroprotection," reviews major injury mechanisms that have formed the basis for many past and present clinical trials conducted around the world. Dr. KIM and colleagues, Washington University School of Medicine, review the status of blocking excitotoxicity as an approach to CNS neuroprotection. Dr. WANG, Pfizer Global Research and Development, Ann Arbor Laboratories, outlines evidence supporting a contribution of apoptosis to pathological neuronal or glial cell loss. Drs. BECKER and HALLENBECK, University of Washington and the National Institute on Neurological Diseases and Stroke, respectively, argue that inhibiting inflammatory pathways in the brain can be neuroprotective. Dr. BEAL, Cornell University, reviews mitochondria-directed approaches to neuroprotection. Dr. MATTSON, National Institute on Aging, summarizes the current understanding of how disturbances in calcium homeostasis and intracellular signaling pathways participate in neurodegeneration; Dr. SASAKI and colleagues, Johns Hopkins University, discuss the neuroprotective effects of inhibiting nitric oxide toxicity. Taken together, the chapters in Sect. I provide expert commentary on well-studied mechanistic approaches to CNS neuroprotection.

Section II, "Neuroprotective Agents," reviews specific pharmacological strategies for CNS neuroprotection. Dr. FOSTER and colleagues, Neurocrine

Biosciences, review adenosine-related approaches to neuroprotection. Dr. TAYLOR, Pfizer Global Research and Development, Ann Arbor Laboratories, argues that sodium and calcium ion channels are targets for neuroprotective drugs. Dr. PHILLIS, Wayne State University, reviews the status of free radical scavengers and other antioxidants as neuroprotective agents; and Drs. FEUERSTEIN and WANG, DuPont Pharmaceuticals, discuss ways of manipulating chemokines and chemokine receptors to reduce neurodegeneration. The chapters in Sect. II cover specific molecular targets for neuroprotection and provide examples of pharmacological agents of current interest.

Section III, "CNS Delivery of Neuroprotective Therapies," highlights gene-based neuroprotection strategies. Drs. BLESCH and TUSZYNSKI, University of California at San Diego, review ex vivo approaches to delivering new genes to the CNS. Drs. HALTERMAN and FEDEROFF, University of Rochester, discuss viral vector-based neuroprotective approaches.

Section IV, "Disease Targeting of Neuroprotective Therapies," presents clinical perspectives on bringing neuroprotective approaches to bear upon specific neurological disorders. Dr. GOLDBERG, Washington University, summarizes some of the progress and challenges in developing neuroprotective drugs to treat stroke. Dr. POOLE, Pfizer Global Research and Development, Ann Arbor Laboratories, discusses approaches to neuroprotective therapy in the traumatically injured CNS. Dr. GANDY, New York University, discusses using neuroprotective approaches to treat patients with Alzheimer's disease; Drs. JACKSON and ROTHSTEIN, Johns Hopkins University, do the same for patients with amyotrophic lateral sclerosis; while Drs. ROSAS and KOROSHETZ, Massachusetts General Hospital, focus on patients with Huntington's disease. Taken together, these chapters yield insights into present and future opportunities for neuroprotective therapies in specific neurodegenerative conditions.

CNS Neuroprotection provides a pharmacological view of neuroprotective therapies. It is not intended to be exhaustive, but rather highlights major mechanistic targets and therapeutic approaches, as well as related challenges and opportunities in the clinic. The editors wish to thank all of the authors for their excellent contributions to this volume and, moreover, to this rapidly developing area of pharmacological research.

FRANK W. MARCOUX
DENNIS W. CHOI

List of Contributors

BEAL, M.F., Department of Neurology, Cornell University Medical College,
525 East 68th Street, New York, NY 10021, USA
e-mail: fbeal@med.cornell.edu

BECKER, K.J., University of Washington School of Medicine, Box 359775
Harborview Medical Center, 325 Ninth Avenue, Seattle, WA 98104-2499,
USA
e-mail: kjb@u.washington.edu

BLESCH, A., Department of Neuroscience, University of California,
San Diego, 9500 Gilman Dr., La Jolla, CA 92093-0626, USA
e-mail: ablesch@ucsd.edu

CHOI, D.W., Washington University School of Medicine, Department of
Neurology, Box 8111, 660 South Euclid Avenue, St. Louis, MO 63110,
USA
e-mail: choid@neuro.wustl.edu

DAWSON, T.M., Johns Hopkins University School of Medicine, Department
of Neurology, 600 North Wolfe Street, Carnegie 214, Baltimore,
MD 21287, USA
e-mail: tdawson@jhmi.edu

DAWSON, V.L., Johns Hopkins University School of Medicine, Department of
Neurology, 600 North Wolfe Street, Carnegie 214, Baltimore, MD 21287,
USA
e-mail: vdawson@jhmi.edu

FEDEROFF, H.J., Center for Aging and Developmental Biology, Aab Institute
of Biomedical Sciences, Departments of Micriobiology and Immunology
and Neurology, University of Rochester School of Medicine and
Dentistry, 601 Elmwood Avenue, Box 603, Rochester, NY 14642, USA
e-mail: howard_federoff@urmc.rochester.edu

FEUERSTEIN, G.Z., Cardiovascular Disease Research, Bristol-Myers Squibb,
Experimental Station, Bldg. 400/3352, Route 141 and Henry Clay Roads,
Wilmington, DE 19880-0400, USA
e-mail: giora.feuerstein@bms.com

FOSTER, A.C., Neurocrine Biosciences Inc., Department of Neuroscience, 10555 Science Center Drive, San Diego, CA 92121, USA
e-mail: afoster@neurocrine.com

GANDY, S., Department of Psychiatry, New York University, The Nathan S. Kline Institute for Psychiatric Research, 140 Old Orangeburg Road, Orangeburg, NY 10962, USA, and
Fisher Center for Alzheimer Research, The Rockefeller University, New York, NY 10021, USA
e-mail: gandy@nki.rfmh.org

GOLDBERG, M.P., Department of Neurology, Washington University School of Medicine, St. Louis, Missouri, USA
e-mail: goldberg@neuro.wustl.edu

HALLENBECK, J.M., Department of Health & Human Services, Public Health Service, National Institutes of Health, Building 36, Room 4A-03, Bethesda, Maryland 20892, USA

HALTERMAN, M.W., University of Rochester School of Medicine & Dentistry, Department of Microbiology and Immunology, 601 Elmwood Ave, Box 672, Rochester, NY 14642, USA
e-mail: mhalterm@rochester.rr.com

JACKSON, M., Johns Hopkins University, School of Medicine, Department Neurology and Neuroscience, Baltimore, MD 21287, USA

KERCHNER, G.A., Washington University School of Medicine, Department of Neurology, Box 8111, 660 South Euclid Avenue, St. Louis, MO 63110, USA

KIM, A.H., Washington University School of Medicine, Department of Neurology, Box 8111, 660 South Euclid Avenue, St. Louis, MO 63110, USA

KOROSHETZ, W.J., Stroke and Clinical Neurology Service, Department of Neurology – VBK 915, Massachusetts General Hospital, 55 Fruit Street, Boston, MA 02114, USA

MATTSON, M.P., Laboratory of Neurosciences, National Institute on Aging, 5600 Nathan Shock Drive, Baltimore, MD 21224, USA
e-mail: mattsonm@grc.nfh.gov

MILLER, L.P., San Diego, CA, USA
e-mail: miller4@aol.com

List of Contributors

PHILLIS, J.W., Department of Physiology, Wayne State University School of Medicine, 540 E. Canfield, Detroit, MI 48201, USA
e-mail: jphillis@med.wayne.edu

POOLE, R.M., Worldwide Development, Pfizer Inc., 50 Peguot Avenue, MS-6025-B2237, New London, CT 06320
e-mail: robert-michael-poole@groton.pfizer.com

ROSAS, H.D., Massachusetts General Hospital, Department of Neurology, Warren 408, Fruit Street, Boston, MA 02114, USA
e-mail: rosas@helix.mgh.harvard.edu

ROTHSTEIN, J.D., Johns Hopkins University, School of Medicine, Department Neurology and Neuroscience, Baltimore, MD 21287, USA
e-mail: jrothste@welchlink.welch.jhu.edu

SASAKI, M., Johns Hopkins University School of Medicine, Department of Neurology, 600 North Wolfe Street, Carnegie 214, Baltimore, MD 21287, USA
e-mail: masasaki@jhmi.edu

TAYLOR, C.P., Department of CNS Pharmacology, Pfizer Global Research and Development, 2800 Plymouth Road, Ann Arbor, MI 48105, USA
e-mail: charles.taylor@pfizer.com

TUSZYNSKI, M.H., Department of Neurosciences-0626, University of California-San Diego, 9500 Gilman Drive, La Jolla, CA 92093-0626, USA
e-mail: mtuszyns@ucsd.edu

WANG, K.K.W., Laboratory of Neuro-Biochemistry, Department of Neuroscience Therapeutics, Parke-Davis Pharmaceutical Research, 2800 Plymouth Road, Ann Arbor, MI 48105, USA
e-mail: kevin.wang@WL.com

WANG, X., Cardiovascular Sciences, Bristol-Myers Squibb Company, Experimental Station, E400/3418, Wilmington, DE 19880, USA

WIESNER, J.B., Ligand Pharmaceuticals, 10275, Science Center Drive, San Diego, CA 92121, USA
e-mail: jwiesner@ligand.com

Contents

Section I: Mechanic Approaches to CNS Neuroprotection

CHAPTER 1

Blocking Excitotoxicity
A.H. KIM, G.A. KERCHNER, and D.W. CHOI 3

A. Introduction 3
B. Contributions to Disease 3
C. Excitotoxicity in Brief 4
D. Extending Excitotoxicity to Glia 6
E. Points of Intervention 7
    I. Reducing Extracellular Glutamate 7
       1. Circuit Activity and Glutamate Release 7
          a) Hypothermia 8
          b) Increasing GABAergic Tone 8
          c) Opening K^+ Channels 9
          d) Modulating Adenosine Receptors 9
          e) Blocking Voltage-Gated Na^+ Channels 10
          f) Blocking Voltage-Gated Ca^{2+} Channels 11
       2. Glutamate Transport 12
    II. Manipulating Glutamate Receptors 12
       1. NMDA Antagonists 12
       2. AMPA/Kainate Antagonists 16
       3. Metabotropic Glutamate Receptors 17
    III. Blocking Downstream Mediators 18
       1. Downstream Effects of Cellular Ca^{2+} Overload 18
       2. Free Radical Formation 19
       3. The Role of PARP 20
F. A Cautionary Note for Antiexcitotoxic Strategies:
   Enhanced Apoptosis? 21
References 22

CHAPTER 2

Limiting Apoptosis as a Strategy for CNS Neuroprotection
K.K.W. WANG. With 4 Figures 37

A. Definition of Necrosis and Apoptosis 37
B. Scheduled Apoptosis in the Development of the CNS 39
C. Unscheduled Apoptosis in Various Neurodegenerative
   Conditions 40
   I. Criteria Used to Identify Neuronal Apoptosis 40
   II. Various Neurodegenerative Disorders with Evidence of
       Apoptosis 40
D. Key Pathways and Components Relevant to Neuronal
   Apoptosis 42
   I. Overview of Apoptosis-Linked Factors 42
   II. Bcl-2/Bax Family 42
   III. Mitochondria and Caspase-9 Mediated Pathway 43
   IV. Death Receptor-Mediated Caspase-8/10 Pathway 44
   V. TNF Receptor 1-Mediated p38 Kinase, JNK and
      NF-κB Activation 45
   VI. Akt-IP3 Kinase-Mediated Antiapoptosis Pathway for
       NGF and Other Survival Factors 48
   VII. Effector Caspases 49
   VIII. Participation of Other Proteases 51
   IX. Reactive Oxygen Species 51
   X. Neuronal-Specific Apoptosis Factors (cdk5, NAIP) 52
E. Comparison of the Potential of Neuronal Apoptosis Factors as
   Neuroprotective Drug Targets 53
References 54

CHAPTER 3

Reducing Neuroinflammation
K.J. BECKER and J.M. HALLENBECK 65

A. Introduction 65
B. The Central Nervous System Inflammatory Response 66
   I. Initiation of the Inflammatory Response 66
   II. Consequences of Leukocyte Activation 67
   III. Cytokines in Brain Injury 69
      1. Interleukin-1β 69
      2. Tumor Necrosis Factor α 71
      3. Interleukin-6 72
      4. Interleukin-8 73
   IV. Nuclear Transcription Factor-κB 73
   V. Leukocyte Adhesion 74

Contents

VI.	Temporal Profile of the Postischemic Inflammatory Response	76
	1. Experimental Models	76
	2. Clinical Data	77
VII.	Endogenous Modulators of Inflammation	78
C. Ischemic Tolerance		79
D. Chronic Inflammatory Disease		80
E. Conclusions		81
References		81

CHAPTER 4

Mitochondrial Approaches to Neuroprotection
M.F. Beal 95

A. Introduction	95
B. Approaches to Neuroprotection	96
I. Mitochondrial Oxidative Interactions	96
II. Mitochondrial Permeability Transition	97
III. Therapeutic Approaches for Neurodegenerative Disease	99
IV. Neuroprotective Effects of Coenzyme Q_{10}	99
V. Neuroprotective Effects of Nicotinamide	100
VI. Neuroprotective Effects of Creatine	101
VII. Creatine and the MPT	102
VIII. Other Potential Neuroprotective Mechanisms of Creatine	102
IX. Creatine Uptake	103
X. Studies in Man	105
C. Conclusions	106
References	106

CHAPTER 5

Stabilizing Calcium Homeostasis
M.P. Mattson. With 12 Figures 115

A. Introduction	115
B. Regulation of Calcium Homeostasis in Neurons	116
I. Plasma Membrane Systems	117
II. Endoplasmic Reticulum	118
III. Mitochondria	121
IV. Other Proteins Involved in Calcium Regulation	123
C. Factors That Destabilize Calcium Homeostasis	126
I. Oxidative Stress	126
II. Metabolic Impairment	130

	III. Excitatory Imbalances	131
	IV. Glucocorticoids	131
	V. Apoptotic Mechanisms	132
D.	Factors That Stabilize Calcium Homeostasis	135
	I. Neurotrophic Factors	135
	II. Cytokines	136
	III. Inhibitory Neurotransmitters	138
	IV. Antioxidants	138
	V. Calcium-Binding Proteins	139
	VI. Heat Shock Proteins	139
	VII. Cytoskeletal Dynamics	139
E.	Strategies to Preserve and Restore Neuronal Calcium Homeostasis	143
References	144

CHAPTER 6

Blocking Nitric Oxide Toxicity
M. Sasaki, T.M. Dawson, and V.L. Dawson 155

A.	Introduction	155
B.	Nitric Oxide Synthase	155
	I. Neuronal Nitric Oxide Synthase	156
	II. Endothelial Nitric Oxide Synthase	156
	III. Immunologic Nitric Oxide Synthase	157
C.	Inhibitors of Nitric Oxide Synthase	157
	I. L-Arginine Site	158
	1. Thiocitrullines	158
	2. Nitro-Indazoles	158
	3. Substituted Guanidines	160
	4. Isothioureas	161
	5. Imidazole	161
	II. Tetrahydrobiopterin	161
	III. Flavoprotein Inhibition	162
	IV. NADPH Oxidase Inhibition	162
	V. Calmodulin Inhibition	162
	VI. Phosphorylation Sites	163
	VII. Miscellaneous Inhibitors of NOS	163
D.	Biochemical Regulation of NOS Catalytic Activity	164
E.	Nitric Oxide and Neurologic Disease	164
	I. Excitotoxicity and Experimental Stroke	164
	II. Parkinson's Disease	165
	III. HIV Dementia	165
	IV. Multiple Sclerosis	166
F.	Conclusions	166
References	167

Section II: Neuroprotective Agents

CHAPTER 7

Adenosine-Based Approaches to the Treatment of Neurodegenerative Disease
A.C. Foster, L.P. Miller, and J.B. Wiesner. With 1 Figure 177

A. Introduction	177
B. The Adenosine System in the Central Nervous System	178
I. Adenosine Receptor Subtypes	180
1. A_1 Receptors	180
2. A_3 Receptors	181
3. A_{2A} and A_{2B} Receptors	181
II. Adenosine Transporters	182
C. The Role of Adenosine in Central Nervous System Physiology	182
D. The Role of Adenosine in Central Nervous System Pathophysiology	185
E. Therapeutic Approaches to Adenosine-Based Neuroprotection in Cerebral Ischemia	187
I. Potential Therapeutic Mechanisms	187
II. Receptor-Based Approaches	187
1. A_1 Receptor Agonists	187
2. A_3 Receptor Agonists/Antagonists	189
3. A_{2A} Receptor Antagonists	189
III. Modulation of Endogenous Adenosine	189
1. Adenosine Transport Inhibitors	190
2. Adenosine Deaminase Inhibitors	191
3. Adenosine Kinase Inhibitors	192
4. Other Adenosine Modulators	193
5. A_1 Receptor Modulators	193
F. Adenosine-Based Therapeutics in Parkinson's Disease	194
I. A_{2A} Antagonists	194
II. Adenosine Modulators	196
G. Adenosine-Based Therapeutics in Epilepsy	197
I. A_1 Receptor Agonists	197
II. Adenosine Modulators	198
H. Conclusions	199
Abbreviations	199
References	200

CHAPTER 8

Sodium and Calcium Channel Blockers
C.P. TAYLOR. With 3 Figures 209

A. Introduction 209
B. Structure and Function of Voltage-Gated Cation
   Channels 210
      I. Molecular Biology and Protein Structure 210
          1. Basic Structure of Channel Proteins 210
          2. Auxiliary Ion Channel Subunit Proteins 215
      II. Activation and Inactivation 216
          1. Activation Gating 216
          2. Inactivation Processes 216
      III. Binding Sites for Blockers and Modulators 217
          1. Tetrodotoxin and Conopeptides 217
          2. Voltage-Dependent Blockers 217
C. Peptides as Specific Ca^{2+} Channel Probes 228
D. Mechanisms of Ischemic Neurotransmitter (Glutamate)
   Release 229
      I. Ca^{2+}-Dependent Release 229
      II. Ca^{2+}-Independent Release 230
E. Neuroprotection with Na^+ Channel Blockers 231
F. Neuroprotection with Small Molecule Ca^{2+} Channel
   Blockers 231
G. Neuroprotection with Conopeptides 232
H. Mixed Na^+ and Ca^{2+} Channel Blockers 233
I. Conclusions 233
References 234

CHAPTER 9

**Neuroprotection by Free Radical Scavengers and
Other Antioxidants**
J.W. PHILLIS. With 8 Figures 245

A. Introduction 245
B. Free Radical Formation in the CNS 246
      I. Reactive Oxygen Species 246
          1. Mitochondrial Activity 246
          2. Excitotoxic Amino Acids 246
          3. Arachidonic Acid Metabolism 247
          4. Purine Catabolism 247
          5. Neutrophils 248
          6. Auto-oxidation of Catecholamines 249
      II. Reactive Nitrogen Species 249
C. Oxidative Stress in the CNS 250
D. Evidence of Free Radical Generation in the CNS 250

E. Antioxidant Defenses	251
F. Cerebroprotective Effects of Administered Antioxidants	254
I. Free Radical Scavengers	254
1. Spin-Trapping Agents	254
2. Mannitol, Dimethylthiourea and Dimethylsulfoxide	258
3. α-Lipoic Acid	259
4. α-Tocopherol	259
5. Superoxide Dismutase and Catalase	260
II. Inhibition of ROS/RNS Formation	261
1. Xanthine Oxidase Inhibitors	261
2. Inhibition of Phospholipases and Arachidonic Acid Metabolism	262
3. Inhibitors of Nitric Oxide Synthases	263
III. Chelation of Metal Ions	264
G. Oxidative Injury and Age-Related Neurodegeneration	265
H. Oxidative Stress and Neurogenerative Diseases	266
I. Amyotrophic Lateral Sclerosis	266
II. Alzheimer's Disease	266
III. Parkinson's Disease	267
IV. Huntington's Disease	268
I. Conclusions	268
Abbreviations	269
References	269

CHAPTER 10

Chemokines and Chemokine Receptors in the Central Nervous System: New Opportunities for Novel Therapeutics in Brain Ischemia and Trauma

G.Z. FEUERSTEIN and X. WANG. With 6 Figures	281
A. Introduction	281
I. The Chemokine Family of Polypeptides and Their Receptors	281
B. Chemokine and the Central Nervous System	285
C. Chemokine Expression in Brain Trauma	286
D. Chemokines and Cerebral Ischemia	287
I. Cytokine-Induced Neutrophil Chemoattractant	287
II. Monocyte Chemoattractant Protein 1	287
III. Monocyte Chemottractant Protein 3	289
IV. Interferon-γ-Inducible Protein 10	289
E. Interleukin-8 in Human Brain Injury	291
F. Chemokines and Neuro-Acquired Immune Deficiency Syndrome	293
G. Chemokines in Neurodegenerative Disorders	293
H. Summary and Conclusion	294
References	295

Section III: CNS Delivery of Neuroprotective Therapies

CHAPTER 11

Ex Vivo Gene Therapy in the Central Nervous System
A. BLESCH and M.H. TUSZYNSKI. With 5 Figures 301

A. Introduction 301
B. Gene Therapy Versus Conventional Drug Delivery 302
C. Practical Considerations 302
    I. Cell Types Suitable for Gene Transfer and Grafting to the CNS 302
    II. Methods of Ex Vivo Gene Transfer 304
    III. Transplantation of Genetically Modified Cells to In Vivo Models of CNS Disease 307
D. Experimental Therapies in Animal Models of Neurodegenerative Diseases 307
    I. Alzheimer's Disease 308
        1. Nerve Growth Factor 309
        2. Biochemical Modulation of Acetylcholine Levels 312
    II. Parkinson's Disease 312
        1. Neuroprotective Strategies 313
        2. Biochemical Modulation of Dopamine Levels 314
    III. Huntington's Disease 314
    IV. Amyotrophic Lateral Sclerosis 317
E. Experimental Therapies in Animal Models of Spinal Cord Injury 319
F. Experimental Therapies of Brain Tumors 323
G. Future Perspectives 324
References 324

CHAPTER 12

Developing Gene-Based Neuroprotection Strategies Using Herpes Amplicon Vectors
M.W. HALTERMAN and H.J. FEDEROFF. With 3 Figures 335

A. Introduction 335
B. Central Nervous System Ischemia Induces Distinct Patterns of Neuron Loss 336
C. A Cell-Autonomous Program Activates the Delayed Form of Neuronal Death 337
    I. Mitochondria Can Sense and Activate Delayed Death 338
D. Herpes Virus-Mediated Gene Delivery 340
    I. The Herpes Amplicon Vector System 340

Contents   XIX

II. Application of Amplicon Vectors in Neuroscience Research	341
E. Developing Neuroprotective Viral Vector Strategies	342
I. Targeting the Late Stage of Delayed Death: IL-1β and Caspase Activity	342
II. Targeting the Intermediate Stage of Delayed Death: bcl-2	344
1. Viral Delivery of bcl-2 Protects Ischemic Neurons	345
F. Using Viral Vectors to Regulate Early Ischemic Transcriptional Responses	347
I. HIF-1α and p53 Regulate Ischemic Gene Expression	348
II. Disruption of HIF-1α Signaling Protects Against Ischemia-Induced Delayed Neuronal Death	349
G. Conclusions	351
References	351

Section IV: Disease Targeting of Neuroprotective Therapies

CHAPTER 13

Stroke

M.P. GOLDBERG	361
A. Introduction	361
I. Cerebrovascular Disorders Defined	361
II. Scope of the Problem	363
III. Rationale for Development of Neuroprotective Agents	364
B. Clinical Trials of Neuroprotective Agents	365
I. Progress in Clinical Trials and Trial Designs	365
II. Glutamate Antagonists	365
1. NMDA Receptor Antagonists	368
2. AMPA Receptor Antagonists	369
III. Other Receptor Agents	370
IV. Ion Channel Blockers	370
V. Antioxidants and Lipid Peroxidation Inhibitors	372
VI. Inhibitors of Inflammation or Leukocyte Activation	372
VII. Agents with Other Mechanisms of Action	372
C. Issues Regarding Preclinical Stroke Research	373
D. Issues Regarding Clinical Stroke Research	375
I. Drug Dosing	375
II. Time of Drug Delivery	377
III. Clinical Trial Design and Outcomes	378
IV. Publication of Trial Results	380

D. Conclusions 381
References 381

CHAPTER 14

Traumatic Central Nervous System Injury
R.M. POOLE. With 1 Figure 387

A. Introduction 387
B. Clinical Trial Experience in Traumatic Brain Injury 387
    I. Adequate Characterization of Compounds in Preclinical
       Experiments 388
    II. Characterization of Compounds in Early Clinical
        Development 390
    III. Special Challenges of Trial Design and Conduct
         for TBI 393
         1. Short Enrollment Time Window 393
         2. Characterization and Stratification of the Study
           Population 394
         3. Standardization of Care 396
         4. Outcome Measures 397
C. Spinal Cord Injury Clinical Trials 403
D. Conclusions 404
References 404

CHAPTER 15

Neurohormonal Signalling Pathways and the Regulation of Alzheimer β-amyloid Metabolism
Integrating Basic and Clinical Observations for Developing Successful Strategies to Delay, Retard or Prevent Cerebral Amyloidosis in Alzheimer's Disease
S. GANDY. With 1 Figure 409

A. Alzheimer's Disease Is Associated with an Intracranial
   Amyloidosis 409
B. Aβ Is a Catabolite of an Integral Precursor 410
C. Pathogenic APP Mutations Occur Within or Near the
   Aβ Domain 413
D. Signal Transduction Regulates the Relative Utilization of APP
   Processing Pathways in Cultured Cell Lines, in Primary Neurons,
   and in Rodents In Vivo 414
E. Insights into Mechanism(s) of Regulated APP Processing 416
F. Therapeutic Manipulation of Aβ Generation Via Ligand or
   Hormonal Manipulation 417
References 418

Contents   XXI

CHAPTER 16

Amyotrophic Lateral Sclerosis
M. Jackson and J.D. Rothstein 423

A. Introduction 423
B. Pathological Mechanisms 424
      I. Free Radicals and Oxidative Stress 424
      II. Cytoskeletal Abnormalities 424
      III. Autoimmune Factors 426
      IV. Excitotoxicity 426
C. Mechanism of Motor Neuron Degeneration 428
D. Clinical Trials 429
      I. Neurotrophic Factors 430
      II. Antioxidant Drugs 430
      III. Antiexcitotoxic Agents 431
E. Development of Preclinical Screening Models 432
F. Future Therapies 436
G. Conclusions 438
References 439

CHAPTER 17

Therapeutics in Huntington's Disease
H.D. Rosas and W. Koroshetz. With 5 Figures 447

A. Background 447
B. The Pathology of Huntington's Disease 448
C. Molecular Basis of Huntington's Disease 451
      I. Wild-Type Huntingtin 451
      II. Huntingtin with Expanded Polyglutamine Repeat 452
         1. Huntingtin and Altered Gene Transcription 454
         2. Pathologic Intracellular Inclusions of Huntingtin ... 455
      III. Reversibility of Molecular Pathology 456
      IV. Apoptosis 456
D. Genetic Animal Models of HD 457
E. Potential Therapeutic Targets 459
      I. Glutamate-Mediated Excitotoxicity in HD 460
      II. NMDA Receptor 461
      III. Selective Neuronal Loss 463
      IV. Selective Vulnerability: Toxins 464
F. Potential Therapeutics Agents 464
      I. Glutamate Release Inhibitors 465
      II. NMDA Channel Blockers 465
      III. Metabotropic Receptor Modulators 466
      IV. Free Radical Inhibitors 466
      V. Nitric Oxide Synthase Inhibitors 466

G.	Energy Metabolism and Its Relationship to Neurodegeneration	467
	I. Mitochondrial Permeability Transition Pore	469
H.	Neuroprotective Strategies	470
	I. Antioxidants	471
	II. Free Radical Spin Traps	471
	III. Energy Boosters	472
	IV. Targets Against the Mitochondrial Permeability Transition Pore	473
	V. Neurotrophic Factors	473
I.	Future Potential Therapeutics	475
	I. Caspase Inhibitors	475
	II. Bcl-2	476
	III. Gene Transcription and Histone Deacetylation Inhibitors	477
	IV. Neural Transplantation	477
J.	Challenges for Clinical Research	477
K.	Conclusions	480
References		480
Subject Index		493

Section I
Mechanic Approaches to CNS Neuroprotection

CHAPTER 1
Blocking Excitotoxicity

A.H. KIM, G.A. KERCHNER, and D.W. CHOI

A. Introduction

The major excitatory transmitter in the mammalian central nervous system is glutamate, which exerts its signaling actions through the stimulation of ionotropic and metabotropic receptors (WATKINS et al. 1981; MAYER and WESTBROOK 1987; NAKANISHI and MASU 1994). Under pathological conditions, glutamate receptor overactivation can trigger neuronal death, a phenomenon known as excitotoxicity (LUCAS and NEWHOUSE 1957; OLNEY 1969). Incentive for developing practical methods for blocking excitotoxicity arises from its implication in several acute and chronic neurological disease states. While recent clinical trials aimed at blocking excitotoxicity in stroke patients have been disappointing, there are several plausible reasons for these trial failures, including specific study design issues, treatment side effects, and a need to achieve concurrent block of parallel injury pathways. In our view, the case for antiexcitotoxic approaches in stroke remains open, and there are other possible disease targets yet to be explored. Ongoing delineation of the cellular and molecular underpinnings of excitotoxicity has led to the progressive unveiling of countermeasures, aimed at attenuating presynaptic glutamate release, postsynaptic receptor activation, the movement or action of cation second messengers, or downstream intracellular injury cascades. The excitotoxicity concept itself may need to be expanded, to encompass the death of oligodendrocytes as well as neurons, and ionic derangements besides Ca^{2+} overload.

B. Contributions to Disease

The ability of glutamate receptor overactivation to cause neuronal death in humans is demonstrated most directly by the toxicity of several naturally occurring glutamate receptor agonists (LUDOLPH et al. 2000), including domoic acid, produced by a phytoplankton that occasionally contaminates blue mussels (PERL et al. 1990; TEITELBAUM et al. 1990), the amino acid β-oxalyl-amino-L-alanine from seeds of the chickling pea (LUDOLPH et al. 1987), and the mushroom poisons acromelic acid and ibotenic acid (LEONHARDT 1949). These toxins all activate ionotropic glutamate receptors and induce a variety of disturbances, including seizures and cognitive alterations, as well as neuronal death.

Beyond dietary exposure to exogenous excitotoxins, excitotoxicity may also be induced by the endogenous neurotransmitter glutamate. Endogenous glutamate-mediated excitotoxicity has been hypothesized to play a fundamental pathogenic role in the neuronal death associated with a wide variety of acute neurological insults, including brain ischemia (both the transient, global interruption of blood supply experienced during cardiac arrest with resuscitation, as well as the focal ischemia associated with thromboembolic stroke), seizures, mechanical trauma, and isolated hypoxia or hypoglycemia (COYLE et al. 1981; ROTHMAN and OLNEY 1986; CHOI 1988b). Elevations in extracellular glutamate concentrations have been observed in the context of ischemia (BENVENISTE et al. 1984), seizures (MELDRUM 1994), and head trauma (KATAYAMA et al. 1990). In the context of excessive extracellular accumulation of glutamate, the movement of cations, including Ca^{2+}, through overactivated glutamate receptors can lead to multiple toxic consequences (see Sect. E.III.).

Glutamate may become lethal even when its synaptic release and extracellular concentration are not especially elevated, in settings where the ability of postsynaptic neurons to maintain homeostasis is compromised by energy depletion (NOVELLI et al. 1988), for example, due to mitochondrial dysfunction (BEAL 2000; NICHOLLS and WARD 2000). It is thus plausible that excitotoxicity may contribute, at least in a secondary fashion, to some of the neuronal loss associated with certain neurodegenerative diseases such as Huntington's disease (COYLE and SCHWARCZ 1976; McGGEER and McGEER 1978), Alzheimer's disease, or Parkinson's disease. In amyotrophic lateral sclerosis, loss of transporter-mediated glutamate uptake has been postulated to induce the excitotoxic death of motor neurons (ROTHSTEIN et al. 1992).

C. Excitotoxicity in Brief

Glutamate kills central neurons by activating several subtypes of ionotropic receptors: N-methyl-D-aspartate (NMDA), α-amino-3-hydroxy-5-methyl-4-isoxazolepropionic acid (AMPA), and kainate (OLVERMAN et al. 1984; WISDEN and SEEBURG 1993; HOLLMANN and HEINEMANN 1994; KERCHNER et al. 1999). Much has been learned about the mechanisms underlying excitotoxic death through studies performed in vitro over the past 15 years. Intense exposure to either glutamate or NMDA for only a few minutes is sufficient to trigger widespread necrosis of cultured cortical neurons over the next hours, a phenomenon we have called "rapidly-triggered excitotoxicity" (CHOI 1992). Neurons swell acutely, due to the massive influx of Na^+ (through NMDA or AMPA/kainate receptors) followed by Cl^- and water, and then undergo delayed neurodegeneration several hours later. This latter component is dependent upon NMDA receptor activation and the presence of extracellular Ca^{2+} and is associated with a massive increase in cytoplasmic free Ca^{2+} concentrations (OGURA et al. 1988; CHENG et al. 1999). The importance of elevations in intracellular Ca^{2+} in mediating excitotoxicity is underscored by the ability of cell-permeable Ca^{2+} chelators to attenuate glutamate-mediated cell death in neuronal culture as well as to decrease injury induced by experimental focal ischemia in rodents (TYMIANSKI et al. 1993).

Neuronal swelling consequent to Na^+, Cl^-, and water influx is not always lethal. Brief exposure to kainate induces marked cortical neuronal swelling in vitro but is followed by little delayed neurodegeneration. The high permeability of NMDA receptors to Ca^{2+} is crucial to the ability of brief glutamate exposure to trigger widespread cortical neuronal death, a hypothesis strengthened by the observation that a subset of cortical neurons containing Ca^{2+}-permeable AMPA receptors, formed when the critical mRNA-edited GluR2 (GluR-B) subunit is absent from heteromeric AMPA receptor complexes (BURNASHEV et al. 1992), is selectively vulnerable to brief, intense activation of those receptors (KOH and CHOI 1988; TURETSKY et al. 1994).

On the other hand, if the exposure time is lengthened from minutes to hours, AMPA/kainate receptor agonists can destroy most cortical neurons ("slowly-triggered excitotoxicity") (CHOI 1992; GWAG et al. 1997). AMPA or kainate toxicity in cultured hippocampal or cerebellar neurons is also dependent upon the presence of extracellular Ca^{2+} (GARTHWAITE and GARTHWAITE 1986; ROTHMAN et al. 1987). Prolonged activation of AMPA or kainate receptors will induce Na^+ influx and sustained depolarization, promoting Ca^{2+} entry via voltage-gated Ca^{2+} channels and reverse operation of the Na^+/Ca^{2+} exchanger (CHOI 1988a; YU and CHOI 1997). Additionally, Ca^{2+} release from intracellular stores may contribute to cytoplasmic Ca^{2+} accumulation (FRANDSEN and SCHOUSBOE 1991).

There are probably many potentially lethal derangements in cellular processes induced by profound elevations in cytoplasmic Ca^{2+}, but work in recent years has assigned particular responsibility for ensuing cellular necrosis to the activation of catabolic enzymes, generation of free radicals including nitric oxide, impairment of mitochondrial energy production, and excessive utilization of energy by the DNA repair enzyme poly(ADP-ribose) polymerase-1 (PARP-1). These downstream steps are discussed further in Sect. E.III.1.

Recent studies have also suggested that Na^+ and Ca^{2+} may not be the only cations whose excessive movement across neuronal membranes can mediate cell death; in particular, movements of Zn^{2+} or K^+ may also contribute. Concentrated in synaptic vesicles at excitatory terminals throughout the forebrain and in some other locations is a chelatable pool of Zn^{2+} (TIMM and NETH 1959; PEREZ-CLAUSELL and DANSCHER 1985; FREDERICKSON 1989). During transient global ischemia, this synaptic Zn^{2+} appears to translocate into postsynaptic neurons that later go on to die (TONDER et al. 1990). Preventing this translocation with an extracellular Zn^{2+} chelator reduced neuronal death (KOH et al. 1996). The ability of excessive exposure to extracellular Zn^{2+} to induce neuronal death has been demonstrated directly in neuronal cell cultures (YOKOYAMA et al. 1986; CHOI et al. 1988; MANEV et al. 1997; AIZENMAN et al. 2000; LOBNER et al. 2000). Zn^{2+}-induced death is potentiated by membrane depolarization (induced by glutamate receptor agonists or elevated extracellular K^+ concentrations), likely reflecting enhancement of toxic Zn^{2+} entry via voltage-gated Ca^{2+} channels and the Na^+/Ca^{2+} exchanger (WEISS et al. 1993; YIN and WEISS 1995). The ability of Zn^{2+} to enter cortical neurons through voltage-gated Ca^{2+} channels was demonstrated directly in electrophysiological experiments, which also revealed a potentiation of Zn^{2+} permeation in conditions of lowered extracellular pH, as may be present during ischemia (KERCHNER et al. 2000).

Neuronal intracellular Zn^{2+} concentration attained during a toxic Zn^{2+} exposure was correlated to the extent of subsequent cell death, with substantial death occurring at intracellular Zn^{2+} concentrations exceeding 250–300 nM (CANZONIERO et al. 1999); astrocytes are more resistant than neurons to death induced by comparable elevations of $[Zn^{2+}]_i$ (DINELEY et al. 2000). At such elevated Zn^{2+} levels, many alterations in intracellular biology can be expected. One consequence may be disruption of glycolysis due to nicotinamide adenine dinucleotide (NAD^+) depletion and consequent secondary inhibition of glyceraldehyde-3-phosphate dehydrogenase, as cortical neurons exposed to toxic levels of extracellular Zn^{2+} exhibited loss of ATP and elevation of the upstream glycolytic substrates dihydroxy-acetone phosphate and fructose 1,6-bisphosphate (SHELINE et al. 2000). While neurons may normally have limited dependence upon glycolysis for energy production (MAGISTRETTI 2000), a crucial role during pathophysiological conditions such as ischemia is not implausible. Mitochondrial disturbances and free radical production may also contribute to Zn^{2+}-induced death (MANEV et al. 1997; KIM et al. 1999a,b).

While little attention has been paid historically to the functional importance of the K^+ permeability of glutamate receptor-gated channels, K^+ efflux has been identified as a potentially important component of the sequence of events leading to programmed cell death. Delayed rectifier K^+ channel current (I_K) is enhanced in neurons undergoing apoptosis (YU et al. 1997b), and blockade of these channels by TEA or clofilium attenuated neuronal death induced by oxygen-glucose deprivation in vitro or by transient focal ischemia in vivo (CHOI et al. 1998). In lymphocytes, loss of intracellular K^+ may be a critical step in the apoptotic cascade, perhaps because DNA fragmentation and proteolytic activation of caspase-3 are inhibited at normal levels of intracellular free K^+ (BORTNER et al. 1997; BORTNER and CIDLOWSKI 1998). While NMDA receptor overactivation typically induces neuronal necrosis mediated by Na^+ and Ca^{2+} influx, NMDA receptor activation could induce apoptosis dependent upon K^+ efflux when the extracellular concentrations of Na^+ and Ca^{2+} were reduced, as occurs in the ischemic brain (YU et al. 1999). Even in the presence of normal extracellular Na^+ and Ca^{2+}, the K^+ efflux mediated by glutamate receptor-gated channels may enhance the propensity of neurons to undergo apoptosis.

D. Extending Excitotoxicity to Glia

Excitotoxicity has conventionally been considered to be specific to neurons. Although Type I astrocytes express functional AMPA receptors (CONDORELLI et al. 1993; MARTIN et al. 1993; MATUTE et al. 1994), they are highly resistant to death upon activation of those receptors by agonist exposure (COYLE et al. 1981; ROTHMAN 1984; CHOI et al. 1987). An ability of glutamate to kill immature oligodendrocytes in vitro was demonstrated by VOLPE and colleagues, but this toxicity appeared to be dependent upon interference with cellular cysteine uptake and consequent glutathione depletion, rather than upon glutamate receptor activation (OKA et al. 1993).

However, recent evidence has suggested that more mature oligodendrocytes may also be vulnerable to a true excitotoxic death mediated by glutamate receptor overactivation. Oligodendroglial lineage cells and oligodendrocytes cultured from rat optic nerve express multiple AMPA and kainate receptor subunits, and exposure to kainate or AMPA plus cyclothiazide (to inhibit AMPA receptor desensitization) can destroy these cells in a Ca^{2+}-dependent manner (YOSHIOKA et al. 1995; MATUTE et al. 1997). Differentiated forebrain oligodendrocytes appear even more sensitive to excitotoxicity, as 100–300 µM AMPA alone can trigger widespread cell death within 24h (MCDONALD et al. 1998a). In vivo, injection of AMPA or kainate into white matter killed oligodendrocytes near the injection site (MATUTE et al. 1997), in a manner sensitive to coinjection of AMPA and kainate receptor antagonists (MCDONALD et al. 1998a).

While further studies are needed to determine why oligodendrocytes are far more vulnerable to AMPA and kainate receptor-mediated toxicity than astrocytes, one possible explanation might be the expression of Ca^{2+}-permeable AMPA receptors in the former cell type (HOLZWARTH et al. 1994; PUCHALSKI et al. 1994). Compared to cortical neurons bearing Ca^{2+}-permeable AMPA receptors, however, oligodendrocytes are killed by longer agonist exposure times, at least 2–3h (MCDONALD et al. 1998a). This agonist exposure time appears intermediate between rapidly and slowly triggered excitotoxicity in neurons. Possibly, differences in AMPA receptor expression or behavior, means of buffering internal Ca^{2+}, or intrinsic differences in vulnerability to Ca^{2+} overload may account for this difference in susceptibility.

E. Points of Intervention
I. Reducing Extracellular Glutamate
1. Circuit Activity and Glutamate Release

One approach to decreasing excitotoxic injury may be to inhibit neuronal circuit activity and, therefore, to reduce vesicular glutamate release from presynaptic terminals. This might be accomplished by several means, including: (a) hypothermia; (b) increasing GABAergic tone; (c) K^+ channel openers; (d) modulating adenosine receptors; (e) blocking voltage-gated Na^+ channels, or (f) blocking voltage-gated Ca^{2+} channels. In the context of decreased energy substrate availability, as during ischemia, these strategies would have the additional benefit of reducing energy demand. Another effect of measures directed at decreasing glutamate release would likely be reduction of synaptic Zn^{2+} release from the same nerve terminals (ASSAF and CHUNG 1984; MARTINEZ-GUIJARRO et al. 1991), although the suggestion that Zn^{2+} may be localized to a subset of synaptic vesicles raises the possibility of a differential modulation of glutamate and Zn^{2+} release (PEREZ-CLAUSELL and DANSCHER 1985).

a) Hypothermia

Well recognized as a neuroprotective maneuver for decades, hypothermia has been proposed to be a "gold standard" against which other interventions should be measured (BUCHAN 1992). Both intra- and postischemic hypothermia produce lasting neuroprotective effects in animal cerebral ischemia studies (BARONE et al. 1997), in large part due to inhibition of glutamate release (BUSTO et al. 1989). Neuroprotective effects of hypothermia can also be demonstrated in neuronal cell cultures, again reflecting a reduction in endogenous glutamate release, as well as probably other actions (BRUNO et al. 1994).

At present, the clinical use of hypothermia is limited to surgical procedures that require concomitant cardiac arrest and neurosurgical procedures such as cerebral aneurysm clipping (TOMMASINO and PICOZZI 1998). Although some benefits of moderate hypothermia have been demonstrated for traumatic brain injury (MARION et al. 1997), testing of hypothermic therapy in human stroke has been slowed by concerns of potential complications such as coagulopathies, arrhythmias, and myocardial infarction (STEEN et al. 1979, 1980). However, hypothermia remains a promising therapeutic approach, especially if methods can be employed to localize cooling to the brain.

b) Increasing GABAergic Tone

GABA, the major inhibitory neurotransmitter in the mammalian brain, mediates its neuronal effects through three receptor subtypes, $GABA_A$, $GABA_B$, and $GABA_C$, all presumably pentameric complexes (BORMANN 2000). $GABA_A$ and $GABA_C$ receptors are ligand-gated chloride channels, while $GABA_B$ receptors are coupled to G-proteins, usually in presynaptic terminals, where they mediate an increase in K^+ conductance and downmodulation of transmitter release (KARLSSON and OLPE 1989; GAGE 1992). $GABA_A$ receptor agonists, such as muscimol or benzodiazepines reduced brain injury following rodent cerebral (STERNAU et al. 1989; LYDEN and HEDGES 1992; SHUAIB et al. 1993; SCHWARTZ-BLOOM et al. 1998) or spinal cord (MADDEN 1994) ischemia. $GABA_A$ receptor stimulation by muscimol also reduced excitotoxicity in neuronal culture, presumably by hyperpolarizing membranes and reducing activation of voltage-gated Ca^{2+} channels, as well as enhancing the voltage-dependent Mg^{2+} block of NMDA receptors (MUIR et al. 1996; c.f. ERDO and MICHLER 1990). However, a cautionary note was raised by the observation that $GABA_A$ receptor agonists paradoxically enhanced excitotoxicity induced by oxygen-glucose deprivation in vitro, possibly due to a contravening effect of maintaining the driving force for Ca^{2+} in energy-depleted and depolarized neurons, thereby outbalancing neuroprotective actions (MUIR et al. 1996).

In contrast to $GABA_A$ receptor agonists, $GABA_B$ receptor agonists like baclofen have provided inconclusive results in animal cerebral ischemia, perhaps in part due to complications such as postischemic hypertension and hemorrhage (STERNAU et al. 1989; ROSENBAUM et al. 1990; JACKSON-FRIEDMAN et al. 1997); they have also been ineffective against excitotoxicity in cell culture (MUIR et al. 1996). $GABA_C$ receptors, which exhibit a predominantly retinal distribution in vertebrates, have not been exploited for antiexcitotoxic purposes (JOHNSTON 1996). Other approaches to increasing GABAergic tone for antiexcitotoxic effects in vivo include the use of GABA reuptake inhibitors such as tiagabine (SUZDAK and JANSEN 1995) and CI-966 (PHILLIS 1995), or blockers of GABA metabolism such as the GABA transaminase inhibitor vigabatrin (SHUAIB et al. 1992).

c) Opening K^+ Channels

Membrane excitability might also be reduced by increasing the opening of K^+ channels other than those gated after $GABA_B$ receptor activation. Mammalian neurons express multiple K^+ channel subtypes, including channels that are voltage-gated (BROWN 1993), ATP-sensitive (HADDAD and JIANG 1994), and Ca^{2+}- or Na^+-activated (DRYER 1994; SAH 1996). Various K^+ channel openers reduced endogenous glutamate release following brief ischemia in hippocampal slices (ZINI et al. 1993). Activators of ATP-sensitive K^+ channels attenuated excitotoxic death in neuronal cultures, at least in part by decreasing the magnitude of intracellular Ca^{2+} elevation (ABELE and MILLER 1990). Similar agents, including Y-26763 (TAKABA et al. 1997), cromakalim (HEURTEAUX et al. 1993), and nicorandil, have exhibited therapeutic value in animal cerebral ischemia studies. The identification of pharmacological openers of large-conductance Ca^{2+}-activated K^+ channels (BK channels) such as BMS-204352 may provide another potentially neuroprotective means of hyperpolarizing neuronal membranes during an excitotoxic insult; phase III clinical trials are underway for this drug as an acute treatment for stroke (BOZIK et al. 2000). However, a note of caution regarding this approach is raised by the potential of K^+ efflux to promote apoptosis (see Sect. C.).

d) Modulating Adenosine Receptors

Adenosine acts as an agonist at three major receptor subtypes, A_1, A_2 (A_{2A} and A_{2B}), and A_3, each of which transduces its signals through coupled G-proteins (OLAH et al. 1995). Stimulation of A_1 receptors leads to multiple circuit depressing effects, including enhancement of K^+ and non-GABAergic Cl^- conductances and reduction of pre- and postsynaptic Ca^{2+} conductances (MAGER et al. 1990; RIBEIRO 1995). Adenosine protected cortical neurons in vitro from oxygen-glucose deprivation (GOLDBERG et al. 1988), presumably through activation of presynaptic A_1 receptors and a subsequent decrease in vesicular glutamate release (CORRADETTI et al. 1984). But results with A_1 receptor agonists in animal studies have been inconsistent, perhaps due to drug-induced bradycardia and hypotension resulting from activation of A_1 receptors in cardiovascular tissues (VON LUBITZ et al. 1995).

Work on the role of A_2 receptors in excitotoxic damage has focused mostly on the A_{2A} subtype, which enhances glutamate release when activated and reduces release when antagonized in ischemic cortex (O'REGAN et al. 1992; SIMPSON et al. 1992). However, available evidence suggests a more complicated role for this receptor in modulating excitotoxicity. Moderately selective A_{2A} receptor antagonists reduced injury subsequent to cerebral ischemia in the sensitive gerbil model (GAO and PHILLIS 1994; VON LUBITZ et al. 1995), and A_{2A} gene deletion provided moderate protection against injury induced by focal cerebral ischemia in mice (CHEN et al. 1999). On the other hand, in contrast to the predicted ability of A_{2A} receptor stimulation to augment excitotoxicity, the selective A_{2A} receptor agonist CGS21680 inhibited hippocampal injury induced by systemic kainate injection (JONES et al. 1998). Similarly, the potential contribution of A_3 receptors to excitotoxicity appears complex, especially given the limitations of current pharmacology and highly species-dependent patterns of expression (VON LUBITZ 1999; KLOTZ 2000). Since a major obstacle for antiexcitotoxic drugs targeting adenosine receptors is the presence of these receptors in nonneural tissues, adenosine analogs with fewer cardiovascular effects have recently been developed. One such A_1 receptor-specific compound, adenosine amine congener (ADAC), reduced injury and improved functional recovery following rodent cerebral ischemia when administered as late as 6h postischemia (VON LUBITZ et al. 1996a; VON LUBITZ et al. 1996b).

Like the adenosine and GABA receptor systems, stimulation of the group II and group III subtypes of metabotropic glutamate receptors may offer yet another modulator-based approach to attenuating transmitter glutamate release (see Sect. II.3.).

e) Blocking Voltage-Gated Na^+ Channels

Na^+ influx through voltage-gated Na^+ channels provides the electrical force for action potential generation and circuit excitation; furthermore, as noted above, intracellular Na^+ accumulation promotes Ca^{2+} influx via voltage-gated Ca^{2+} channels and reverse operation of the Na^+/Ca^{2+} exchanger. A favorable characteristic to consider when designing or screening Na^+ channel blockers for antiexcitotoxic potential may be the use-dependence of the agent since drugs with this property would be predicted to inhibit the most active neurons preferentially. The Na^+ channel blockers tetrodotoxin (YAMASAKI et al. 1991; LYSKO et al. 1994), phenytoin (CULLEN et al. 1979; TAFT et al. 1989), and riluzole (PRATT et al. 1992) decreased neuronal injury following cerebral ischemia in rodents. In culture, the antiexcitotoxic effect of local anesthetics and anticonvulsants alone, which exert their effects predominantly through use-dependent Na^+

channel blockade, has been variable (KOH and CHOI 1987; OGURA et al. 1988; MATTSON and KATER 1989).

White matter represents a particularly important therapeutic target for Na$^+$ channel blocking drugs. A series of elegant studies using the rat optic nerve has suggested that activation of voltage-gated Na$^+$ channels in the setting of anoxic injury is responsible for triggering toxic Ca^{2+} influx in axons, mainly through the reverse activity of the Na$^+$/Ca^{2+} exchanger (STYS et al. 1991). The prominence of this dependence of Ca^{2+} influx upon Na$^+$ channels in white matter may reflect the absence of glutamate receptor-mediated entry routes available in gray matter. Indeed, in cultured cortical neurons, once NMDA and AMPA receptors are pharmacologically blocked, an additional neuroprotective effect of tetrodotoxin or phenytoin against neuronal death triggered by oxygen-glucose deprivation can be unmasked (LYNCH et al. 1995). These findings suggest that combined therapy targeting both postsynaptic glutamate receptors and axonal Na$^+$ channels may provide more effective neuroprotection than either alone.

f) Blocking Voltage-Gated Ca^{2+} Channels

At least seven subtypes of voltage-gated Ca^{2+} channels, each with distinct electrophysiological properties and cellular localization, have been identified in mammalian neurons (MILJANICH and RAMACHANDRAN 1995; PEREZ-REYES and Schneider 1995). Presynaptic N-type Ca^{2+} channels play a crucial role in vesicular neurotransmitter release (KAMIYA et al. 1988; DUTAR et al. 1989), and drugs selective for this channel subtype attenuated neuronal injury in rodent cerebral ischemia studies (VALENTINO et al. 1993; BUCHAN et al. 1994). P- and Q-type channels, also expressed presynaptically, regulate physiological glutamate transmission in hippocampal slices (WHEELER et al. 1994), and a peptide inhibitor of these channels reduced infarct volume following rodent cerebral ischemia (ASAKURA et al. 1997). While L-type Ca^{2+} channels are predominantly located on postsynaptic cell bodies, activation of these channels can in certain cases enhance neurotransmitter release, so that antagonists may offer the dual benefit of reducing postsynaptic Ca^{2+} influx as well as presynaptic glutamate release during an excitotoxic insult (MIDDLEMISS and SPEDDING 1985). Consistent with these predictions, dihydropyridine antagonists reduced excitotoxic death in neuronal culture (ABELE et al. 1990; WEISS et al. 1990). However, as with N-type channel blockers, the therapeutic value of neuronal L-type channel blockade in cerebral ischemia remains ambiguous due to complicating cardiovascular effects in vivo (KOBAYASHI and MORI 1998). Additionally, clinical trials with dihydropyridines in the context of subarachnoid hemorrhage or ischemic stroke have yielded inconsistent or disappointing results (ROSENBAUM et al. 1991; AMERICAN NIMODIPINE STUDY GROUP 1992; MURPHY 1992; KASTE et al. 1994). More recently, broad-spectrum neuronal voltage-gated Ca^{2+} channel antagonists with minimal cardiovascular side effects, such as SB 201823A, have shown some promise in rodent cerebral ischemia studies (BARONE et al. 1995).

2. Glutamate Transport

The Ca^{2+}-dependent, vesicular release of glutamate does not wholly account for the rise in extracellular glutamate concentrations during brain ischemia. A substantial contribution is also made in a Ca^{2+}-independent fashion, by reverse operation of glutamate transporters in both neurons and glia (NICHOLLS and ATTWELL 1990; SZATKOWSKI et al. 1990; ATTWELL et al. 1993; JABAUDON et al. 2000; ROSSI et al. 2000). These transporters normally function to remove glutamate from synapses and thus to terminate a synaptic signaling event. However, since the direction in which a glutamate transporter operates is governed by the gradients of the other ions that are co- or countertransported, perturbations in intra- and extracellular ionic conditions can induce release of glutamate from the cytoplasm of astrocytes and neurons into the extracellular space.

The transport of one glutamate anion is coupled to cotransport of three Na^+ ions, countertransport of one K^+ ion, and cotransport of one proton or counter-transport of one hydroxyl ion (KANNER and SHARON 1978; BARBOUR et al. 1988; BOUVIER et al. 1992; ZERANGUE and KAVANAUGH 1996). During brain ischemia, cells experience a shortage of high-energy phosphates; the Na^+-K^+ ATPase is inhibited, and extracellular K^+ and intracellular Na^+ concentrations rise. The magnitude of this run-down in ionic gradients predicts that glutamate transporters would operate in reverse until a new equilibrium is reached, with extracellular glutamate concentrations reaching beyond $100\,\mu M$ (SZATKOWSKI and ATTWELL 1994), levels that are potentially neurotoxic. Therefore, glutamate transporters may represent a useful pharmacological target in attenuating excitotoxic damage (VANDENBERG 1998), whether or not impairment of transporter function is involved in disease pathogenesis as has been suggested in the case of amyotrophic lateral sclerosis (see Sect. B). There might be particular value in developing agents that selectively inhibit reverse transport, in analogy with the recent development of selective blockers of reverse Na^+/Ca^{2+} exchange (IWAMOTO et al. 1996; HOYT et al. 1998). In addition, glutamate transporters activate a Cl^- conductance that is not directly coupled to glutamate transport (FAIRMAN et al. 1995; WADICHE et al. 1995). That the Cl^- conductance is decoupled from glutamate uptake is supported by the ability of Zn^{2+} to modulate these two activities differentially in certain transporter subtypes (SPIRIDON et al. 1998; VANDENBERG et al. 1998). By developing an agent that enhances this hyperpolarizing flow of Cl^-, it may be possible to decrease the magnitude of ischemic depolarization in cells that express transporters, thus favoring forward operation.

II. Manipulating Glutamate Receptors

1. NMDA Antagonists

Consistent with the prominent role of NMDA receptors in mediating glutamate-induced Ca^{2+} overload and rapidly-triggered excitotoxic neurodegeneration in vitro, NMDA antagonists can reduce the death of cultured cortical neurons induced by hypoxia, glucose deprivation, and trauma (CHOI 1992), and a substantial literature indicates that NMDA antagonists can reduce neuronal death in multiple models of brain injury in vivo. These include animal models of ischemia (SIMON et al. 1984; MCCULLOCH 1992), hypoglycemia (WIELOCH 1985), sustained seizures (MELDRUM 1994), and trauma (MCINTOSH et al. 1989). Unfortunately, several recent clinical trials of NMDA antagonists in stroke patients have been disappointing; side effects including hallucinations, ataxia, or hypotension were prominent with several drugs (KEMP et al. 1999; READ et al. 1999). It remains to be seen whether efficacy can be established with this strategy, perhaps with the aid of enhancements in dosage regimens or drug characteristics, or whether utility in human stroke will prove to be fundamentally constrained (see below). Nonetheless, considering the high potential of the NMDA receptor system to contribute to excitotoxic neuronal death, we think it likely that NMDA antagonists will eventually find use as neuroprotective agents in one or another disease setting.

NMDA receptor blockade can be achieved in a variety of ways, using agents that act at distinct molecular sites within the heteromeric receptor complex. Competitive antagonists bind the glutamate recognition site; channel blockers, also termed uncompetitive antagonists, bind sites within the channel pore; glycine antagonists bind the glycine recognition site; and noncompetitive antagonists bind other sites on NMDA receptors, downmodulating receptor activation via remote actions, for example, via allosteric changes. The latter modulatory sites include those responding to polyamines (RANSOM and STEC 1988), redox potential (AIZENMAN et al. 1989), hydrogen ions (TANG et al. 1990; TRAYNELIS and CULL-CANDY 1990), and Zn^{2+} (PETERS et al. 1987; WESTBROOK and MAYER 1987; CHRISTINE and CHOI 1990; LEGENDRE and WESTBROOK 1990). Whereas NMDA receptor activation would be reduced by the free radicals and lactic acid produced by ischemia, the ischemic release of polyamines, including putrescine, spermine, and spermidine would be expected to upmodulate NMDA receptor activity (PASCHEN et al. 1992; KERCHNER et al. 1999). Zn^{2+} effects might be complex, as acute direct NMDA receptor inhibition might be followed by more lasting Src kinase-mediated upregulation (MANZERRA et al. 2001).

A well-recognized theoretical limitation of competitive NMDA receptor antagonists is that they are more effective when ambient glutamate concentrations are low and hence may be more effective against receptors operating physiologically than at the overactivated receptors contributing to acute excitotoxic damage. Channel blockers, glycine antagonists, and noncompetitive antagonists would not have this difficulty, but all of these antagonists are at risk for evoking what are probably mechanism-driven cognitive and motor side effects. NMDA antagonists also have the potential for inducing vacuolization or even death in small numbers of neurons in the cingulate or retrosplenial cortex, perhaps mediated by the paradoxical release of excitation in specific circuits (OLNEY et al. 1989).

How might the therapeutic index of NMDA antagonists be improved? Three approaches are currently being explored: (a) preferentially blocking overactivated NMDA receptors relative to physiologically-activated receptors, (b) limiting antagonism with partial or weak antagonists; and (c) enhancing target specificity with subtype selective antagonists. The first approach might be achieved by using activity-dependent channel blocking compounds that not only require channel opening to reach their binding site, but also exhibit a greater degree of blockade at higher levels of receptor activity. Memantine is such a compound (and it also has low affinity for its channel binding site – see below, this section); it has has shown promise in attenuating excitotoxic neuronal loss in vitro as well as brain damage in a rodent model of stroke, at concentrations that might permit near-normal levels of physiological NMDA receptor-mediated synaptic transmission (CHEN et al. 1992). In addition, the apparent affinity of ifenprodil and related antagonists for NMDA receptors increases at higher agonist concentrations (KEW et al. 1996), which, in addition to other useful properties (see below, this section), may contribute to a reduced side effect profile.

Low affinity channel blockers may represent one means to achieve moderate levels of NMDA receptor antagonism. Interestingly, despite the value generally attached to potency in drug development, an inverse relationship between drug affinity and toxicity is apparent for many NMDA channel blockers (ROGAWSKI 1993; PALMER and WIDZOWSKI 2000). In principle, because lower affinity compounds typically exhibit faster unblocking rates and require a higher concentration to achieve a given level of blockade, they equilibrate with their receptors more quickly, resulting in faster termination of NMDA receptor gating than is achieved by equieffective doses of higher affinity agents. At the same time, the faster unbinding of low affinity channel blockers should lead to less trapping of antagonist as receptor activity falls off and channels close. Such properties are attractive and may underlie reduced side effects.

A practical method for achieving limited antagonism of the NMDA receptor may be through the use of glycine site antagonists. While complete glycine site antagonism would be expected to generate a set of mechanism-driven side effects comparable to those produced by glutamate site antagonists and channel blockers, partial glycine site agonists, such as cycloserine (HOOD et al. 1989), by producing limited-efficacy blockade of NMDA receptor activity, may be able to strike an attractive balance between reduction of excitotoxicity and the downsides associated with high-level receptor blockade. Alternatively, levels of the endogenous glycine-site agonist, D-serine (SCHELL et al. 1995; SNYDER and KIM 2000), might be therapeutically reduced, hopefully still leaving enough ambient D-serine or glycine to keep receptors from shutting down completely. D-serine is synthesized by the enzyme serine racemase (WOLOSKER et al. 1999) within a discrete population of protoplasmic astrocytes that ensheath synapses (SCHELL et al. 1997); degradation of D-serine with exogenously applied D-amino acid oxidase inhibits NMDA receptor-mediated synaptic transmission in hippocampal slices (MOTHET et al. 2000).

Another promising approach involves the use of NMDA receptor subunit-selective antagonists. Ifenprodil, originally recognized as an NMDA antagonist that interacts with the polyamine binding site (CARTER et al. 1989; REYNOLDS and MILLER 1989), acts as a noncompetitive NMDA receptor antagonist and reduces excitotoxic neurodegeneration following glutamate or NMDA exposure in vitro (GRAHAM et al. 1992) and focal ischemia in vivo (GOTTI et al. 1988). It turned out to be approximately 400-fold more potent at NMDA receptor complexes containing the subunit NR2B than those containing NR2A (WILLIAMS 1993), NR2C, or NR2D (WILLIAMS 1995). Presumably reflecting this subtype specificity – and thus regional specificity, as NR2B-containing NMDA receptors are expressed preferentially in the adult forebrain, in a nonuniform distribution between various forebrain structures and neuronal populations (WATANABE et al. 1993; MONYER et al. 1994) – ifenprodil and related compounds appear to exhibit less side effects than broad spectrum NMDA antagonists (KEMP et al. 1999).

Even while efforts are underway to improve the molecular profile of NMDA antagonist drugs, it is worth noting that the simple physiological channel blocker, Mg^{2+}, responsible for conferring voltage sensitivity to NMDA receptors (Nowak et al. 1984), has shown promise as a therapeutic agent in animal models of stroke, as well as traumatic brain injury (VINK and CERNAK 2000). It also has been used extensively in humans for the prevention of seizures associated with preeclampsia and eclampsia (MASON et al. 1994; ANTHONY et al. 1996) and has been suggested to reduce the risk of cerebral palsy in human infants born to preeclamptic mothers (NELSON and GRETHER 1995). Of course, the beneficial effects of Mg^{2+} may not be limited to NMDA receptor antagonism. To the extent that it penetrates into the CNS in a given disease setting, it would likely reduce glutamate release, and inhibit voltage-gated Ca^{2+} channel-mediated Ca^{2+} entry into neurons and vascular smooth muscle (the latter effect leading to enhancements of cerebral blood flow).

Finally, there may be some settings where NMDA antagonists, regardless of molecular mechanism of action, may not be beneficial. As noted above, NMDA receptor overactivation may already be limited by endogenous tissue factors such as lowered extracellular pH, Zn^{2+}, and oxygen free radicals. In addition, there has been recent recognition that the ability of NMDA antagonists to reduce Ca^{2+} influx may concurrently increase the likelihood of apoptosis for neurons that are in a state of relative Ca^{2+} starvation versus Ca^{2+} overload (LEE et al. 1999). In the developing rat brain, brief administration of NMDA antagonists has been shown to induce widespread apoptotic neuronal death (IKONOMIDOU et al. 1999); in agreement with that observation, the effect of ethanol to promote massive programmed cell death of central neurons in immature rat brains may reflect its ability to inhibit NMDA receptors (IKONOMIDOU et al. 2000).

2. AMPA/Kainate Antagonists

As discussed already, AMPA/kainate receptors can directly mediate excitotoxic cell death, albeit less powerfully than NMDA receptors. The competitive AMPA/kainate receptor antagonist NBQX is effective in reducing neuronal loss following both global (SHEARDOWN et al. 1990) and focal (BUCHAN et al. 1991) cerebral ischemia, spinal cord ischemia (XU et al. 1993), and brain trauma (WRATHALL et al. 1992), although the possibility of a contribution from cerebral hypothermia has been raised (COLBOURNE et al. 1997). The noncompetitive AMPA receptor antagonist GYKI-52466 has also exhibited neuroprotective effects in studies of global (LE PEILLET et al. 1992) or focal (SMITH and MELDRUM 1992; XUE et al. 1994) ischemia.

In addition, AMPA/kainate receptor antagonists may be of special value in certain settings. Although the death of most cortical neurons induced by brief glutamate exposure at neutral pH is AMPA/kainate receptor antagonist-insensitive (KOH and CHOI 1991), lowering pH selectively enhanced AMPA/kainate receptor-mediated neurotoxicity, perhaps by delaying recovery of intracellular Ca^{2+} homeostasis (MCDONALD et al. 1998b). In addition, a small subpopulation of neurons, largely GABAergic, that express Ca^{2+}-permeable AMPA receptors exhibits prominent vulnerability to AMPA receptor-mediated excitotoxicity (KOH and CHOI 1988; JONAS et al. 1994; TURETSKY et al. 1994). Brief glutamate exposure raises intracellular Ca^{2+} and destroys these cells even when NMDA receptors are blocked. Ca^{2+}-permeable AMPA receptors are likely also permeable to Zn^{2+} and hence confer vulnerability to Zn^{2+} neurotoxicity (SENSI et al. 1997; WEISS and SENSI 2000). Besides protecting neuronal subpopulations expressing Ca^{2+}-permeable AMPA receptors, AMPA antagonists could have value in reducing the excitotoxic loss of oligodendrocytes, which likely also express Ca^{2+}-permeable AMPA receptors (see above).

The prevalence of Ca^{2+}-permeable AMPA receptors in populations of selectively vulnerable neurons in certain disease settings highlights a potentially important therapeutic role for AMPA/kainate antagonists. In amyotrophic lateral sclerosis, the motor neurons that undergo selective degeneration express AMPA receptors that are Ca^{2+}-permeable due to low levels of GluR2 expression (SHAW and INCE 1997). Indeed, brief kainate exposure induced a Ca^{2+}-dependent and Ca^{2+}-permeable AMPA receptor antagonist-sensitive death in spinal motor neurons but not dorsal horn neurons (VAN DEN BOSCH and ROBBERECHT 2000). In the context of transient global cerebral ischemia, the prevalence of Ca^{2+}-permeable AMPA receptors may rise among hippocampal CA1 neurons, a population of cells particularly vulnerable to this type of insult, due to a downregulation in expression of GluR2 relative to other AMPA receptor subunits (PELLEGRINI-GIAMPIETRO et al. 1992).

Historically, the roles of AMPA and kainate receptors in neuronal physiology have been difficult to distinguish, due to insufficient pharmacology. With the development of selective, noncompetitive AMPA receptor antagonists

(PELLETIER et al. 1996), it has become possible to discriminate between the relative contributions of AMPA and kainate receptors to several phenomena. Experiments with cultured cortical neurons have suggested that slowly-triggered excitotoxicity, induced by prolonged exposure to kainate (see above) is mediated predominantly by AMPA receptors, suggesting that activation of cortical neuronal kainate receptors alone may not suffice to induce cell death (TURETSKY et al. 1998). Moreover, there is reason to consider that selective kainate receptor antagonism could potentially be counterproductive, as activation of presynaptic kainate receptors by synaptically released glutamate inhibits excitatory transmission in the hippocampus (SCHMITZ et al. 2000), a phenomenon that may reflect a direct negative-feedback pathway for glutamate release. Of note, the AMPA-selective antagonists developed to date exhibit noncompetitive kinetics, providing effective blockade even in the context of excess extracellular glutamate.

3. Metabotropic Glutamate Receptors

Eight metabotropic glutamate receptors (mGluRs), which are linked to G-proteins rather than ion channels, have been identified and segregated into three groups based on sequence similarity and mechanisms of signal transduction (NAKANISHI and MASU 1994; CONN and PIN 1997). Group I mGluRs (mGluR1 and -5) couple via phospholipase C to phosphoinositide turnover and Ca^{2+} release from intracellular stores, whereas group II (mGluR2 and -3) and III (mGluR4, -6, -7, and -8) receptors couple to the inhibition of adenyl cyclase and reduction in cAMP levels. Although mGluRs do not directly mediate excitotoxicity, they can modify excitotoxicity and thus may be useful targets for therapeutic manipulation. The first clue to neuroprotective actions was the demonstration that the nonselective mGluR agonist, trans-1-aminocyclopentane-1,3-dicarboxylic acid (tACPD), could attenuate glutamate-induced neuronal death (KOH et al. 1991); nonselective activation of mGluRs also reduced infarct volume in vivo after focal ischemia (CHIAMULERA et al. 1992).

Since mGluR group II and III receptors typically have inhibitory effects on circuit excitation and glutamate release, whereas group I receptors are typically proexcitatory (CONN and PIN 1997; CARTMELL and SCHOEPP 2000), it is plausible that agonists selective for group II or III mGluRs would have more powerful antiexcitotoxic effects than nonselective agonists. The mechanisms by which group II/III mGluRs downregulate transmitter release are not entirely understood but likely involve inhibition of presynaptic voltage-gated Ca^{2+} channels (STEFANI et al. 1996) and activation of presynaptic K^+ conductances (SLADECZEK et al. 1993). Initial studies with group II agonists suggested antiexcitotoxic actions against NMDA-induced degeneration in vitro (BRUNO et al. 1995a; PIZZI et al. 1996), although available drugs had confounding weak agonist/antagonist activity at NMDA receptors (BUISSON et al. 1996; CONTRACTOR et al. 1998). The more selective group II mGluR agonist, (+)-2-

aminobicyclo[3.1.0]hexane-2,6-dicarboxylic acid (LY354740), was subsequently found surprisingly to lack antiexcitotoxic effects either in vitro or in vivo (LAM et al. 1998; BEHRENS et al. 1999). However, the group III mGluR agonists L-(+)-2-amino-4-phosphonobutyric acid and L-serine-*O*-phosphate did reduce trauma-induced neuronal death in vitro, adding to the protective effects of an NMDA antagonist (FADEN et al. 1997). Recently, it was demonstrated that the selective group III mGluR agonist (+)-4-phosphonophenylglycine attenuated NMDA-induced excitotoxic neuronal death both in cortical cultures and in vivo; these effects were completely abolished in mice lacking the mGluR4 gene (BRUNO et al. 2000).

Agonists at group I mGluRs enhance neuronal excitability through several mechanisms, including regulation of Ca^{2+} and K^+ channels (CONN and PIN 1997). In addition, these mGluRs have complex modulatory effects upon NMDA receptors, inducing both a rapid, membrane-delimited reduction of NMDA receptor currents (YU et al. 1997a) and a long-term NMDA receptor upregulation via Src family kinase-mediated phosphorylation of NR2 receptors (BEHRENS et al. 2000); the latter effect may be more important from the standpoint of excitotoxicity. Consistent with proexcitatory and NMDA receptor-enhancing actions, activation of group I mGluRs generally potentiates excitotoxicity. In cell culture studies, neuronal death secondary to NMDA or kainate exposure (BRUNO et al. 1995b; BUISSON and CHOI 1995; STRASSER et al. 1998) or traumatic injury (MUKHIN et al. 1996) was potentiated by group I mGluR agonists and attenuated by antagonists.

III. Blocking Downstream Mediators

1. Downstream Effects of Cellular Ca^{2+} Overload

Many enzymes, including proteases, lipases, endonucleases, kinases, and phosphatases are activated directly or indirectly by increases in intracellular Ca^{2+} concentration and may contribute to cellular damage after excitotoxic receptor activation. Calpain inhibition attenuated neuronal death triggered by exogenous excitotoxins in vitro (BRORSON et al. 1994) and following transient global ischemia in rodents (LEE et al. 1991). More recently, MDL 28,170, a potent inhibitor of calpains, decreased infarct volume after focal ischemia when administered even 6h postocclusion (MARKGRAF et al. 1998). Ca^{2+}-activated cytoplasmic phospholipase A_2 ($cPLA_2$) can catabolize phospholipids to liberate arachidonic acid (DUMUIS et al. 1988), which may augment excitotoxicity by reducing glutamate reuptake (YU et al. 1986), promoting glutamate release (FREEMAN et al. 1990), and producing free radicals in the process of downstream metabolism (see next section). $cPLA_2$ gene deletion increased the resistance of mice to focal ischemia-induced brain injury (BONVENTRE et al. 1997).

Although Ca^{2+}-activated endonucleases have also been suggested to contribute to excitotoxic death, the poor selectivity of available endonuclease inhibitors needs to be kept in mind (KURE et al. 1991; ROBERTS-LEWIS et al. 1993; ZEEVALK et al. 1993; POSNER et al. 1995).

2. Free Radical Formation

Cytosolic Ca^{2+} loading consequent to glutamate receptor stimulation triggers the formation of multiple free radical species, which have deleterious effects on proteins, DNA, and lipids. Antioxidants can reduce neuronal death induced by exogenous excitotoxins in culture or by intrastriatal injection of excitotoxins in vivo (DYKENS et al. 1987; MIYAMOTO and COYLE 1990; MONYER et al. 1990).

At least four pathways may link an excitotoxic increase in intracellular free Ca^{2+} to free radical overproduction: xanthine dehydrogenase, cyclooxygenases, nitric oxide synthases, or mitochondrial electron transport. Elevated intracellular Ca^{2+} indirectly converts xanthine dehydrogenase into xanthine oxidase, which can produce superoxide radicals ($\cdot O_2^-$) (DYKENS et al. 1987; ATLANTE et al. 1997). However, due to the low expression of this enzyme in the human brain, the pathophysiological relevance of this free radical pathway in humans remains uncertain (SARNESTO et al. 1996). Cyclooxygenase (COX)-mediated metabolism of arachidonic acid to a prostaglandin intermediate can also lead to the production of toxic $\cdot O_2^-$ (CHAN et al. 1985; WEI et al. 1986). Fenamate derivatives, which inhibit both COX-1 and -2, decreased the cell death induced by either NMDA or kainate in isolated chick retina (CHEN et al. 1998). In cortical neuronal cultures, NMDA-induced excitotoxicity was decreased by a specific COX-2 inhibitor, NS-398 (HEWETT et al. 2000). Consistent with a role for COX proteins in brain injury subsequent to focal ischemia, prostaglandin production was observed to increase early (15min) following focal ischemia in rats and was attenuated by pretreatment with the fenamate derivative meclofenamate (BUCCI et al. 1990). Significantly, COX-2 inhibition afforded neuroprotection against focal ischemia when administered to rats postocclusion (NOGAWA et al. 1997).

Rises in intracellular Ca^{2+} concentration can also activate neuronal nitric oxide synthase (nNOS), which forms the weak oxidant, nitric oxide (DAWSON and SNYDER 1994). In the presence of $\cdot O_2^-$, however, nitric oxide can be converted to peroxynitrite, a powerfully destructive free radical (BECKMAN and KOPPENOL 1996). Thus, nNOS plays a central role in mediating cell death induced by the overactivation of NMDA receptors in neuronal culture (DAWSON et al. 1991; DAWSON et al. 1996) and in mouse striatum in vivo (AYATA et al. 1997), as well as in rodents following focal ischemia (HUANG et al. 1994). An inducible form of nitric oxide synthase, iNOS, which is expressed in inflammatory (IADECOLA et al. 1995a), vascular (IADECOLA et al. 1996), and glial (ENDOH et al. 1994) cells after cytokine exposure in culture (HEWETT et al. 1994) or after the onset of ischemia in vivo, can also contribute to excitotoxic damage. In vitro, cytokine-dependent induction of iNOS in astrocytes potentiated NMDA-mediated neuronal death (HEWETT et al. 1994), and aminoguanidine, an inhibitor of iNOS, reduced infarct volume following focal ischemia, even when administered 24h following occlusion (Iadecola et al. 1995b; Zhang et al. 1996).

While mitochondria have the ability to buffer elevations in intracellular Ca^{2+} (Gunter and Pfeiffer 1990; Wang and Thayer 1996), excessive Ca^{2+} accumulation in mitochondria may lead to the uncoupling of energy production from electron transport and the formation of toxic levels of free radicals (Dugan et al. 1995; Reynolds and Hastings 1995; Schinder et al. 1996; White and Reynolds 1996). Pharmacological blockade of mitochondrial Ca^{2+} uptake substantially decreased glutamate-mediated neuronal death in culture (Stout et al. 1998). Furthermore, additional increases in free radical production may occur if mitochondria release their Ca^{2+} stores into the cytoplasm, amplifying the Ca^{2+}-dependent free radical cascades mentioned above (White and Reynolds 1996).

Beneficial results have been obtained with several free radical scavenger drugs in animal studies of ischemic or traumatic brain injury (Clemens and Panetta 1994), although the magnitude of neuroprotection observed has typically not been very large. Additionally, recent clinical trial experience with the antioxidant tirilazad mesylate in subarachnoid hemorrhage was not especially encouraging (Ranttas Investigators 1996; Kassell et al. 1996; Haley et al. 1997). It is possible that more powerful antioxidant agents may yield greater therapeutic benefits. The spin trapping agent, α-phenyl-N-tert-butyl nitrone (PBN) reduced infarct volume following focal ischemia (Cao and Phillis 1994) when administered up to 3h after ischemia onset (Zhao et al. 1994), perhaps reflecting an ability of its breakdown product, N-t-butyl hydroxylamine, to inhibit mitochondrial superoxide production (Atamna et al., 2000).

Recent reports have suggested that Zn^{2+}-induced neuronal death may also in part be mediated by an increase in oxidative stress. In neuronal cultures, Zn^{2+} exposure caused an early increase in reactive oxygen species production and lipid peroxidation, and antioxidants attenuated neuronal death triggered by Zn^{2+} (Kim et al. 1999a,b; Sensi et al. 1999); however, in other studies a relatively lower prominence of free radical-mediated injury after Zn^{2+} exposure was observed (L.L. Dugan and D.W. Choi, unpublished results).

3. The Role of PARP

A particularly damaging consequence of reactive oxygen species formation may be single-stranded DNA breakage, leading to activation of the repair enzyme, poly(ADP-ribose) polymerase-1 (PARP-1), and consequent depletion of cellular NAD^+ and energy stores (Szabo and Dawson 1998). Consistent with the idea that PARP-1 activation leads to lethal energy depletion under excitotoxic conditions, pharmacological inhibition or gene deletion of PARP-1 attenuated neuronal death induced by glutamate receptor agonists in vitro (Zhang et al. 1994; Eliasson et al. 1997). PARP-1 knockout mice also exhibited increased resistance to focal ischemia (Zhang et al. 1994; Eliasson et al. 1997; ENDRES et al. 1997) as well as to damage induced by intrastriatal NMDA injection (MANDIR et al. 2000). Several PARP inhibitors have demonstrated neuroprotective effects in rodent focal ischemia studies (ENDRES et al. 1997; TAKAHASHI et al. 1997).

F. A Cautionary Note for Antiexcitotoxic Strategies: Enhanced Apoptosis?

The basic nature of excitotoxicity – influx of cations into cells through overactivated glutamate receptors, leading to acute cell swelling and subsequent death – is suggestive of necrosis. Indeed, multiple studies support the notion that excitotoxic glutamate receptor overactivation in vitro typically induces necrosis (GOTTRON et al. 1997; GWAG et al. 1997; CHIHAB et al. 1998). However, any insult can probably induce programmed cell death in certain circumstances, and excitotoxicity is no exception, particularly when excitotoxic conditions are mild or when target neurons are immature (BONFOCO et al. 1995; McDONALD et al. 1997). Apoptosis after excitotoxic insults may also be favored by reductions in extracellular Na^+ and Ca^{2+}, as occur in ischemic tissue; these disturbances alter the ionic driving forces governing NMDA receptor currents, increasing, in particular, the relative contribution of proapoptotic K^+ efflux (Yu et al. 1997b, 1999). In vivo, intrastriatal injection of excitotoxins (FERRER et al. 1995; PORTERA-CAILLIAU et al. 1995; QIN et al. 1996) and cerebral ischemia (LINNIK et al. 1993; MACMANUS et al. 1993) lead to neuronal death outcomes that lie on a spectrum of morphological and biochemical phenotypes, ranging from necrosis to classic apoptosis, with many cells exhibiting a mixture of markers. The greater prominence of apoptosis after excitotoxin administration in vivo compared to in vitro, particularly at sites remote from an injection site in vivo, may reflect in part loss of innervation or trophic support originally provided by destroyed injection-site neurons.

The idea that certain insults may drive neurons simultaneously towards excitotoxic necrosis and apoptosis argues for caution in selecting neuroprotective strategies, since maneuvers that attenuate one type of death may have little effect or even a deleterious influence on the other. For instance, upstream antiexcitotoxic approaches reducing glutamate release or glutamate receptor activation may have a general tendency to reduce excitotoxic necrosis but enhance apoptosis triggered by other independent events. The converse may also be true. One might conceivably use mild proexcitotoxic manipulations such as the activation of group I mGluRs to raise intracellular Ca^{2+} and attenuate neuronal apoptosis (ALLEN et al. 2000).

Compared to upstream antiexcitotoxic approaches, downstream strategies aimed at blocking intracellular injury events may afford more opportunity for blocking excitotoxic necrosis without promoting apoptosis. Indeed, some strategies, such as free radical scavengers, may be effective against both excitotoxic necrosis and apoptosis. Besides a reduced risk of enhancing apoptosis, downstream approaches may generally offer a longer therapeutic window of opportunity than upstream approaches. However, since glutamate receptor overactivation triggers multiple parallel injury cascades, downstream approaches may be unlikely to achieve the neuroprotective efficacy of upstream approaches, unless several pathways are simultaneously targeted.

If the contribution of excitotoxic necrosis to injury is large enough, upstream antiexcitotoxic approaches alone may be of value, but if apoptosis is prominent, it may be necessary to add concurrent blockers of apoptosis. Alternatively, it may be possible to separate a necrotic phase of injury from an apoptotic phase in time and/or space. For example, after ischemic insults, excitotoxic necrosis may be most prominent near the ischemic core and at early time points, whereas apoptosis may be more prominent in penumbral regions and at later time points.

References

Abele AE, Miller RJ (1990) Potassium channel activators abolish excitotoxicity in cultured hippocampal pyramidal neurons. Neurosci Lett 115:195–200

Abele AE, Scholz KP, Scholz WK, Miller RJ (1990) Excitotoxicity induced by enhanced excitatory neurotransmission in cultured hippocampal pyramidal neurons. Neuron 4:413–419

Aizenman E, Lipton SA, Loring RH (1989) Selective modulation of NMDA responses by reduction and oxidation. Neuron 2:1257–1263

Aizenman E, Stout AK, Hartnett KA, Dineley KE, McLaughlin B, Reynolds IJ (2000) Induction of neuronal apoptosis by thiol oxidation: putative role of intracellular zinc release. J Neurochem 75:1878–1888

Allen JW, Knoblach SM, Faden AI (2000) Activation of group I metabotropic glutamate receptors reduces neuronal apoptosis but increases necrotic cell death in vitro. Cell Death Differ 7:470–476

American Nimodipine Study Group (1992) Clinical trial of nimodipine in acute ischemic stroke. Stroke 23:3–8

Anthony J, Johanson RB, Duley L (1996) Role of magnesium sulfate in seizure prevention in patients with eclampsia and pre-eclampsia. Drug Saf 15:188–199

Asakura K, Matsuo Y, Kanemasa T, Ninomiya M (1997) P/Q-type Ca2+ channel blocker omega-agatoxin IVA protects against brain injury after focal ischemia in rats. Brain Res 776:140–145

Assaf SY, Chung SH (1984) Release of endogenous Zn2+ from brain tissue during activity. Nature 308:734–736

Atamna H, Martínez-Paler A, Ames BN (2000) N-t-butyl hydroxylamine, a hydrolysis product of α-phenyl-N-t-butyl nitrone, is more potent in delaying senescence in human lung fibroblasts. J Biol Chem 275:6741–6748

Atlante A, Gagliardi S, Minervini GM, Ciotti MT, Marra E, Calissano P (1997) Glutamate neurotoxicity in rat cerebellar granule cells: a major role for xanthine oxidase in oxygen radical formation. J Neurochem 68:2038–2045

Attwell D, Barbour B, Szatkowski M (1993) Nonvesicular release of neurotransmitter. Neuron 11:401–407

Ayata C, Ayata G, Hara H, Matthews RT, Beal MF, Ferrante RJ, Endres M, Kim A, Christie RH, Waeber C, Huang PL, Hyman BT, Moskowitz MA (1997) Mechanisms of reduced striatal NMDA excitotoxicity in type I nitric oxide synthase knock-out mice. J Neurosci 17:6908–6917

Barbour B, Brew H, Attwell D (1988) Electrogenic glutamate uptake in glial cells is activated by intracellular potassium. Nature 335:433–435

Barone FC, Feuerstein GZ, White RF (1997) Brain cooling during transient focal ischemia provides complete neuroprotection. Neurosci Biobehav Rev 21:31–44

Barone FC, Lysko PG, Price WJ, Feuerstein G, al-Baracanji KA, Benham CD, Harrison DC, Harries MH, Bailey SJ, Hunter AJ (1995) SB 201823-A antagonizes calcium currents in central neurons and reduces the effects of focal ischemia in rats and mice. Stroke 26:1683–1689

Beal MF (2000) Energetics in the pathogenesis of neurodegenerative diseases. Trends Neurosci 23:298–304

Beckman JS, Koppenol WH (1996) Nitric oxide, superoxide, and peroxynitrite: the good, the bad, and ugly. Am J Physiol 271:C1424–1437

Behrens MM, Heidinger V, Manzerra P, Ichinose T, Yu SP, Choi DW (2000) Neuronal mGluR1 activation enhances NMDA receptor function via src-family kinase-mediated phosphorylation. Soc Neurosci Abstr 26:1412

Behrens MM, Strasser U, Heidinger V, Lobner D, Yu SP, McDonald JW, Won M, Choi DW (1999) Selective activation of group II mGluRs with LY354740 does not prevent neuronal excitotoxicity. Neuropharmacology 38:1621–1630

Benveniste H, Drejer J, Schousboe A, Diemer NH (1984) Elevation of the extracellular concentrations of glutamate and aspartate in rat hippocampus during transient cerebral ischemia monitored by intracerebral microdialysis. J Neurochem 43: 1369–1374

Bonfoco E, Krainc D, Ankarcrona M, Nicotera P, Lipton SA (1995) Apoptosis and necrosis: two distinct events induced, respectively, by mild and intense insults with N-methyl-D-aspartate or nitric oxide/superoxide in cortical cell cultures. Proc Natl Acad Sci USA 92:7162–7166

Bonventre JV, Huang Z, Taheri MR, O'Leary E, Li E, Moskowitz MA, Sapirstein A (1997) Reduced fertility and postischemic brain injury in mice deficient in cytosolic phospholipase A2. Nature 390:622–625

Bormann J (2000) The "ABC" of GABA receptors. Trends Pharmacol Sci 21:16–19

Bortner CD, Cidlowski JA (1998) A necessary role for cell shrinkage in apoptosis. Biochem Pharmacol 56:1549–1559

Bortner CD, Hughes FM Jr, Cidlowski JA (1997) A primary role for K+ and Na+ efflux in the activation of apoptosis. J Biol Chem 272:32436–32442

Bouvier M, Szatkowski M, Amato A, Attwell D (1992) The glial cell glutamate uptake carrier countertransports pH-changing anions. Nature 360:471–474

Bozik ME, Smith JM, Douglass A, Caplik J, Sullivan MA, Fisher M, Fayad P (2000) Double-Blind, Placebo Controlled, Safety and Efficacy Trials of Intravenous BMS-204352 in Patients with Acute Stroke. On-going Clinical Trials Session, 25th International Stroke Conference.

Brorson JR, Manzolillo PA, Miller RJ (1994) Ca2+ entry via AMPA/KA receptors and excitotoxicity in cultured cerebellar Purkinje cells. J Neurosci 14:187–197

Brown AM (1993) Functional bases for interpreting amino acid sequences of voltage-dependent K+ channels. Annu Rev Biophys Biomol Struct 22:173–198

Bruno V, Battaglia G, Copani A, Giffard RG, Raciti G, Raffaele R, Shinozaki H, Nicoletti F (1995a) Activation of class II or III metabotropic glutamate receptors protects cultured cortical neurons against excitotoxic degeneration. Eur J Neurosci 7:1906–1913

Bruno V, Battaglia G, Ksiazek I, van der Putten H, Catania MV, Giuffrida R, Lukic S, Leonhardt T, Inderbitzin W, Gasparini F, Kuhn R, Hampson DR, Nicoletti F, Flor PJ (2000) Selective activation of mGlu4 metabotropic glutamate receptors is protective against excitotoxic neuronal death. J Neurosci 20:6413–6420

Bruno V, Copani A, Knopfel T, Kuhn R, Casabona G, Dell'Albani P, Condorelli DF, Nicoletti F (1995b) Activation of metabotropic glutamate receptors coupled to inositol phospholipid hydrolysis amplifies NMDA-induced neuronal degeneration in cultured cortical cells. Neuropharmacology 34:1089–1098

Bruno VM, Goldberg MP, Dugan LL, Giffard RG, Choi DW (1994) Neuroprotective effect of hypothermia in cortical cultures exposed to oxygen-glucose deprivation or excitatory amino acids. J Neurochem 63:1398–1406

Bucci MN, Black KL, Hoff JT (1990) Arachidonic acid metabolite production following focal cerebral ischemia: time course and effect of meclofenamate. Surg Neurol 33:12–14

Buchan A (1992) Advances in cerebral ischemia: experimental approaches. Neurol Clin 10:49–61

Buchan AM, Gertler SZ, Li H, Xue D, Huang ZG, Chaundy KE, Barnes K, Lesiuk HJ (1994) A selective N-type Ca(2+)-channel blocker prevents CA1 injury 24h following severe forebrain ischemia and reduces infarction following focal ischemia. J Cereb Blood Flow Metab 14:903–910

Buchan AM, Xue D, Huang ZG, Smith KH, Lesiuk H (1991) Delayed AMPA receptor blockade reduces cerebral infarction induced by focal ischemia. Neuroreport 2:473–476

Buisson A, Choi DW (1995) The inhibitory mGluR agonist, S-4-carboxy-3-hydroxy-phenylglycine selectively attenuates NMDA neurotoxicity and oxygen-glucose deprivation-induced neuronal death. Neuropharmacology 34:1081–1087

Buisson A, Yu SP, Choi DW (1996) DCG-IV selectively attenuates rapidly triggered NMDA-induced neurotoxicity in cortical neurons. Eur J Neurosci 8:138–143

Burnashev N, Monyer H, Seeburg PH, Sakmann B (1992) Divalent ion permeability of AMPA receptor channels is dominated by the edited form of a single subunit. Neuron 8:189–198

Busto R, Globus MY, Dietrich WD, Martinez E, Valdes I, Ginsberg MD (1989) Effect of mild hypothermia on ischemia-induced release of neurotransmitters and free fatty acids in rat brain. Stroke 20:904–910

Canzoniero LM, Turetsky DM, Choi DW (1999) Measurement of intracellular free zinc concentrations accompanying zinc- induced neuronal death. J Neurosci (Online) 19:RC31

Cao X, Phillis JW (1994) alpha-Phenyl-tert-butyl-nitrone reduces cortical infarct and edema in rats subjected to focal ischemia. Brain Res 644:267–272

Carter C, Rivy JP, Scatton B (1989) Ifenprodil and SL 82.0715 are antagonists at the polyamine site of the N-methyl-D-aspartate (NMDA) receptor. Eur J Pharmacol 164:611–612

Cartmell J, Schoepp DD (2000) Regulation of neurotransmitter release by metabotropic glutamate receptors. J Neurochem 75:889–907

Chan PH, Fishman RA, Longar S, Chen S, Yu A (1985) Cellular and molecular effects of polyunsaturated fatty acids in brain ischemia and injury. Prog Brain Res 63:227–235

Chen HS, Pellegrini JW, Aggarwal SK, Lei SZ, Warach S, Jensen FE, Lipton SA (1992) Open-channel block of N-methyl-D-aspartate (NMDA) responses by memantine: therapeutic advantage against NMDA receptor-mediated neurotoxicity. J Neurosci 12:4427–4436

Chen JF, Huang Z, Ma J, Zhu J, Moratalla R, Standaert D, Moskowitz MA, Fink JS, Schwarzschild MA (1999) A(2 A) adenosine receptor deficiency attenuates brain injury induced by transient focal ischemia in mice. J Neurosci 19:9192–9200

Chen Q, Olney JW, Lukasiewicz PD, Almli T, Romano C (1998) Fenamates protect neurons against ischemic and excitotoxic injury in chick embryo retina. Neurosci Lett 242:163–166

Cheng C, Fass DM, Reynolds IJ (1999) Emergence of excitotoxicity in cultured forebrain neurons coincides with larger glutamate-stimulated $[Ca(2+)](i)$ increases and NMDA receptor mRNA levels. Brain Res 849:97–108

Chiamulera C, Albertini P, Valerio E, Reggiani A (1992) Activation of metabotropic receptors has a neuroprotective effect in a rodent model of focal ischaemia. Eur J Pharmacol 216:335–336

Chihab R, Oillet J, Bossenmeyer C, Daval JL (1998) Glutamate triggers cell death specifically in mature central neurons through a necrotic process. Mol Genet Metab 63:142–147

Choi DW (1988a) Calcium-mediated neurotoxicity: relationship to specific channel types and role in ischemic damage. Trends Neurosci 11:465–469

Choi DW (1988b) Glutamate neurotoxicity and diseases of the nervous system. Neuron 1:623–634
Choi DW (1992) Excitotoxic cell death. J Neurobiol 23:1261–1276
Choi DW, Maulucci-Gedde M, Kriegstein AR (1987) Glutamate neurotoxicity in cortical cell culture. J Neurosci 7:357–368
Choi DW, Yokoyama M, Koh J (1988) Zinc neurotoxicity in cortical cell culture. Neuroscience 24:67–79
Choi DW, Yu SP, Wei L, Gottron F (1998) Potassium channel blockers attenuate neuronal deaths induced by hypoxic insults in cortical culture and by transient focal ischemia in the rat. Society for Neuroscience Abstracts 24:1226
Christine CW, Choi DW (1990) Effect of zinc on NMDA receptor-mediated channel currents in cortical neurons. J Neurosci 10:108–116
Clemens JA, Panetta JA (1994) Neuroprotection by antioxidants in models of global and focal ischemia. Ann N Y Acad Sci 738:250–256
Colbourne F, Sutherland G, Corbett D (1997) Postischemic hypothermia. A critical appraisal with implications for clinical treatment. Mol Neurobiol 14:171–201
Condorelli DF, Dell'Albani P, Corsaro M, Barresi V, Giuffrida Stella AM (1993) AMPA-selective glutamate receptor subunits in astroglial cultures. J Neurosci Res 36:344–356
Conn PJ, Pin JP (1997) Pharmacology and functions of metabotropic glutamate receptors. Annu Rev Pharmacol Toxicol 37:205–237
Contractor A, Gereau RWT, Green T, Heinemann SF (1998) Direct effects of metabotropic glutamate receptor compounds on native and recombinant N-methyl-D-aspartate receptors. Proc Natl Acad Sci USA 95:8969–8974
Corradetti R, Lo Conte G, Moroni F, Passani MB, Pepeu G (1984) Adenosine decreases aspartate and glutamate release from rat hippocampal slices. Eur J Pharmacol 104:19–26
Coyle JT, Bird SJ, Evans RH, Gulley RL, Nadler JV, Nicklas WJ, Olney JW (1981) Excitatory amino acid neurotoxins: selectivity, specificity, and mechanisms of action. Neurosci Res Program Bull 19:1–427
Coyle JT, Schwarcz R (1976) Lesion of striatal neurones with kainic acid provides a model for Huntington's chorea. Nature 263:244–246
Cullen JP, Aldrete JA, Jankovsky L, Romo-Salas F (1979) Protective action of phenytoin in cerebral ischemia. Anesth Analg 58:165–169
Dawson TM, Snyder SH (1994) Gases as biological messengers: nitric oxide and carbon monoxide in the brain. J Neurosci 14:5147–5159
Dawson VL, Dawson TM, London ED, Bredt DS, Snyder SH (1991) Nitric oxide mediates glutamate neurotoxicity in primary cortical cultures. Proc Natl Acad Sci USA 88:6368–6371
Dawson VL, Kizushi VM, Huang PL, Snyder SH, Dawson TM (1996) Resistance to neurotoxicity in cortical cultures from neuronal nitric oxide synthase-deficient mice. J Neurosci 16:2479–2487
Dineley KE, Scanlon JM, Kress GJ, Stout AK, Reynolds IJ (2000) Astrocytes are more resistant than neurons to the cytotoxic effects of increased $[Zn^{2+}]_i$. Neurobiol Disease 7:310–320
Dryer SE (1994) Na(+)-activated K+ channels: a new family of large-conductance ion channels. Trends Neurosci 17:155–160
Dugan LL, Sensi SL, Canzoniero LM, Handran SD, Rothman SM, Lin TS, Goldberg MP, Choi DW (1995) Mitochondrial production of reactive oxygen species in cortical neurons following exposure to N-methyl-D-aspartate. J Neurosci 15:6377–6388
Dumuis A, Sebben M, Haynes L, Pin JP, Bockaert J (1988) NMDA receptors activate the arachidonic acid cascade system in striatal neurons. Nature 336:68–70
Dutar P, Rascol O, Lamour Y (1989) Omega-conotoxin GVIA blocks synaptic transmission in the CA1 field of the hippocampus. Eur J Pharmacol 174:261–266

Dykens JA, Stern A, Trenkner E (1987) Mechanism of kainate toxicity to cerebellar neurons in vitro is analogous to reperfusion tissue injury. J Neurochem 49: 1222–1228

Eliasson MJ, Sampei K, Mandir AS, Hurn PD, Traystman RJ, Bao J, Pieper A, Wang ZQ, Dawson TM, Snyder SH, Dawson VL (1997) Poly(ADP-ribose) polymerase gene disruption renders mice resistant to cerebral ischemia. Nat Med 3:1089–1095

Endoh M, Maiese K, Wagner J (1994) Expression of the inducible form of nitric oxide synthase by reactive astrocytes after transient global ischemia. Brain Res 651: 92–100

Endres M, Wang ZQ, Namura S, Waeber C, Moskowitz MA (1997) Ischemic brain injury is mediated by the activation of poly(ADP-ribose) polymerase. J Cereb Blood Flow Metab 17:1143–1151

Erdo SL, Michler A (1990) GABA does not protect cerebro-cortical cultures against excitotoxic cell death. Eur J Pharmacol 182:203–206

Faden AI, Ivanova SA, Yakovlev AG, Mukhin AG (1997) Neuroprotective effects of group III mGluR in traumatic neuronal injury. J Neurotrauma 14:885–895

Fairman WA, Vandenberg RJ, Arriza JL, Kavanaugh MP, Amara SG (1995) An excitatory amino acid transporter with properties of a ligand-gated chloride channel. Nature 375:599–603

Ferrer I, Martin F, Serrano T, Reiriz J, Perez-Navarro E, Alberch J, Macaya A, Planas AM (1995) Both apoptosis and necrosis occur following intrastriatal administration of excitotoxins. Acta Neuropathol 90:504–510

Frandsen A, Schousboe A (1991) Dantrolene prevents glutamate cytotoxicity and Ca2+ release from intracellular stores in cultured cerebral cortical neurons. J Neurochem 56:1075–1078

Frederickson CJ (1989) Neurobiology of zinc and zinc-containing neurons. Int Rev Neurobiol 31:145–238

Freeman EJ, Terrian DM, Dorman RV (1990) Presynaptic facilitation of glutamate release from isolated hippocampal mossy fiber nerve endings by arachidonic acid. Neurochem Res 15:743–750

Gage PW (1992) Activation and modulation of neuronal K+ channels by GABA. Trends Neurosci 15:46–51

Gao Y, Phillis JW (1994) CGS 15943, an adenosine A2 receptor antagonist, reduces cerebral ischemic injury in the Mongolian gerbil. Life Sci 55:L61–65

Garthwaite G, Garthwaite J (1986) Neurotoxicity of excitatory amino acid receptor agonists in rat cerebellar slices: dependence on calcium concentration. Neurosci Lett 66:193–198

Goldberg MP, Monyer H, Weiss JH, Choi DW (1988) Adenosine reduces cortical neuronal injury induced by oxygen or glucose deprivation in vitro. Neurosci Lett 89:323–327

Gotti B, Duverger D, Bertin J, Carter C, Dupont R, Frost J, Gaudilliere B, MacKenzie ET, Rousseau J, Scatton B et al. (1988) Ifenprodil and SL 82.0715 as cerebral anti-ischemic agents. I. Evidence for efficacy in models of focal cerebral ischemia. J Pharmacol Exp Ther 247:1211–1221

Gottron FJ, Ying HS, Choi DW (1997) Caspase inhibition selectively reduces the apoptotic component of oxygen-glucose deprivation-induced cortical neuronal cell death. Mol Cell Neurosci 9:159–169

Graham D, Darles G, Langer SZ (1992) The neuroprotective properties of ifenprodil, a novel NMDA receptor antagonist, in neuronal cell culture toxicity studies. Eur J Pharmacol 226:373–376

Gunter TE, Pfeiffer DR (1990) Mechanisms by which mitochondria transport calcium. Am J Physiol 258:C755–786

Gwag BJ, Koh JY, DeMaro JA, Ying HS, Jacquin M, Choi DW (1997) Slowly triggered excitotoxicity occurs by necrosis in cortical cultures. Neuroscience 77:393–401

Haddad GG, Jiang C (1994) Mechanisms of neuronal survival during hypoxia: ATP-sensitive K+ channels. Biol Neonate 65:160–165

Haley EC Jr, Kassell NF, Apperson-Hansen C, Maile MH, Alves WM (1997) A randomized, double-blind, vehicle-controlled trial of tirilazad mesylate in patients with aneurysmal subarachnoid hemorrhage: a cooperative study in North America. J Neurosurg 86:467–744

Heurteaux C, Bertaina V, Widmann C, Lazdunski M (1993) K+ channel openers prevent global ischemia-induced expression of c-fos, c-jun, heat shock protein, and amyloid beta-protein precursor genes and neuronal death in rat hippocampus. Proc Natl Acad Sci USA 90:9431–9435

Hewett SJ, Csernansky CA, Choi DW (1994) Selective potentiation of NMDA-induced neuronal injury following induction of astrocytic iNOS. Neuron 13:487–494

Hewett SJ, Uliasz TF, Vidwans AS, Hewett JA (2000) Cyclooxygenase-2 contributes to N-methyl-D-aspartate-mediated neuronal cell death in primary cortical cell culture. J Pharmacol Exp Ther 293:417–425

Hollmann M, Heinemann S (1994) Cloned glutamate receptors. Annu Rev Neurosci 17:31–108

Holzwarth JA, Gibbons SJ, Brorson JR, Philipson LH, Miller RJ (1994) Glutamate receptor agonists stimulate diverse calcium responses in different types of cultured rat cortical glial cells. J Neurosci 14:1879–1891

Hood WF, Compton RP, Monahan JB (1989) D-cycloserine: a ligand for the N-methyl-D-aspartate coupled glycine receptor has partial agonist characteristics. Neurosci Lett 98:91–95

Hoyt KR, Arden SR, Aizenman E, Reynolds IJ (1998) Reverse Na+/Ca2+ exchange contributes to glutamate-induced intracellular Ca2+ concentration increases in cultured rat forebrain neurons. Mol Pharmacol 53:742–749

Huang Z, Huang PL, Panahian N, Dalkara T, Fishman MC, Moskowitz MA (1994) Effects of cerebral ischemia in mice deficient in neuronal nitric oxide synthase. Science 265:1883–1885

Iadecola C, Zhang F, Casey R, Clark HB, Ross ME (1996) Inducible nitric oxide synthase gene expression in vascular cells after transient focal cerebral ischemia. Stroke 27:1373–1380

Iadecola C, Zhang F, Xu S, Casey R, Ross ME (1995a) Inducible nitric oxide synthase gene expression in brain following cerebral ischemia. J Cereb Blood Flow Metab 15:378–384

Iadecola C, Zhang F, Xu X (1995b) Inhibition of inducible nitric oxide synthase ameliorates cerebral ischemic damage. Am J Physiol 268:R286–292

Ikonomidou C, Bittigau P, Ishimaru MJ, Wozniak DF, Koch C, Genz K, Price MT, Stefovska V, Horster F, Tenkova T, Dikranian K, Olney JW (2000) Ethanol-induced apoptotic neurodegeneration and fetal alcohol syndrome. Science 287:1056–1060

Ikonomidou C, Bosch F, Miksa M, Bittigau P, Vockler J, Dikranian K, Tenkova TI, Stefovska V, Turski L, Olney JW (1999) Blockade of NMDA receptors and apoptotic neurodegeneration in the developing brain. Science 283:70–74

Iwamoto T, Watano T, Shigekawa M (1996) A novel isothiourea derivative selectively inhibits the reverse mode of Na+/Ca2+ exchange in cells expressing NCX1. J Biol Chem 271:22391–22397

Jabaudon D, Scanziani M, Gahwiler BH, Gerber U (2000) Acute decrease in net glutamate uptake during energy deprivation. Proc Natl Acad Sci USA 97:5610–5615

Jackson-Friedman C, Lyden PD, Nunez J, Jin A, Zweifler R (1997) High dose baclofen is neuroprotective but also causes intracerebral hemorrhage: a quantal bioassay study using the intraluminal suture occlusion method. Exp Neurol 147:346–352

Johnston GA (1996) GABAc receptors: relatively simple transmitter -gated ion channels? Trends Pharmacol Sci 17:319–323

Jonas P, Racca C, Sakmann B, Seeburg PH, Monyer H (1994) Differences in Ca2+ permeability of AMPA-type glutamate receptor channels in neocortical neurons caused by differential GluR-B subunit expression. Neuron 12:1281–1289

Jones PA, Smith RA, Stone TW (1998) Protection against hippocampal kainate excitotoxicity by intracerebral administration of an adenosine A2A receptor antagonist. Brain Res 800:328–335

Kamiya H, Sawada S, Yamamoto C (1988) Synthetic omega-conotoxin blocks synaptic transmission in the hippocampus in vitro. Neurosci Lett 91:84–88

Kanner BI, Sharon I (1978) Solubilization and reconstitution of the L-glutamic acid transporter from rat brain. FEBS Lett 94:245–248

Karlsson G, Olpe HR (1989) Late inhibitory postsynaptic potentials in rat prefrontal cortex may be mediated by GABAB receptors. Experientia 45:157–158

Kassell NF, Haley EC Jr, Apperson-Hansen C, Alves WM (1996) Randomized, double-blind, vehicle-controlled trial of tirilazad mesylate in patients with aneurysmal subarachnoid hemorrhage: a cooperative study in Europe, Australia, and New Zealand. J Neurosurg 84:221–228

Kaste M, Fogelholm R, Erila T, Palomaki H, Murros K, Rissanen A, Sarna S (1994) A randomized, double-blind, placebo-controlled trial of nimodipine in acute ischemic hemispheric stroke. Stroke 25:1348–1353

Katayama Y, Becker DP, Tamura T, Hovda DA (1990) Massive increases in extracellular potassium and the indiscriminate release of glutamate following concussive brain injury. J Neurosurg 73:889–900

Kemp JA, Kew JNC, Gill R (1999) NMDA receptor antagonists and their potential as neuroprotective agents. In: Jonas P, Monyer H (eds) Ionotropic glutamate receptors in the CNS. Springer, Berlin Heidelberg New York, pp 495–527

Kerchner GA, Canzoniero LM, Yu SP, Ling C, Choi DW (2000) Zn2+ current is mediated by voltage-gated Ca2+ channels and enhanced by extracellular acidity in mouse cortical neurones. J Physiol 528:39–52

Kerchner GA, Kim AH, Choi DW (1999) Glutamate-mediated excitotoxicity. In: Jonas P, Monyer H (eds) Ionotropic glutamate receptors in the CNS. Springer, Berlin Heidelberg New York, pp 443–469

Kew JN, Trube G, Kemp JA (1996) A novel mechanism of activity-dependent NMDA receptor antagonism describes the effect of ifenprodil in rat cultured cortical neurones. J Physiol (Lond) 497:761–772

Kim EY, Koh JY, Kim YH, Sohn S, Joe E, Gwag BJ (1999a) Zn2+ entry produces oxidative neuronal necrosis in cortical cell cultures. Eur J Neurosci 11:327–334

Kim YH, Kim EY, Gwag BJ, Sohn S, Koh JY (1999b) Zinc-induced cortical neuronal death with features of apoptosis and necrosis: mediation by free radicals. Neuroscience 89:175–182

Klotz KN (2000) Adenosine receptors and their ligands. Naunyn Schmiedebergs Arch Pharmacol 362:382–391

Kobayashi T, Mori Y (1998) Ca2+ channel antagonists and neuroprotection from cerebral ischemia. Eur J Pharmacol 363:1–15

Koh JY, Choi DW (1987) Effect of anticonvulsant drugs on glutamate neurotoxicity in cortical cell culture. Neurology 37:319–322

Koh JY, Choi DW (1988) Vulnerability of cultured cortical neurons to damage by excitotoxins: differential susceptibility of neurons containing NADPH-diaphorase. J Neurosci 8:2153–2163

Koh JY, Choi DW (1991) Selective blockade of non-NMDA receptors does not block rapidly triggered glutamate-induced neuronal death. Brain Res 548:318–321

Koh JY, Palmer E, Cotman CW (1991) Activation of the metabotropic glutamate receptor attenuates N-methyl-D-aspartate neurotoxicity in cortical cultures. Proc Natl Acad Sci USA 88:9431–9435

Koh JY, Suh SW, Gwag BJ, He YY, Hsu CY, Choi DW (1996) The role of zinc in selective neuronal death after transient global cerebral ischemia. Science 272:1013–1016

Kure S, Tominaga T, Yoshimoto T, Tada K, Narisawa K (1991) Glutamate triggers internucleosomal DNA cleavage in neuronal cells. Biochem Biophys Res Commun 179:39–45

Lam AG, Soriano MA, Monn JA, Schoepp DD, Lodge D, McCulloch J (1998) Effects of the selective metabotropic glutamate agonist LY354740 in a rat model of permanent ischaemia. Neurosci Lett 254:121–123

Le Peillet E, Arvin B, Moncada C, Meldrum BS (1992) The non-NMDA antagonists, NBQX and GYKI 52466, protect against cortical and striatal cell loss following transient global ischaemia in the rat. Brain Res 571:115–120

Lee JM, Zipfel GJ, Choi DW (1999) The changing landscape of ischaemic brain injury mechanisms. Nature 399:A7–14

Lee KS, Frank S, Vanderklish P, Arai A, Lynch G (1991) Inhibition of proteolysis protects hippocampal neurons from ischemia. Proc Natl Acad Sci USA 88:7233–7237

Legendre P, Westbrook GL (1990) The inhibition of single N-methyl-D-aspartate-activated channels by zinc ions on cultured rat neurones. J Physiol (Lond) 429:429–449

Leonhardt W (1949) Über Rauschzustände bei Patherpilzvergiftung. Nervenarzt 20: 181–185

Linnik MD, Zobrist RH, Hatfield MD (1993) Evidence supporting a role for programmed cell death in focal cerebral ischemia in rats. Stroke 24:2002–2008

Lobner D, Canzoniero LM, Manzerra P, Gottron F, Ying H, Knudson M, Tian M, Dugan LL, Kerchner GA, Sheline CT, Korsmeyer SJ, Choi DW (2000) Zinc-induced neuronal death in cortical neurons. Cell Mol Biol (Noisy-Le-Grand) 46:797–806

Lucas DR, Newhouse JP (1957) The toxic effects of sodium L-glutamate on the inner layers of the retina. Arch Ophthalmol 58:193–201

Ludolph AC, Hugon J, Dwivedi MP, Schaumburg HH, Spencer PS (1987) Studies on the aetiology and pathogenesis of motor neuron diseases. 1. Lathyrism: clinical findings in established cases. Brain 110:149–615

Ludolph AC, Meyer T, Riepe MW (2000) The role of excitotoxicity in ALS–what is the evidence? J Neurol 247 Suppl 1:I7–16

Lyden PD, Hedges B (1992) Protective effect of synaptic inhibition during cerebral ischemia in rats and rabbits. Stroke 23:1463–1469

Lynch JJ, 3rd Yu SP, Canzoniero LM, Sensi SL, Choi DW (1995) Sodium channel blockers reduce oxygen-glucose deprivation-induced cortical neuronal injury when combined with glutamate receptor antagonists. J Pharmacol Exp Ther 273:554–560

Lysko PG, Webb CL, Yue TL, Gu JL, Feuerstein G (1994) Neuroprotective effects of tetrodotoxin as a Na+ channel modulator and glutamate release inhibitor in cultured rat cerebellar neurons and in gerbil global brain ischemia. Stroke 25: 2476–2482

MacManus JP, Buchan AM, Hill IE, Rasquinha I, Preston E (1993) Global ischemia can cause DNA fragmentation indicative of apoptosis in rat brain. Neurosci Lett 164:89–92

Madden KP (1994) Effect of gamma-aminobutyric acid modulation on neuronal ischemia in rabbits. Stroke 25:2271–2274

Magistretti PJ (2000) Cellular basis of functional brain imaging: insights from neuron-glia metabolic coupling. Brain Res 886:108–112

Mager R, Ferroni S, Schubert P (1990) Adenosine modulates a voltage-dependent chloride conductance in cultured hippocampal neurons. Brain Res 532:58–62

Mandir AS, Poitras MF, Berliner AR, Herring WJ, Guastella DB, Feldman A, Poirier GG, Wang ZQ, Dawson TM, Dawson VL (2000) NMDA but not non-NMDA excitotoxicity is mediated by Poly(ADP-ribose) polymerase. J Neurosci 20:8005–8011

Manev H, Kharlamov E, Uz T, Mason RP, Cagnoli CM (1997) Characterization of zinc-induced neuronal death in primary cultures of rat cerebellar granule cells. Exp Neurol 146:171–178

Manzerra P, Behrens MM, Canzoniero LMT, Wang XQ, Heidinger V, Ichinose T, Yu SP, Choi DW (2001) Zinc induces a Src family kinase-mediated upregulation of NMDA receptor activity and excitotoxicity. Proc Natl Acad Sci 98:11055–11061

Marion DW, Penrod LE, Kelsey SF, Obrist WD, Kochanek PM, Palmer AM, Wisniewski SR, DeKosky ST (1997) Treatment of traumatic brain injury with moderate hypothermia. N Engl J Med 336:540–546

Markgraf CG, Velayo NL, Johnson MP, McCarty DR, Medhi S, Koehl JR, Chmielewski PA, Linnik MD (1998) Six-hour window of opportunity for calpain inhibition in focal cerebral ischemia in rats. Stroke 29:152–158

Martin LJ, Blackstone CD, Levey AI, Huganir RL, Price DL (1993) AMPA glutamate receptor subunits are differentially distributed in rat brain. Neuroscience 53: 327–358

Martinez-Guijarro FJ, Soriano E, Del Rio JA, Lopez-Garcia C (1991) Zinc-positive boutons in the cerebral cortex of lizards show glutamate immunoreactivity. J Neurocytol 20:834–843

Mason BA, Standley CA, Irtenkauf SM, Bardicef M, Cotton DB (1994) Magnesium is more efficacious than phenytoin in reducing N-methyl-D-aspartate seizures in rats. Am J Obstet Gynecol 171:999–1002

Mattson MP, Kater SB (1989) Excitatory and inhibitory neurotransmitters in the generation and degeneration of hippocampal neuroarchitecture. Brain Res 478: 337–348

Matute C, Gutierrez-Igarza K, Rio C, Miledi R (1994) Glutamate receptors in astrocytic end-feet. Neuroreport 5:1205–1208

Matute C, Sanchez-Gomez MV, Martinez-Millan L, Miledi R (1997) Glutamate receptor-mediated toxicity in optic nerve oligodendrocytes. Proc Natl Acad Sci USA 94: 8830–8835

Mayer ML, Westbrook GL (1987) The physiology of excitatory amino acids in the vertebrate central nervous system. Prog Neurobiol 28:197–276

McCulloch J (1992) Excitatory amino acid antagonists and their potential for the treatment of ischaemic brain damage in man. Br J Clin Pharmacol 34:106–114

McDonald JW, Althomsons SP, Hyrc KL, Choi DW, Goldberg MP (1998a) Oligodendrocytes from forebrain are highly vulnerable to AMPA/kainate receptor-mediated excitotoxicity. Nat Med 4:291–297

McDonald JW, Behrens MI, Chung C, Bhattacharyya T, Choi DW (1997) Susceptibility to apoptosis is enhanced in immature cortical neurons. Brain Res 759:228–232

McDonald JW, Bhattacharyya T, Sensi SL, Lobner D, Ying HS, Canzoniero LM, Choi DW (1998b) Extracellular acidity potentiates AMPA receptor-mediated cortical neuronal death. J Neurosci 18:6290–6299

McGeer PL, McGeer EG (1978) Intracerebral injections of kainic acid and tetanus toxin: possible models for the signs of chorea and dystonia. Adv Neurol 21:331–338

McIntosh TK, Vink R, Soares H, Hayes R, Simon R (1989) Effects of the N-methyl-D-aspartate receptor blocker MK-801 on neurologic function after experimental brain injury. J Neurotrauma 6:247–259

Meldrum BS (1994) The role of glutamate in epilepsy and other CNS disorders. Neurology 44:S14–23

Middlemiss DN, Spedding M (1985) A functional correlate for the dihydropyridine binding site in rat brain. Nature 314:94–96

Miljanich GP, Ramachandran J (1995) Antagonists of neuronal calcium channels: structure, function, and therapeutic implications. Annu Rev Pharmacol Toxicol 35:707–734

Miyamoto M, Coyle JT (1990) Idebenone attenuates neuronal degeneration induced by intrastriatal injection of excitotoxins. Exp Neurol 108:38–45

Monyer H, Burnashev N, Laurie DJ, Sakmann B, Seeburg PH (1994) Developmental and regional expression in the rat brain and functional properties of four NMDA receptors. Neuron 12:529–540

Monyer H, Hartley DM, Choi DW (1990) 21-Aminosteroids attenuate excitotoxic neuronal injury in cortical cell cultures. Neuron 5:121–126

Mothet JP, Parent AT, Wolosker H, Brady RO Jr, Linden DJ, Ferris CD, Rogawski MA, Snyder SH (2000) D-serine is an endogenous ligand for the glycine site of the N-methyl-D-aspartate receptor. Proc Natl Acad Sci USA 97:4926–4931

Muir JK, Lobner D, Monyer H, Choi DW (1996) GABAA receptor activation attenuates excitotoxicity but exacerbates oxygen-glucose deprivation-induced neuronal injury in vitro. J Cereb Blood Flow Metab 16:1211–1218

Mukhin A, Fan L, Faden AI (1996) Activation of metabotropic glutamate receptor subtype mGluR1 contributes to post-traumatic neuronal injury. J Neurosci 16: 6012–6020

Murphy JJ (1992) The role of calcium antagonists in the treatment of cerebrovascular disease. Drugs Aging 2:1–6

Nakanishi S, Masu M (1994) Molecular diversity and functions of glutamate receptors. Annu Rev Biophys Biomol Struct 23:319–348

Nelson KB, Grether JK (1995) Can magnesium sulfate reduce the risk of cerebral palsy in very low birthweight infants? Pediatrics 95:263–269

Nicholls D, Attwell D (1990) The release and uptake of excitatory amino acids. Trends Pharmacol Sci 11:462–468

Nicholls DG, Ward MW (2000) Mitochondrial membrane potential and neuronal glutamate excitotoxicity: mortality and millivolts. Trends Neurosci 23:166–174

Nogawa S, Zhang F, Ross ME, Iadecola C (1997) Cyclo-oxygenase-2 gene expression in neurons contributes to ischemic brain damage. J Neurosci 17:2746–2755

Novelli A, Reilly JA, Lysko PG, Henneberry RC (1988) Glutamate becomes neurotoxic via the N-methyl-D-aspartate receptor when intracellular energy levels are reduced. Brain Res 451:205–212

Nowak L, Bregestovski P, Ascher P, Herbet A, Prochiantz A (1984) Magnesium gates glutamate-activated channels in mouse central neurones. Nature 307:462–465

O'Regan MH, Simpson RE, Perkins LM, Phillis JW (1992) The selective A2 adenosine receptor agonist CGS 21680 enhances excitatory transmitter amino acid release from the ischemic rat cerebral cortex. Neurosci Lett 138:169–172

Ogura A, Miyamoto M, Kudo Y (1988) Neuronal death in vitro: parallelism between survivability of hippocampal neurones and sustained elevation of cytosolic Ca2+ after exposure to glutamate receptor agonist. Exp Brain Res 73:447–458

Oka A, Belliveau MJ, Rosenberg PA, Volpe JJ (1993) Vulnerability of oligodendroglia to glutamate: pharmacology, mechanisms, and prevention. J Neurosci 13:1441–1453

Olah ME, Ren H, Stiles GL (1995) Adenosine receptors: protein and gene structure. Arch Int Pharmacodyn Ther 329:135–150

Olney JW (1969) Brain lesion, obesity and other disturbances in mice treated with monosodium glutamate. Science 164:719–721

Olney JW, Labruyere J, Price MT (1989) Pathological changes induced in cerebrocortical neurons by phencyclidine and related drugs. Science 244:1360–1362

Olverman HJ, Jones AW, Watkins JC (1984) L-glutamate has higher affinity than other amino acids for [3H]-D-AP5 binding sites in rat brain membranes. Nature 307:460–462

Palmer GC, Widzowski D (2000) Low affinity use-dependent NMDA receptor antagonists show promise for clinical development. Amino Acids 19:151–155

Paschen W, Widmann R, Weber C (1992) Changes in regional polyamine profiles in rat brains after transient cerebral ischemia (single versus repetitive ischemia): evidence for release of polyamines from injured neurons. Neurosci Lett 135:121–124

Pellegrini-Giampietro DE, Zukin RS, Bennett MV, Cho S, Pulsinelli WA (1992) Switch in glutamate receptor subunit gene expression in CA1 subfield of hippocampus following global ischemia in rats. Proc Natl Acad Sci USA 89:10499–10503

Pelletier JC, Hesson DP, Jones KA, Costa AM (1996) Substituted 1,2-dihydrophthalazines: potent, selective, and noncompetitive inhibitors of the AMPA receptor. J Med Chem 39:343–346

Perez-Clausell J, Danscher G (1985) Intravesicular localization of zinc in rat telencephalic boutons. A histochemical study. Brain Res 337:91–98

Perez-Reyes E, Schneider T (1995) Molecular biology of calcium channels. Kidney Int 48:1111–1124

Perl TM, Bedard L, Kosatsky T, Hockin JC, Todd EC, Remis RS (1990) An outbreak of toxic encephalopathy caused by eating mussels contaminated with domoic acid. N Engl J Med 322:1775–1780

Peters S, Koh J, Choi DW (1987) Zinc selectively blocks the action of N-methyl-D-aspartate on cortical neurons. Science 236:589–593

Phillis JW (1995) CI-966, a GABA uptake inhibitor, antagonizes ischemia-induced neuronal degeneration in the gerbil. Gen Pharmacol 26:1061–1064

Pizzi M, Consolandi O, Memo M, Spano PF (1996) Activation of multiple metabotropic glutamate receptor subtypes prevents NMDA-induced excitotoxicity in rat hippocampal slices. Eur J Neurosci 8:1516–1521

Portera-Cailliau C, Hedreen JC, Price DL, Koliatsos VE (1995) Evidence for apoptotic cell death in Huntington disease and excitotoxic animal models. J Neurosci 15: 3775–3787

Posner A, Raser KJ, Hajimohammadreza I, Yuen PW, Wang KK (1995) Aurintricarboxylic acid is an inhibitor of mu- and m-calpain. Biochem Mol Biol Int 36:291–299

Pratt J, Rataud J, Bardot F, Roux M, Blanchard JC, Laduron PM, Stutzmann JM (1992) Neuroprotective actions of riluzole in rodent models of global and focal cerebral ischaemia. Neurosci Lett 140:225–230

Puchalski RB, Louis JC, Brose N, Traynelis SF, Egebjerg J, Kukekov V, Wenthold RJ, Rogers SW, Lin F, Moran T et al. (1994) Selective RNA editing and subunit assembly of native glutamate receptors. Neuron 13:131–147

Qin ZH, Wang Y, Chase TN (1996) Stimulation of N-methyl-D-aspartate receptors induces apoptosis in rat brain. Brain Res 725:166–176

Ransom RW, Stec NL (1988) Cooperative modulation of [3H]MK-801 binding to the N-methyl-D- aspartate receptor-ion channel complex by L-glutamate, glycine, and polyamines. J Neurochem 51:830–836

RANTTAS Investigators (1996) A randomized trial of tirilazad mesylate in patients with acute stroke (RANTTAS). Stroke 27:1453–8

Read SJ, Hirano T, Davis SM, Donnan GA (1999) Limiting neurological damage after stroke: a review of pharmacological treatment options. Drugs Aging 14:11–39

Reynolds IJ, Hastings TG (1995) Glutamate induces the production of reactive oxygen species in cultured forebrain neurons following NMDA receptor activation. J Neurosci 15:3318–3327

Reynolds IJ, Miller RJ (1989) Ifenprodil is a novel type of N-methyl-D-aspartate receptor antagonist: interaction with polyamines. Mol Pharmacol 36:758–765

Ribeiro JA (1995) Purinergic inhibition of neurotransmitter release in the central nervous system. Pharmacol Toxicol 77:299–305

Roberts-Lewis JM, Marcy VR, Zhao Y, Vaught JL, Siman R, Lewis ME (1993) Aurintricarboxylic acid protects hippocampal neurons from NMDA- and ischemia-induced toxicity in vivo. J Neurochem 61:378–381

Rogawski MA (1993) Therapeutic potential of excitatory amino acid antagonists: channel blockers and 2,3-benzodiazepines. Trends Pharmacol Sci 14:325–331

Rosenbaum D, Zabramski J, Frey J, Yatsu F, Marler J, Spetzler R, Grotta J (1991) Early treatment of ischemic stroke with a calcium antagonist. Stroke 22:437–441

Rosenbaum DM, Grotta JC, Pettigrew LC, Ostrow P, Strong R, Rhoades H, Picone CM, Grotta AT (1990) Baclofen does not protect against cerebral ischemia in rats. Stroke 21:138–140

Rossi DJ, Oshima T, Attwell D (2000) Glutamate release in severe brain ischaemia is mainly by reversed uptake. Nature 403:316–321

Rothman S (1984) Synaptic release of excitatory amino acid neurotransmitter mediates anoxic neuronal death. J Neurosci 4:1884–1891

Rothman SM, Olney JW (1986) Glutamate and the pathophysiology of hypoxic–ischemic brain damage. Ann Neurol 19:105–111

Rothman SM, Thurston JH, Hauhart RE (1987) Delayed neurotoxicity of excitatory amino acids in vitro. Neuroscience 22:471–480

Rothstein JD, Martin LJ, Kuncl RW (1992) Decreased glutamate transport by the brain and spinal cord in amyotrophic lateral sclerosis. N Engl J Med 326:1464–1468

Sah P (1996) Ca(2+)-activated K+ currents in neurones: types, physiological roles and modulation. Trends Neurosci 19:150–154

Sarnesto A, Linder N, Raivio KO (1996) Organ distribution and molecular forms of human xanthine dehydrogenase/xanthine oxidase protein. Lab Invest 74:48–56

Schell MJ, Brady RO Jr, Molliver ME, Snyder SH (1997) D-serine as a neuromodulator: regional and developmental localizations in rat brain glia resemble NMDA receptors. J Neurosci 17:1604–1615

Schell MJ, Molliver ME, Snyder SH (1995) D-serine, an endogenous synaptic modulator: localization to astrocytes and glutamate-stimulated release. Proc Natl Acad Sci USA 92:3948–3952

Schinder AF, Olson EC, Spitzer NC, Montal M (1996) Mitochondrial dysfunction is a primary event in glutamate neurotoxicity. J Neurosci 16:6125–6133

Schmitz D, Frerking M, Nicoll RA (2000) Synaptic activation of presynaptic kainate receptors on hippocampal mossy fiber synapses. Neuron 27:327–338

Schwartz-Bloom RD, McDonough KJ, Chase PJ, Chadwick LE, Inglefield JR, Levin ED (1998) Long-term neuroprotection by benzodiazepine full versus partial agonists after transient cerebral ischemia in the gerbil. J Cereb Blood Flow Metab 18:548–558

Sensi SL, Canzoniero LM, Yu SP, Ying HS, Koh JY, Kerchner GA, Choi DW (1997) Measurement of intracellular free zinc in living cortical neurons: routes of entry. J Neurosci 17:9554–9564

Sensi SL, Yin HZ, Carriedo SG, Rao SS, Weiss JH (1999) Preferential Zn^{2+} influx through Ca^{2+}-permeable AMPA/kainate channels triggers prolonged mitochondrial superoxide production. Proc Natl Acad Sci USA 96:2414–2419

Shaw PJ, Ince PG (1997) Glutamate, excitotoxicity and amyotrophic lateral sclerosis. J Neurol 244 Suppl 2:S3–14

Sheardown MJ, Nielsen EO, Hansen AJ, Jacobsen P, Honore T (1990) 2,3-Dihydroxy-6-nitro-7-sulfamoyl-benzo(F)quinoxaline: a neuroprotectant for cerebral ischemia. Science 247:571–574

Sheline CT, Behrens MM, Choi DW (2000) Zinc-induced cortical neuronal death: contribution of energy failure attributable to loss of NAD(+) and inhibition of glycolysis. J Neurosci 20:3139–3146

Shuaib A, Ijaz S, Hasan S, Kalra J (1992) Gamma-vinyl GABA prevents hippocampal and substantia nigra reticulata damage in repetitive transient forebrain ischemia. Brain Res 590:13–17

Shuaib A, Mazagri R, Ijaz S (1993) GABA agonist "muscimol" is neuroprotective in repetitive transient forebrain ischemia in gerbils. Exp Neurol 123:284–288

Simon RP, Swan JH, Griffiths T, Meldrum BS (1984) Blockade of N-methyl-D-aspartate receptors may protect against ischemic damage in the brain. Science 226:850–852

Simpson RE, O'Regan MH, Perkins LM, Phillis JW (1992) Excitatory transmitter amino acid release from the ischemic rat cerebral cortex: effects of adenosine receptor agonists and antagonists. J Neurochem 58:1683–1690

Sladeczek F, Momiyama A, Takahashi T (1993) Presynaptic inhibitory action of a metabotropic glutamate receptor agonist on excitatory transmission in visual cortical neurons. Proc R Soc Lond B Biol Sci 253:297–303

Smith SE, Meldrum BS (1992) Cerebroprotective effect of a non-N-methyl-D-aspartate antagonist, GYKI 52466, after focal ischemia in the rat. Stroke 23:861–864

Snyder, SH Kim, PM (2000) D-amino acids as putative neurotransmitters: focus on D-serine. Neurochem Res 25:553–560

Spiridon M, Kamm D, Billups B, Mobbs P, Attwell D (1998) Modulation by zinc of the glutamate transporters in glial cells and cones isolated from the tiger salamander retina. J Physiol (Lond) 506:363–376

Steen PA, Milde JH, Michenfelder JD (1980) The detrimental effects of prolonged hypothermia and rewarming in the dog. Anesthesiology 52:224–230

Steen PA, Soule FH, Michenfelder JD (1979) Deterimental effect of prolonged hypothermia in cats and monkeys with and without regional cerebral ischemia. Stroke 10:522–529

Stefani A, Pisani A, Mercuri NB, Calabresi P (1996) The modulation of calcium currents by the activation of mGluRs. Functional implications. Mol Neurobiol 13:81–95

Sternau LL, Lust WD, Ricci AJ, Ratcheson R (1989) Role for gamma-aminobutyric acid in selective vulnerability in gerbils. Stroke 20:281–287

Stout AK, Raphael HM, Kanterewicz BI, Klann E, Reynolds IJ (1998) Glutamate-induced neuron death requires mitochondrial calcium uptake. Nat Neurosci 1: 366–373

Strasser U, Lobner D, Behrens MM, Canzoniero LM, Choi DW (1998) Antagonists for group I mGluRs attenuate excitotoxic neuronal death in cortical cultures. Eur J Neurosci 10:2848–2855

Stys PK, Waxman SG, Ransom BR (1991) Reverse operation of the Na(+)-Ca2+ exchanger mediates Ca2+ influx during anoxia in mammalian CNS white matter. Ann NY Acad Sci 639:328–332

Suzdak PD, Jansen JA (1995) A review of the preclinical pharmacology of tiagabine: a potent and selective anticonvulsant GABA uptake inhibitor. Epilepsia 36:612–626

Szabo C, Dawson VL (1998) Role of poly(ADP-ribose) synthetase in inflammation and ischaemia-reperfusion. Trends Pharmacol Sci 19:287–298

Szatkowski M, Attwell D (1994) Triggering and execution of neuronal death in brain ischaemia: two phases of glutamate release by different mechanisms. Trends Neurosci 17:359–365

Szatkowski M, Barbour B, Attwell D (1990) Non-vesicular release of glutamate from glial cells by reversed electrogenic glutamate uptake. Nature 348:443–446

Taft WC, Clifton GL, Blair RE, DeLorenzo RJ (1989) Phenytoin protects against ischemia-produced neuronal cell death. Brain Res 483:143–148

Takaba H, Nagao T, Yao H, Kitazono T, Ibayashi S, Fujishima M (1997) An ATP-sensitive potassium channel activator reduces infarct volume in focal cerebral ischemia in rats. Am J Physiol 273:R583–586

Takahashi K, Greenberg JH, Jackson P, Maclin K, Zhang J (1997) Neuroprotective effects of inhibiting poly(ADP-ribose) synthetase on focal cerebral ischemia in rats. J Cereb Blood Flow Metab 17:1137–1142

Tang CM, Dichter M, Morad M (1990) Modulation of the N-methyl-D-aspartate channel by extracellular H+. Proc Natl Acad Sci USA 87:6445–6449

Teitelbaum JS, Zatorre RJ, Carpenter S, Gendron D, Evans AC, Gjedde A, Cashman NR (1990) Neurologic sequelae of domoic acid intoxication due to the ingestion of contaminated mussels. N Engl J Med 322:1781–1787

Timm F, Neth R (1959) Die normalen Schwermetalle der niere. Histochemie 1:403–419

Tommasino C, Picozzi P (1998) Mild hypothermia. J Neurosurg Sci 42:37–38

Tonder N, Johansen FF, Frederickson CJ, Zimmer J, Diemer NH (1990) Possible role of zinc in the selective degeneration of dentate hilar neurons after cerebral ischemia in the adult rat. Neurosci Lett 109:247–252

Traynelis SF, Cull-Candy SG (1990) Proton inhibition of N-methyl-D-aspartate receptors in cerebellar neurons. Nature 345:347–350

Turetsky DM, Canzoniero LMT, Choi DW (1998) Kainate-induced toxicity in cultured neocortical neurons is reduced by the AMPA receptor selective antagonist SYM 2206. Soc Neurosci Abstr 24:578

Turetsky DM, Canzoniero LM, Sensi SL, Weiss JH, Goldberg MP, Choi DW (1994) Cortical neurones exhibiting kainate-activated Co2+ uptake are selectively vulnerable to AMPA/kainate receptor-mediated toxicity. Neurobiol Dis 1:101–110

Tymianski M, Wallace MC, Spigelman I, Uno M, Carlen PL, Tator CH, Charlton MP (1993) Cell-permeant Ca2+ chelators reduce early excitotoxic and ischemic neuronal injury in vitro and in vivo. Neuron 11:221–235

Valentino K, Newcomb R, Gadbois T, Singh T, Bowersox S, Bitner S, Justice A, Yamashiro D, Hoffman BB, Ciaranello R et al. (1993) A selective N-type calcium channel antagonist protects against neuronal loss after global cerebral ischemia. Proc Natl Acad Sci USA 90:7894–7897

Van Den Bosch L, Robberecht W (2000) Different receptors mediate motor neuron death induced by short and long exposures to excitotoxicity. Brain Res Bull 53: 383–388

Vandenberg RJ (1998) Molecular pharmacology and physiology of glutamate transporters in the central nervous system. Clin Exp Pharmacol Physiol 25:393–400

Vandenberg RJ, Mitrovic AD, Johnston GA (1998) Molecular basis for differential inhibition of glutamate transporter subtypes by zinc ions. Mol Pharmacol 54: 189–196

Vink R, Cernak I (2000) Regulation of intracellular free magnesium in central nervous system injury. Front Biosci 5:D656–665

Von Lubitz DK (1999) Adenosine and cerebral ischemia: therapeutic future or death of a brave concept? Eur J Pharmacol 365:9–25

Von Lubitz DK, Beenhakker M, Lin RC, Carter MF, Paul IA, Bischofberger N, Jacobson KA (1996a) Reduction of postischemic brain damage and memory deficits following treatment with the selective adenosine A1 receptor agonist. Eur J Pharmacol 302:43–48

von Lubitz DK, Carter MF, Beenhakker M, Lin RC, Jacobson KA (1995) Adenosine: a prototherapeutic concept in neurodegeneration. Ann NY Acad Sci 765:163–178

Von Lubitz DK, Lin RC, Jacobson KA (1995) Cerebral ischemia in gerbils: effects of acute and chronic treatment with adenosine A2 A receptor agonist and antagonist. Eur J Pharmacol 287:295–302

Von Lubitz DK, Lin RC, Paul IA, Beenhakker M, Boyd M, Bischofberger N, Jacobson KA (1996b) Postischemic administration of adenosine amine congener (ADAC): analysis of recovery in gerbils. Eur J Pharmacol 316:171–179

Wadiche JI, Amara SG, Kavanaugh MP (1995) Ion fluxes associated with excitatory amino transport. Neuron 15:721–728

Wang GJ, Thayer SA (1996) Sequestration of glutamate-induced Ca2+ loads by mitochondria in cultured rat hippocampal neurons. J Neurophysiol 76:1611–1621

Watanabe M, Inoue Y, Sakimura K, Mishina M (1993) Distinct distributions of five N-methyl-D-aspartate receptor channel subunit mRNAs in the forebrain. J Comp Neurol 338:377–390

Watkins JC, Davies J, Evans RH, Francis AA, Jones AW (1981) Pharmacology of receptors for excitatory amino acids. Adv Biochem Psychopharmacol 27:263–273

Wei EP, Ellison MD, Kontos HA, Povlishock JT (1986) O2 radicals in arachidonate-induced increased blood-brain barrier permeability to proteins. Am J Physiol 251: H693–699

Weiss JH, Hartley DM, Koh J, Choi DW (1990) The calcium channel blocker nifedipine attenuates slow excitatory amino acid neurotoxicity. Science 247:1474–1477

Weiss JH, Hartley DM, Koh JY, Choi DW (1993) AMPA receptor activation potentiates zinc neurotoxicity. Neuron 10:43–49

Weiss JH, Sensi SL (2000) Ca2+-Zn2+ permeable AMPA or kainate receptors: possible key factors in selective neurodegeneration. Trends Neurosci 23:365–371

Westbrook GL, Mayer ML (1987) Micromolar concentrations of Zn2+ antagonize NMDA and GABA responses of hippocampal neurons. Nature 328:640–643

Wheeler DB, Randall A, Tsien RW (1994) Roles of N-type and Q-type Ca2+ channels in supporting hippocampal synaptic transmission. Science 264:107–111

White RJ, Reynolds IJ (1996) Mitochondrial depolarization in glutamate-stimulated neurons: an early signal specific to excitotoxin exposure. J Neurosci 16:5688–5697

Wieloch T (1985) Hypoglycemia-induced neuronal damage prevented by an N-methyl-D-aspartate antagonist. Science 230:681–683

Williams K (1993) Ifenprodil discriminates subtypes of the N-methyl-D-aspartate receptor: selectivity and mechanisms at recombinant heteromeric receptors. Mol Pharmacol 44:851–859

Williams K (1995) Pharmacological properties of recombinant N-methyl-D-aspartate (NMDA) receptors containing the epsilon 4 (NR2D) subunit. Neurosci Lett 184: 181–184

Wisden W, Seeburg PH (1993) Mammalian ionotropic glutamate receptors. Curr Opin Neurobiol 3:291–298

Wolosker H, Blackshaw S, Snyder SH (1999) Serine racemase: a glial enzyme synthesizing D-serine to regulate glutamate-N-methyl-D-aspartate neurotransmission. Proc Natl Acad Sci USA 96:13409–13414

Wrathall JR, Teng YD, Choiniere D, Mundt DJ (1992) Evidence that local non-NMDA receptors contribute to functional deficits in contusive spinal cord injury. Brain Res 586:140–143

Xu XJ, Hao JX, Seiger A, Wiesenfeld-Hallin Z (1993) Systemic excitatory amino acid receptor antagonists of the alpha-amino-3-hydroxy-5-methyl-4-isoxazolepropionic acid (AMPA) receptor and of the N-methyl-D-aspartate (NMDA) receptor relieve mechanical hypersensitivity after transient spinal cord ischemia in rats. J Pharmacol Exp Ther 267:140–144

Xue D, Huang ZG, Barnes K, Lesiuk HJ, Smith KE, Buchan AM (1994) Delayed treatment with AMPA, but not NMDA, antagonists reduces neocortical infarction. J Cereb Blood Flow Metab 14:251–261

Yamasaki Y, Kogure K, Hara H, Ban H, Akaike N (1991) The possible involvement of tetrodotoxin-sensitive ion channels in ischemic neuronal damage in the rat hippocampus. Neurosci Lett 121:251–254

Yin HZ, Weiss JH (1995) Zn(2+) permeates Ca(2+) permeable AMPA/kainate channels and triggers selective neural injury. Neuroreport 6:2553–2556

Yokoyama M, Koh J, Choi DW (1986) Brief exposure to zinc is toxic to cortical neurons. Neurosci Lett 71:351–355

Yoshioka A, Hardy M, Younkin DP, Grinspan JB, Stern JL, Pleasure D (1995) Alpha-amino-3-hydroxy-5-methyl-4-isoxazolepropionate (AMPA) receptors mediate excitotoxicity in the oligodendroglial lineage. J Neurochem 64:2442–2448

Yu AC, Chan PH, Fishman RA (1986) Effects of arachidonic acid on glutamate and gamma-aminobutyric acid uptake in primary cultures of rat cerebral cortical astrocytes and neurons. J Neurochem 47:1181–1189

Yu SP, Choi DW (1997) Na(+)-Ca2+ exchange currents in cortical neurons: concomitant forward and reverse operation and effect of glutamate. Eur J Neurosci 9:1273–1281

Yu SP, Sensi SL, Canzoniero LM, Buisson A, Choi DW (1997a) Membrane-delimited modulation of NMDA currents by metabotropic glutamate receptor subtypes 1/5 in cultured mouse cortical neurons. J Physiol (Lond) 499:721–732

Yu SP, Yeh C, Strasser U, Tian M, Choi DW (1999) NMDA receptor-mediated K+ efflux and neuronal apoptosis. Science 284:336–339

Yu SP, Yeh CH, Sensi SL, Gwag BJ, Canzoniero LM, Farhangrazi ZS, Ying HS, Tian M, Dugan LL, Choi DW (1997b) Mediation of neuronal apoptosis by enhancement of outward potassium current. Science 278:114–117

Zeevalk GD, Schoepp D, Nicklas WJ (1993) Aurintricarboxylic acid prevents NMDA-mediated excitotoxicity: evidence for its action as an NMDA receptor antagonist. J Neurochem 61:386–389

Zerangue N, Kavanaugh MP (1996) Flux coupling in a neuronal glutamate transporter. Nature 383:634–637

Zhang F, Casey RM, Ross ME, Iadecola C (1996) Aminoguanidine ameliorates and L-arginine worsens brain damage from intraluminal middle cerebral artery occlusion. Stroke 27:317–323

Zhang J, Dawson VL, Dawson TM, Snyder SH (1994) Nitric oxide activation of poly(ADP-ribose) synthetase in neurotoxicity. Science 263:687–689

Zhao Q, Pahlmark K, Smith ML, Siesjo BK (1994) Delayed treatment with the spin trap alpha-phenyl-N-tert-butyl nitrone (PBN) reduces infarct size following transient middle cerebral artery occlusion in rats. Acta Physiol Scand 152:349–350

Zini S, Roisin MP, Armengaud C, Ben-Ari Y (1993) Effect of potassium channel modulators on the release of glutamate induced by ischaemic-like conditions in rat hippocampal slices. Neurosci Lett 153:202–205

Ueda Y, Obrenovitch TP, Lok SY, Sarna GS, Symon L (1992) Changes in extracellular glutamate concentration produced in the rat striatum by repeated ischemia. Stroke 23:1125–1130

CHAPTER 2
Limiting Apoptosis as a Strategy for CNS Neuroprotection

K.K.W. Wang

A. Definition of Necrosis and Apoptosis

The term necrosis is a Greek word meaning "deadness." A contemporary definition of necrosis is the sum of morphological changes indicative of cell death (Majno and Joris 1995). Necrosis usually applies to cell death that occurs to a group of cells or part of an organ in vivo, but the term has more recently been applied to cells in culture. Necrosis is the form of cell death which usually occurs when cells were injured by extreme physical stress or chemical challenges to the point that is beyond repair. As a result of the presence of massive ion influx (e.g., Na^+, Ca^{2+}), early and marked mitochondria swelling and cell swelling (oncosis) characterize necrosis (Majno and Joris 1995; Trump et al. 1997) (Fig. 1). Nonspecific DNA breakage results in the formation of chromatin fragments (in a punctuate fashion) all over the nuclei. Eventually, the nuclei become leaky and ultimately the plasma membrane ruptures. At least under in vitro conditions, necrosis is rapid and the time course of cell death is usually within 1–5-h death (Majno and Joris 1995).

On the other hand, the term apoptosis is derived from a Greek word meaning "falling leaves." The contemporary criteria for apoptosis are morphological; for example, early presence of condensed chromatin around the inner surface of the nuclear envelope. This is then followed by shrinkage of both nucleus volume and cytoplasm volume (Fig. 1). These are then followed by the breakdown of the whole cells into plasma membrane-bound apoptotic bodies, which contain cytosolic volume, and certain organelles as well as nuclear and/or chromatin fragments (like "falling leaves"). Needless to say, the strictest identification is made using high-power microscopy to electron microscopy. It was later discovered that concomitant with DNA condensation in the nuclei, DNA fragmentation at the nucleosome linkage regions form a series of DNA fragments of 180-bp intervals apart when extracted chromosomal DNA were subjected to agarose gel electrophoresis.

Apoptosis is also called programmed cell death (PCD) because it is a form of desired and programmed physiological cell death, critical for developing organisms as well as allowing a mature organism to rid itself of unwanted cells (McConkey 1988). More recently, it has been shown that unscheduled apop-

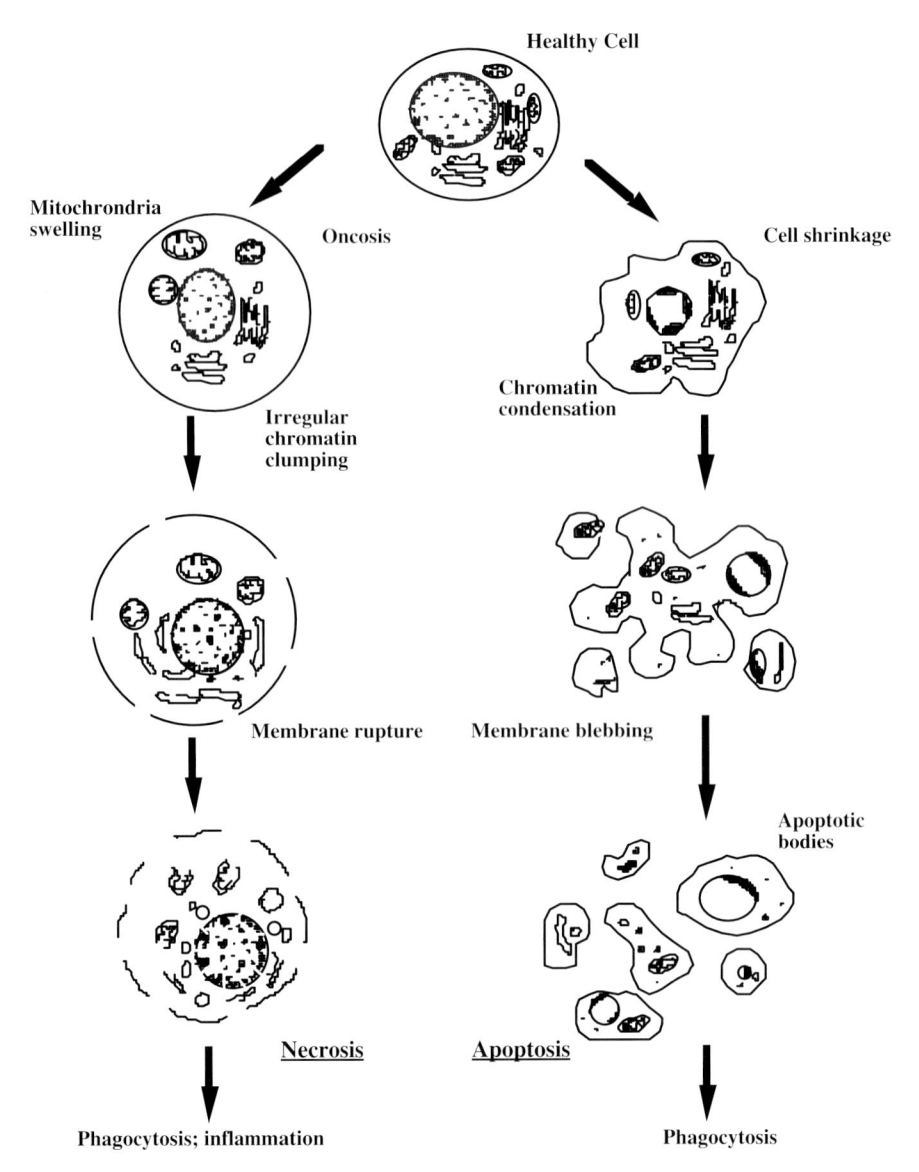

Fig. 1. Distinct features of necrosis and apoptosis. Morphological and anatomical characteristics of necrosis (*left*) and apoptosis (*right*) are illustrated

tosis could also be triggered when various mammalian cell types are under certain physical (e.g., hypoxia) or chemical challenges (e.g., toxins). Cell shrinkage occurs, resulting in the ultimate formation of apoptotic bodies. In keeping with the term PCD, in most (but not all) forms of apoptosis, new gene induction and new protein synthesis appear to be necessary. In fact, the use of transcription inhibitors (e.g., actinomycin D) and protein synthesis inhibitors (e.g., cylcohexamide) can retard or reduce apoptotic death in various situations (MCCONKEY 1988). It is important to stress, however, that this is not a fail-proof test for diagnosis of apoptosis (see Sect. II.). Under in vitro conditions, apoptosis is more delayed and the time course of cell death is usually 8–16h. There are two well-written review articles on apoptosis versus necrosis (MAJNO and JORIS 1995; MCCONKEY 1988), as well as two well-written general review articles on neuronal apoptosis (PETTMANN and HENDERSON 1998; LEIST and NICOTERA 1998).

B. Scheduled Apoptosis in the Development of the CNS

It is important to stress that apoptosis is a form of biological programming aimed at the elimination of unwanted cells at various stages of the life span of an organism. In mammals, one of the most notable manifestations of "physiological apoptosis" is in fact the development of the central nervous system (OPPENHEIM 1991). If one looks at the complexity of the various synaptic and other neural connections and the anatomical organization of the CNS, one could not help marvelat it. Apparently, it is now understood that during development there are far more neurons than the final number that makes up a mature brain. It seems that redundant and adjacent neurons would compete for proper synaptic formation and cell–cell networking with the microgila cells. It is this local electro-stimulation, cell-association, and growth factor-based microenvironment that allow the lucky neurons (about 50%) to continue to survive and flourish, while those that lack such support system are destined to be eliminated by apoptosis. This notion is particularly illustrated by the recent reports that show both the apoptosis-required caspase-3 and caspase-9 knockout mice have abnormally developed brain and a larger number of neurons than that seen in normal mice (KUIDA et al. 1996, 1998). Thus, in these cases, it appears that the lack of physiologically-timed neuronal apoptosis in the rat brain leads to a failure to eliminate redundant central neurons.

Apoptosis is also important in the normal development of other organs (e.g., liver, kidney, etc.) and other body structures such as digit formation (MAJNO and JORIS 1995). Yet, the occurrence of apoptosis is not limited to immature animals. Even in adult mammals, skin epithelial cells are turned over constantly, as well as a variety of erthropoetic cells (e.g., T-cells). Another key function of apoptosis is to eliminate cells that are damaged or malfunctioning as a result of spontaneous or induced DNA mutations, and breakage or other damage, chemical toxin-poisoning (e.g., a side effect from chemotherapeutic drugs or environmental poisons) or microbial infection (e.g., bacteria, virus). For example, with DNA damage, when the cell attempts unsuccessfully to repair its DNA, it will trigger the induction of p53 gene. The overexpression of cell-cycle checkpoint protein p53 has two important functions: (a) to allow the cell to enter cell cycle arrest at G1-S transition, and (b) in some cases, to trigger the cell to go into apoptosis (EVAM and LITTLEWOOD 1998). Both events have a common goal of preventing manifestation of this genetic mistake. Indeed, the reverse is also true, i.e., the failure of needed apoptosis could result in tumor formation in various tissues. To date, an increasing number of oncogenes have now been linked to the very cascade of apoptosis, including bcl-2, c-myc, caspase-3 (EVAM and LITTLEWOOD 1998). These are important issues to bear in mind as we start to address suppressing apoptosis as a strategy for neuroprotection under neurodegenerative conditions (see Sect. D.).

C. Unscheduled Apoptosis in Various Neurodegenerative Conditions

I. Criteria Used to Identify Neuronal Apoptosis

Apoptosis is most strictly defined by the presence of condensed chromatin, which sometimes form margination of the inside surface of the nuclear envelope. Obviously, this morphology has to be confirmed by high-resolution microscopy or even electron microscopy. Other supportive evidence includes TUNEL and Hoechst staining of the nuclei. Apoptosis is also associated with the shrinkage of cytoplasm volume, plasma membrane blebbing, and formation of apoptotic bodies. However, most of these features might not be readily observed in vivo due to the tissue organization. When analyzing apoptosis in the brain, one also needs to confirm that the apoptotic cells are indeed neurons rather than astrocytes using the immunostaining method. DNA fragmentation is also a test but again not necessary fail-proof (CHARRIAUT-MARLANGUE 1995). In addition, some forms of apoptosis are inhibitable by cyclohexamide or actinomycin D, but there are exceptions. In recent years, the emergence of caspase activation as a unique feature in apoptosis but not necrosis also offers hope that biochemical markers for apoptosis might become available, such as those that monitored the caspase-specific protein fragments (see Sect. VII.). Also, to date, most forms of apoptosis are inhibited or at least retarded by caspase inhibitors. Thus, these pharmacological agents are useful in identifying and studying neuronal apoptosis (THORNBERRY and LAZEBNIK 1998).

II. Various Neurodegenerative Disorders with Evidence of Apoptosis

LINNIK and colleagues (LINNIK et al. 1993) was the first group to report evidence for apoptosis in a rat focal ischemia model by showing the presence of DNA laddering. Shortly after, MACMANUS and colleagues (MACMANUS et al. 1993) reported similar findings in a four-vessel-occlusion global ischemia model. Since then numerous reports appeared documenting the presence of apoptosis in other focal ischemia or excitotoxicity models (HERON et al. 1993; FILIPKOWSKI et al. 1994; LI et al. 1995; CHEN et al. 1997; POLLARD et al. 1994;

PULERA et al. 1998). In general, electron micrographic analysis has been employed to identify apoptotic neurons. However, several research groups found little or no evidence of classic apoptosis (DESSI et al. 1993; SCOTT and HEGYI 1997). Terminal deoxyribonucleotidyl transferase (TdT)-mediated biotin-16-dUTP nick-end labeling (TUNEL) is one method that has been misused to positively identify apoptosis as it detects only the increase of fragmented DNA, which occurs in both apoptosis and necrosis (CHARRIAUT-MARLANGUE et al. 1995). Regarding excitotoxicity, BONFOCO et al. (1995) and ANKARCRONA (1995) reported that under specific conditions, excitotoxin challenge could lead to apoptosis in cultured CNS neurons (based on nuclear morphology). Yet, other research groups found little evidence for apoptosis when cerebrocortical cultures or cerebellar granule neurons (CGNs) were challenged with excitotoxins (DESSI et al. 1993; MACMANUS et al. 1993; PORTERA-CAILLIAU et al. 1997).

Recently, neuronal apoptosis has also been identified in spinal cord injury and traumatic models [based on evidence of DNA laddering and apoptotic nuclei morphology (EMERY et al. 1998; LOU et al. 1998; YAKOVLEV et al. 1997; RINK et al. 1995)]. Apoptotic neuronal death has also been identified in human Alzheimer's disease (COTMAN and PERSON 1995; SU et al. 1997), Parkinson's disease (ANGLADE et al. 1997; RUBERG et al. 1997; ZHANG et al. 1998), amyotrophic lateral sclerosis (TROOST et al. 1995), and spinal cord motor neurodegeneration (ROY et al. 1995). Again, it is worth-noting that these disorders have been traditionally viewed as leading to necrosis exclusively.

In fact, while most researchers agree that there is a subpopulation of neurons that are typically apoptotic in acute neuronal injury (especially in the cerebral ischemia field), many disagree on the relative proportion of apoptosis that would account for total neuronal loss. Some of these discrepancies in views and experimental data could probably be explained by the notion that injury-induced neuronal apoptosis, when it occurs in vivo, could be atypical. CHOI (1996) in fact, proposed that rather than obligatorily forcing neuronal cell death into either apoptosis or necrosis, there is a continuum between these two extreme forms of cell death. Further, his group showed that when cerebrocortical neurons were challenged with glucose-oxygen deprivation alone, an overwhelming calcium influx through ionotropic glutamate receptors (NMDA, AMPA and kainate receptors) produce a rapid and dominating necrotic cell death pathway. Yet, if the deprivation was done in saturating concentrations of NMDA receptor antagonist MK-801 and AMPA receptor antagonist NBQX, the previously hidden cyclohexamide-sensitive apoptotic component came to the surface (GWAG et al. 1995). ANKARCRONA et al. (1995) also argued that the intracellular ATP level might be a critical factor in dictating where the cells undergo apoptosis or necrosis. Similarly, LIPTON's group argued that high glutamate concentrations lead to necrosis, while milder lower concentrations of glutamate challenge yield apoptosis (BONFOCO et al. 1995). Also, TAN et al. (TAN et al. 1998a) reported that oxidative stress causes hippocampal cell line (HT-22) to experience a form of cell death that is neither apoptosis nor necrosis by their strict and classic morphological and anatomical definitions. In fact, glutamate and oxidative stress-induced neuronal death might be a continuum of necrosis to apoptosis (PORTERA-CAILLIAU et al. 1997; TAN et al. 1998b; NICOTERA et al. 1997). Also, LEMAIRE et al. (1998) have shown that caspase inhibitor can convert cell death from an apoptotic phenotype to one that resembles necrosis. Thus, it is becoming clear that this area of research will benefit from a biochemical and molecular rather than anatomical approach to define the forms of neuronal cell death.

D. Key Pathways and Components Relevant to Neuronal Apoptosis

I. Overview of Apoptosis-Linked Factors

During the past 7–8 years, there have been several breakthrough discoveries that move forward our understanding of the fundamental mechanism of apoptosis. These are: (1) the identification of nematode *C. elegans* antiapoptotic protein CED-9 and the homologous mammalian Bcl-2/Bax (ADAMS and CORY 1998); (2) the cloning of the *C. elegans* gene product CED-3, which is absolutely required for apoptosis and the discovery that it is highly homologous to the mammalian interleukin-1β converting enzyme (ICE) and the subsequent identification of a large subfamily of mammalian ICE-like proteases (caspases) that are apoptosis-linked (THORNBERRY and LAZEBNIK 1998); (3) the discovery of various apoptosis-mediated receptors containing the so-called death domain and their interactions with cytosolic adapter molecules that also contain death domains (ASHKENAZI and DIXIT 1998); (4) the discovery that cytochrome C release from mitochondria is a key step in many forms of apoptosis; and (5) the cloning of apoptosis-linked CED-4 protein in *C. elegans* (GREEN and REED 1998) and subsequently the mammalian homologue Apaf-1 (HU et al. 1998). We will go into some of the details in Sects. III and IV., especially those pertinent to neuronal apoptosis.

II. Bcl-2/Bax Family

The bcl-2/Bax family members can be roughly divided into two groups: (a) those that promote survival against apoptosis (Blc-2, Bcl-XL), and (b) those that promote apoptosis typified by Bax and Bid. Both families of proteins have been recently reviewed (ADAMS and CORY 1998).

Antiapoptotic family members such as Bcl-2, Bcl-XL and Bcl-w (about 30–35 kDa) all have (a) three helical domains BH1, BH2, and BH3, which are important for homo- and hetero-dimerization, and (b) BH4 domain near the N-terminal, which is prosurvival and is also essential for interacting with Apf-1 protein (in the case of Bcl-XL). In contrast, There are two subfamilies of proapoptotic members (the Bax family and the BH3 family) (ADAMS and CORY

1998). Bax family members such as Bax, Bak, and Bok all have BH1 and BH2 domains but with an alternated BH3 domain. Thus, if there were dimerized with bcl-2 via BH3 domain interaction, they might produce a different conformation to the complex. Also, this group of protein lacks BH4 domain and thus is usually smaller in size (20–25 kDa). The BH3 family includes Blk, Bik, Bid and Bad. They are nonhomologous to other bcl-2 family members except for their BH3 region. Like the Bax subfamily, the BH3 subfamily members are viewed perhaps as the direct antagonists of Bcl-XL or Bcl-2 by BH3-homophilic hetero-dimerization (HEGDE et al. 1998).

III. Mitochondria and Caspase-9 Mediated Pathway

Mitochondria appears to be an important organelle in many pathways of apoptosis (GREEN and REED 1998). Typically, Bcl-2, which resides primarily on the outer membrane of mitochondria, exists as a homodimer and serves as apoptosis-suppressor. On the other hand, the proapoptotic protein Bax, which normally resides in the cytosol, will translocate to the outer membrane of mitochondria upon apoptotic signal, through an as yet unidentified mechanism,. Indeed, Bax is shown to be a required factor in at least some forms of neuronal apoptosis (DECKWERTH et al. 1996). Also, Bax is overexpressed in the cerebellum of mutant lurcher and weaver mice where apoptosis occurs during development (WULLNER et al. 1998). There, it has two potential effects: it can form an ion channel and thus might alter mitochondria ion permeability (GREEN and REED 1998). It can also heterodimerize with Bcl-2 and by doing so negate the antiapoptotic effects of Bcl-2. Regardless of whether either hypothesis is true in vivo, the release of loosely-associated cytochrome C from mitochondria apparently follows, as has been shown in a number of neuronal injury systems (DU et al. 1997; GLEICHMANN et al. 1998; MCGINNIS et al. 1999). Cytosolic cytochrome C, in addition to the presence of dATP, will associate with a protein factor called Apaf-1. The cytochrom-C/Apaf-1 complex then associates and activates procaspase-9. Procaspase-9 (49 kDa) processing to its processed form (40 kDa) is predicted to be autolytic (LI et al. 1997; SRINIVASULA et al. 1998; PAN 1998a). It is now well documented that caspase-3 can be activated by caspase-9 (PAN 1998b). Very recently, two studies demonstrated that caspase-9 knockout mice experience severe impairment in neural development as a direct result of the lack of proper neuronal apoptosis (KUIDA et al. 1998; HAKEM et al. 1998). Predictably, caspase-3 inhibitors (such as Z-D-DCB, Z-VAD-fmk, and Ac-DEVD-CHO) or the viral protein Crm-A are potent inhibitors of this apoptosis pathway (GARCIA-CALVO et al. 1998).

Interestingly, Bcl-XL can act downstream from cytochrome C release to inhibit apoptosis (LI et al. 1997). Apparently Bcl-XL interacts with cytochrome C/Apaf-1 complex and prevents its association to procaspase-9 (HU et al. 1998). Again, BH3 family members such as Bad could sequester Bcl-XL and thus prevent its protective effects.

Mitochondria permeability transition (MPT) has also been proposed to play an important role in the initiation of apoptosis cascade. MPT can indeed be observed in a wide range of apoptosis models and also in necrosis models (LEMASTERS et al. 1998). Bax also binds to the MPT pore and induces MPT directly (MARZO et al. 1998). However, sometimes the occurrence of MPT appears to be a late event during apoptosis. MPT blockers, such as bongkrekic acid, cyclosporin A (CyA) in combination with the phospholipase A2 inhibitor aristolochic acid are effective in some forms of apoptosis (MARZO et al. 1998; HORTELANO et al. 1998; BRADHAM et al. 1998). Cyclosporin A was also shown to protect against hypoxic/hypoglycemic neuronal injury (LI et al. 1997; FRIBERG et al. 1998).

IV. Death Receptor-Mediated Caspase-8/10 Pathway

Alternative to the cytochrome-C and mitochondria-mediated pathway, caspase-3 activation can also be triggered by a receptor-mediated pathway. Death receptors include TNF-α receptor 1, NGF p75 receptor, etc. (ASHKENAZI and DIXIT 1998). Most cell types, including neural cells, express TNF-receptor 1 (TNF-R1; 55kDa), which responds to TNF-α. TNF-α also binds to TNF-R2 (75kDa), which seems to have antagonistic properties and is not essential for TNF-α action (PESCHON et al. 1998; DECKERT-SCHLUTER et al. 1998). TNF-R1, when activated, couples to the TNF-R1-associated death domain protein (TRADD) (Fig. 2). TRADD can interact with TRAF2 (TNF-R1-associated factor 2) or with another adapter protein called FADD (Fas-associated death domain protein). FADD then triggers apoptosis by association with procaspase-8 (or procaspase-10) (VILLALBA et al. 1997). Upon binding to FADD (Fig. 2), procaspase-8 (55kDa) is predicted to autolytically process to its activated form (48kDa) (HAKEM et al. 1998l; STENNICKE et al. 1998). Processing and activation of caspase-8 has recently been reported in a rat model of traumatic brain injury (BEER et al. 2001). Again, like caspase-9, TRADD-FADD-caspase-8/10 interaction induces the autolytic activation of caspase-8/10, which in turns processes and activates the common apoptosis mediator caspase-3 (Fig. 2). The FADD-caspase-8-caspase-3 interaction is essential for TNF-α-mediated apoptosis (MUZIO et al. 1996). Caspase inhibitors such as Z-D-DCB or Z-VAD-fmk, are capable of inhibiting TNF-α-mediated apoptosis in a number of cell types (SIDOTI-DE FRAISSE et al. 1998; JAESCHKE et al. 1998). Interestingly, receptor-mediated caspase-8 activation could be amplified by mitochondrial pathway of cytochrome C and caspase-9 activation (KUWANA et al. 1998), via two pathways. Firstly, Bid is a caspase-8 substrate, which, when truncated, translocates to mitochondria and releases cytochrome C (LUO et al. 1998; LI et al. 1998). Secondly, caspase-8/10 activated-caspase-3 can process Bcl2 to a Bax-like molecule, activating the mitochondria-mediated pathway (CHENG et al. 1997).

On the other hand, activation of caspase-2 (Nedd-2) takes on a different pathway: activated TNF-R1 could associate with adapter protein RAIDD

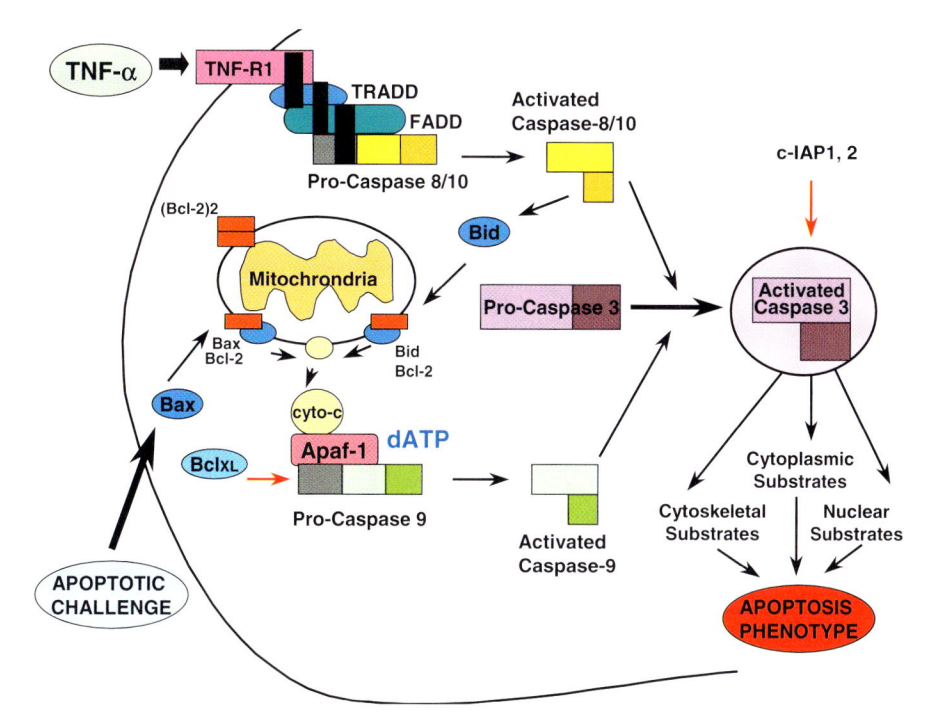

Fig. 2. Schematic for major neuronal apoptosis cascades. Both the mitochondria-caspase-9- and the TNF-α-receptor-caspase-8/10-mediated pathways are illustrated. For details, see Sects. III and IV

which associates with the extended N-terminal prodomain of caspase-2 (DUAN and DIXIT 1997). Caspase-2 m-RNA has been reported to be elevated in a global ischemia model (KINOSHITA et al. 1997) and is activated in PC12 apoptosis (STEFANIS et al. 1998). However, unlike caspase-3, the biochemical steps immediately downstream to caspase-2 activation and the relative role of caspase-2 in neuronal apoptosis are still unclear. Lastly, Fas ligand, when bound to Fas receptor, can also trigger apoptosis via the caspase-8 pathway. It was been shown that both Fas ligand and Fas are upregulated in neurons in regions subjected to traumatic brain injury (BEER et al. 2000).

V. TNF Receptor 1-Mediated p38 Kinase, JNK and NF-κB Activation

In addition to activation of caspase-8 via the TRADD–FADD interaction, the TNF-R1/TRADD association with TRAF2 also could mediate activation of three transcription cascades (NF-κB, JUN, and ATF2 activation) (Hsu et al. 1995; EDER 1997) (Fig. 3). For FOS activation, TRAF2 activates the apoptosis signal-regulating kinase 1 (ASK1), which activates MKK3/6, which in turn acti-

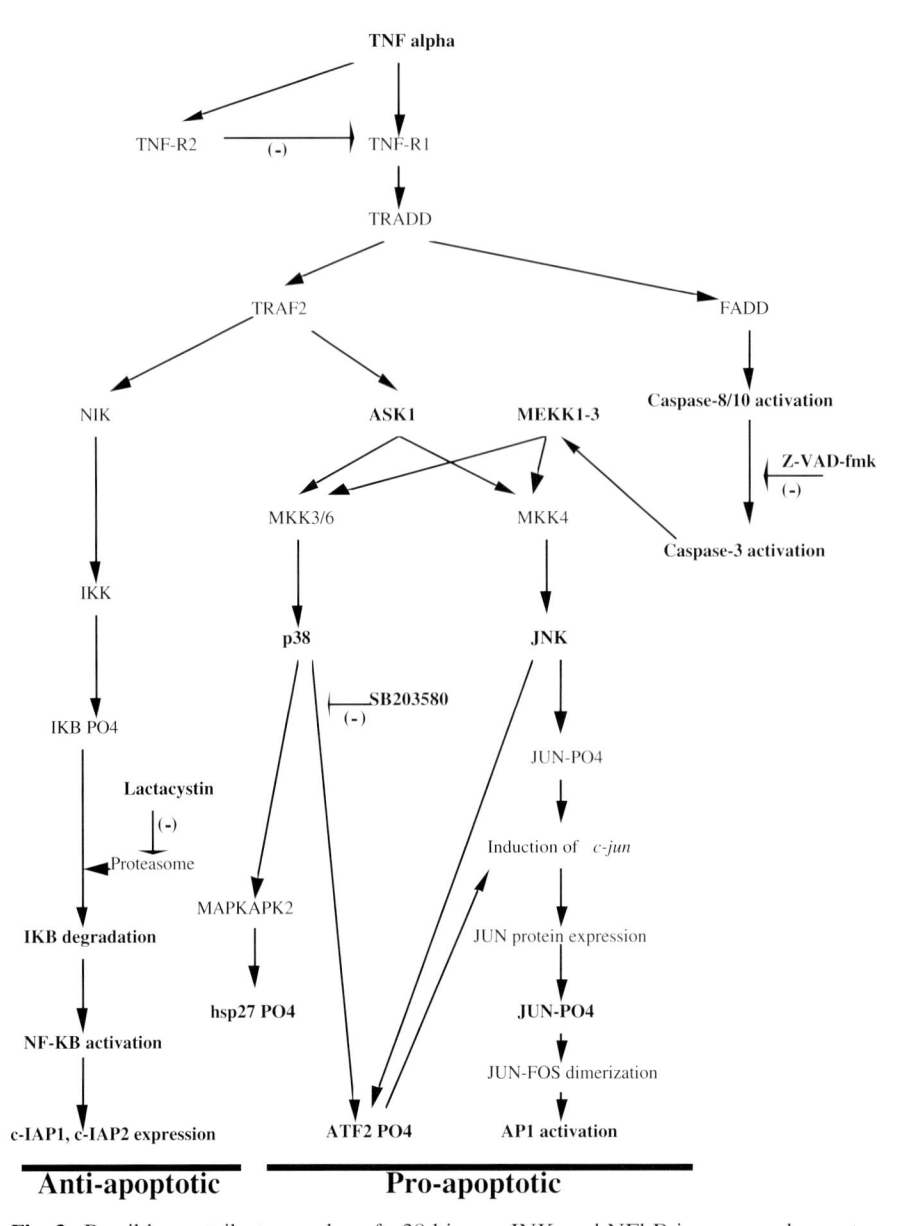

Fig. 3. Possible contributory roles of p38 kinase, JNK and NFkB in neuronal apoptosis. PO4 indicates phosphorylation. For details, see Sect. V

vates p38 kinase by phosphorylation (Fig. 3). p38 then activates MAPK-activated protein kinase 2 (MAPKAPK2), which phosphorylated heat shock protein hsp27 (GUAY et al. 1997). P38 also phosphorylates activating transcription factor 2 (ATF2) (JIANG et al. 1996). In parallel, for JUN activation,

TRAF2 activates MEKK1–3, which activates MKK4. MKK4 then activates JUN N-terminal kinase (JNK). JNK activates JUN and ATF2 by phosphorylation to induce more c-jun expression. Newly synthesized JUN is further activated by JNK. Then convergently activated FOS and JUN dimerize and activate AP1 (Fig. 3). Regarding the potential gene that might be turned on by these pathways, the author also refer to a review article on specific gene induction during neuronal apoptosis (ESTUS 1998). Importantly, MEKK1 is recently shown to be proteolytically activated by caspase-3 (CARDONE et al. 1998), thus caspase-3 activation could further feed forward into JUN activation (Fig. 3). It has been documented that both FOS and JUN activation is associated with NGF withdrawal-induced apoptosis in sympathetic neurons (ESTUS 1998; SILOS-SANTIAGO et al. 1995). In addition, JUN phosphorylation was found to be essential to low potassium-mediated apoptosis in cerebellar granule neurons (WATSON et al. 1998). Furthermore, both specific p38 kinase inhibitor SB203580 as well as olomoucine, which inhibits JNK (and cyclin-dependent kinase 2 and 4), have been shown to retard apoptosis in neurotrophin-withdrawal-mediated CNS/PNS neurons and glutamate-treated cerebellar granule neurons (MAAS et al. 1998; KAWASAKI et al. 1997). However, a recent study showed that JNK activation is not linked to TNF-mediated apoptosis (LIU et al. 1996).

For the NF-κB pathway, TRAF2 apparently binds to and activates an NF-κB inducing kinase (NIK). It in turn activates I-κB kinase (IκK) by phosphorylation. IκK then phosphorylates I-κB, priming it for degradation by proteasome (EDER 1997). The cytosolic p50+p65 NF-κB complex is thus released from the degraded I-κB and translocated into nucleus (Fig. 3). Several research groups have now shown that NF-κB activation apparently suppresses apoptosis (WANG et al. 1996; VAN ANTWERP et al. 1996; WU et al. 1996). Most recently, WANG et al. (1998) showed that NF-κB transcriptional activity induced the expression of endogenous caspase inhibitor proteins (c-IAP1 and c-IAP-2) which exert antiapoptotic effects. It is intriguing that TNF-R1 mediates two opposing pathways, one antiapoptotic while the other proapoptotic. Since NF-κB inhibition is apparently antiapoptotic, it is likely that the suppression of the NF-κB pathway will shift the equilibrium towards apoptosis. This could explain why, as indicated in several studies, the addition of TNF-α alone is not sufficient to promote apoptosis in cultured hippocampal neurons (VIVIANI et al. 1998) and murine neuroblastoma line, N1E-115 (SIPE et al. 1996). Recently, an important study shows that proteasome inhibitor lactacystin sensitizes TNF-R1-mediated apoptosis in chronic leukemia lymphocytes (DELIC et al. 1998). It is believed that lactacystin inhibits the degradation of I-κB, thus preventing NF-κB translocation to the nucleus (Fig. 3). It is likely that the suppression of the NF-κB pathway will shift the equilibrium towards apoptosis. Indeed, proteasome inhibitors lactacystin, IGAL-CHO, and Cbz-LLL-CHO all induce apoptosis in tumor cell lines (SHINOHARA et al. 1996; DREXLER 1997). Alternatively, transcription inhibitor actinomycin D also enhances TNF-α-induced apoptosis. It is assumed that it might do so by inhibiting the NF-κB-induced gene expression. Since most cell types express TNF-R1 including neural cells (Sipe et al. 1996) neurons could be sensitized to become vulnerable to TNF-α-mediated apoptosis. An important determinant is the NF-κB activation in defining whether apoptosis occurs (Clemens et al. 1997). It is well documented that in excitotoxin-, ischemia-, or trauma-injured brains, TNF-α levels are elevated (Botchkina et al. 1997; Uno et al. 1997; Kita et al. 1997). If, under the same conditions, the NF-κB activation is impaired, TNF-α would push the system into apoptosis by both the JUN-FOS pathway as well as the caspase-8/10 pathway (Figs. 2, 3).

VI. Akt-IP3 Kinase-Mediated Antiapoptosis Pathway for NGF and Other Survival Factors

The involvement of Akt protein kinase and phosphatidylinositol 3-kinase in apoptosis suppression by nerve growth factor (NGF) is one of the better understood neuronal survival pathways. It is well known that NFG binding to the high affinity receptor TrkA is a survival signal for sympathetic neurons such as rat superior cervical ganglion (SCG) (Deckwerth and Johnson 1993), as well as NGF-differentiated rat pheochromocytoma PC-12 cells (Yao and Cooper 1995). However, the signaling pathway that promotes cell survival was not identified until Yao and Copper (1995) reported that NGF's ability to prevent apoptosis in differentiated PC-12 cells is abolished by two specific PI-3 kinase inhibitors: wortmannin and LY294002. More recently, essentially the same effects of PI-3 kinase inhibitors were confirmed in SCG neurons in culture (Crowder and Freeman 1998). Upon NGF binding to TrkA, the receptor-associated tyrosine kinase autophosphorylates itself, thus allowing binding to the SH2 domain in PI-3 kinase (PI-3K) (Coffer et al. 1998). PI-3 kinase mediates the 3′-phosphorylation of PIP2 to PIP3, which recruits the plecstrin homology (PH) domain-containing PDK1 and PDK2. PDK1/2, in turn, phosphorylates and thus activates Akt protein kinase (also called PKB) (Coffer et al. 1998) (Fig. 4). Consistent with that, Akt was identified as a downstream effector of PI-3 kinase in signaling cell survival (Dudek et al. 1997; Kauffmann-Zeh et al. 1997; Kennedy et al. 1997). Akt can also be alternatively phosphorylated by CaM-dependent kinase kinase (CaMKK) (Yano et al. 1998). Interestingly, insulin receptor activation also leads to inactivation of p38 kinase and p38 kinase-specific inhibitors (SB203580, PD169316) mimic the survival signal of trophic factors (Heidenreich and Kummer 1996; Kummer et al. 1997) (Fig. 4). The PI-3K-Akt pathway is particularly important in CGN cells, as high extracellular K$^+$ (25mM) is needed to suppress spontaneous apoptosis in culture. It is now shown that the slight depolarization elevated intracellular calcium and thus calmodulin and CaMKK, leading to phosphorylation of Akt (Yano et al. 1998) (Fig. 4). One of the principal substrates of Akt is glycogen synthase kinase-3β (GSK3β), which is inactivated upon phosphorylation by Akt. This leads to the hypothesis that GSK3 might also be

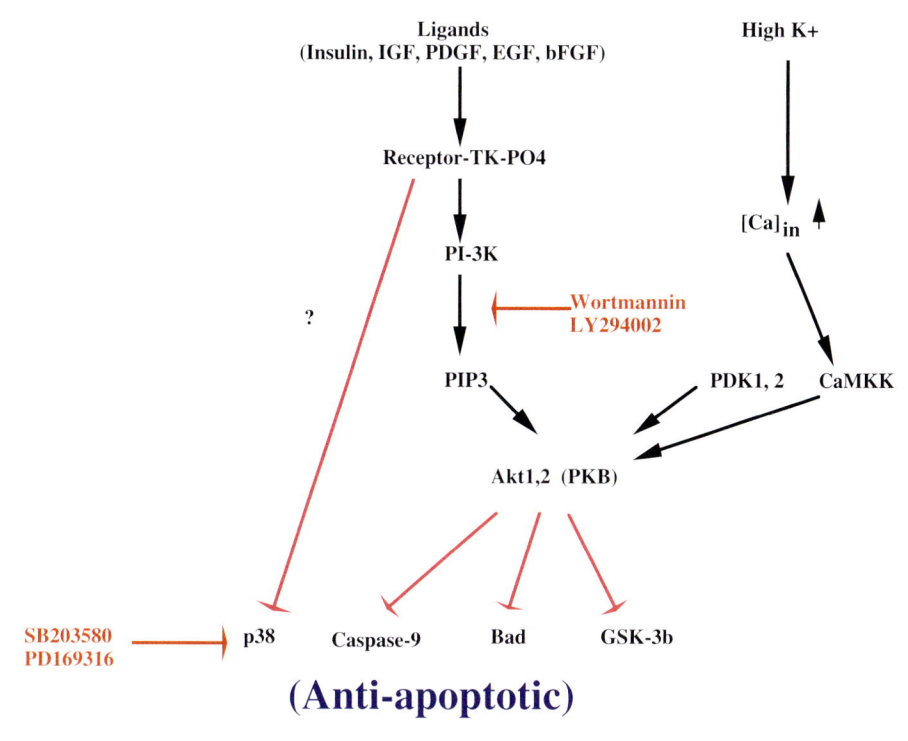

Fig. 4. Cell survival signal transduction pathways. Survival signal ligands activate their respective receptor-linked tyrosine kinase autophosphorylation (*TK-PO4*) which recruits PI-3 kinase and activating Akt1,2 inducing an antiapoptotic pathway. Similarly, high extracellular K$^+$ also triggers calcium influx and CaMKK activation which also activates Akt1,2

involved in cell survival. Using this argument, active GSK3β might promote apoptosis. In fact, overexpression of catalytically active GSK3β did just that (Pap and Cooper 1998). In addition, Akt activation also directly and/or indirectly leads to phosphorylation of proapoptotic protein Bad (on Ser112 and S136) (Datta et al. 1997; del Peso et al. 1997). Phosphorylated Bad, instead of binding to Bcl-XL, binds 14–3-3, which abolishes its apoptotic ability (Yaffe et al. 1997). Lastly, Akt can also inactivate the proteolytic activity of caspase-9 by direct phosphorylation (Cardone et al. 1998). The importance of the PI3K-Akt signaling pathway in neuronal survival was recently reviewed by Brunet et al. (2001).

VII. Effector Caspases

Caspase-3 and its highly homologous cousin caspase-7 are generally regarded as the effector caspases that target a wide range but selecte subgroup of protein substrates (THORNBERRY and LAZEBNIK 1998) (Fig. 1). The exception is the CNS, where no mRNA for caspase-7 was detected (JUAN et al. 1997). The determinant with the most specificity is the Asp (D) in the P1 and P4 positions, identified using synthetic peptides as substrates. Again, like calpain, caspase-3 tends to produce limit fragments of its substrates, leaving them as a fingerprint for caspase-3 activation. The first ever identified caspase-3 substrate was PARP; the major cleavage site was found to be DEVD*G, which conforms to the DXXD-specificity. Other substrates can be categorized into (a) cytoskeletal proteins (such as actin, α-, β-spectrin II, GAS-2, vimentin), (b) signal transduction enzymes (protein kinase C δ and θ isoforms CaMPK-II and IV, FAK (focal adhesion kinase), phospholipase C, p21-activated kinase and MEKK1), and (c) cell cycle proteins (PITSLRE kinase; Rb, MDM-2 and cdk inhibitors p27, p21). Caspase-3 also degrades a number of nuclear substrates, such as DNA-PKcs, PARP, U1–70K and NuMA, many of which participate in DNA repair (THORNBERRY and LAZEBNIK 1998). Caspase-3 is also believed to activate caspase-6, which is responsible for degrading lamin A, B1, B2, and C (Fig. 2). Another important substrate for caspases is the cytosolic inhibitor of the caspase-activated DNAase (ICAD). Upon cleavage, it is released from the caspase-activated DNAase (CAD). CAD then enters the nucleus and starts the process of DNA fragmentation (Fig. 2).

Recently we have identified a caspase-3 cleavage site in αII-spectrin (DETD1185*SKTAS) that produces a 150-kDa SBDP (termed SBDP150i) which occurs only in neuronal apoptosis (NATH et al. 1996). This 150kDa SBDP, although similar in size, is nonidentical to the SBDP150 generated by calpain (at VY1176*GMMPR). In addition, a 120-kDa fragment (SBDP120) is generated solely by caspase-3 (DEVD1478*SVEAL) (WANG et al. 1998; JANICKE et al. 1998). In fact, using SBDP120 as a marker, we identified the presence of caspase-3 activation in both NMDA/kainate and oxygen-glucose-deprivation challenged cerebrocortical neurons and in an in vivo model of TBI (NATH et al. 1998; PIKE et al. 1998). Caspase activation is also involved in glutamate and 1-Methyl-4-phenylpyridinium (MPP$^+$) toxicity to CGN neurons (DU et al. 1997a,b; NI et al. 1996; LEIST et al. 1998).

Caspase-3 expression has recently been shown to be induced under various injury-related conditions: in rat kidney after ischemia-reperfusion injury (KAUSHAL et al. 1998), in the CA1 region in a rat global forebrain ischemia model (NI et al. 1998), and in the hippocampus after transient cerebral (CHEN et al. 1998). The increased caspase-3 protein, while probably nonessential, could accelerate apoptosis.

A number of pharmacological inhibitors of caspase are available: general caspase inhibitors such as Z-D-DCB, caspase-3/7 selective Ac-DEVD-CHO, and Z-VAD-fmk and Z-VAD-DCB which act on caspase-8/9. The cowpox vial protein Crm-A is potent against caspase-1, 8, and 9 but not caspase-2, 3, or 7 (GARCIA-CALVO et al. 1998). There is a group of endogenous caspase inhibitor proteins called IAP (inhibitors of apoptosis proteins) present in insect (Op-IAP) to human (cIAP1, cIAP2) that apparently directly inhibit caspases

(TAMM et al. 1998). Caspase inhibitors (Z-VAD-fmk, Z-D-DCB) are found to protect against excitotoxin- and hypoxia-induced neuronal injury (NATH et al. 1998; HARA et al. 1997; GOTTRON et al. 1997) as well as against several focal ischemic brain injury models in vivo (using Z-VAD-DCB, Z-VAD-fmk, Z-DEVD-fmk, and Z-D-DCB) (ENDRES et al. 1998; MA et al. 1998; HIMI et al. 1998).

VIII. Participation of Other Proteases

Other proteases might also participate in neuronal apoptosis. For example, calpain activation was first demonstrated in thymocyte and T-cell apoptosis under certain conditions (SQUIER et al. 1994; SARIN et al. 1995; NATH 1996b). Our laboratories then extended the finding to show that calpain is indeed activated in staurosporine-treated neuroblastoma SH-SY5Y cells, in NGF-deprived rat PC-12 cells, and in low potassium-treated rat cerebellar granule and hippocampal neurons (NATH 1996a,b; PIKE et al. 1998). In fact, calpain and caspase seem to simultaneously attack a number of cellular substrates, such as αII- and βII-spectrin (fodrins) (WANG et al. 1998; LITERSKY and JOHNSON 1995; CANU et al. 1998) and CaMPK-II and IV (HAJIMOHAMMADREZA et al. 1997; MCGINNIS et al. 1998). Thus, the calpain system can be viewed as an auxiliary unit that furthers proteolytic destruction of the cell (WANG et al. 2000). Calpain inhibitors have been found to protect against certain forms of neuronal apoptosis (NATH et al. 1996b; JORDAN et al. 1997).

IX. Reactive Oxygen Species

A reactive oxygen species (ROS) requirement for neuronal apoptosis was first documented by JOHNSON's group using the NGF-dependent supercervical ganglion (SCG) neurons (DECKWERTH and JOHNSON 1993). Subsequently, they showed that micro-injection of the reactive oxygen scavenger superoxide dismutase protein (SOD) into cultured SCG cells rescues them from NGF-deprivation-mediated apoptosis (GREENLUND et al. 1995). In the CGN culture model, potassium deprivation-induced apoptosis again is found to require reactive oxygen species (ROS) (SCHULZ et al. 1996). ROS was detected by an oxidation-sensitive indicator dihydrorhodamine 123 and showed that it occurred after caspase activation. They also found that antioxidants SOD, N-acetyl-L-cysteine, catalase, vitamin E, and free radical spin-trap N-tert-butyl-α-phenylnitrone (PBN), when added to culture medium, all protect against apoptotic death. Similarly, TAN et al. found glutamate toxicity in an immortalized mouse hippocampal cell line (HT22) is dependent on ROS formation (TAN et al. 1998a). Along the same concept, nitric oxide is shown to induce neuronal apoptosis in CGN via NMDA-receptor activation (BONFOCO et al. 1996). In addition, peroxynitrite (ONOO$^-$) and lipid peroxidation is associated with apoptosis in pheochromocytoma PC6 cells induced by Fe^{2+}, amyloid β peptide, or NO donors (KELLER et al. 1998). They also showed that overexpression of mitochondria manganese SOD prevents these forms of apoptosis. Similarly, ONOO⁻ was shown to induce apoptosis in NSC34 neuroblastoma-spinal cord cell line (COOKSON et al. 1998) and in leukemic HL-60 cells (LIN et al. 1998; KELLER et al. 1998) further showed that SOD transgenic mice are protected against ischemic brain injury in an vivo focal ischemia model.

X. Neuronal-Specific Apoptosis Factors (cdk5, NAIP)

One mammalian gene encoding for a neuronal apoptosis inhibitory protein (NAIP) is homologous to the IAP family of antiapoptosis proteins (LISTON et al. 1996). One important genetic observation is that the NAIP gene is partially deleted in individuals with spinal muscular atrophy (SMA) (ROY et al. 1995; VELASCO et al. 1996). SMA is manifested by an inappropriate persistence of normally occurring motor neuron apoptosis. This points to the important apoptosis nature of NAIP. More recently, it was shown that elevation of neuronal expression reduces ischemic damage in the rat hippocampus (XU et al. 1997).

Cyclin dependent kinase 5 (Cdk5) is another protein that has been linked to apoptosis. Cdk5 is highly expressed in the CNS, but is also present in other peripheral tissues. However, the two known activator proteins (p35 and p39) essential for cd5 activation are only present in the nervous system (LEE et al. 1997). Induction of the cyclin dependent kinase Cdk5 protein appears to coincide with the time point when neurons were irreversibly committed to die in vitro (SHIRVAN et al. 1998). Cdk5 is associated with apoptotic cell death in the CNS during development and tissue remodeling (ZHANG et al. 1997). Consistent with that, the expression of Cdk5 and its activator p35 are associated with natural apoptotic cell death in both adult and embryonic tissues in vivo (AHUJA et al. 1997). Within spinal cord in sporadic and two superoxide dismutase type 1 (SOD1) familial cases of ALS, intense cdk-5 immunoreactivity was observed in perikarya of degenerating neurons. Furthermore, cdk-5 colocalized with lipofuscin, which is linked to an oxidative cascade that could result in apoptosis (BAJAJ et al. 1998). Also, cdk5 accumulates in neurons with early stages of Alzheimer's disease (PEI et al. 1998). Pharmacologically, roscovitine and olomoucine are the only two available potent and selective inhibitors of the cyclin-dependent kinases cdc2, cdk2, and cdk5. Both compounds are reported to inhibit neurotrophic factor withdrawal-induced apoptosis in rat retinal ganglion cells, but their lack of total selectivity among cdks makes the assessment of the role of cdk5 ambivalent (MAAS et al. 1998). Recently, p35 have been shown to be processed to p25 by calpain (NATH et al. 2000; LEE et al. 2000; KUSAKAWA et al. 2000). P25-associated cdk5 is considered to be more readily activated. Furthermore, cdk5 activation has been found to be tied to a β-induced τ hyperphosphorylation and the associated neuronal death (LEE et al. 2000).

E. Comparison of the Potential of Neuronal Apoptosis Factors as Neuroprotective Drug Targets

Under the assumption that apoptosis is a significant component of neurodegeneration, there are a number of approaches to reduce unscheduled neuronal apoptosis (Table 1). These strategies are reviewed in the following paragraphs.

Overexpression or delivery of natural antiapoptotic proteins. Bcl-2 is a universal apoptosis inhibitor that appears to act upstream in the cascade. Given its low endogenous levels in mature CNS neurons, Bcl-2 offers good potential as a drug. In fact, several biotechnology companies are currently exploiting its potential for gene therapy. Bcl-XL would be an alternative approach to Bcl-2. NIAP would potentially be an even more attractive drug as it is a naturally occurring neuronal-specific protein. If gene therapy is used, it will be advisable to develop a CNS-specific delivery system to minimize the exposure of peripheral tissues to antiapoptosis proteins as these proteins may have tumor-promoting properties.

Another interesting approach would be to mimic the natural cell survival signal pathway. In this case, small molecule IP3-kinase activator or Akt activator will be desirable. Alternatively, p38 inhibitor (such as prototype compounds SB203580 and PD169316) and to a lesser extent, JNK inhibitor, might have some potential. In addition, cdk5 inhibitor might offer the advantage of targeting a neuron-specific pathway. However, further validations of these mechanisms in in vivo neurodegeneration models are needed.

Table 1. Neuronal apoptosis-linked proteins as potential drug targets

Drug target	Drug	Potential
Bcl2	Bcl2 protein or gene therapy	+++
Bcl-XL	Bcl-XL protein or gene therapy	++
NIAP	NIAP protein or gene therapy	+++
PI3-kinase	PI3-kinase activator	+
Akt	Akt activator	+
P38 kinase	p38 inhibitor	+++
JNK	JNK inhibitor	+
GSK3b	GSK3b inhibitor	+
Cdk5	cdk5 inhibitor	++
MPT	MPT inhibitor	++
TNF-α-R1	TNF-α-R1 antagonist	?
Fas	Receptor antagonist	?
ROS	Reactive oxygen scavenger	++
Caspase-9	Caspase-9 inhibitor	+++
Caspase-8	Caspase-8 inhibitor	+++
Caspase-3	Caspase-3 inhibitor	+++
Caspase-2	Caspase-2 inhibitor	?
μ-, m-Calpain	Calpain inhibitor	++
CAD	DNAase inhibitor	+

Significant literature data have pointed to the cytochrome C-caspase-9–caspase-3 pathway as relevant in neuronal apoptosis. This opens up several therapeutic opportunities. Mitochondria permeability transition might be involved in cytochrome C release, then MPT inhibitors would have potential neuroprotective value. MPT inhibitors which are more selective than cyclosporine A would be desirable, as they would not have immuno-suppressive side effects. Additionally, caspase-8, caspase-9, and caspase-3 inhibitors are attractive, in light of existing neuroprotection studies using caspase inhibitors (ELDADAH BA and FADEN 2000). As calpain and caspase-3 have synergic actions of cellular protein fragmentation, the use of calpain inhibitor in addition to caspase inhibitor should further protect against apoptosis.

For the receptor-mediated pathway to apoptosis, more direct evidence is needed to establish a link of the TNF-α receptor system to neuronal apoptosis. If indeed this pathway is involved, then TNF-α-R1 antagonist should be neuroprotective (YANG et al. 1998; NAWASHIRO et al. 1997). Similarly, the caspase-8/10 and caspase-3 inhibitors should provide a downstream inhibition of this pathway. The Ca^{2+}/Mg^{2+}-activated endonuclease CAD, which is indirectly activated by caspase-3, mediates the DNA fragmentation characteristics of apoptosis. It is conceivable to develop inhibitors to this DNAase. However, as DNA fragmentation is a late event in the apoptosis cascade, inhibition of it is unlikely to rescue the dying neurons.

Lastly, we have discussed that ROS seems to be a mediator of various forms of neuronal apoptosis. With the other potential beneficial effects such as antioxidation, ROS scavengers would appear to be of potential therapeutic value in treating neurodegenerative diseases.

In summary, there are a number of strategies to prevent or suppress unscheduled neuronal apoptosis. However, one should bear in mind that neuronal apoptosis occurs also physiologically, particularly in the immature brain. In addition, chronic suppression of apoptosis posts unknown risk as it could promote tumorigenesis. Thus apoptosis inhibitors will be better suited to use against acute neurodegenerative conditions such as cerebral ischemia, where the antiapoptosis agent is only given to patients for a short treatment period. Alternatively, it would be desirable to develop a neuronal apoptosis-specific inhibitor to avoid side effects as a result of inhibition of physiological apoptosis in peripheral tissues.

Acknowledgements. I want to thank the contributions of Ms. Rathna Nath, Dr. Kim McGinnis, Dr. Rand Posmantur. I also thank Dr. Ronald Hayes, Dr. Margaret Gnegy, Dr. Po-wai Yuen and Dr. Richard Gilbertsen for stimulating and critical discussions.

References

Adams JM, Cory S (1998) The Bcl-2 protein family: arbiters of cell survival. Science 281:1322–1326
Ahuja HS, Zhu Y, Zakeri Z (1997) Association of cyclin-dependent kinase 5, its activator p35 with apoptotic cell death. Dev Genet 21:258–267

Anglade P, Vyas S, Javoy-Agid F, Herrero MT, Michel PP, Marquez J, Mouatt-Prigent A, Ruberg M, Hirsch EC, Agid Y (1997) Apoptosis, autophagy in nigral neurons of patients with Parkinson's disease. Histol Histopathol 12:25–31

Ankarcrona M, Dypbukt JM, Bonfoco E, Zhivotovsky B, Orrenius S, Lipton SA, Nicotera, P (1995) Glutamate-induced neuronal death: a succession of necrosis or apoptosis depending on mitochondrial function. Neuron 15:961–973

Ashkenazi A, Dixit VM (1998) Death receptors: signaling, modulation. Science 281: 1305–1308

Bajaj NP, Al-Sarraj ST, erson V, Kibble M, Leigh N, Miller CC (1998) Cyclin dependent kinase-5 is associated with lipofuscin in motor neurones in amyotrophic lateral sclerosis. Neurosci Lett 245:45–48

Beer R, Franz G, Krajewski S, Pike BR, Hayes RL, Reed JC, Wang KK, Klimmer C, Schmutzhard E, Poewe W, Kampfl A (2001) Temporal and spatial profile of caspase 8 expression and proteolysis after experimental traumatic brain injury. J Neurochem 78:862–873

Beer R, Franz G, Schopf M, Reindl M, Zelger B, Schmutzhard E, Poewe W, Kampfl A (2000) Expression of Fas and Fas ligand after experimental traumatic brain injury in the rat. J Cereb Blood Flow Metab 20:669–677

Bonfoco E, Krainc D, Ankarcrona M, Nicotera P, Lipton SA (1995) Apoptosis, necrosis: two distinct events induced, respectively, by mild, intense insults with N-methyl-D-aspartate or nitric oxide/superoxide in cortical cell cultures. Proc Natl Acad Sci USA 92:7162–7166

Bonfoco E, Leist M, Zhivotovsky B, Orrenius S, Lipton SA, Nicotera P (1996) Cytoskeletal breakdown, apoptosis elicited by NO donors in cerebellar granule cells require NMDA receptor activation. J Neurochem 67:2484–2493

Botchkina GI, Meistrell M, Botchkina IL, Tracey KJ (1997) Expression of TNF, TNF receptors (p55, p75) in the rat brain after focal cerebral ischemia Mol Med 3:765–781

Bradham CA, Qian T, Streetz K, Trautwein C, Brenner DA, Lemasters JJ (1998) The mitochondrial permeability transition is required for tumor necrosis factor α-mediated apoptosis, cytochrome c release. Mol Cell Biol 18:6353–6364

Brunet A, Datta SR, Greenberg ME (2001)Transcription-dependent and -independent control of neuronal survival by the PI3K-Akt signaling pathway. Curr Opin Neurobiol 11:297–305

Canu N, Dus L, Barbato C, Ciotti MT, Brancolini C, Rinaldi AM, Novak M, Cattaneo A, Bradbury A, Calissano P (1998) τ cleavage, dephosphorylation in cerebellar granule neurons undergoing apoptosis J Neurosci 18:7061–7074

Cardone MH, Roy N, Stennicke HR, Salvesen GS, Franke TF, Stanbridge E, Frisch S, Reed JC (1998) Regulation of cell death protease caspase-9 by phosphorylation. Science 282:1318–1321

Charriaut-Marlangue C, Margaill I Plotkine M, Ben-Ari Y (1995) Early endonuclease activation following reversible focal ischemia in the rat brain J Cereb Blood Flow Metab 15:385–388

Chen J, Jin K, Chen M Pei W, Kawaguchi K, Greenberg DA, Simon RP (1997) Early detection of DNA strand breaks in the brain after transient focal ischemia: implications for the role of DNA damage in apoptosis, neuronal cell death. J Neurochem 69:232–245

Chen J, Nagayama T, Jin K, Stetler RA, Zhu RL, Graham SH, Simon RP (1998) Induction of caspase-3-like protease may mediate delayed neuronal death in the hippocampus after transient cerebral ischemia. J Neurosci 18:4914–4928

Cheng EH, Kirsch DG, Clem RJ, Ravi R, Kastan MB, Bedi A, Ueno K, Hardwick JM (1997) Conversion of Bcl-2 to a Bax-like death effector by caspases. Science 278: 1966–1968

Choi DW (1996) Ischemia-induced neuronal apoptosis. Curr Opin Neurobiol 6:667–72

Clemens JA, Stephenson DT, Smalstig EB, Dixon EP, Little SP (1997) Global ischemia activates nuclear factor-κ B in forebrain neurons of rats. Stroke 28:1073–80; discussion 1080–1081

Coffer PJ, Jin J, Woodgett JR (1998) Protein kinase B (c-Akt): a multifunctional mediator of phosphatidylinositol 3-kinase activation. Biochem J 335:1–13

Cookson MR, Ince PG, Shaw PJ (1998) Peroxynitrite, hydrogen peroxide induced cell death in the NSC34 neuroblastoma x spinal cord cell line: role of poly (ADP-ribose) polymerase. J Neurochem 70:501–508

Cotman CW, Person AJ (1995) A potential role for apoptosis in neurodegeneration, Alzheimer's disease. Mol Neurobiol 10:19–45

Crowder RJ, Freeman RS (1998) Phosphatidylinositol 3-kinase, Akt protein kinase are necessary, sufficient for the survival of nerve growth factor-dependent sympathetic neurons. J Neurosci 18:2933–2943

Datta SR, Dudek H, Tao X, Masters S, Fu H, Gotoh Y, Greenberg ME (1997) Akt phosphorylation of BAD couples survival signals to the cell- intrinsic death machinery. Cell 91:231–241

Deckert-Schluter M, Bluethmann H, Rang A, Hof H, Schluter D (1998) Crucial role of TNF receptor type 1 (p55), but not of TNF receptor type 2 (p75), in murine toxoplasmosis. J Immunol 160:3427–3436

Deckwerth TL, Elliott JL, Knudson CM, Johnson E, JR Snider WD, Korsmeyer SJ (1996) BAX is required for neuronal death after trophic factor deprivation, during development. Neuron 17:401–411

Deckwerth TL, Johnson E, JR (1993) Temporal analysis of events associated with programmed cell death (apoptosis) of sympathetic neurons deprived of nerve growth factor. J Cell Biol 123:1207–1222

del Peso L, Gonzalez-Garcia M, Page C, Herrera R, Nunez G (1997) Interleukin-3 induced phosphorylation of BAD through the protein kinase Akt. Science 278:687–689

Delic J, Masdehors P, Omura S, Cosset JM, Dumont J, Binet JL, Magdelenat H (1998) The proteasome inhibitor lactacystin induces apoptosis, sensitizes chemo-, radioresistant human chronic lymphocytic leukaemia lymphocytes to TNF-α-initiated apoptosis. Br J Cancer 77:1103–1107

Dessi F, Charriaut-Marlangue C, Khrestchatisky M, Ben-Ari Y (1993) Glutamate-induced neuronal death is not a programmed cell death in cerebellar culture. J Neurochem 60:1953–1955

Drexler HC (1997) Activation of the cell death program by inhibition of proteasome function. Proc Natl Acad Sci USA 94:855–860

Du Y, Dodel RC, Bales KR, Jemmerson R, Hamilton-Byrd E, Paul SM (1997a) Involvement of a caspase-3-like cysteine protease in 1-methyl-4-phenylpyridinium-mediated apoptosis of cultured cerebellar granule neurons. J Neurochem 69:1382–1388

Du Y, Bales KR, Dodel RC, Hamilton-Byrd E, Horn JW, Czilli DL, Simmons LK, Ni B, Paul SM (1997b) Activation of a caspase 3-related cysteine protease is required for glutamate-mediated apoptosis of cultured cerebellar granule neurons. Proc Natl Acad Sci USA 94:11657–11662

Duan H, Dixit VM (1997) RAIDD is a new "death" adaptor molecule. Nature 385:86–89

Dudek H, Datta SR, Franke TF, Birnbaum MJ, Yao R, Cooper GM, Segal RA, Kaplan DR, Greenberg ME (1997) Regulation of neuronal survival by the serine-threonine protein kinase Akt (see comments). Science 275:661–665

Eder J (1997) Tumour necrosis factor α, interleukin 1 signalling: do MAPKK kinases connect it all? Trends Pharmacol Sci 18:319–322

Eldadah BA, Faden AI (2000) Caspase pathways, neuronal apoptosis, and CNS injury. J Neurotrauma 17:811–17829

Emery E, Aldana P, Bunge MB, Puckett W, Srinivasan A, Keane RW, Bethea J, Levi AD (1998) Apoptosis after traumatic human spinal cord injury. J Neurosurg 89:911–920

Endres M, Namura S, Shimizu-Sasamata M, Waeber C, Zhang L, Gomez-Isla T, Hyman BT, Moskowitz MA (1998) Attenuation of delayed neuronal death after mild focal ischemia in mice by inhibition of the caspase family. J Cereb Blood Flow Metab 18:238–247

Estus S (1998) Gene induction, neuronal apoptosis In: Mattson EMP (ed) Neuroprotective Signal Transduction. Humana Press, Totowa, NJ, pp 84–94

Evan G, Littlewood T (1998) A matter of life, cell death Science 281:1317–1322

Filipkowski RK, Hetman M, Kaminska B, Kaczmarek L (1994) DNA fragmentation in rat brain after intraperitoneal administration of kainate Neuroreport 5:1538–1540

Friberg H, Ferrand-Drake M, Bengtsson F, Halestrap AP, Wieloch T (1998) Cyclosporin A, but not FK 506, protects mitochondria, neurons against hypoglycemic damage, implicates the mitochondrial permeability transition in cell death. J Neurosci 18:5151–5159

Garcia-Calvo M, Peterson EP, Leiting B, Ruel R, Nicholson DW, Thornberry NA (1998) Inhibition of human caspases by peptide-based, macromolecular inhibitors. J Biol Chem 273:32608–32613

Gleichmann M, Beinroth S, Reed JC, Krajewski S, Schulz JB, Wullner U, Klockgether T, Weller M (1998) Potassium deprivation-induced apoptosis of cerebellar granule neurons: cytochrome c release in the absence of altered expression of Bcl-2 family proteins. Cell Physiol Biochem 8:194–201

Gottron FJ, Ying HS, Choi DW (1997) Caspase inhibition selectively reduces the apoptotic component of oxygen-glucose deprivation-induced cortical neuronal cell death Mol Cell Neurosci 9:159–169

Green DR, Reed JC (1998) Mitochondria, apoptosis. Science 281:1309–1312

Greenlund LJ, Deckwerth TL, Johnson E, JR (1995) Superoxide dismutase delays neuronal apoptosis: a role for reactive oxygen species in programmed neuronal death. Neuron 14:303–315

Guay J, Lambert H, Gingras-Breton G, Lavoie JN, Huot J, Landry J (1997) Regulation of actin filament dynamics by p38 map kinase-mediated phosphorylation of heat shock protein 27. J Cell Sci 110:357–368

Gwag BJ, Lobner D, Koh JY, Wie MB, Choi DW (1995) Blockade of glutamate receptors unmasks neuronal apoptosis after oxygen-glucose deprivation in vitro. Neuroscience 68:615–619

Hajimohammadreza I, Raser KJ, Nath R, Nadimpalli R, Scott M, Wang KK (1997) Neuronal nitric oxide synthase, calmodulin-dependent protein kinase IIα undergo neurotoxin induced proteolysis. J Neurochem 69:1006–1013

Hakem R, Hakem A, Duncan GS, Henderson JT, Woo M, Soengas MS, Elia A, de la Pompa JL, Kagi D, Khoo W, Potter J, Yoshida R, Kaufman SA, Lowe SW, Penninger JM, Mak TW (1998) Differential requirement for caspase 9 in apoptotic pathways in vivo. Cell 94:339–352

Hara H, Friedlander RM, Gagliardini V, Ayata C, Fink K, Huang Z, Shimizu-Sasamata M, Yuan J, Moskowitz MA (1997) Inhibition of interleukin 1β converting enzyme family proteases reduces ischemic, excitotoxic neuronal damage. Proc Natl Acad Sci USA 94:2007–2012

Hegde R, Srinivasula SM, Ahmad M, Fernandes-Alnemri T, Alnemri ES (1998) Blk, a BH3-containing mouse protein that interacts with Bcl-2, Bcl-xL, is a potent death agonist. J Biol Chem 273:7783–7786

Heidenreich KA, Kummer JL (1996) Inhibition of p38 mitogen-activated protein kinase by insulin in cultured fetal neuron. J Biol Chem 271:9891–9894

Heron A, Pollard H, Dessi F, Moreau F, Lasbennes F, Ben-Ari Y, Charriaut-Marlangue C (1993) Regional variability in DNA fragmentation after global ischemia evidenced by combined histological, gel electrophoresis observations in the rat brain. J Neurochem 61:1973–6

Himi T, Ishizaki Y, Murota S (1998) A caspase inhibitor blocks ischaemia-induced delayed neuronal death in the gerbil. Eur J Neurosci 10:777–781

Hortelano S, Dallaporta B, Zamzami N, Hirsch T, Susin SA, Marzo I, Bosca L, Kroemer G (1997) Nitric oxide induces apoptosis via triggering mitochondrial permeability transition. Febs Lett 410:373–377

Hsu H, Xiong J, Goeddel DV (1995) The TNF receptor 1-associated protein TRADD signals cell death, NF-κB activation. Cell 81:495–504

Hu Y, Benedict MA, Wu D, Inohara N, Nunez G (1998) Bcl-XL interacts with Apaf-1, inhibits Apaf-1-dependent caspase-9 activation. Proc Natl Acad Sci USA 95: 4386–4391

Jaeschke H, Fisher MA, Lawson JA, Simmons CA, Farhood A, Jones DA (1998) Activation of caspase 3 (CPP32)-like proteases is essential for TNF- α-induced hepatic parenchymal cell apoptosis, neutrophil-mediated necrosis in a murine endotoxin shock model. J Immunol 160:3480–3486

Janicke RU, Ng P, Sprengart ML, Porter AG (1998) Caspase-3 is required for α fodrin cleavage but dispensable for cleavage of other death substrates in apoptosis. J Biol Chem 273:15540–15545

Jiang Y, Chen C, Li Z, Guo W, Gegner JA, Lin S, Han J (1996) Characterization of the structure, function of a new mitogen- activated protein kinase (p38β). J Biol Chem 271:17920–17926

Jordan J, Galindo MF, Miller RJ (1997) Role of calpain-, interleukin-1 β converting enzyme-like proteases in the β-amyloid-induced death of rat hippocampal neurons in culture. J Neurochem 68:1612–1621

Juan TS, McNiece IK, Argento JM, Jenkins NA, Gilbert DJ, Copeland NG, Fletcher FA (1997) Identification, mapping of Casp7, a cysteine protease resembling CPP32 β, interleukin-1 β converting enzyme, CED-3. Genomics 40:86–93

Kauffmann-Zeh A, Rodriguez-Viciana P, Ulrich E, Gilbert C, Coffer P, Downward J, Evan G (1997) Suppression of c-Myc-induced apoptosis by Ras signalling through PI(3)K. PKB Nature 385:544–548

Kaushal GP, Singh AB, Shah SV (1998) Identification of gene family of caspases in rat kidney, altered expression in ischemia-reperfusion injury. Am J Physiol 274:F587–F595

Kawasaki H, Morooka T, Shimohama S, Kimura J, Hirano T, Gotoh Y, Nishida E (1997) Activation, involvement of p38 mitogen-activated protein kinase in glutamate-induced apoptosis in rat cerebellar granule cells. J Biol Chem 272:18518–18521

Keller JN, Kindy MS, Holtsberg FW, St Clair DK, Yen HC, Germeyer A, Steiner SM, Bruce-Keller AJ, Hutchins JB, Mattson MP (1998) Mitochondrial manganese superoxide dismutase prevents neural apoptosis, reduces ischemic brain injury: suppression of peroxynitrite production, lipid peroxidation, mitochondrial dysfunction. J Neurosci 18:687–697

Kennedy SG, Wagner AJ, Conzen SD, Jordan J, Bellacosa A, Tsichlis PN, Hay N (1997) The PI 3-kinase/Akt signaling pathway delivers an antiapoptotic signal. Genes Dev 11:701–713

Kinoshita M, Tomimoto H, Kinoshita A, Kumar S, Noda M (1997) Up-regulation of the Nedd2 gene encoding an ICE/Ced-3-like cysteine protease in the gerbil brain after transient global ischemia. J Cereb Blood Flow Metab 17:507–514

Kita T, Liu L, Tanaka N, Kinoshita Y (1997) The expression of tumor necrosis factor α in the rat brain after fluid percussive injury. Int J Legal Med 110:305–311

Kuida K, Haydar TF, Kuan CY, Gu Y, Taya C, Karasuyama H, Su MS, Rakic P, Flavell RA (1998) Reduced apoptosis, cytochrome c-mediated caspase activation in mice. lacking caspase 9. Cell 94:325–337

Kuida K, Zheng TS, Na S, Kuan C, Yang D, Karasuyama H, Rakic P, Flavell RA (1996) Decreased apoptosis in the brain, premature lethality in CPP32- deficient mice. Nature 384:368–372

Kummer JL, Rao PK, Heidenreich KA (1997) Apoptosis induced by withdrawal of trophic factors is mediated by p38 mitogen-activated protein kinase. J Biol Chem 272:20490–20494

Kusakawa G, Saito T, Onuki R, Ishiguro K, Kishimoto T, Hisanaga S (2000) Calpain-dependent proteolytic cleavage of the p35 cyclin-dependent kinase 5 activator to p25. J Biol Chem 2;275:17166–17172

Kuwana T, Smith JJ, Muzio M, Dixit V, Newmeyer DD, Kornbluth S (1998) Apoptosis induction by caspase-8 is amplified through the mitochondrial release of cytochrome C. J Biol Chem 273:16589–16594

Lee KY, Qi Z, Yu YP, Wang JH (1997) Neuronal Cdc2-like kinases: neuron-specific forms of Cdk5. Int J Biochem Cell Biol 29:951–958

Lee MS, Kwon YT, Li M, Peng J, Friedlander RM, Tsai LH (2000). Neurotoxicity induces cleavage of p35 to p25 by calpain. Nature 405(6784):360–364

Leist M, Nicotera P (1998) Apoptosis, excitotoxicity, neuropathology. Exp Cell Res 239:183–201

Leist M, Volbracht C, Fava E, Nicotera P (1998) 1-Methyl-4-phenylpyridinium induces autocrine excitotoxicity, protease activation, neuronal apoptosis. Mol Pharmacol 54:789–801

Lemaire C, reau K, Souvannavong V, Adam A (1998) Inhibition of caspase activity induces a switch from apoptosis to necrosis. Febs Lett 425:266–270

Lemasters JJ, Nieminen AL, Qian T, Trost LC, Elmore SP, Nishimura Y, Crowe RA, Cascio WE, Bradham CA, Brenner DA, Herman B (1998) The mitochondrial permeability transition in cell death: a common mechanism in necrosis, apoptosis, autophagy. Biochim Biophys Acta 1366:177–196

Li Y, Chopp M, Jiang N, Zaloga C (1995) In situ detection of DNA fragmentation after focal cerebral ischemia in mice. Brain Res Mol Brain Res 28:164–168

Li H, Zhu H, Xu CJ, Yuan J (1998) Cleavage of BID by caspase 8 mediates the mitochondrial damage in the Fas pathway of apoptosis. Cell 94:491–501

Li P, Nijhawan D, Budihardjo I, Srinivasula SM, Ahmad M, Alnemri ES, Wang X (1997) Cytochrome c, dATP-dependent formation of Apaf-1/caspase-9 complex initiates an apoptotic protease cascade. Cell 91:479–489

Li PA, Uchino H, Elmer E, Siesjo BK (1997) Amelioration by cyclosporin A of brain damage following 5 or 10min of ischemia in rats subjected to preischemic hyperglycemia. Brain Res 753:133–210

Lin KT, Xue JY, Lin MC, Spokas EG, Sun FF, Wong PY (1998) Peroxynitrite induces apoptosis of HL-60 cells by activation of a caspase-3 family protease. Am J Physiol 274:C855–C860

Linnik MD, Zobrist RH, Hatfield MD (1993) Evidence supporting a role for programmed cell death in focal cerebral ischemia in rats. Stroke 24:2002–2008; discussion 2008–2009

Liston P, Roy N, Tamai K, Lefebvre C, Baird S, Cherton-Horvat G, Farahani R, McLean M, Ikeda JE, MacKenzie A, Korneluk, RG (1996) Suppression of apoptosis in mammalian cells by NAIP, a related family of IAP genes. Nature 379:349–353

Litersky JM, Johnson GV (1995) Phosphorylation of τ in situ: inhibition of calcium dependent proteolysis. J Neurochem 65:903–911

Liu ZG, Hsu H, Goeddel DV, Karin M (1996) Dissection of TNF receptor 1 effector functions: JNK activation is not linked to apoptosis while NF-κB activation prevents cell death. Cell 87:565–576

Loddick SA, MacKenzie A, Rothwell NJ (1996) An ICE inhibitor, z-VAD-DCB attenuates ischaemic brain damage in the rat. Neuroreport 7:1465–1468

Lou J, Lenke LG, Ludwig FJ, O'Brien MF (1998) Apoptosis as a mechanism of neuronal cell death following acute experimental spinal cord injury. Spinal Cord 36:683–690

Luo X, Budihardjo I, Zou H, Slaughter C, Wang X (1998) Bid, a Bcl2 interacting protein, mediates cytochrome c release from mitochondria in response to activation of cell surface death receptors. Cell 94:481–490

Ma J, Endres M, Moskowitz MA (1998) Synergistic effects of caspase inhibitors, MK 801 in brain injury after transient focal cerebral ischaemia in mice. Br J Pharmacol 124:756–762

Maas J JR, Horstmann S, Borasio GD, Anneser JM, Shooter EM, Kahle PJ (1998) Apoptosis of central, peripheral neurons can be prevented with cyclin-dependent kinase/mitogen-activated protein kinase inhibitors. J Neurochem 70:1401–1410

MacManus JP, Buchan AM, Hill IE, Rasquinha I, Preston E (1993) Global ischemia can cause DNA fragmentation indicative of apoptosis in rat brain. Neurosci Lett 164:89–92

MacManus JP, Rasquinha I, Black MA, Laferriere NB, Monette R, Walker T, Morley P (1997) Glutamate-treated rat cortical neuronal cultures die in a way different from the classical apoptosis induced by staurosporine. Exp Cell Res 233:310–320

Majno G, Joris I (1995) Apoptosis, oncosis, necrosis An overview of cell death. Am J Pathol 146:3–15

Marzo I, Brenner C, Zamzami N, Jurgensmeier JM, Susin SA, Vieira HL, Prevost MC, Xie Z, Matsuyama S, Reed JC, Kroemer G (1998) Bax, adenine nucleotide translocator cooperate in the mitochondrial control of apoptosis. Science 281:2027–2031

McConkey DJ (1998) Biochemical determinants of apoptosis, necrosis [In Process Citation]. Toxicol Lett 99:157–168

McGinnis KM, Gnegy ME, Wang KK (1999) Endogenous Bax translocation in SH-SY5Y human neuroblastoma cells and cerebellar granule neurons undergoing apoptosis. J Neurochem 72:1899–1906

McGinnis KM, Whitton MM, Gnegy ME, Wang KK (1998) Calcium/calmodulin dependent protein kinase IV is cleaved by caspase-3, calpain in SH-SY5Y human neuroblastoma cells undergoing apoptosis. J Biol Chem 273:19993–20000

Muzio M, Chinnaiyan AM, Kischkel FC, O'Rourke K, Shevchenko A, Ni J, Scaffidi C, Bretz JD, Zhang M, Gentz R, Mann M, Krammer PH, Peter ME, Dixit VM (1996) FLICE, a novel FADD-homologous ICE/CED-3-like protease, is recruited to the CD95 (Fas/APO-1) death – inducing signaling complex. Cell 85:817–827

Nath R, Davis M, Probert AW, Kupina NC, Ren X, Schielke GP, Wang KK (2000) Processing of cdk5 activator p35 to its truncated form (p25) by calpain in acutely injured neuronal cells. Biochem Biophys Res Commun 274:16–21

Nath R, Mcginnis K, Dutta S, Shivers B, Wang KKW (2001) Inhibition of p38 kinase mimics survival signal-linked protection against apoptosis in rat cerebellar granule neurons. Cell Mol Biol Lett 6:173–184

Nath R, Probert A, JR, McGinnis KM, Wang KK (1998) Evidence for activation of caspase-3-like protease in excitotoxin-, hypoxia/hypoglycemia-injured neurons. J Neurochem 71:186–195

Nath R, Raser KJ, McGinnis K, Nadimpalli R, Stafford D, Wang KK (1996b) Effects of ICE-like protease, calpain inhibitors on neuronal apoptosis. Neuroreport 8: 249–255

Nath R, Raser KJ, Stafford D, Hajimohammadreza I, Posner A, Allen H, Talanian RV, Yuen P, Gilbertsen RB, Wang KK (1996a) Nonerythroid α-spectrin breakdown by calpain, interleukin 1 β-converting-enzyme-like protease(s) in apoptotic cells: contributory roles of both protease families in neuronal apoptosis. Biochem J 319:683–690

Nawashir H, Martin D, Hallenbeck JM (1997) Neuroprotective effects of TNF binding protein in focal cerebral ischemia. Brain Res 778:265–271

Ni B, Wu X, Du Y, Su Y, Hamilton-Byrd E, Rockey PK, Rosteck P JR, Poirier GG, Paul SM (1996) Cloning, expression of a rat brain interleukin-1β-converting enzyme (ICE)-related protease (IRP), its possible role in apoptosis of cultured cerebellar granule neurons. 1561–1569

Ni B, Wu X, Su Y, Stephenson D, Smalstig EB, Clemens J, Paul SM (1998) Transient global forebrain ischemia induces a prolonged expression of the caspase-3 mRNA in rat hippocampal CA1 pyramidal neurons. J Cereb Blood Flow Metab 18: 248–256

Nicotera P, Ankarcrona M, Bonfoco E, Orrenius S, Lipton SA (1997) Neuronal necrosis, apoptosis: two distinct events induced by exposure to glutamate or oxidative stress. Adv Neurol 72:95–101

Oppenheim RW (1991) Cell death during development of the nervous system. Annu Rev Neurosci 14:453–501

Pan G, Humke EW, Dixit VM (1998a) Activation of caspases triggered by cytochrome c. Febs Lett 426:151–154

Pan G, O'Rourke K, Dixit VM (1998b) Caspase-9, Bcl-XL, Apaf-1 form a ternary complex. J Biol Chem 273:5841–5845

Pap M, Cooper GM (1998) Role of glycogen synthase kinase-3 in the phosphatidylinositol 3- Kinase/Akt cell survival pathway. J Biol Chem 273:19929–19932

Pastorino JG, Chen ST, Tafani M, Snyder JW, Farber JL (1998) The overexpression of Bax produces cell death upon induction of the mitochondrial permeability transition. J Biol Chem 273:7770–7775

Pei JJ, Grundke-Iqbal I, Iqbal K, Bogdanovic N, Winblad B, Cowburn RF (1998) Accumulation of cyclin-dependent kinase 5 (cdk5) in neurons with early stages of Alzheimer's disease neurofibrillary degeneration. Brain Res 797:267–277

Peschon JJ, Torrance DS, Stocking KL, Glaccum MB, Otten C, Willis CR, Charrier K, Morrissey PJ, Ware CB, Mohler KM (1998) TNF receptor-deficient mice reveal divergent roles for p55, p75 in several models of inflammation. J Immunol 160: 943–952

Pettmann B, Henderson CE (1998) Neuronal cell death. Neuron 20:633–647

Pike BR, Zhao X, Newcomb JK, Posmantur RM, Wang KK, Hayes RL (1998) Regional calpain, caspase-3 proteolysis of α-spectrin after traumatic brain injury. Neuroreport 9:2437–2442

Pike BR, Zhao X, Newcomb JK, Wang KK, Posmantur RM, Hayes RL (1998) Temporal relationships between de novo protein synthesis, calpain, caspase 3-like protease activation, DNA fragmentation during apoptosis in septo-hippocampal cultures. J Neurosci Res 52:505–520

Pollard H, Charriaut-Marlangue C, Cantagrel S, Represa A, Robain O, Moreau J, Ben-Ari Y (1994) Kainate-induced apoptotic cell death in hippocampal neurons. Neuroscience 63:7–18

Portera-Cailliau C, Price DL, Martin LJ (1997) Excitotoxic neuronal death in the immature brain is an apoptosis- necrosis morphological continuum. J Comp Neurol 378:70–87

Portera-Cailliau C, Price DL, Martin LJ (1997) NonNMDA, NMDA receptor-mediated excitotoxic neuronal deaths in adult brain are morphologically distinct: further evidence for an apoptosis-necrosis continuum. J Comp Neurol 378:88–104

Pulera MR, Adams LM, Liu H, Santos DG, Nishimura RN, Yang F, Cole GM, Wasterlain CG, del Zoppo GJ (1998) Apoptosis in a neonatal rat model of cerebral hypoxia-ischemia. Stroke 29:2622–2630

Rink A, Fung KM, Trojanowski JQ, Lee VM, Neugebauer E, McIntosh TK (1995) Evidence of apoptotic cell death after experimental traumatic brain injury in the rat. Am J Pathol 147:1575–1583

Roy N, Mahadevan MS, McLean M, Shutler G, Yaraghi Z, Farahani R, Baird S, Besner-Johnston A, Lefebvre C, Kang X, et al. (1995) The gene for neuronal apoptosis inhibitory protein is partially deleted in individuals with spinal muscular atrophy. Cell 80:167–178

Ruberg M, France-Lanord V, Brugg B, Lambeng N, Michel PP, Anglade P, Hunot S, Damier P, Faucheux B, Hirsch E, Agid Y (1997) [Neuronal death caused by apoptosis in Parkinson disease]. Rev Neurol 153:499–508

Sarin A, Nakajima H, Henkart PA (1995) A protease-dependent TCR-induced death pathway in mature lymphocytes. J Immunol 154:5806–5812

Schulz JB, Weller M, Klockgether T (1996) Potassium deprivation-induced apoptosis of cerebellar granule neurons: a sequential requirement for new mRNA, protein synthesis, ICE-like protease activity, reactive oxygen species. J Neurosci 16:4696–4706

Scott RJ, Hegyi L (1997) Cell death in perinatal hypoxic-ischaemic brain injury. Neuropathol. Appl Neurobiol 23:307–314

Shinohara K, Tomioka M, Nakano H, Tone S, Ito H, Kawashima S (1996) Apoptosis induction resulting from proteasome inhibition. Biochem J 317:385–388

Shirvan A, Ziv I, Zilkha-Falb R, Machlyn T, Barzilai A, Melamed E (1998) Expression of cell cycle-related genes during neuronal apoptosis: is there a distinct pattern? Neurochem Res 23:767–777

Sidoti-de Fraisse C, Rincheval V, Risler Y, Mignotte B, Vayssiere JL (1998) TNF-α activates at least two apoptotic signaling cascades. Oncogene 17:1639–1651

Silos-Santiago I, Greenlund LJ, Johnson E JR, Snider WD (1995) Molecular genetics of neuronal survival. Curr Opin Neurobiol 5:42–49

Sipe KJ, Srisawasdi D, Dantzer R, Kelley KW, Weyhenmeyer JA (1996) An endogenous 55 kDa TNF receptor mediates cell death in a neural cell line. Brain Res Mol Brain Res 38:222–232

Squier MK, Miller AC, Malkinson AM, Cohen JJ (1994) Calpain activation in apoptosis. J Cell Physiol 159:229–237

Srinivasula SM, Ahmad M, Fernandes-Alnemri T, Alnemri ES (1998) Autoactivation of procaspase-9 by Apaf-1-mediated oligomerization. Mol Cell 1:949–957

Stefanis L, Troy CM, Qi H, Shelanski ML, Greene LA (1998) Caspase-2 (Nedd-2) processing, death of trophic factor-deprived PC12 cells, sympathetic neurons occur independently of caspase-3 (CPP32)- like activity. J Neurosci 18:9204–9215

Stennicke HR, Jurgensmeier JM, Shin H, Deveraux Q, Wolf BB, Yang X, Zhou Q, Ellerby HM, Ellerby LM, Bredesen D, Green DR, Reed JC, Froelich CJ, Salvesen GS (1998) Procaspase-3 is a major physiologic target of caspase-8. J Biol Chem 273:27084–27090

Su JH, Deng G, Cotman CW (1997) Bax protein expression is increased in Alzheimer's brain: correlations with DNA damage, Bcl-2 expression, brain pathology. J Neuropathol Exp Neurol 56:86–93

Tamm I, Wang Y, Sausville E, Scudiero DA, Vigna N, Oltersdorf T, Reed JC (1998) IAP-family protein survivin inhibits caspase activity, apoptosis induced by Fas (CD95), Bax, caspases, anticancer drugs. Cancer Res 58:5315–5320

Tan S, Sagara Y, Liu Y, Maher P, Schubert D (1998a) The regulation of reactive oxygen species production during programmed cell death. J Cell Biol 141:1423–1432

Tan S, Wood M, Maher P (1998b) Oxidative stress induces a form of programmed cell death with characteristics of both apoptosis, necrosis in neuronal cells. J Neurochem 71:95–105

Thornberry NA, Lazebnik Y (1998) Caspases: enemies within. Science 281:1312–1316

Troost D, Aten J, Morsink F, de Jong JM (1995) Apoptosis in amyotrophic lateral sclerosis is not restricted to motor neurons. Bcl-2 expression is increased in unaffected post-central gyrus. Neuropathol Appl Neurobiol 21:498–504

Trump BF, Berezesky IK, Chang SH, Phelps PC (1997) The pathways of cell death: oncosis, apoptosis, and necrosis. Toxicol Pathol 25:82–88.

Uno H, Matsuyama T, Akita H, Nishimura H, Sugita M (1997) Induction of tumor necrosis factor-α in the mouse hippocampus following transient forebrain ischemia. J Cereb Blood Flow Metab 17:491–499

Van Antwerp DJ, Martin SJ, Kafri T, Green DR, Verma IM (1996) Suppression of TNF-α-induced apoptosis by NF-κB. Science 274:787–789

Vanags DM, Porn-Ares MI, Coppola S, Burgess DH, Orrenius S (1996) Protease involvement in fodrin cleavage, phosphatidylserine exposure in apoptosis. J Biol Chem 271:31075–31085

Velasco E, Valero C, Valero A, Moreno F Hernandez-Chico C (1996) Molecular analysis of the SMN, NAIP genes in Spanish spinal muscular atrophy (SMA) families, correlation between number of copies of cBCD541, SMA phenotype. Hum Mol Genet 5:257–263

Villalba M, Bockaert J, Journot L (1997) Concomitant induction of apoptosis, necrosis in cerebellar granule cells following serum, potassium withdrawal. Neuroreport 8:981–985

Viviani B, Corsini E, Galli CL, Marinovich M (1998) Glia increase degeneration of hippocampal neurons through release of tumor necrosis factor-α. Toxicol Appl Pharmacol 150:271–276

Wang CY, Mayo MW, Baldwin A, JR (1996) TNF-, cancer therapy-induced apoptosis: potentiation by inhibition of NF-κB. Science 274:784–787

Wang CY, Mayo MW, Korneluk RG, Goeddel DV, Baldwin A, JR (1998) NF κB anti-apoptosis: induction of TRAF1, TRAF2, c-IAP1, c- IAP2 to suppress caspase 8 activation. Science 281:1680–1683

Wang KKW (2000) Calpain and Caspase: Can You Tell the Difference. Trends Neurosci 23:20–26

Wang KK, Posmantur R, Nath R, McGinnis K, Whitton M, Talanian RV, Glantz SB, Morrow JS (1998) Simultaneous degradation of αII-, βII-spectrin by caspase 3 (CPP32) in apoptotic cells. J Biol Chem 273:22490–22497

Watson A, Eilers A, Lallemand D, Kyriakis J, Rubin LL, Ham J (1998) Phosphorylation of c-Jun is necessary for apoptosis induced by survival signal withdrawal in cerebellar granule neurons. J Neurosci 18:751–762

Wu M, Lee H, Bellas RE, Schauer SL, Arsura M, Katz D, FitzGerald MJ, Rothstein TL, Sherr DH, Sonenshein GE (1996) Inhibition of NF-κB/Rel induces apoptosis of murine B cells. Embo J 15:4682–4690

Wullner U, Weller M, Schulz JB, Krajewski S, Reed JC, Klockgether T (1998) Bcl-2, Bax, Bcl-x expression in neuronal apoptosis: a study of mutant weaver, lurcher mice. Acta Neuropathol 96:233–238

Xu DG, Crocker SJ, Doucet JP, St-Jean M, Tamai K, Hakim AM, Iked JE, Liston P, Thompson CS, Korneluk RG, MacKenzie A, Robertson GS (1997) Elevation of neuronal expression of NAIP reduces ischemic damage in the rat hippocampus. Nat Med 3:997–1004

Yaffe MB, Rittinger K, Volinia S, Caron PR, Aitken A, Leffers H, Gamblin SJ, Smerdon SJ, Cantley LC (1997) The structural basis for 14-3-3: phosphopeptide binding specificity. Cell 91:961–971

Yakovlev AG, Knoblach SM, Fan L, Fox GB, Goodnight R, Faden AI (1997) Activation of CPP32-like caspases contributes to neuronal apoptosis, neurological dysfunction after traumatic brain injury. J Neurosci 17:7415–7424

Yang GY, Gong C, Qin Z, Ye W, Mao Y, Bertz AL (1998) Inhibition of TNFα attenuates infarct volume, ICAM-1 expression in ischemic mouse brain. Neuroreport 9:2131–2134

Yano S, Tokumitsu H, Soderling TR (1998) Calcium promotes cell survival through CaM K kinase activation of the protein-kinase-B pathway. Nature 396:584–587

Yao R, Cooper GM (1995) Requirement for phosphatidylinositol-3 kinase in the prevention of apoptosis by nerve growth factor. Science 267:2003–2006

Zhang Q, Ahuja HS, Zakeri ZF, Wolgemuth DJ (1997) Cyclin-dependent kinase 5 is associated with apoptotic cell death during development, tissue remodeling. Dev Biol 183:222–233

Zhang J, Price JO, Graham DG, Montine TJ (1998) Secondary excitotoxicity contributes to dopamine-induced apoptosis of dopaminergic neuronal cultures. Biochem Biophys Res Commun 248:812–816

CHAPTER 3
Reducing Neuroinflammation

K.J. Becker and J.M. Hallenbeck

A. Introduction

The response of living tissue to injury is characterized by an increase in blood flow and a local influx of leukocytes. These events, mediated by either the upregulation or induction of adhesion molecules and the release of biologically active substances, are collectively termed *inflammation*. The inflammatory response is not antigen-specific and occurs following all modes of tissue injury, including traumatic, ischemic, and infectious insults. In infection, the inflammatory response is usually accompanied by an immune response, which is antigen specific. The brain has traditionally been regarded as an immunologically privileged organ, but it is subject to continuous lymphocyte surveillance and inflammatory and immunological reactions do occur in the brain. The character of the inflammatory response in the brain, however, differs from that of other organs. These differences result from the fact that under normal circumstances, the blood–brain barrier (BBB) exists and the cerebral microvessels express few of the adhesion molecules required for leukocyte trafficking into the brain. But when the brain is injured, the BBB becomes compromised (Belayev et al. 1996) and there is an upregulation of leukocyte adhesion molecules, both of which allow for an inflammatory response, and potentially an immune response, to occur. Because neural tissue does not express major histocompatibility (MHC) antigens, effective antigen presenting cells (APCs) and adequate costimulation are lacking in the brain (Perry 1998), and components of the neuronal cell membranes are immunosuppressive (Irani et al. 1996), initiation of immunological responses in the brain tends to be limited. From a teleological perspective, the purpose of the inflammatory response is to help destroy invading pathogens, which would be important in an infectious insult. In the case of ischemic or traumatic brain injury, however, there are no pathogens to destroy. The purpose of an inflammatory response in these instances is unclear, but may be related to a need for phagocytosis of cellular debris in order to promote wound healing. In the process, however, the inflammatory response may enhance ischemic brain injury.

    This chapter will explore how the initiation and the execution of the inflammatory response are affected by the unique microenvironment of the brain. It is impossible to consider the actions and consequences of one biological mediator of inflammation, such as a cytokine, in isolation from others, as these mediators are part of a network that induce or inhibit each other as the inflammatory response evolves. Current strategies aimed at limiting leukocyte-mediated brain injury, however, are usually quite specific and target just one inflammatory mediator at a time. Because of the redundancy and overlap in function of these mediators, the potential clinical efficacy of a single intervention is probably limited. Despite these limitations, experience with anti-inflammatory therapy for central nervous system injury is increasing, and the therapeutic utility of intervening with inflammatory pathways in the brain, particularly in the context of acute ischemic stroke, will be discussed.

B. The Central Nervous System Inflammatory Response

I. Initiation of the Inflammatory Response

The signals responsible for inciting the initial inflammatory response after tissue injury are not clear, but the signals that maintain the response are well characterized. Bacterial specific peptides, or *n*-formylated methionyl peptides, are potent chemotaxins for leukocytes. Mitochondria, which are presumably of prokaryotic origin, also synthesize *n*-formylated methionyl peptides and release them upon injury (CARP 1982). Whether or not *n*-formylated methionyl peptides are released by injured brain has never been studied, but these peptides can stimulate neutrophil chemotaxis across the BBB (CORVIN et al. 1990). Complement activation is central to the inflammatory process. For a complete discussion of the role of complement in inflammation, see the review by Frank and Fries (FRANK and FRIES 1991). Complement is an acute phase reactant synthesized by the liver. Extrahepatic production of complement has been documented and there is evidence that complement is synthesized in the brain (BARNUM 1995) and activated following traumatic brain injury (BELLANDER et al. 1996) and ischemic stroke (LINDSBERG et al. 1996b). C5a, or anaphylatoxin, is an extremely potent chemoattractant for neutrophils and macrophages. Both C3a and C5a cause smooth muscle contraction, vasodilation, and enhance vascular permeability. Receptors for C3a and C5a are abundant within the brain and can be found on astrocytes and microglia, as well as on invading cells (GASQUE et al. 1995; GASQUE et al. 1998).

Phospholipase A_2 (PLA_2) is a cytosolic enzyme that is activated by a number of stimuli, including excitotoxic amino acids (BONVENTRE 1997) and ischemia (CLEMENS et al. 1996). PLA_2 degrades membrane phospholipids to eicosanoids and arachidonic acid. Arachidonic acid can then be metabolized by lipoxygenase or cyclooxygenase (COX) to generate either leukotrienes or prostaglandins, respectively. Leukotriene B_4 (LTB_4) is one of the most potent leukocyte chemoattractants known. Production of both LTB_4 and leukotriene C_4 (LTC_4) is stimulated by cerebral ischemia and reperfusion (NAMURA et al. 1994), and antagonism of either decreases postreperfusion cerebral edema

(DEMPSEY et al. 1986; MINAMISAWA et al. 1988). Prostaglandin products include thromboxanes and prostacyclins. Thromboxane A_2 (TxA_2) is a vasoconstrictor and stimulates platelet aggregation. The stable metabolite of TxA_2 is thromboxane B_2 (TxB_2). In models of transient cerebral ischemia, brain (COLE et al. 1993) and plasma (SADOSHIMA et al. 1989) levels of TxB_2 rise during reperfusion. The major source of brain TxA_2 is microglia (GIULIAN et al. 1996). Inhibition of TxA_2 synthesis during ischemia decreases neuronal death (IIJIMA et al. 1996; IMURA et al. 1995) and improves postischemic cerebral blood flow (CBF) (PETTIGREW et al. 1989; SADOSHIMA et al. 1989). PGD_2 and PGE_2, the major prostaglandin products in ischemic brain (KEMPSKI et al. 1987), also increase vascular permeability but prevent platelet aggregation.

Platelet activating factors, or PAFs, are biologically active lipid derivatives of the phospholipid membrane produced by activated platelets and leukocytes. PAFs cause platelet aggregation, are potent chemoattractants, and activate macrophages and polymorphonuclear cells (PMNs). PAFs also enhance excitatory amino acid release and activate the expression of matrix metalloproteinases (MMPs) (BAZAN and ALLAN 1996). PAF is produced in ischemic brain (NISHIDA and MARKEY 1996; PETTIGREW et al. 1995) and PAF antagonists improve outcome from both ischemic (LIU et al. 1996) and traumatic (TOKUTOMI et al. 1994) brain injury.

II. Consequences of Leukocyte Activation

Most phagocytic and cytotoxic leukocytes possess the necessary cellular machinery to produce a variety of toxic oxygen species. These cells include PMNs, macrophages, monocytes, microglia, and subsets of lymphocytes. Upon activation, these cells undergo a "respiratory burst", during which time they consume large amounts of oxygen and produce a number of toxic oxygen compounds. For a complete review of the subject, see ROSEN et al. (ROSEN et al. 1995). Briefly, during the respiratory burst, oxygen is reduced to either superoxide anion ($O_2^{\bullet-}$) or the perhydroxyl radical ($HO_2^{\bullet}$).

$$O_2 + e^- \rightarrow \underset{\text{perhydroxyl radical}}{HO_2^{\bullet}} \leftrightarrow H^+ + \underset{\text{superoxide anion}}{O_2^{\bullet-}}$$

The superoxide anion is toxic by virtue of production of a number of other toxic byproducts, including hydrogen peroxide (H_2O_2). Hydrogen peroxide is formed through a reaction catalyzed by superoxide dismustase (SOD).

$$HO_2^{\bullet} + O_2^{\bullet-} + H^+ \xrightarrow{SOD} O_2 + H_2O_2$$

Hydrogen peroxide can be detoxified by catalase or glutathione peroxidase, but it can also be converted to hypochlorous acid (HOCl) by myeloperoxidase (MPO), an enzyme stored in the azurophilic granules of neutrophils. Hypochlorous acid is metabolized to hypochlorite (OCl^-) and chlorine. Hypochlorite, and chlorine reaction products such as the chloramines, are extremely cytotoxic, either directly or in conjunction with granule-based proteinases (WEISS 1989).

$$H_2O_2 + Cl^- + H^+ \xrightarrow{MPO} H_2O + \underset{\text{hypochlorous acid}}{HOCl}$$

$$HOCl \leftrightarrow H^+ + \underset{\text{hypochlorite}}{OCl^-}$$

$$HOCl + Cl^- \rightarrow \underset{\text{chlorine}}{Cl_2} + OH^-$$

The hydroxyl radical (·OH), another cytotoxic byproduct of H_2O_2 metabolism, can be produced by several different reactions, including the Haber-Weiss reaction:

$$Fe^{2+} + H_2O_2 \rightarrow Fe^{3+} + OH^- + \text{·}OH$$

The hydroxyl radical is a powerful oxidant and can extract electrons from a number of different compounds to form new radicals that then oxidize other substances:

$$\text{·}OH + R \rightarrow OH^- + R\text{·}$$

These toxic oxygen metabolites damage cell membranes, proteins and DNA and can trigger PMN activation and adhesion to vascular endothelium (FRATICELLI et al. 1996). Antagonists of PAF and TxA_2 can decrease the generation of these oxygen free radicals (MATSUO et al. 1996).

In addition to the toxic oxygen radicals produced by activated leukocytes, a number of potentially injurious proteases are secreted from the lysosomal granules of these cells. The proteases, which include elastase, collagenase, gelatinase, cathepsin D, and cathepsin G, aid in tissue destruction and allow further leukocyte infiltration, inflammation, and injury to occur. The endogenous inhibitor of neutrophil elastase, α_1-antitrypsin, is inhibited by HOCl, therefore the respiratory burst enhances the process of tissue degradation (WEISS 1989). Tissue destruction is also aided by a variety of MMPs that are activated in ischemia (HEO et al. 1999; ROMANIC et al. 1998). The loss of vascular integrity due to disruption of the basal lamina is associated with hemorrhagic transformation of ischemic infarcts (HAMANN et al. 1996; HEO et al. 1999).

Nonsteroidal anti-inflammatory drugs (NSAIDs) inhibit the synthesis of thromboxanes and prostacyclins by inhibiting COX activity. Ibuprofen decreases brain levels of TxB_2, improves postischemic CBF (GRICE et al. 1987), and decreases neuronal injury (PATEL et al. 1993). Indomethacin decreases cerebral edema following transient ischemia (DELUGA et al. 1991) and, like ibuprofen, improves postischemic blood flow (HALLENBECK and FURLOW 1979; JOHSHITA et al. 1989). Aspirin, the most widely used NSAID, has been shown to improve outcome from stroke in clinical trials (INTERNATIONAL STROKE TRIAL COLLABORATIVE GROUP 1997; CAST COLLABORATIVE GROUP 1997). The sooner aspirin is administered after stroke onset, the more robust its effect (CAST

COLLABORATIVE GROUP 1997). Tetracycline derivatives can inhibit microglial activation and are neuroprotective in animal models of cerebral ischemia (CLARK et al. 1994; YRJANHEIKKI et al. 1998).

Leukocytes can directly injure brain tissue through the release of toxic oxygen metabolites and proteases. Activated leukocytes also indirectly injure brain tissue by secreting substances like cytokines that are directly cytotoxic or that amplify the inflammatory process. Interleukin (IL)-1, tumor necrosis factor (TNF), and IL-8 are proinflammatory cytokines. This chapter will focus on the contribution of the cytokines to ischemia-induced inflammation, but the noninflammatory and nonimmunologic effects of cytokines will be mentioned for the sake of completeness. It is also important to realize that not all cytokines promote inflammation or brain injury. In fact, cytokines like IL-1 receptor antagonist (IL-1ra) and transforming growth factor (TGF)-β, produced as the inflammatory response evolves, can limit the extent of the inflammatory response and be neuroprotective. The time course of cytokine expression and leukocyte invasion in the brain is quite stereotyped and largely independent of the mode of injury. It is important to realize, however, that the species, strain, and sex of the animals studied can influence experimental results. While the ensuing discussion focuses on cerebral ischemia, the response to traumatic brain injury is similar and will therefore not be addressed separately.

III. Cytokines in Brain Injury

1. Interleukin-1β

Leukocytes are the primary source of most cytokines, but many of the cytokines are also produced within the brain tissue itself. Genes that code for interleukin-1 converting enzyme (ICE), IL-1α, IL-1β, and IL-1ra are present within the cerebral microvasculature (WONG et al. 1995). In normal brain, there is either little or no expression of IL-1 (MINAMI et al. 1992; SAIRANEN et al. 1997; ZHANG et al. 1998c), but within hours after ischemia there is induction of IL-1β expression (DAVIES et al. 1999; MINAMI et al. 1992; SAITO et al. 1996; WANG et al. 1994b; ZHAI et al. 1997). Most of the expression is localized to the area of ischemia and is seen within the vasculature and in microglial cells (DAVIES et al. 1999; GIULIAN et al. 1986; ZHANG et al. 1998c). Astroglia and oligodendroglia, however, can also produce IL-1β (SAIRANEN et al. 1997) and IL-1β seen in the contralateral hemisphere within 48h after ischemia (DAVIES et al. 1999).

Interleukin-1β is synthesized as an inactive precursor that is converted to its active form by a highly selective protease called interleukin-1β converting enzyme (ICE). ICE, also known as caspase I, causes apoptosis (MILLER et al. 1997b). The role of ICE in neuronal apoptosis following cerebral ischemia has been a focus of intense research over recent years. While the subject of apoptosis is beyond the scope of this chapter, it is important to mention that given the same degree of ischemia, ICE deficient animals have smaller strokes than their wild type counterparts (SCHIELKE et al. 1998). The mechanism of this protection, however, is unclear.

IL-1β can promote brain injury by disrupting the BBB (QUAGLIARELLO et al. 1991), which may occur by virtue of neutrophil recruitment (BOLTON et al. 1998). IL-1β also enhances the release of arachidonic acid from astrocytes (STELLA et al. 1997) and induces COX-2 mRNA expression within the cerebral microvasculature (CAO et al. 1997). Furthermore, IL-1 induces the expression of L-selectin ligands (KANDA et al. 1995) and intercellular adhesion molecule (ICAM)-1 (WONG and DOROVINI-ZIS 1992) on microvascular endothelial cells. Ultimately, these effects allow for leukocyte entry into the brain, and thus, leukocyte-mediated brain damage. Consistent with these observations is the fact that injection of IL-1 into the striatum during stroke increases brain injury and infarct size (STROEMER and ROTHWELL 1998).

Interleukin-1β may act as a neurotransmitter as well as a cytokine. IL-1β has been shown to promote slow-wave sleep (GEMMA et al. 1997) and enhance long-term potentiation (SCHNEIDER et al. 1998). IL-1β has profound effects on the hypothalamus and stimulates the release of corticotropin-releasing factor (CRF), which leads to glucocorticoid secretion (SAPOLSKY et al. 1987) and, potentially, hyperglycemia. Both hypercortisolemia (O'NEILL et al. 1991) and hyperglycemia (TRACEY et al. 1993) are associated with a worse outcome following stroke. IL-1β, in concert with IL-6, also acts on the hypothalamus to cause fever (KLIR et al. 1993; MILLER et al. 1997a). Hyperthermia enhances leukocyte function and potentiates the inflammatory response (HUANG et al. 1996; WANG et al. 1998) and is associated with increased morbidity and mortality in the poststroke period (AZZIMONDI et al. 1995; CASTILLO et al. 1998; REITH et al. 1996).

For a cytokine to have a direct effect on the brain, receptors for that cytokine must be present in the brain. IL-1 receptors were first detected in the brain in the late 1980s (FARRAR et al. 1987). Since that time it has become apparent that there are two distinct types of IL-1 receptors. The distribution, expression, and function of these receptors differ. True signal transduction takes place at the type I receptor (SIMS et al. 1994). Type I receptors are expressed at high levels throughout the normal brain and their expression is further upregulated following ischemia (WANG et al. 1997). These receptors are located predominantly on barrier-related cells, including the leptomeninges, portions of the ependyma, the choroid plexus, and vascular endothelium (ERICSSON et al. 1995; VAN DAM et al. 1996), but they are also found on neurons (PARNET et al. 1994; YABUUCHI et al. 1994). The type II receptors are felt to be decoy receptors in that they bind IL-1 but do not transduce signal (SIMS et al. 1994). As such, type II receptors serve as endogenous antagonists of IL-1. The type II receptors are present at very low levels in normal brain, but their expression is dramatically increased shortly (6h) after ischemia (WANG et al. 1997). IL-1ra is another endogenous antagonist of IL-1. It is also expressed at very low levels in the normal brain (MATSUKAWA et al. 1997) and upregulated during focal ischemia (LODDICK et al. 1997; WANG et al. 1997).

For some, or perhaps all, of the reasons listed above, inhibition of IL-1 activity at the time of stroke improves outcome. When inhibitors of ICE are injected into the ventricles of animals prior to ischemia, IL-1β expression and infarct size are decreased (Hara et al. 1997). Neutralization of endogenous IL-1ra worsens outcome from stroke (Loddick et al. 1997) and exogenous administration of IL-1ra improves outcome (Garcia et al. 1995; Relton et al. 1996). Animals that overexpress IL-1ra in the brain as a result of virally mediated transfection also experience smaller infarcts (Betz et al. 1995; Yang et al. 1997), and inhibition of IL-1 activity decreases postischemic cerebral edema (Yamasaki et al. 1994).

2. Tumor Necrosis Factor α

There is a fair amount of redundancy between the actions of TNF-α and IL-1β, and, like IL-1β, TNF-α expression in the brain is upregulated shortly after ischemia (Saito et al. 1996; Wang et al. 1994b; Yoshimoto et al. 1997). The TNF-α within the brain is produced largely by infiltrating macrophages, PMNs, and microglia (Sebire et al. 1993; Uno et al. 1997), but endothelial cells, neurons and astrocytes are also capable of producing TNF-α (Botchkina et al. 1997; Liu et al. 1994). There is a corresponding increase in TNF-α levels in the peripheral circulation following stroke (Lavine et al. 1998). This cytokine is produced as an inactive membrane-bound cytokine that is converted to its soluble active form by MMPs (Gearing et al. 1994; McGeehan et al. 1994). Ischemia (Heo et al. 1999; Romanic et al. 1998) and cytokines (Zhang et al. 1998b) induce the production of MMPs. Inhibitors of metalloproteinases prevent the release of TNF-α into the circulation (Amour et al. 1998; Gearing et al. 1994; McGeehan et al. 1994), and inhibition of MMPs (MMP-9 in particular) decreases infarct size after cerebral ischemia (Romanic et al. 1998).

Similar to IL-1β, TNF-α can impair the integrity of the BBB, although it is not as effective in doing so (Quagliarello et al. 1991). However, unlike IL-1β, for which there is little disagreement about its detrimental effects in stroke, there is some controversy surrounding the role of TNF-α in cerebral ischemia. Much of the controversy arises from the fact that animals lacking TNF receptors experience larger infarcts than their wild type counterparts (Bruce et al. 1996; Gary et al. 1998). Because TNF has such pleiotropic effects, however, it is difficult to imagine that an animal could develop normally without the ability to respond to this cytokine. Thus, it is possible that the TNF receptor "knock out" animal has developed alternative pathways to respond to TNF and the true pathophysiology of cerebral ischemic injury is not adequately modeled. Alternatively, since TNF-α seems to play a role in promoting cell growth under normal conditions (Cheng et al. 1994), animals lacking the ability to respond to TNF may be chronically "stressed" or subject to a "deprivation state" and therefore more susceptible to injury. In animals replete with TNF receptors, intracerebroventricular injection of TNF-α immediately prior to MCA occlusion significantly increases infarct size (BARONE et al. 1997) and injection of TNF-α into the cortex of the brain increases neutrophil accumulation within the microvasculature (LIU et al. 1994). Transgenic animals that overexpress TNF-α in their astrocytes, but not their neurons, develop neurologic disease with evidence of chronic inflammation and degeneration (AKASSOGLOU et al. 1997).

Like the receptors for IL-1, there are distinct forms of the TNF receptor with differing functions. These receptors are referred to as the 55-kd or p55 (TNF-RI) and the 75-kd or p75 (TNF-RII) receptors. Engagement of either receptor can lead to apoptotic cell death, although the p55 receptor is more potent than the p75 receptor in this regard (BAZZONI and BEUTLER 1996). The p55 receptor is also largely responsible for TNF-mediated activation of the hypothalamic-pituitary-adrenal axis and for the central induction of peripheral IL-6 by TNF (BENIGNI et al. 1996). MCA occlusion induces significant upregulation of both TNF receptors. The p55 receptor appears within 6h of ischemia while the p75 receptor does not appear until 24h after the onset of ischemia (BOTCHKINA et al. 1997). Animals that are depleted of the p55 receptor, but not the p75 receptor, are more susceptible to ischemic injury (GARY et al. 1998).

Evidence to support the detrimental effects of TNF in stroke comes from the fact that antagonism of TNF-α activity decreases infarct size. Intracerebroventricular injection of either soluble TNF-RI (sTNF-RI), which binds to and immobilizes TNF, or TNF neutralizing antibodies prior to focal ischemia decreases stroke size (BARONE et al. 1997; YANG et al. 1998). Similarly, topical cortical administration of TNF binding protein (TNF-bp), which also binds to and immobilizes TNF, can significantly decrease infarct size (NAWASHIRO et al. 1997a). Finally, systemic administration of either TNF-bp or TNF neutralizing antibodies improves microvascular perfusion and decreases infarct size (DAWSON et al. 1996a; LAVINE et al. 1998).

3. Interleukin-6

Interleukin-6 is a pivotal cytokine in many ways. It is often equated with an acute phase reactant and its presence is indicative of ongoing inflammation, yet it also induces the production of cytokines that serve to limit the inflammatory response (MANTOVANI 1997). It has been shown to have both neuroprotective and neurodestructive properties (GADIENT and OTTEN 1997). IL-1 potently induces IL-6 production (SEBIRE et al. 1993), and intracerebroventricular administration of IL-1 leads to increases in peripheral IL-6 (DE SIMONI et al. 1993; KITAMURA et al. 1998). IL-6 is also felt to mediate the pyrogenic effects of IL-1β (LUHESHI et al. 1997).

Animal studies show that IL-6 production is induced in cerebral ischemia (HAGBERG et al. 1996; LODDICK et al. 1998). Its levels peak at a later time than those of TNF-α and IL-1β (SAITO et al. 1996; WANG et al. 1995b; YOSHIMOTO et al. 1997). Astrocytes and neurons can produce IL-6, but microglia are the main source of IL-6 in the brain (SEBIRE et al. 1993). Cells that produce IL-6 also express receptors for IL-6, suggesting that the cytokine has autocrine activity (SCHOBITZ et al. 1993). Transgenic animals that over-express IL-6 in their astrocytes develop extensive breakdown of the BBB (BRETT et al. 1995), produce large amounts of C3 (BARNUM et al. 1996), and develop progressive neurodegenerative disease (HERNANDEZ et al. 1997). If the IL-6 production is targeted to neurons, however, there is no neuronal damage despite astrogliosis and an increase in the number of ramified microglia (FATTORI et al. 1995). In clinical studies, CSF levels of interleukin-6 following stroke are predictive of infarct size (TARKOWSKI et al. 1995). Plasma levels of IL-6 are also elevated in acute stroke (KIM et al. 1996) and are higher in those patients with the largest strokes (BEAMER et al. 1995). In traumatic brain injury, jugular levels of IL-6 are higher than systemic levels, indicating CNS production of the cytokine (MCKEATING et al. 1997). In animal studies, infusion of IL-6 into the ventricles during permanent focal ischemia seems to be neuroprotective (LODDICK et al. 1998). Studies of IL-6 antagonism in acute stroke have not been done.

While a number of other cytokines can be found in the brain, none have been studied as extensively as IL-1β, TNF-α, and IL-6. These cytokines have redundant, overlapping, and pleiotropic effects, which likely reflects how integral they are to our survival. That IL-6 and TNF-α have been implicated to have both beneficial and deleterious effects highlights the complexities of the inflammatory response and the cytokine network. These are essential cytokines and sustained blockade of their activity is probably not desirable. Thus, from a pharmacotherapeutic perspective, it is important to determine when cytokine antagonism should be instituted in the time period following brain injury, as well as how long it should be maintained.

4. Interleukin-8

Interleukin-8 is an important chemoattractant for and activator of neutrophils. IL-8 production in extravascular tissue promotes leukocyte accumulation, while high concentrations of intravascular IL-8 tend to inhibit leukocyte emigration (LEY et al. 1993). IL-8 production is stimulated by IL-1, TNF-α, and C5a (EMBER et al. 1994). Much less is known about the role of IL-8 in stroke as compared to IL-1β, TNF-α, and IL-6. IL-8 levels are elevated in the brains of animals following transient focal cerebral ischemia (MATSUMOTO et al. 1997) and there is an increase in intrathecal IL-8 following stroke (TARKOWSKI et al. 1997). As for the other cytokines, microglia are the primary source of brain IL-8 (EHRLICH et al. 1998) and neutralization of IL-8 following stroke decreases brain edema and infarct size (MATSUMOTO et al. 1997).

IV. Nuclear Transcription Factor-κB

Nuclear factor-κB (NF-κB) is a transcription factor that is responsible for the transcription of genes coding for chemotaxins, proinflammatory cytokines, and adhesion molecules. In normal brain, the role of NF-κB is probably limited by the ubiquitous presence of gangliosides, which prevents its activation in leukocytes (IRANI et al. 1996). Oxygen radicals, TNF-α, IL-1β and PAF are all known to activate NF-κB in vitro (BALDWIN 1996; YE et al. 1996). Hypoxia and reoxygenation activate NF-κB in human brain endothelial cells (HOWARD et al. 1998), while excitotoxic amino acids (O'NEILL and KALTSCHMIDT 1997) and transient cerebral ischemia activate NF-κB in neurons (CLEMENS et al. 1997; SALMINEN et al. 1995; TERAI et al. 1996). As mentioned previously, aspirin improves outcome in stroke (INTERNATIONAL STROKE TRIAL COLLABORATIVE GROUP 1997; CAST COLLABORATIVE GROUP 1997). Part of this benefit may be attributable to the fact that, at least at high doses, aspirin can inhibit NF-κB mobilization (WEBER et al. 1995) and thereby limit glutamate mediated neurotoxicity (GRILLI et al. 1996).

V. Leukocyte Adhesion

For leukocytes to travel to an area of inflammation, a well-orchestrated series of events must first occur. In addition to the chemotactic signals mentioned above, leukocyte adhesion molecules need to be expressed on the inflamed microvasculature so that the cells can bind to the vessel and emigrate into the tissue. For a thorough review of this topic, see the review by FRENETTE and WAGNER (FRENETTE and WAGNER 1996a and 1996b). The selectins are adhesion molecules that mediate the initial interaction between the leukocytes and the endothelium. L-selectin is constitutively expressed on most white blood cells (WBCs). E-selectin is expressed only on cytokine-activated endothelial cells. P-selectin is stored in granules of platelets and endothelial cells and is rapidly mobilized to the cell surface by a number of cytokines and inflammatory mediators. Both P- and E-selectin recognize mucin-like sialylated glycoproteins on the surface of leukocytes while L-selectin recognizes similarly glycosylated mucins on the surface of endothelial cells. In addition to the cytokines that induce P- and E-selectin expression, toxic oxygen radicals can also stimulate expression (MEHTA et al. 1995; NORGARD et al. 1993). The interaction of selectins with their ligands is weak and results in a rapid on-off rate and *rolling* of the leukocytes along the endothelium. The process of rolling allows for the interaction of other leukocyte and endothelial adhesion molecules that ultimately result in firm adherence of the leukocytes to the endothelium.

In addition to the selectins, leukocytes express surface glycoproteins known as integrins. The integrins are heterodimers composed of an α and β chain. The β_2 integrins share a common β chain (CD18) and include CD11a/CD18 (LFA-1) and CD11b/CD18 (Mac-1), which are the most important molecules involved in neutrophil and monocyte adhesion. These molecules bind to members of the immunoglobulin (Ig) superfamily, including intercellular adhesion molecule (ICAM)-1 and ICAM-2, which are expressed on vascular endothelium. The process of leukocyte rolling not only brings these molecules into apposition, but it increases the expression and avidity of the β_2 integrins for their ligands. The subsequent interaction of the β_2 integrins with their ligands results in firm cellular adhesion.

The brain undergoes constant immune surveillance by T cells (LASSMANN 1997) and these cells express another class of integrins referred to as very late activation (VLA) antigens. Resting T lymphocytes express VLA-4, VLA-5, and VLA-6, which mediate binding of the cells to the vascular endothelium as well as to the extracellular matrix. The most important of the VLA molecules is VLA-4. VLA-4 binds to fibronectin and endothelial vascular cell adhesion molecule (VCAM)-1, another member of the Ig superfamily. In contrast to neutrophils and monocytes, T cells have unique antigen specificity. The antigen specificity of the T cell, however, is not important for entry into the brain during an inflammatory response, but it is dependent upon the complementary expression of VLA-4 and VCAM-1 (IRANI and GRIFFIN 1996). The antigen specificity of the cells determines whether they will be retained in the brain after the initial inflammatory response subsides (IRANI and GRIFFIN 1996).

Most of the described adhesion molecules are either not expressed on normal brain microvessels or are expressed only at very low levels. Brain microvessels constitutively express large amounts of ICAM-2, but only minimal ICAM-1 (NAVRATIL et al. 1997) and no P-selectin (NAVRATIL et al. 1997; BARKALOW et al. 1996; SUZUKI et al. 1998; ZHANG et al. 1998a), E-selectin, (HARING et al. 1996; ZHANG et al. 1996), or VCAM-1 (MAENPAA et al. 1997). In vitro, the inflammatory cytokines IL-1β and TNF-α, as well as hypoxia-reoxygenation, can stimulate the surface expression of ICAM-1 and VCAM-1 in human cerebral microvascular endothelial cells (HESS et al. 1994; STANIMIROVIC et al. 1997; STINS et al. 1997). E-selectin expression is induced to a lesser extent (STINS et al. 1997). In vivo, endothelial expression of ICAM-1 (CLARK et al. 1995; LINDSBERG et al. 1996a; OKADA et al. 1994; WANG et al. 1994a; ZHANG et al. 1995b), VCAM-1 (JANDER et al. 1996; KRUPINSKI et al. 1993), P-selectin (OKADA et al. 1994), and E-selectin is induced by ischemia (HARING et al. 1996; ZHANG et al. 1996). VCAM-1 may also be expressed on activated astrocytes (ROSENMAN et al. 1995).

In addition to leukocyte rolling and adhesion, the process of leukocyte trafficking involves transmigration of the leukocytes out of the blood vessels and into the tissue. This process involves platelet/endothelial cell adhesion molecule (PECAM)-1 (LEY 1996). PECAM-1 is a member of the Ig superfamily and is expressed on endothelium, leukocytes, and platelets. PECAM-1 is expressed constitutively and, unlike ICAM-1, its levels are not affected by cytokine stimulation (BRAYTON et al. 1998; WONG and DOROVINI-ZIS 1996) or tissue injury (BELL and PERRY 1995; CARLOS et al. 1997). The endothelial expression of PECAM-1 is highest at cell junctions, and because it displays strong homotypic interactions (MULLER 1995), PECAM-1 is probably important for maintaining vascular integrity. These homotypic interactions also allow for leukocyte transmigration through the vasculature via leukocyte PECAM-1 and endothelial PECAM-1 interactions; an observation supported by the fact that antibodies directed against PECAM-1 prevent leukocyte transmigration without affecting cell adhesion (MULLER 1995). Once a leukocyte has passed the vessel wall, its movement through tissue is facilitated by the proteases released from the lysosomal granules and by MMPs. MMP-2 and MMP-9 can be detected in ischemic brain and seem to be localized to infiltrating cells (ROMANIC et al. 1998). MMP-2, for instance, is induced in T cells upon interaction of VLA-4 and VCAM-1 (MADRI et al. 1996).

VI. Temporal Profile of the Postischemic Inflammatory Response

1. Experimental Models

Summarizing the data outlined above, one can construct a model with a fairly consistent sequence of events that occur after the onset of cerebral injury. Microglia, the endogenous macrophages of the central nervous system, are activated almost immediately after ischemia (RUPALLA et al. 1998; ZHANG et al. 1997). Message for TNF-α and IL-1β increases within 30–60 min after the onset of ischemia (MINAMI et al. 1992; WANG et al. 1994b) and levels peak at 3–6 h (WANG et al. 1994b; ZHAI et al. 1997). The p55 TNF receptor (BOTCHKINA et al. 1997) and the type II IL-1R (WANG et al. 1997) appear within 6 h of ischemia, while the p75 TNF receptor and the type I IL-1R do not peak until 24 h (BOTCHKINA et al. 1997) and 5 days (WANG et al. 1997), respectively. ICAM-1 mRNA is detected 1 h after the onset of ischemia; its expression reaches a maximum at 10 h of reperfusion and persists for 1 week (CLARK et al. 1995; WANG et al. 1994a; ZHANG et al. 1995b). E-selectin (ZHANG et al. 1996) and P-selectin (OKADA et al. 1994; SUZUKI et al. 1998; ZHANG et al. 1998a) expression is induced in the ischemic tissue within 2 h of reperfusion. E-selectin is also expressed in the nonischemic hemisphere by 24 h after reperfusion (HARING et al. 1996). Neutrophils appear in the microvessels of the ischemic hemisphere within 30 min after arterial occlusion and reach a maximum 12 h later; the number of extravasated intraparenchymal neutrophils peaks at 24 h (GARCIA et al. 1994). Monocytes first appear in the ischemic microvessels approximately 4–6 h after the onset of ischemia (GARCIA et al. 1994). IL-6 expression peaks 12–24 h after ischemia (WANG et al. 1995b; YOSHIMOTO et al. 1997), while IL-1ra expression increases within 12 h and remains elevated for days (WANG et al. 1997). Macrophages begin to infiltrate the core of the infarct within 24 h (SCHROETER et al. 1994). T cells can be detected on the pial surface by day 1 and at the edge of the infarct by day 3; they persist in the infarcted hemisphere for up to 1 week (SCHROETER et al. 1994).

Given the above data, it makes intuitive sense that interfering with the process of leukocyte adhesion would affect the outcome after stroke. In animal models, this has been shown to be the case in transient, but not permanent, cerebral ischemia (MORIKAWA et al. 1996; ZHANG et al. 1995a). Blocking the interactions of the β_2 integrins with ICAM-1 decreases infarct size and improves outcome (CHEN et al. 1994; CHOPP et al. 1994; MORI et al. 1992; ZHANG et al. 1995c). Inhibition of E-selectin-mediated binding also improves outcome from stroke (MORIKAWA et al. 1996; ZHANG et al. 1996), while neutralization of L-selectin is not convincingly beneficial (BEDNAR et al. 1998). Interfering with VCAM-1-mediated adhesion also appears to limit infarct damage, suggesting a role for lymphocytes in cerebral ischemic injury (BECKER et al. 2001; RELTON et al. 2001). Again, part of the benefit of aspirin in treating acute stroke may be related to its ability to inhibit adhesion molecule expression through inhibition of NF-κB mobilization (WEBER et al. 1995). In order for leukocytes to injure brain tissue, they must first gain access to that tissue. If ischemia occurs without reperfusion, leukocytes are prevented from reaching the ischemic tissue. Leukocyte-mediated injury would therefore only be an issue in instances where blood flow has been restored. Reperfusion can be hastened through the use of thrombolytics, but it usually occurs spontaneously through the activity of endogenous fibrinolytic systems. However, despite recanalization of large conducting vessels, microcirculatory impairments of blood flow may persist. This is referred to as the "no-reflow" phenomenon and is attributed, in part, to sludging caused by activated leukocytes that become strongly adherent to the microvasculature. Because leukocytes are larger and less deformable than erythrocytes, microvascular perfusion is impaired. Sentinel studies of leukocyte-mediated microvascular impairment in cerebral ischemia were done in the late 1980s and early 1990s (DEL ZOPPO et al. 1991). Both neutrophil depletion (BEDNAR et al. 1991; DAWSON et al. 1996b; GROGAARD et al. 1989) and blockade of leukocyte adherence to the endothelium (MORI et al. 1992; MORIKAWA et al. 1996) improve microvascular patency following cerebral ischemia and reperfusion.

2. Clinical Data

Clinical information about the role of inflammation in stroke has been harder to collect and interpret than animal data. The clinical data, by necessity, rely on surrogate measures of brain inflammation, such as circulating levels of the cytokines and soluble adhesion molecules, and the activity of peripheral leukocytes. Much of the data is conflicting, and reflects differences in the composition of patient and control populations, differences in the time of assay after stroke, and differences in specimen preparation. For instance, in patients studied within 24h of stroke onset, sICAM-1 levels were found to be elevated by some (SHYU et al. 1997), but not others (CLARK et al. 1993; DE BLEECKER et al. 1998). Some investigators have found elevations in soluble E-selectin up to 24h after stroke onset (FASSBENDER et al. 1995), while others report no changes (SHYU et al. 1997), or even a delayed decrease (DE BLEECKER et al. 1998). In rare instances where autopsy specimens were examined after acute stroke, ICAM-1 expression was increased in brain microvessels (LINDSBERG et al. 1996a). Levels of sVCAM-1 are elevated within 4h after the onset of stroke and remain elevated for at least 5 days (FASSBENDER et al. 1995).

Plasma PAF levels are elevated in stroke patients (SATOH et al. 1992) and both serum and CSF levels of IL-6 increase acutely following stroke and remain elevated for days (FASSBENDER et al. 1994; KIM et al. 1996); IL-6 levels correlate to infarct size (BEAMER et al. 1995; TARKOWSKI et al. 1995). Elevations in serum IL-1 and TNF have been more difficult to document (FASSBENDER et al. 1994). This may be due to the fact that the elevations are transient and occur very early, or the fact that soluble receptors interfere with the assay of the cytokines. Indeed, plasma levels of the soluble TNF-RI (ELNEIHOUM et al. 1996) are elevated in stroke patients, as are IL-1ra levels (BEAMER et al. 1995). Intrathecal levels of IL-8 are elevated at day 2 after stroke (TARKOWSKI et al. 1997), and monocytes from stroke patients can be stimulated to express higher levels of IL-8 than age-matched controls (KOSTULAS et al. 1998).

Transient polymorphonuclear responses have been documented in the CSF of stroke patients for nearly a century (SORNAS et al. 1972). Using positron emission tomography (PET) imaging, macrophages can be seen to accumulate in infarcted brain approximately 2 weeks after the event and persist up to 3 weeks (RAMSAY et al. 1992). Neutrophils isolated from patients with acute stroke have increased adhesive properties (CLARK et al. 1993) and bind more tightly to laminin and fibronectin compared to age- and risk factor-matched controls (GRAU et al. 1992). This may be due to increases in CD11a expression (KIM et al. 1995) or CD18 expression (FISZER et al. 1998), which are seen after stroke onset. Increased neutrophil activation in acute stroke is reflected by elevated plasma levels of neutrophil gelatinase-associated lipocalin (NGAL) and neutrophil proteinase 4 (NP4) (ELNEIHOUM et al. 1996).

Given the experimental data, clinical trials of anti-adhesion therapy in acute stroke have been undertaken. In the first study, a monoclonal antibody, Enlimomab, directed against ICAM-1 was administered to patients within 6h of acute stroke (THE ENLIMOMAB ACUTE STROKE TRIAL INVESTIGATORS 1997). There was increased morbidity and mortality among the patients who received the antibody. The poor outcome has been attributed to the fact that the antibody was of murine origin and, as a foreign protein, incited an inflammatory response that contributed to the injury process, although this has not been proven. A second study investigated the efficacy of a humanized monoclonal antibody, LeukArrest or Hu23F2G, directed to CD18. The trial was stopped early because it was claimed to show no chance of being effective (www.ICOS.com).

VII. Endogenous Modulators of Inflammation

Inflammation is a process that evolves over time. In most situations, the inflammatory response is self-limited because the body produces endogenous modulators of that response. IL-1ra and the IL-1 type II receptor limit the effects of IL-1; the sTNF-R, and potentially the p75 receptor, limit the effects of TNF; IL-6 induces production of immunosuppressive cytokines, and so on. Of these immunosuppressive cytokines, TGF-β1 is the best characterized. The brain expresses high endogenous levels of TGF-β1, which may contribute to its relative immunoprivilege (TAYLOR and STREILEIN 1996). TGF-β1 expression is induced after stroke; it peaks at 2 days but remains elevated for up to 2 weeks (WANG et al. 1995a). TGF-β1 is produced by lymphocytes, but is also produced by neurons and glia (BECKER et al. 1997). Receptors for TGF-β1 are expressed on both neurons and astrocytes (VIVIEN et al. 1998). Plasma levels of TGF-β1 are inversely correlated to infarct size (K.J. BECKER, unpublished data), and exogenous TGF-β1 decreases neuronal injury following focal (GROSS et al. 1993) and global cerebral ischemia (HENRICH-NOACK et al. 1996). Infarct size can also be decreased by enhancing local TGF-β1 production (BECKER et al. 1997). TGF-β1 may be neuroprotective by virtue of its ability to modulate the immune response (KEHRL 1991); it can prevent the development of cytotoxic T lymphocytes (GORDON et al. 1998; KING et al. 1998) and inhibit the expression of VCAM-1 on astrocytes (WINKLER and BEVENISTE 1998). TGF-β1 also seems to have a direct protective effect on neurons (FLANDERS et al. 1998; HENRICH-NOACK et al. 1994) that is independent of its effect on the immune system. In addition to the high endogenous levels of TGF-β1, other aspects of the brain's microenvironment can affect initiation and propagation of the inflammatory response. For instance, brain-derived gangliosides inhibit activation of NF-κB and decrease lymphocyte proliferation and (IRANI et al. 1996).

C. Ischemic Tolerance

It has been appreciated for quite some time that if an animal experiences brief transient episodes of sublethal cerebral ischemia prior to a level of ischemia that induces infarction, the ultimate degree of tissue injury is decreased. This phenomenon is referred to as *ischemic preconditioning* and animals are said to have *ischemic tolerance*. There are a number of hypotheses, along with supporting data, that attempt to explain why ischemic preconditioning is beneficial. Most of these hypotheses focus on changes in gene expression that are induced by the brief episodes of ischemia and the likelihood that these changes render the cell more resistant to the final episode of ischemia. There are also data to support modulation of the ischemic inflammatory response as the mechanism by which ischemic preconditioning produces ischemic tolerance.

As mentioned previously, inflammation is a self-limited process because the response is regulated through feedback inhibition and production of endogenous antagonists. If ischemic preconditioning induces cytokine production to a lesser degree than would a more prolonged ischemic insult, it could render the animal tolerant to the detrimental effects those cytokines at a later point in time. For instance, administration of IL-1 at the time of stroke enhances brain injury (STROEMER and ROTHWELL 1998), but systemic administration of IL-1 several days prior to an ischemic insult is neuroprotective (OHTSUKI et al. 1996). The benefit of preischemic IL-1 administration can be abrogated by concurrent administration of IL-1ra (OHTSUKI et al. 1996). Classic ischemic tolerance, induced by brief periods of transient ischemia, is associated with increased production of IL-1ra (BARONE et al. 1998); administration of IL-1ra can abrogate the benefit of the preconditioning ischemia (OHTSUKI et al. 1996). These data suggest that IL-1 may be involved in the process of ischemic tolerance. Similar data exists for TNF; intracisternal administration of TNF-α 2 days prior to MCA occlusion results in significant neuroprotection (NAWASHIRO et al. 1997b). Finally, injection of lipopolysaccharide (LPS), which induces secretion of both TNF and IL-1, several days prior to stroke lessens ischemic injury. This benefit can be nullified by concurrent administration of TNF-bp (TASAKI et al. 1997).

D. Chronic Inflammatory Disease

There is no question that inflammation contributes to the acute injury process in stroke, but there is also an increasing number of reports that suggest systemic inflammatory states may increase the risk of stroke (GRAU et al. 1996; MACKO et al. 1996). The Physician's Health Study showed that C-reactive protein (CRP) levels were highest among males who went on to experience myocardial infarction or stroke, and that aspirin was only effective in decreasing this risk in patients with the highest CRP levels (RIDKER et al. 1997). Atherosclerosis is caused by an inflammatory reaction within the vascular wall (BERLINER et al. 1995), and individuals with classic stroke risk factors have persistent elevations of sICAM-1 (HWANG et al. 1997) and L-selectin (FASSBENDER et al. 1995). Independent of risk factor profile, circulating levels of sICAM-1 and sVCAM-1 positively correlate with the degree of carotid atherosclerosis (ROHDE et al. 1998). In carotid plaques from patients with symptoms of cerebral ischemia, higher levels of TNF-α and ICAM-1 are expressed than in plaques from patients without symptoms (DEGRABA 1997). Symptomatic plaques are also rich in macrophages and T cells, which may destabilize the plaque and lead to thrombosis (JANDER et al. 1998).

Multiple sclerosis (MS) is the prototypic inflammatory disease of the CNS, but unlike the inflammation induced by ischemia and trauma, the pathogenesis of MS is felt to involve antigen specific immune responses. However, in any situation where inflammation becomes chronic, either due to a persistence of the inflammatory stimulus or a failure of the endogenous modulatory systems, immune responses can be evoked. A history of traumatic brain injury has been associated with increased risk of both MS and Alzheimer's disease (AD), and could contribute to the pathogenesis of the disease by provoking the initial inflammatory response. Discussion of the factors involved in the development of MS is beyond the scope of this chapter, but evidence is mounting to support a role for inflammation in AD. IL-6 levels are elevated in patients with AD (GRUOL and NELSON 1997; KALMAN et al. 1997), and both amyloid precursor protein (BARGER and HARMON 1997) and β-amyloid protein (MEDA et al. 1995) activate microglia. Multiple epidemiological studies suggest a protective role for NSAIDs in AD (McGeer et al. 1996; Stewart et al. 1997). This benefit has been attributed, in part, to inhibition of microglial activation (Mackenzie and Munoz 1998). Pentoxifylline and related compounds that inhibit TNF production also show clinical benefit in trials for AD and multi-infarct dementia (Rother et al. 1998).

E. Conclusions

An inflammatory response accompanies brain injury and contributes to the injury process. There exist various methods to interfere with this response and there are a number of points in time at which to do so. The most appropriate interventions can be determined based on the time that they are to be instituted. For instance, inhibition of the initiation and amplification stages of inflammation, such as blockade of complement activation, eicosanoid production, or cytokine expression would need to be instituted soon after the injury in order to be protective. Because of the redundancy and overlap of effects of these mediators of inflammation, it is unlikely that any inhibitor of a single mediator would be of clinical significance. More directed blockade of leukocyte recruitment and activation through antagonism of adhesion molecule interactions, for example, could potentially be accomplished with benefit throughout the period of inflammation. Clinical therapies are currently delivered systemically, which increases the risk of side effects, including those related to immunosuppression. Ultimately, the most effective time window for institution of each therapy and the most efficacious combinations of therapies will need to be determined, realizing that ill-timed delivery or prolonged administration of anti-inflammatory agents could potentially worsen outcome.

References

Akassoglou K, Probert L, Kontogeorgos, G, Kollias G (1997) Astrocyte-specific but not neuron-specific transmembrane TNF triggers inflammation and degeneration in the central nervous system of transgenic mice. J Immunol 158:438–445

Amour A, Slocombe PM, Webster A, Butler M, Knight CG, Smith BJ, Stephens PE, Shelley C, Hutton M, Knauper V, Docherty AJ, Murphy G (1998) TNF-α converting enzyme (TACE) is inhibited by TIMP-3. FEBS Lett 435:39–44

Azzimondi G, Bassein L, Nonino F, Fiorani L, Vignatelli L, Re G, D'Alessandro R (1995) Fever in acute stroke worsens prognosis. A prospective study. Stroke 26: 2040–2043

Baldwin AS Jr (1996) The NF-κB and I κB proteins: new discoveries and insights. Annu Rev Immunol 14:649–683

Barger SW, Harmon AD (1997) Microglial activation by Alzheimer amyloid precursor protein and modulation by apolipoprotein E. Nature 388:878–881

Barkalow FJ, Goodman MJ, Gerritsen ME, Mayadas TN (1996) Brain endothelium lack one of two pathways of P-selectin-mediated neutrophil adhesion. Blood 88: 4585–4593

Barnum SR (1995) Complement biosynthesis in the central nervous system. Crit Rev Oral Biol Med 6:132–146

Barnum SR, Jones JL, Muller-Ladner U, Samimi A, Campbell IL (1996) Chronic complement C3 gene expression in the CNS of transgenic mice with astrocyte-targeted interleukin-6 expression. Glia 18:107–117

Barone FC, Arvin B, White RF, Miller A, Webb CL, Willette RN, Lysko PG, Feuerstein GZ (1997) Tumor necrosis factor-α. A mediator of focal ischemic brain injury. Stroke 28:1233–1244

Barone FC, White RF, Spera, PA, Ellison J, Currie RW, Wang X, Feuerstein GZ (1998) Ischemic preconditioning and brain tolerance: temporal histological and functional outcomes, protein synthesis requirement, and interleukin-1 receptor antagonist and early gene expression. Stroke 29:1937–1950

Bazan NG, Allan G (1996) Platelet-activating factor in the modulation of excitatory amino acid neurotransmitter release and of gene expression. J Lipid Mediat Cell Signal 14:321–330

Bazzoni F, Beutler B (1996) The tumor necrosis factor ligand and receptor families. N Engl J Med 334:1717–1725

Beamer NB, Coull BM, Clark WM, Hazel JS, Silberger JR (1995) Interleukin-6 and interleukin-1 receptor antagonist in acute stroke. Ann Neurol 37:800–805

Becker K, Kindrick D, Relton J, Harlan J, Winn R (2001) Antibody to the alpha 4 integrin decreases infarct size in transient focal cerebral ischemia in rats. Stroke 32:206–211

Becker KJ, McCarron RM, Ruetzler C, Laban O, Sternberg E, Flanders KC, Hallenbeck JM (1997) Immunologic tolerance to myelin basic protein decreases stroke size after transient focal cerebral ischemia. Proc Natl Acad Sci USA 94:10873–10978

Bednar MM, Gross CE, Russell SR, Fuller SP, Ellenberger CL, Schindler E, Klingbeil C, Vexler V (1998) Humanized anti-L-selectin monoclonal antibody DREG200 therapy in acute thromboembolic stroke. Neurol Res 20:403–408

Bednar MM, Raymond S, McAuliffe T, Lodge PA, Gross CE (1991) The role of neutrophils and platelets in a rabbit model of thromboembolic stroke. Stroke 22:44–50

Belayev L, Busto R, Zhao W, Ginsberg MD (1996) Quantitative evaluation of blood-brain barrier permeability following middle cerebral artery occlusion in rats. Brain Res 739:88–96

Bell MD, Perry VH (1995) Adhesion molecule expression on murine cerebral endothelium following the injection of a proinflammagen or during acute neuronal degeneration. J Neurocytol 24:695–710

Bellander BM, von Holst H, Fredman P, Svensson M (1996) Activation of the complement cascade and increase of clusterin in the brain following a cortical contusion in the adult rat. J Neurosurg 85:468–475

Benigni F, Faggioni R, Sironi M, Fantuzzi G, Vandenabeele P, Takahashi N, Sacco S, Fiers W, Buurman WA, Ghezzi P (1996) TNF receptor p55 plays a major role in centrally mediated increases of serum IL-6 and corticosterone after intracerebroventricular injection of TNF. J Immunol 157:5563–5568

Berliner JA, Navab M, Fogelman AM, Frank JS, Demer LL, Edwards PA, Watson AD, Lusis AJ (1995) Atherosclerosis: basic mechanisms. Oxidation, inflammation, and genetics. Circulation 91:2488–2496

Betz AL, Yang GY, Davidson BL (1995) Attenuation of stroke size in rats using an adenoviral vector to induce overexpression of interleukin-1 receptor antagonist in brain. J Cereb Blood Flow Metab 15:547–551

Bolton SJ, Anthony DC, Perry VH (1998) Loss of the tight junction proteins occludin and zonula occludens-1 from cerebral vascular endothelium during neutrophil-induced blood-brain barrier breakdown in vivo. Neuroscience 86:1245–1257

Bonventre JV (1997) Roles of phospholipases A2 in brain cell and tissue injury associated with ischemia and excitotoxicity. J Lipid Mediat Cell Signal 16:199–208

Botchkina GI, Meistrell ME 3rd, Botchkina IL, Tracey KJ (1997) Expression of TNF and TNF receptors (p55 and p75) in the rat brain after focal cerebral ischemia. Mol Med 3:765–781

Brayton J, Qing Z, Hart MN, VanGilder JC, Fabry Z (1998) Influence of adhesion molecule expression by human brain microvessel endothelium on cancer cell adhesion. J Neuroimmunol 89:104–112

Brett FM, Mizisin AP, Powell HC, Campbell IL (1995) Evolution of neuropathologic abnormalities associated with blood-brain barrier breakdown in transgenic mice expressing interleukin-6 in astrocytes. J Neuropathol Exp Neurol 54:766–775

Bruce AJ, Boling W, Kindy MS, Peschon J, Kraemer PJ, Carpenter MK, Holtsberg FW, Mattson MP (1996) Altered neuronal and microglial responses to excitotoxic and ischemic brain injury in mice lacking TNF receptors. Nat Med 2:788–794

Cao C, Matsumura K, Watanabe Y (1997). Induction of cyclooxygenase-2 in the brain by cytokines. Ann NY Acad Sci 813:307–309

Carlos TM, Clark RS, Franicola-Higgins D, Schiding JK, Kochanek PM (1997) Expression of endothelial adhesion molecules and recruitment of neutrophils after traumatic brain injury in rats. J Leukoc Biol 61:279–285

Carp H (1982) Mitochondrial N-formylmethionyl proteins as chemoattractants for neutrophils. J Exp Med 155:264–275

CAST Collaborative Group (1997) CAST: randomised placebo-controlled trial of early aspirin use in 20 000 patients with acute ischemic stroke. Lancet 349:1641–1649

Castillo J, Davalos A, Marrugat J, Noya M (1998) Timing for fever-related brain damage in acute ischemic stroke. Stroke 29:2455–2460

Chen H, Chopp M, Zhang RL, Bodzin G, Chen Q, Rusche JR, Todd RF 3rd (1994) Anti-CD11b monoclonal antibody reduces ischemic cell damage after transient focal cerebral ischemia in rat. Ann Neurol 35:458–463

Cheng B, Christakos S, Mattson MP (1994) Tumor necrosis factors protect neurons against metabolic-excitotoxic insults and promote maintenance of calcium homeostasis. Neuron 12:139–153

Chopp M, Zhang RL, Chen H, Li Y, Jiang N, Rusche JR (1994) Postischemic administration of an anti-Mac-1 antibody reduces ischemic cell damage after transient middle cerebral artery occlusion in rats. Stroke 25:869–875

Clark WM, Calcagno FA, Gabler WL, Smith JR, Coull BM (1994) Reduction of central nervous system reperfusion injury in rabbits using doxycycline treatment. Stroke 25:1411–1415

Clark WM, Coull BM, Briley DP, Mainolfi E, Rothlein R (1993) Circulating intercellular adhesion molecule-1 levels and neutrophil adhesion in stroke. J Neuroimmunol 44:123–125

Clark WM, Lauten JD, Lessov N, Woodward W, Coull BM (1995) Time course of ICAM-1 expression and leukocyte subset infiltration in rat forebrain ischemia. Mol Chem Neuropathol 26:213–230

Clemens JA, Stephenson DT, Smalstig EB, Dixon EP, Little SP (1997) Global ischemia activates nuclear factor-κB in forebrain neurons of rats. Stroke 28:1073–1080

Clemens JA, Stephenson DT, Smalstig EB, Roberts EF, Johnstone EM, Sharp JD, Little SP, Kramer RM (1996) Reactive glia express cytosolic phospholipase A2 after transient global forebrain ischemia in the rat. Stroke 27:527–535

Cole DJ, Patel PM, Schell RM, Drummond JC, Osborne TN (1993) Brain eicosanoid levels during temporary focal cerebral ischemia in rats: a microdialysis study. J Neurosurg Anesthesiol 5:41–47

Corvin S, Schurer L, Abels C, Kempski O, Baethmann A (1990) Effect of stimulation of leukocyte chemotaxis by fMLP on white blood cell behaviour in the microcirculation of rat brain. Acta Neurochir Suppl 51:55–57

Davies C, Loddick S, Toulmond S, Stroemer R, Hunt J, Rothwell N (1999) The progression and topographic distribution of interleukin-1β expression after permanent middle cerebral artery occlusion in the rat. J Cereb Blood Flow Metab 19:87–98

Dawson DA, Martin D, Hallenbeck JM (1996a) Inhibition of tumor necrosis factor-α reduces focal cerebral ischemic injury in the spontaneously hypertensive rat. Neurosci Lett 218:41–44

Dawson DA, Ruetzler CA, Carlos TM, Kochanek PM, Hallenbeck JM (1996b) Polymorphonuclear leukocytes and microcirculatory perfusion in acute stroke in the SHR. Keio J Med 45:248–252

De Bleecker J, Coulier I, Fleurinck C, De Reuck J (1998) Circulating intercellular adhesion molecule-1 and E-selectin in acute ischemic stroke. J Stroke Cerebrovasc Dis 7:192–195

De Simoni MG, De Luigi A, Gemma L, Sironi M, Manfridi A, Ghezzi P (1993) Modulation of systemic interleukin-6 induction by central interleukin-1. Am J Physiol 265:R739–R742

DeGraba TJ (1997) Expression of inflammatory mediators and adhesion molecules in human atherosclerotic plaque. Neurology 49:S15–S19

del Zoppo GJ, Schmid-Schonbein GW, Mori E, Copeland BR, Chang CM (1991) Polymorphonuclear leukocytes occlude capillaries following middle cerebral artery occlusion and reperfusion in baboons. Stroke 22:1276–1283

Deluga KS, Plotz FB, Betz AL (1991) Effect of indomethacin on edema following single and repetitive cerebral ischemia in the gerbil. Stroke 22:1259–1264

Dempsey RJ, Roy MW, Cowen DE, Maley ME (1986) Lipoxygenase metabolites of arachidonic acid and the development of ischaemic cerebral oedema. Neurol Res 8:53–56

Ehrlich LC, Hu S, Sheng WS, Sutton RL, Rockswold GL, Peterson PK, Chao CC (1998) Cytokine regulation of human microglial cell IL-8 production. J Immunol 160:1944–1948

Elneihoum AM, Falke P, Axelsson L, Lundberg E, Lindgarde F, Ohlsson K (1996) Leukocyte activation detected by increased plasma levels of inflammatory mediators in patients with ischemic cerebrovascular diseases. Stroke 27:1734–1738

Ember JA, Sanderson SD, Hugli TE, Morgan EL (1994) Induction of interleukin-8 synthesis from monocytes by human C5a anaphylatoxin. Am J Pathol 144:393–403

Ericsson A, Liu C, Hart RP, Sawchenko PE (1995) Type 1 interleukin-1 receptor in the rat brain: distribution, regulation, and relationship to sites of IL-1-induced cellular activation. J Comp Neurol 361:681–698

Farrar WL, Kilian PL, Ruff MR, Hill JM, Pert CB (1987) Visualization and characterization of interleukin 1 receptors in brain. J Immunol 139:459–463

Fassbender K, Mossner R, Motsch L, Kischka U, Grau A, Hennerici M (1995) Circulating selectin- and immunoglobulin-type adhesion molecules in acute ischemic stroke. Stroke 26:1361–1364

Fassbender K, Rossol S, Kammer T, Daffertshofer M, Wirth S, Dollman M, Hennerici M (1994) Proinflammatory cytokines in serum of patients with acute cerebral ischemia: kinetics of secretion and relation to the extent of brain damage and outcome of disease. J Neurol Sci 122:135–139

Fattori E, Lazzaro D, Musiani P, Modesti A, Alonzi T, Ciliberto G (1995) IL-6 expression in neurons of transgenic mice causes reactive astrocytosis and increase in ramified microglial cells but no neuronal damage. Eur J Neurosci 7:2441–2449

Fiszer U, Korczak-Kowalska G, Palasik W, Korlak J, Gorski A, Czlonkowska A (1998) Increased expression of adhesion molecule CD18 (LFA-1β) on the leukocytes of peripheral blood in patients with acute ischemic stroke. Acta Neurol Scand 97:221–224

Flanders KC, Ren RF, Lippa CF (1998). Transforming growth factor-βs in neurodegenerative disease. Prog Neurobiol 54:71–85

Frank MM, Fries LF (1991) The role of complement in inflammation and phagocytosis. Immunol Today 12:322–326

Fraticelli A, Serrano CV Jr, Bochner BS, Capogrossi MC, Zweier JL (1996) Hydrogen peroxide and superoxide modulate leukocyte adhesion molecule expression and leukocyte endothelial adhesion. Biochim Biophys Acta 1310:251–259

Frenette PS, Wagner DD (1996a) Adhesion Molecules–Part I. N Engl J Med 334: 1526–1529

Frenette PS, Wagner DD (1996b) Adhesion molecules–Part II: Blood vessels and blood cells. N Engl J Med 335:43–45

Gadient RA, Otten UH (1997) Interleukin-6 (IL-6) – a molecule with both beneficial and destructive potentials. Prog Neurobiol 52:379–390

Garcia JH, Liu KF, Relton JK (1995) Interleukin-1 receptor antagonist decreases the number of necrotic neurons in rats with middle cerebral artery occlusion. Am J Pathol 147:1477–1486

Garcia JH, Liu KF, Yoshida Y, Lian J, Chen S, del Zoppo GJ (1994) Influx of leukocytes and platelets in an evolving brain infarct (Wistar rat). Am J Pathol 144: 188–199

Gary DS, Bruce-Keller AJ, Kindy MS, Mattson MP (1998) Ischemic and excitotoxic brain injury is enhanced in mice lacking the p55 tumor necrosis factor receptor. J Cereb Blood Flow Metab 18:1283–1287

Gasque P, Chan P, Fontaine M, Ischenko A, Lamacz M, Gotze O, Morgan BP (1995) Identification and characterization of the complement C5a anaphylatoxin receptor on human astrocytes. J Immunol 155:4882–4889

Gasque P, Singhrao SK, Neal JW, Wang P, Sayah S, Fontaine M, Morgan BP (1998) The receptor for complement anaphylatoxin C3a is expressed by myeloid cells and nonmyeloid cells in inflamed human central nervous system: analysis in multiple sclerosis and bacterial meningitis. J Immunol 160:3543–3554

Gearing AJ, Beckett P, Christodoulou M, Churchill M, Clements J, Davidson AH, Drummond AH, Galloway WA, Gilbert R, Gordon JL et al. (1994) Processing of tumour necrosis factor-α precursor by metalloproteinases. Nature 370:555–557

Gemma C, Imeri L, de Simoni MG, Mancia M (1997) Interleukin-1 induces changes in sleep, brain temperature, and serotonergic metabolism. Am J Physiol 272:R601–R606

Giulian D, Baker TJ, Shih LC, Lachman LB (1986) Interleukin 1 of the central nervous system is produced by ameboid microglia. J Exp Med 164:594–604

Giulian D, Corpuz M, Richmond B, Wendt E, Hall ER (1996) Activated microglia are the principal glial source of thromboxane in the central nervous system. Neurochem Int 29:65–76

Gordon LB, Nolan SC, Ksander BR, Knopf PM, Harling-Berg CJ (1998) Normal cerebrospinal fluid suppresses the in vitro development of cytotoxic T cells: role of the brain microenvironment in CNS immune regulation. J Neuroimmunol 88: 77–84

Grau AJ, Berger E, Sung KL, Schmid-Schonbein GW (1992) Granulocyte adhesion, deformability, and superoxide formation in acute stroke. Stroke 23:33–39

Grau AJ, Buggle F, Becher H, Werle E, Hacke W (1996) The association of leukocyte count, fibrinogen and C-reactive protein with vascular risk factors and ischemic vascular diseases. Thromb Res 82:245–255

Grice SC, Chappell ET, Prough DS, Whitley JM, Su M, Watkins WD (1987) Ibuprofen improves cerebral blood flow after global cerebral ischemia in dogs. Stroke 18: 787–791

Grilli M, Pizzi M, Memo M, Spano P (1996) Neuroprotection by aspirin and sodium salicylate through blockade of NF-κB activation. Science 274:1383–1385

Grogaard B, Schurer L, Gerdin B, Arfors KE (1989) Delayed hypoperfusion after incomplete forebrain ischemia in the rat. The role of polymorphonuclear leukocytes. J Cereb Blood Flow Metab 9:500–505

Gross CE, Bednar MM, Howard DB, Sporn MB (1993) Transforming growth factor-β1 reduces infarct size after experimental cerebral ischemia in a rabbit model. Stroke 24:558–562

Gruol DL, Nelson TE (1997) Physiological and pathological roles of interleukin-6 in the central nervous system. Mol Neurobiol 15:307–339

Hagberg H, Gilland E, Bona E, Hanson LA, Hahin-Zoric M, Blennow M, Holst M, McRae A, Soder O (1996) Enhanced expression of interleukin (IL)-1 and IL-6 messenger RNA and bioactive protein after hypoxia-ischemia in neonatal rats. Pediatr Res 40:603–609

Hallenbeck JM, Furlow TW Jr (1979) Prostaglandin I2 and indomethacin prevent impairment of post-ischemic brain reperfusion in the dog. Stroke 10:629–637

Hamann GF, Okada Y, del Zoppo GJ (1996) Hemorrhagic transformation and microvascular integrity during focal cerebral ischemia/reperfusion. J Cereb Blood Flow Metab 16:1373–1378

Hara H, Friedlander RM, Gagliardini V, Ayata C, Fink K, Huang Z, Shimizu-Sasamata M, Yuan J, Moskowitz MA (1997) Inhibition of interleukin 1β converting enzyme family proteases reduces ischemic and excitotoxic neuronal damage. Proc Natl Acad Sci USA 94:2007–2012

Haring HP, Berg EL, Tsurushita N, Tagaya M, del Zoppo GJ (1996) E-selectin appears in nonischemic tissue during experimental focal cerebral ischemia. Stroke 27:1386–1391

Henrich-Noack P, Prehn JH, Krieglstein J (1994) Neuroprotective effects of TGF-β1. J Neural Transm [Suppl] 43:33–45

Henrich-Noack P, Prehn JHM, Krieglstein J (1996) TGF-β1 protects hippocampal neurons against degeneration caused by transient global ischemia. Dose-response relationship and potential neuroprotective mechanisms. Stroke 27:1609–1614

Heo J, Lucero J, Abumiya T, Koziol J, Copeland B, del Zoppo GJ (1999) Matrix metalloproteinases increase very early during experimental focal cerebral ischemia. J Cereb Blood Flow Metab 19:624–633

Hernandez J, Molinero A, Campbell IL, Hidalgo J (1997) Transgenic expression of interleukin 6 in the central nervous system regulates brain metallothionein-I and -III expression in mice. Brain Res Mol Brain Res 48:125–131

Hess DC, Zhao W, Carroll J, McEachin M, Buchanan K (1994) Increased expression of ICAM-1 during reoxygenation in brain endothelial cells. Stroke 25:1463–1467

Howard EF, Chen Q, Cheng C, Carroll JE, Hess D (1998) NF-κB is activated and ICAM-1 gene expression is upregulated during reoxygenation of human brain endothelial cells. Neurosci Lett 248:199–203

Huang YH, Haegerstrand A, Frostegard J (1996) Effects of in vitro hyperthermia on proliferative responses and lymphocyte activity. Clin Exp Immunol 103:61–66

Hwang SJ, Ballantyne CM, Sharrett AR, Smith LC, Davis CE, Gotto AM Jr, Boerwinkle E (1997) Circulating adhesion molecules VCAM-1, ICAM-1, and E-selectin in carotid atherosclerosis and incident coronary heart disease cases: the Atherosclerosis Risk In Communities (ARIC) study. Circulation 96:4219–4225

Iijima T, Sawa H, Shiokawa Y, Saito I, Ishii H, Nakamura Z, Sankawa H (1996) Thromboxane synthetase inhibitor ameliorates delayed neuronal death in the CA1 subfield of the hippocampus after transient global ischemia in gerbils. J Neurosurg Anesthesiol 8:237–242

Imura Y, Kiyota Y, Nagai Y, Nishikawa K, Terashita Z (1995) Beneficial effect of CV-4151 (Isbogrel), a thromboxane A2 synthase inhibitor, in a rat middle cerebral artery thrombosis model. Thromb Res 79:95–107

International Stroke Trial Collaborative Group (1997) The International Stroke Trial (IST): a randomised trial of aspirin, subcutaneous heparin, both, or neither among 19,435 patients with acute ischemic stroke. Lancet 349:1569–1581

Irani DN, Griffin DE (1996) Regulation of lymphocyte homing into the brain during viral encephalitis at various stages of infection. J Immunol 156:3850–3857

Irani DN, Lin KI, Griffin DE (1996) Brain-derived gangliosides regulate the cytokine production and proliferation of activated T cells. J Immunol 157:4333–4340

Jander S, Pohl J, Gillen C, Schroeter M, Stoll G (1996) Vascular cell adhesion molecule-1 mRNA is expressed in immune-mediated and ischemic injury of the rat nervous system. J Neuroimmunol 70:75–80

Jander S, Sitzer M, Schumann R, Schroeter M, Siebler M, Steinmetz H, Stoll G (1998) Inflammation in high-grade carotid stenosis: a possible role for macrophages and T cells in plaque destabilization. Stroke 29:1625–1630

Johshita H, Asano T, Hanamura T, Takakura K (1989) Effect of indomethacin and a free radical scavenger on cerebral blood flow and edema after cerebral artery occlusion in cats. Stroke 20:788–794

Kalman J, Juhasz A, Laird G, Dickens P, Jardanhazy T, Rimanoczy A, Boncz I, Parry-Jones WL, Janka Z (1997) Serum interleukin-6 levels correlate with the severity of dementia in Down syndrome and in Alzheimer's disease. Acta Neurol Scand 96:236–240

Kanda T, Yamawaki M, Ariga T, Yu RK (1995) Interleukin 1 β upregulates the expression of sulfoglucuronosyl paragloboside, a ligand for L-selectin, in brain microvascular endothelial cells. Proc Natl Acad Sci USA 92:7897–7901

Kehrl J (1991) Transforming growth factor-β: an important mediator of immunoregulation. Int J Cell Cloning 9:438–450

Kempski O, Shohami E, von Lubitz D, Hallenbeck JM, Feuerstein G (1987) Postischemic production of eicosanoids in gerbil brain. Stroke 18:111–119

Kim JS, Chopp M, Chen H, Levine SR, Carey JL, Welch KM (1995) Adhesive glycoproteins CD11a and CD18 are upregulated in the leukocytes from patients with ischemic stroke and transient ischemic attacks. J Neurol Sci 128:45–50

Kim JS, Yoon SS, Kim YH, Ryu JS (1996) Serial measurement of interleukin-6, transforming growth factor-β, and S-100 protein in patients with acute stroke. Stroke 27:1553–1557

King C, Davies J, Mueller R, Lee MS, Krahl T, Yeung B, O'Connor E, Sarvetnick N (1998) TGF-β1 alters APC preference, polarizing islet antigen responses toward a Th2 phenotype. Immunity 8:601–613

Kitamura H, Okamoto S, Shimamoto Y, Morimatsu M, Terao A, Saito M (1998) Central IL-1 differentially regulates peripheral IL-6 and TNF synthesis. Cell Mol Life Sci 54:282–287

Klir JJ, Roth J, Szelenyi Z, McClellan JL, Kluger MJ (1993) Role of hypothalamic interleukin-6 and tumor necrosis factor-alpha in LPS fever in rat. Am J Physiol 265:R512–517

Kostulas N, Kivisakk P, Huang Y, Matusevicius D, Kostulas V, Link H (1998) Ischemic stroke is associated with a systemic increase of blood mononuclear cells expressing interleukin-8 mRNA. Stroke 29:462–466

Krupinski J, Kaluza J, Kumar P, Kumar S, Wang JM (1993) Some remarks on the growth-rate and angiogenesis of microvessels in ischemic stroke. Morphometric and immunocytochemical studies. Patol Pol 44:203–209

Lassmann H (1997) Basic mechanisms of brain inflammation. J Neural Transm [Suppl] 50:183–190

Lavine SD, Hofman FM, Zlokovic BV (1998) Circulating antibody against tumor necrosis factor-α protects rat brain from reperfusion injury. J Cereb Blood Flow Metab 18:52–58

Ley K (1996) Molecular mechanisms of leukocyte recruitment in the inflammatory process. Cardiovasc Res 32:733–742

Ley K, Baker JB, Cybulsky MI, Gimbrone MA Jr, Luscinskas FW (1993) Intravenous interleukin-8 inhibits granulocyte emigration from rabbit mesenteric venules without altering L-selectin expression or leukocyte rolling. J Immunol 151:6347–6357

Lindsberg PJ, Carpen O, Paetau A, Karjalainen-Lindsberg M-L, Kaste M (1996a) Endothelial ICAM-1 expression associated with inflammatory cell response in human ischemic stroke. Stroke 94:939–945

Lindsberg PJ, Ohman J, Lehto T, Karjalainen-Lindsberg ML, Paetau A, Wuorimaa T, Carpen O, Kaste M, Meri S (1996b) Complement activation in the central nervous system following blood-brain barrier damage in man. Ann Neurol 40:587–596

Liu T, Clark RK, McDonnell PC, Young PR, White RF, Barone FC, Feuerstein GZ (1994) Tumor necrosis factor-α expression in ischemic neurons. Stroke 25:1481–1488

Liu XH, Eun BL, Silverstein FS, Barks JD (1996) The platelet-activating factor antagonist BN 52021 attenuates hypoxic-ischemic brain injury in the immature rat. Pediatr Res 40:797–803

Loddick SA, Turnbull AV, Rothwell NJ (1998) Cerebral interleukin-6 is neuroprotective during permanent focal cerebral ischemia in the rat. J Cereb Blood Flow Metab 18:176–179

Loddick SA, Wong ML, Bongiorno PB, Gold PW, Licinio J, Rothwell NJ (1997) Endogenous interleukin-1 receptor antagonist is neuroprotective. Biochem Biophys Res Commun 234:211–215

Luheshi GN, Stefferl A, Turnbull AV, Dascombe MJ, Brouwer S, Hopkins SJ, Rothwell NJ (1997) Febrile response to tissue inflammation involves both peripheral and brain IL-1 and TNF-α in the rat. Am J Physiol 272:R862–R868

Mackenzie IR, Munoz DG (1998) Nonsteroidal anti-inflammatory drug use and Alzheimer-type pathology in aging. Neurology 50:986–990

Macko R, Ameriso S, Barndt R, Clough W, Weiner J, Fisher M (1996) Precipitants of brain infarction. Roles of preceding infection/inflammation and recent psychological stress. Stroke 27:1999–2004

Madri JA, Graesser D, Haas T (1996) The roles of adhesion molecules and proteinases in lymphocyte transendothelial migration. Biochem Cell Biol 74:749–757

Maenpaa A, Kovanen PE, Paetau A, Jaaskelainen J, Timonen T (1997) Lymphocyte adhesion molecule ligands and extracellular matrix proteins in gliomas and normal brain: expression of VCAM-1 in gliomas. Acta Neuropathol (Berl) 94:216–225

Mantovani A (1997) The interplay between primary and secondary cytokines. Cytokines involved in the regulation of monocyte recruitment. Drugs 54 [Suppl 1]:15–23

Matsukawa A, Fukumoto T, Maeda T, Ohkawara S, Yoshinaga M (1997) Detection and characterization of IL-1 receptor antagonist in tissues from healthy rabbits: IL-1 receptor antagonist is probably involved in health. Cytokine 9:307–315

Matsumoto T, Ikeda K, Mukaida N, Harada A, Matsumoto Y, Yamashita J, Matsushima K (1997) Prevention of cerebral edema and infarct in cerebral reperfusion injury by an antibody to interleukin-8. Lab Invest 77:119–125

Matsuo Y, Kihara T, Ikeda M, Ninomiya M, Onodera H, Kogure K (1996) Role of platelet-activating factor and thromboxane A2 in radical production during ischemia and reperfusion of the rat brain. Brain Res 709:296–302

McGeehan GM, Becherer JD, Bast RC Jr, Boyer CM, Champion B, Connolly KM, Conway JG, Furdon P, Karp S, Kidao S et al. (1994) Regulation of tumour necrosis factor-α processing by a metalloproteinase inhibitor. Nature 370:558–561

McGeer PL, Schulzer M, McGeer EG (1996) Arthritis and anti-inflammatory agents as possible protective factors for Alzheimer's disease: a review of 17 epidemiologic studies. Neurology 47:425–432

McKeating EG, Andrews PJ, Signorini DF, Mascia L (1997) Transcranial cytokine gradients in patients requiring intensive care after acute brain injury. Br J Anaesth 78:520–523

Meda L, Cassatella MA, Szendrei GI, Otvos L Jr, Baron P, Villalba M, Ferrari D, Rossi F (1995) Activation of microglial cells by β-amyloid protein and interferon-γ. Nature 374:647–650

Mehta A, Yang B, Khan S, Hendricks JB, Stephen C, Mehta JL (1995) Oxidized low-density lipoproteins facilitate leukocyte adhesion to aortic intima without affecting endothelium-dependent relaxation. Role of P-selectin. Arterioscler Thromb Vasc Biol 15:2076–2083

Miller AJ, Hopkins SJ, Luheshi GN (1997a) Sites of action of IL-1 in the development of fever and cytokine responses to tissue inflammation in the rat. Br J Pharmacol 120:1274–1279

Miller DK, Myerson J, Becker JW (1997b) The interleukin-1 β converting enzyme family of cysteine proteases. J Cell Biochem 64:2–10

Minami M, Kuraishi Y, Yabuuchi K, Yamazaki A, Satoh M (1992) Induction of interleukin-1 β mRNA in rat brain after transient forebrain ischemia. J Neurochem 58:390–392

Minamisawa H, Terashi A, Katayama Y, Kanda Y, Shimizu J, Shiratori T, Inamura K, Kaseki H, Yoshino Y (1988) Brain eicosanoid levels in spontaneously hypertensive rats after ischemia with reperfusion: leukotriene C4 as a possible cause of cerebral edema. Stroke 19:372–377

Mori E, del Zoppo GJ, Chambers JD, Copeland BR, Arfors KE (1992) Inhibition of polymorphonuclear leukocyte adherence suppresses no-reflow after focal cerebral ischemia in baboons. Stroke 23:712–718

Morikawa E, Zhang SM, Seko Y, Toyoda T, Kirino T (1996) Treatment of focal cerebral ischemia with synthetic oligopeptide corresponding to lectin domain of selectin. Stroke 27:951–955

Muller WA (1995) The role of PECAM-1 (CD31) in leukocyte emigration: studies in vitro and in vivo. J Leukoc Biol 57:523–528

Namura Y, Shio H, Kimura J (1994) LTC4/LTB4 alterations in rat forebrain ischemia and reperfusion and effects of AA-861, CV-3988. Acta Neurochir [Suppl] 60:296–299

Navratil E, Couvelard A, Rey A, Henin D, Scoazec JY (1997) Expression of cell adhesion molecules by microvascular endothelial cells in the cortical and subcortical regions of the normal human brain: an immunohistochemical analysis. Neuropathol Appl Neurobiol 23:68–80

Nawashiro H, Martin D, Hallenbeck JM (1997a) Inhibition of tumor necrosis factor and amelioration of brain infarction in mice. J Cereb Blood Flow Metab 17:229–232

Nawashiro H, Tasaki K, Ruetzler CA, Hallenbeck JM (1997b) TNF-α pretreatment induces protective effects against focal cerebral ischemia in mice. J Cereb Blood Flow Metab 17:483–490

Nishida K, Markey SP (1996) Platelet-activating factor in brain regions after transient ischemia in gerbils. Stroke 27:514–518

Norgard KE, Han H, Powell L, Kriegler M, Varki A, Varki NM (1993) Enhanced interaction of L- selectin with the high endothelial venule ligand via selectively oxidized sialic acids. Proc Natl Acad Sci USA 90:1068–1072

Ohtsuki T, Ruetzler CA, Tasaki K, Hallenbeck JM (1996) Interleukin-1 mediates induction of tolerance to global ischemia in gerbil hippocampal CA1 neurons. J Cereb Blood Flow Metab 16:1137–1142

Okada Y, Copeland BR, Mori E, Tung MM, Thomas WS, del Zoppo GJ (1994) P-selectin and intercellular adhesion molecule-1 expression after focal brain ischemia and reperfusion. Stroke 25:202–211

O'Neill LA, Kaltschmidt C (1997) NF-κB: a crucial transcription factor for glial and neuronal cell function. Trends Neurosci 20:252–258

O'Neill PA, Davies I, Fullerton KJ, Bennett D (1991) Stress hormone and blood glucose response following acute stroke in the elderly. Stroke 22:842–847

Parnet P, Amindari S, Wu C, Brunke-Reese D, Goujon E, Weyhenmeyer JA, Dantzer R, Kelley KW (1994) Expression of type I and type II interleukin-1 receptors in mouse brain. Brain Res Mol Brain Res 27:63–70

Patel PM, Drummond JC, Sano T, Cole DJ, Kalkman CJ, Yaksh TL (1993) Effect of ibuprofen on regional eicosanoid production and neuronal injury after forebrain ischemia in rats. Brain Res 614:315–324

Perry VH (1998) A revised view of the central nervous system microenvironment and major histocompatibility complex class II antigen presentation. J Neuroimmunol 90:113–121

Pettigrew LC, Grotta JC, Rhoades HM, Wu KK (1989) Effect of thromboxane synthase inhibition on eicosanoid levels and blood flow in ischemic rat brain. Stroke 20:627–632

Pettigrew LC, Meyer JJ, Craddock SD, Butler SM, Tai HH, Yokel RA (1995) Delayed elevation of platelet activating factor in ischemic hippocampus. Brain Res 691:243–247

Quagliarello VJ, Wispelwey B, Long WJ Jr, Scheld WM (1991) Recombinant human interleukin-1 induces meningitis and blood-brain barrier injury in the rat. Characterization and comparison with tumor necrosis factor. J Clin Invest 87:1360–1366

Ramsay SC, Weiller C, Myers R, Cremer JE, Luthra SK, Lammertsma AA, Frackowiak RS (1992) Monitoring by PET of macrophage accumulation in brain after ischaemic stroke. Lancet 339:1054–1055

Reith J, Jorgensen H, Pedersen P, Nakayama H, Raaschou H, Jeppesen L, Olsen T (1996) Body temperature in acute stroke: relation to stroke severity, infarct size, and mortality. Lancet 347:422–425

Relton J, Sloan, KE, Frew EM, Whalley ET, Adams SP, Lobb RR (2001) Stroke 32:199–205

Relton JK, Martin D, Thompson RC, Russell DA (1996) Peripheral administration of Interleukin-1 Receptor antagonist inhibits brain damage after focal cerebral ischemia in the rat. Exp Neurol 138:206–213

Ridker PM, Cushman M, Stampfer MJ, Tracy RP, Hennekens CH (1997) Inflammation, aspirin, and the risk of cardiovascular disease in apparently healthy men. N Engl J Med 336:973–979

Rohde LE, Lee RT, Rivero J, Jamacochian M, Arroyo LH, Briggs W, Rifai N, Libby P, Creager MA, Ridker PM (1998) Circulating cell adhesion molecules are correlated with ultrasound-based assessment of carotid atherosclerosis. Arterioscler Thromb Vasc Biol 18:1765–1770

Romanic AM, White RF, Arleth AJ, Ohlstein EH, Barone FC (1998) Matrix metalloproteinase expression increases after cerebral focal ischemia in rats: inhibition of matrix metalloproteinase-9 reduces infarct size. Stroke 29:1020–1030

Rosen GM, Pou S, Ramos CL, Cohen MS, Britigan BE (1995) Free radicals and phagocytic cells. FASEB J 9:200–209

Rosenman SJ, Shrikant P, Dubb L, Benveniste EN, Ransohoff RM (1995) Cytokine-induced expression of vascular cell adhesion molecule-1 (VCAM-1) by astrocytes and astrocytoma cell lines. J Immunol 154:1888–1899

Rother M, Erkinjuntti T, Roessner M, Kittner B, Marcusson J, Karlsson I (1998) Propentofylline in the treatment of Alzheimer's disease and vascular dementia: a review of phase III trials. Dement Geriatr Cogn Disord 9 [Suppl 1]: 36–43

Rupalla K, Allegrini PR, Sauer D, Wiessner C (1998) Time course of microglia activation and apoptosis in various brain regions after permanent focal cerebral ischemia in mice. Acta Neuropathol (Berl) 96:172–178

Sadoshima S, Ooboshi H, Okada Y, Yao H, Ishitsuka T, Fujishima M (1989) Effect of thromboxane synthetase inhibitor on cerebral circulation and metabolism during experimental cerebral ischemia in spontaneously hypertensive rats. Eur J Pharmacol 169:75–83

Sairanen TR, Lindsberg PJ, Brenner M, Siren AL (1997) Global forebrain ischemia results in differential cellular expression of interleukin-1β (IL-1β) and its receptor at mRNA and protein level. J Cereb Blood Flow Metab 17:1107–1120

Saito K, Suyama K, Nishida K, Sei Y, Basile AS (1996) Early increases in TNF-α, IL-6 and IL-1 β levels following transient cerebral ischemia in gerbil brain. Neurosci Lett 206:149–152

Salminen A, Liu PK, Hsu CY (1995) Alteration of transcription factor binding activities in the ischemic rat brain. Biochem Biophys Res Commun 212:939–944

Sapolsky R, Rivier C, Yamamoto G, Plotsky P, Vale W (1987) Interleukin-1 stimulates the secretion of hypothalamic corticotropin-releasing factor. Science 238:522–524

Satoh K, Imaizumi T, Yoshida H, Hiramoto M, Takamatsu S (1992) Increased levels of blood platelet-activating factor (PAF) and PAF-like lipids in patients with ischemic stroke. Acta Neurol Scand 85:122–127

Schielke GP, Yang GY, Shivers BD, Betz AL (1998) Reduced ischemic brain injury in interleukin-1 β converting enzyme-deficient mice. J Cereb Blood Flow Metab 18:180–185

Schneider H, Pitossi F, Balschun D, Wagner A, del Rey A, Besedovsky HO (1998) A neuromodulatory role of interleukin-1β in the hippocampus. Proc Natl Acad Sci USA 95:7778–7783

Schobitz B, de Kloet ER, Sutanto W, Holsboer F (1993) Cellular localization of interleukin 6 mRNA and interleukin 6 receptor mRNA in rat brain. Eur J Neurosci, 5:1426–1435

Schroeter M, Jander S, Witte OW, Stoll G (1994) Local immune responses in the rat cerebral cortex after middle cerebral artery occlusion. J Neuroimmunol 55:195–203

Sebire G, Emilie D, Wallon C, Hery C, Devergne O, Delfraissy JF, Galanaud P, Tardieu M (1993) In vitro production of IL-6, IL-1 β, and tumor necrosis factor-α by human embryonic microglial and neural cells. J Immunol 150:1517–1523

Shyu KG, Chang H, Lin CC (1997) Serum levels of intercellular adhesion molecule-1 and E-selectin in patients with acute ischaemic stroke. J Neurol 244:90–93

Sims JE, Giri JG, Dower SK (1994) The two interleukin-1 receptors play different roles in IL-1 actions. Clin Immunol Immunopathol 72:9–14

Sornas R, Ostlund H, Muller R (1972). Cerebrospinal fluid cytology after stroke. Arch Neurol 26:489–501

Stanimirovic DB, Wong J, Shapiro A, Durkin JP (1997) Increase in surface expression of ICAM-1, VCAM-1 and E-selectin in human cerebromicrovascular endothelial cells subjected to ischemia-like insults. Acta Neurochir Suppl (Wien) 70:12–16

Stella N, Estelles A, Siciliano J, Tence M, Desagher S, Piomelli D, Glowinski J, Premont J (1997) Interleukin-1 enhances the ATP-evoked release of arachidonic acid from mouse astrocytes. J Neurosci 17:2939–2946

Stewart WF, Kawas C, Corrada M, Metter EJ (1997) Risk of Alzheimer's disease and duration of NSAID use. Neurology 48:626–632

Stins MF, Gilles F, Kim KS (1997) Selective expression of adhesion molecules on human brain microvascular endothelial cells. J Neuroimmunol 76:81–90

Stroemer RP, Rothwell NJ (1998) Exacerbation of ischemic brain damage by localized striatal injection of interleukin-1β in the rat. J Cereb Blood Flow Metab 18:833–839

Suzuki H, Abe K, Tojo S, Kimura K, Mizugaki M, Itoyama Y (1998). A change of P-selectin immunoreactivity in rat brain after transient and permanent middle cerebral artery occlusion. Neurol Res 20:463–469

Tarkowski E, Rosengren L, Blomstrand C, Wikkelso C, Jensen C, Ekholm S, Tarkowski A (1995) Early intrathecal production of interleukin-6 predicts the size of brain lesion in stroke. Stroke 26:1393–1398

Tarkowski E, Rosengren L, Blomstrand C, Wikkelso C, Jensen C, Ekholm S, Tarkowski A (1997) Intrathecal release of pro- and anti-inflammatory cytokines during stroke. Clin Exp Immunol 110:492–499

Tasaki K, Ruetzler CA, Ohtsuki T, Martin D, Nawashiro H, Hallenbeck JM (1997) Lipopolysaccharide pre-treatment induces resistance against subsequent focal cerebral ischemic damage in spontaneously hypertensive rats. Brain Res 748:267–270

Taylor AW, Streilein JW (1996) Inhibition of antigen-stimulated effector T cells by human cerebrospinal fluid. Neuroimmunomodulation 3:112–118

Terai K, Matsuo A, McGeer EG, McGeer PL (1996) Enhancement of immunoreactivity for NF-κB in human cerebral infarctions. Brain Res 739:343–349

The Enlimomab Acute Stroke Trial Investigators (1997) The Enlimomab acute stroke trial: final results. Neurology 48:A270:S33.001

Tokutomi T, Sigemori M, Kikuchi T, Hirohata M (1994) Effect of platelet-activating factor antagonist on brain injury in rats. Acta Neurochir [Suppl] 60:508–510

Tracey F, Crawford VL, Lawson JT, Buchanan KD, Stout RW (1993) Hyperglycaemia and mortality from acute stroke. Q J Med 86:439–446

Uno H, Matsuyama T, Akita H, Nishimura H, Sugita M (1997) Induction of tumor necrosis factor-α in the mouse hippocampus following transient forebrain ischemia. J Cereb Blood Flow Metab 17:491–499

Van Dam AM, De Vries HE, Kuiper J, Zijlstra FJ, De Boer AG, Tilders FJ, Berkenbosch F (1996) Interleukin-1 receptors on rat brain endothelial cells: a role in neuroimmune interaction? FASEB J 10:351–356

Vivien D, Bernaudin M, Buisson A, Divoux D, MacKenzie ET, Nouvelot A (1998) Evidence of type I and type II transforming growth factor-beta receptors in central nervous tissues: changes induced by focal cerebral ischemia. J Neurochem 70:2296–2304

Wang WC, Goldman LM, Schleider DM, Appenheimer MM, Subjeck JR, Repasky EA, Evans SS (1998) Fever-range hyperthermia enhances L-selectin-dependent adhesion of lymphocytes to vascular endothelium. J Immunol 160:961–969

Wang X, Barone FC, Aiyar NV, Feuerstein GZ (1997) Interleukin-1 receptor and receptor antagonist gene expression after focal stroke in rats. Stroke 28:155–161

Wang X, Siren AL, Liu Y, Yue TL, Barone FC, Feuerstein GZ (1994a) Upregulation of intercellular adhesion molecule 1 (ICAM-1) on brain microvascular endothelial cells in rat ischemic cortex. Brain Res Mol Brain Res 26:61–68

Wang X, Yue TL, Barone FC, White RF, Gagnon RC, Feuerstein GZ (1994b) Concomitant cortical expression of TNF-α and IL-1 β mRNAs follows early response gene expression in transient focal ischemia. Mol Chem Neuropathol 23:103–114

Wang X, Yue TL, White RF, Barone FC, Feuerstein GZ (1995a) Transforming growth factor-β1 exhibits delayed gene expression following focal cerebral ischemia. Brain Res Bull 36:607–609

Wang X, Yue TL, Young PR, Barone FC, Feuerstein GZ (1995b) Expression of interleukin-6, c-fos, and zif268 mRNAs in rat ischemic cortex. J Cereb Blood Flow Metab 15:166–171

Weber C, Erl W, Pietsch A, Weber PC (1995) Aspirin inhibits nuclear factor-κB mobilization and monocyte adhesion in stimulated human endothelial cells. Circulation 91:1914–1917

Weiss SJ (1989) Tissue destruction by neutrophils. N Engl J Med 320:365–376

Winkler MK, Beveniste EN (1998) Transforming growth factor-β inhibition of cytokine-induced vascular cell adhesion molecule-1 expression in human astrocytes. Glia 22:171–179

Wong D, Dorovini-Zis K (1992) Upregulation of intercellular adhesion molecule-1 (ICAM-1) expression in primary cultures of human brain microvessel endothelial cells by cytokines and lipopolysaccharide. J Neuroimmunol 39:11–21

Wong D, Dorovini-Zis K (1996) Platelet/endothelial cell adhesion molecule-1 (PECAM-1) expression by human brain microvessel endothelial cells in primary culture. Brain Res 731:217–220

Wong ML, Bongiorno PB, Gold PW, Licinio J (1995) Localization of interleukin-1β converting enzyme mRNA in rat brain vasculature: evidence that the genes encoding the interleukin-1 system are constitutively expressed in brain blood vessels. Pathophysiological implications. Neuroimmunomodulation 2:141–148

Yabuuchi K, Minami M, Katsumata S, Satoh M (1994) Localization of type I interleukin-1 receptor mRNA in the rat brain. Brain Res Mol Brain Res 27:27–36

Yamasaki Y, Shozuhara H, Onodera H, Kogure K (1994) Blocking of interleukin-1 activity is a beneficial approach to ischemia brain edema formation. Acta Neurochir Suppl 60:300–302

Yang GY, Gong C, Qin Z, Ye W, Mao Y, Bertz AL (1998) Inhibition of TNFα attenuates infarct volume and ICAM-1 expression in ischemic mouse brain. Neuroreport 9:2131–2134

Yang GY, Zhao YJ, Davidson BL, Betz AL (1997). Overexpression of interleukin-1 receptor antagonist in the mouse brain reduces ischemic brain injury. Brain Res 751:181–188

Ye RD, Kravchenko VV, Pan Z, Feng L (1996) Stimulation of NF-κB activation and gene expression by platelet-activating factor. Adv Exp Med Biol 416:143–151

Yoshimoto T, Houkin K, Tada M, Abe H (1997) Induction of cytokines, chemokines and adhesion molecule mRNA in a rat forebrain reperfusion model. Acta Neuropathol (Berl) 93:154–158

Yrjanheikki J, Keinanen R, Pellikka M, Hokfelt T, Koistinaho J (1998) Tetracyclines inhibit microglial activation and are neuroprotective in global brain ischemia. Proc Natl Acad Sci USA 95:15769–15774

Zhai QH, Futrell N, Chen FJ (1997) Gene expression of IL-10 in relationship to TNF-α, IL-1β and IL- 2 in the rat brain following middle cerebral artery occlusion. J Neurol Sci 152:119–124

Zhang R, Chopp M, Zhang Z, Jiang N, Powers C (1998a) The expression of P- and E-selectins in three models of middle cerebral artery occlusion. Brain Res 785:207–214

Zhang RL, Chopp M, Jiang N, Tang WX, Prostak J, Manning AM, Anderson DC (1995a) Anti-intercellular adhesion molecule-1 antibody reduces ischemic cell damage after transient but not permanent middle cerebral artery occlusion in the Wistar rat. Stroke 27:1438–1442

Zhang RL, Chopp M, Zaloga C, Zhang ZG, Jiang N, Gautam SC, Tang WX, Tsang W, Anderson DC, Manning AM (1995b) The temporal profiles of ICAM-1 protein and mRNA expression after transient MCA occlusion in the rat. Brain Res 682:182–188

Zhang RL, Chopp M, Zhang ZG, Phillips ML, Rosenbloom CL, Cruz R, Manning A (1996) E-selectin in focal cerebral ischemia and reperfusion in the rat. J Cereb Blood Flow Metab 16:1126–1136

Zhang Y, McCluskey K, Fujii K, Wahl LM (1998b) Differential regulation of monocyte matrix metalloproteinase and TIMP-1 production by TNF-α, granulocyte-macrophage CSF, and IL-1β through prostaglandin-dependent and -independent mechanisms. J Immunol 161:3071–3076

Zhang Z, Chopp M, Goussev A, Powers C (1998c) Cerebral vessels express interleukin 1β after focal cerebral ischemia. Brain Res 784:210–217

Zhang Z, Chopp M, Powers C (1997) Temporal profile of microglial response following transient (2 h) middle cerebral artery occlusion. Brain Res 744:189–198

Zhang ZG, Chopp M, Tang WX, Jiang N, Zhang RL (1995c) Postischemic treatment (2–4 h) with anti-CD11b and anti-CD18 monoclonal antibodies are neuroprotective after transient (2 h) focal cerebral ischemia in the rat. Brain Res 698:79–85

CHAPTER 4
Mitochondrial Approaches to Neuroprotection

M.F. BEAL

A. Introduction

A potential critical role of mitochondrial dysfunction in neurodegenerative diseases is becoming increasingly compelling. Mitochondrial dysfunction leads to a number of deleterious consequences for the cell including impaired calcium buffering, generation of free radicals, activation of nitric oxide synthase, activation of the mitochondrial permeability transition, and secondary excitotoxicity (BEAL 1992, 1995). This can lead to both apoptotic and necrotic cell death depending on the severity of the insult. Neurodegenerative diseases have widely disparate etiologies but may share mitochondrial dysfunction as a final common pathway.

Perhaps the best evidence for a role of mitochondrial dysfunction in neurodegenerative diseases comes from recent discoveries in Friedrich's ataxia. Friedrich's ataxia is characterized by neurodegeneration involving the spinocerebellar pathways as well as a cardiomyopathy. The gene product has been designated frataxin. The disease is caused by an expansion of a polymorphic GAA in trinucleotide repeat situated in the first intron of the corresponding gene on chromosome 25 (CAMPUZANO et al. 1996). Studies in yeast have shown that there is a gene homologous to the human frataxin protein which encodes a mitochondrial protein involved in iron homeostasis and respiratory function (WILSON and ROOF 1997; KOUTNIKOVA et al. 1997; BABCOCK et al. 1997). Both antibody staining and studies of human frataxin linked to green fluorescent protein have shown that the frataxin protein is localized to mitochondria (KOUTNIKOVA et al. 1997; BABCOCK et al. 1997; PRILLER et al. 1997). Disruption of the homologous yeast protein leads to respiratory insufficiency with an inability to carry out oxidative phosphorylation and a loss of mitochondrial DNA. Yeast with a disruption of the frataxin homologue also show markedly increased iron transport and hypersensitivity to oxidative stress mediated by hydrogen peroxide (BABCOCK et al. 1997). Consistent with this, endomyocardial biopsies of two patients with Friedrich's ataxia showed marked deficiencies of aconitase and complexes I and II-III of the electron transport chains, which are iron-sulfer clusters containing enzymes and are particularly susceptible to oxidative stress (ROTIG et al. 1997).

Familial amyotrophic lateral sclerosis is associated with point mutations in superoxide dismutase which may lead to increased generation of free radicals and thereby contribute to mitochondrial dysfunction. Mitochondria are particularly vulnerable to oxidative stress, and mitochondrial swelling and vacuolization are amongst the earliest pathologic features of two strains of transgenic amyotrophic lateral sclerosis (ALS) mice with copper/zinc (Cu, Zn) superoxide dismutase mutations (WONG et al. 1995; GURNEY et al. 1994). Mice with the G93A human SOD-1 mutations show altered electron transport enzymes, and expression of the mutant enzyme in vitro results in a loss of mitochondrial membrane potential, as well as elevated cytosolic calcium concentrations (CARRI et al. 1997; BROWNE et al. 1998). Recent work has demonstrated that there is a marked increase in mitochondrial swelling and vacuolization which immediately precedes the rapid phase of motor weakness in the transgenic ALS mice (KONG and XU 1998).

Recent studies have shown that Huntington's disease is associated with the development of intranuclear inclusions (DIFIGLIA et al. 1997). The disease is caused by an expansion of a CAG repeat in an unknown protein termed huntingtin. The means by which this leads to neuronal degeneration, however, remains obscure. A number of lines of evidence suggest that energetic defects are directly involved in pathogenesis. We demonstrated that there are elevations in lactate concentrations in both the cerebral cortex and basal ganglia of Huntington's disease patients (JENKINS et al. 1993, 1998). We also demonstrated that there was an elevation in phosphocreatine to inorganic phosphate ratio in resting gastrocnemius muscle (KOROSHETZ et al. 1997). Other evidence is the increased caloric intake despite weight loss in these patients (OBRIEN et al. 1990). We have also carried out a number of studies with mitochondrial toxins such as 3-nitropropionic acid which show that it can replicate both the histologic and neurochemical changes of Huntington's disease in rodents and primates (BEAL et al. 1993; BROUILLET et al. 1995).

In Alzheimer's disease there is evidence for deficiencies of both α-ketoglutarate dehydrogenase as well as cytochrome oxidase. Recent studies using cybrid cell lines have shown that the cytochrome oxidase defects can be transferred into mitochondrial deficient cell lines from patients platelets (DAVIS et al. 1997). Similarly, in Parkinson's disease there is evidence for a complex I defect in the substantia nigra (SCHAPIRA et al. 1990), which has also been demonstrated in platelets in several studies (PARKER et al. 1989; BENECKE et al. 1992; HAAS et al. 1995). The complex I defect in the patients' platelets has been shown to be transferable into mitochondria deficient cell lines (SWERDLOW et al. 1996).

B. Approaches to Neuroprotection
I. Mitochondrial Oxidative Interactions

Mitochondria are both important targets, and important sources, of reactive oxygen species. Because of the potential for univalent transfer of electrons from the electron transport chain to oxygen (electron leak) mitochondria are likely to be the major source of reactive species in eukaryotes (NOHL et al. 1993; PACKER et al. 1996; DYKENS 1994). Generation of reactive oxidants, including reactive oxygen species (ROS), appears to be increased in damaged mitochondria and in cells with compromised mitochondrial function. Conversely, mitochondrial structure and function are extremely sensitive to oxidants. For example, as described in more detail below in Sect. VII., acute exposure to relatively high levels of oxidants, especially in the presence of calcium, can induce the mitochondrial permeability transition, uncouple oxidative phosphorylation, and contribute to cytotoxicity via necrosis and/or apoptosis (through release of cytochrome c). The chemical composition of the mitochondria and its constituents contributes to its oxidant sensitivity. Mitochondrial membranes are highly polyunsaturated, making them excellent targets for peroxidation. Iron-sulfur proteins (e.g., aconitase) are abundant, essential, and highly susceptible to oxidant-mediated damage. The potential combination of free iron or copper and hydrogen peroxide formed from the dismutation of $O_2^{\bullet}$ increases the odds of production of $^{\bullet}OH$ through Fenton chemistry. Potential byproducts of $^{\bullet}OH$-mediated damage are indeed observed. These damage byproducts include the aromatic hydroxylation products o- and m-tyrosine, which are seen in some but not all mitochondrial preparations (KRISTAL et al. 1998), and the DNA damage product 8-hydroxydeoxyguanosine, which may reach concentrations approximately 15 times that found in nuclear DNA (RICHTER et al. 1988).

Mitochondrial resistance to oxidants is primarily mediated by a combination of enzymatic free radical scavengers (glutathione peroxidase, glutathione reductase, Mn-superoxide dismutase, and, in heart mitochondria, catalase) and low molecular weight antioxidants (e.g., tocopherols, lipoates, uric acid).

II. Mitochondrial Permeability Transition

A central role for the mitochondrial permeability transition (MPT) in both necrotic and apoptotic cell death is starting to emerge (ZORATTI and SZABO 1995; BERNARDI 1996). The MPT is attributed to a voltage-gated, cation-permeable channel, whose opening is favored by a number of factors including depolarization, intramatrix Ca^{2+} and oxidizing agents, and whose closing is favored by protons (low matrix pH) and adenine nucleotides. Oxidation of pyridine nucleotides favors pore opening while reduction favors pore closing (RIGOBELLO et al. 1995). Cyclosporin A is a good blocker of the channel that appears to require an interaction with mitochondrial cyclophilin (ANDREEVA et al. 1995). The channel functions in both high- and low-conductance modes. Induction involves the calcium-mediated opening of a pore in the inner mitochondrial membrane that allows free diffusion of all solutes smaller than 1,500 Da in and out of the mitochondrial matrix. Induction therefore leads to loss of the proton gradient and inability to conduct oxidative phosphorylation.

Exposure of mitochondria to supraphysiological levels of calcium, particularly in the presence of inorganic phosphate or oxidants, leads to mitochondrial dysfunction. At least two partially interactive and partially independent biochemical mechanisms mediate this loss of mitochondrial function (ZORATTI and SZABO 1995; BERNARDI 1996; KRISTAL and DUBINSKY 1997). The first of these is termed calcium cycling; the second is the MPT. "Calcium-cycling" refers to the depletion of mitochondrial energy stores, such as membrane potential, by the futile shuttling of calcium in and out of the mitochondrial matrix by the mitochondrial calcium transporters. In some models of calcium and/or oxidant mediated injury, calcium cycling appears to explain most or all of the mitochondrial injury observed (SCHLEGEL et al. 1992). In other models, calcium cycling appears to play a minor role, and the dominant mechanism seems to be the induction of a proteinaceous pore (ZORATTI and SZABO 1995). Furthermore, in some models of mitochondrial injury, calcium cycling contributes to (or perhaps is the primary cause of) induction of an MPT (SCHLEGEL et al. 1992). It is therefore not surprising that in many cases both cycling-dependent and -independent processes are likely involved in MPT induction (KRISTAL and DUBINSKY 1997).

Opening the MPT in the high conductance mode appears to be irreversible and is linked to cell death. It results in osmotic changes and mitochondrial swelling leading to straightening of the mitochondrial cristae. This in turn leads to a rupture of the outer mitochondrial membrane since the inner membrane is larger than the outer membrane. Rupture of the outer membrane releases apoptosis inducing factor (AIF) and cytochrome c into the cytoplasm (KROEMER et al. 1997). The release of cytochrome c appears to be due to a mechanically damaged outer mitochondrial membrane, rather than a specific release mechanism (PETIT et al. 1998). These factors then initiate apoptotic cell death by activation of a cascade of caspases. Activation of the MPT is enhanced by several caspases and is inhibited by the antiapoptotic proteins Bcl-2 and Bcl-X (MARZO et al. 1998). Activation of the MPT also plays an important role in excitotoxic cell death in vitro since NMDA induced mitochondrial depolarization and cell death are antagonized by cyclosporin A (SCHINDER et al. 1996; WHITE and REYNOLDS 1996). A role in vivo is suggested by the finding that cyclosporin A can inhibit ischemia-induced cell death in the hippocampus (UCHINO et al. 1995). Cyclosporin A also attenuates the degeneration of dopamingeric neurons induced by 6-hydroxydopamine in mouse brain (MATSUURA et al. 1996). Furthermore bcl-2, which inhibits apoptosis, inhibits opening of the MPT and release of cytochrome c from mitochondria (KLUCK et al. 1997). Bcl-2 resides in the outer mitochondrial membrane and at contact points between the inner and outer membranes (KRAJEWSKI et al. 1993).

Proposed components of the MPT are hexokinase 1, voltage dependent anion channel (porin), the peripheral benzodiazepine receptor and bcl-2 in the outer membrane, mitochondrial creatine kinase in the intermembrane space, and the adenine nucleotide transporter in the inner membrane. Drugs that bind to the translocator such as atractyloside and bongkrekic acid are respectively activators and inhibitors of the MPT. Complexes of hexokinase, mitochondrial creatine kinase, porin and the adenine nucleotide translocator can be isolated from rat brain (BEUTNER et al. 1997, 1998).

III. Therapeutic Approaches for Neurodegenerative Disease

If mitochondrial dysfunction plays a key role in neurodegenerative diseases then agents which improve mitochondrial energy production or which protect mitochondria from damage should be neuroprotective. As discussed below there are a number of approaches which could lead to both these goals. Several agents have been reported to show some efficacy in human mitochondrial diseases. Amongst those are coenzyme Q_{10}, nicotinamide, and creatine. Each of these agents has been shown to have some efficacy in human diseases associated with mitochondrial defects.

IV. Neuroprotective Effects of Coenzyme Q_{10}

There has been considerable interest in the use of coenzyme Q_{10} for treatment of mitochondrial disorders. The release of cytochrome c appears to be due to a mechanically damaged outer mitochondrial membrane, rather than a specific release mechanism (PETIT et al. 1998). Several reports found both clinical and biochemical improvement in patients with mitochondrial diseases. A recent report showed improvement of brain PCr levels in patients with mitochondrial cytopathies (BARBIROLI et al. 1997). Coenzyme Q_{10} has been reported to improve adenosine triphosphate (ATP) generation in vitro (BEYER 1992).

Coenzyme Q is an essential cofactor in the electron transport chain where it accepts electrons from complex I and II (BEYER 1992; EARNEST et al. 1996). The predominant form of coenzyme Q in man is coenzyme Q_{10}, which contains 10 isoprenoid units in the tail, while the predominant form in rodent is coenzyme Q_9, which has 9 isoprenoid units in the tail. Coenzyme Q is soluble and mobile in the hydrophobic core of the phospholipid bilayer of the inner membrane of the mitochondria, where it transfers electrons one at a time to complex III of the electron transport chain. Coenzyme Q also serves as an important antioxidant in both mitochondria and lipid membranes (NOACK et al. 1994; FORSMARK-ANDREE et al. 1997). Coenzyme Q, which is also known as ubiquinone, is a lipid soluble compound composed of a redox active quinoid moiety and a hydrophobic tail. If defects in energy metabolism and oxidative damage play a role in the pathogenesis of neurodegenerative diseases, then treatment with coenzyme Q_{10} could exert beneficial therapeutic effects.

We showed that oral administration of coenzyme Q_{10} significantly attenuated lesions produced by intrastriatal administration of malonate in rats, as well as malonate-induced depletions of ATP and increases in lactate concentrations (BEAL et al. 1994). An important issue, however, is whether oral administration of coenzyme Q_{10} can increase brain concentrations. We recently found that feeding with coenzyme Q_{10} increases cerebral cortex concentrations in 12- and 24-month-olds (MATTHEWS et al. 1998a).

Coenzyme Q levels are known to decrease with aging in several animal species and in man. We administered coenzyme Q_{10} to 12- and 24-month-old rats (MATTHEWS et al. 1998a). Oral supplementation with coenzyme Q_{10} at a dose of 200mg/kg to 12-month-old Sprague-Dawley rats produced significant increases in coenzyme Q_{10} levels in the cerebral cortex in the 30% range. Similarly, there was significant increases in 24-month-old Fisher 344 rats. We also demonstrated that oral supplementation with coenzyme Q_{10} resulted in significant time-dependent increases in brain mitochondrial concentrations of coenzyme Q_{10} at 7, 30, and 60 days with increases at 60 days being significant as compared to controls. Coenzyme Q_{10} supplementation for 1 week prior to administration of 3-nitropropionic (3-NP) acid dramatically reduced the size of the lesions in the treated animals in the range of approximately 85%. Oral supplementation with coenzyme Q_{10} also significantly attenuated reductions in reduced coenzyme Q_9 and reduced coenzyme Q_{10} following 3-NP administration in the same animals. We also examined the effects of coenzyme Q_{10} in the G93A transgenic model of familial ALS. The G93A mice were placed on diet supplemented with 200mg/kg of coenzyme Q_{10} or on a normal diet starting at 70 days of age. There was a significant increase in survival from a mean of 135 to 141 days of age (MATTHEWS et al. 1998a). In Huntington's disease patients, we found that administration of coenzyme Q_{10} at a dose of 360mg daily resulted in a significant decrease in elevated occipital cortex lactate concentrations, which was reversible upon withdrawal of coenzyme Q_{10} (KOROSHETZ et al. 1997).

V. Neuroprotective Effects of Nicotinamide

Nicotinamide is a precursor of nicotinamide adenine dinucleotide (NADH), which is an essential cofactor of several enzymes involved in energy production, including NADH-ubiquinone oxidoreductase and α-ketoglutarate dehydrogenase. Nicotinamide has been utilized as a treatment for mitochondrial disorders with limited success. Nicotinamide is also an inhibitor of poly (ADP-ribose) polymerase. This enzyme is activated after damage to DNA, such as that produced by peroxynitrite (ZHANG et al. 1994). Activation of the enzyme leads to a depletion of cellular levels of ATP and NADH, since the transfer of ADP-ribose is energy dependent (4 ATP molecules/generation of 1 NADH molecule). Nicotinamide protects against cell death produced by NO and peroxynitrite in vitro (ZHANG et al. 1994). We found that nicotinamide protects against malonate neurotoxicity in vivo and this protection is associated with protection against ATP depletion (BEAL et al. 1994). Both we and others found that nicotinamide protects against γ1-methyl-4-phenyl-1,2,3,6-tetrahydropyridine (MPTP) neurotoxicity (COSI et al. 1996; SCHULZ et al. 1995a). Recent work showed that poly (ADP-ribose) polymerase inhibitors are effective in animal models of stroke (TAKAHASHI et al. 1997; ENDRES et al. 1997). Studies in transgenic mice with a knockout of poly (ADP-ribose) polymerase also showed a marked resistance to ischemic lesions (ENDRES et al. 1997; ELIASSON et al. 1997).

VI. Neuroprotective Effects of Creatine

If the initiating step in the pathologic cascade is a depletion of cellular energy stores, then agents which can buffer cellular energy stores may be neuroprotective. The brain isoform of creatine kinase, along with the mitochondrial isoform and the substances (Cr) and (PCr), constitutes a system that seems to be critical in regulating energy homeostasis in the brain and other organs with high and fluctuating energy demands (HEMMER and WALLIMANN 1993). The mitochondrial isoform is part of a complex of proteins which form an efficient tightly coupled multienzyme energy channel which generates and transports energy in the form of PCr from the mitochondrial matrix to the cytoplasm. Creatine is an excellent stimulant for mitochondrial respiration resulting in the generation of PCr (O'GORMAN et al. 1996; KERNEC et al. 1996).

Substantial evidence supports a direct functional coupling of creatine kinase with Na^+/K^+ ATPase, neurotransmitter release, and the maintenance of membrane potentials, and restoration of ion gradients before and after depolarization (HEMMER and WALLIMANN 1993; DUNANT et al. 1988). High energy turnover and high creatine kinase concentrations have been found in those regions of the brain that are rich in synaptic connections, e.g., molecular layer of the cerebellum, glomerular structures of the granule layer, and the hippocampus (KALDIS et al. 1996). An important role of creatine kinase in the adult brain is supported by in vivo ^{31}P-NMR transfer measurements showing that the pseudo first-order rate constant of the creatine kinase reaction (in the direction of ATP synthesis) (CORBETT and LAPTOOK 1994). Creatine kinase flux correlates with brain activity as measured by EEG, and the amount of 2-deoxyglucose phosphate found in the brain (SAUTER and RUDIN 1993).

A strategy to improve brain energy stores is therefore to administer either creatine, or a creatine analogue such as cyclocreatine, to increase high energy phosphagens which could potentially sustain ATP production by the creatine kinase reaction. Prior studies in both the heart and in skeletal muscle showed that cyclocreatine administration resulted in increased tissue levels of cyclocreatine and phosphocyclocreatine, delayed depletion of ATP levels under ischemic conditions, and delayed onset of ischemia induced rigor (ANNESLEY and WALKER 1980; ELGEBALY et al. 1994; GRIFFITHS and WALKER 1976; ROBERTS and WALKER 1982; TURNER and WALKER 1987). Creatine has no effects on ischemia induced ATP depletion in the heart, but it protects against ATP depletion produced by arterial hypertension (TURNER and WALKER 1985; OSBAKKEN et al. 1992; CONSTANTIN-TEODOSIU et al. 1995). In hippocampal slices creatine supplementation increases PCr levels, delays synaptic failure, and ameliorates neuronal damage produced by anoxia (WHITTINGHAM and LIPTON

1981; CARTER et al. 1995). Cyclocreatine administration in vivo increased brain concentrations of PCCr and appeared to buffer ATP stores (WOZNICKI and WALKER 1980).

VII. Creatine and the MPT

Mitochondrial creatine kinase lies between the inner and outer membrane, and the hexokinase and porin on the outer membrane. Porin complexes with Mi-CK in vitro and induces its octamerization (BRDICZKA et al. 1994). Recent studies showed that complexes between hexokinase and creatine kinase can be isolated from rat brain homogenates (BEUTER et al. 1996). These complexes also contained porin and the adenine nucleotide translocator. When the complexes were reconstituted in phospholipid vesicles they showed cyclosporin A inhibitable release of both malate and ATP in response to Ca^{2+}, suggesting that they were major components of the MPT (BEUTNER et al. 1998). The hexokinase complex, porin, and the adenine nucleotide carrier form a channel which couples the enzyme directly to internal ATP (BEUTNER et al. 1997).

Recent work has shown that creatine and cyclocreatine can inhibit activation of the MPT in vitro. MiCK is believed to occur as an octomer in vivo, and has preferential access to ATP exported from the mitochondrial matrix by the adenine nucleotide transporter. Normally creatine kinase is not expressed in the liver but transgenic mice have been produced in which the human Mi-CK gene is specifically expressed in liver cells and is successfully imported into their mitochondria. Both calcium and atractyloside induce cyclosporin A sensitive swelling in the liver mitochondria of these mice, which is accompanied by a 2.5-fold increase in state 4 respiration rates (O'GORMAN et al. 1997). Creatine and cyclocreatine inhibited the increases in respiration rates. It was proposed that creatine and cyclocreatine stabilizes MiCk in an octomeric form which is resistant to activation in the MPT. In contrast, dimerization of MiCK in complexes resulted in cyclosporin A sensitive activation of the MPT in the presence of Ca^{2+}. Only the octomer form of creatine kinase is able to interact with the two boundary membranes. A plausible mechanism for the neuroprotective effects of creatine is therefore stabilization of the MPT.

VIII. Other Potential Neuroprotective Mechanisms of Creatine

Another potential neuroprotective mechanism of creatine is that it can increase PCr which is a direct substrate for glutamate uptake into synaptic vesicles (XU et al. 1997). XU et al. reported a PCr-dependent glutamate uptake system in addition to the ATP-dependent glutamate uptake system (XU et al. 1996). There is substantial evidence for an impairment of L-glutamate uptake in ALS (ROTHSTEIN et al. 1992).

Another potential neuroprotective mechanism of creatine is altering Ca^{2+} responses. Recent work in transgenic mice with mutations in both mitochondrial and cytosolic creatine kinase showed only small alterations in PCr (STEEGHS et al. 1997). The plateau levels of Ca^{2+} in myotubes of these mice were significantly increased after stimulation with acetylcholine or KCl. This suggests that PCr is intimately connected to the processes of Ca^{2+} release and storage in skeletal muscle. It was suggested that creatine kinase may therefore preferentially provide energy for Ca^{2+} uptake into the smooth endoplasmic reticulum. There is also substantial evidence that Ca^{2+} uptake in mitochondria plays a key role in neurotoxic mechanisms, such as the MPT, and impaired MiCK may therefore also potentially contribute to neurotoxicity.

IX. Creatine Uptake

Creatine is taken up by the choline transporter (GUIMBAL and KILIMANN 1993). This transporter is highly expressed in brain and skeletal muscle (SCHLOSS et al. 1994). In situ hybridization showed that the transporter is widely expressed in the central nervous system, but there are high levels of expression in the cerebellum (Purkinje and granule cell layers), choroid plexus, medial habenula, pontine nuclei, several brainstem nuclei, and hippocampus. Moderate signal is detected in the cortex, globus pallidus, and white matter while there is low level expression in the basal ganglia (HAPPE and MURRIN 1995).

We found that oral administration of both creatine and cyclocreatine produced dose-dependent neuroprotective effects against malonate lesions (MATTHEWS et al. 1998b). We also found protection following subacute administration of 3-NP. A significant 83% reduction in lesion volume was observed in animals fed 1% creatine.

The mechanism of neuroprotection involves protection against depletions of both PCr and ATP. We found that administration of either creatine or cyclocreatine results in increased brain concentrations of PCr or PCCr respectively. Administration of 3-NP produced significant decreases in creatine, PCr, GDP, AMP, NAD, ADP/GTP, and ATP. These energy metabolites are also decreased by cerebral ischemia (LAZZARINO et al. 1992), and by 3-NP in vitro (ERECINSKA and NELSON 1994), and in vivo (BROUILLET et al. 1993; TSAI et al. 1997). Creatine administration significantly protected against the decreases, and protected against 3-NP-induced increases of lactate as assessed by ^1H magnetic resonance spectroscopy in vivo (MATTHEWS et al. 1998b).

A secondary consequence of energy impairment is increased intracellular concentrations of calcium (BEAL 1995; BEAL 1992). This leads to increased free radical production by the mitochondria as well as activation of neuronal nitric oxide synthase, which is calmodulin dependent. This can lead to the generation of peroxynitrite, formed by the interaction of $O_2^{\bullet}$ with $NO^{\bullet}$. Peroxynitrite can oxidize intracellular molecules by a "hydroxyl radical" like activity and it can also lead to nitration of tyrosines (BECKMAN et al. 1990, 1992). We found that both malonate and 3-NP result in increases in both "hydroxyl radical activity," as well as 3-nitrotyrosine (SCHULZ et al. 1995a,b).

Creatine pretreatment delayed increases in intracellular calcium produced by 3-NP in cortical and striatal astrocytes in vitro (DESHPANDE et al. 1997). We found that administration of both creatine and cyclocreatine significantly attenuated malonate-induced increases in 2,3 and 2,5 DHBA/salicylate, markers of hydroxyl radical generation (FLOYD et al. 1984). Increased 3-nitrotyrosine levels following malonate and 3-NP were also attenuated. These findings therefore suggest that improved energy buffering can act upstream to attenuate free radical generation associated with cell death produced by mitochondrial toxins.

The development of transgenic mouse models of motor neuron degeneration associated with SOD1 mutations has further implicated mitochondrial dysfunction in ALS (WONG et al. 1995; GURNEY et al. 1994; RIPPS et al. 1995). These mice develop progressive weakness and degeneration of motor neurons leading to death. Two strains of these mice show early mitochondrial swelling and vacuolization as a prominent pathologic feature which precedes motor neuron loss (WONG et al. 1995; GURNEY et al. 1994; BRUIJN et al. 1997; DAL CANTO and GURNEY 1994). Transgenic mice with the G93A SOD1 mutation show increased mitochondrial complex I activity in cerebral cortex similar to observations in patients with the A4V SOD1 mutation (BROWNE et al. 1998), and expression of the G93A SOD1 mutation in neuroblastoma cells causes a loss of mitochondrial membrane potential (CARRI et al. 1997).

We examined the effects of creatine in G93A transgenic ALS mice. Dietary supplementation with creatine dose-dependently extended survival and preserved motor function in transgenic ALS mice (Klevenyi et al., unpublished data). Creatine administration resulted in improved maximal motor performance on the rotorod test from 116 to 136 days of age. Survival was extended by 13 days with 1% creatine and by 26 days with 2% creatine, which is superior to improvement with riluzole, which extended survival by 13 days in this model (GURNEY et al. 1996). Histologic evaluation at 120 days of age showed a 50% depletion of motor neurons in animals on normal diets and creatine administration resulted in significant protection against motor neuron loss.

We investigated whether neuroprotection in the G93A mice is associated with protection against oxidative stress. We found that creatine administration blocked increases in spinal cord 3-nitrotyrosine levels which were observed in FALS mice on normal diets (FERRANTE et al. 1997). We also utilized in vivo microdialysis to evaluate the conversion of 4HBA to 3,4DHBA following systemic administration of the mitochondrial toxin 3-nitropropionic acid to stress mitochondria (BEAL 1995). A significant increase in 3,4DHBA/4HBA was seen in FALS mice on control diets, which was significantly attenuated in mice with 1% creatine diets. Lastly, we found that creatine supplementation increased spinal cord creatine levels in FALS mice.

The therapeutic benefit of creatine on survival and motor function in G93A SOD1 transgenic mice provides further evidence that mitochondrial dysfunction and oxidative damage may play a role in disease pathogenesis, and suggest that creatine supplementation might be a novel therapeutic strategy for the treatment of ALS.

X. Studies in Man

A number of studies have examined the effects of creatine loading on creatine and PCr concentrations in muscle. In normal individuals ingestion of 20 g of creatine each day for 5 days leads on average to a more than 20% increase in muscle total creatine and an approximately 10% increase in PCr (HARRIS et al. 1992). Creatine is taken up by a specific transporter which is highly expressed in muscle and nervous tissue (SCHLOSS et al. 1994). The majority of creatine uptake occurs within the first few days of ingestion. One study examined ingestion at 20g per day for 6 days either alone or followed by maintenance with 2g per day (HULTMAN et al. 1996). The ingestion of 20g per day resulted in significant increases in muscle creatine levels at day 7, which then declined towards normal on days 21 and 35; however, the ingestion of 2g per day after the initial loading dose maintained elevated concentrations of creatine at days 21 and 35. Following ingestion of 3g per day significant increases in creatine and PCr were found at days 14 and 28, with the levels achieved on day 28 being comparable to those observed with 20g per day for 6 days. There were no significant changes in ATP concentrations. In another study 1gm of creatine produced increases in plasma creatine which did not exceed 100 μmol/l, while a dose of 5g resulted in a mean peak concentration of 295 μmol/l 1h after ingestion (HARRIS et al. 1992). Supplementation with 20–30g for 2 or more days resulted in significant increases in creatine in the quadriceps muscle, raising levels to the upper limit of normal found in normal subjects. An increase in PCr amounting to 20% of the creatine increase was observed, but there were no changes in ATP. No side effects of creatine administration were encountered. Carbohydrate ingestion has been shown to augment skeletal muscle creatine accumulation during creatine supplementation in man (GREEN et al. 1996).

Subsequent work has shown that muscle creatine accumulation is enhanced when creatine ingestion occurs in combination with simple carbohydrates (GREEN et al. 1996). Creatine ingestion enhanced PCr resynthesis post exercise (GREENHAFF et al. 1994). In one study creatine ingestion of 20g daily was reported to have no effects on muscle performance or muscle creatine (ODLAND et al. 1997). In another study, however, creatine administration at 20g daily for 5 days improved muscle performance during maximal exercise, and improved ATP resynthesis after exercise (CASEY et al. 1996). Elevated extracellular creatine leads to downregulation of creatine transport into human muscle cells, which serves to maintain creatine homeostasis (LOIKE et al. 1988).

There are a limited number of studies of creatine as a therapeutic in man. In 1981 SIPILA et al. reported that patients with gyrate atrophy (a condition in which creatine synthesis is impaired) treated with 1.5g of creatine per day resulted in subjective improvements in strength and it reversed the Type II muscle fiber atrophy associated with the illness (Sipila et al. 1981). Creatine administration for 1 year was well tolerated without adverse effects. A recent randomized controlled trial examined the effects of creatine in 6 patients with mitochondrial encephalopathy lactic acidosis and strokes (MELAS) and a seventh with a mitochondrial myopathy. Creatine was administered at 5 g po bid for 14 days followed by 2 g po bid for 7 days (Tarnopolsky et al. 1997). Creatine administration resulted in significant increases in hand grip strength, dorsiflexion torque and postexercise lactate. A prior report also noted improvement in a MELAS patient (Hagenfeldt et al. 1994). Gordon and colleagues found that 10 days of creatine at 20 g/day in patients with heart failure increased muscle PCr by 12% and improved one-leg knee extension ergometry by 21% (Gordon et al. 1995).

Administration of creatine at a dose of 20 g/d for 5 days followed by 10 g/day for 51 days had no adverse effects on hepatorenal function or clinical biochemistry (Greenhaff 1997). Patients with inborn errors in creatine synthesis have developmental delay, extrapyramidal symptoms, and seizures (Stockler et al. 1996). Oral supplementation with 4–8 g creatine per day or 2 g/kg per day normalizes brain concentrations and produces clinical improvement (Stockler et al. 1996; Ganesan et al. 1997). Although there are some anecdotal and press reports of adverse effects of creatine supplementation, there are no published data showing health risks associated with creatine supplementation. Creatine administration at 5 g per day for 56 days reduced plasma cholesterol and other lipids (Earnest et al. 1996). We have used creatine at a dose of 4.2 g daily for 1–2 months in preliminary studies in Huntington's disease patients and we have not encountered any adverse effects (Rosas et al., unpublished data).

C. Conclusions

There is substantial evidence that coenzyme Q_{10}, nicotinamide, and creatine can exert neuroprotective effects in experimental animal models. The prospects for using these compounds in man appear promising and coenzyme Q_{10} is presently in clinical trials in HD. It is being tested in a double-blind randomized clinical trial of 300 patients over 30 months either alone or in combination with the *N*-methyl-D-aspartate antagonist remacemide (CARE-HD Trial). Our recent results with creatine suggest that it is a promising agent for therapeutic studies in both HD and ALS.

References

Andreeva L, Tanveer A, Crompton M (1995) Evidence for the involvement of a membrane-associated cyclosporin-A-binding protein in the Ca^{2+}-activated inner membrane pore of heart mitochondria. Eur J Biochem 230:1125–2232

Annesley TM, Walker JB (1980) Energy metabolism of skeletal muscle containing cyclocreatine phosphate. J Biol Chem 255:3924–3930

Babcock M, de Silva D, Oaks R, Davis-Kaplan S, Jiralerspong S, Montermini L, Pandolfo M, Kaplan J (1997) Regulation of mitochondrial iron accumulation by Yfh1p, a putative homolog of frataxin. Science 276:1709–1712

Barbiroli B, Frassineti C, Martinelli P, Iotti S, Lodi R, Cortelli P, Montagna P (1997) Coenzyme Q_{10} improves mitochondrial respiration in patients with mitochondrial cytopathies. An in vivo study on brain and skeletal muscle by phosphorus magnetic resonance spectroscopy. Cell Molec Biol 43:741–749

Beal MF (1992) Does impairment of energy metabolism result in excitotoxic neuronal death in neurodegenerative illnesses? Ann Neurol 31:119–130

Beal MF (1995) Aging, energy and oxidative stress in neurodegenerative diseases. Ann Neurol 38:357–366

Beal MF, Brouillet E, Jenkins BG, Ferrante RJ, Kowall NW, Miller JM, Storey E, Srivastava R, Rosen BR, Hyman B (1993) Neurochemical and histologic characterization of excitotoxic lesions produced by the mitochondrial toxin 3-nitropropionic acid. J Neurosci 13:4181–4191

Beal MF, Henshaw R, Jenkins BG, Rosen BR, Schulz JB (1994) Coenzyme Q_{10} and nicotinamide block striatal lesions produced by the mitochondrial toxin malonate. Ann Neurol 36:882–888

Beckman JS, Beckman TW, Chen J, marshall PA, Freeman BA (1990) Apparent hydroxyl radical production by peroxynitrite: implications for endothelial injury from nitric oxide and superoxide. Proc Natl Acad Sci USA 87:1620–1624

Beckman JS, Ischiropoulos H, Zhu L, van der Woerd M, Smith C, Chen J, Harrison J, Martin JC, Tsai M (1992) Kinetics of superoxide dismutase- and iron-catalyzed nitration of phenolics by peroxynitrite. Arch Biochem Biophys 298:438–445

Benecke R, Strumper P, Weiss H (1992) Electron transfer complex I defect in idiopathic dystonia, Ann Neurol 32:683–686

Bernardi P (1996) The permeability transition pore. Control points of a cyclosporin A-sensitive mitochondrial channel involved in cell death. Biochim Biophys Acta 1275:5–9

Beuter G, Ruck A, Riede B, Welte W, Brdiczka D (1996) Complexes between kinases, mitochondrial porin and adenylate translocator in rat brain resemble the permeability transition pore. FEBS Lett 396:189–195

Beutner G, Ruck A, Riede B, Brdiczka D (1997) Complexes between hexokinase, mitochondrial porin and adenylate translocator in brain: regulation of hexokinase, oxidative phosphorylation and permeability transition pore. Biochem Soc Trans 25:151–157

Beutner G, Ruck A, Riede B, Brdiczka D (1998) Complexes between porin, hexokinase, mitochondrial creatine kinase and adenylate translocator display properties of the permeability transition pore. Implication for regulation of permeability transition by the kinases. Biochim Biophys Acta 1368:7–18

Beyer RE (1992) An analysis of the role of coenzyme Q in free radical generation and as an antioxidant. Biochem Cell Biol 70:390–403

Brdiczka D, Kaldis P, Wallimann T (1994) In vitro complex formation between the octamer of mitochondrial creatine kinase and porin. J Biol Chem 269:27640–27644

Brouillet E, Hantraye P, Ferrante RJ, Dolan R, Leroy-Willig A, Kowall NW, Beal MF (1995) Chronic mitochondrial energy impairment produces selective striatal degeneration and abnormal choreiform movements in primates. Proc Natl Acad Sci USA 92:7105–7109

Brouillet E, Jenkins BG, Hyman BT, Ferrante RJ, Kowall NW, Srivastava R, Roy DS, Rosen BR, Beal MF (1993) Age-dependent vulnerability of the striatum to the mitochondrial toxin 3-nitropropionic acid. J Neurochem 60:356–359

Browne SE, Bowling AC, Baik MJ, Gurney M, Brown JRH, Beal MF (1998) Metabolic dysfunction in familial, but not sporadic, amyotrophic lateral sclerosis. J Neurochem 71:281–287

Bruijn LI, Becher MW, Lee MK, Anderson KL, Jenkins NA, Copeland NG, Sisodia SS, Rothstein JD, Borchelt DR, Price DL, Cleveland DW (1997) ALS-linked SOD1 mutant G85R mediates damage to astrocytes and promotes rapidly progressive disease and SOD1-containing inclusions. Neuron 18:327–338

Campuzano V, Montermini L, Molto MD, Pianese L, Cossee M, Cavalcanti F, Monros E, Rodius F, Duclos F, Monticelli A, Zara F, Canizares J, Koutnikova H, Bidichandani SI, Gellera C, Brice A, Trouillas P, De Michele G, Filla A, De Frutos R, Palau F, Patel PI, Di Donato S, Mandel J-L, Cocozza S, Koenig M, Pandolfo M (1996) Friedrich's ataxia: autosomal recessive disease caused by an intronic GAA triplet repeat expansion. Science 271:1423–1427

Carri MT, Ferri A, Battistoni A, Famhy L, Gabbianelli R, Poccia F, Rotilio G (1997) Expression of a Cu,Zn superoxide dismutase typical of familial amyotrophic lateral sclerosis induces mitochondrial alteration and increase of cytosolic Ca^{2+} concentration in transfected neuroblastoma SH-SY5Y cells. FEBS Lett 414: 365–368

Carter AJ, Muller RE, Pschorn U, Stransky W (1995) Preincubation with creatine enhances levels of creatine phosphate and prevents anoxic damage in rat hippocampal slices. J Neurochem 64:2691–2699

Casey A, Constantin-Teodosiu D, Howell S, Hultman E, Greenhaff PL (1996) Creatine ingestion favorably affects performance and muscle metabolism during maximal exercise in humans. Am J Physiol 271:E31–E37

Constantin-Teodosiu D, Greenhaff PL, Gardiner SM, Randall MD, March JE, Bennett T (1995) Attenuation by creatine of myocardial metabolic stress in Brattleboro rats caused by chronic inhibition of nitric oxide synthase. Br J Pharmacol 116: 3288–3292

Corbett RJT, Laptook AR (1994) Age-related changes in swine brain creatine kinase-catalyzed ^{31}P exchange measured in vivo using ^{31}P NMR magnetization transfer. J Cereb Blood Flow Metab 14:1070–1077

Cosi C, Colpaert F, Koek W, Degryse A, Marien M (1996) Poly(ADP-ribose) polymerase inhibitors protect against MPTP-induced depletions of striatal dopamine and cortical noradrenaline in C57B1/6 mice. Brain Res 729:264–269

Dal Canto MC, Gurney ME (1994) Development of central nervous system pathology in a murine transgenic model of human amyotrophic lateral sclerosis. Am J Pathol 145:1271–1280

Davis RE, Miller S, Hermstadt C, Ghosh SS, Fahy E, Shinobu L, Galasko D, Thal LJ, Beal MF, Howell N, Parker JWD (1997) Mutations in mitochondrial cytochrome c oxidase genes segregate with late-onset Alzheimer's disease. Proc Natl Acad Sci USA 94:4526–4531

Deshpande SB, Fukuda A, Nishino H (1997) 3-Nitropropionic acid increases the intracellular Ca^{2+} in cultured astrocytes by reverse operation of the Na^{+}-Ca^{2+} exchanger. Exp Neurol 145:38–45

DiFiglia M, Sapp E, Chase KO, Davies SW, Bates GP, Vonsattel JP, Aronin N (1997) Aggregation of huntingtin in neuronal intranuclear inclusions and dystrophic neurites in brain. Science 277:1990–1993

Dunant Y, Loctin F, Marsal J, Muller D, Parducz A, Rabasseda X (1988) Energy metabolism, quantal acetylcholine release: effects of botulinum toxin, 1-fluoro-2,4-dinitrobenzene, and diamide in the *Torpedo* electric organ. J Neurochem 50:431–439

Dykens JA (1994) Isolated cerebral and cerebellar mitochondria produce free radicals when exposed to elevated Ca^{+} and Na^{+}: implications for neurodegeneration. J Neurochem 63:584–591

Earnest CP, Almada AL, Mitchell TL (1996) High-performance capillary electrophoresis-pure creatine monohydrate reduces blood lipids in men and women. Clin Sci 91:113–118

Elgebaly SA, Wei Z, Tyles E, Elkerm AF, Houser SL, Gillies C, Kaddurah-Daouk R (1994) Enhancement of the recovery of rat hearts after prolonged cold storage by cyclocreatine phosphate. Transplantation 57:803–806

Eliasson MJL, Sampei K, Mandir AS, Hurn PD, Traystman RJ, Bao J, Pieper A, Wang Z-Q, Dawson TM, Snyder SH, Dawson FL (1997) Poly(ADP-ribose) polymerase gene disruption renders mice resistent to cerebral ischemia. Nature Med 3: 1089–1095

Endres M, Wang Z-Q, Namura S, Waeber C, Moskowitz MA (1997) Ischemic brain injury is mediated by the activation of poly(ADP-ribose)polymerase. J Cereb Blood Flow Metab 17:1143–1151

Erecinska M, Nelson D (1994) Effects of 3-nitropropionic acid on synaptosomal energy and transmitter metabolism: relevance to neurodegenerative brain disease. J Neurochem 63:1033–1041

Ferrante RJ, Browne SE, Shinobu LA, Bowling AC, Baik MJ, MacGarvey U, Kowall NW, Brown JRH, Beal MF (1997) Evidence of increased oxidative damage in both sporadic and familial amyotrophic lateral sclerosis. J Neurochem 69: 2064–2074

Floyd RA, Watson JJ, Wong PK (1984) Sensitive assay of hydroxyl radical formation utilizing high pressure liquid chromatography with electrochemical detection of phenol and salicylate hydroxylation products. J Biochem Biophys Methods 10: 221–235

Forsmark-Andree P, Lee C-P, Dallner G, Ernster L (1997) Lipid peroxidation and changes in the ubiquinone content and the respiratory chain enzymes of submitochondrial particles. Free Radic Biol Med 22:391–400

Ganesan V, Johnson A, Connelly A, Eckhardt S, Surtees RA (1997) Guanidinoacetate methyltransferase deficiency: new clinical features. Pediatr Neurol 17:155–157

Gordon A, Hultman E, Kijser L, Kristjansson S, Rolf CJ, Nyquist O, Sylven C (1995) Creatine supplementation in chronic heart failure increases skeletal muscle creatine phosphate and muscle performance. Cardiovasc Res 30:413–418

Green AL, Hultman E, MacDonald IA, Sewell DA, Greenhaff PL (1996) Carbohydrate ingestion augments skeletal muscle creatine accumulation during creatine supplementation in humans. Am J Physiol 271:E821–E826

Greenhaff PL (1997) The nutritional biochemistry of creatine, J Nutr Biochem 8: 610–618

Greenhaff PL, Bodin K, Soderlund K, Hultman E (1994) Effect of oral creatine supplementation on skeletal muscle phosphocreatine resynthesis. Am J Physiol 266:E725–E730

Griffiths GR, Walker JB (1976) Accumulation of analog of phosphocreatine in muscle of chicks fed 1-carboxymethyl-2-iminoimidazolidine (cyclocreatine). J Biol Chem 251:2049–2054

Guimbal C, Kilimann MA (1993) A Na(+)-dependent creatine transporter in rabbit brain, muscle, heart, and kidney. cDNA cloning and functional expression. J Biol Chem 268:8418–8421

Gurney ME, Cutting FB, Zhai P, Dubocovich ML, Doble A, Taylor CP, Andrus PK, Hall ED (1996) Antioxidants and inhibitors of glutamatergic transmission have therapeutic benefit in a transgenic model of familial amyotrophic lateral sclerosis. Ann Neurol 39:147–157

Gurney ME, Pu H, Chiu AY, Dal Canto MC, Polchow CY, Alexander DD, Caliendo J, Hentati A, Kwon YW, Deng H-X, Chen W, Zhai P, Sufit RL, Siddique T (1994) Motor neuron degeneration in mice that express a human Cu,Zn superoxide dismutase mutation. Science 264:1772–1775

Haas RH, Nasirian F, Nakano K, Ward D, Pay M, Hill R, Shults CW (1995) Low platelet mitochondrial complex I and complex II/III activity in early untreated Parkinson's disease. Ann Neurol 37:714–722

Hagenfeldt L, von Dobeln U, Solders G, Kaijser L (1994) Creatine treatment in MELAS. Muscle Nerve Oct:1236–1237

Happe HK, Murrin LC (1995) In situ hybridization analysis of CHOT1, a creatine transporter, in the rat central nervous system. J Comp Neurol 35:94–103

Harris RC, Soderlund K, Hultman E (1992) Elevation of creatine in resting and exercised muscle of normal subjects by creatine supplementation, Clin Sci 83:367–374

Hemmer W, Wallimann T (1993) Functional aspects of creatine kinase in brain. Dev Neurosci 15:249–260

Hultman E, Soderlund K, Timmons JA, Cederblad TG, Greenhaff PL (1996) Muscle creatine loading men. J Appl Physiol 81:232–237

Jenkins B, Koroshetz W, Beal MF, Rosen B (1993) Evidence for an energy metabolism defect in Huntington's disease using localized proton spectroscopy. Neurology 43: 2689–2695

Jenkins BG, Rosas HD, Chen Y-C, Makabe T, Myers R, MacDonald M, Rosen BR, Beal MF, Koroshetz WJ (1998) ^{1}H NMR spectroscopy in Huntington's. Striatal asymmetries and correlations with CAG repeats. Neurology 50:1357–1365

Kaldis P, Hemmer W, Zanolla E, Holtzman D, Wallimann T (1996) "Hot spots" of creatine kinase localization in brain: cerebellum, hippocampus and choroid plexus. Develop Neurosci 18:542–554

Kernec F, Le Tallec N, Nadal L, Begue J-M, Le Rumeur E (1996) Phosphocreatine synthesis by isolated rat skeletal muscle mitochondria is not dependent upon external ADP: a 31P NMR study. Biochem Biophys Res Commun 225:819–825

Kluck RM, Bossy-Wetzel E, Green DR, Newmeyer DD (1997) The release of cytochrome c from mitochondria: a primary site for Bcl-2 regulation of apoptosis. Science 275:1132–1136

Kong J, Xu Z (1998) Massive mitochondrial degeneration in motor neurons triggers the onset of amyotrophic lateral sclerosis in mice expressing a mutant SOD1. J Neurosci 18:3241–3250

Koroshetz WJ, Jenkins BG, Rosen BR, Beal MF (1997) Energy metabolism defects in Huntington's disease and possible therapy with coenzyme Q_{10}. Ann Neurol 41: 160–165

Koutnikova H, Campuzano V, Foury F, Dolle P, Cazzalini O, Koenig M (1997) Studies of human, mouse and yeast homologues indicate a mitochondrial function for frataxin. Nature Genet 16:345–351

Krajewski S, Tanaka S, Takayama S, Schibler MJ, Fenton W, Reed JC (1993) Investigation of the subcellular distribution of the *bcl*-2 oncoprotein: residence in the nuclear envelope, endoplasmic reticulum, and outer mitochondrial membranes. Cancer Res 53:4701–4714

Kristal BS, Dubinsky JM (1997) Mitochondrial permeability transition in the central nervous system: induction by calcium cycling-dependent and independent pathways. J Neurochem 69:524–538

Kristal BS, Vigneau-Callahan KE, Matson WR (1998) Modulated purine catabolism augments mitochondrial antioxidant defenses, in press

Kroemer G, Zamzami N, Susin SA (1997) Mitochondrial control of apoptosis. Trends Immunol 18:44–51

Lazzarino G, Vagnozzi R, Tavazzi B, Pastore FS, Di Pierro D, Siragusa P, Belli A, Giuffre R, Giardina B (1992) MDA, oxypurines, and nucleosides relate to reperfusion in short-term incomplete cerebral ischemia in the rat. Free Radic Biol Med 13: 489–498

Loike JD, Zalutsky DL, Kaback E, Miranda AF, Silverstein SC (1988) Extracellular creatine regulates creatine transport in rat and human muscle cell. Proc Natl Acad Sci USA 85:807–811

Marzo I, Brenner C, Zamzami N, Susin SA, Beutner G, Brdiczka D, Remy R, Xie Z-H, Reed JC, Kroemer G (1998) The permeability transition pore complex: A target for apoptosis regulation by caspases and Bcl-2-related proteins. J Exp Med 187:1261–1271

Matsuura K, Kabuto H, Makino H, Ogawa N (1996) Cyclosporin A attenuates degeneration of dopaminergic neurons induced by 6-hydroxydopamine in the mouse brain. Brain Res 733:101–104

Matthews RT, Yang L, Jenkins BG, Ferrante RJ, Rosen BR, Kaddurah-Daouk R, Beal MF (1998a) Neuroprotective effects of creatine and cyclocreatine in animal models of Huntington's disease. J Neurosci 18:156–163

Matthews RT, Yang S, Browne S, Baik M, Beal MF (1998b) Coenzyme Q_{10} administration increases brain mitochondrial concentrations and exerts neuroprotective effects. Proc Natl Acad Sci USA in press

Noack H, Kube U, Augustin W (1994) Relations between tocopherol depletion and coenzyme Q during lipid peroxidation in rat liver mitochondria. Free Radic Res 20:375–386

Nohl H, Koltover V, Stolze K (1993) Ischemia/reperfusion impairs mitochondrial energy conservation and triggers $O_2^{\cdot-}$-release as a byproduct of respiration. Free Radic Res Comm 18:127–137

O'Gorman E, Beutner G, Dolder M, Koretsky AP, Brdiczka D, Wallimann T (1997) The role of creatine kinase inhibition of mitochondrial permeability transition. FEBS Lett 414:253–257

O'Gorman E, Beutner G, Wallimann T, Brdiczka D (1996) Differential effects of creatine depletion on the regulation of enzyme activities and on creatine-stimulated mitochondrial respiration in skeletal muscle, heart, and brain. Biochim Biophys Acta 1276:161–170

Obrien CF, Miller C, Goldblatt D, Welle S, Forbes G, Lipinski B, Panzik J, Peck R, Plumb S, Oakes D, Kurlan R, Shoulson I (1990) Extraneural metabolism in early Huntington's disease. Ann Neurol 28:300–301

Odland LM, MacDougall JD, Tarnopolsky MA, Elorriaga A, Borgmann A (1997) Effect of oral creatine supplmentation on muscle [PCr] and short-term maxium power ouput. Med Sci Sports Exerc 29:216–219

Osbakken M, Ito K, Zhang D, Ponomarenko I, Ivanics T, Jahngen EGE, Cohn M (1992) Creatine and cyclocreatine effects on ischemic myocardium: ^{31}P nuclear magnetic resonance evaluation of intact heart. Cardiology 80:184–195

Packer MA, Porteous CM, Murphy MP (1996) Superoxide production by mitochondria in the presence of nitric oxide forms peroxynitrite. Biochem Molec Biol Intl 40:527–534

Parker WD, Jr, Boyson SJ, Parks JK (1989) Abnormalities of the electron transport chain in idiopathic Parkinson's disease. Ann Neurol 26:719–723

Petit PX, Goubern M, Diolez P, Susin SA, Zamzami N, Kroemer G (1998) Disruption of the outer mitochondrial membrane as a result of large amplitude swelling: the impact of irreversible permeability transition. FEBS Lett 426:111–116

Priller J, Scherzer CR, Faber PW, MacDonald ME, Young AB (1997) Frataxin gene of Friedreich's ataxia is targeted to mitochondria. Ann Neurol 42:265–269

Richter C, Park J-W, Ames B (1988) Normal oxidative damage to mitochondrial and nuclear DNA is extensive. Proc Natl Acad Sci USA 52:515–520

Rigobello MP, Turcato F, Bindoli A (1995) Inhibition of rat liver mitochondrial permeability transition by respiratory substrates. Arch Biochem Biophys 319:225–230

Ripps ME, Huntley GW, Hof PR, Morrison JH, Gordon JW (1995) Transgenic mice expressing an altered murine superoxide dismutase gene provide an animal model of amyotrophic lateral sclerosis. Proc Natl Acad Sci USA 92:689–693

Roberts JJ, Walker JB (1982) Feeding a creatine analogue delays ATP depletion and onset of rigor in ischemic heart. Am J Physiol 243:H911–H916

Rothstein JD, Martin LJ, Kuncl RW (1992) Decreased glutamate transport by the brain and spinal cord in amyotrophic lateral sclerosis. N Engl J Med 236:1464–1468

Rotig A, de Lonlay P, Chretien D, Foury F, Koenig M, Sidi D, Munnich A, Rustin P (1997) Aconitase and mitochondrial iron-sulphur protein deficiency in Friedreich ataxia. Nature Gene 17:215–217

Sauter A, Rudin M (1993) Determination of creatine kinase parameters in rat brain by NMR magnetization transfer: correlation with brain function. J Biol Chem 268:13166–13171

Schapira AHV, Cooper JM, Dexter D, Clark JB, Jenner P, Marsden CD (1990) Mitochondrial complex I deficiency in Parkinson's disease. J Neurochem 54:823–827

Schinder AF, Olson EC, Spitzer NC, Montal M (1996) Mitochondrial dysfunction is a primary event in glutamat neurotoxicity. J Neurosci 16:6125–6133

Schlegel J, Schweizer M, Richter C (1992) "Pore" formation is not required for the hydroperoxide-induced Ca^{2+} release from rat liver mitochondria. Biochem J 285:65–69

Schloss P, Mayser W, Betz H (1994) The putative rat choline transporter CHOT1 transports creatine and is highly expressed in neural and muscle-rich tissues. Biochem Biophys Res Commun 198:637–645

Schulz JB, Henshaw DR, Matthews RT, Beal MF (1995a) Coenzyme Q_{10} and nicotinamide and a free radical spin trap protect against MPTP neurotoxicity. Exp Neurol 132:279–283

Schulz JB, Henshaw DR, Siwek D, Jenkins BG, Ferrante RJ, Cipolloni PB, Kowall NW, Rosen BR, Beal MF (1995b) Involvement of free radicals in excitotoxicity in vivo. J Neurochem 64:2239–2247

Sipila I, Rapola J, Simell O, Vannas A (1981) Supplementary creatine as a treatment for gyrate atrophy of the choroid and retina. N Engl J Med 304:867–870

Steeghs K, Benders A, Oerlemans F, de Haan A, Heerschap A, Ruitenbeek W, Jost C, van Deursen J, Perryman B, Pette D, Bruckwilder M, Koudijs J, Jap P, Veerkamp J and Wieringa B (1997) Altered Ca^{2+} responses in muscles with combined mitochondrial and cytosolic creatine kinase deficiencies. Cell 89:93–103

Stockler S, Isbrandt D, Hanefeld F, Schmidt B, von Figura K (1996) Guanidinoacetate methyltransferase deficiency: the first inborn error of creatine metabolism in man. Am J Hum Genet 58:914–922

Swerdlow RH, Parks JK, Miller SW, Tuttle JB, Trimmer PA, Sheehan JP, Bennett JP, Jr, Davis RE, Parker WD, Jr (1996) Origin and functional consequences of the complex I defect in Parkinson's disease. Ann Neurol 40:663–671

Takahashi K, Greenberg JH, Jackson P, Maclin K, Zhang J (1997) Neuroprotective effects of inhibiting poly(ADP-ribose) synthetase on focal cerebral ischemia in rats. J Cereb Blood Flow Metab 17:1137–1142

Tarnopolsky MA, Roy BD, MacDonald JR (1997) A randomized, controlled trial of creatine monohydrate in patients with mitochondrial cytopathies. Muscle Nerve 1502–1509

Tsai MJ, Goh CC, Wan YL, Chang C (1997) Metabolic alterations produced by 3-nitropropionic acid in rat striata and cultured astrocytes: quantitative *in vitro* ^{1}H nuclear magnetic resonance spectroscopy and biochemical characterization. Neuroscience 79:819–826

Turner DM, Walker JB (1985) Relative abilities of phosphagens with different thermodynamic or kinetic properties to help sustain ATP and total adenylate pools in heart during ischemia. Arch Biochem Biophys 238:642–651

Turner DM, Walker JB (1987) Enhanced ability of skeletal muscle containing cyclocreatine phosphate to sustain ATP levels during ischemia following β-adrenergic stimulation. J Biol Chem 262:6605–6609

Uchino H, Elmer E, Uchino K, Lindvall O, Siesjo BK (1995) Cyclosporin A dramatically ameliorates CA1 hippocampal damage following transient forebrain ischaemia in the rat. Acta Physiol Scand 155:469–471

White RJ, Reynolds IJ (1996) Mitochondrial depolarization in glutamate-stimulated neurons: an early signal specific to excitotoxin exposure. J Neurosci 16:5688–5697

Whittingham TS, Lipton P (1981) Cerebral synaptic transmission during anoxia is protected by creatine. J Neurochem 37:1618–1621

Wilson RB, Roof DM (1997) Respiratory deficiency due to loss of mitochondrial DNA in yeast lacking the frataxin homologue. Nature Genet 16:352–357

Wong PC, Pardo CA, Borchelt DR, Lee MK, Copeland NG, Jenkins NA, Sisodia SS, Cleveland DW, Price DL (1995) An adverse property of a familial ALS-linked SOD1 mutation causes motor neuron disease characterized by vacuolar degeneration of mitochondria. Neuron 14:1105–1116

Woznicki DT, Walker JB (1980) Utilization of cyclocreatine phosphate, an analogue of creatine phosphate, by mouse brain during ischemia and its sparing action on brain energy reserves. J Neurochem 34:1247–1253

Xu CJ, Kanfer JN, Klunk WE, Xiong Q, McClure RJ, Pettegrew JW (1997) Effect of phosphomonoesters, phosphodiesters, and phosphocreatine on glutamate uptake by synaptic vesicles. Molec Chem Neuropathol 32:89–99

Xu CJ, Klunk WE, Kanfer JN, Xiong Q, Miller G, Pettegrew JW (1996) Phosphocreatine-dependent glutamate uptake by synaptic vesicles. J Biol Chem 271:13435–13440

Zhang J, Dawson VL, Dawson TM, Snyder SH (1994) Nitric oxide activation of poly(ADP-ribose) synthetase in neurotoxicity. Science 263:687–689

Zoratti M, Szabo I (1995) The mitochondrial permeability transition. Biochim Biophys Acta 1241:139–176

CHAPTER 5
Stabilizing Calcium Homeostasis

M.P. MATTSON

A. Introduction

The premise of this chapter is that disruption of cellular calcium homeostasis plays a pivotal role in the neurodegenerative process that occurs in many different neurological disorders, with a focus on cerebral ischemia. I will not attempt to review the extensive literature supporting the involvement of calcium in ischemic injury to neurons (see CHOI 1995; MATTSON and MARK 1996 for review), but rather will provide selected examples of such evidence, together with a more thorough account of preventative and therapeutic approaches that target systems involved in either disruption or stabilization of neuronal calcium homeostasis. A variety of cell culture and animal models of neurodegenerative processes relevant to ischemic brain injury have been developed. Such experimental models have allowed the manipulation and monitoring of Ca^{2+} movements in living neurons exposed to conditions relevant to humans suffering from stroke and other neurodegenerative conditions, thus clarifying roles of calcium in neuronal cell death processes. Other chapters in this volume expand upon some of the mechanisms that contribute to the sustained elevations of intracellular free calcium level ($[Ca^{2+}]_i$) that occur in neurons following ischemic or traumatic insults – these include energy failure, overstimulation of glutamate receptors, and oxidative stress. Recent progress in understanding mechanisms of programmed cell death (apoptosis) has provided additional information concerning the role calcium in cell death, and has emphasized its involvement in both early and late events in the cell death process.

In the present chapter I will emphasize the importance of neuroprotective signal transduction pathways that are activated when CNS tissue is injured (see MATTSON 1997 for review), including recently discovered and quite surprising mechanisms. The concept of "programmed cell life" (MATTSON and FURUKAWA 1996) is perhaps more parsimonious with evolutionary theory than is "programmed cell death" in that mechanisms designed to prevent neuronal death following injury will enhance the chances of survival of the individual and the species. It is very clear that injury to the nervous system, as occurs following stroke or trauma, results in increased expression of a myriad of trophic factors and cytokines. Studies have documented that many of these injury-induced factors (e.g., bFGF, NGF, BDNF, TGF-β, sAPPα, TNF-α) can protect neurons against the ischemic insult. As will be detailed later in this chapter, these neuroprotective factors activate receptors linked to intracellular signaling pathways that regulate cellular calcium homeostasis, either by modulating gene expression or by inducing postranslational modifications (e.g., phosphorylation) of calcium-regulating proteins. In addition to intercellular signaling factors, increased levels of intracellular Ca^{2+} and reactive oxygen species (ROS) that occur as the result of ischemic and traumatic injury, can activate cytoprotective mechanisms within neurons. Such intrinsic neuroprotective pathways have evolved to include a diverse array of proteins, ranging from protein kinases to transcription factors to cytoskeletal proteins.

B. Regulation of Calcium Homeostasis in Neurons

The plasma membrane, and membranes of the endoplasmic reticulum and mitochondria, contain proteins that play central roles in regulation of $[Ca^{2+}]_i$ under basal conditions and following physiological and cytotoxic elevations of $[Ca^{2+}]_i$. In addition, there are many different calcium-binding proteins located in cytosolic and/or organellar compartments that serve as transducers of calcium signals or as calcium buffers. The importance of calcium in nervous system function is highlighted by the variety of enzymes and structural proteins that are regulated, either directly or indirectly, by calcium (Table 1). Many of these calcium-responsive systems are intimately involved in the regulation of neuronal survival and plasticity, and are therefore of considerable interest with respect to the adverse consequences of cerebral ischemia on neuronal circuits. A better understanding of the various systems that regulate and/or respond to changes in cytoplasmic and intraorganellar changes in calcium levels is required to develop therapeutic approaches that preserve and restore functional neuronal circuits following ischemic brain injury.

Table 1. Examples of regulatory systems responsive to calcium

Protein	Effect
Calmodulin	Activation
CaM kinases	Activation
Phospholipase C	Enhanced activation
Protein kinase C	Enhanced activation
Transcription factors	Increased or decreased gene expression
Nitric oxide synthase	Activation
Synaptic vesicle-associated proteins	Neurotransmitter release
Microtubule-associated proteins	Increased microtubule binding
Actin-binding proteins	Actin polymerization and depolymerization
Caspases	Activation and cleavage of various substrates

I. Plasma Membrane Systems

Several different types of integral membrane proteins mediate movements of Ca^{2+} into or out of neurons (Table 2). The 10000-fold concentration gradient of calcium ions across the plasma membrane of neurons (approximately 1 mM outside the cell and 100 nM in the cytoplasm) is maintained largely by the constitutive activity of the plasma membrane Ca^{2+}-ATPase (PMCA) of which there are several isoforms. Activity of the PMCA can be regulated by intracellular signaling pathways, with calmodulin playing a role in stimulation of its activity (CARAFOLI et al. 1996). The PMCA has a high affinity for Ca^{2+} but such a low capacity that it is relatively ineffective in rapidly restoring $[Ca^{2+}]_i$ following a large influx of Ca^{2+}, as can occur following an action potential. Rapid removal of Ca^{2+} from the cytoplasm can be accomplished via the activation of the plasma membrane Na^+/Ca^{2+} exchanger (LEDERER et al. 1996). The Na^+/K^+-ATPase plays an indirect role in the regulation of Ca^{2+} movement across the plasma membrane by maintaining resting membrane potential. Depolarization of the plasma membrane results in the opening of voltage-dependent Ca^{2+} channels (VDCC) which provide conduits for influx of Ca^{2+} down its concentration gradient. Membrane depolarization can also result in Ca^{2+} influx through the N-methyl-D-aspartate (NMDA) subtype of glutamate receptor as the result of relief of the voltage-dependent block of the receptor channel by Mg^{2+}. Certain subtypes of AMPA (α-amino-3-hydroxy-5-methylisoxazole-4-propionate) receptors, particularly those lacking the GluR2 subunit, are also permeable to calcium.

Table 2. Plasma membrane proteins involved in regulation of neuronal calcium homeostasis

Protein	Function/Characteristics
Ca^{2+}-ATPase (PMCA1–4)	Maintenance of rest $[Ca^{2+}]$C/high affinity–low capacity
Na^+/Ca^{2+} exchanger	Rapid restoration of $[Ca^{2+}]$C/low affinity–high capacity
Voltage-dependent Ca^{2+} channels	Mediate depolarization-induced Ca^{2+} influx
L, N, T and P subtypes	Differ in current characteristics and pharmacological properties
NMDA receptors (NR1, NR2A-C)	Flux large amounts of Ca^{2+} in response to glutamate
AMPA/kainate receptors (GluR1–4)	Membrane depolarization in response to glutamate
Na^+/K^+-ATPase	Maintenance of membrane potential – indirect control of VDCC
Transmitter receptors (e.g., GABA, acetylcholine, norepinephrine)	Modulate Ca^{2+} channels via second messenger pathways or changes in membrane potential
Glucose transporter (GLUT3)	Glucose uptake – maintenance of ATP levels

There is no doubt that plasma membrane proteins involved in calcium regulation play important roles in promoting or suppressing the excessive accumulation of $[Ca^{2+}]_i$ that occurs following ischemic, excitotoxic and traumatic insults. Activation of glutamate receptors can kill neurons, and such cell death is mediated by calcium influx (MATTSON et al. 1989; CHOI 1995). Glutamate receptor antagonists can protect neurons in cell culture against insults relevant to stroke, such as hypoxia and glucose deprivation (MATTSON and MARK 1996). Ouabain, an inhibitor of the Na^+/K^+-ATPase, promotes opening of VDCC and NMDA receptors, and greatly increases neuronal vulnerability to glutamate-induced calcium overload and cell death (Fig. 1). Activation of gamma-aminobutyric acid (GABA) receptors stabilizes membrane potential and can thereby reduce calcium influx and protect against calcium overload (MATTSON and KATER 1988). In addition to membrane proteins that directly modulate ion fluxes, many different plasma membrane proteins indirectly influence calcium homeostasis. For example, the glucose transporter plays a major role in moving glucose into neurons, and this glucose is critical for maintenance of ATP levels. Impairment of glucose transport results in a dramatic increase in the vulnerability of neurons to glutamate-induced calcium overload and cell death (Fig. 1).

II. Endoplasmic Reticulum

The endoplasmic reticulum (ER) is an organelle that is intimately involved in calcium signaling in neurons (Fig. 2). The concentration of Ca^{2+} in the ER lumen is at least 100-fold higher than in the cytoplasm; this Ca^{2+} gradient is maintained by the activity of the smooth endoplasmic reticulum Ca^{2+}-ATPase (SERCA), an enzyme that is distinct from the PMCA. The ER membrane also houses Ca^{2+} channels, of which there are several different types including those responsive to inositol triphosphate (IP_3) and those responsive to ryanodine. IP_3 is released when membrane-associated phosphatidyl inositol bisphosphate (PIP_2) is cleaved by phospholipase C, whereas activation of ryanodine receptors is initiated by the opening of surface membrane Ca^{2+} channels (BOOTMAN et al. 1997; FRANZINI-ARMSTRONG and PROTASI 1997; BENNETT et al. 1998). Emerging data suggest that in many cell types ER and plasma membranes are in close apposition, and that molecules associated with the inner surface of the plasma membrane can interact with IP_3 and ryanodine receptors. For example, PIP_2 was recently shown to negatively modulate opening of IP_3 receptors, apparently by competing with IP_3 for binding to the channel protein (LUPU et al. 1998). The ER also contains several Ca^{2+}-binding proteins including calreticulin and calretinin (BAIMBRIDGE et al. 1992; KRAUSE and MICHALAK 1997). In addition to these established ER calcium-regulating proteins, there are likely to be other ER-resident proteins that play important roles in calcium homeostasis. For example, the glucose-regulated proteins, GRP78 and GRP94, were recently shown to play important roles in protecting cells against apoptosis by a mechanism that appears to involve stabilization of calcium home-

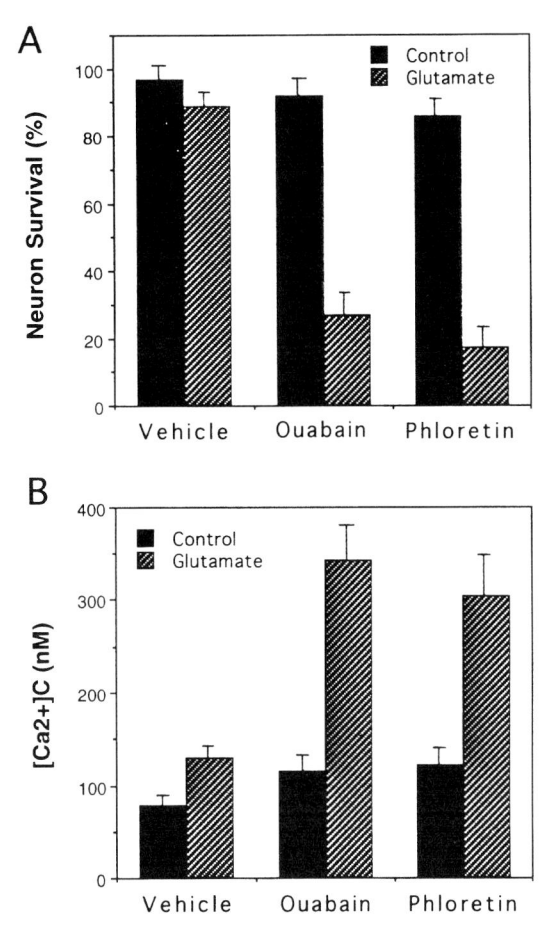

Fig. 1A,B. Evidence for pivotal roles of the plasma membrane Na$^+$/K$^+$-ATPase and glucose transporter in preventing glutamate receptor-mediated calcium overload. **A** Rat hippocampal cell cultures were exposed to ouabain (100 μM; an inhibitor of the Na$^+$/K$^+$-ATPase), phloretin (5 μM, an inhibitor of the glucose transporter), glutamate (5 μM), or combinations of ouabain or phloretin plus glutamate. Neuronal survival was quantified 24 h following treatment. Values are the mean and SE of determinations made in four cultures. **B** Rat hippocampal cell cultures were exposed to ouabain (100 μM), phloretin (5 μM), glutamate (5 μM), or combinations of ouabain or phloretin plus glutamate. The [Ca^{2+}]C was quantified 20 min following treatment by fluorescence ratio imaging of the calcium indicator dye fura-2. Values are the mean and SE of determinations made in four cultures (12–18 neurons analyzed/culture)

ostasis. (LIU et al. 1997). Supporting a role for GRP78 in stabilizing neuronal calcium homeostasis are data showing that suppression of GRP78 expression in cultured rat hippocampal neurons using antisense technology results in increased vulnerability of these cells to excitotoxic and oxidative insults (Fig. 3; YU et al. 1999).

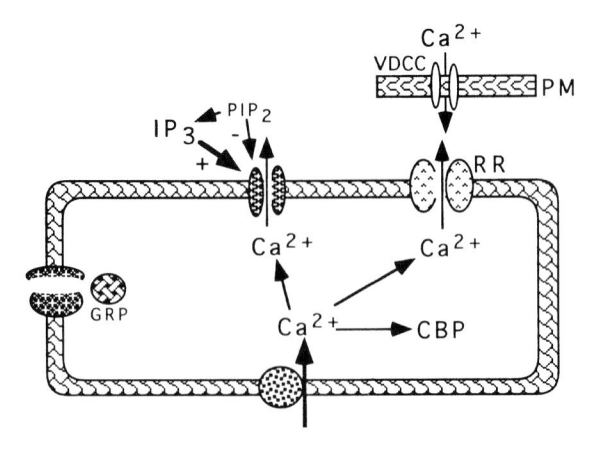

Fig. 2. Endoplasmic reticulum systems involved in modulation of cellular calcium homeostasis. Under basal conditions Ca^{2+} levels are typically 100-fold greater in the ER lumen than in the cytoplasm. Two different types of ER ligand-gated Ca^{2+} channels mediate release of Ca^{2+} from the ER. IP_3 receptors are activated in response to agonists that cause hydrolysis of plasma membrane - associated phosphatidylinositol bisphosphate (PIP_2), resulting in the release of IP_3; there are three main types of IP_3 receptors, with the type 1 IP_3 receptor being the most prominent in neurons. PIP_2 may negatively regulate IP_3 receptor channels when plasma and ER membranes are in close apposition. The second type of ER Ca^{2+} channel is the ryanodine receptor (*RR*), which is activated in response to Ca^{2+} influx through plasma membrane channels. Ca^{2+} is actively "pumped" into the ER via activity of the smooth endoplasmic reticulum Ca^{2+} ATPase (*SERCA*). The ER contains several calcium-binding proteins (*CBP*), such as calreticulin and calretinin, that likely serve as Ca^{2+} buffers analogous to the cytosolic CBP calbindin and parvalbumin. ER "stress proteins" such as the glucose-regulated proteins (*GRP*) may also modulate ER Ca^{2+} homeostasis, and play a role in protecting cells against apoptosis

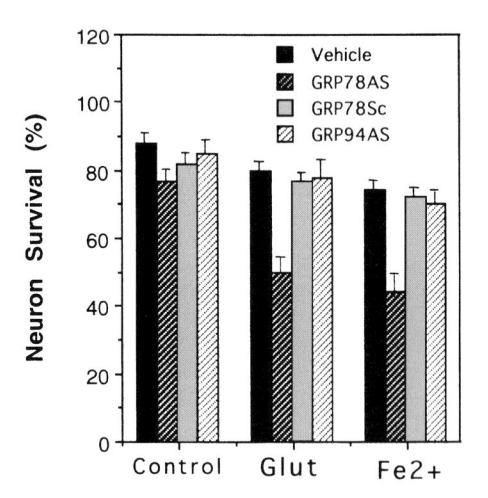

Fig. 3. Evidence that the ER stress-responsive chaperone protein GRP78 protects hippocampal neurons against excitotoxic and oxidative injury. Hippocampal cultures were exposed for 24 h to saline (*control*), $20 \mu M$ GRP78 antisense DNA (*AS*), $20 \mu M$ scrambled DNA (*Sc*), or $20 \mu M$ GRP94 antisense DNA. Cultures were then exposed for 20 h to vehicle, $10 \mu M$ glutamate or $5 \mu M$ $FeSO_4$ and neuronal survival was quantified. Values are the mean and SE of determinations made in four to six separate cultures. (Modified from Yu et al. 1999)

Increasing evidence suggests that perturbed ER Ca^{2+} homeostasis contributes to the neurodegenerative process in cerebral ischemia, not least of which is the ability of agents that block Ca^{2+} release from ER to prevent cell death following exposure of neurons to metabolic and excitotoxic insults (FRANDSEN et al. 1991; GUO et al. 1998; Fig. 4). Moreover, exposure of cultured hippocampal neurons to thapsigargin, an inhibitor of the ER Ca^{2+}-ATPase, greatly increases neuronal vulnerability to excitotoxic and metabolic insults (Fig. 4). Indirect evidence that calcium release from ER contributes to excitotoxic injury includes data showing enhancement of glutamate toxicity in cultured neurons exposed to agonists of muscarinic cholinergic (IP_3-linked) receptors (MATTSON 1989), and reduced neuronal degeneration following treatments with muscarinic receptor antagonists in in vivo brain injury models (HAYES et al. 1992). A general role for ER calcium release in neuronal cell death is suggested by studies implicating IP_3 receptors in apoptosis (KHAN et al. 1996). Interestingly, mutations in a gene called presenilin-1, which are linked to many cases of autosomal dominant early-onset inherited Alzheimer's disease, have been shown to result in perturbed ER calcium regulation (GUO et al. 1996; 1998). Enhanced calcium release in neurons expressing mutant presenilin-1 results in increased vulnerability to apoptosis (GUO et al. 1997; 1998) and to excitotoxicity (GUO et al. 1999).

III. Mitochondria

The importance of mitochondria in regulation of $[Ca^{2+}]_i$ in both physiological and pathophysiological settings is increasingly appreciated (Fig. 5, RICHTER 1993; WERTH and THAYER 1994). By producing the vast majority of ATP generated in neurons, mitochondria provide the energy that drives ion-motive ATPases in the plasma membrane and ER. Beyond the latter indirect role in maintenance of calcium homeostasis, mitochondria are capable of sequestering large quantities of Ca^{2+}. Mitochondria possess a Ca^{2+} uniporter responsible for Ca^{2+} uptake from the cytosol (GUNTER et al. 1994), which may play a major role in rapidly effecting local decreases in cytoplasmic Ca^{2+} levels (BABCOCK et al. 1997). Mitochondria can release large amounts of Ca^{2+}, and such release may play roles in both normal physiological modulation of Ca^{2+} homeostasis and in neurodegenerative processes. Ca^{2+} release from mitochondria occurs in response to a decrease in mitochondrial transmembrane potential, and appears to involve the formation of pores with two different states. One state is tightly coupled to the H^+ gradient and involves opening of somewhat selective Ca^{2+} channels in the inner membrane, while the second state involves formation of very large conductance nonselective "megapores" that span both the inner and outer membranes (BERNARDI 1996; KROEMER et al. 1997). The megachannels are called permeability transition pores (PTP), and formation of such pores is now known to play a key role in apoptosis. The quite remarkable ability of cyclosporin A to prevent neuronal apoptosis and stabilize calcium homeostasis appears to result from blockade of the

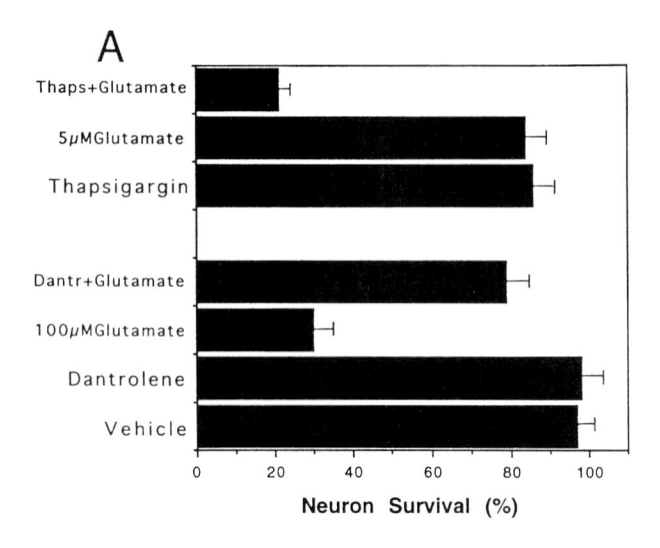

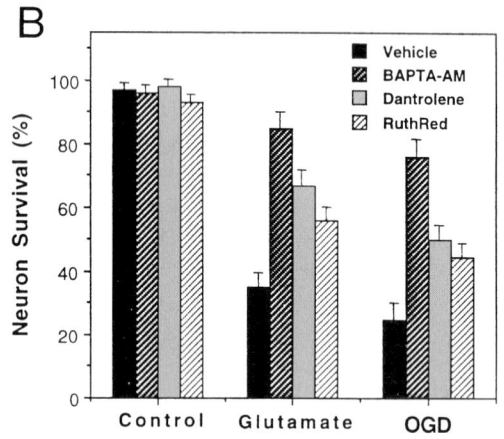

Fig. 4A,B. Blockade of ER calcium release decreases, and stimulation of ER calcium release increases, neuronal vulnerability to excitotoxicity. **A** Neuronal survival in rat hippocampal cell cultures was quantified 24h following exposure to vehicle (0.2% dimethylsulfoxide), 10 μM dantrolene, 100 μM glutamate, dantrolene+100 μM glutamate, 100 nM thapsigargin, 5 μM glutamate, and 100 nM thapsigargin+ 5 μM glutamate. Values are the mean and SE of determinations made in four separate cultures. **B** Rat hippocampal cell cultures were pretreatred for 1h with vehicle, 5 μM BAPTA-AM, 100 μM dantrolene or 25 μM ruthenium red. Cultures were then exposed for 24h to 10 μM glutamate or combined oxygen/glucose deprivation (*OGD*), and neuronal survival was quantified. Values are the mean and SE of determinations made in four separate cultures

mitochondrial permeability transition (ZAMZAMI et al. 1996; KELLER et al. 1998).

Roles for mitochondrial perturbations in the neuronal injury and death following stroke is strongly suggested by studies demonstrating compromise

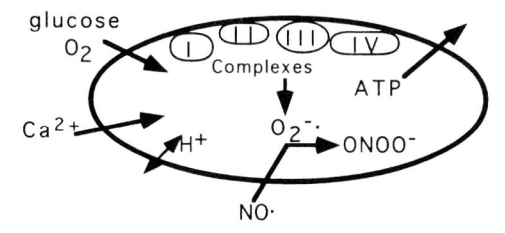

Fig. 5. Mitochondrial systems involved in modulation of cellular calcium homeostasis. Mitochondrial electron transport, in which glucose and oxygen are used to generate ATP, involves the activities of several enzyme complexes (I-IV). Both ATP and oxyradicals (principally $O_2^{-\bullet}$ and $ONOO^-$) generated during the electron transport process play important indirect roles in regulating neuronal calcium homeostasis. ATP is used to drive ion-motive ATPases and in phosphorylation reactions that modulate functions of various calcium-regulating proteins (e.g., glutamate receptors and voltage-dependent calcium channels). Oxyradicals can impair the function of ion-motive ATPases, glucose transporters, and GTP-binding proteins. Mitochondria possess a membrane Ca^{2+} uniporter that moves Ca^{2+} from the cytoplasm into the mitochondrial matrix. Changes in mitochondrial transmembrane potential influence Ca^{2+} movements into and out of the mitochondria. In cells undergoing apoptosis, large pores are formed that span both the inner and outer mitochondrial membranes and mediate dumping of Ca^{2+} into the cytoplasm

of various mitochondrial parameters in rodent stroke models (see SIESJO 1992 for review). Increased oxidative stress appears to play a major role in mitochondrial dysfunction following ischemia-reperfusion. As evidence, overexpression of mitochondrial Mn-SOD in transgenic mice results in reduced brain damage in an ischemia-reperfusion model of focal ischemic brain injury (Fig. 6; KELLER et al. 1998). The latter study also showed that overexpression of Mn-SOD in cultured neural cells suppresses peroxynitrite production and greatly increases their resistance to apoptosis induced by various insults, including exposure to NO donors.

IV. Other Proteins Involved in Calcium Regulation

There are several cytosolic Ca^{2+}-binding proteins that can act to reduce $[Ca^{2+}]_i$, prominent among which are calbindin D28k, parvalbumin and calmodulin. Calbindin and parvalbumin are present at very high (mM) levels in some neurons, and may reduce $[Ca^{2+}]_i$ by directly sequestering Ca^{2+} (MATTSON et al. 1991; CHARD et al. 1993). Calmodulin, on the other hand, indirectly regulates $[Ca^{2+}]_i$ by modifying the activity of various Ca^{2+}-regulating proteins. For example, calmodulin has been shown to activate the PMCA (MORGAN et al. 1986) and thereby stimulates calcium removal. Calmodulin is also known to activate nitric oxide synthase, resulting in the generation of nitric oxide (BREDT and SNYDER 1992). Nitric oxide activates soluble guanylate cyclase which, in turn, leads to phosphorylation of various proteins involved in regulation of Ca^{2+} homeostasis including potassium ion channels (FURUKAWA et al. 1996).

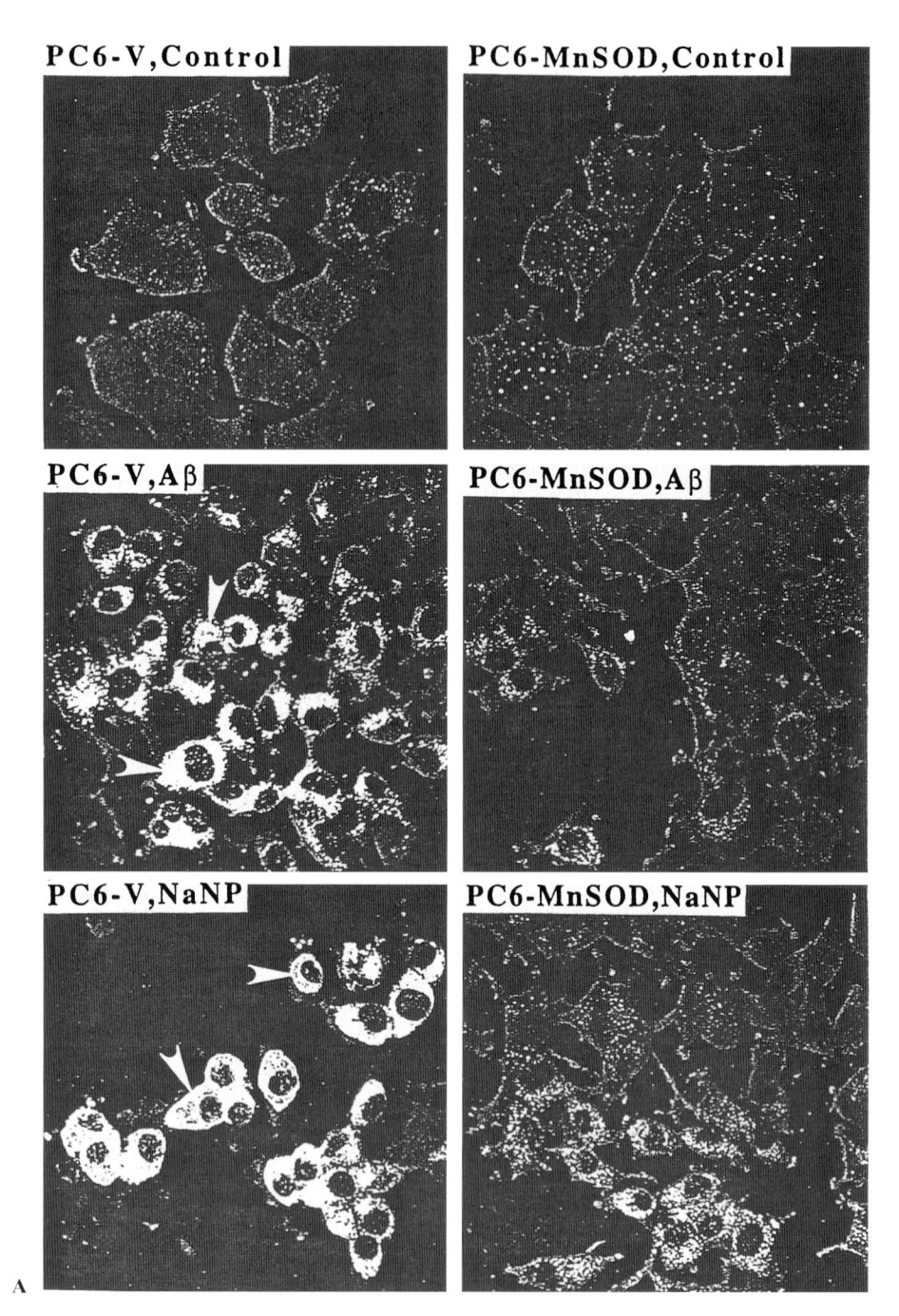

PC6-V,Control	PC6-MnSOD,Control
PC6-V,Aβ	PC6-MnSOD,Aβ
PC6-V,NaNP	PC6-MnSOD,NaNP

A

Fig. 6A–C. Overexpression of mitochondrial Mn-SOD suppresses peroxynitrite production, and reduces ischemic brain injury. **A** Confocal laser scanning microscope images of nitrotyrosine immunoreactivity in PC6-V and PC6-MnSOD cells exposed for 6h to either vehicle (*control*), 50 μM Aβ, or 100 μM sodium nitroprusside (*NaNP*). **B**, **C** Cortical infarct volumes and TBARS fluorescence (measure of membrane lipid peroxidation) were quantified 24h following MCA occlusion in WT and Mn-SOD-overexpressing transgenic mice. Values are the mean and SE (six mice in each group). (Modified from KELLER et al. 1998)

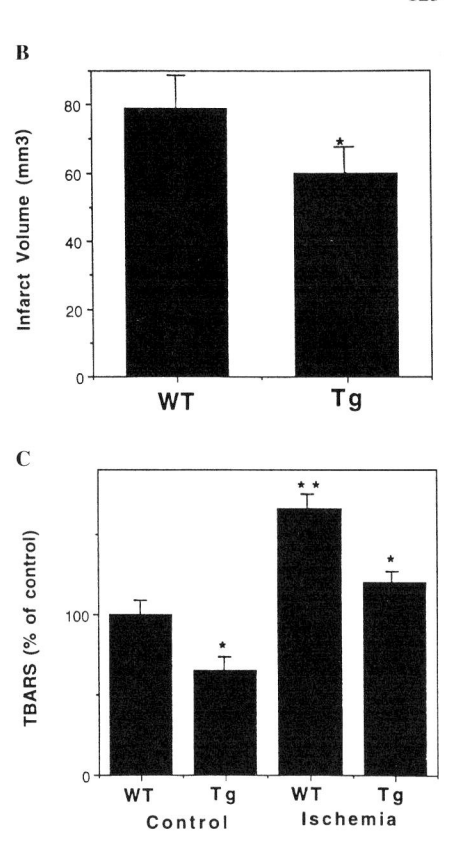

Activation of K^+ channels in response to cGMP production hyperpolarizes the membrane and protects neurons against excitotoxicity. Nitric oxide can also alter Ca^{2+} homeostasis by modulating the activities of NMDA receptor channels and other ion channels via a redox-based mechanism (LEVY et al. 1990).

Various second messengers, kinases, and phosphatases are involved in regulation of $[Ca^{2+}]_i$ by virtue of their ability to alter the function of ion channels, ion-motive ATPases, and other $[Ca^{2+}]_i$-regulating proteins. For example, cyclic AMP activates cyclic AMP-dependent protein kinase which, in turn, phosphorylates protein substrates resulting in increased opening of the NMDA receptor and enhanced Ca^{2+} influx (RAMAN et al. 1996). Protein phosphatases such as calcineurin and protein phosphatase-2A have been shown to alter activities of K^+ and Ca^{2+} channels and modulate synaptic plasticity (MULKEY et al. 1994; FURUKAWA et al. 1996). GTP-binding proteins may directly modulate ion channel function or may, via stimulation of effector proteins (e.g., phospholipase C), promote activation of kinases (e.g., protein kinase C) (NELSON and ALKON 1991).

Finally, cellular systems involved in controlling cell shape and adhesion are increasingly recognized as playing important roles in regulation of $[Ca^{2+}]_i$. For example, actin filaments regulate functions of NMDA receptors and VDCC in cultured hippocampal neurons; depolymerization of actin filaments enhances rundown of the channels resulting in reduced Ca^{2+} influx (Rosenmund and Westbrook 1993; Furukawa et al. 1995). The Ca^{2+}-activated actin-severing protein gelsolin appears to play a key role in depolymerizing actin, and thereby suppressing Ca^{2+} influx in response to sustained elevations of $[Ca^{2+}]_i$ (Furukawa et al. 1997). Cell adhesion molecules such as integrins and cadherins have been shown to influence the activities of ion channels in nonneuronal cells, and may also modulate calcium homeostasis in neurons (Schwartz et al. 1993; Gary and Mattson 2001).

C. Factors That Destabilize Calcium Homeostasis

A complex array of intra- and intercellular alterations contribute to perturbed neuronal calcium homeostasis following cerebral ischemia. Cerebral ischemia results in profound disturbances in the function of the different plasma membrane systems involved in regulation of neuronal Ca^{2+} homeostasis. In addition to the compromise of ion-motive ATPase activities as the result of ATP depletion, changes in cellular redox state can have quite striking effects on the function of membrane Ca^{2+}-regulating proteins in the plasma membrane, as well as in mitochondria and ER (Lipton et al. 1993; Keller et al. 1997). Understanding the various cascades of events leading to disruption of neuronal ion homeostasis following cerebral ischemia is viewed as a critical step towards developing effective therapeutic approaches aimed at preventing neuronal injury and death, and restoring cell functions.

I. Oxidative Stress

Studies of mechanisms of ischemic brain injury in both cell culture and animal models have documented major contributions of oxidative stress to the neurodegenerative process. Oxidative stress is defined as increased accumulation of various reactive oxygen species (ROS) within cells that results in damage to proteins, nucleic acids, and lipids. Major pathways that lead to oxidative stress in neurons have been identified (Fig. 7) and include: mitochondrial production of superoxide anion radical ($O_2^{-\bullet}$), which occurs during the electron transport process; production of hydrogen peroxide (H_2O_2), which is formed primarily as a result of the activities of superoxide dismutases (Mn-SOD and Cu/Zn-SOD), which convert $O_2^{-\bullet}$ to H_2O_2; hydroxyl radical ($OH^\bullet$) production via the Fenton reaction in which Fe^{2+} and Cu^+ catalyze the conversion of H_2O_2 to $OH^\bullet$; and peroxynitrite ($ONOO^-$) formation, which occurs when nitric oxide (NO) interacts with $O_2^{-\bullet}$. While $OH^\bullet$ and $ONOO^-$ can cause direct

Stabilizing Calcium Homeostasis

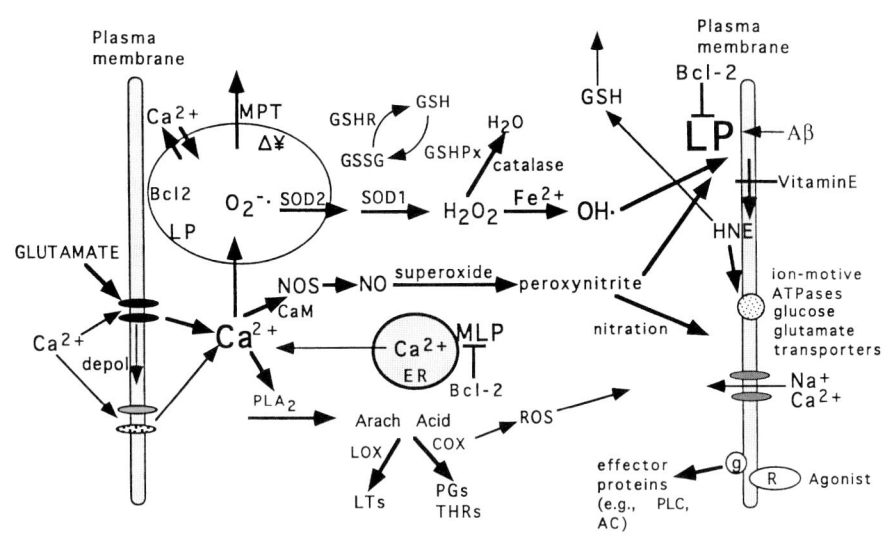

Fig. 7. Interactions between reactive oxygen species and calcium-regulating systems in neurons. The major source of oxyradicals is mitochondria wherein $O_2^{-\bullet}$ is generated during the electron transport process. Superoxide dismutases (*SOD*) convert $O_2^{-\bullet}$ to H_2O_2 which, in the presence of Fe^{2+}, generates $OH^{\bullet}$. $O_2^{-\bullet}$ can also interact with nitric oxide (*NO*) to form peroxynitrite. Both $OH^{\bullet}$ and peroxynitrite induce membrane lipid peroxidation (*LP*), which may occur in the plasma membrane, mitochondrial membranes, and endoplasmic reticulum (*ER*) membranes. Additionally, exogenous agents such as amyloid β-peptide (*Aβ*) can induce LP. LP liberates 4-hydroxynonenal (*HNE*), which binds to membrane transporters and ion channels, and thereby alters their activities. Impairment of the Na^+/K^+-ATPase, glucose transporter, and glutamate transporters results in membrane depolarization and excessive activation of glutamate receptors, resulting in excessive calcium influx and cell injury. LP also perturbs ion homeostasis in ER and mitochondria, and thereby compromises their important Ca^{2+} sequestration functions. The antiapoptotic gene product Bcl-2 may act, in part, by suppressing LP in plasma, mitochondrial and ER membranes. It should be noted that not only does LP lead to an elevation of $[Ca^{2+}]_i$ but, conversely, elevation of $[Ca^{2+}]_i$ promotes LP by inducing NO and $O_2^{-\bullet}$ production; NO and $O_2^{-\bullet}$ interact to form peroxynitrite, a potent inducer of LP. In addition, calcium promotes activation of phospholipases, resulting in production of arachidonic acid, which is then acted upon by cyclooxygenases (*COX*) and lipoxygenases (*LOX*) with resultant generation of $O_2^{-\bullet}$. *AC*, adenylate cyclase; *CaM*, calmodulin; *LT*, leukotrienes; *NOS*, nitric oxide synthase; *PG*, prostaglandins; *PLC*, phospholipase C; *THRs*, thromboxanes

damage to proteins and DNA, a major mechanism whereby they damage cells involves induction of membrane lipid peroxidation (MLP), a process involving free radical attack on double bonds of unsaturated fatty acids that comprise membrane phospholipids (MATTSON 1998a).

In rodent models of stroke, including middle cerebral artery occlusion (MCAO) and transient global forebrain ischemia, there is clear evidence for ROS production and free radical damage to cellular components. For example, increased MLP, protein oxidation, and DNA damage have been documented within ischemic brain tissue (OLIVER et al. 1990; JESBERGER and RICHARDSON 1991; CHEN et al. 1997; YOSHINO et al. 1997; KELLER et al. 1998). Moreover, studies have shown that levels of $O_2^{-\bullet}$, $OH^{\bullet}$, and $ONOO^{-\bullet}$ are increased within neurons in ischemic rodent brain tissue (PIANTADOSI and ZHANG 1996; YOSHINO et al. 1997; FUKUYAMA et al. 1998; KELLER et al. 1998). The events that occur during ischemia that lead to oxidative stress in neurons likely include mitochondrial dysfunction, increased Fe^{2+} accumulation, and Ca^{2+} influx. Indeed, a major stimulus for production of both $O_2^{-\bullet}$ and NO is elevation of $[Ca^{2+}]_i$, which means that oxidative stress is one consequence of activation of glutamate receptors and VDCC (Fig. 7). The impact of NO on neuronal calcium homeostasis is complex, with our current level of understanding indicating that NO can disrupt calcium homeostasis via a peroxynitrite-mediated mechanism (BRORSON et al. 1997), while it can stabilize calcium homeostasis via a cGMP-mediated mechanism (Fig. 8; BARGER et al. 1995; FURUKAWA et al. 1996). Exposure of cultured primary hippocampal neurons to excitotoxic and metabolic insults results in generation of $O_2^{-\bullet}$, $OH^{\bullet}$ and $ONOO^{-}$, and MLP (LAFON-CAZAL et al. 1993; MATTSON et al. 1995; DUGAN et al. 1995; KELLER et al. 1998). Importantly, various antioxidants (e.g., vitamin E, 21-aminosteroids, glutathione, uric acid, and estrogen) and overexpression of antioxidant enzymes (Mn-SOD, Cu/Zn-SOD and glutathione peroxidase) have been reported to be effective in preventing neuronal degeneration in cell culture and animal models of cerebral ischemia (YANG et al. 1994; GOODMAN et al. 1996; MARK et al. 1997a; KELLER et al. 1998; WEISBROT-LEFKOWITZ et al. 1998; YU et al. 1998).

Additional evidence for a major role for oxidative stress in disruption of calcium homeostasis and neuronal death following cerebral ischemia comes from studies of transgenic and knockout mice. Transgenic mice in which Cu/Zn-SOD is overexpressed in neural cells exhibit reduced brain injury following focal cerebral ischemia-reperfusion (YANG et al. 1994). Conversely, mice lacking Cu/Zn-SOD exhibit increased neuronal injury following focal cerebral ischemia-reperfusion (KONDO et al. 1997). Mitochondrial Mn-SOD appears to be a particularly important neuroprotective antioxidant enzyme. Transgenic mice overexpressing Mn-SOD exhibit reduced infarct volume and improved behavioral outcome following focal cerebral ischemia-reperfusion (KELLER et al. 1998). On the other hand, ischemic brain injury is increased in Mn-SOD knockout mice (MURAKAMI et al. 1998). Cell culture studies have shown that peroxynitrite production and membrane lipid peroxidation are suppressed in neurons overexpressing Mn-SOD, and that this reduced oxidative stress is correlated with resistance to apoptosis induced by a variety of insults (KELLER et al. 1998). Neural cells overexpressing Mn-SOD exhibit improved mitochondrial calcium handling, and are able to maintain calcium homeostasis following exposure to apoptotic insults (BRUCE-KELLER et al. 1998).

ROS production has profound adverse effects on neuronal Ca^{2+} homeostasis. Particularly important targets of oxyradical attack are membrane

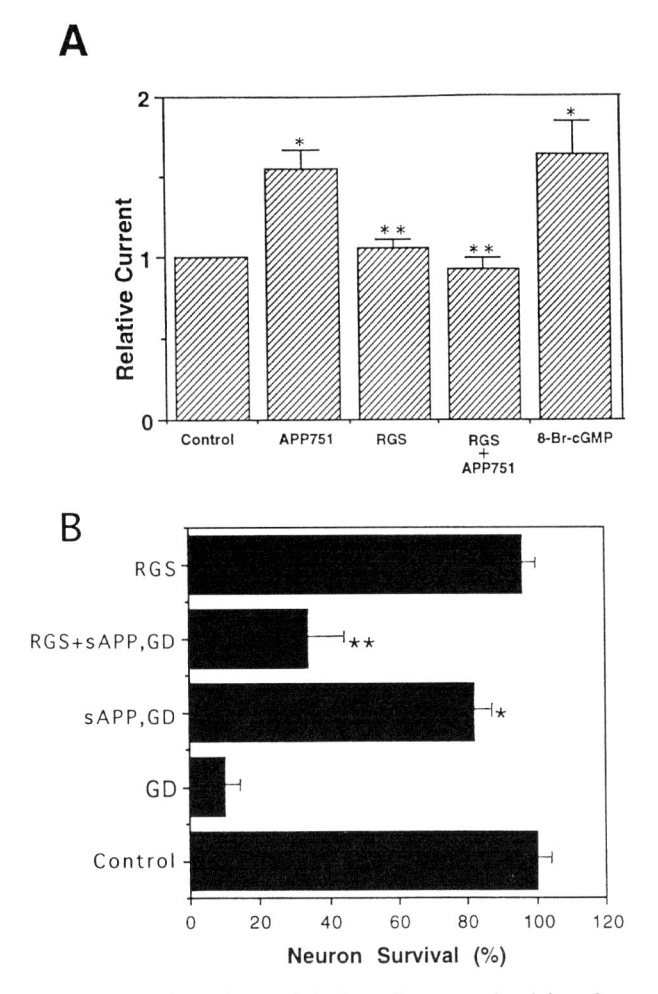

Fig. 8A,B. Role of cyclic GMP in modulation of neuronal calcium homeostasis and protection against metabolic/excitotoxic insults. **A** Whole-cell K$^+$ currents were measured in cultured hippocampal neurons under basal conditions (*control*) and in neurons exposed to sAPPα (*APP751*), RGS (an inhibitor of cGMP-dependent protein kinase), sAPPα plus RGS, or 8-bromo-cGMP. **B** Hippocampal cultures were pretreated with sAPP, RGS, or RGS+sAPP. Cultures were then subjected to glucose deprivation (*GD*) for 24h, and neuronal survival was quantified. Values are the mean and SE of determinations made in three separate cultures. *$P < 0.05$ compared to GD value; **$P < 0.05$ compared to sAPP, GD value. (Modified from BARGER et al. 1995 and FURUKAWA et al. 1996)

transport systems involved in maintenance of Ca^{2+} homeostasis. Oxidative insults can lead to impairment of function of the plasma membrane Na$^+$/K$^+$- and Ca^{2+}-ATPases (MARK et al. 1995; KELLER et al. 1997), plasma membrane glucose transporters (MARK et al. 1997), and glutamate transporters (VOLTERA et al. 1994; KELLER et al. 1997). While direct oxidation of the transport proteins may contribute to ischemia-induced impairment of their function, recent data suggest a more indirect mechanism involving MLP. When MLP occurs, lipid hydroperoxides and an array of aldehydes of different carbon chain lengths (e.g., hexanal, heptanal, octanal, etc.) are produced (ESTERBAUER et al. 1991). One aldehydic production of MLP, 4-hydroxynonenal (HNE), appears to play a pivotal role in the cell death cascade resulting from MLP (Fig. 7). HNE covalently modifies proteins by binding to cysteine, lysine, and histidine residues, and can also interact with proteins via Schiff-base reactions (ESTERBAUER et al. 1991; MARK et al. 1997). HNE production is increased in ischemic brain tissue (YOSHINO et al. 1997) and following traumatic injury to the brain and spinal cord (SPRINGER et al. 1997; and M.P. Mattson unpublished data). HNE can disrupt neuronal calcium homeostasis at concentrations generated following exposure to insults relevant to cerebral ischemia and traumatic brain and spinal cord injury (MARK et al. 1997a). HNE disrupts calcium homeostasis by impairing the function of the Na^+/K^+-ATPase (MARK et al. 1997a) and glucose transporter (MARK et al. 1997b) in neurons, and glutamate transporters in astrocytes (KELLER et al. 1997a,b; SPRINGER et al. 1997; BLANC et al. 1998). In this way MLP greatly increases neuronal vulnerability to excitotoxicity (MARK et al. 1997a). MLP in the ER and mitochondria (KRISTAL et al. 1996; BRUCE et al. 1998) are also likely to contribute to excitotoxic cascades in many different neurodegenerative conditions.

II. Metabolic Impairment

Cerebral ischemia, by definition, involves reduced availability of oxygen and glucose to brain cells. As a consequence there is a marked depletion of ATP levels in neurons (GINSBERG 1990). Moreover, the initial metabolic insult adversely affects mitochondrial function, thus exacerbating oxyradical production during the reperfusion period, resulting in a further compromise of mitochondrial function and continued ATP depletion. ATP depletion endangers neurons by compromising the function of ion-motive ATPases in the plasma membrane and organellar membranes. Impairment of the Na^+/K^+-ATPase results in membrane depolarization which, in turn, relieves the voltage-dependent Mg^{2+} block of the NMDA receptor thereby leading to calcium influx through the NMDA receptor channel (REYNOLDS 1990). Membrane depolarization also activates VDCC of various subtypes (e.g., L, N, and P channels) which accentuates calcium influx. Activation of both NMDA receptors and VDCC apparently play important roles in the neuronal calcium overload associated with ischemic brain injury because antagonists of NMDA receptors (e.g., MK801 and APV) and VDCC (e.g., nifedipine) can protect neurons against ischemic conditions (PERUCHE and KRIEGLSTEIN 1993). ATP depletion impairs the function of the Ca^{2+}-ATPases in the plasma membrane and ER resulting in dumping of ER Ca^{2+} stores and a failure of the cells to remove Ca^{2+} from the cytoplasm.

III. Excitatory Imbalances

There is evidence for increased levels of extracellular glutamate in ischemic brain tissue (HOSSMANN 1994). The increased glutamate levels may result from a combination of enhanced glutamate release from presynaptic terminals and impaired uptake into astrocytes. The increased extracellular glutamate levels result in overactivation of AMPA and kainate receptors which depolarize the membrane, resulting in activation of VDCC and NMDA receptors. Alterations in other transmitter systems during ischemia may also contribute to an excitotoxic process. As evidence, increased activation of muscarinic cholinergic receptors may promote excitotoxicity (MATTSON 1989), while activation of GABA receptors may attenuate the excitotoxic process (MATTSON and KATER 1989). GABA receptor agonists have been shown to reduce seizure-induced brain injury in rodent models of temporal lobe epilepsy and in some stroke models (GREEN et al. 2000). Modulation of neuronal vulnerability to ischemic brain injury by different neurotransmitters is tightly correlated with effects of those transmitters on $[Ca^{2+}]_i$.

IV. Glucocorticoids

An increasing body of evidence indicates that the neuroendocrine stress response involving the hypothalamic–pituitary–adrenal axis can exacerbate brain damage in stroke and epilepsy patients (SAPOLSKY 1997). Damage to hippocampal neurons induced by the excitotoxin kainic acid was reduced in adrenalectomized rats, and exacerbated in rats administered corticosterone (ELLIOTT et al. 1993). Suppression of endogenous glucocorticoid production, effected by administration of the 11-β-hydroxylase inhibitor metyrapone, significantly reduced brain damage in two different rat models of cerebral ischemia, a middle cerebral artery occlusion focal ischemia model and a transient global forebrain ischemia paradigm, which results in selective loss of CA1 hippocampal neurons (SMITH-SWINTOSKY et al. 1996). Thus, ischemia results in a massive stress response accompanied by increased levels of circulating glucocorticoids (OLSSON 1990; SMITH-SWINTOSKY et al. 1996), which likely exacerbates neuronal damage. Indeed, physiological stress (e.g., exposure to cold temperature, painful stimuli, and psychological stressors such as crowding) can cause atrophy of apical dendrites of hippocampal CA3 pyramidal neurons (WATANABE et al. 1992), and can increase the vulnerability of hippocampal neurons to excitotoxic injury (STEIN-BEHRENS et al. 1994).

The mechanism whereby glucocorticoids endanger neurons may involve metabolic compromise and disruption of calcium homeostasis. Exposure of cultured hippocampal neurons to corticosterone resulted in a reduction in the level of glucose transport (HORNER et al. 1990), and enhanced elevations of $[Ca^{2+}]_i$ following exposure to glutamate and kainate (ELLIOTT and SAPOLSKY 1992; GOODMAN et al. 1996). Additional mechanisms of glucocorticoid endangerment include suppression of production of neurotrophic factors (COSI et al. 1993), and impaired antioxidant defense mechanisms (McIntosh et al. 1998). Indeed, glucocorticoids increase the vulnerability of cultured hippocampal neurons to oxidative insults (Goodman et al. 1996), while neurotrophic factors protect neurons against such insults (Mattson and Lindvall 1997).

V. Apoptotic Mechanisms

Apoptosis is characterized by cell shrinkage and nuclear DNA condensation and fragmentation, and by the ability of macromolecular synthesis inhibitors and inhibitors of caspases to prevent cell death (Bredesen 1996). Apoptosis of neurons, which clearly occurs during development of the nervous system, is increasingly recognized as playing a role in ischemic brain injury (Choi 1996; Chalmers-Redmond et al. 1997). Morphological changes and DNA fragmentation suggestive of apoptosis occur in the ischemic penumbra in focal ischemia models, and in hippocampal CA1 neurons in transient global forebrain ischemia models (Linnik et al. 1993; MacManus et al. 1993; Nitatori et al. 1995). Moreover, overexpression of the antiapoptotic gene product Bcl-2 can reduce ischemic brain injury in rodent models (Martinou et al. 1994; Linnik et al. 1995). In most cases examined to date, calcium and ROS have been shown to play central roles in neuronal apoptosis. As evidence, studies of cultured neural cells have shown that apoptosis induced by many different insults (e.g., exposure to staurosporine, oxidative insults and trophic factor withdrawal) is associated with a relatively early increase of $[Ca^{2+}]C$ that precedes mitochondrial and nuclear events in the apoptotic cascade (Kruman et al. 1997; Prehn et al. 1997; Kruman et al. 1998). Moreover, agents that suppress $[Ca^{2+}]C$ increases (e.g., BAPTA-AM and overexpression of the calcium-binding protein calbindin) prevent apoptosis in many different paradigms (Kruman et al. 1998; Guo et al. 1998). Calcium influx through plasma membrane channels and release from ER contribute to neuronal apoptosis in many cases because agents that suppress calcium influx or release can prevent apoptosis (Guo et al. 1996; Keller et al. 1998). Caspase activation appears to contribute to disruption of cellular calcium homeostasis in neurons undergoing apoptosis because caspase inhibitors can stabilize $[Ca^{2+}]C$ and preserve mitochondrial function following challenges with apoptotic insults (Kruman et al. 1998; Keller et al. 1998). The caspase substrates that influence neuronal calcium homeostasis remain to be identified, but certain protein kinase C isoforms and actin filaments are candidates (Furukawa et al. 1995).

Although there are differences in the early initiation phases of apoptosis depending upon the insult (e.g., apoptosis induced by trophic factor withdrawal versus that induced by staurosporine or metabolic insults), the events involved in the subsequent "effector" phase of apoptosis appear to be shared. Events occurring in the mitochondria appear to play a pivotal role in apoptosis by transducing the apoptosis initiation signal into an irreversible cascade that leads to nuclear disintegration and cell death (KROEMER et al. 1997). The mitochondrial alterations include depolarization of the inner membrane, PTP formation, and release of "suicide" proteins which may include cytochrome C and an apoptosis-inducing factor. Oxyradical production and MLP play important roles in apoptotic mitochondrial alterations because MLP leads to a decrease in potential across the inner mitochondrial membrane, followed by elevation of intramitochondrial calcium levels and PTP pore formation. Indeed, HNE causes decreases in mitochondrial transmembrane potential and energy charge, and induces apoptosis in cultured neurons (KRUMAN et al. 1997), and induces PTP formation in isolated mitochondria (KRISTAL et al. 1996). As in the case of MLP-induced impairment of plasma membrane transporters, MLP impairs the function of proteins involved in maintenance of mitochondrial transmembrane ion gradients. A pivotal role for HNE in oxidative stress-induced neuronal apoptosis is further suggested by data showing that glutathione, an endogenous detoxifier of HNE, can prevent cellular accumulation of HNE and neuronal apoptosis following exposure of cells to Fe^{2+} and amyloid β-peptide (MARK et al. 1997; KRUMAN et al. 1997).

The major sites of calcium influx during cerebral ischemia are postsynaptic spines in dendrites at glutamatergic synapses (DENK et al. 1996). Interestingly, recent findings suggest that biochemical events characteristic of apoptosis occur locally in synapses (MATTSON et al. 1998a). Loss of membrane phospholipid asymmetry, caspase activation, and mitochondrial alterations (membrane depolarization, calcium overload, and oxyradical accumulation) characteristic of apoptosis occurred in cortical synaptosomes exposed to staurosporine. Extracts from synaptosomes exposed to apoptotic insults induced chromatin condensation and fragmentation in isolated nuclei indicating that signals capable of inducing nuclear apoptosis are generated locally in synapses (MATTSON et al. 1998a 1998b). The identity of the apoptotic factor(s) released from mitochondria is not firmly established, but may include cytochrome c (KLUCK et al. 1997; YANG et al. 1997) and a 50-kDa apoptosis-inducing protein (SUSIN et al. 1996). Exposure of cultured hippocampal neurons to glutamate resulted in caspase activation and mitochondrial membrane depolarization in distal dendrites (MATTSON et al. 1998a). The possibility that calcium influx can trigger apoptotic events locally in synapses and dendrites is supported by cell culture and in vivo data suggesting that activation of glutamate receptors can induce neuronal apoptosis (ANKARCRONA et al. 1995; DU et al. 1997), and by studies suggesting that calcium and reactive oxygen species may induce caspase activation and mitochondrial membrane depolarization during apoptosis (RICHTER 1993; KROEMER et al. 1997; KRUMAN et al. 1998; NICOTERA et al. 1997). Thus, apoptotic biochemical cascades can be activated locally in synapses and likely initiate an apoptotic cascade that eventually leads to cell death (Fig. 9).

Activation of caspases and release of apoptotic factors from mitochondria in synapses might lead to nuclear apoptosis by movement of the caspases and apoptotic factors to the cell body, or by initiating a degenerative cascade

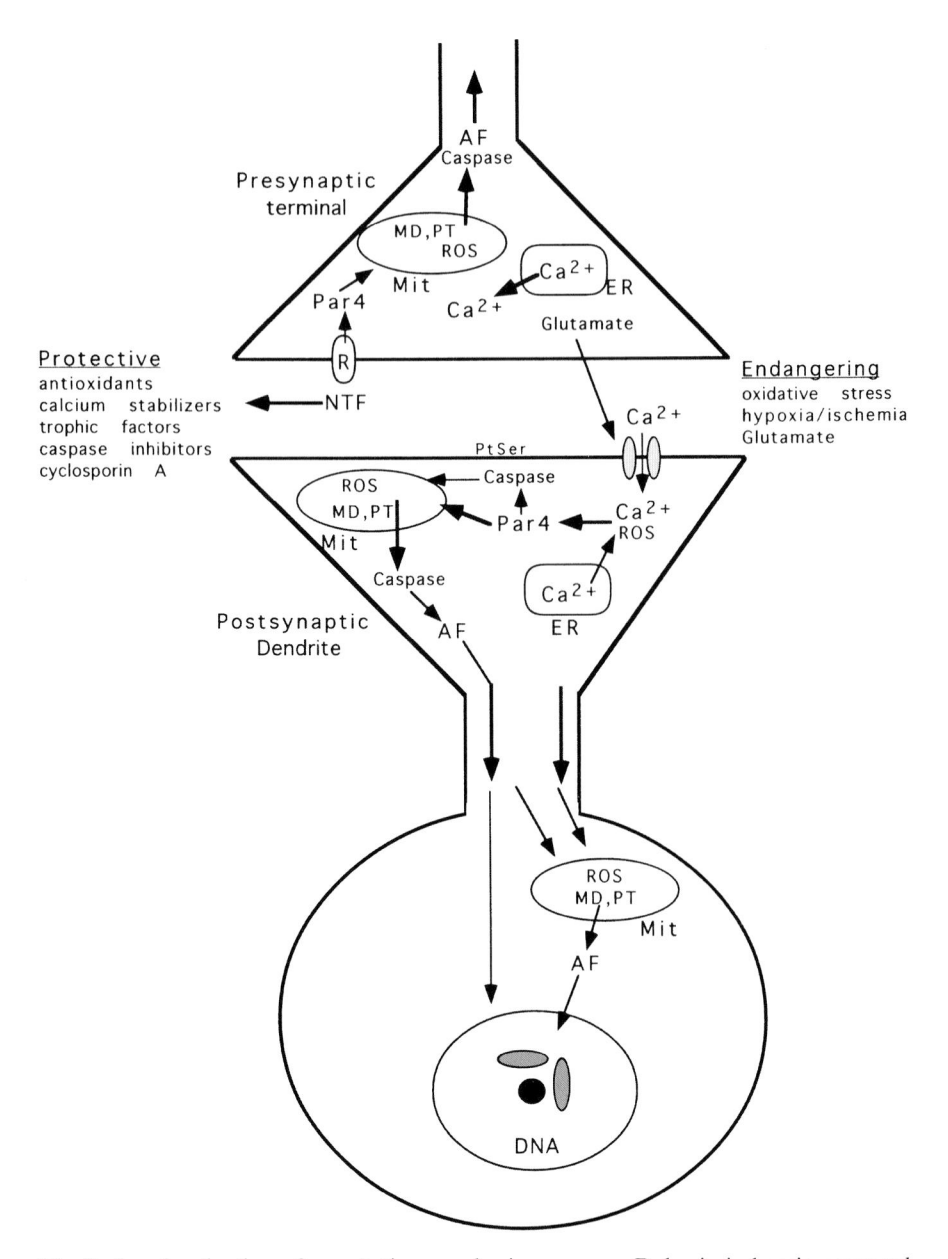

Fig. 9. Local activation of apoptotic cascades in synapses: Roles in ischemic neuronal cell death. Excessive glutamate receptor activation results in local calcium influx and ROS production in postsynaptic regions of dendrites which, in turn, induce caspase activation, mitochondrial membrane depolarization (*MD*), and permeability transition (*PT*). Apoptotic factors (*AF*) are released from the mitochondria locally in synapses. These local apoptotic cascades may propagate to the cell body and elicit mitochondrial and/or nuclear apoptotic events. Reduced levels of neurotrophic factors (*NTF*) input may also elicit synaptic apoptosis. In addition, synaptic apoptosis may occur in the absence of cell death resulting in, for example, a local microglial response to exposure of phosphatidylserine (*PS*) on the synaptic membrane surface. (Modified from MATTSON et al. 1998a)

involving reactive oxygen species and calcium that activates caspases and induces mitochondrial alterations in the cell body. Local inflammatory responses to apoptosing synapses might play a role in the synapse removal and remodeling that occurs following cerebral ischemic insults. For example, exposure of phosphatidylserine on the surface of synaptic membranes (MATTSON et al. 1998a) may induce a local microglial response that results in removal of synapses.

D. Factors That Stabilize Calcium Homeostasis

The vast majority of basic research in the fields of stroke, epilepsy, and traumatic brain and spinal cord injury has focused on the biochemical and molecular cascades that promote neuronal death. Less studied, but perhaps of equal importance, are the events that occur in injured neural tissue that represent attempts by the cells to prevent their demise (see MATTSON and FURUKAWA 1996 for review). It seems intuitively obvious that evolution would favor stress-responsive cellular mechanisms that increase neuronal resistance to cell death, rather than mechanisms the promote neuronal death (and hence death of the organism). Recent findings in many different laboratories suggest that, indeed, an array of different neuroprotective signal transduction pathways are activated following injury to the nervous system. The signaling pathways include both intercellular signals such as neurotrophic factors and cytokines, and intracellular cascades involving both transcription-dependent and transcription-independent processes (MATTSON and FURUKAWA 1996). It is of considerable interest that essentially all of the endogenous protective mechanisms thus far identified ultimately act by stabilizing cellular calcium homeostasis and/or suppressing ROS accumulation (MATTSON and LINDVALL 1997).

I. Neurotrophic Factors

The expression of a remarkable array of neurotrophic factors occurs following ischemic and traumatic brain injury, and following severe epileptic seizures (MATTSON and LINDVALL 1997). In many cases the neurotrophic factors have been shown to protect cultured neurons against excitotoxic, metabolic, and oxidative insults. For example, bFGF, NGF, BDNF, NT-4/5, IGF-1, and PDGF each protected cultured hippocampal neurons against death induced by glutamate, glucose deprivation, and oxidative insults (MATTSON et al. 1989; CHENG and MATTSON 1991, 1992, 1994, 1995; CHENG et al. 1994; ZHANG et al. 1993). In several cases the same neurotrophic factors have proven effective in reducing the extent of brain injury in rodent stroke models. For example, bFGF was effective in reducing brain damage following hypoxia in neonatal rats (NOZAKI et al. 1993) and in reducing infarct size and behavioral deficits following middle cerebral artery occlusion in rats (KOKETSU et al. 1994). This preclinical research has led to a clinical trial of bFGF in human stroke patients, which is currently in progress.

Measurements of $[Ca^{2+}]_i$ in cultured hippocampal neurons have shown that the elevations of $[Ca^{2+}]_i$ induced by glutamate and glucose deprivation are significantly attenuated in cultures pretreated with bFGF, BDNF, and IGF-1 (MATTSON et al. 1989, 1993; CHENG and MATTSON 1991; CHENG and MATTSON 1992). In the case of bFGF, evidence suggests that the mechanism underlying the calcium-stabilizing effect involves modulation of the expression of the calcium-binding protein calbindin D28k and suppression of the expression of an NMDA receptor protein (COLLAZO et al. 1992; MATTSON et al. 1993). Interestingly, bFGF and NGF may increase expression of VDCC (CHENG et al. 1993), and bFGF can increase the expression of the AMPA subunit GluR1 (CHENG et al. 1995). Neurotrophic factors may also indirectly stabilize neuronal calcium homeostasis by enhancing antioxidant defense mechanisms. For example, bFGF, NGF, and BDNF were reported to induce the expression of one or more antioxidant enzymes (Cu/Zn-SOD, catalase, and glutathione peroxidase) in hippocampal cell cultures (MATTSON et al. 1995).

II. Cytokines

Cytokines have, perhaps, received a bad rap when it comes to their roles in cellular responses to brain injury. The two most heavily studied cytokines with respect to brain injury are tumor necrosis factor-α (TNF-α) and interleukin-1β (see ROTHWELL and HOPKINS 1995; MATTSON et al. 1997 for review). IL-1β appears to promote degeneration of some populations of neurons following cerebral ischemia (see ROTHWELL and HOPKINS 1995 for review). As evidence, an IL-1β receptor antagonist protects neurons against focal cerebral ischemic injury and administration of IL-1β worsens ischemic injury. Links between IL-1β production and caspase activation are intriguing (IL-1β converting enzyme, also known as caspase 1, cleaves the IL-1β precursor), although the specific mechanisms involved remain elusive.

Based upon the fact that TNF-α can induce apoptosis of several types of mitotic cells (e.g., tumor cells, lymphocytes, and hepatocytes), it had long been assumed that large and rapid increases in expression of TNF-α contributed to neuronal damage in the brain following ischemic, traumatic, and excitotoxic insults (BARONE et al. 1997). However, when the first experiments were performed to directly test the latter hypothesis, the results were quite surprising: TNF-α protected cultured hippocampal neurons against excitotoxic, metabolic, and oxidative insults (CHENG et al. 1994; BARGER et al. 1995). Calcium responses to glutamate were suppressed in hippocampal neurons pretreated with TNF-α (CHENG et al. 1994). Subsequent findings suggest that the mechanism whereby TNF-α prevents excitotoxicity and neuronal apoptosis involves activation of the transcription factor NF-κB. Treatment of hippocampal neurons with TNF-α results in an increased production of the antioxidant enzyme Mn-SOD and the calcium-binding protein calbindin, which is abolished by co-treatment with κB decoy DNA (Fig. 10; CHENG et al. 1994; MATTSON et al. 1995; BRUCE et al. 1996; MATTSON et al. 1997). Whole-cell patch clamp analyses of currents through VDCC, and NMDA and kainate receptor channels, have shown that TNF-α can enhance calcium currents while suppressing NMDA- and kainate-induced currents (FURUKAWA et al. 1998). The

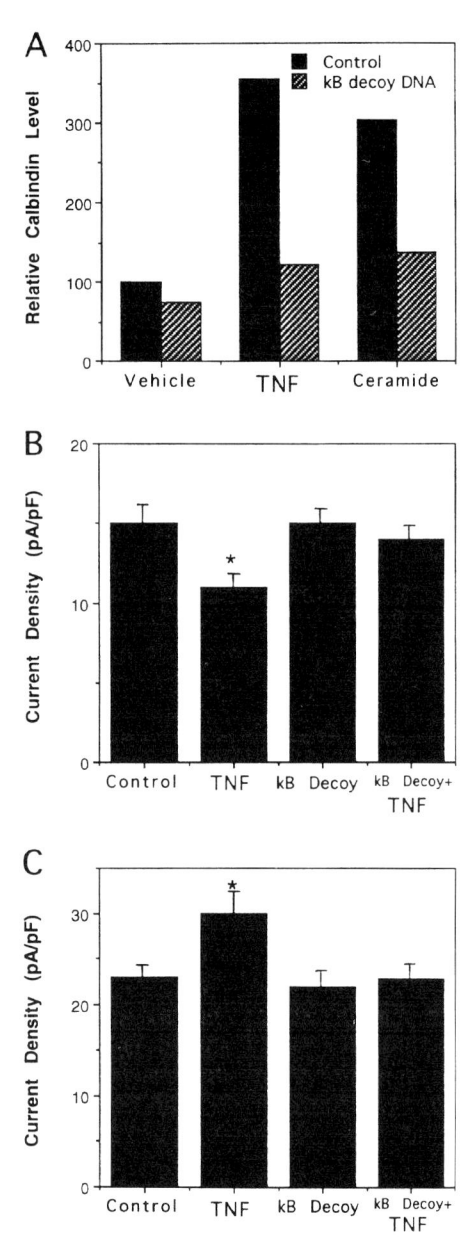

Fig. 10A–C. The transcription factor NF-κB modulates expression/activity of calcium-regulating proteins. A Hippocampal cultures were exposed to vehicle, TNF, ceramide (an activator of NF-κB), κB decoy DNA, κB decoy DNA plus TNF, or κB decoy DNA plus ceramide. Relative levels of calbindin were assessed by Western blot analysis. Values are expressed as percentage of the level in vehicle-treated cultures and represent the mean of two separate experiments. B, C Whole cell NMDA-induced currents (B) or Ca^{2+} currents (C) were recorded in hippocampal neurons that had been treated for 24h with vehicle (control), 100ng/ml TNF, 25μM κB decoy DNA, or κB decoy DNA plus TNF. *$P < 0.05$ compared to each of the other values. (Modified from FURUKAWA et al. 1998)

latter study showed that the effects of TNF-α were blocked by κB decoy DNA and mimicked by ceramide (an agent that activates NF-κB), strongly suggesting a role for NF-κB in modulation of the expression and/or function of neuronal calcium channels and glutamate receptors. Studies of cellular responses to ischemic and excitotoxic brain injury in mice lacking TNF-α receptors provided direct evidence that endogenous injury-induced TNF-α plays a neuroprotective role (BRUCE et al. 1996). The neuroprotective action of TNF-α against focal cerebral ischemia-reperfusion appears to be mediated by the p55 TNF-α receptor (GARY et al. 1998).

III. Inhibitory Neurotransmitters

GABA is the preeminent inhibitory neurotransmitter in the CNS. Cell culture and in vivo studies have provided compelling evidence that activation of GABA receptors can protect neurons against excitotoxic and metabolic insults. For example, GABA suppressed dendritic degeneration and cell death in cultured hippocampal neurons exposed to glutamate (MATTSON and KATER 1989). Pharmacological blockade of GABA receptors can induce seizures and exacerbate neuronal injury in vivo. GABA receptor agonists such as diazepam have proven effective in stabilizing calcium homeostasis and protecting neurons against a variety of insults including exposure to glutamate and oxidative insults (MATTSON and KATER 1989; MARK et al. 1995). Adenosine is another inhibitory transmitter reported to protect neurons against insults relevant to ischemic brain injury (SWEENEY 1997).

IV. Antioxidants

Changes in levels of several different endogenous antioxidants and antioxidant enzymes have been shown to occur following various insults to the CNS, including cerebral ischemia. Increases in expression of Cu/Zn-SOD and Mn-SOD following excitotoxic and ischemic brain insults have been documented (OHTSUKI et al. 1993; LIU et al. 1994; BRUCE et al. 1996). Glutathione is a critical antioxidant that, by detoxifying HNE, protects cells against injury resulting from MLP (MARK et al. 1997a). Glutathione levels in neurons decrease following cerebral ischemia (KRAMER et al. 1992). On the other hand, levels of uric acid (the major antioxidant in human plasma and CSF) are increased following ischemia-reperfusion in adult rats (KANEMITSU et al. 1989; UEMURA et al. 1991). Cell culture studies have shown that uric acid is a potent antioxidant that can protect neurons against excitotoxic and oxidative insults (MATTSON et al. 1997; KELLER et al. 1998; KRUMAN et al. 1998). Uric acid appears to act by suppressing accumulation of both hydroxyl radical and peroxynitrite (MATTSON et al. 1997; KELLER et al. 1998; YU et al. 1998). We recently found that intraperitoneal administration of uric acid to adult rats prior to, or intravenous administration 1h following, middle cerebral artery occlusion results in a marked reduction in infarct volume in an ischemia-reperfusion stroke model in rats (Yu et al. 1998). The latter study also showed that uric acid suppresses elevations of $[Ca^{2+}]C$ following exposure of cultured hippocampal neurons to glutamate and metabolic toxins.

V. Calcium-Binding Proteins

Considerable data suggest that several different calcium-binding proteins can protect neurons against excitotoxic, ischemic, and traumatic injury. The distribution of calbindin D28k in hippocampus is consistent with a neuroprotective role in that dentate granule cells, which express high levels of calbindin D28k (Fig. 11), are resistant to degeneration in ischemic paradigms (SMITH-SWINTOSKY et al. 1996). Cultured hippocampal neurons expressing calbindin D28k are relatively resistant to death induced by glutamate, glucose deprivation, and exposure to oxidative insults (MATTSON et al. 1991; CHENG et al. 1994). Overexpression of calbindin D28k in cultured neural cells confers resistance of the cells to apoptosis induced by trophic factor withdrawal and oxidative insults (Fig. 11; GUO et al. 1998). Calbindin expression is increased in some neurons and astrocytes following excitotoxic, metabolic, and traumatic brain insults in vivo (MATTSON et al. 1995; LEE et al. 1997).

VI. Heat Shock Proteins

Cerebral ischemia, severe epileptic seizures, and traumatic brain injury are accompanied by increased levels of expression of several different "heat shock" proteins, including the cytosolic proteins HSP-70 and HSP-60, the mitochondrial protein GRP75, and the ER-resident proteins GRP78 and GRP94 (see WELCH 1993; LIANG and MACRAE 1997 for review). There is evidence that HSP-70 can protect neurons against death induced by excitotoxic and metabolic insults (LOWENSTEIN et al. 1991). GRP78 and GRP 94 are known to act as molecular chaperones that regulate protein folding and translocation into the ER, and protein secretion (LEE 1992). Recent findings suggest that GRP78 can protect cultured hippocampal neurons against excitotoxic and oxidative insults, in part by stabilizing calcium homeostasis (Yu et al. 1998).

VII. Cytoskeletal Dynamics

It has been known for decades that cytoskeletal elements, particularly actin filaments and microtubules, depolymerize in cells dying of apoptosis and necrosis. Excessive elevation of $[Ca^{2+}]_i$ plays a prominent role in such cytoskeletal disruption in many cases, including when neurons are exposed to excitotoxic and metabolic insults. It had been widely assumed that such cytoskeletal alterations contribute to the cell injury and death process by disrupting cell structure and function. Surprisingly, however, at least some of the cytoskeletal "derangements" appear to promote stabilization of cellular calcium homeostasis and resistance to cell death. Pharmacological manipula-

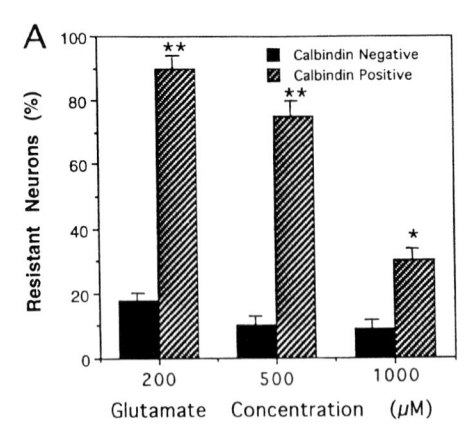

Fig. 11A–D. Antiexcitotoxic action of the calcium-binding protein calbindin D28k. **A** Hippocampal neurons expressing calbindin are relatively resistant to glutamate toxicity. (Modified from MATTSON et al. 1991). **B, C** Calbindin expression is increased in response to TNF treatment in cultured hippocampal neurons (**B**) and following kainate-induced seizures in hippocampus in vivo (**C**). (Modified from CHENG et al. 1994 and LEE et al. 1997). **D** Calbindin immunoreactivity in hippocampus of an adult mouse. Note expression in all dentate granule neurons and a subset of CA1 neurons, and lack of calbindin in CA3 neurons

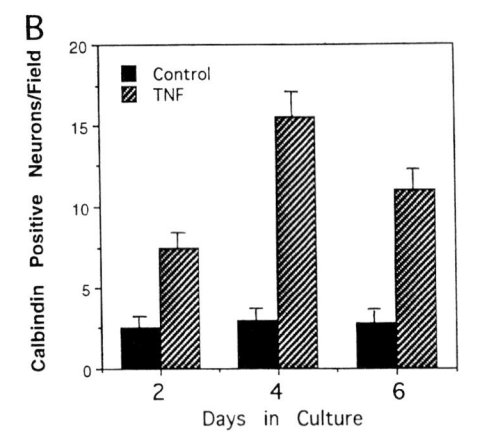

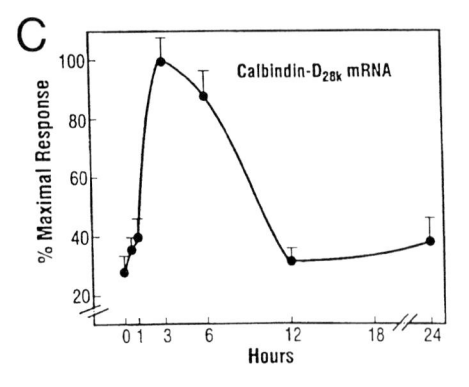

tions of actin polymerization revealed a role for actin filaments in the regulation of calcium influx through NMDA receptor channels and VDCC (ROSENMUND and WESTBROOK 1993; JOHNSON and BYERLY 1993). Treatment of cultured rat hippocampal neurons with the actin-depolymerizing agent

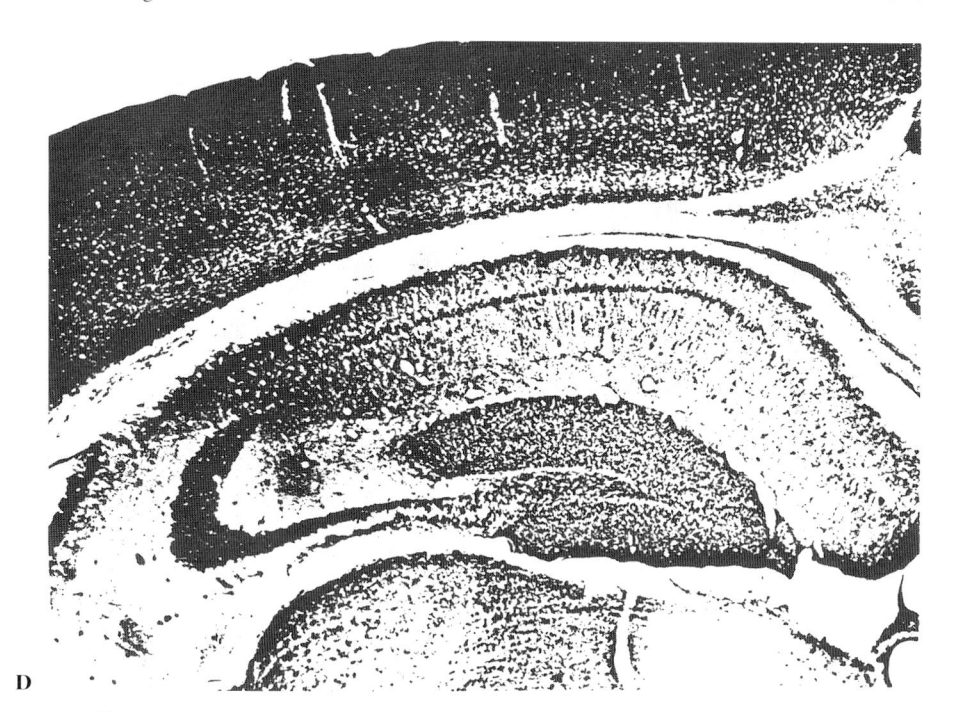

Fig. 11A–D. *Continued*

cytochalasin D resulted in attenuated calcium responses to glutamate and increased resistance to excitotoxicity (FURUKAWA et al. 1995). Administration of cytochalasin D to adult rats reduced seizure-induced damage to hippocampal pyramidal neurons (FURUKAWA et al. 1995) and reduced infarct volume following focal cerebral ischemia-reperfusion in adult mice (ENDRES et al. 1998).

Gelsolin is a calcium-activated actin-severing protein that is expressed in neurons. Studies of gelsolin knockout mice have revealed an important role for this protein in modulating neuronal calcium homeostasis and vulnerability to excitotoxic and ischemic injury. Cultured hippocampal neurons lacking gelsolin exhibit decreased actin filament depolymerization and enhanced calcium influx following exposure to glutamate (FURUKAWA et al. 1997). Measurements of whole-cell currents through NMDA receptors and VDCC revealed enhancement of currents through both types of channels in gelsolin-deficient neurons, which was due to decreased current run-down (Fig. 12). Vulnerability of cultured hippocampal neurons to glutamate toxicity, and of hippocampal neurons to seizure-induced injury in vivo, was greater in mice lacking gelsolin (FURUKAWA et al. 1997). Infarct volume and calcium-mediated proteolysis were significantly increased following focal ischemia-reperfusion in adult gelsolin knockout mice (ENDRES et al. 1998). These findings identify

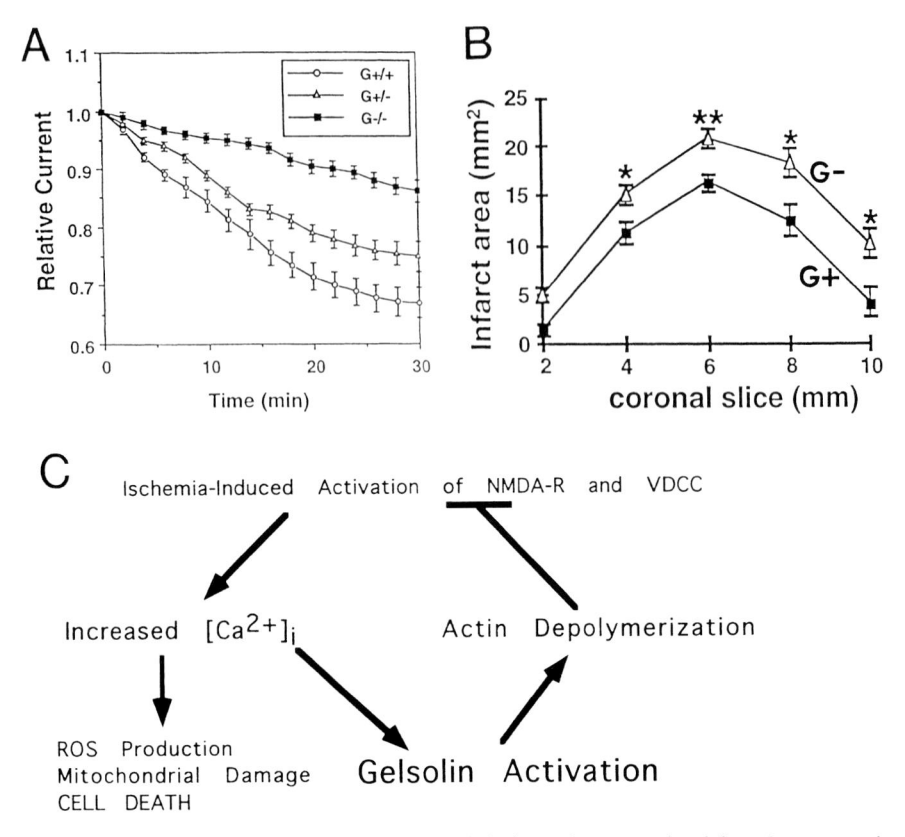

Fig. 12A–C. Role of actin filaments in modulation of neuronal calcium homeostasis and vulnerability to excitotoxic and ischemic insults. **A** Modulation of currents through NMDA receptors and voltage-dependent calcium channels by actin filaments. Rundown of whole-cell calcium currents and NMDA-induced currents is reduced in hippocampal neurons lacking the calcium-activated actin-severing protein gelsolin. (Modified from FURUKAWA et al. 1997). **B** Focal ischemic brain injury is enhanced in mice lacking gelsolin. G^+, wild-type mice; G^-, gelsolin knockout mice. (Modified from ENDRES et al. 1999). **C** Model for regulation of neuronal calcium homeostasis by actin filaments

novel roles for gelsolin and actin filaments in controlling neuronal calcium homeostasis and vulnerability to ischemic injury.

Microtubules are very sensitive to elevations of $[Ca^{2+}]C$, and rapidly depolymerize with relatively modest (subtoxic) event levels of calcium influx (MATTSON 1992). In contrast to actin filament depolymerization, microtubule depolymerization appears to destabilize neuronal calcium homeostasis. As evidence, treatment of cultured rat hippocampal neurons with colchicine, an agent that disrupts microtubules, enhances vulnerability of the neurons to glutamate toxicity (FURUKAWA and MATTSON 1995). Conversely, taxol, an agent that promotes microtubule polymerization, stabilizes cellular calcium homeostasis and protects hippocampal neurons against excitotoxicity (FURUKAWA and MATTSON 1995).

E. Strategies to Preserve and Restore Neuronal Calcium Homeostasis

This brief review of mechanisms of neuronal degeneration and neuroprotection relevant to ischemic brain injury identifies numerous avenues that may lead to improved outcome in human stroke patients (Table 3). The available data suggest that the majority of these interventions stabilize neuronal calcium homeostasis, either directly or indirectly.

The old adage "an ounce of prevention is worth a pound of cure" is certainly true of neurodegenerative disorders. While the bulk of current research dollars are being used to identify mechanisms leading to neuronal injury and death once the degenerative process has begun (e.g., following a stroke or following the onset of symptoms in Alzheimer's and Parkinson's patients), little time and effort is being placed on developing approaches that prevent the initiation of the neurodegenerative cascades (Table 3). In the case of stroke there are clear risk factors (e.g., high blood pressure, a high cholesterol level, obesity, smoking) that can be reduced through dietary and behavioral modification. Moreover, recent findings suggest that diet can greatly influence outcome following acute brain insults relevant to stroke. Thus, rats maintained for 2–4

Table 3. Approaches aimed at stabilizing neuronal calcium homeostasis

Approach	Examples	Mechanism of Action
VDCC blockers (t)	Nifedipine, nimodipine	Suppress calcium influx
NMDA receptor antagonists (t)	MK801, memantine	Suppress calcium influx
Antagonists of ER channels (t)	Dantrolene	Suppress calcium release from ER
Block PTP formation (t)	Cyclosporin A	Prevents PTP protein aggregation
Antioxidants (p, t)	Vitamin E, uric acid, GSH	Suppress protein and lipid damage
Neurotrophic factors (t)	bFGF, NGF, BDNF	Induction of "life" genes
Cytokines (t)	TNF	Activation of NF-κB; gene expression
Actin depolymerizing agents (t)	Cytochalasin D	Suppress calcium influx
Caspase inhibitors (t)	zVAD-fmk, DEVD-CHO	Stabilize mitochondrial function
Food restriction (p)	Reduced caloric intake	Reduced ROS, increase stress proteins

t, Treatment; p, prevention.

months on a food restricted diet (alternate day feeding regimen) exhibit markedly reduced damage to hippocampal and striatal neurons following administration of the excitotoxin kainate and the mitochondrial toxin 3-nitropropionic acid, respectively (BRUCE-KELLER et al. 1999).

Given that ischemic stroke will continue to be a prominent health problem in our society, therapeutic interventions that reduce brain damage and promote recovery are sorely needed. The basic research described above supports efforts to identify treatments that will stabilize neuronal calcium homeostasis in the hours to days following stroke. Table 3 provides a short list of the kinds of agents that have proven effective in reducing neuronal injury in cell culture and animal models of stroke. They target a variety of calcium-regulating systems including glutamate receptors, VDCC, ER calcium channels, and mitochondrial calcium-regulating systems. In addition, agents such as antioxidants and growth factors can stabilize calcium homeostasis through a variety of mechanisms.

Acknowledgements. I thank Y. Goodman, W. Fu and H. Luo for technical assistance. The invaluable contributions of past and present lab members, as well as collaborators, to original research from this laboratory are greatly appreciated.

References

Abele A, Miller R (1990) Potassium channel activators abolish excitotoxicity in cultured hippocampal pyramidal neurons. Neurosci Lett 115:195–200

Ankarcrona M, Dypbukt JM, Bonfoco E, Zhivotovsky B, Orrenius S, Lipton SA, Nicotera P (1995) Glutamate-induced neuronal death: a succession of necrosis or apoptosis depending on mitochondrial function. Neuron 15:961–973

Babcock DF, Herrington J, Goodwin PC, Park YB, Hille B (1997) Mitochondrial participation in the intracellular Ca^{2+} network. J Cell Biol 136:833–844

Baimbridge KG, Celio MR, Rogers JH (1992) Calcium-binding proteins in the nervous system. Trends Neurosci 15:303–308

Barger SW, Fiscus RR, Ruth P, Hofmann F, Mattson MP (1995) Role of cyclic GMP in the regulation of neuronal calcium and survival by secreted forms of β amyloid precursor. J Neurochem 64:2087–2096

Barger SW, Horster D, Furukawa K, Goodman Y, Krieglstein J, Mattson MP (1995) Tumor necrosis factors α and β protect neurons against amyloid β-peptide toxicity: evidence for involvement of a κB-binding factor and attenuation of peroxide and Ca^{2+} accumulation. Proc Natl Acad Sci USA 92:9328–9332

Barone FC, Arvin B, White RF, Miller A, Webb CL, Willette RN, Lysko PG, Feuerstein GZ (1997) Tumor necrosis factor-α. A mediator of focal ischemic brain injury. Stroke 28:1233–1244

Bennett DL, Bootman MD, Berridge MJ, Cheek TR (1998) Ca^{2+} entry into PC12 cells initiated by ryanodine receptors or inositol 1,4,5-trisphosphate receptors. Biochem J 329:349–357

Bernardi P (1996) The permeability transition pore. Control points of a cyclosporin A-sensitive mitochondrial channel involved in cell death. Biochim Biophys Acta 1275:5–9

Bootman MD, Berridge MJ, Lipp P (1997) Cooking with calcium: the recipes for composing global signals from elementary events. Cell 91:367–373

Bredt DS, Snyder SH (1992) Nitric oxide, a novel neuronal messenger. Neuron 8:3–11
Brorson JR, Sulit RA, Zhang H (1997) Nitric oxide disrupts Ca^{2+} homeostasis in hippocampal neurons. J Neurochem 68:95–105
Bruce AJ, Boling W, Kindy MS, Peschon J, Kraemer PJ, Carpenter MK, Holtsberg FW, Mattson MP (1996) Altered neuronal and microglial responses to brain injury in mice lacking TNF receptors. Nature Medicine 2:788–794
Bruce-Keller AJ, Begley JG, Fu W, Butterfield DA, Bredesen DE, Hutchins JB, Hensley K, Mattson MP (1998) Bcl-2 protects isolated plasma and mitochondrial membranes against lipid peroxidation induced by hydrogen peroxide and amyloid β-peptide. J Neurochem 70:31–39
Bruce-Keller AJ, Geddes JW, Knapp PE, Keller JN, Holtsberg FW, Steiner SM, Mattson MP (1998) Antiapoptotic effects of TNF against metabolic poisoning: mitochondrial stabilization by MnSOD. J Neuroimmunol In press
Bruce-Keller AJ, Umberger G, McFall R, Mattson MP (1999) Food restriction reduces brain damage and improves behavioral outcome following excitotoxic and metabolic insults. Ann Neurol 45:8–15
Carafoli E, Garcia-Martin E, Guerini D (1996) The plasma membrane calcium pump: recent developments and future perspectives. Experientia 52:1091–1100
Chalmers-Redman RM, Fraser AD, Ju WY, Wadia J, Tatton NA, Tatton WG (1997) Mechanisms of nerve cell death: apoptosis or necrosis after cerebral ischaemia. Int Rev Neurobiol 40:1–25
Chard PS, Bleakman D, Christakos S, Fullmer CS, Miller RJ (1993) Calcium buffering properties of calbindin D28k and parvalbumin in rat sensory neurones. J Physiol 472:341–357
Chen J, Jin K, Chen M, Pei W, Kawaguchi K, Greenberg DA, Simon RP (1997) Early detection of DNA strand breaks in the brain after transient focal ischemia: implications for the role of DNA damage in apoptosis and neuronal death. J Neurochem 69:232–245
Cheng B, Mattson MP (1991) NGF and bFGF protect rat and human central neurons against hypoglycemic damage by stabilizing calcium homeostasis. Neuron 7:1031–1041
Cheng B, Mattson MP (1992) IGF-I and IGF-II protect cultured hippocampal and septal neurons against calcium-mediated hypoglycemic damage. J Neurosci 12:1558–1566
Cheng B, McMahon D, Mattson MP (1993) Modulation of calcium current, intracellular calcium levels and cell survival by hypoglycemia, NGF and bFGF in cultured hippocampal neurons. Brain Res 607:275–285
Cheng B, Mattson MP (1994) NT-3 and BDNF protect CNS neurons against metabolic/excitotoxic insults. Brain Res 640:56–67
Cheng B, Goodman Y, Begley JG, Mattson MP (1994) Neurotrophin 4/5 protects hippocampal and cortical neurons against energy deprivation- and excitatory amino acid-induced injury. Brain Res 650:331–335
Cheng B, Christakos S, Mattson MP (1994) Tumor necrosis factors protect neurons against excitotoxic/metabolic insults and promote maintenance of calcium homeostasis. Neuron 12:139–153
Cheng B, Mattson MP (1995) PDGFs protect hippocampal neurons against energy deprivation and oxidative injury: evidence for induction of antioxidant pathways. J Neurosci 15:7095–7104
Cheng B, Furukawa K, O'Keefe JA, Goodman Y, Kihiko M, Fabian T, Mattson MP (1995) Basic fibroblast growth factor selectively increases AMPA-receptor subunit GluR1 protein level and differentially modulates Ca^{2+} responses to AMPA and NMDA in hippocampal neurons. J Neurochem 65:2525–2536
Choi DW (1995) Calcium: still center stage in hypoxic-ischemic neuronal death. Trends Neurosci 18:58–60
Choi DW (1996) Ischemia-induced neuronal apoptosis. Curr Opin Neurobiol 6:667–672

Cosi C, Spoerri PE, Comelli MC, Guidolin D, Skaper SD (1993) Glucocorticoids depress activity-dependent expression of BDNF mRNA in hippocampal neurones. Neuroreport 4:527–530

Denk W, Yuste R, Svoboda K, Tank DW (1996) Imaging calcium dynamics in dendritic spines. Curr Opin Neurobiol 6:372–378

Du Y, Bales KR, Dodel RC, Hamilton-Byrd E, Horn JW, Czilli DL, Simmons LK, Ni B, Paul SM (1997) Activation of a caspase 3-related cysteine protease is required for glutamate-mediated apoptosis of cultured cerebellar granule neurons. Proc Natl Acad Sci USA 94:11657–11662

Dubinsky J, Rothman S (1991) Intracellular calcium concentrations during "chemical hypoxia" and excitotoxic neuronal injury. J Neurosci 11:2545–2551

Dugan LL, Sensi SL, Canzoniero LM, Handran SD, Rothman SM, Goldberg MP, Choi DW (1996) Mitochondrial production of reactive oxygen species in cortical neurons following exposure to N-methyl-D-aspartate. J Neurosci 15:6377–6388

Elliott E, Sapolsky RM (1992) Corticosterone enhances kainic acid-induced calcium mobilization in cultured hippocampal neurons. J Neurochem 59:1033–1038

Elliott E, Mattson MP, Vanderklish P, Lynch G, Chang I, Sapolsky RM (1993) Corticosterone exacerbates kainate-induced alterations in hippocampal tau immunoreactivity and spectrin proteolysis in vivo. J Neurochem 61:57–67

Endres M, Fink K, Zhu J, Stagliano NE, Bondada V, Geddes JW, Azuma T, Mattson MP, Kwiatkowski DJ, Moskowitz MA (1999) Neuroprotective effects of gelsolin and cytochalasin D during murine stroke. J Clin Invest In press

Frandsen A, Schousboe A (1991) Dantrolene prevents glutamate cytotoxicity and calcium release from intracellular stores in cultured cerebral cortical neurons. J Neurochem 56:1075–1078

Franzini-Armstrong C, Protasi F (1997) Ryanodine receptors of striated muscles: a complex channel capable of multiple interactions. Physiol Rev 77:699–729

Fukuyama N, Takizawa S, Ishida H, Hoshiai K, Shinohara Y, Nakazawa H (1998) Peroxynitrite formation in focal cerebral ischemia-reperfusion in rats occurs predominantly in the peri-infarct region. J Cereb Blood Flow Metab 18:123–129

Furukawa K, Mattson MP (1995) Taxol stabilizes $[Ca^{2+}]_i$ and protects hippocampal neurons against excitotoxicity. Brain Res 689:141–146

Furukawa K, Smith-Swintosky VL, Mattson MP (1995) Evidence that actin depolymerization protects hippocampal neurons against excitotoxicity by stabilizing $[Ca^{2+}]_i$. Exp Neurol 133:153–163

Furukawa K, Barger SW, Blalock E, Mattson MP (1996) Activation of K^+ channels and suppression of neuronal activity by secreted β-amyloid precursor protein. Nature 379:74–78

Furukawa K, Fu W, Li Y, Witke W, Kwiatkowski DJ, Mattson MP (1997) The actin-severing protein gelsolin modulates calcium channel and NMDA receptor activities and vulnerability to excitotoxicity in hippocampal neurons. J Neurosci 17: 8178–8186

Furukawa K, Mattson MP (1998) The transcription factor NF-κB mediates increases in calcium currents and decreases in NMDA and AMPA/kainate-induced currents in response to TNF-α in hippocampal neurons. J Neurochem 70:1876–1886

Gary DS, Bruce-Keller AJ, Kindy MS, Mattson MP (1998) Ischemic and excitotoxic brain injury is enhanced in mice lacking the p55 tumor necrosis factor receptor. J Cereb Blood Flow Metab 18:1283–1287

Gary DS, Mattson MP (2001) Integrin signaling via the PI3-kinase-Akt pathway increases neuronal resistance to glutamate-induced apoptosis. J Neurochem 76: 1485–1496

Gilman SC, Bonner MJ, Pellmar TC (1992) Peroxide effects on [^{3}H]-glutamate release by synaptosomes isolated from the cerebral cortex. Neurosci Lett 140:157–160

Ginsberg MD (1990) Local metabolic responses to cerebral ischemia. Cerebrovasc Brain Metab Rev 2:58–93

Goldberg MP, Choi DW (1993) Combined oxygen and glucose deprivation in cortical cell culture: calcium-dependent and calcium-independent mechanisms of neuronal injury. J Neurosci 13:3510–3524

Goodman Y, Steiner MR, Steiner SM, Mattson MP (1994) Nordihydroguaiaretic acid protects hippocampal neurons against amyloid β-peptide toxicity, and attenuates free radical and calcium accumulation. Brain Res 654:171–176

Goodman Y, Mattson MP (1996) K$^+$ channel openers protect hippocampal neurons against oxidative injury and amyloid β-peptide toxicity. Brain Res 706:328–332

Goodman Y, Mattson MP (1996) Ceramide protects hippocampal neurons against excitotoxic and oxidative insults, and amyloid β-peptide toxicity. J Neurochem 66:869–872

Goodman Y, Bruce AJ, Cheng B, Mattson MP (1996) Estrogens attenuate and corticosterone exacerbates excitotoxicity, oxidative injury and amyloid β-peptide toxicity in hippocampal neurons. J Neurochem 66:1836–1844

Green AR, Hainsworth AH, Jackson DM (2000) GABA potentiation: A logical pharmacological approach for the treatment of acute ischaemic stroke. Neuropharmacology 39:1483–1494

Gunter TE, Gunter KK, Sheu SS, Gavin CE (1994) Mitochondrial calcium transport: physiological and pathological relevance. Am J Physiol 267:C313–C339

Guo Q, Furukawa K, Sopher BL, Pham DG, Robinson N, Martin GM, Mattson MP (1996) Alzheimer's PS-1 mutation perturbs calcium homeostasis and sensitizes PC12 cells to death induced by amyloid β-peptide. NeuroReport 8:379–383

Guo Q, Sopher BL, Pham DG, Furukawa K, Robinson N, Martin GM, Mattson MP (1997) Alzheimer's presenilin mutation sensitizes neural cells to apoptosis induced by trophic factor withdrawal and amyloid β-peptide: involvement of calcium and oxyradicals. J Neurosci 17:4212–4222

Guo Q, Christakos S, Robinson N, Mattson MP (1998) Calbindin blocks the proapoptotic actions of mutant presenilin-1: reduced oxidative stress and preserved mitochondrial function. Proc Natl Acad Sci USA 95:3227–3232

Guo Q, Fu W, Sopher BL, Miller MW, Ware CB, Martin GM, Mattson MP (1999) Increased vulnerability of hippocampal neurons to excitotoxic necrosis in presenilin-1 mutant knock-in mice. Nature Med 5:101–107

Hara H, Nagasawa H, Kogure K (1990): Nimodipine prevents postischemic brain damage in the early phase of focal cerebral ischemia. Stroke 21:IV102–104

Hartley DM, Kurth MC, Bjerkness L, Weiss JH, Choi DW (1993) Glutamate receptor-induced ^{45}Ca^{2+} accumulation in cortical cell culture correlates with subsequent neuronal degeneration. J Neurosci 13:1993–2000

Hayes RL, Jenkins LW, Lyeth BG (1992) Neurotransmitter-mediated mechanisms of traumatic brain injury: acetylcholine and excitatory amino acids. J Neurotrauma 9:S173–S187

He H, Lam M, McCormick TS, Distelhorst CW (1997) Maintenance of calcium homeostasis in the endoplasmic reticulum by Bcl-2. J Cell Biol 138:1219–1228

Holtsberg FW, Steiner MR, Furukawa K, Keller JN, Mattson MP, Steiner SM (1997) Lysophosphatidic acid induces a sustained elevation of neuronal intracellular calcium. J Neurochem 69:68–75

Horner H, Packan D, Sapolsky RM (1990) Glucocorticoids inhibit glucose transport in cultured hippocampal neurons and glia. Neuroendocrinology 52:57–64

Hossmann KA (1994) Glutamate-mediated injury in focal cerebral ischemia: the excitotoxin hypothesis revised. Brain Pathol 4:23–36

Hoth M, Fanger CM, Lewis RS (1997) Mitochondrial regulation of store-operated calcium signaling in T lymphocytes. J Cell Biol 137:633–648

Jesberger JA, Richardson JS (1991) Oxygen free radicals and brain dysfunction. Int J Neurosci 57:1–17

Johnson B, Byerly L (1993) A cytoskeletal mechanism for Ca^{2+} channel metabolic dependence and inactivation by intracellular Ca^{2+}. Neuron 10:797–804

Kanemitsu H, Tamura A, Kirino T, Oka H, Sano K, Iwamoto T, Yoshiura M, Iriyama K (1989) Allopurinol inhibits uric acid accumulation in the rat brain following focal cerebral ischemia. Brain Res 499:367–370

Keller JN, Mark RJ, Bruce AJ, Blanc EM, Rothstein JD, Uchida K, Mattson MP (1997a) 4-hydroxynonenal, an aldehydic product of membrane lipid peroxidation, impairs

glutamate transport and mitochondrial function in synaptosomes. Neuroscience 80:685–696
Keller JN, Pang Z, Geddes JW, Begley JG, Germeyer A, Waeg G, Mattson MP (1997b) Impairment of glucose and glutamate transport and induction of mitochondrial oxidative stress and dysfunction in synaptosomes by amyloid β-peptide: role of the lipid peroxidation product 4-hydroxynonenal. J Neurochem 69:273–284
Keller JN, Mattson MP (1997) 17β-estradiol attenuates oxidative impairment of synaptic Na$^+$/K$^+$-ATPase activity, glucose transport and glutamate transport induced by amyloid β-peptide and iron. J Neurosci Res 50:522–530
Keller JN, Kindy MS, Holtsberg FW, St Clair DK, Yen H-C, Germeyer A, Steiner SM, Bruce-Keller AJ, Hutchins JB, Mattson MP (1998) Mitochondrial MnSOD prevents neural apoptosis and reduces ischemic brain injury: suppression of peroxynitrite production, lipid peroxidation and mitochondrial dysfunction. J Neurosci 18:687–697
Khan AA, Soloski MJ, Sharp AH, Schillin G, Sabatini DM, Li SH, Ross CA, Snyder SH (1996) Lymphocyte apoptosis: mediation by increased type 3 inositol 1,4,5-trisphosphate receptor. Science 273:503–507
Kluck RM, Bossy-Wetzel E, Green DR, Newmeyer DD (1997) The release of cytochrome c from mitochondria: a primary site for Bcl-2 regulation of apoptosis. Science 275:1132–1136
Knopfel T, Spuler A, Grafe P, Gahwiler BH (1990) Cytosolic calcium during glucose deprivation in hippocampal pyramidal cells of rats. Neurosci Lett 117:295–299
Koketsu N, Berlove DJ, Moskowitz MA, Kowall NW, Caday CG, Finklestein SP Pretreatment with intraventricular basic fibroblast growth factor (bFGF) decreases infarct size following focal cerebral ischemia in rats. Ann Neurol 35:451–457
Kondo T, Reaume AG, Huang TT, Carlson E, Murakami K, Chen SF, Hoffman EK, Scott RW, Epstein CJ, Chan PH (1997) Reduction of CuZn-superoxide dismutase activity exacerbates neuronal cell injury and edema formation after transient focal cerebral ischemia. J Neurosci 17:4180–4189
Kramer K, Voss HP, Grimbergen JA, Smink C, Timmerman H, Bast A (1992) Glutathione mobilization during cerebral ischemia and reperfusion in the rat. Gen Pharmacol 23:105–108
Krause KH, Michalak M (1997) Calreticulin. Cell 88:439–443
Kristal BS, Park BK, Yu BP (1996) 4-Hydroxyhexenal is a potent inducer of the mitochondrial permeability transition. J Biol Chem 271:6033–6038
Kroemer G, Zamzami N, Susin SA (1997) Mitochondrial control of apoptosis. Immunol Today 18:44–51
Kruman I, Bruce-Keller AJ, Bredesen DE, Waeg G, Mattson MP (1997) Evidence that 4-hydroxynonenal mediates oxidative stress-induced neuronal apoptosis. J Neurosci 17:5089–5100
Kruman I, Guo Q, Mattson MP (1998) Calcium and reactive oxygen species mediate staurosporine-induced mitochondrial dysfunction and apoptosis in PC12 cells. J Neurosci Res 51:293–308
Kruman I, Pang Z, Geddes JW, Mattson MP (1998) Pivotal role of mitochondrial calcium uptake in neural cell apoptosis and necrosis. J Neurochem Submitted
Lafon-Cazal M, Pietri S, Culcasi M, Bockaert J (1993) NMDA-dependent superoxide production and neurotoxicity. Nature 364:535–537
Lederer WJ, He S, Luo S, duBell W, Kofuji P, Kieval R, Neubauer CF, Ruknudin A, Cheng H, Cannell MB, Rogers TB, Schulze DH (1996) The molecular biology of the Na$^+$-Ca^{2+} exchanger and its functional roles in heart, smooth muscle cells, neurons, glia, lymphocytes, and nonexcitable cells. Ann N Y Acad Sci 779: 7–17
Lee AS (1992) Mammalian stress response: induction of the glucose regulated protein family. Curr Opin Cell Biol 4:267–273
Lee S, Williams J, Lothman EW, Szele FG, Chesselet MF, Sapolsky RM, Mattson MP, Christakos S (1997) Early induction of mRNA for calbindin-D28k and BDNF but

not NT-3 in rat hippocampus after kainic acid treatment. Mol Brain Res 47: 183–194

Levy DI, Sucher NJ, Lipton SA (1990) Redox modulation of NMDA receptor-mediated toxicity in mammalian central neurons. Neurosci Lett 110:291–296

Levy DI, Sucher NJ, Lipton SA (1991) Glutathione prevents N-methyl-D-aspartate receptor-mediated neurotoxicity. Neuropharmacol Neurotoxicol 2:345–348

Liang P, MacRae TH (1997) Molecular chaperones and the cytoskeleton. J Cell Sci 110:1431–1440

Linnik MD, Zobrist RH, Hatfield MD (1993) Evidence supporting a role for programmed cell death in focal cerebral ischemia in rats. Stroke 24:2002–2008

Linnik MD, Zahos P, Geschwind MD, Federoff HJ (1995) Expression of bcl-2 from a defective herpes simplex virus-1 vector limits neuronal death in focal cerebral ischemia. Stroke 26:1670–1674

Lipton SA, Choi YB, Pan ZH, Lei SZ, Chen HSV, Sucher NJ, Loscaizo J, Singel DJ, Stamler JS (1993) A redox-based mechanism for the neuroprotective and neurodestructive effects of nitric oxide and related nitroso-compounds. Nature 364:626–632

Little E, Tocco G, Baudry M, Lee AS, Schreiber SS (1996) Induction of glucose-regulated protein (glucose-regulated protein 78/BiP and glucose-regulated protein 94) and heat shock protein 70 transcripts in the immature rat brain following status epilepticus. Neuroscience 75:209–219

Liu X, Kato H, Araki T, Itoyama Y, Kato K, Kogure K (1994) An immunohistochemical study of copper / zinc super oxide dismutase and manganese superoxide dismutase following focal cerebral ischemia in the rat. Brain Res 644: 257–266

Louis JC, Magal E, Takayama S, Varon S (1993) CNTF protection of oligodendrocytes against natural and tumor necrosis factor-induced death. Science 259:689–692

Lowenstein DH, Chan P, Miles M (1991) The stress protein response in cultured neurons: characterization and evidence for a protective role in excitotoxicity. Neuron 7:1053–1060

Lowenstein DH, Gwinn RP, Seren MS, Simon RP, McIntosh TK (1994) Increased expression of mRNA encoding calbindin-D28k, the glucose-regulated proteins, or the 72 heat-shock protein in three models of acute CNS injury. Mol Brain Res 22:299–308

Lupu VD, Kaznacheyeva E, Krishna UM, Falck JR, Bezprozvanny I (1998) Functional coupling of phosphatidylinositol 4,5-bisphosphate to inositol 1,4,5-trisphosphate receptor. J Biol Chem 273:14067–14070

MacManus JP, Buchan AM, Hill IE, Rasquinha I, Preston E (1993) Global ischemia can cause DNA fragmentation indicative of apoptosis in rat brain. Neurosci Lett 164:89–92

Marchetti P, Castedo M, Susin SA, Zamzami N, Hirsch T, Macho A, Haeffner A, Hirsch F, Geuskens M, Kroemer G (1996a) Mitochondrial permeability transition is a central coordinating event of apoptosis. J Exp Med 184:1155–1160

Marin MC, Fernandez A, Bick RJ, Brisbay S, Buja LM, Snuggs M, McConkey DJ, von Eschenbach AC, Keating MJ, McDonnell TJ (1996) Apoptosis suppression by bcl-2 is correlated with the regulation of nuclear and cytosolic Ca^{2+}. Oncogene 12: 2259–2266

Mark RJ, Ashford JW, Mattson MP (1995) Anticonvulsants attenuate amyloid β-peptide neurotoxicity and promote maintenance of calcium homeostasis. Neurobiol Aging 16:187–198

Mark RJ, Hensley K, Butterfield DA, Mattson MP (1995) Amyloid β-peptide impairs ion-motive ATPase activities: evidence for a role in loss of neuronal Ca^{2+} homeostasis and cell death. J Neurosci 15:6239–6249

Mark RJ, Lovell MA, Markesbery WR, Uchida K, Mattson MP (1997a) A role for 4-hydroxynonenal in disruption of ion homeostasis and neuronal death induced by amyloid β-peptide. J Neurochem 68:255–264

Mark RJ, Pang Z, Geddes JW, Uchida K, Mattson MP (1997b) Amyloid β-peptide impairs glucose uptake in hippocampal and cortical neurons: involvement of membrane lipid peroxidation. J Neurosci 17:1046–1054

Martinou JC, Dubois-Dauphin M, Staple JK, Rodriguez I, Frankowski H, Missotten M, Albertini P, Talabot D, Catsicas S, Pietra C (1994) Overexpression of BCL-2 in transgenic mice protects neurons from naturally occurring cell death and experimental ischemia. Neuron 13:1017–1030

Mattson, M. P. (1989) Acetylcholine potentiates glutamate-induced neurodegeneration in cultured hippocampal neurons. Brain Res 497:402–406

Mattson MP, Kater SB (1989) Excitatory and inhibitory neurotransmitters in the generation and degeneration of hippocampal neuroarchitecture. Brain Res 478:337–348

Mattson MP, Murrain M, Guthrie PB, Kater SB (1989) Fibroblast growth factor and glutamate: Opposing actions in the generation and degeneration of hippocampal neuroarchitecture. J Neurosci 9:3728–3740

Mattson MP, Guthrie PB, Kater SB (1989) A role for Na^+-dependent Ca^{2+} extrusion in protection against neuronal excitotoxicity. FASEB J 3:2519–2526

Mattson MP, Rychlik B, Chu C, Christakos S (1991) Evidence for calcium-reducing and excitoprotective roles for the calcium binding protein (calbindin-D28k) in cultured hippocampal neurons. Neuron 6:41–51

Mattson MP (1992) Calcium as sculptor and destroyer of neural circuitry. Exp Gerontol 27:29–49

Mattson MP, Cheng B, Culwell A, Esch F, Lieberburg I, Rydel RE (1993) Evidence for excitoprotective and intraneuronal calcium-regulating roles for secreted forms of β-amyloid precursor protein. Neuron 10:243–254

Mattson MP, Zhang Y, Bose S (1993) Growth factors prevent mitochondrial dysfunction, loss of calcium homeostasis and cell injury, but not ATP depletion in hippocampal neurons deprived of glucose. Exp Neurol 121:1–13

Mattson MP, Kumar K, Cheng B, Wang H, Michaelis EK (1993) Basic FGF regulates the expression of a functional 71 kDa NMDA receptor protein that mediates calcium influx and neurotoxicity in cultured hippocampal neurons. J Neurosci 13:4575–4588

Mattson MP (1994) Secreted forms of β-amyloid precursor protein modulate dendrite outgrowth and calcium responses to glutamate in cultured embryonic hippocampal neurons. J Neurobiol 25:439–450

Mattson MP, Cheng B, Baldwin S, Smith-Swintosky VL, Keller JN, Geddes JW, Scheff SW, Christakos S (1995) Brain injury and tumor necrosis factors induce expression of calbindin D-28k in astrocytes: a cytoprotective response. J Neurosci Res 42:357–370

Mattson MP, Lovell MA, Furukawa K, Markesbery WR (1995) Neurotrophic factors attenuate glutamate-induced accumulation of peroxides, elevation of $[Ca^{2+}]_i$ and neurotoxicity, and increase antioxidant enzyme activities in hippocampal neurons. J Neurochem 65:1740–1751

Mattson MP, Mark RJ (1996) Excitotoxicity and excitoprotection in vitro. In: Siesjo B, Wielock T (eds) Cellular and Molecular Mechanisms of Ischemic Brain Damage Adv Neurol Lippincott-Raven, New York, 71:1–37

Mattson MP, Lindvall O (1997) Neurotrophic factor and cytokine signaling in the aging brain. In: Mattson MP, Geddes JW (eds) The Aging Brain, Adv Cell Aging Gerontol 2:299–345

Mattson MP, Goodman Y, Luo H, Fu W, Furukawa K (1997) Activation of NF-κB protects hippocampal neurons against oxidative stress-induced apoptosis: evidence for induction of Mn-SOD and suppression of peroxynitrite production and protein tyrosine nitration. J Neurosci Res 49:681–697

Mattson MP, Barger SW, Furukawa K, Bruce AJ, Wyss-Coray T, Mark RJ, Mucke L (1997) Cellular signaling roles of TGFβ, TNFα and βAPP in brain injury responses and Alzheimer's disease. Brain Res Rev 23:47–61

Mattson MP (1998a) Modification of ion homeostasis by lipid peroxidation: roles in neuronal degeneration and adaptive plasticity. Trends Neurosci 21:53–57

Mattson MP (1998b) Free radicals, calcium, and the synaptic plasticity – cell death continuum: emerging roles of the transcription factor NF-κB. Int Rev Neurobiol 42:103–168

Mattson MP, Keller JN, Begley JG (1998a) Evidence for synaptic apoptosis. Exp Neurol 153:35–48

Mattson MP, Partin J, Begley JG (1998b) Amyloid β-peptide induces apoptosis-related events in synapses and dendrites. Brain Res 807:167–176

McIntosh LJ, Hong KE, Sapolsky RM (1998) Glucocorticoids may alter antioxidant enzyme capacity in the brain: baseline studies. Brain Res 791:209–214

Meyer EM, Judkins JH (1993) Effects of membrane peroxidation on [^{3}H]-acetylcholine release in rat cerebral cortical synaptosomes. Neurochem Res 18:1047–1050

Michaels RL, Rothman SM (1990) Glutamate neurotoxicity in vitro: antagonist pharmacology and intracellular calcium concentrations. J Neurosci 10:283–292

Morgan DW, Welton AF, Heick AE, Christakos S (1986) Specific in vitro activation of Ca, Mg-ATPase by vitamin D-dependent rat renal calcium binding protein (calbindin D28K). Biochem Biophys Res Commun 138:547–553

Mossakowski MJ, Gadamski R (1990) Nimodipine prevents delayed neuronal death of sector CA1 pyramidal cells in short-term forebrain ischemia in mongolian gerbils. Stroke 21:IV120–122

Mulkey RM, Endo S, Shenolikar S, Malenka RC (1994) Involvement of a calcineurin/inhibitor-1 phosphatase cascade in hippocampal long-term depression. Nature 369:486–488

Muller U, Krieglstein J (1995) Prolonged pretreatment with alpha-lipoic acid protects cultured neurons against hypoxic, glutamate-, or iron-induced injury. J Cereb Blood Flow Metab 5:624–630

Murakami K, Kondo T, Kawase M, Li Y, Sato S, Chen SF, Chan PH (1998) Mitochondrial susceptibility to oxidative stress exacerbates cerebral infarction that follows permanent focal cerebral ischemia in mutant mice with manganese superoxide dismutase deficiency. J Neurosci 18:205-213

Myers KM, Fiskum G, Liu Y, Simmens SJ, Bredesen DE, Murphy AN (1995) Bcl-2 protects neural cells from cyanide/aglycemia-induced lipid oxidation, mitochondrial injury, and loss of viability. J Neurochem 65:2432–2440

Nelson TJ, Alkon DL (1991) GTP-binding proteins and potassium channels involved in synaptic plasticity and learning. Mol Neurobiol 5:315–328

Nicotera P, Ankarcrona M, Bonfoco E, Orrenius S, Lipton SA (1997) Neuronal necrosis and apoptosis: two distinct events induced by exposure to glutamate or oxidative stress. Adv Neurol 72:95–101

Nitatori T, Sato N, Waguri S, Karasawa Y, Araki H, Shibanai K, Kominami E, Uchiyama Y (1995) Delayed neuronal death in the CA1 pyramidal cell layer of the gerbil hippocampus following transient ischemia is apoptosis. J Neurosci 15:1001–1011

Novelli A, Reilly JA, Lysko PG, Henneberry RC (1988) Glutamate becomes neurotoxic via the N-methyl-D-aspartate receptor when intracellular energy levels are reduced. Brain Res 451:205–212

Nozaki K, Finklestein SP, Beal MF (1993) Basic fibroblast growth factor protects against hypoxia-ischemia and NMDA neurotoxicity in neonatal rats. J Cereb Blood Flow Metab 13:221–228

Ohtsuki T, Matsumoto M, Suzuki K, Taniguchi N, Kamada T (1993) Effect of transient forebrain ischemia on superoxide dismutases in gerbil hippocampus. Brain Res 620:305–309

Oliver CN, Starke-Reed PE, Stadtman ER, Liu GJ, Carney JM, Floyd RA (1990) Oxidative damage to brain proteins, loss of glutamine synthetase activity, and production of free radicals during ischemia/reperfusion-induced injury to gerbil brain. Proc Natl Acad Sci USA 87:5144–5147

Olsson T (1990) Urinary free cortisol excretion shortly after ischaemic stroke. J Intern Med 228:177–181

Peruche B, Krieglstein J (1993) Mechanisms of drug actions against neuronal damage caused by ischemia–an overview. Prog Neuropsychopharmacol Biol Psychiatry 17:21–70

Piantadosi CA, Zhang J (1996) Mitochondrial generation of reactive oxygen species after brain ischemia in the rat. Stroke 27:327–331

Prehn JHM, Backhaub C, Krieglstein J (1993) Transforming growth factor-β prevents glutamate toxicity in rat neocortical cell cultures and protects mouse from ischemic injury in vivo. J Cereb Blood Flow Metab 13:521–525

Prehn JHM, Jordan J, Ghadge GD, Preis E, Galindo MF, Roos RP, Krieglstein J, Miller RJ (1997) Ca^{2+} and reactive oxygen species in staurosporine-induced neuronal apoptosis. J Neurochem 68:1679–1685

Racay P, Bezakova G, Kaplan P, Lehotsky J, Mezesova V (1994) Alteration in rabbit brain endoplasmic reticulum Ca^{2+} transport by free oxygen radicals in vitro. Biochem Biophys Res Commun 199:63–69

Raman IM, Tong G, Jahr CE (1996) Beta-adrenergic regulation of synaptic NMDA receptors by cAMP-dependent protein kinase. Neuron 16:415–421

Reynolds IJ (1990) Modulation of NMDA receptor responsiveness by neurotransmitters, drugs and chemical modification. Life Sci 47:1785–1792

Richter C, Kass GEN (1991) Oxidative stress in mitochondria: its relationship to cellular Ca^{2+} homeostasis, cell death, proliferation, and differentiation. Chem Biol Interactions 77:1–23

Richter C (1993) Pro-oxidants and mitochondrial calcium: their relationship to apoptosis and oncogenesis. FEBS Lett 325:104–107

Rosenberg PA, Aizenman E (1989) Hundred-fold increase in neuronal vulnerability to glutamate toxicity in astrocyte-poor cultures of rat cerebral cortex. Neuroscience 103:162–168

Rosenmund C, Westbrook G (1993) Calcium-induced actin depolymerization reduces NMDA channel activity. Neuron 10:805–814

Rothwell NJ, Hopkins SJ (1995) Cytokines and the nervous system II: actions and mechanisms of action. Trends Neurosci 18:130–136

Sapolsky RM (1997) Deleterious and salutary effects of steroid hormones in the nervous system. In: Mattson MP (ed) Neuroprotective Signal Transduction, Humana Press, Totowa, NJ. pp 259–283

Scharfman HE, Schwartzkroin PA (1989) Protection of dentate hilar cells from prolonged stimulation by intracellular calcium chelation. Science 246:257–260

Schwartz MA, Brown EJ, Fazeli B (1993) A 50-kDa integrin-associated protein is required for intergrin-regulated calcium entry in endothelial cells. J Biol Chem 268:19931–19934

Siesjo BK, Bengtsson G, Grampp W, Theander S (1989) Calcium, Exitotoxins, and neuronal death in brain. Ann NY Acad Sci 568:234–251

Siesjo BK (1992) Pathophysiology and treatment of focal cerebral ischemia. Part I. pathophysiology. J Neurosurg 77:169–184

Smith-Swintosky VL, Pettigrew LC, Craddock SD, Culwell AR, Rydel RE, Mattson MP (1994) Secreted forms of β-amyloid precursor protein protect against ischemic brain injury. J Neurochem 63:781–784

Smith-Swintosky VL, Zimmer S, Fenton JW, Mattson MP (1995) Protease nexin-I and thrombin modulate neuronal Ca^{2+} homeostasis and sensitivity to glucose deprivation-induced injury. J Neurosci 15:5840–5850

Smith-Swintosky VL, Pettigrew LC, Sapolsky RM, Phares C, Craddock SD, Brooke SM, Mattson MP (1996) Metyrapone, an inhibitor of glucocorticoid production, reduces brain injury induced by focal and global ischemia and seizures. J Cerebr Blood Flow Metab 16:585–598

Stein-Behrens B, Mattson MP, Chang I, Yeh M, Sapolsky RM (1994) Stress exacerbates neuron loss and cytoskeletal pathology in the hippocampus. J Neurosci 14:5373–5380

Susin SA, Zamzami N, Castedo M, Hirsch T, Marchetti P, Macho A, Kroemer G (1996) Bcl-2 inhibits the mitochondrial release of apoptogenic protease. J Exp Med 184: 1331–1341

Sweeney MI (1997) Neuroprotective effects of adenosine in cerebral ischemia: Window of opportunity. Neurosci Biobehav Rev 21:207-217

Tymianski M, Wallace MC, Spigelman I, Uno M, Carlen PL, Tator CH, Charlton MP (1993) Cell-permeant Ca^{2+} chelators reduce early excitotoxic and ischemic neuronal injury in vitro and in vivo. Neuron 11:221–235

Uemura Y, Miller JM, Matson WR, Beal MF (1991) Neurochemical analysis of focal ischemia in rats. Stroke 22:1548–1553

Vibulsreth S, Hefti F, Ginsberg MD, Dietrich WD, Busto R (1987) Astrocytes protect cultured neurons from degeneration induced by anoxia. Brain Res 422:303–311

Volterra A, Trotti D, Tromba C, Floridi S, Racagni G (1994) Glutamate uptake inhibition by oxygen free radicals in rat cortical astrocytes. J Neurosci 14:2924–2932

Watanabe Y, Gould E, McEwen BS (1992) Stress induces atrophy of apical dendrites of hippocampal CA3 pyramidal neurons. Brain Res 588:341–345

Weisbrot-Lefkowitz M, Reuhl K, Perry B, Chan PH, Inouye M, Mirochnitchenko O (1998) Overexpression of human glutathione peroxidase protects transgenic mice against focal cerebral ischemia/reperfusion damage. Mol Brain Res 53:334–339

Weiss JH, Hartley DM, Koh J, Choi DW (1990) The calcium channel blocker nifedipine attenuates slow excitatory amino acid neurotoxicity. Science 247:1474–1477

Welch WJ (1993) Heat shock proteins functioning as molecular chaperones: their roles in normal and stressed cells. Philos Trans R Soc Lond B Biol Sci 339:327–333

Werth JL, Thayer SA (1994) Mitochondria buffer physiological calcium loads in cultured rat dorsal root ganglion neurons. J Neurosci 14:348–356

Yang G, Chan PH, Chen J, Carlson E, Chen SF, Weinstein P, Epstein CJ, Kamii H (1994) Human copper-zinc superoxide dismutase transgenic mice are highly resistant to reperfusion injury after focal cerebral ischemia. Stroke 25:165–170

Yang CS, Lin NN, Tsai PJ, Liu L, Kuo JS (1996) In vivo evidence of hydroxyl radical formation induced by elevation of extracellular glutamate after cerebral ischemia in the cortex of anesthetized rats. Free Radic Biol Med 20:245–250

Yang J, Liu X, Bhalla K, Kim CN, Ibrado AM, Cai J, Peng TI, Jones DP, Wang X (1997) Prevention of apoptosis by Bcl-2: release of cytochrome C from mitochondria blocked. Science 275:1129–1132

Yoshino H, Hattori N, Urabe T, Uchida K, Tanaka M, Mizuno Y (1997) Postischemic accumulation of lipid peroxidation products in the rat brain: immunohistochemical detection of 4-hydroxy-2-nonenal modified proteins. Brain Res 767:81–86

Yu ZF, Bruce-Keller AJ, Goodman Y, Mattson MP (1998) Uric acid protects neurons against excitotoxic and metabolic insults in cell culture, and against focal ischemic brain injury in vivo. J Neurosci Res 53:613–625

Yu ZF, Guo Q, Luo H, Fu W, Mattson MP (1999) The endoplasmic reticulum protein GRP78 protects neurons against excitotoxicity and apoptosis: suppression of oxidative stress and stabilization of calcium homeostasis. Exp Neurol In press

Zamzami N, Marchetti P, Castedo M, Hirsch T, Susin SA, Masse B, Kroemer G (1996) Inhibitors of permeability transition interfere with the disruption of the mitochondrial transmembrane potential during apoptosis. FEBS Lett 384:53–57

Zhang Y, Tatsuno T, Carney J, Mattson MP (1993) Basic FGF, NGF, and IGFs protect hippocampal neurons against iron-induced degeneration. J Cerebral Blood Flow Metab 13:378–388

CHAPTER 6
Blocking Nitric Oxide Toxicity

M. Sasaki, T.M. Dawson, and V.L. Dawson

A. Introduction

Nitric oxide (NO) is a novel neuronal messenger molecule which can mediate rapid signaling, diffusing freely in three dimensions to act throughout local regions of neural tissue (Garthwaite and Boulton 1995; Yun et al. 1996). NO and other nitrogen oxides are derived from chemical reactions following NOS activation. NO production can regulate ion channel activity, neurotransmitter release, cerebral blood flow, synaptic plasticity, growth cone structure, and gene expression in the nervous system. NO can elicit such diverse cellular signaling due to a wide variety of molecular targets including soluble guanylate cyclase, other heme-containing enzymes including cyclooxygenase, thiol moieties and tyrosine residues on proteins, iron sulfur-containing proteins, and superoxide anion. Thus, NO is important in normal neuronal signaling; however, when NO production is unregulated or excessive, NO can mediate neuronal degeneration. NO has been implicated in the neuropathology of stroke, trauma, AIDS dementia, Alzheimer's disease, multiple sclerosis, as well as bacterial and viral encephalitis (Dawson and Snyder 1994; Meldrum 1996). Limiting NO production under pathologic conditions has become a potentially important therapeutic target for the treatment of various neurologic diseases and disorders.

B. Nitric Oxide Synthase

Nitric oxide synthase enzymatically converts l-arginine to l-citrulline and NO. This reaction requires flavoproteins, flavin adenine dinucleotide (FAD) and flavin mononucleotide (FNN), and tetrahydrobiopterin as cofactors (Bredt and Snyder 1994; Mayer et al. 1991). The reaction proceeds in the presence of l-arginine, oxygen, and nicotinamide adenine dinucleotide phosphate (NADPH). Although all three known isoforms of NOS require the same cofactors and substrates and dimerization for activity, there are significant differences in the regulatory mechanisms, biophysical properties (Table 1), and localization of each of the NOS isoforms.

Table 1. Physiologic roles for nitric oxide synthase/nitric oxide

Neuronal NOS	Endothelial NOS	Immunologic NOS
Neurotransmission	Vasodilation	Antimicrobial
Brain Development	Maintenance of cerebral blood flow	Antitumorigenic
Neuronal Plasticity (LTP, LTD)	Inibition of: Platelet Aggregation	Septic shock – contributes to severe vasodilation
Excitotoxicity	Mast Cell Degranulation	

I. Neuronal Nitric Oxide Synthase

Neuronal NOS (nNOS; Type 1) was first purified and cloned from rat brain (BREDT et al. 1990). With the exception of the cerebellum, where most granule cells express nNOS, in most brain regions nNOS is expressed in only 1-2% of the total neuronal population (DAWSON et al. 1991). nNOS accounts for the histochemical stain, NADPH diaphorase, when tissue is fixed in 4% paraformaldehyde. nNOS/NADPH diaphorase neurons are unique in that they are selectively resistant to a variety of neurodegenerative, neurologic, and excitotoxic injuries including stroke, Alzheimer's disease, Parkinson's disease, and glutamate and NMDA neurotoxicity (CHOI 1988; FERRANTE et al. 1985; HYMAN et al. 1992; KOH and CHOI 1988; KOH et al. 1986; UEMURA et al. 1990). Additionally, these neurons are unique in the density and diversity of their neuronal projections. Even though nNOS neurons comprise only 1%–2% of the total neuronal population, the extent of neuronal projections coupled with the diffusibility of NO has led to the suggestion that every neuron in the central nervous system could be affected and modulated by NO production. Following protein synthesis, nNOS monomers loosely dimerize. Upon the rapid influx of calcium and the subsequent engagement of calcium-calmodulin to nNOS, nNOS tightly dimerizes, and, in the presence of cofactors and substrates, NO is formed (MARLETTA 1994). In neurons, nNOS is predominantly activated by calcium influx through the N-methyl-D-aspartate (NMDA) receptor channel (CRAVEN and BREDT 1998). nNOS contains a PDZ-binding domain which allows it to couple to the postsynaptic density protein-95 (PSD-95) and PSD-93, which tightly couples nNOS to the NMDA receptor (BRENMAN et al. 1996). This physical coupling of nNOS to NMDA receptors likely accounts for the functional coupling of NMDA receptor activation with nNOS.

II. Endothelial Nitric Oxide Synthase

Endothelial NOS (eNOS, Type III) was first cloned from endothelial cells from the vasculature (MARSDEN et al. 1992). Additionally, eNOS is expressed in a small population of neurons in the central nervous system (CNS), including hippocampal pyramidal cells and dentate gyrus granule cells (DINERMAN et al.

1994; DOYLE and SLATER 1997). Vascular eNOS is the major regulator of blood flow (HUANG et al. 1995). eNOS, like nNOS is constitutively expressed and is a calcium dependent isoform which transiently produces NO following a rapid increase of intracellular calcium. The role of neuronally-expressed eNOS in the CNS is not entirely understood, but it may play an essential role in the induction and maintenance of long-term potentiation (KANTOR et al. 1996; SON et al. 1996). Vascular eNOS is critical in maintaining cerebral blood flow (DREIER et al. 1998). Inhibition of vascular eNOS, particularly under neuropathologic conditions, can have dramatic and drastic effects on neuronal survival.

III. Immunologic Nitric Oxide Synthase

Immunologic or inducible NOS (iNOS, Type II) was first cloned from macrophages and then hepatocytes (LOWENSTEIN et al. 1992; LYONS et al. 1992; XIE et al. 1992). iNOS message and protein are not detected in the healthy brain; however, following pathologic insults, iNOS can be expressed in the brain. iNOS has been detected following ischemic insult (COEROLI et al. 1998; GRANDATI et al. 1997; HIGUCHI et al. 1998), trauma (HANDY and MOORE 1998), viral infection (FENYK-MELODY et al. 1998), bacterial infection (ARKOVITZ et al. 1996; BUNE et al. 1996), and other immunologic challenges. Expression of iNOS is observed predominantly in microglia and astrocytes (MURPHY et al. 1993). iNOS is considered the high-output isoform of NO synthase. Following protein synthesis, iNOS monomers assemble to form a functional dimer. Sustained catalysis results from binding calmodulin without dependence on elevated calcium levels. This ability to bind calmodulin independently of calcium requires a unique protein domain between amino acids 484 to 726, in addition to the canonical calmodulin binding domain present in all NOS isoforms (RUAN et al. 1996). Regulation of iNOS occurs primarily at the transcriptional and translational level (LOWENSTEIN et al. 1993; XIE et al. 1993). Once the transcriptional signal has been eliminated, iNOS message and protein are rapidly degraded.

C. Inhibitors of Nitric Oxide Synthase

Coordination of both heme moieties in each NOS monomer is required for appropriate and functional binding of L-arginine and catalysis, which leads to an overall 5-electron oxidation of the guanidinium nitrogen of L-arginine, and production of L-citrulline and NO (ABU-SOUD et al. 1994). A common strategy to inhibit NOS catalytic activity is to design chemical reagents which can effectively compete with L-arginine for the arginine-binding site in NOS. Chemical classes of such inhibitors include the arginine analogs, thiocitrullines, indazoles, substituted guanidoamines, and thioureas. These agents are effective inhibitors of NOS and are competitive because excess L-arginine can effectively compete with them and activate NOS. Unfortunately, the prototype NOS inhibitors are nonselective; they inhibit all three NOS isoforms effectively, which has rendered them essentially useless for in vivo studies and most clinical applications. Efforts have been devoted to generating more specific and selective NOS inhibitors (BERKA et al. 1998) (Table 2).

I. L-Arginine Site

The best known and most widely used NOS inhibitors are the L-arginine analogs which include, but are not limited to, nitro-L-arginine, methyl-arginine, iminoethyl-L-ornithine, nitro-L-arginine methyl ester, and methyl-L-arginine acetate salt (BABU and GRIFFITH 1998; RESINK et al. 1996). Most arginine analog inhibitors of NOS are generated by substituting the guanidino nitrogen of arginine. Two arginine analogs which do not have this guanidino modification are α-guanidinoglutaric acid and arcaine, which inhibit NOS activity at equivalent potency to monomethyl-L-arginine (KABUTO et al. 1995; YOKOI et al. 1994). All these agents are general NOS inhibitors with minimal specificity for the three isoforms. The administration of these agents can result in systemic hypertension and potent cerebrovasoconstriction. Even low concentrations of these agents, which do not induce general systemic hypertension, can reduce cerebral blood flow by over 80%. These dramatic effects on cerebral blood flow effectively limit the possibility of using these agents for the therapeutic treatment of human diseases. Additionally, these deleterious effects on cerebral blood flow also limit the use of these agents in vivo. Decreased cerebral blood flow will dramatically affect most experimental outcome measures of neuronal function. Therefore, when designing experiments to investigate the role of NO in CNS function, one should use isoform selective NOS inhibitors.

1. Thiocitrullines

Another class of agents which occupy the L-arginine binding site are the thiocitrullines. These agents are less well-known and not used as widely as the arginine analogs, but have many of the same features. These agents nonspecifically inhibit all isoforms of NOS, and this NOS inhibition is reversible by excess L-arginine. The prototype agent is L-thiocitrulline, which inhibits NOS, in part, by blocking electron flow through the heme-iron (ABU-SOUD et al. 1994). Other thiocitrullines inhibit NOS simply by occupying the L-arginine catalytic site. The most potent thiocitrulline currently reported is S-alkyl-L-thiocitrulline (ARKOVITZ et al. 1996; NARAYANAN et al. 1995). The proposed use of these agents is for treatment of hypotension due to the overproduction of nitric oxide in conditions such as septic shock (ROSSELET et al. 1998).

2. Nitro-Indazole

A considerable amount of resources and effort have gone into the development of isoform-selective NOS inhibitors. 7-Nitro-indazole (7-NI) was the first

Table 2. NOS inhibitors

Compound	Selectivity
Arginine analogs	
Nitro-L-arginine	nNOS = eNOS = iNOS
Methyl-arginine	nNOS = eNOS = iNOS
Nitro-L-arginine methyl ester	nNOS = eNOS = iNOS
Methyl-L-arginine acetate salt	nNOS = eNOS = iNOS
Nitro-diamino-L-arginine	nNOS = eNOS = iNOS
Nitro-dimethyl-L-ariginine	nNOS = eNOS = iNOS
Iminoethyl-L-ornithine (NIO)	nNOS = eNOS = iNOS
Methyl-L-Iminoethyl-L-ornithine	nNOS = eNOS = iNOS
Ethyl-L-Iminoethyl-L-ornithine	nNOS = eNOS = iNOS
Vinyl-iminoethyl-L-ornithine	nNOS > eNOS >> iNOS
Iminoethyl-L-lysine	iNOS > nNOS, eNOS
Thiocitrullines	
L-Thiocitrulline	nNOS = eNOS = iNOS
S-Methyl-thiocitrulline	nNOS = eNOS = iNOS
S-Ethyl-thiocitrulline	nNOS = eNOS = iNOS
Indazoles	
7-Nitro-indazole	nNOS > eNOS, iNOS
6-Nitro-indazole	nNOS > eNOS, iNOS
5-Nitro-indazole	nNOS > eNOS, iNOS
3-Chloro-indazole	nNOS > eNOS, iNOS
3-Chloro 5-nitro-indazole	nNOS > eNOS, iNOS
Indazole	nNOS > eNOS, iNOS
Substituted guanidines	
Aminoguanidine	iNOS > eNOS, nNOS
2-Methyl-aminoguanidine	iNOS > eNOS, nNOS
2-Ethyl-aminoguanidine	iNOS > eNOS, nNOS
2-Hydroxy-aminoguanidine	iNOS > eNOS, nNOS
Thioureas	
S-Methyl-isothiourea	iNOS > eNOS, nNOS
S-(2-aminoethyl)-isothiourea	iNOS > eNOS, nNOS
S-Ethyl-isothiourea	iNOS > eNOS, nNOS
S-Isopropyl-isothiourea	iNOS > eNOS, nNOS
SS-(phenylenebis (1,2-ethanediyl))-bisisothiourea	iNOS > nNOS >>> eNOS
S-Ethyl-N-phenyl-isothiourea	nNOS = eNOS = iNOS
2-Aminoethylisothiouronium	nNOS = eNOS = iNOS
N-Phenyl-isothiourea 39	nNOS > eNOS, iNOS
Imidazole	
1-(2-trifluoromethylphenyl) imidazole (TRIM)	nNOS, iNOS >> eNOS
Noncatalytic site inhibitors	
Methotrexate	nNOS = eNOS = iNOS
4-Amino tetrahydrobiopterin	nNOS = eNOS = iNOS
Epigallocatechin gallate	nNOS = eNOS = iNOS
Diphenyleneiodonium	nNOS = eNOS = iNOS
Calmidazolium	nNOS = eNOS
W7	nNOS = eNOS
FK506	nNOS = eNOS
ARL-17477	NNOS >> iNOS = eNOS
A-84643	nNOS
2-Iminopiperidine	iNOS >>eNOS
1400W	iNOS >>> nNOS >>>> eNOS

relatively selective nNOS inhibitor developed (BABBEDGE et al. 1993). Although in vitro 7-NI inhibits all three isoforms with equal affinity, it has a marked selectivity for inhibition of nNOS over eNOS in vivo and thus has minimal vascular effects. 7-NI has been extremely useful in investigating the role of nNOS in a variety of experimental paradigms (TOKIME et al. 1998). 7-NI provides neuroprotection in in vivo models of neurotoxicity. In experimental stroke models of permanent focal ischemia (DALKARA et al. 1994; NANRI et al. 1998), transient focal ischemia (CHALIMONIUK et al. 1996; CHALIMONIUK and STROSZNAJDER 1998; COERT et al. 1999; KAMII et al. 1996), and thromboembolic stroke (STAGLIANO et al. 1997), administration of 7-NI reduces infarct volume. 7-NI also reduces the contusion volume injury following experimental traumatic brain injury induced by parasagittal fluid-percussion. Lesion volume induced by intrastriatal administration of excitotoxins and mitochondrial toxins, and 1-methyl-4-phenyl-1,2,3,6-tetra-hydropyridine (MPTP)-induced parkinsonism (PRZEDBORSKI et al. 1996; SCHULZ et al. 1995; SCHULZ et al. 1995) are also reduced with administration of 7-NI. The effects of 7-NI can be competed, in part, by excess L-arginine, indicating that its effects are in part through disruption of arginine binding. Additionally, 7-NI interferes with the binding of the essential cofactor, tetrahydrobiopterin (MAYER et al. 1994). Other indazole analogs also inhibit nNOS in vitro and include in rank order of potency: 6-nitroindazole >5-nitroindazole >3-chloroindazole >3-chloro 5-nitroindazole >, and indazole (BABBEDGE et al. 1993). In vivo, 7-NI, 6-nitroindazole and indazole produce a dose-dependent antinociception and fail to influence mean arterial pressure (HANDY and MOORE 1998; MOORE et al. 1993). These observations are consistent with inhibition of nNOS and a lack of effect on eNOS activity.

3. Substituted Guanidines

The substituted guanidines have been investigated for their potential specific activity on the iNOS isoform over nNOS or eNOS. Aminoguanidine is a prototype agent in this class. Aminoguanidine inhibition of iNOS can be completely reversed with L-arginine. Since the discovery that aminoguanidine has a relative selectivity for iNOS, it has been used extensively in vivo (ARKOVITZ et al. 1996; HIGUCHI et al. 1998; ZHANG and IADECOLA 1998). However, caution must be used when designing experiments using aminoguanidine. Before its discovery as an iNOS inhibitor, aminoguanidine was used as a diamine oxidase inhibitor, polyamine oxidase inhibitor, general oxidase inhibitor, ribonucleotide reductase inhibitor, decarboxylase inhibitor, catalase inhibitor, and an inhibitor of advanced glycosylation end products. Other substituted guanidines are also under development and have been investigated (SZABO et al. 1996; WOLFF and LUBESKIE 1996). The most potent of these are 1-amino-2-hydroxy-guanidine (RUETTEN et al. 1996) and 2-ethyl-amino guanidine (WOLFF et al. 1997). The potential metabolic effects of these agents are currently unknown.

4. Isothioureas

The isothioureas in general are potent and selective inhibitors of iNOS. These novel, nonarginine-based compounds have been used extensively to study the role of iNOS produced NO in the pathophysiology of inflammation (JANG et al. 1996) and circulatory shock (ARKOVITZ et al. 1996; ROSSELET et al. 1998; RUETTEN et al. 1996; SAETRE et al. 1998; THIEMERMANN et al. 1995). The S-substituted isothioureas are 8–24 times more potent than *N*-methyl-L-arginine and 200 times more potent than nitro-L-arginine in inhibiting iNOS activity in vitro (HRABAK et al. 1997). This may be due in part to their lack of arginase inhibition. Nonspecific arginine analog NOS inhibitors also have the capacity to inhibit the enzyme arginase, which degrades L-arginine. Inhibition of arginine degradation would effectively increase arginine concentrations which could compete for the NOS catalytic site. Isothioureas are not arginase inhibitors and therefore, do not affect cellular concentrations of L-arginine. These compounds may be of more practical use for inhibiting NO production in cells which simultaneously express iNOS and arginase. While investigating the isothiourea compounds, a novel *S*-ethyl-*N*-phenyl-isothiourea was found to be a potent inhibitor of all NOS isoforms. Further investigation identified an analog with selective nNOS activity, *N*-phenyl-isothiourea-39 (SHEARER et al. 1997). This agent competes with L-arginine to elicit nNOS inhibition. These studies have been conducted in vitro and the usefulness of this agent in vivo has yet to be determined.

5. Imidazole

Imidazoles can interact with heme proteins. A novel imidazole, 1-(2-trifluoromethylphenyl) imidazole (TRIM) is a potent inhibitor of both neuronal and inducible NOS. In contrast, TRIM is a poor inhibitor of eNOS (HANDY et al. 1996). TRIM competes with L-arginine substrate binding on NOS and interferes with the binding of tetrahydrobiopterin to NOS (HANDY and MOORE 1997). TRIM has been shown to have antinociceptive effects in vivo without affecting systemic blood pressure (HANDY et al. 1995). Additionally, TRIM reduces infarct volume after transient focal cerebral ischemia in the rat (ESCOTT et al. 1998).

II. Tetrahydrobiopterin

Tetrahydrobiopterin is best known as a cofactor of enzymes which hydroxylate aromatic amino acids (MAYER and WERNER 1995). However, tetrahydrobiopterin is also an essential cofactor in the biosynthesis of NO from NOS. The role of tetrahydrobiopterin in NO synthesis has been elusive, but current experimental evidence suggests that tetrahydrobiopterin is involved both as an allosteric modulator of NOS, as well as a reactant in the L-arginine oxidation reaction. Competition for the biopterin binding site on NOS can afford effective inhibition of NOS catalytic activity (PFEIFFER et al. 1997). Agents such as methotrexate (ROBBINS et al. 1998), 4-amino tetrahydrobiopterin (an analog of tetrahydrobiopterin and epigallocatechin gallate) compete for the biopterin binding site (CHAN et al. 1997). Additionally, many agents which compete for the arginine binding site such as imidazole (BERKA et al. 1998), indazoles and substituted aminoguanidines also prevent tetrahydrobiopterin binding (DICK and LEFEBVRE 1997; KLATT et al. 1994). It is not clear whether these latter agents directly compete for the tetrahydrobiopterin binding site or whether the site is structurally located near the arginine binding site and so these agents functionally inhibit both sites. Additionally, NOS catalytic activity can be prevented by inhibition of tetrahydrobiopterin's synthesis thereby limiting the availability of this essential cofactor for NOS catalytic activity (BUNE et al. 1996; KILBOURN and GRIFFITH 1992; KINOSHITA et al. 1997).

III. Flavoprotein Inhibition

NOS stoichiometrically binds flavin mononucleotide (FMN) and flavin adenine dinucleotide (FAD). The flavoproteins are essential in the shuttling of the electrons necessary for the oxidation of L-arginine. Flavoprotein inhibitors can block this electron shuttling and functionally inhibit NOS catalytic activity. Diphenyleneiodonium (DPI) is a prototype flavoprotein inhibitor which effectively inhibits NOS catalytic activity both in vivo and in vitro (STUEHR et al. 1991). DPI is effective in reducing glutamate and NMDA neurotoxicity in vitro, DPI inhibits relaxation of aortic smooth muscle in vitro (DAWSON et al. 1993; DODD-O et al. 1997), and produces pressor and tachycardic effects in vivo (WANG and PANG 1993). Although potent and effective, DPI is not a candidate for therapeutic development because inhibition of flavoproteins would eventually inhibit mitochondrial respiration, resulting in a negative effect in vivo.

IV. NADPH Oxidase Inhibition

NADPH is an essential cofactor for NOS catalytic activity (MAYER et al. 1991). It is an essential source of electrons necessary for the conversion of L-arginine to NO. Agents which interfere with the NADPH oxidase activity of nNOS are effective in inhibiting NOS catalytic activity (KUMAGAI et al. 1998; MAYER and WERNER 1995). The class of porphyrins are effective inhibitors of NOS at the NADPH oxidase site. Agents in this class include protoporphyrin IX and zinc protoporphyrin IX (WOLFF et al. 1996).

V. Calmodulin Inhibition

Calcium/calmodulin binding to NOS is essential for tight dimerization in eNOS and nNOS, and for shuttling of electrons in all NOS isoforms. Calmodulin inhibitors, such as calmidazolium and W-7, effectively inhibit NOS catalytic activity (WIN et al. 1996). These agents are protective against glutamate and NMDA-mediated neurotoxicity (DAWSON et al. 1993).

VI. Phosphorylation Sites

NOS has multiple consensus sites for phosphorylation. In vitro, NOS can be phosphorylated and catalytic activity reduced by protein kinase C, calcium calmodulin kinase 2, protein kinase A, and protein kinase G (BREDT et al. 1992; DINERMAN et al. 1994). NOS is de-phosphorylated by calcineurin. Inhibition of calcineurin can enhance phosphorylation of NOS, maintaining NOS in a catalytically inactive state (DAWSON et al. 1993). The immunosuppressant FK506 (tacrolimus) binds to FKBP, which subsequently binds to calcineurin and inhibits the phosphatase activity. In vitro and in cell culture, FK506 is effective in enhancing phosphorylation of nNOS and inhibiting its catalytic activity. In in vitro primary neuronal cell culture, FK506 is a potent inhibitor of glutamate and NMDA neurotoxicity, in part through inhibition of nNOS (DAWSON et al. 1993). However, the degree of neuroprotection afforded by FK506 in this excitotoxic model exceeded that observed with NOS inhibitors and with genetic deletion of nNOS. In cultures from nNOS knockout mice, FK506 still elicits neuroprotection, indicating that there are other cellular targets of FK506 which mediate neuroprotection against excitotoxicity (DAWSON et al. 1996). In vivo, FK506 is potently protective against various models of cerebral ischemia (SHARKEY and BUTCHER 1994). Originally, this neuroprotection was attributed to inhibition of nNOS; however, recent studies indicate that FK506 does not effectively inhibit nNOS in vivo (TOUNG et al. 1999). Therefore, the reductions in infarct volume observed in vivo in various models of cerebral ischemia must be due to inhibition of other calcineurin targets or other targets of the FK506/FKBP12 complex.

VII. Miscellaneous Inhibitors of NOS

There are numerous reports in the literature of agents which inhibit NOS catalytic activity but for which the mechanism is not yet described (IADECOLA 1997; MONCADA and HIGGS 1995). These include the selective nNOS inhibitors ARL-17477 and A-84643 (ZHANG et al. 1996). These agents potently inhibit nNOS catalytic activity and dramatically decrease infarct volume in animal models of transient focal cerebral ischemia. The rat hypothalamus secretes a substance in response to increased salt, termed hypothalamic hypertensive factor, which shares common chemical features to the substituted guanidines. Hypothalamic hypertensive factor is an effective endogenous inhibitor of NOS catalytic activity. It may act in opposition to vascular NO and be involved in the rise of arterial pressure (MORRIS et al. 1997). Analogs of 2-iminopiperidine are reported as potent and selective inhibitors of iNOS catalytic activity with respect to eNOS catalytic activity (HANSEN et al. 1998; WEBBER et al. 1998). Their effect on nNOS is not known.

D. Biochemical Regulators of NOS Catalytic Activity

nNOS activation is coupled to NMDA receptor mediated calcium influx and is thought to be due to a physical coupling of nNOS and the receptor by an intermediary adaptor protein, PSD95, through interaction of the PDZ domains present on the NMDA receptor, PSD95, and nNOS (CRAVEN and BREDT 1998). Disruption of the nNOS association with PSD95 by small peptides prevents NMDA receptor mediated activation of nNOS. Under these conditions calcium influx through the NMDA receptor is not altered. Recently, a novel nNOS-associated protein, which is highly enriched in brain and colocalizes with nNOS, was determined to interact with the nNOS/PDZ domain (JAFFREY et al. 1998). This protein, termed CAPON, competes with PSD95 for interaction with nNOS. Overexpression of CAPON results in a loss of PSD95/nNOS complexes. Since disruption of the nNOS PSD95 interaction results in inactivation of NOS in response to NMDA receptor-mediated calcium influx, it is likely that CAPON regulates the activity of nNOS in response to NMDA receptor activation by regulating its ability to associate with the NMDA receptor via PSD95 (JAFFREY et al. 1998).

Another protein inhibitor of nNOS has been identified and designated protein inhibitor of NOS (PIN) (JAFFREY and SNYDER 1996). PIN is one of the most conserved proteins in nature with a 92% amino acid identity with the nematode and rat homologs. Binding of PIN destabilizes the nNOS dimer, resulting in inactivation of nNOS catalytic activity. The physiologic activity of PIN is not yet known, but it is possible that it regulates numerous biologic processes by regulating nNOS catalytic activity (GILLARDON et al. 1998).

E. Nitric Oxide and Neurologic Disease

I. Excitotoxicity and Experimental Stroke

A role for NO in neurotoxicity was first described in an in vitro model of focal ischemia (DAWSON et al. 1991). In primary cortical cultures, a 5min exposure to glutamate or NMDA induces a series of events which results in neuronal cell death 12–18h later. Nonselective and selective NOS inhibitors provide potent neuroprotection in a dose-dependent manner. Additionally, cultures generated from nNOS knockout mice are resistant to the neurotoxic effects of glutamate and NMDA (DAWSON et al. 1996). NO donors elicit neurotoxicity, which develops over a similar time course as that of glutamate and NMDA neurotoxicity. These initial observations have been replicated in in vivo models of cerebral ischemia (CAGGIANO and KRAIG 1998; NANRI et al. 1998; NOWICKI et al. 1991; ONITSUKA et al. 1998; WEI and QUAST 1998). Confusion and controversy was generated in the literature through the use of nonselective NOS inhibitors, which effectively inhibited cerebral blood flow in addition to inhibiting nNOS. Through the generation of selective NOS inhibitors and the generation of knockout mice to nNOS, eNOS or iNOS, the role of NO generated from each of these isoforms in the pathogenesis of cerebral ischemia has been revealed. NO generated from nNOS immediately triggered by the ischemic event sets in motion a neurotoxic cascade through the generation of peroxynitrite. NO derived from eNOS is critical in maintaining cerebral blood flow (ENDRES et al. 1998; HUANG et al. 1995). Any inhibition of eNOS results in dramatic increases in infarct volume (DREIER et al. 1998). Twenty-four hours after the ischemic insult, iNOS is induced and mediates secondary neuronal damage (GRANDATI et al. 1997). Inhibition of iNOS or genetic knockout of iNOS results in decreased infarct volume following cerebral ischemia (NAGAYAMA et al. 1998). Therefore, in stroke nNOS mediates early neuronal toxicity, iNOS mediates late phase neurotoxicity, and eNOS is neuroprotective by maintaining cerebral blood flow.

II. Parkinson's Disease

Parkinson's disease is a movement disorder characterized by the selective loss of dopamine neurons in the substantia nigra which project to the striatum. Parkinsonism can be induced in mice by the toxin, 1-methyl-4-phenyl-1,2,3,6-tetrahydropyridine (MPTP). MPTP is converted to MPP$^+$ and taken up through the dopamine transporter into substantia nigra neurons. MPP$^+$ inhibits mitochondrial respiration resulting in decreased ATP production and increased superoxide anion formation (PRZEDBORSKI et al. 1992). Under this setting the increased superoxide anion formation is permissive for the formation of peroxynitrite from endogenously produced NO (PRZEDBORSKI et al. 1996). Peroxynitrite can elicit a feed forward cycle by damaging MnSOD activity and further inhibiting other enzymes in the mitochondrial respiratory chain thus increasing the production of superoxide anion and therefore, the production of peroxynitrite (GONZALEZ-ZULUETA et al. 1998; MACMILLAN-CROW et al. 1996; RADI et al. 1991). In the mouse model of parkinsonism, the selective nNOS inhibitor 7-NI prevents the loss of dopamine and its metabolites (SCHULZ et al. 1995). These results are also observed in the primate model exposed to MPTP (HANTRAYE et al. 1996) and have been confirmed in the nNOS knockout mice, which are also resistant to the effects of MPTP (PRZEDBORSKI et al. 1996).

III. HIV Dementia

It is estimated that 15% of human immunodeficiency virus (HIV)-infected patients who have progressed to acquired immunodeficiency syndrome (AIDS) are affected by a progressive dementing illness termed HIV dementia (NAVIA et al. 1986; PRICE 1996; WILEY and ACHIM 1995). Since HIV does not directly infect or damage neurons, alternative mechanisms have been sought to explain HIV dementia (EPSTEIN and GENDELMAN 1993; GENIS et al. 1992). Recently expression of iNOS was found to correlate with the severity and the rate of progression of HIV dementia (ADAMSON et al. 1999; ADAMSON et al.

1996). Additionally, expression of the HIV coat protein, gp41, also correlates with the severity and the rate of progression of HIV dementia (ADAMSON et al. 1999; ADAMSON et al. 1996). In in vitro rodent cortical cultures, HIV gp41 induces expression of iNOS over the course of 72 h, which results in NO mediated neurotoxicity observed five days following exposure to gp41, reaching maximal neurotoxicity at seven days postexposure (ADAMSON et al. 1999; ADAMSON et al. 1996). It is possible that gp41 expressed in HIV-infected cells stimulates local microglia and astrocytes to express iNOS and that iNOS production of NO contributes significantly to the neuronal damage and severity of HIV dementia. In vitro nonspecific NOS inhibitors can completely inhibit gp41/iNOS-induced neurotoxicity (ADAMSON et al. 1999; ADAMSON et al. 1996). However, in vivo iNOS is a potent inhibitor of viral replication. In patients infected with HIV, inhibition of iNOS may not be the first choice of therapeutic strategies as this may result in increased viral replication and an overall deleterious effect.

IV. Multiple Sclerosis

In multiple sclerosis, expression of iNOS can be found in active lesions (Bo et al. 1994). Since NO can injure oligodendrocytes and result in demyelination, it is thought that expression of iNOS may significantly contribute to the pathogenesis of multiple sclerosis (MERRILL et al. 1993). In an animal model of MS, experimental allergic encephalomyelitis, inhibitors of iNOS, as well as iNOS knockout mice, have been examined (FENYK-MELODY et al. 1998). In this animal model, inhibition of iNOS or genetic knockout of iNOS greatly exacerbates the progression of experimental autoimmune encephalomyelitis. It appears that this effect is, in part, due to timing. Production of NO plays a protective role in the induction of an autoimmune response to the injection of myelin basic protein (COWDEN et al. 1998). However, inhibition of iNOS, both pharmacologically and through antisense strategies during the progressive stage of EAE, can inhibit the progression (OKUDA et al. 1998). Therefore, timing of administration of iNOS inhibitors is important in EAE and may very well be important in the treatment of multiple sclerosis.

F. Conclusions

Nitric oxide is an important neuronal messenger molecule. It regulates synaptic transmission, neurobehavior, and synaptic plasticity. However, when produced in excess under oxidizing conditions, NO, largely through the production of peroxynitrite, elicits neurotoxicity. Pathologic conditions which overstimulate nNOS or induce expression of iNOS in the CNS can result in significant neuronal damage. Strategies which target limiting the production of NO under these conditions have been shown to be neuroprotective both in vitro and in vivo. The limitation in carrying these agents forward to therapeutic use in humans revolves around the need for isoform specific NOS inhibitors which do not have deleterious side effects.

Acknowledgements. This work was supported by US PHS Grants NS33277 and NS37090 (TMD) and NS33142 and NS37460 (VLD). TMD is an Established Investigator of the American Heart Association and is supported by the Paul Beeson Faculty Scholar Award in Aging Research. VLD is a Virginia Davids Scholar, and a Staglin Music Festival NARSAD Investigator. Under an agreement between the John Hopkins University and Guilford Pharmaceuticals, TMD and VLD are entitled to a share of sales royalty received by the University from Guilford. TMD and the University also own Guilford stock, and the University stock is subject to certain restrictions under University policy. The terms of this arrangement are being managed by the University in accordance with its conflict of interest policies.

References

Abu-Soud HM, Feldman PL, Clark P, Stuehr DJ (1994) Electron transfer in the nitric-oxide synthases. Characterization of L-arginine analogs that block heme iron reduction. J Biol Chem 269:32318–32326

Adamson DC, Kopnisky KL, Dawson TM, Dawson VL (1999) Mechanisms and Structural Determinants of HIV-1 Coat Protein gp41-Induced Neurotoxicity. J Neurosci 19:64–71

Adamson DC, McArthur JC, Dawson TM, Dawson VL (1999) Course of HIV-Associated Dementia: Correlations with gp41, iNOS and Macrophage/Microglial Activation. Mol Med in press:

Adamson DC, Wildemann B, Sasaki M, Glass JD, McArthur JC, Christov VI, Dawson TM, Dawson VL (1996) Immunologic NO synthase: elevation in severe AIDS dementia and induction by HIV-1 gp41. Science 274:1917–1921

Arkovitz MS, Wispe JR, Garcia VF, Szabo C (1996) Selective inhibition of the inducible isoform of nitric oxide synthase prevents pulmonary transvascular flux during acute endotoxemia. J Pediatr Surg 31:1009–1015

Babbedge RC, Bland-Ward PA, Hart SL, Moore PK (1993) Inhibition of rat cerebellar nitric oxide synthase by 7-nitro indazole and related substituted indazoles. Br J Pharmacol 110:225–228

Babu BR, Griffith OW (1998) N5-(1-Imino-3-butenyl)-L-ornithine. A neuronal isoform selective mechanism-based inactivator of nitric oxide synthase. J Biol Chem 273:8882–8889

Berka V, Palmer G, Chen PF, Tsai AL (1998) Effects of various imidazole ligands on heme conformation in endothelial nitric oxide synthase. Biochemistry 37:6136–6144

Bo L, Dawson TM, Wesselingh S, Mork S, Choi S, Kong PA, Hanley D, Trapp BD (1994) Induction of nitric oxide synthase in demyelinating regions of multiple sclerosis brains. Ann Neurol 36:778–786

Bredt DS, Ferris CD, Snyder SH (1992) Nitric oxide synthase regulatory sites. Phosphorylation by cyclic AMP-dependent protein kinase, protein kinase C, and calcium/calmodulin protein kinase; identification of flavin and calmodulin binding sites. J Biol Chem 267:10976–10981

Bredt DS, Hwang PM, Snyder SH (1990) Localization of nitric oxide synthase indicating a neural role for nitric oxide. Nature 347:768–770

Bredt DS, Snyder SH (1994) Nitric oxide: a physiologic messenger molecule. Annu Rev Biochem 63:175–195

Brenman JE, Christopherson KS, Craven SE, McGee AW, Bredt DS (1996) Cloning and characterization of postsynaptic density 93, a nitric oxide synthase interacting protein. J Neurosci 16:7407–7415

Bune AJ, Brand MP, Heales SJ, Shergill JK, Cammack R, Cook HT (1996) Inhibition of tetrahydrobiopterin synthesis reduces in vivo nitric oxide production in experimental endotoxic shock. Biochem Biophys Res Commun 220:13–19

Caggiano AO, Kraig RP (1998) Neuronal nitric oxide synthase expression is induced in neocortical astrocytes after spreading depression. J Cereb Blood Flow Metab 18:75–87

Chalimoniuk M, Glód B, Strosznajder J (1996) NMDA receptor mediated nitric oxide dependent cGMP synthesis in brain cortex and hippocampus. Effect of ischemia on NO related biochemical processes during reperfusion. Neurologia I Neurochirurgia Polska 30 Suppl 2:65–84

Chalimoniuk M, Strosznajder J (1998) NMDA receptor-dependent nitric oxide and cGMP synthesis in brain hemispheres and cerebellum during reperfusion after transient forebrain ischemia in gerbils: effect of 7-Nitroindazole. Journal Of Neuroscience Research 54:681–690

Chan MM, Fong D, Ho CT, Huang HI (1997) Inhibition of inducible nitric oxide synthase gene expression and enzyme activity by epigallocatechin gallate, a natural product from green tea. Biochem Pharmacol 54:1281–1286

Choi DW (1988) Glutamate neurotoxicity and diseases of the nervous system. Neuron 1:623–634

Coeroli L, Renolleau S, Arnaud S, Plotkine D, Cachin N, Plotkine M, Ben-Ari Y, Charriaut-Marlangue C (1998) Nitric oxide production and perivascular tyrosine nitration following focal ischemia in neonatal rat. J Neurochem 70:2516–2525

Coert BA, Anderson RE, Meyer FB (1999) A comparative study of the effects of two nitric oxide synthase inhibitors and two nitric oxide donors on temporary focal cerebral ischemia in the Wistar rat. Journal Of Neurosurgery 90:332–338

Cowden WB, Cullen FA, Staykova MA, Willenborg DO (1998) Nitric oxide is a potential down-regulating molecule in autoimmune disease: inhibition of nitric oxide production renders PVG rats highly susceptible to EAE. J Neuroimmunol 88:1–8

Craven SE, Bredt DS (1998) PDZ proteins organize synaptic signaling pathways. Cell 93:495–498

Dalkara T, Yoshida T, Irikura K, Moskowitz MA (1994) Dual role of nitric oxide in focal cerebral ischemia. Neuropharmacology 33:1447–1452

Dawson TM, Bredt DS, Fotuhi M, Hwang PM, Snyder SH (1991) Nitric oxide synthase and neuronal NADPH diaphorase are identical in brain and peripheral tissues. Proc Natl Acad Sci USA 88:7797–7801

Dawson TM, Snyder SH (1994) Gases as biological messengers: nitric oxide and carbon monoxide in the brain. J Neurosci 14:5147–5159

Dawson TM, Steiner JP, Dawson VL, Dinerman JL, Uhl GR, Snyder SH (1993) Immunosuppressant FK506 enhances phosphorylation of nitric oxide synthase and protects against glutamate neurotoxicity. Proc Natl Acad Sci. USA 90:9808–9812

Dawson VL, Dawson TM, Bartley DA, Uhl GR, Snyder SH (1993) Mechanisms of nitric oxide-mediated neurotoxicity in primary brain cultures. J Neurosci 13:2651–2661

Dawson VL, Dawson TM, London ED, Bredt DS, Snyder SH (1991) Nitric oxide mediates glutamate neurotoxicity in primary cortical cultures. Proceedings of the National Academy of Sciences of the United States of America 88:6368–6371

Dawson VL, Kizushi VM, Huang PL, Snyder SH, Dawson TM (1996) Resistance to neurotoxicity in cortical cultures from neuronal nitric oxide synthase-deficient mice. J Neurosci 16:2479–2487

Dick JM, Lefebvre RA (1997) Influence of different classes of NO synthase inhibitors in the pig gastric fundus. Naunyn Schmiedebergs Arch Pharmacol 356:488–494

Dinerman JL, Dawson TM, Schell MJ, Snowman A, Snyder SH (1994) Endothelial nitric oxide synthase localized to hippocampal pyramidal cells: implications for synaptic plasticity. Proc Natl Acad Sci USA 91:4214–4218

Dinerman JL, Steiner JP, Dawson TM, Dawson V, Snyder SH (1994) Cyclic nucleotide dependent phosphorylation of neuronal nitric oxide synthase inhibits catalytic activity. Neuropharmacology 33:1245–1251

Dodd-o JM, Zheng G, Silverman HS, Lakatta EG, Ziegelstein RC (1997) Endothelium-independent relaxation of aortic rings by the nitric oxide synthase inhibitor diphenyleneiodonium. Br J Pharmacol 120:857–864

Doyle CA, Slater P (1997) Localization of neuronal and endothelial nitric oxide synthase isoforms in human hippocampus. Neuroscience 76:387–395

Dreier JP, Korner K, Ebert N, Gorner A, Rubin I, Back T, Lindauer U, Wolf T, Villringer A, Einhaupl KM, Lauritzen M, Dirnagl U (1998) Nitric oxide scavenging by hemoglobin or nitric oxide synthase inhibition by N-nitro-L-arginine induces cortical spreading ischemia when K+ is increased in the subarachnoid space. J Cereb Blood Flow Metab 18:978–990

Endres M, Laufs U, Huang Z, Nakamura T, Huang P, Moskowitz MA, Liao JK (1998) Stroke protection by 3-hydroxy-3-methylglutaryl (HMG)-CoA reductase inhibitors mediated by endothelial nitric oxide synthase. Proc Natl Acad Sci USA 95:8880–8885

Epstein LG, Gendelman HE (1993) Human immunodeficiency virus type 1 infection of the nervous system: pathogenetic mechanisms. Ann Neurol 33:429–436

Escott KJ, Beech JS, Haga KK, Williams SC, Meldrum BS, Bath PM (1998) Cerebroprotective effect of the nitric oxide synthase inhibitors, 1-(2-trifluoromethylphenyl) imidazole and 7-nitro indazole, after transient focal cerebral ischemia in the rat. J Cereb Blood Flow Metab 18:281–287

Fenyk-Melody JE, Garrison AE, Brunnert SR, Weidner JR, Shen F, Shelton BA, Mudgett JS (1998) Experimental autoimmune encephalomyelitis is exacerbated in mice lacking the NOS2 gene. J Immunol 160:2940–2946

Ferrante RJ, Kowall NW, Beal MF, Richardson EP, Jr, Bird ED, Martin JB (1985) Selective sparing of a class of striatal neurons in Huntington's disease. Science 230:561–563

Garthwaite J, Boulton CL (1995) Nitric oxide signaling in the central nervous system. Annu. Rev. Physiol. 57:683–706

Genis P, Jett M, Bernton EW, Boyle T, Gelbard HA, Dzenko K, Keane RW, Resnick L, Mizrachi Y, Volsky DJ, et al. (1992) Cytokines and arachidonic metabolites produced during human immunodeficiency virus (HIV)-infected macrophage-astroglia interactions: implications for the neuropathogenesis of HIV disease. J Exp Med 176:1703–1718

Gillardon F, Krep H, Brinker G, Lenz C, Bottiger B, Hossmann KA (1998) Induction of protein inhibitor of neuronal nitric oxide synthase/cytoplasmic dynein light chain following cerebral ischemia. Neuroscience 84:81–88

Gonzalez-Zulueta M, Ensz LM, Mukhina G, Lebovitz RM, Zwacka RM, Engelhardt JF, Oberley LW, Dawson VL, Dawson TM (1998) Manganese superoxide dismutase protects nNOS neurons from NMDA and nitric oxide-mediated neurotoxicity. J Neurosci 18:2040–2055

Grandati M, Verrecchia C, Revaud ML, Allix M, Boulu RG, Plotkine M (1997) Calcium-independent NO-synthase activity and nitrites/nitrates production in transient focal cerebral ischaemia in mice. Br J Pharmacol 122:625–630

Handy RL, Harb HL, Wallace P, Gaffen Z, Whitehead KJ, Moore PK (1996) Inhibition of nitric oxide synthase by 1-(2-trifluoromethylphenyl) imidazole (TRIM) in vitro: antinociceptive and cardiovascular effects. Br J Pharmacol 119:423–431

Handy RL, Moore PK (1997) Mechanism of the inhibition of neuronal nitric oxide synthase by 1-(2- trifluoromethylphenyl) imidazole (TRIM). Life Sci 60:PL389–394

Handy RL, Moore PK (1998) A comparison of the effects of L-NAME, 7-NI and L-NIL on carrageenan-induced hindpaw oedema and NOS activity. Br J Pharmacol 123:1119–1126

Handy RL, Wallace P, Gaffen ZA, Whitehead KJ, Moore PK (1995) The antinociceptive effect of 1-(2-trifluoromethylphenyl) imidazole (TRIM), a potent inhibitor of neuronal nitric oxide synthase in vitro, in the mouse. Br J Pharmacol 116:2349–2350

Hansen DW, Jr, Peterson KB, Trivedi M, Kramer SW, Webber RK, Tjoeng FS, Moore WM, Jerome GM, Kornmeier CM, Manning PT, Connor JR, Misko TP, Currie MG,

Pitzele BS (1998) 2-Iminohomopiperidinium salts as selective inhibitors of inducible nitric oxide synthase (iNOS). J Med Chem 41:1361–1366

Hantraye P, Brouillet E, Ferrante R, Palfi S, Dolan R, Matthews RT, Beal MF (1996) Inhibition of neuronal nitric oxide synthase prevents MPTP-induced parkinsonism in baboons. Nature Medicine 2:1017–1021

Higuchi Y, Hattori H, Kume T, Tsuji M, Akaike A, Furusho K (1998) Increase in nitric oxide in the hypoxic-ischemic neonatal rat brain and suppression by 7-nitroindazole and aminoguanidine. Eur J Pharmacol 342:47–49

Hrabak A, Bajor T, Southan GJ, Salzman AL, Szabo C (1997) Comparison of the inhibitory effect of isothiourea and mercapto- alkylguanidine derivatives on the alternative pathways of arginine metabolism in macrophages. Life Sci 60: PL395–401

Huang PL, Huang Z, Mashimo H, Bloch KD, Moskowitz MA, Bevan JA, Fishman MC (1995) Hypertension in mice lacking the gene for endothelial nitric oxide synthase . Nature 377:239–242

Hyman BT, Marzloff K, Wenniger JJ, Dawson TM, Bredt DS, Snyder SH (1992) Relative sparing of nitric oxide synthase-containing neurons in the hippocampal formation in Alzheimer's disease. Ann Neurol 32:818–820

Iadecola C (1997) Bright and dark sides of nitric oxide in ischemic brain injury . Trends Neurosci 20:132–139

Jaffrey SR, Snowman AM, Eliasson MJ, Cohen NA, Snyder SH (1998) CAPON: a protein associated with neuronal nitric oxide synthase that regulates its interactions with PSD95. Neuron 20:115–124

Jaffrey SR, Snyder SH (1996) PIN: an associated protein inhibitor of neuronal nitric oxide synthase. Science 274:774–777

Jang D, Szabo C, Murrell GA (1996) S-substituted isothioureas are potent inhibitors of nitric oxide biosynthesis in cartilage. Eur J Pharmacol 312:341–347

Kabuto H, Yokoi I, Habu H, Asahara H, Mori A (1995) Inhibitory effect of arcaine on nitric oxide synthase in the rat brain. Neuroreport 6:554–556

Kamii H, Mikawa S, Murakami K, Kinouchi H, Yoshimoto T, Reola L, Carlson E, Epstein CJ, Chan PH (1996) Effects of nitric oxide synthase inhibition on brain infarction in SOD-1-transgenic mice following transient focal cerebral ischemia. Journal Of Cerebral Blood Flow And Metabolism 16:1153–1157

Kantor DB, Lanzrein M, Stary SJ, Sandoval GM, Smith WB, Sullivan BM, Davidson N, Schuman EM (1996) A role for endothelial NO synthase in LTP revealed by adenovirus-mediated inhibition and rescue. Science 274:1744–1748

Kilbourn RG, Griffith OW (1992) Inhibition of inducible nitric oxide synthase with inhibitors of tetrahydrobiopterin biosynthesis J Natl Cancer Inst 84:1672

Kinoshita H, Milstien S, Wambi C, Katusic ZS (1997) Inhibition of tetrahydrobiopterin biosynthesis impairs endothelium-dependent relaxations in canine basilar artery. Am J Physiol 273:H718–H724

Klatt P, Schmid M, Leopold E, Schmidt K, Werner ER, Mayer B (1994) The pteridine binding site of brain nitric oxide synthase. Tetrahydrobiopterin binding kinetics, specificity, allosteric interaction with the substrate domain. J Biol Chem 269: 13861–13866

Koh JY, Choi DW (1988) Cultured striatal neurons containing NADPH-diaphorase or acetylcholinesterase are selectively resistant to injury by NMDA receptor agonists. Brain Res 446:374–378

Koh JY, Peters S, Choi DW (1986) Neurons containing NADPH-diaphorase are selectively resistant to quinolinate toxicity. Science 234:73–76

Kumagai Y, Nakajima H, Midorikawa K, Homma-Takeda S, Shimojo N (1998) Inhibition of nitric oxide formation by neuronal nitric oxide synthase by quinones: nitric oxide synthase as a quinone reductase. Chem Res Toxicol 11:608–613

Lowenstein CJ, Alley EW, Raval P, Snowman AM, Snyder SH, Russell SW, Murphy WJ (1993) Macrophage nitric oxide synthase gene: two upstream regions mediate induction by interferon gamma and lipopolysaccharide. Proc Natl Acad Sci USA 90:9730–9734

Lowenstein CJ, Glatt CS, Bredt DS, Snyder SH (1992) Cloned and expressed macrophage nitric oxide synthase contrasts with the brain enzyme. Proceedings of the National Academy of Sciences of the United States of America 89:6711–6715

Lyons CR, Orloff GJ, Cunningham JM (1992) Molecular cloning and functional expression of an inducible nitric oxide synthase from a murine macrophage cell line. J Biol Chem 267:6370–6374

MacMillan-Crow LA, Crow JP, Kerby JD, Beckman JS, Thompson JA (1996) Nitration and inactivation of manganese superoxide dismutase in chronic rejection of human renal allografts. Proc Natl Acad Sci USA 93:11853–11858

Marletta MA (1994) Nitric oxide synthase: aspects concerning structure and catalysis. Cell 78:927–930

Marsden PA, Schappert KT, Chen HS, Flowers M, Sundell CL, Wilcox JN, Lamas S, Michel T (1992) Molecular cloning and characterization of human endothelial nitric oxide synthase. FEBS Lett 307:287–293

Mayer B, John M, Heinzel B, Werner ER, Wachter H, Schultz G, Bohme E (1991) Brain nitric oxide synthase is a biopterin- and flavin-containing multi functional oxidoreductase. FEBS Lett 288:187–191

Mayer B, Klatt P, Werner ER, Schmidt K (1994) Molecular mechanisms of inhibition of porcine brain nitric oxide synthase by the antinociceptive drug 7-nitro-indazole Neuropharmacology 33:1253–1259

Mayer B, Werner ER (1995) In search of a function for tetrahydrobiopterin in the biosynthesis of nitric oxide. Naunyn Schmiedebergs Arch Pharmacol 351:453–463

Meldrum BS (1996) The role of nitric oxide in ischemic damage. Adv Neurol 71:355–363

Merrill JE, Ignarro LJ, Sherman MP, Melinek J, Lane TE (1993) Microglial cell cytotoxicity of oligodendrocytes is mediated through nitric oxide. J Immunol 151:2132–2141

Moncada S, Higgs EA (1995) Molecular mechanisms and therapeutic strategies related to nitric oxide . Faseb J 9:1319–1330

Moore PK, Wallace P, Gaffen Z, Hart SL, Babbedge RC (1993) Characterization of the novel nitric oxide synthase inhibitor 7-nitro indazole and related indazoles: antinociceptive and cardiovascular effects. Br J Pharmacol 110:219–224

Morris HR, Etienne AT, Panico M, Tippins JR, Alaghband-Zadeh J, Holland SM, Mehdizadeh S, de Belleroche J, Das I, Khan NS, de Wardener HE (1997) Hypothalamic hypertensive factor: an inhibitor of nitric oxide synthase activity. Hypertension 30:1493–1498

Murphy S, Simmons ML, Agullo L, Garcia A, Feinstein DL, Galea E, Reis DJ, Minc-Golomb D, Schwartz JP (1993) Synthesis of nitric oxide in CNS glial cells. Trends Neurosci 16:323–328

Nagayama M, Zhang F, Iadecola C (1998) Delayed treatment with aminoguanidine decreases focal cerebral ischemic damage and enhances neurologic recovery in rats J Cereb Blood Flow Metab 18:1107–1113

Nanri K, Montecot C, Springhetti V, Seylaz J, Pinard E (1998) The selective inhibitor of neuronal nitric oxide synthase, 7-nitroindazole, reduces the delayed neuronal damage due to forebrain ischemia in rats. Stroke 29:1248–53; discussion 1253–1254

Narayanan K, Spack L, McMillan K, Kilbourn RG, Hayward MA, Masters BS, Griffith OW (1995) S-alkyl-L-thiocitrullines. Potent stereoselective inhibitors of nitric oxide synthase with strong pressor activity in vivo. J Biol Chem 270:11103–11110

Navia BA, Jordan BD, Price RW (1986) The AIDS dementia complex: I. Clinical features. Ann Neurol 19:517–524

Nowicki JP, Duval D, Poignet H, Scatton B (1991) Nitric oxide mediates neuronal death after focal cerebral ischemia in the mouse. Eur J Pharmacol 204:339–340

Okuda Y, Sakoda S, Fujimura H, Yanagihara T (1998) Aminoguanidine, a selective inhibitor of the inducible nitric oxide synthase, has different effects on experimental allergic encephalomyelitis in the induction and progression phase. J Neuroimmunol 81:201–210

Onitsuka M, Mihara S, Yamamoto S, Shigemori M, Higashi H (1998) Nitric oxide contributes to irreversible membrane dysfunction caused by experimental ischemia in rat hippocampal CA1 neurons. Neurosci Res 30:7–12

Pfeiffer S, Gorren AC, Pitters E, Schmidt K, Werner ER, Mayer B (1997) Allosteric modulation of rat brain nitric oxide synthase by the pterin- site enzyme inhibitor 4-aminotetrahydrobiopterin. Biochem J 328:349–352

Price RW (1996) Neurological complications of HIV infection. Lancet 348:445–452

Przedborski S, Jackson-Lewis V, Yokoyama R, Shibata T, Dawson VL, Dawson TM (1996) Role of neuronal nitric oxide in 1-methyl-4-phenyl-1,2,3,6-tetrahydropyridine (MPTP)-induced dopaminergic neurotoxicity. Proc Natl Acad Sci USA 93:4565–4571

Przedborski S, Kostic V, Jackson-Lewis V, Naini AB, Simonetti S, Fahn S, Carlson E, Epstein CJ, Cadet JL (1992) Transgenic mice with increased Cu/Zn-superoxide dismutase activity are resistant to N-methyl-4-phenyl-1,2,3,6-tetrahydropyridine-induced neurotoxicity. J Neurosci 12:1658–1667

Radi R, Beckman JS, Bush KM, Freeman BA (1991) Peroxynitrite-induced membrane lipid peroxidation: the cytotoxic potential of superoxide and nitric oxide. Arch Biochem Biophys 288:481–487

Resink AM, Dawson VL, Dawson TM (1996) Nitric Oxide Inhibitors: Future Therapies for CNS Disorders. CNS Drugs Review 6:351–357

Robbins RA, Jinkins PA, Bryan TW, Prado SC, Milligan SA (1998) Methotrexate inhibition of inducible nitric oxide synthase in murine lung epithelial cells in vitro. Am J Respir Cell Mol Biol 18:853–859

Rosselet A, Feihl F, Markert M, Gnaegi A, Perret C, Liaudet L (1998) Selective iNOS inhibition is superior to norepinephrine in the treatment of rat endotoxic shock. Am J Respir Crit Care Med 157:162–170

Ruan J, Xie Q, Hutchinson N, Cho H, Wolfe GC, Nathan C (1996) Inducible nitric oxide synthase requires both the canonical calmodulin- binding domain and additional sequences in order to bind calmodulin and produce nitric oxide in the absence of free Ca2+. J Biol Chem 271:22679–22686

Ruetten H, Southan GJ, Abate A, Thiemermann C (1996) Attenuation of endotoxin-induced multiple organ dysfunction by 1-amino- 2-hydroxy-guanidine, a potent inhibitor of inducible nitric oxide synthase. Br J Pharmacol 118:261–270

Saetre T, Gundersen Y, Thiemermann C, Lilleaasen P, Aasen AO (1998) Aminoethyl-isothiourea, a selective inhibitor of inducible nitric oxide synthase activity, improves liver circulation and oxygen metabolism in a porcine model of endotoxemia. Shock 9:109–115

Schulz JB, Matthews RT, Jenkins BG, Ferrante RJ, Siwek D, Henshaw DR, Cipolloni PB, Mecocci P, Kowall NW, Rosen BR et al. (1995) Blockade of neuronal nitric oxide synthase protects against excitotoxicity in vivo. J Neurosci 15:8419–8429

Schulz JB, Matthews RT, Muqit MM, Browne SE, Beal MF (1995) Inhibition of neuronal nitric oxide synthase by 7-nitroindazole protects against MPTP-induced neurotoxicity in mice. Journal of Neurochemistry 64:936–939

Sharkey J, Butcher SP (1994) Immunophilins mediate the neuroprotective effects of FK506 in focal cerebral ischaemia. Nature 371:336–339

Shearer BG, Lee S, Oplinger JA, Frick LW, Garvey EP, Furfine ES (1997) Substituted N-phenylisothioureas: potent inhibitors of human nitric oxide synthase with neuronal isoform selectivity. J Med Chem 40:1901–1105

Son H, Hawkins RD, Martin K, Kiebler M, Huang PL, Fishman MC, Kandel ER (1996) Long-term potentiation is reduced in mice that are doubly mutant in endothelial and neuronal nitric oxide synthase. Cell 87:1015–1023

Stagliano NE, Dietrich WD, Prado R, Green EJ, Busto R (1997) The role of nitric oxide in the pathophysiology of thromboembolic stroke in the rat. Brain Research 759:32–40

Stuehr DJ, Fasehun OA, Kwon NS, Gross SS, Gonzalez JA, Levi R, Nathan CF (1991) Inhibition of macrophage and endothelial cell nitric oxide synthase by diphenyleneiodonium and its analogs. Faseb J 5:98–103

Szabo C, Bryk R, Zingarelli B, Southan GJ, Gahman TC, Bhat V, Salzman AL, Wolff DJ (1996) Pharmacological characterization of guanidinoethyldisulphide (GED), a novel inhibitor of nitric oxide synthase with selectivity towards the inducible isoform. Br J Pharmacol 118:1659–1668

Thiemermann C, Ruetten H, Wu CC, Vane JR (1995) The multiple organ dysfunction syndrome caused by endotoxin in the rat: attenuation of liver dysfunction by inhibitors of nitric oxide synthase. Br J Pharmacol 116:2845–2851

Tokime T, Nozaki K, Sugino T, Kikuchi H, Hashimoto N, Ueda K (1998) Enhanced poly(ADP-ribosyl)ation after focal ischemia in rat brain. J Cereb Blood Flow Metab 18:991–997

Toung TJ, Bhardwaj A, Dawson VL, Dawson TM, Traystman RJ, Hurn PD (1999) Immunosuppressent FK506 Affords Focal Ischemic Neuroprotection without Altering Nitric Oxide Production In Vivo. Stroke in press

Uemura Y, Kowall NW, Beal MF (1990) Selective sparing of NADPH-diaphorase-somatostatin-neuropeptide Y neurons in ischemic gerbil striatum. Ann Neurol 27:620–625

Wang YX, Pang CC (1993) Functional integrity of the central and sympathetic nervous systems is a prerequisite for pressor and tachycardic effects of diphenyleneiodonium, a novel inhibitor of nitric oxide synthase. J Pharmacol Exp Ther 265:263–272

Webber RK, Metz S, Moore WM, Connor JR, Currie MG, Fok KF, Hagen TJ, Hansen DW, Jr, Jerome GM, Manning PT, Pitzele BS, Toth MV, Trivedi M, Zupec ME, Tjoeng FS (1998) Substituted 2-iminopiperidines as inhibitors of human nitric oxide synthase isoforms. J Med Chem 41:96–101

Wei J, Quast MJ (1998) Effect of nitric oxide synthase inhibitor on a hyperglycemic rat model of reversible focal ischemia: detection of excitatory amino acids release and hydroxyl radical formation. Brain Res 791:146–156

Wiley CA, Achim CL (1995) Human immunodeficiency virus encephalitis and dementia. Ann Neurol 38:559–560

Win NH, Ishikawa T, Saito N, Kato M, Yokokura H, Watanabe Y, Iida Y, Hidaka H (1996) A new and potent calmodulin antagonist, HF-2035; which inhibits vascular relaxation induced by nitric oxide synthase. Eur J Pharmacol 299:119–126

Wolff DJ, Gauld DS, Neulander MJ, Southan G (1997) Inactivation of nitric oxide synthase by substituted aminoguanidines and aminoisothioureas. J Pharmacol Exp Ther 283:265–273

Wolff DJ, Lubeskie A (1996) Inactivation of nitric oxide synthase isoforms by diaminoguanidine and NG-amino-L-arginine. Arch Biochem Biophys 325:227–234

Wolff DJ, Naddelman RA, Lubeskie A, Saks DA (1996) Inhibition of nitric oxide synthase isoforms by porphyrins. Arch Biochem Biophys 333:27–34

Xie QW, Cho HJ, Calaycay J, Mumford RA, Swiderek KM, Lee TD, Ding A, Troso T, Nathan C (1992) Cloning and characterization of inducible nitric oxide synthase from mouse macrophages. Science 256:225–228

Xie QW, Whisnant R, Nathan C (1993) Promoter of the mouse gene encoding calcium-independent nitric oxide synthase confers inducibility by interferon gamma and bacterial lipopolysaccharide. J Exp Med 177:1779–1784

Yokoi I, Kabuto H, Habu H, Inada K, Toma J, Mori A (1994) Structure-activity relationships of arginine analogues on nitric oxide synthase activity in the rat brain. Neuropharmacology 33:1261–1265

Yun HY, Dawson VL, Dawson TM (1996) Neurobiology of nitric oxide. Crit Rev Neurobiol 10:291–316

Zhang F, Iadecola C (1998) Temporal characteristics of the protective effect of aminoguanidine on cerebral ischemic damage. Brain Res 802:104–110

Zhang ZG, Reif D, Macdonald J, Tang WX, Kamp DK, Gentile RJ, Shakespeare WC, Murray RJ, Chopp M (1996) ARL 17477; a potent and selective neuronal NOS inhibitor decreases infarct volume after transient middle cerebral artery occlusion in rats. J Cereb Blood Flow Metab 16:599–604

Section II
Neuroprotective Agents

CHAPTER 7
Adenosine-Based Approaches to the Treatment of Neurodegenerative Disease

A.C. FOSTER, L.P. MILLER, and J.B. WIESNER

A. Introduction

Adenosine is arguably the prototypic "neuromodulator" and has been described as the brain's natural anticonvulsant (DRUGANOW 1986) and neuroprotective agent (RUDOLPHI et al. 1992). The ability of adenosine to inhibit neuronal activity has been repeatedly demonstrated over the years in many kinds of in vitro and in vivo experiments. The stimulant effects of the adenosine receptor antagonists, caffeine and theophylline, are due to a reversal of adenosine's inhibitory tone in the central nervous system (CNS), a phenomenon that has been experienced by virtually every human being on the planet! Because of these well-documented effects on the animal and human CNS, efforts have been made over the past several decades to harness the neuronal depressant activities of adenosine for therapeutic advantage, including the treatment of neurodegenerative diseases. Despite these efforts, no drugs are currently in clinical use for CNS disorders that act specifically on the adenosine system. Perhaps the major reason for this has been concerns over the side effects which are observed when adenosine receptors are activated. The principle worries are the general sedative effects and alterations in cardiovascular function caused by agents which activate adenosine receptors, since adenosine also has potent depressant effects on the heart and is an effective vasodilator. However, some recent strategies have been devised either to minimize or avoid these potential side-effect concerns.

In this chapter we review the molecular mechanisms by which adenosine can provide neuroprotective effects and describe the efficacy and side effects which occur in animal models. In addition, we discuss the relative merits of both direct and indirect approaches to harness the potential therapeutic benefits of adenosine and minimize the potential side effects. For some recent reviews of adenosine physiology and pharmacology in the CNS, see JACOBSON et al. (1992), BRUNDEGE and DUNWIDDIE (1997), GUIEU et al. (1998), VON LUBITZ (1999) and MILLER (1999).

B. The Adenosine System in the Central Nervous System

Adenosine is not a neurotransmitter in the classical sense (i.e., released by one neuron to pass a signal to a receptive neuron), but appears to work as a local modulator of neuronal activity. As shown in Fig. 1, intracellular adenosine is formed from the breakdown of ATP through ADP and AMP. In the CNS, this

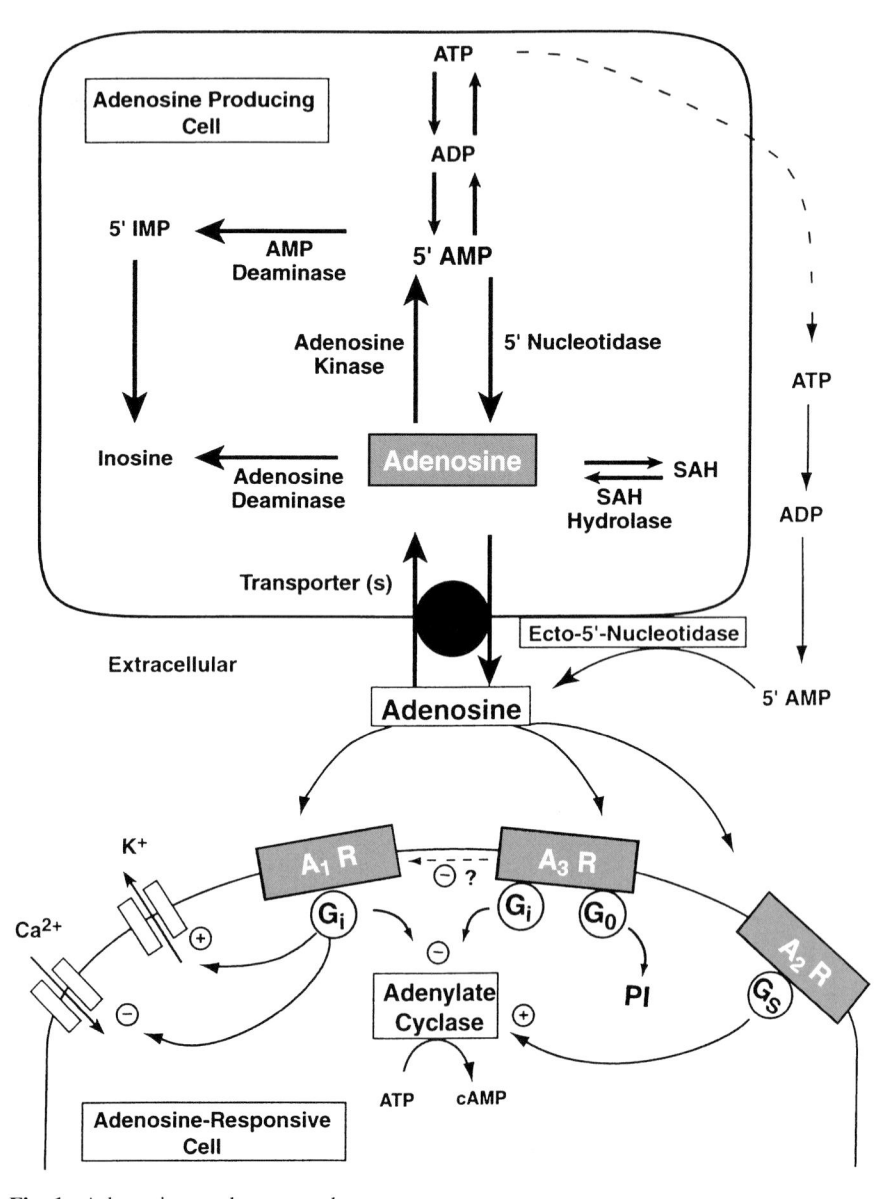

Fig. 1. Adenosine pathways and receptor systems

occurs continually, but is accelerated during periods of energy failure, such as the loss of blood supply during stroke, or when energy demand outstrips supply, such as the excessive neuronal firing which occurs during seizures. Adenosine is then catabolized principally by two enzymes, adenosine kinase and adenosine deaminase, or leaves the cells via specific transporters which deliver it to the extracellular space. The same transporters can also function in the opposite direction to allow cells to take up adenosine. On the cell surface, four adenosine receptor subtypes are known to exist: A_1, A_{2A}, A_{2B}, and A_3; these are G-protein coupled receptors, and differ in their structure, pharmacology, localization, and second messenger responses (Table 1). An alternative source of extracellular adenosine is by synthesis from extra-

Table 1. Adenosine receptor subtypes

Subtype	A_1	A_{2a}	A_{2b}	A_3
Amino acids (human)	326	412	332	318
G protein	$G_{i(1-3)}$	G_s	G_s	G_i/G_o
Effectors	↓ cAMP ↑ IP_3 ↑ K channel ↓ Calcium channel	↑ cAMP	↑ cAMP	↓ cAMP ↑ IP_3
Agonists	Adenosine (nM) NECA CHA[a] CPA[a]	Adenosine (nM) NECA CGS 21680[a]	Adenosine (μM) NECA	Adenosine (μM) NECA 2-Cl-IB-MECA[a]
Antagonists	Theophylline XAC CPT[a] DPCPX[a]	Theophylline XAC CSC[a] SCH-58261[a] KF 17837[a] ZM 241385	Theophylline XAC	MRS 1191[b] MRS1220[b] L-249313[b]
Localization	Widespread in CNS, including cerebral cortex, basal ganglia, hippocampus, cerebellum, spinal cord Peripheral nervous system Heart	Caudate nucleus, putamen, olfactory tubercle, nucleus accumbens Cerebral blood vessels Peripheral blood vessels	Diffuse, low level of expression in CNS	Diffuse, low level of expression in CNS Heart Immune cells

[a] Selective for subtype.
[b] High affinity for human A_3 receptor, but not subtype selective in rat.

cellular nucleotides through the action of ecto-5′-nucleotidases (Fig. 1). Application of ATP, ADP, AMP, (and to a lesser extent) cyclic AMP produces adenosine-like inhibitory effects on neurons which are blocked by adenosine receptor antagonists and by inclusion of adenosine deaminase, indicating that these nucleotides are converted to adenosine which then acts on adenosine receptors.

A recent study by DUNWIDDIE et al. (1997b) provided an elegant demonstration of this phenomenon in rat hippocampal slices, and indicated that the conversion of these nucleosides to adenosine is very rapid, the rate-limiting step being the conversion of AMP to adenosine by 5′-nucleotidase. Since ATP has been shown to be co-released with classical neurotransmitters (Jo and SCHLICHTER 1999), this may be an important pathway for the regulation of neurotransmission.

I. Adenosine Receptor Subtypes

1. A_1 Receptors

When applied to CNS tissue, the general effect of adenosine is to inhibit neuronal function. Activation of the A_1 receptor subtype is primarily responsible for this. A_1 receptors are coupled to G_i, with a resulting inhibition of adenylate cyclase. Despite this link to cyclic AMP, two of the most powerful inhibitory mechanisms activated by activation of the A_1 receptor are independent of this second messenger, namely, inhibition of neuronal calcium channels (DOLPHIN and ARCHER 1983; THOMPSON et al. 1992; WU and SAGGAU 1994) and activation of potassium channels (GREENE and HAAS 1991). Principle targets of adenosine, through the A_1 receptor, are glutamate-releasing pathways. Glutamate is the major excitatory neurotransmitter in the CNS and its dysfunction has been implicated in a number of neurodegenerative conditions, including epilepsy, stroke, and Alzheimer's disease. One of the best-studied glutamate-releasing pathways is the Schaeffer-collateral input to the CA1 pyramidal neurons of the hippocampus. A_1 receptors are present both pre- and postsynaptically in this pathway and act to inhibit glutamate's actions (THOMPSON et al. 1992). Activation of postsynaptic A_1 receptors increases potassium channel activity, producing a hyperpolarization which counteracts the depolarizing effects of postsynaptic ionotropic glutamate receptors. Activation of presynaptic A_1 receptors inhibits neuronal calcium channels, which are required for the entry of calcium into nerve terminals and trigger the release of glutamate (WU and SAGGAU 1994). Consequently, adenosine, through the A_1 receptor, acts to inhibit release of the excitatory neurotransmitter and also to dampen its effects on the postsynaptic cell. This dual action, a highly efficient way of down-tuning this major excitatory neurotransmitter, may be in operation during physiological glutamate pathway activity (MITCHELL et al. 1993; MANZONI et al. 1994), but is certainly in evidence during pathological states such as hypoxia, glucose deprivation, or ischemia. For example, when hippocampal slices are subjected to hypoxia, an abrupt inhibition of Schaeffer-collateral-CA1 pathway activity occurs which can be largely prevented or delayed by A_1-selective adenosine receptor antagonists (ZENG et al. 1992; KATCHMAN and HERSHOWITZ 1993; CRONING et al. 1995), indicating that endogenous adenosine is released under these conditions. This response of the tissue to a hypoxic insult may serve as a protective mechanism, to shut down activity and conserve energy expenditure during the insult to increase the chances of cell survival.

2. A_3 Receptors

Like A_1 receptors, A_3 receptors are linked to G_i, but have also been demonstrated to activate phospholipase C and the inositol phosphate pathway (ABBRACCHIO et al. 1995). A_3 receptors are expressed at lower levels than other adenosine receptors in the mammalian CNS, and their functional role there is not clearly defined. Studies in rat hippocampal slices suggest that A_3 receptors have little influence on synaptic responses, but may act to influence the activity of A_1 receptors selectively (DUNWIDDIE et al. 1997a). More recent studies have indicated an inhibitory influence of A_3 receptor activation, through protein kinase C on presynaptic type II G-protein-coupled glutamate receptors (MACEK et al. 1998). A number of antagonists with selectivity for A_3 versus A_1 receptors have been identified (JACOBSON 1998). The designation of this selectivity, however, has mainly been based on radioligand binding studies using human A_3 receptors versus rat A_1 receptors. This selectivity is less apparent when rat A_3 and A_1 receptors are compared directly. Consequently, the effects of A_3 receptor antagonists in rat studies must be interpreted carefully, in terms of receptor-specific effects. The lack of selective, bioavailable pharmacological tools has been the major roadblock to understanding A_3 receptor function in the CNS.

3. A_{2A} and A_{2B} Receptors

A_{2A} and A_{2B} receptors are linked to G_o/G_q and increase cAMP levels in cells. A_{2B} receptors have a widespread CNS distribution, including localization on glial cells, whereas, in all species studied, A_{2A} receptors are largely confined to the striatum, nucleus accumbens, and olfactory tubercle. Little is known about the function of A_{2B} receptors, primarily due to the lack of selective pharmacological agents; however, the recent development of selective A_{2A} receptor antagonists has revealed some interesting effects of A_{2A} receptor activation on neuronal activity. In situ hybridization studies have shown that A_{2A} receptors are expressed in the medium spiny striatal γ-aminobutyric acidergic (GABAergic) neurons in the rat which also contain enkephalin and project to the external globus pallidus (SCHIFFMAN et al. 1991a). These neurons also contain dopamine D_2 receptors and several studies have indicated that A_{2A} and D_2 receptors act in a mutually antagonistic fashion, in terms of cellular responses (see Sect. F.I.). MORI et al. (1996) demonstrated that synaptic

GABAergic inhibition of the medium spiny neurons is modulated by A_{2A} receptor activation, in that the selective agonist, CGS-21680, reduced ipsp's and the selective antagonist, KF17837, enhanced ipsp's, indicating that in these experiments synaptic GABA responses were under the tonic control of endogenous adenosine. Further experiments indicated that the A_{2A} receptors were presynaptic. Whether the GABAergic input in question arose from GABAergic interneurons or from collaterals of the medium spiny neurons themselves is not clear. Nevertheless, these studies clearly indicate that A_{2A} receptors, like A_1 receptors, can exist on presynaptic terminals and, in this case, control the release of an inhibitory neurotransmitter.

II. Adenosine Transporters

Two types of adenosine transporters have been identified in mammalian tissues, "concentrative" and "equilibrative" (for a review see THORN and JARVIS 1996). These nucleoside transporters are not selective for adenosine as a substrate, and are also subject to inhibition by several classes of compounds. The concentrative transporters provide a means for cells to accumulate adenosine and are sodium-dependent, the direction of transport being driven by the sodium gradient. Five subtypes have been identified from pharmacological and molecular cloning experiments (WANG et al. 1997). In the brain, the equilibrative transporters appear to dominate (JONES and HAMMOND 1995), and these fall into two subtypes, es and ei, depending on their sensitivity to the inhibitor nitrobenzylthioinosine (NBMPR), which have recently been cloned (GRIFFITHS et al. 1997a,b). The equilibrative transporters facilitate the transfer of adenosine across cell membranes, the direction being dependent on the adenosine concentration gradient. NBMPR has low nanomolar affinity at es, and micromolar affinity for ei. Compounds such as dipyridamole, dilazep, soluflazine, and propentofylline are inhibitors of es and ei to differing degrees, a property which is proposed as the primary basis of their pharmacological actions (see Sect. E.III.1.). There are also species differences with respect to the sensitivity of es and ei to some of these inhibitors. In rat tissues, dipyridamole and dilazep have equal affinity for es and ei, whereas in guinea pig, mouse, and man, es is approximately 1000-fold less sensitive than ei to these inhibitors (THORN and JARVIS 1996).

C. The Role of Adenosine in Central Nervous System Physiology

The fact that adenosine has an important role to play in the normal physiological functioning of the CNS is most clearly evident in the stimulant effects of adenosine receptor antagonists such as theophylline and caffeine. A role for adenosine in the control of arousal has been demonstrated in experimental animals. A compelling series of experiments in rats and cats has provided strong evidence that endogenous adenosine exerts a tonic inhibitory control over basal forebrain cholinergic neurons, whose activation produces electroencephalograph (EEG) arousal (RAINNIE et al. 1994; PORKKA-HEISKANEN et al. 1997; PORTAS et al. 1997). This line of work implicates endogenous adenosine as a key inducer of sleep induced by prolonged wakefulness. Other brain regions also show changes in extracellular adenosine in relation to the sleep–wake cycle. In the hippocampus of rats, extracellular adenosine levels rise towards the end of the active (dark) period and are associated with "sleep-like" behavior, reduced activity, and consummatory activities (HUSTON et al. 1996). Electrophysiological studies in the guinea pig hippocampus have demonstrated that synaptic activity which activates the NMDA subtype of glutamate receptors produces adenosine release and local inhibition of glutamate release through activation of presynaptic A_1 receptors (MANZONI et al. 1994). Consequently, adenosine appears to be a tonically-active neuromodulator in the CNS which participates in normal physiological processes. Even so, under normal conditions, the receptors responsible for these actions of adenosine (principally the A_1 and A_{2A} subtype) are clearly not saturated since agonists which cross the blood–brain barrier will produce profound sedative effects in animals (MARSTON et al. 1998). Opportunities exist, therefore, to either up or downregulate adenosine receptor activity by changing extracellular levels of endogenous adenosine, or with synthetic receptor agonists or antagonists.

What is the source of extracellular adenosine under physiological conditions and how is it regulated? At the present time, we have few answers to these questions. The classical view is that intracellular adenosine levels rise in response to increased neuronal activity through the breakdown of ATP, as this energy substrate is consumed, and adenosine is then extruded from the cell by facilitated diffusion through adenosine transporters. However, cells strive to maintain their ATP levels and so it seems unlikely that ATP breakdown is a major source of adenosine when energy metabolism is not severely compromised. Adenosine kinase is an important enzyme in this respect, since it phosphorylates adenosine to produce AMP, and so it works to maintain nucleotide pools. The K_m of adenosine kinase for adenosine is approximately $1\,\mu M$, in the range of "physiological" adenosine levels. Consequently, this enzyme may be a key regulator of adenosine tone. In rat hippocampal slices, iodotubercidin, an inhibitor of adenosine kinase, increases adenosine release and produces an inhibition of glutamate-mediated synaptic responses, an effect which is inhibited by an A_1 receptor antagonist (PAK et al. 1994; DOOLETTE 1997). Inhibitors of the other adenosine metabolizing enzymes, adenosine deaminase and s-adenosylhomocysteine hydrolase have little or no effect (PAK et al. 1994). Consequently, the endogenous tone of adenosine is under the control of adenosine kinase. Since the predominant adenosine transporters are not "concentrative," but are facilitators of adenosine flux across cell membranes in the direction of the adenosine concentration gradient, intracellular adenosine kinase activity may provide the driving force to maintain an extracellular-to-intracellular flow of adenosine, which controls the degree of adenosine receptor activation. An alternative source of adenosine is an extracellular pool of nucleotides, which, as discussed above, could come from ATP released as a neurotransmitter in its own right or along with other neurotransmitters, or from cyclic AMP. In either case, adenosine levels would be responsive to neuronal activity without the need to compromise the major pools of nucleosides available for energy metabolism.

When adenosine is infused into pyramidal neurons in hippocampal slices by means of a recording patch electrode, glutamate-mediated excitatory inputs to that neuron are selectively inhibited via presynaptic A_1 receptors (BRUNDEGE and DUNWIDDIE 1996). This demonstrates that adenosine is capable of acting as a retrograde messenger in such a way that changes in the postsynaptic neuron could feed back to modify excitatory inputs. However, when inhibitors of adenosine kinase, adenosine deaminase, or nucleoside transport were introduced into pyramidal neurons in the same way, no effect was observed, suggesting that insufficient endogenous intracellular adenosine is generated under these experimental conditions to provide an inhibitory tone (BRUNDEGE and DUNWIDDIE 1998). However, a recent paper by MASINO and DUNWIDDIE (1999) provided convincing evidence that the inhibition of excitatory synaptic responses in pyramidal neurons in hippocampal slices which results from elevating the temperature from 33°C (as used by BRUNDEGE and DUNWIDDIE 1996; 1998) to 38.5°C, was entirely due to increased extracellular adenosine. Consequently, it would be interesting to know if inhibitors of adenosine metabolism or transport applied inside pyramidal neurons have effects at more physiological temperatures. If so, postsynaptic neurons may indeed be a source of the adenosine which provides an inhibitory tone under normal conditions.

The discussion so far has centered on the neuronal depressant effects of adenosine. At the neuronal level, the A_1 receptor subtype appears to be primarily responsible, although at the behavioral level, both selective A_1 or A_{2A} receptor agonists produce locomotor depression in rodents (MARSTON et al. 1998). A_{2A} receptors are also located on cerebral blood vessels and produce vasodilation (SCIOTTI and VAN WYLEN 1993). It has been speculated that interstitial changes in adenosine produced by increased neuronal activity could activate these receptors to produce a local vasodilation and increase cerebral blood flow. Is there a physiological role for the A_{2B} and A_3 receptor subtypes in the CNS? A lack of selective pharmacological tools has not allowed a characterization of the function of A_{2B} receptors. In a carefully conducted study, DUNWIDDIE et al. (1997a) showed that an A_3 receptor agonist, chloro-N^6-(3-iodobenzyl)-5'-N-methyluronamide (Cl-IB-MECA), produced a desensitization of A_1 receptors in the rat hippocampus, which was prevented by an A_3 receptor antagonist, MRS 1191. Since the A_3 receptor has a lower affinity for adenosine compared to the A_1 receptor, the authors proposed that in situations where adenosine levels are elevated, such as ischemia, A_3 receptor activation may downregulate the A_1 receptor and limit the protective effects which are mediated by this receptor subtype. This would suggest that an A_3 receptor antagonist would potentiate the neuroprotective effects of endogenous adenosine. Electrophysiological studies in the hippocampus have also revealed an interaction between A_3 receptors and G-protein-coupled glutamate receptors, whereby A_3 receptor activation leads to downregulation of presynaptic type III mGluRs via inhibition of protein kinase C (MACEK et al. 1998). If this effect is important either physiologically, or under ischemic conditions, an A_3 agonist would, through downregulation of glutamate autoreceptors, potentiate glutamate-mediated neurotransmission and neurotoxicity.

D. The Role of Adenosine in Central Nervous System Pathophysiology

It is clear that in the CNS, as in other systems, adenosine levels are elevated in response to potentially injurious conditions. In the brain, large local elevations of extracellular adenosine are evoked by a period of ischemia, brain injury or by epileptiform activity. For example, in cats with occlusion of the middle cerebral artery, a 54-fold increase in microdialysate levels of adenosine occurs within 30min after occlusion (MATSUMOTO et al. 1993). In patients with temporal lobe epilepsy who are hospitalized for seizure focus localization prior to surgery, microdialysis in the hippocampus has revealed large elevations of adenosine during, and for a period after, spontaneous seizures (DURING and SPENCER 1992). In animal models of head trauma, a transient increase in extracellular adenosine occurs soon after the insult (NILSSON et al. 1990; HEADRICK et al. 1994; BELL et al. 1998), and in man cerebrospinal fluid adenosine levels are elevated (KOCHANEK et al. 1997; CLARK et al. 1997). The origin of the extracellular adenosine has not been examined in detail, but is thought to come from intracellular pools following the breakdown of ATP.

Elevated levels of endogenous adenosine during seizures or cerebral ischemia are presumed to play a protective role, helping to reduce neuronal degeneration and epileptiform activity. If this is the case, adenosine receptor antagonists would be predicted to worsen the pathology by preventing the actions of endogenous adenosine. Theophylline has been shown to increase the incidence and duration of epileptiform events both in vitro and in vivo (DRAGANOW et al. 1986; KNUTSEN and MURRAY 1997). In vitro, application of the A_1 receptor antagonist, 8-cyclopentyltheophylline (8-CPT), led to increased glutamate release and neuronal degeneration in cortical neuronal cultures exposed to oxygen and glucose deprivation (LOBNER and CHOI 1994; LYNCH et al. 1998), and 8-CPT was also shown to reduce the synaptic depression in hippocampal slices during hypoxia (GRIBKOFF et al. 1990), indicating a protective effect of endogenous adenosine. Similarly, the degree of recovery of injury following anoxic injury to rat optic nerve was reduced by theophylline, suggesting a protective effect of endogenous adenosine towards white matter (FERN et al. 1994). The situation in in vivo models of neurodegeneration is less clear. In global ischemia models in the rat, theophylline was reported to have either no effect (LEKIEFFRE et al. 1991) or to increase the percentage of degenerated neurons (RUDOLPHI et al. 1987; ZHOU et al. 1994). Theophylline was shown to increase hippocampal neurodegeneration in the gerbil global ischemia model (DELEO et al. 1988; DUX et al. 1990). Microdialysis studies indicated that 8-(p-sulfophenyl)theophylline (8-SPT) delivered via the microdialysis probe augmented the ischemia-induced elevations of extracellular glutamate and aspartate (SCIOTTI et al. 1992). In the gerbil global ischemia model, acute administration of caffeine (SUTHERLAND et al. 1991) or the A_1-selective antagonist 8-cyclopentyl-1,3-dipropylxanthine (DPCPX; VON LUBITZ et al. 1994) worsened outcome, although following subchronic administration outcome was improved (RUDOLPHI et al. 1989; SUTHERLAND et al. 1991; VON LUBITZ et al. 1994). This improvement may be due to upregulation of A_1 receptors in response to the continued presence of the antagonist. In neonatal rats subjected to hypoxia/ischemia, theophylline reduced neuronal degeneration, an effect that was partly reproduced by an A_{2A}-selective antagonist (SCH 58261), but not by an A_1-selective antagonist (DPCPX) (BONA et al. 1997). In gerbils, the outcome of global ischemia was made more severe by the A_1 receptor-selective antagonist DPCPX, and improved with the selective A_{2A} receptor antagonists CGS 15943, 8-(3-chlorostyryl) caffeine (CSC) and CP 66,713 (PHILLIS 1995; VON LUBITZ et al. 1995; GAO and PHILLIS 1994). Following 1 h of middle cerebral artery occlusion in the cat, theophylline reduced brain tissue damage, hyperemia, edema and blood–brain barrier breakdown (SEIDA et al. 1988) and in the pig, theophylline prevented hypoxia-induced cerebral vessel dilatation (BARI et al. 1998). In the mouse, the A_{2A} receptor-selective antagonist SCH 58261 reduced infarct size after middle cerebral artery occlusion (MONOPOLI et al. 1998).

Consequently, no clear picture emerges from the in vivo studies of the beneficial or deleterious role played by endogenous adenosine during cerebral ischemia. Experiments using the nonselective, but bioavailable, adenosine receptor antagonist theophylline are subject to the concerns that blocking both A_1 and A_{2A} receptors may have opposing effects on the outcome of cerebral ischemia. A_{2A} receptor antagonists appear to be consistently protective. This may be due to effects on the vasculature, if a general cerebral vasodilation mediated by adenosine removed blood from the ischemic region (a "steal" effect), or due to excitatory effects on neurons. Unfortunately, no systematic study of an A_1 receptor-selective antagonist in a focal ischemia model appears to be available in the literature, consequently it is not possible to determine the effects of this receptor subtype under true simulated stroke conditions independently.

E. Therapeutic Approaches to Adenosine-Based Neuroprotection in Cerebral Ischemia

I. Potential Therapeutic Mechanisms

There are multiple mechanisms through which adenosine has the potential to provide therapeutic benefit in the setting of cerebral ischemia. First, adenosine can provide direct neuroprotection through activation of the A_1 receptor subtype, present on neurons. Adenosine's actions here are largely due to a reduction of the overactivation of glutamate receptors which occurs during cerebral ischemia, by inhibiting neuronal calcium channels to reduce the release of glutamate from presynaptic terminals, and by opening potassium channels on the postsynaptic neurons to dampen cellular depolarization (see reviews by RUDOLPHI et al. 1992; MILLER and HSU 1992). Second, adenosine can protect white matter, such as axons, from the consequences of ischemia (FERN et al. 1994). Third, activation of adenosine receptors on neurons has been demonstrated to produce "ischemic preconditioning" in the brain. This is a phenomenon, originally characterized in the heart, where a brief, noninjurious period of ischemia gives rise to a subsequent period of time (which can last over several days) during which protection towards neurodegeneration caused by longer ischemic periods occurs. Evidence for an A_1 receptor involvement in this phenomenon has been provided both in vitro (PEREZ-PINZON et al. 1996) and in vivo (HEURTEAUX et al. 1995). The downstream events which underlie the phenomenon are unknown. Fourth, A_{2A} receptors in the cerebral vasculature produce vasodilation which, if local, can restore blood supply to ischemic tissue. Fifth, activation of A_{2A} receptors on neutrophils and platelets reduces activation of these blood-born cells, preventing the plugging of arterioles and the formation of platelet clots and emboli. Thus, the adenosine system provides a unique and wide-ranging assortment of potentially powerful mechanisms which may provide benefit in cerebral ischemia. Some of the therapeutic strategies based on adenosine focus on one or two of these mechanisms, and some attempt to capture most, if not all.

II. Receptor-Based Approaches

1. A_1 Receptor Agonists

The data reviewed so far indicate that A_1 receptor activation may be a good rationale for providing protection of neurons during cerebral ischemia. In vitro, selective A_1 receptor agonists such as N^6-cyclohexyladenosine (CHA) and R-N^6-(2-phenylisopropyl)adenosine (R-PIA) are effective neuroprotectants in primary neuronal cultures exposed to simulated ischemia (GOLDBERG et al. 1988; LOBNER and CHOI 1994; DAVAL and NICOLAS 1994; LOGAN and SWEENEY 1997) or trauma (MITCHELL et al. 1995), although one report in hippocampal slice cultures indicated that under certain conditions both adenosine and A_1 receptor agonists potentiate the neurotoxic effects of mild ischemia (BARTH et al. 1997). Demonstration of the neuroprotective effects of A_1 receptor agonists in vivo has been made difficult due to the profound cardiovascular and hypothermic effects which result from systemic administration of such compounds. Reports of protective effects of A_1 receptor agonists in gerbil global ischemia models (DAVAL et al. 1989; VON LUBITZ et al. 1994; 1996a,b) have not taken these important confounding factors into account (see MILLER and HSU 1992). Indeed, when ROUSSEL et al. (1991) tested R-PIA in a middle cerebral artery occlusion model in spontaneously-hypertensive rats at a dose which produced minimal hemodynamic effects, no reduction in infarct size was observed. A potential strategy to avoid these concerns is to administer A_1 agonists intracerebroventricularly, although it is not necessarily the case that the cardiovascular and hypothermic effects are exclusively mediated by peripheral A_1 receptors. When this has been done, neuroprotective effects have been observed in both global (EVANS et al. 1987; ZHOU et al. 1994) and focal ischemia models (L.P. Miller, unpublished observations). In addition, R-PIA and 2-chloroadenosine protect against excitotoxin-induced neuronal degeneration when injected directly into the rat striatum (ARVIN et al. 1988; FINN et al. 1991).

To date, A_1 receptor agonists have not been taken into the clinic for examination of their effects in stroke patients, and it is unlikely that this will happen, given the concerns over side effects, particularly the cardiovascular changes which would be prohibitive in such a cardiovascularly-compromised patient population. (Note that the hypothermia which results from A_1 receptor activation could be advantageous from a cerebral protection point of view, provided that it is mild and could be controlled; profound sedative effects in stroke patients, however, would also be problematic.) Alternative strategies are to find agonists which have a selective effect on central, rather than peripheral, A_1 receptors. However, at present, there are no indications that subtypes of A_1 receptors exist, and native A_1 receptors in the heart and brain have identical pharmacologies. Another strategy is to develop partial agonists for the A_1 receptor, and hope that the relative receptor reserve in the heart and brain would lead to a selective activation of central receptors. At present, there are no indications that this is feasible. A further strategy is to develop compounds which allosterically enhance A_1 receptor function. Such compounds have been discovered (BRUNS and FERGUS 1990; BRUNS et al. 1990), but work equally well on the heart and brain A_1 receptors (MIZUMURA et al. 1996); however, they offer the potential for a site-selective enhancement of A_1 receptor function, which is reviewed in Sect. E.III.5. Finally, the preconditioning effects of A_1 receptor activation, reviewed in Sect. I, could conceivably be utilized to advantage in patients who are due to undergo surgical procedures where a high risk of stroke and/or neurological deficits from focal cerebral ischemia is anticipated, e.g., coronary artery graft bypass surgery; although, again, tolerability to the side effects of A_1 receptor activation would need to be assessed carefully in the chosen patient population.

2. A_3 Receptor Agonists/Antagonists

Few data are available about the potential therapeutic effects of A_3 receptor ligands. VON LUBITZ et al. (1994) reported that the acute administration of the A_3 receptor agonist, N^6-(3-iodobenzyl)-5'-N-methyluronamide (IB-MECA), increased neuronal degeneration in the gerbil global ischemia model, whereas chronic administration provided protection. The effects of acute administration would be consistent with the observations that A_3 receptor activation downregulates both A_1 receptor function and presynaptic type-III G-protein glutamate receptor function in the hippocampus (see Sect. C.), in that increased glutamate-mediated neurotoxicity would be expected. The protective outcome of chronic IB-MECA administration could conceivably result from A_3 receptor desensitization, although no direct evidence of this has been provided. One could conclude, based on the discussion above, that a selective A_3 receptor antagonist could be neuroprotective; however, there are no reports to date of A_3 receptor antagonists in relevant cerebral ischemia models.

3. A_{2A} Receptor Antagonists

Several reports have now indicated that the selective A_{2A} receptor antagonists which have been developed recently have neuroprotective effects in ischemia models (see review by ONGINI et al. 1997). Thus, CSC, CP 66,713, and CGS 15943 are protective in the gerbil global ischemia model (GAO and PHILLIS 1994; PHILLIS 1995; VON LUBITZ et al. 1995), CSC protects retinal function and structure following 60 min of ischemia (LI et al. 1999), and SCH 58261 reduces infarct size in neonatal rats subjected to hypoxia/ischemia (BONA et al. 1997) and in both normotensive and hypertensive adult rats following middle cerebral artery occlusion (MONOPOLI et al. 1998). The mechanism of this protective action is unknown. Given the low doses of these agents which are effective following systemic application, it would appear that an effect primarily on the vasculature is the most likely. There are no reports of neuroprotective effects of A_{2A} receptor antagonists in in vitro brain slice or cultured neuron preparations, consequently potential direct neuroprotective effects have yet to be assessed. However, JONES et al. (1998) reported that injection of the A_{2A} receptor antagonist, ZM 241385, into the rat hippocampus reduced neuronal degeneration caused by intrahippocampal kainate injection, suggesting a direct neuroprotective effect in this model. Given the interest in A_{2A} receptor antagonists for symptomatic relief in Parkinson's disease (see Sect. F.I.), a detailed characterization of their potential neuroprotective effects would seem warranted.

III. Modulation of Endogenous Adenosine

As outlined in the sections above, extracellular levels of endogenous adenosine rise markedly during cerebral ischemia, and activation of adenosine receptors on the cell surface can give rise to a number of potentially beneficial effects which can counteract the ongoing pathology. Although studies with adenosine receptor antagonists have yielded inconclusive results in the setting of cerebral ischemia (see Sect. D.), positive results have been obtained with drugs designed to potentiate either the levels of adenosine or adenosine receptor activation. A potential advantage of this approach is that the drugs employed may be relatively silent under physiological conditions, but have their greatest effects when ischemia induces a rise in endogenous adenosine levels. Consequently, the compounds would be less susceptible to the side effect concerns which have plagued the direct-acting adenosine receptor agonists. Data with the different classes of agents, namely adenosine transport inhibitors, adenosine deaminase inhibitors, adenosine kinase inhibitors, other adenosine modulators and allosteric modulators of A_1 receptors, are described in the following sections.

1. Adenosine Transport Inhibitors

Propentofylline (HWA 285) is the adenosine transport inhibitor that has received the most extensive profiling in animal models of cerebral ischemia, and has also entered clinical studies for vascular dementia and Alzheimer's disease (KITTNER et al. 1997). This compound inhibits adenosine transport ($K_i = 9\mu M$ for es; PARKINSON et al. 1993), but also is an inhibitor of cAMP-dependent phosphodiesterase (K_i = approx. $100\mu M$; NAGATA et al. 1985). Propentofylline is neuroprotective in the rat neonatal hypoxia/ischemia model (GIDDAY et al. 1995), in the gerbil model of global ischemia (DE LEO et al. 1988; DUX et al. 1990), and in a rat middle cerebral artery occlusion model (PARK and RUDOLPHI 1994). In a rat middle cerebral artery occlusion model with reperfusion, propentofylline reduced infarct size when administration was started at 30 or 60 min, but not 3h after artery occlusion (JOHNSON et al. 1998). Propentofylline was also reported to potentiate ischemic preconditioning in the gerbil global ischemia model (KAWAHARA et al. 1998). Evidence for an adenosine-mediated mechanism was provided by ANDINE et al. (1990), who demonstrated that the increased extracellular levels of adenosine in the hippocampus in response to global ischemia were further elevated by propentofylline, with a concomitant decrease in extracellular glutamate. Nevertheless, the relative contributions of adenosine transport inhibition and inhibition of cAMP-dependent phosphodiesterase, which would also be expected to potentiate A_2 receptor responses, is unclear. Indeed, the protective effects of propentofylline reported by DE LEO et al. (1988) were only partially reversed by co-administration of theophylline, which might suggests suggest mechanisms other than adenosine transport inhibition. A further issue is whether the effects of this compound are primarily within neuronal parenchyma to elevate local adenosine in close apposition to the vulnerable neurons, or in the cerebral vasculature to produce a local vasodilation. Propentofylline itself is a peripheral and central vasodilator and the adenosine transport inhibitor dipyridamole has been shown to potentiate pial arteriole diameter during hypoxia along with increased adenosine levels (MENO et al. 1993).

One way to differentiate these mechanisms would be to assess the neuroprotective effects of adenosine transport inhibitors in neuronal cultures. This has been done by LOBNER and CHOI (1994), who reported that dipyridamole increased the neuronal degeneration induced by oxygen and glucose deprivation, in parallel with a decrease in extracellular adenosine levels. These observations could be explained by assuming that the source of the increased extracellular adenosine in this paradigm was from intracellular stores and was released through the adenosine transporter, the inhibition of which (by dipyridamole) was deleterious. This would suggest that the likely site of action of adenosine transporter inhibitors in in vivo stroke models is the cerebral vasculature. It is interesting to note that dipyridamole, which is an antiplatelet agent, is effective in stroke prevention, an effect which has been speculated to also involve an adenosine component (PICANO and ABBRACCHIO 1998).

Soluflazine is one of a series of additional adenosine transport inhibitors which have been investigated for their potential neuroprotective effects. Soluflazine is a potent inhibitor of NBMPR-insensitive nucleoside transport (IC_{50} approximately 100nM; GRIFFITH et al. 1990). In rat hippocampal slices, soluflazine by itself produced an inhibition of excitatory postsynaptic potentials which was reversed by the A_1 receptor antagonist 8-CPT, and delayed hypoxia-induced depolarization (BOISSARD and GRIBKOFF 1993). In guinea pig hippocampal slices, soluflazine potentiated the inhibitory effects of adenosine (ASHTON et al. 1987) and inhibited epileptiform activity (ASHTON et al. 1988). Unfortunately, there are few studies of the in vivo effects of soluflazine. PHILLIS et al. (1989) found no effect of the transport inhibitor to alter hypoxia/ischemia-induced elevations of adenosine from rat cerebral cortex, and O'CONNOR et al. (1991) demonstrated a sleep-promoting effect of soluflazine after intracerebroventricular administration to rats. Consequently, the potential neuroprotective effects of soluflazine remain largely untested.

2. Adenosine Deaminase Inhibitors

Deoxycoformycin is a potent and selective adenosine deaminase inhibitor (ROGLER-BROWN et al. 1978). Deoxycoformycin was reported to reduce hippocampal neurodegeneration in the gerbil global ischemia model (PHILLIS and O'REGAN 1989), although DELANEY et al. (1993) found no effect in a rat global ischemia model. Deoxycoformycin was also effective in reducing infarct size in the neonatal rat hypoxia/ischemia model (GIDDAY et al. 1995). In the one report using a focal ischemia model (middle cerebral artery occlusion coupled with right carotid artery occlusion) LIN and PHILLIS (1992) indicated that a protective effect was evident with pretreatment, but not when treatment was delayed for 1h after the occlusion (note that in the same series of experiments, oxypurinol was effective with a delay of 1h). Unfortunately, there are no reports of the effects of deoxycoformycin in the currently favored permanent or transient middle cerebral artery occlusion models which have been used to examine many potential neuroprotective drugs. There are also no reports of the effects of deoxycoformycin in in vitro neurodegeneration models, consequently any direct neuroprotective effects have not been evaluated.

A theoretical advantage of adenosine deaminase inhibition is that, relative to adenosine kinase, the enzyme has a low affinity for adenosine, and might be expected to deal primarily with the excess adenosine which is present during ischemia (or other pathologies). Consequently, an adenosine deaminase inhibitor may have little effect on the normal "adenosine tone" and, therefore, fewer side effects. This does not appear to be the case in epilepsy models, however, where deoxycoformycin is rather ineffective (see Sect. G.II.). Unfortunately, there are too little data to compare the relative efficacies and side-effect ratios of adenosine deaminase and adenosine kinase inhibitors in cerebral ischemia models.

3. Adenosine Kinase Inhibitors

A clearer picture can be established for inhibitors of adenosine kinase. Despite the initial report that a prototype adenosine kinase inhibitor, 5-iodotubercidin, was ineffective in the gerbil global ischemia model (PHILLIS and SMITH-BARBOUR 1993), three studies using well-established middle cerebral artery occlusion models have demonstrated neuroprotective effects with this class of agent. Using the potent adenosine kinase inhibitor, 5′-deoxy-5-iodotubercidin, MILLER et al. (1996) observed a reduction in infarct size in a rat middle cerebral artery occlusion model with reperfusion when drug administration was started either before or during the occlusion period. Similarly, JIANG et al. (1997) observed neuroprotection with the same compound when treatment was initiated prior to middle cerebral artery occlusion and reperfusion. Using the novel, bioavailable adenosine kinase inhibitor, GP683, whose blood–brain barrier penetration characteristics have been well-described (WIESNER et al. 1999), TATLISUMAK et al. (1998) showed a 44% reduction of infarct size following 90min of middle cerebral artery occlusion, with repeated doses of 1 mg/kg, but a higher dose showed no significant effect, indicating a U-shaped dose–response relationship. In this study, some important variables were monitored to ensure that the neuroprotective effects of GP683 were not accompanied by changes in blood pressure, body or brain temperature, blood gases, or changes in cerebral blood flow. It seems likely, therefore, that the reduction of infarct size caused by GP683 was due to a direct neuroprotective effect mediated by the increase of endogenous adenosine, although adenosine-mediated local increases in blood flow to the ischemic region, or effects on neutrophils in this reperfusion model, two documented effects of adenosine kinase inhibitors (SCIOTTI and VAN WYLEN 1993; FIRESTEIN et al. 1995), cannot be entirely ruled out. A recent report by BRITTON et al. (1999) provided evidence for a "pathology-dependent" enhancement of extracellular adenosine concentrations in the rat striatum following systemic administration of 5′-

deoxy-5-iodotubercidin. This suggests that enhancement of adenosine's activities at the neuronal level may underlie the protective effects of adenosine kinase inhibitors.

One report has appeared which examined an adenosine kinase inhibitor in an in vitro neurodegeneration model. LYNCH et al. (1998) found no effect of 5'-iodotubercidin (1 μM) on oxygen-glucose deprivation-induced neuronal degeneration or elevation of adenosine levels in rat cortical neuronal cultures. However, adenosine kinase activity in these cultures was markedly reduced by oxygen-glucose deprivation, which may explain the lack of effect of the inhibitor. In situations like this, where ATP levels are presumed to be severely compromised, the resultant reduction of adenosine kinase activity (since ATP is a substrate for the enzyme) may indeed contribute to the elevated adenosine levels observed. This may be more analogous to the "core" of an ischemic lesion, a region which will not be spared unless blood flow can be restored. Adenosine kinase inhibitors would be expected to be most effective in the "penumbra," where ATP levels are less severely reduced. It is the penumbra which is the target of neuroprotective therapy.

4. Other Adenosine Modulators

Acadesine, an "adenosine regulating agent" which elevates the levels of endogenous adenosine during cardiac ischemia (GRUBER et al. 1989), was examined in a model of embolic stroke where the number of platelet emboli, which can break off from a thrombus and lodge in the cerebral vasculature, was measured (DIETRICH et al. 1995). Acadesine significantly reduced the deposition of these emboli, indicating the potential for adenosine modulators to be effective in embolic stroke or in surgical indications where emboli cause "mini strokes" resulting in neurological deficits. An analogue of acadesine with improved adenosine modulating properties, GP668, was also shown to reduce infarct size in a temporary middle cerebral artery occlusion model (MILLER et al. 1994). Although the mechanism of action of these compounds in elevating endogenous adenosine levels remains unclear, these data strengthen the idea that local elevations in adenosine during cerebral ischemia have beneficial and potentially therapeutic effects.

5. A_1 Receptor Modulators

A series of compounds which allosterically enhance agonist binding to A_1 receptors was first described by BRUNS and colleagues (BRUNS and FERGUS 1990; BRUNS et al. 1990) and is exemplified by PD 81,723. The advantage of this type of compound over A_1 receptor agonists is the potential to enhance A_1 receptor activation only when the endogenous agonist is present, thus providing a local pathology-dependent enhancement of A_1 receptor activation, reducing the potential for side effects. Although several studies have demonstrated the ability of PD 81,723 to enhance A_1 receptor mediated effects in the heart (e.g., MIZUMURA et al. 1996), fewer studies are available in the CNS. In hippocampal slices, PD 81,732 had no affect on CA1 synaptic responses, but potentiated the inhibitory effects of applied adenosine (Janusz et al. 1991). However, epileptiform activity in hippocampal slices evoked by low magnesium perfusion was reversed by PD 81,732, mimicking the action of applied adenosine (Janusz and Berman 1993). Three studies have examined the effects of PD 81,732 in cerebral ischemia. Phillis et al. (1994) showed that the compound reduced the increased glutamate release from rat cerebral cortex which could be measured during global ischemia using the cortical cup technique. However, no effect of PD 81,732 was found in the gerbil model of global ischemia (Cao and Phillis 1995). On the other hand, in the neonatal rat hypoxia model, Halle et al. (1997) demonstrated that PD 81,732 prevented the hemispheric weight loss characteristic of infarction in this model. Although this approach of using an allosteric enhancer to take advantage of the endogenous activation of the A_1 receptor appears attractive, it may be limited by the relatively low affinities of these compounds (in in vitro studies the effects are generally seen in the 10–100M range), and by more recent findings which indicate that they are not selective for the A_1 receptor, and possess additional inhibitory activity towards adenylate cyclase (Musser et al. 1999).

F. Adenosine-Based Therapeutics in Parkinson's Disease

Adenosine has been proposed to provide the potential for both symptomatic relief and prevention of disease progression in Parkinson's disease. The former is based on the effects of selective A_{2A} receptor antagonists in the basal ganglia, and the latter on the neuroprotective effects of A_1 receptor activation. Since the daunting side effect profile and potential for receptor desensitization would preclude long-term treatment with an A_1 receptor agonist in this chronic neurodegenerative disease, approaches based on modulation of endogenous adenosine would seem to be a more viable option to halt disease progression. For some recent reviews see Richardson et al. (1997) and Ferre et al. (1997).

I. A_{2A} Antagonists

In the rat, primate, and human caudate nucleus, A_{2A} receptors are present on the GABA and enkephalin-containing medium spiny neurons, which are the cells of origin of the "indirect" output pathway to the external globus pallidus and contain D_2 dopamine receptors (Schiffmann et al. 1991a,b; Fink et al. 1992; Svenningsson et al. 1998). A_{2A} receptors also appear to be present on striatal cholinergic neurons (Dixon et al. 1996). In both the GABAergic and cholinergic cells, neurochemical evidence suggests that one site of A_{2A} receptor function is on presynaptic nerve terminals, since A_{2A} receptor activation inhibits the release of GABA and increases the release of acetyl choline from striatal synaptosomes (KIRK and RICHARDSON 1994; KUROKAWA et al. 1994). The cellular actions of A_{2A} and D_2 receptor activation are mutually antagonistic (YANG et al. 1995). In vivo, A_{2A} agonists reduce locomotor activity (MARSTON et al. 1998), and an A_{2A} antagonist injected into the rat striatum increased locomotor activity and reversed catalepsy induced by selective D_1 or D_2 receptor antagonists (HAUBER et al. 1998). Given this mutually antagonistic nature, in Parkinson's disease where striatal dopamine receptor activation is diminished due to the degeneration of the nigrostriatal pathway, an A_{2A} antagonist has the potential to relieve the endogenous adenosine tone which may now dominate striatal neuron activity and provide symptomatic relief similar to that of dopamine receptor activation by direct-acting agonists or L-3,4-dihydroxyphenylalanine (L-DOPA). In addition, it would be expected that A_{2A} antagonists would work additively, or synergistically with L-DOPA. In this respect, the methylxanthine adenosine receptor antagonists have been shown to potentiate the actions of dopamine receptor agonists and L-DOPA in the 6-hydroxydopamine-lesioned rat model of Parkinson's disease (FUXE and UNGERSTEDT 1974). In contrast to the well-documented loss of presynaptic dopamine markers in Parkinson's disease patients and alterations in postsynaptic dopamine receptors, caudate nucleus A_{2A} receptors appear to be unchanged in the disease (MARTINEZ-MIR et al. 1991), and in 6-hydroxydopamine-lesioned rats, extracellular striatal adenosine levels are also unchanged (BALLARIN et al. 1987), indicating that the adenosine system is intact.

Three recent reports have indicated the potential of A_{2A} receptor antagonists for relief of parkinsonian symptoms in N-methyl-4-phenyl-1,2,3,6-tetrahydropyridine (MPTP)-treated primate models. KANDA et al. (1998a) reported that oral administration of the A_{2A} receptor antagonist, KW-6002, restored locomotor activity to normal in MPTP-treated marmosets without producing abnormal movements, such as stereotypy or dyskinesias in animals which were primed to show dyskinesias in response to L-DOPA administration. The beneficial effects of KW-6002 were present over 21 days of treatment, without any signs of tolerance. In a further study, the effects of KW-6002 were found to be reversed by the A_{2A} receptor agonist, 2-[p-[2-(2-aminoethylamino)carbonylethyl] phenethylamino]-5'-N-ethylcarboxamidoadenosine (APEC; KANDA et al. 1998b). In MPTP-treated cynamologous monkeys, GRONDIN et al. (1999) reported that KW-6002 reversed parkinsonian symptoms and potentiated the effects of L-DOPA, without increasing the incidence of dyskinesias.

These reports are very encouraging for the prospect of A_{2A} antagonists in the treatment of Parkinson's disease, and it is to be hoped that human trials with this class of compound are forthcoming. In addition to symptomatic relief, A_{2A} antagonists may provide the possibility of disease modification if the neuroprotective effects of this class of compounds described in Sect. E.II.3 is confirmed, and is relevant for the mechanisms of nigro-striatal neurodegeneration in this disease.

II. Adenosine Modulators

Given the neuroprotective effects of adenosine in cerebral ischemia models (see Sect. E.), consideration has been given to the idea that adenosine may provide neuroprotection towards the nigro-striatal dopamine neurons which undergo degeneration in Parkinson's disease. Glutamate-mediated mechanisms of neurodegeneration have received some attention, both from the evidence that glutamate receptor antagonists can prevent chemically-induced nigro-striatal dopamine neuron degeneration (SONSALLA et al. 1998), and that mitochondrial dysfunction may render nigro-striatal neurons more vulnerable to excitotoxic mechanisms (BLANDINI et al. 1996). In addition, there is convincing evidence that reversal of glutamate hyperactivity in the subthalamic nucleus of parkinsonian primates provides marked symptomatic relief (GREENAMYRE et al. 1994). Since adenosine, through the A_1 receptor, has a potent inhibitory effect on glutamate neurotransmission and neurotoxicity, it would seem reasonable to investigate the ability of A_1 receptor activation to both prevent nigro-striatal dopamine neuron loss and reverse parkinsonian symptoms. Only three studies have addressed the potential protective effects of adenosine in models of Parkinson's disease. LAU and MOURADIAN (1993) reported that systemically-applied CHA reduced MPTP-induced striatal dopamine depletion in mice, an effect which was reversed by DPCPX. The effect was also apparent when CHA administration was started 5h after MPTP administration, a time when striatal dopamine already showed a significant reduction. Unfortunately, it is not possible in this experiment to discern whether this was a direct neuroprotective effect of CHA, or if the beneficial effect was the result of the cardiovascular and, in particular, hypothermic effects which would be manifest at the CHA doses used. DELLE DONNE and SONSALLA (1994) also found a protective effect of an A_1 receptor agonist, N^6-cyclopropyladenosine (CPA), towards methamphetamine-induced degeneration of nigro-striatal dopamine neurons in the mouse, which was reversed by CPT. This study is subject to the same concerns noted above, however the authors did additional experiments with caffeine which provide more convincing evidence for a protective effect of adenosine. When they examined the effects of acute caffeine administration, they found a potentiation of the effects of methamphetamine on striatal dopamine markers, whereas chronic caffeine administration, which caused an upregulation of striatal adenosine receptors, reduced the effects of methamphetamine. However, this chronic caffeine regimen failed to prevent MPTP-induced neurochemical changes. Finally, adenosine has been shown to prevent the spontaneous degeneration of dopamine neurons in rat mesencephalic primary cultures, although the receptor specificity of this effect is not clear (MICHEL et al. 1999).

Clearly there is room for further investigation of the potential neuroprotective and potentially symptom-relieving effects of elevating adenosine in Parkinson's disease. From a therapeutic standpoint, drugs which modulate adenosine levels would be the favored approach for the treatment of this chronic disease, given the side-effect concerns and potential desensitizing effects of direct acting A_1 receptor agonists. A systematic study of adenosine transport inhibitors, adenosine kinase inhibitors, and adenosine deaminase inhibitors would be valuable in both in vivo and in vitro models of nigrostriatal dopamine neuron degeneration. In addition, an examination of the effects of these agents on parkinsonian symptoms in MPTP-treated primates would seem warranted. This latter aspect appears contrary to the evidence that an A_{2A} receptor antagonist provides symptomatic relief. However, if local adenosine levels are elevated in response to overactivation of glutamate pathways in regions of the basal ganglia outside of the caudate nucleus, potentiation of this local adenosine release by adenosine-modulating drugs may be beneficial and potentially synergistic with A_{2A} receptor blockade.

G. Adenosine-Based Therapeutics in Epilepsy

Neurodegeneration occurs in epileptic patients and in animals which undergo repetitive seizures. There has been a great deal of interest in the role of adenosine in epilepsy, and the therapeutic potential of adenosine-based therapies in this disorder. Since the focus of this chapter is neuroprotection, we will not address the anticonvulsant potential of adenosine, for which the reader is referred to the following reports: DRUGANOW 1986; KNUTSEN and MURRAY 1997; WIESNER et al. 1999. However, the mechanisms of seizure generation and seizure-related brain damage have relevance for neurodegenerative disorders in general, and the effects of adenosine therapeutics on these aspects will briefly be discussed here.

I. A_1 Receptor Agonists

A_1 receptor agonists exert anticonvulsant effects in a wide variety of seizure models, including audiogenic seizures, chemically-induced seizures, maximum electroshock-induced seizures, and kindling (DRUGANOW 1986; KNUTSEN and MURRAY 1997; WIESNER and ZIMRING 1994), and adenosine receptor antagonists increase the epileptiform events in vitro and in vivo. These robust antiepileptic effects are evidence of the powerful inhibitory effects of adenosine towards abnormally activated neurons; however, the side effect concerns associated with A_1 receptor agonists have so far prevented the consideration of their clinical use.

Epileptic conditions are often associated with neuronal degeneration which has similar characteristics to that which occurs in many neurodegenerative diseases, in terms of selective neuronal vulnerability and involvement of glutamate mechanisms (SCHWARCZ and BEN-ARI 1986). Neuronal degeneration which results from seizures has typically been examined in two types of animal models: systemic administration of the cholinergic agonist pilocarpine to induce repetitive generalized seizures, and systemic administration of kainic acid to induce repetitive limbic seizures. In the pilocarpine model, theophylline was shown to produce status epilepticus in rats treated with a subthreshold convulsive dose of pilocarpine, with resultant neuronal degeneration in the hippocampus, thalamus, amygdala, olfactory cortex, substantia nigra, and neocortex (TURSKI et al. 1985). Systemically-applied 2-chloroadenosine prevented the seizures and brain damage which resulted from a higher dose of pilocarpine (TURSKI et al. 1985). In the kainic acid model, theophylline increased the extent of neuronal degeneration (PINARD et al. 1990). Systemic administration of the A_1 agonist R-PIA reduced seizure-related neuronal degeneration in certain, but not all, regions of the hippocampus, amygdala and cortex (MACGREGOR et al. 1996; 1998). These regional differences could reflect a differentiation between suppression of seizures and direct neuroprotective effects. Complete suppression of seizures in the kainic acid model (with diazepam, for example) leads to protection against all extra-hippocampal neuronal degeneration. The regionally-selective protection by R-PIA may reflect differential distribution of A_1 receptors or suboptimal seizure suppression at the doses used. From the studies reported, it is not possible to dissociate these possibilities, nor is it possible to precisely elucidate the role that A_1 agonist-induced cardiovascular changes and hypothermia play in the protection observed. Again, these issues could be addressed with local injection of A_1 agonists into vulnerable brain regions in these seizure-induced neuronal degeneration models.

II. Adenosine Modulators

The observations that adenosine receptor antagonists are pro-convulsant and exacerbate seizure-related brain damage (see Sect. G.II.) implicate a protective role for endogenous adenosine, and provide firm support for the idea that drugs which can enhance adenosine's actions may have therapeutic potential. Where the different adenosine modulation strategies have been examined, adenosine kinase inhibitors appear to be the most effective as anticonvulsant agents. ZHANG et al. (1993) examined the anticonvulsant activity of the adenosine kinase inhibitors, 5'-amino-5'-deoxyadenosine and 5'-iodotubercidin, the adenosine deaminase inhibitor, deoxycoformycin, and the adenosine transport inhibitors, NBMPR and dilazep, applied locally in the rat prepyriform cortex against bicuculline-induced seizures. Both adenosine kinase inhibitors produced potent antiepileptic effects, as did dilazep, however, NBMPR and deoxycoformycin were ineffective. In the rat maximal electroshock-induced convulsion model, WIESNER et al. (1999) demonstrated potent anticonvulsant activity with several known and novel classes of adenosine kinase inhibitors applied systemically, but reported no anticonvulsant effects of the adenosine deaminase inhibitor, deoxycoformycin or the adenosine transport inhibitors NBMPR and dipyridamole. In the same study, the adenosine kinase inhibitors were effective in reducing epileptiform activity caused by removal of magnesium from the solution bathing rat cortical slices, whereas the adenosine deaminase inhibitors deoxycoformycin and EHNA were ineffective, as was NBMPR, although dipyridamole showed some inhibition.

WIESNER et al. (1999) identified several classes of adenosine kinase inhibitors, one of which, exemplified by GP683, demonstrated anticonvulsant activity by the oral route of administration and possessed good blood–brain barrier penetration. Furthermore, comparison of GP683 with the A_1 receptor agonists CPA and CHA indicated that, at equivalent anticonvulsant doses, GP683 was devoid of the profound reductions of heart rate and blood pressure produced by A_1 receptor agonists, and produced minimal hypothermic effects and less sedative activity compared to these agents. These experiments are a direct demonstration of the advantages of adenosine modulation, compared to direct-acting adenosine receptor agonists in terms of side effect profile. As noted above (Sect. E.III.3.), GP683 has neuroprotective activity in a rat focal stroke model; consequently, the side effect profile described by WIESNER et al. (1999) for this compound provides encouragement that clinical evaluation of adenosine kinase inhibitors may be feasible in the setting of stroke and chronic neurodegenerative disease. An analog with improved oral bioavailability has also been identified (WIESNER et al. 1997).

H. Conclusions

The information reviewed in this chapter constitutes a solid body of evidence for the beneficial effects of adenosine in acute neurodegenerative diseases such as stroke. Unfortunately, information about the potential protective effects of adenosine in chronic neurodegenerative diseases such as Parkinson's disease and Alzheimer's disease is limited. The fact that adenosine is an important brake on the glutamate neurotransmitter system in many areas of the CNS would imply beneficial effects of adenosine where "excitotoxic" mechanisms are important in chronic human neurodegeneration, and the effects of adenosine in the vasculature, white matter, and on neuronal inflammation (SCHUBERT et al. 1998), may also be relevant. The challenge for the development of adenosine-based therapeutics has been delivery of the beneficial effects of adenosine without the unwanted side effects. At the present time, compounds which modulate adenosine levels appear to be the most promising in this respect, particularly adenosine kinase inhibitors. Finally, it is to be hoped that the promising data on A_{2A} receptor antagonists as symptomatic relief in Parkinson's disease models may translate into efficacy in forthcoming clinical studies.

Abbreviations

APEC	2-[p-[2-(2-aminoethylamino)carbonylethyl]phenethylamino]-5′-N-ethylcarboxamidoadenosine
CHA	N^6-cyclohexyladenosine
Cl-IB-MECA	chloro-N^6-(3-iodobenzyl)-5′-N-methyluronamide

CNS	central nervous system
CPA	N^6-cyclopropyladenosine
8-CPT	8-cyclopentyltheophylline
CSC	8-(3-chlorostyryl)caffeine
DPCPX	8-cyclopentyl-1,3-dipropylxanthine
IB-MECA	N^6-(3-iodobenzyl)-5'-N-methyluronamide
L-DOPA	L-3,4-dihydroxyphenylalanine
MPTP	N-methyl-4-phenyl-1,2,3,6-tetrahydropyridine
NECA	5'-N-ethylcarboxamidoadenosine
NBMPR	nitrobenzylthioinosine
8-SPT	8-(p-sulfophenyl)theophylline
R-PIA	R-N^6-(2-phenylisopropyl)adenosine
IB-MECA	N^6-(3-iodobenzyl)-5'-N-methyluronamide
XAC	xanthine amine congener

References

Abbracchio MP, Brambilla R, Ceruti S, Kim HO, von Lubitz DK, Jacobson KA, Cattabeni F (1995) G protein-dependent activation of phospholipase C by adenosine A3 receptors in rat brain. Mol Pharmacol 48:1038–1045

Andine P, Rudolphi KA, Fredholm BB, Hagberg H (1990) Effect of propentofylline (HWA 285) on extracellular purines and excitatory amino acids in CA1 of rat hippocampus during transient ischaemia. Br J Pharmacol 100:814–818

Arvin B, Neville LF, Roberts PJ (1988) 2-Chloroadenosine prevents kainic acid-induced toxicity in rat striatum. Neurosci Lett 93:336–340

Ashton D, De Prins E, Willems R, Van Belle H, Wauquier A (1988) Anticonvulsant action of the nucleoside transport inhibitor, soluflazine, on synaptic and nonsynaptic epileptogenesis in the guinea-pig hippocampus. Epilepsy Res 2:65–71

Ashton D, Willems R, De Prins E, Van Belle H, Wauquier A (1987) The nucleoside-transport inhibitor soluflazine (R 64 719) increases the effects of adenosine in the guinea-pig hippocampal slice and is antagonized by adenosine deaminase. Eur J Pharmacol 142:403–408

Ballarin M, Herrera-Marschitz M, Casas M, Ungerstedt U (1987) Striatal adenosine levels measured "in vivo" by microdialysis in rats with unilateral dopamine denervation. Neurosci Lett 83:338–344

Bari F, Louis TM, Busija DW (1998) Effects of ischemia on cerebral arteriolar dilation to arterial hypoxia in piglets. Stroke 29:222–227

Barth A, Newell DW, Nguyen LB, Winn HR, Wender R, Meno JR, Janigro D (1997) Neurotoxicity in organotypic hippocampal slices mediated by adenosine analogues and nitric oxide. Brain Res 762:79–88

Bell MJ, Kochanek PM, Carcillo JA, Mi Z, Schiding JK, Wisniewski SR, Clark RS, Dixon CE, Marion DW, Jackson E (1998) Interstitial adenosine, inosine, and hypoxanthine are increased after experimental traumatic brain injury in the rat. J Neurotrauma 15:163–170

Blandini F, Porter RH, Greenamyre JT (1996) Glutamate and Parkinson's disease. Mol Neurobiol 12:73–94

Boissard CG, Gribkoff VK (1993) The effects of the adenosine reuptake inhibitor soluflazine on synaptic potentials and population hypoxic depolarizations in area CA1 of rat hippocampus in vitro. Neuropharmacology 32:149–155

Bona E, Aden U, Gilland E, Fredholm BB, Hagberg H (1997) Neonatal cerebral hypoxia-ischemia: the effect of adenosine receptor antagonists. Neuropharmacology 36:1327–1338

Britton DR, Mikusa J, Lee CH, Jarvis MF, Williams M, Kowaluk EA (1999) Site and event specific increase of striatal adenosine release by adenosine kinase inhibition in rats. Neurosci Lett 266:93–96

Brundege JM, Dunwiddie TV (1998) Metabolic regulation of endogenous adenosine release from single neurons. NeuroReport 9:3007–3011

Brundege JM, Dunwiddie TV (1997) Role of adenosine as a modulator of synaptic activity in the central nervous system. Advances in Pharmacology 39:353–391

Brundege JM, Dunwiddie TV (1996) Modulation of excitatory synaptic transmission by adenosine released from single hippocampal pyramidal neurons. J Neuroscience 16:5603–5612

Bruns RF, Fergus JH (1990) Allosteric enhancement of adenosine A1 receptor binding and function by 2-amino-3-benzoylthiophenes. Mol Pharmacol 38:939–949

Bruns RF, Fergus JH, Coughenour LL, Courtland GG, Pugsley TA, Dodd JH, Tinney FJ (1990) Structure-activity relationships for enhancement of adenosine A1 receptor binding by 2-amino-3-benzoylthiophenes. Mol Pharmacol 38:950–958

Clark RS, Carcillo JA, Kochanek PM, Obrist WD, Jackson EK, Mi Z, Wisniewski SR, Bell MJ (1997) Cerebrospinal fluid adenosine concentration and uncoupling of cerebral blood flow and oxidative metabolism after severe head injury in humans. Neurosurgery 41:1284–1292

Cao X, Phillis JW (1995) Adenosine A1 receptor enhancer, PD 81,723, and cerebral ischemia/reperfusion injury in the gerbil. Gen Pharmacol 26:1545–1548

Croning MD, Zetterstrom TS, Grahame-Smith DG, Newberry NR (1995) Action of adenosine receptor antagonists on hypoxia-induced effects in the rat hippocampus in vitro. Br J Pharmacol 116:2113–2119

Daval JL, Nicolas F (1994) Opposite effects of cyclohexyladenosine and theophylline on hypoxic damage in cultured neurons. Neurosci Lett 175:114–116

Duval JL, Von Lubitz DK, Deckert J, Redond DJ, Marangos PJ (1989) Protective effect of cyclohexyladenosine on adenosine A1-receptors, guanine nucleotide and forskolin binding sites following transient brain ischemia: a quantitative autoradiographic study. Brain Res 491:212–226

DeLeo J, Schubert P, Kreutzberg GW (1988) Propentofilline (HWA 285) protects hippocampal neurons of Mongolian gerbils against ischemic damage in the presence of an adenosine antagonist. Neurosci Lett 84:307–311

Delle Donne KT, Sonsalla PK (1994) Protection against methamphetamine-induced neurotoxicity to neostriatal dopaminergic neurons by adenosine receptor activation. J Pharmacol Exp Ther 271:1320–1326

Dietrich WD, Miller LP, Prado R, Dewanjee S, Alex N, Dewanjee MK, Gruber H (1995) Acadesine reduces indium-labeled platelet deposition after photothrombosis of the common carotid artery in rats. Stroke 26:111–116

Dixon AK, Gubitz AK, Sirinathsinghji DJ, Richardson PJ, Freeman TC (1996) Tissue distribution of adenosine receptor mRNAs in the rat. Br J Pharmacol 118:141–148

Dolphin AC, Archer ER (1983) An adenosine agonist inhibits and a cyclic AMP analog enhances the release of glutamate but not GABA from slices of rat dentate gyrus. Neurosci Lett 43:49–54

Doolette DJ (1997) Mechanism of adenosine accumulation in the hippocampal slice during energy deprivation. Neurochem Int 30:211–223

Doolette DJ, Kerr, DI (1995) Hyperexcitability in CA1 of the rat hippocampal slice following hypoxia or adenosine. Brain Res 677:127–137

Dragunow M, Goddard GV, Laverty R (1985) Is adenosine an endogenous anticonvulsant? Epilepsia 26:480–487

Dunwiddie TV, Diao L, Kim HO, Jiang JL, Jacobson KA (1997a) Activation of hippocampal adenosine A_3 receptors produces a desensitization of A_1 receptor-mediated responses in rat hippocampus. J Neuroscience 17:607–614

Dunwiddie TV, Diao L, Proctor WR (1997b) Adenine nucleotides undergo rapid, quantitative conversion to adenosine in the extracellular space in rat hippocampus. J Neuroscience 17:7673–7682

During MJ, Spencer DD (1992) Adenosine: a potential mediator of seizure arrest and postictal refractoriness. Ann Neurol 32:618–624

Dux E, Fastbom J, Ungerstedt U, Rudolphi K, Fredholm BB (1990) Protective effect of adenosine and a novel xanthine derivative propentofyline on the cell damage after bilateral carotid occlusion in the gerbil hippocampus. Brain Res 516:248–256

Evans MC, Swan JH, Meldrum BS (1987) An adenosine analogue, 2-chloroadenosine, protects against long term development of ischaemic cell loss in the rat hippocampus. Neuroscience Lett 83:287–292

Fern R, Waxman SG, Ransom BR (1994) Modulation of anoxic injury in CNS white matter by adenosine and interaction between adenosine and GABA. J Neurophysiol 72:2609–2616

Ferré S, Fredholm BB, Morelli M, Popoli P, Fuxe K (1997) Adenosine-dopamine receptor-receptor interactions as an integrative mechanism in the basal ganglia. TINS 20:482–487

Fink JS, Weaver DR, Rivkees SA, Peterfreund RA, Pollack AE, Adler EM, Reppert SM (1992) Molecular cloning of the rat A2 adenosine receptor: selective co-expression with D2 dopamine receptors in rat striatum. Brain Res Mol Brain Res 14:186–195

Finn SF, Swartz KJ, Beal MF (1991) 2-Chloroadenosine attenuates NMDA, kainate, and quisqualate toxicity. Neurosci Lett 126:191–194

Firestein GS, Bullough DA, Erion MD, Jimenez R, Ramirez-Weinhouse M, Barankiewicz J, Smith CW, Gruber HE, Mullane KM (1995) Inhibition of neutrophil adhesion by adenosine and an adenosine kinase inhibitor. The role of selectins. J Immunol 154:326–334

Fuxe K, Ungerstedt U (1974) Action of caffeine and theophyllamine on supersensitive dopamine receptors: considerable enhancement of receptor response to treatment with DOPA and dopamine receptor agonists. Med Bio 52:48–54

Gao Y, Phillis JW (1994) CGS 15943, an adenosine A2 receptor antagonist, reduces cerebral ischemic injury in the Mongolian gerbil. Life Sci 55:PL61–PL65

Gidday JM, Fitzgibbons JC, Shah AR, Kraujalis MJ, Park TS (1995) Reduction in cerebral ischemic injury in the newborn rat by potentiation of endogenous adenosine. Pediatr Res 38:306–311

Goldberg MP, Moyer H, Weiss JH, Choi DW (1988) Adenosine reduces cortical neuronal injury induced by oxygen or glucose deprivation in vitro. Neuroscience Lett 89:323–327

Greenamyre JT, Eller RV, Zhang Z, Ovadia A, Kurlan R, gash DM (1994) Antiparkinsonian effects of remacemide hydrochloride, a glutamate antagonist, in rodent and primate models of Parkinson's disease. Ann Neurol 35:655–661

Greene RW, Haas HL (1991) The electrophysiology of adenosine in the mammalian central nervous system. Prog Neurobiol 36:329–341

Gribkoff VK, Bauman LA, VanderMaelen CP (1990) The adenosine antagonist 8-cyclopentyltheophylline reduces the depression of hippocampal neuronal responses during hypoxia. Brain Res 512:353–357

Griffith DA, Conant AR, Jarvis SM (1990) Differential inhibition of nucleoside transport systems in mammalian cells by a new series of compounds related to lidoflazine and mioflazine. Biochem Pharmacol 40:2297–303

Griffiths M, Beaumont N, Yao SYM, Sundaram M, Boumah CE, Davies A, Kwong FYP, Coe I, Cass CE, Young JD, Baldwin SA (1997a) Cloning of a human nucleoside transporter implicated in cellular uptake of adenosine and chemotherapeutic drugs. Nature Medicine 3:89–93

Griffiths M, Yao SY, Abidi F, Phillips SE, Cass CE, Young JD, Baldwin SA (1997b) Molecular cloning and characterization of a nitrobenzylthioinosine-insensitive (ei) equilibrative nucleoside transporter from human placenta. Biochem J 328:739–743

Grondin R, Bedard PJ, Hadj Tahar A, Gregoire L, Mori A, Kase H (1999) Antiparkinsonian effect of a new selective adenosine A2A receptor antagonist in MPTP-treated monkeys. Neurology 52:1673–1677

Gruber HE, Hoffer ME, McAllister DR, Laikind PK, Lane TA, Schmid-Schoenbein GW, Engler RL (1989) Increased adenosine concentration in blood from ischemic myocardium by AICA riboside. Effects on flow, granulocytes, and injury. Circulation 80:1400–1411

Guieu R, Dussol B, Halimi G, Bechis G, Sampieri F, Berland Y, Sampol J, Couraud F, Rochat H (1998) Adenosine and the nervous system: pharmacological data and therapeutic perspectives. Gen Pharmacol 31:553–561

Halle JN, Kasper CE, Gidday JM, Koos BJ (1997) Enhancing adenosine A1 receptor binding reduces hypoxic-ischemic brain injury in newborn rats. Brain Res 759: 309–312

Hauber W, Nagel J, Sauer R, Müller CE (1998) Motor effects induced by a blockade of adenosine A_{2A} receptors in the caudate-putamen. NeuroReport 9:1803–1806

Headrick JP, Bendall MR, Faden AI, Vink R (1994) Dissociation of adenosine levels from bioenergetic state in experimental brain trauma: potential role in secondary injury. J Cereb Blood Flow Metab 14:853–861

Heurteaux C, Lauritzen I, Widmann C, Lazdunski M (1995) Essential role of adenosine, adenosine A1 receptors, and ATP-sensitive K+ channels in cerebral ischemic preconditioning. Proc Natl Acad Sci USA 92:4666–4670

Huston JP, Haas HL, Boix F, Pfister M, Decking U, Schrader J, Schwarting RK (1996) Extracellular adenosine levels in neostriatum and hippocampus during rest and activity periods of rats. Neuroscience 73:99–107

Jacobson KA (1998) Adenosine A3 receptors: novel ligands and paradoxical effects. Trends Pharmacol Sci 19:184–191

Jacobson KA, van Galen PJ, Williams M (1992) Adenosine receptors: pharmacology, structure-activity relationships, and therapeutic potential. J Med Chem 35:407–422

Janusz CA, Berman RF (1993) The adenosine binding enhancer, PD 81,723, inhibits epileptiform bursting in the hippocampal brain slice. Brain Res 619:131–136

Janusz CA, Bruns RF, Berman RF (1991) Functional activity of the adenosine binding enhancer, PD 81,723, in the in vitro hippocampal slice. Brain Res 567:181–187

Jiang N, Kowaluk EA, Lee CH, Mazdiyasni H, Chopp M (1997) Adenosine kinase inhibition protects brain against transient focal ischemia in rats. Eur J Pharmacol 320:131–137

Jo Y-H, Schlicter R (1999) Synaptic corelease of ATP and GABA in cultured spinal neurons. Nature Neuroscience 2:241

Johnson MP, McCarty DR, Chmielewski PA (1998) Temporal dependent neuroprotection with propentofylline (HWA 285) in a temporary focal ischemia model. Eur J Pharmacol 356:151–157

Jones PA, Smith RA, Stone TW (1998) Protection against hippocampal kainate excitotoxicity by intracerebral administration of an adenosine A2A receptor antagonist. Brain Res 800:328–335

Jones KW, Hammond JR (1995) Characterization of nucleoside transport activity in rabbit cortical synaptosomes. Can J Physiol Pharmacol 83:1733–1741

Kanda T, Jackson MJ, Smith LA, Pearce RKB, Nakamura J, Kase H, Kuwana Y, Jenner P (1998a) Adenosine A_{2A} antagonist: a novel antiparkinsonian agent that does not provoke dyskinesia in parkinsonian monkeys. Annals of Neurology 43:508–513

Kanda T, Tashiro T, Kuwana Y, Jenner P (1998b) Adenosne A_{2A} receptors modify motor function in MPTP-treated common marmosets. NeuroReport 9:2857–2860

Katchman AN, Hershkowitz N (1993) Adenosine antagonists prevent hypoxia-induced depression of excitatory but not inhibitory synaptic currents. Neurosci Lett 159: 123–126

Kawahara N, Ide T, Saito N, Kawai K, Kirino T (1998) Propentofylline potentiates induced ischemic tolerance in gerbil hippocampal neurons via adenosine receptor. J Cereb Blood Flow Metab 18:472–475

Kirk IP, Richardson PJ (1994) Adenosine A2a receptor-mediated modulation of striatal [3H]GABA and [3H]acetylcholine release. J Neurochem 62:960–966

Kittner B, Rossner M, Rothr M (1997) Clinical trials in dementia with propentofylline. Ann NY Acad Sci 826:307–316

Knutsen LJS, Murray TF (1997) Adenosine and ATP in epilepsy. In: Jacobson KA, Jarvis MF (eds) Purinergic Approaches in Experimental Therapeutics, pp 423–447, Wiley-Liss, New York

Kochanek PM, Clark RS, Obrist WD, Carcillo JA, Jackson EK, Mi Z, Wisniewski SR, Bell MJ, Marion DW (1997) The role of adenosine during the period of delayed cerebral swelling after severe traumatic brain injury in humans. Acta Neurochir Suppl (Wien) 70:109–111

Kurokawa M, Kirk IP, Kirkpatrick KA, Kase H, Richardson PJ (1994) Inhibition by KF17837 of adenosine A2A receptor-mediated modulation of striatal GABA and Ach release. Br J Pharmacol 113:43–48

Lau YS, Mouradian MM (1993) Protection against acute MPTP-induced dopamine depletion in mice by adenosine A1 agonist. J Neurochem 60:768–771

Lekieffre D, Callebert J, Plotkine M, Allix M, Boulu RG (1991) Enhancement of endogenous excitatory amino acids by theophylline does not modify the behavioral and histological consequences of forebrain ischemia. Brain Res 565: 353–357

Lin Y, Phillis JW (1992) Deoxycoformycin and oxypurinol: protection against focal ischemic brain injury in the rat. Brain Res 571:272–280

Lobner D, Choi DW (1994) Dipyridamole increases oxygen-glucose deprivation-induced injury in cortical cell culture. Stroke 25:2085–2089; discussion 2089–2090

Logan M, Sweeney MI (1997) Adenosine A1 receptor activation preferentially protects cultured cerebellar neurons versus astrocytes against hypoxia-induced death. Mol Chem Neuropathol 31:119–133

Lynch JJ 3rd, Alexander KM, Jarvis MF, Kowaluk EA (1998) Inhibition of adenosine kinase during oxygen-glucose deprivation in rat cortical neuronal cultures. Neurosci Lett 252:207–210

Macek TA, Schaffhauser H, Conn PJ (1998) Protein kinase C and A_3 adenosine receptor activation inhibit presynaptic metabotropic glutamate receptor (mGluR) function and uncouple mGluRs from GTP-binding proteins. J Neurosci 18:6138–6146

MacGregor DG, Graham DI, Jones PA, Stone TW (1998) Protection by an adenosine analogue against kainate-induced extrahippocampal neuropathology. Gen Pharmacol 31:233–238

MacGregor DG, Jones PA, Maxwell WL, Graham DI, Stone TW (1996) Prevention by a purine analogue of kainate-induced neuropathology in rat hippocampus. Brain Res 725:115–120

Manzoni OJ, Manabe T, Nicoll RA (1994) Release of adenosine by activation of NMDA receptors in the hippocampus. Science 265:2098–2101

Marston HM, Finlayson K, Maemoto T, Olverman HJ, Akahane, A, Sharkey J, Butcher SP (1998) Pharmacological characterization of a simple behavioral response mediated selectively by central adenosine A_1 receptors, using in vivo and in vitro techniques. J Pharmacology Exp Therapeutics 285:1023–1030

Martinez-Mir MI, Probst A, Palacios JM (1991) Adenosine A2 receptors: selective localization in the human basal ganglia and alterations with disease. Neuroscience 42:697–706

Masino SA, Dunwiddie TV (1999) Temperature-dependent modulation of excitatory transmission in hippocampal slices is mediated by Extracellular adenosine. J Neuroscience 19:1932–1939

Matsumoto K, Graf R, Rosner G, Taguchi J, Heiss WD (1993) Elevation of neuroactive substances in the cortex of cats during prolonged focal ischemia. J Cereb Blood Flow Metab 13:586–594

Meno JR, Ngai AC, Winn HR (1993) Changes in pial arteriolar diameter and CSF adenosine concentrations during hypoxia. J Cerebral Blood Flow and Metabolism 13:214–220

Michel PP, Marien M, Ruberg M, Colpaert F, Agid Y (1999) Adenosine prevents the death of mesencephalic dopaminergic neurons by a mechanism that involves astrocytes. J Neurochem 72:2074–2082

Miller LP (1999) Adenosine in ischemic brain injury. In stroke therapy, basic, preclinical directions, Ed Leonard P. Miller, Wiley-Liss, Inc.

Miller LP, Jelovich LA, Yao L, DaRe J, Ugarkar B, Foster AC (1996) Pre- and peri-stroke treatment with the adenosine kinase inhibitor, 5'-deoxyiodotubercidin, significantly reduces infarct volume after temporary occlusion of the middle cerebral artery in rats. Neurosci Lett 220:73–76

Miller LP, Chiang PC, Carriedo S, Metzner K, Foster AC (1994) The adenosine regulating agent, GP-1–668, reduces infarct volume and neurological deficits in a rat model of focal stroke with reperfusion. Abs. Stroke Council 19th International Joint Converence on Stroke and Cerebral Circulation, February 17–19, 1994, San Diego, CA USA, p 21

Miller LP, Hsu C (1992) Therapeutic potential for adenosine receptor activation in ischemic brain injury. J Neurotrauma 9 [Suppl 2]:S563–S577

Mitchell JB, Lupica CR, Dunwidde TV (1993) Activity-dependent release of endogenous adenosine modulates synaptic responses in the rat hippocampus. J Neurosci 13:3439–3447

Mitchell HL, Frisella WA, Brooker RW, Yoon KW (1995) Attenuation of traumatic cell death by an adenosine A1 agonist in rat hippocampal cells. Neurosurgery 36:1003–1007; discussion 1007–1008

Mizumura T, Auchampach JA, Linden J, Bruns RF, Gross GJ (1996) PD 81,723, an allosteric enhancer of the A1 adenosine receptor, lowers the threshold for ischemic preconditioning in dogs. Circ Res 79:415–423

Monopoli A, Lozza G, Forlani A, Mattavelli A, Ongini E (1998) Blockade of adenosine A2A receptors by SCH 58261 results in neuroprotective effects in cerebral ischaemia in rats. Neuroreport 9:3955–3959

Mori A, Shindou T, Ichimura M, Nonaka H, Kase H (1996) The role of adenosine A2a receptors in regulating GABAergic synaptic transmission in striatal medium spiny neurons. J Neurosci 16:605–611

Musser B, Mudumbi RV, Liu J, Olson RD, Vestal RE (1999) Adenosine A1 receptor-dependent and -independent effects of the allosteric enhancer PD 81,723. J Pharmacol Exp Ther 288:446–454

Nagata K, Ogawa T, Omosu M, Fujimoto K, Hayashi S (1985) In vitro and in vivo inhibitory effects of propentofylline on cyclic AMP phosphodiesterase activity. Arzneimittelforschung 35:1034–1036

Nilsson P, Hillered L, Ponten U, Ungerstedt U (1990) Changes in cortical extracellular levels of energy-related metabolites and amino acids following concussive brain injury in rats. J Cereb Blood Flow Metab 19:631–637

O'Connor SD, Stojanovic M, Radulovacki M (1991) The effect of soluflazine on sleep in rats. Neuropharmacology 30:671–674

Ongini E, Adami M, Ferri C, Bertorelli R (1997) Adenosine A2A receptors and neuroprotection. Ann NY Acad Sci 825:30–48

Pak MA, Haas HL, Decking UK, Schrader J (1994) Inhibition of adenosine kinase increases endogenous adenosine and depresses neuronal activity in hippocampal slices. Neuropharmacology 33:1049–1053

Park CK, Rudolphi KA (1994) Antiischemic effects of propentofylline (HWA 285) against focal cerebral infarction in rats. Neurosci Lett 178:235–238

Perez-Pinzon MA, Mumford PL, Rosenthal M, Sick TJ (1996) Anoxic preconditioning in hippocampal slices: role of adenosine. Neuroscience 75:687–694

Phillis JW (1995) The effects of selective A1 and A2a adenosine receptor antagonists on cerebral ischemic injury in the gerbil. Brain Res 705:79–84

Phillis JW, Smith-Barbour M, Perkins LM, O'Regan MH (1994) Characterization of glutamate, aspartate, and GABA release from ischemic rat cerebral cortex. Brain Res Bull 34:457–466

Phillis JW, Smith-Barbour M (1993) The adenosine kinase inhibitor, 5-iodotubercidin, is not protective against cerebral ischemic injury in the gerbil. Life Sci 53:497–502

Phillis JW, O'Regan MH (1989) Deoxycoformycin antagonizes ischemia-induced neuronal degeneration. Brain Res Bull 22:537–540

Phillis JW, O'Regan MH, Walter GA (1989) Effects of two nucleoside transport inhibitors, dipyridamole and soluflazine, on purine release from the rat cerebral cortex. Brain Res 481:309–316

Picano E, Abbracchio MP (1998) European stroke prevention study-2 results: serendipitous demonstration of neuroprotection induced by endogenous adenosine accumulation. TIPS 19:14–16

Pinard E, Riche D, Puiroud S, Seylaz J (1990) Theophylline reduces cerebral hyperaemia and enhances brain damage induced by seizures. Brain Res 511:303–309

Porkka-Heiskanen T, Strecker RE, Thakkar M, Bjorkum AA, Greene RW, McCarley RW (1997) Adenosine: a mediator of the sleep-inducing effects of prolonged wakefulness. Science 276:1265–1268

Portas CM, Thakkar M, Rainnie DG, Greene RW, McCarley RW (1997) Role of adenosine in behavioral state modulation: a microdialysis study in the freely moving car. Neuroscience 79:225–235

Rainnie DG, Grunze HC, McCarley RW, Greene RW (1994) Adenosine inhibition of mesopontine cholinergic neurons: implications for EEG arousal. Science 263:689–692

Richardson PJ, Kase H, Jenner PG (1997) Adenosine A_{2A} receptor antagonists as new agents for the treatment of Parkinson's disease. TIPS 17:338–344

Rogler-Brown T, Agarwal RP, Parks RE Jr (1978) Tight binding inhibitors–VI. Interactions of deoxycoformycin and adenosine deaminase in intact human erythrocytes and sarcoma 180 cells. Biochem Pharmacol 27:2289–2296

Roussel S, Pinard E, Seylaz J (1991) Focal cerebral ischemia in chronic hypertension: no protection by R-phenylisopropyladenosine. Brain Res 545:171–174

Rudolphi KA, Schubert P, Parkinson FE, Fredholm BB (1992) Neuroprotective role of adenosine in cerebral ischaemia. TIPS 13:439–445

Rudolphi KA, Keil M, Fastbom J, Fredholm BB (1989) Ischaemic damage in gerbil hippocampus is reduced following upregulation of adenosine (A1) receptors by caffeine treatment. Neurosci Lett 103:275–280

Rudolphi KA, Keil M, Hinze HJ (1987) Effect of theophylline on ischemically induced hippocampal damage in Mongolian gerbils: a behavioral and histopathological study. J Cereb Blood Flow Metab 7:74–81

Schiffmann SN, Jacobs O, Vanderhaeghen JJ (1991a) Striatal restricted adenosine A2 receptor (RDC8) is expressed by enkephalin but not by substance P neurons: an in situ hybridization histochemistry study. J Neurochem 57:1062–1067

Schiffmann SN, Libert F, Vassart G, Vanderhaeghen JJ (1991b) Distribution of adenosine A2 receptor mRNA in the human brain. Neurosci Lett 130:177–181

Schubert P, Ogata T, Miyazaki H, Marchini C, Ferroni S, Rudolphi K (1998) Pathological immuno-reactions of glial cells in Alzheimer's disease and possible sites of interference. J Neural Transm Suppl 54:167–174

Schwarcz R, Ben-Ari Y (1986) Acids and epilepsy. Advances in Experimental Medicine and Biology, Vol. 203, Plennum Press, New York

Sciotti VM, Van Wylen DG (1993) Increases in interstitial adenosine and cerebral blood flow with inhibition of adenosine kinase and adenosine deaminase. J Cereb Blood Flow Metab 13:201–207

Sciotti VM, Roche FM, Grabb MC, Van Wylen DG (1992) Adenosine receptor blockade augments interstitial fluid levels of excitatory amino acids during cerebral ischemia. J Cereb Blood Flow Metab 12:646–655

Seida M, Wagner HG, Vass K, Klatzo I (1988) Effect of aminophylline on postischemic edema and brain damage in cats. Stroke 10:1275–1282

Sonsalla PK, Albers DS, Zeevalk GD (1998) Role of glutamate in neurodegeneration of dopamine neurons in several animal models of parkinsonism. Amino Acids 14:69–74

Sutherland GR, Peeling J, Lesiuk HJ, Brownstone RM, Rydzy M, Saunders JK, Geiger JD (1991) The effects of caffeine on ischemic neuronal injury as determined by magnetic resonance imaging and histopathology. Neuroscience 42:171–182

Svenningsson P, Le Moine C, Aubert I, Burbaud P, Fredholm BB, Bloch B (1998) Cellular distribution of adenosine A2A receptor mRNA in the primate striatum. J Comp Neurol 399:229–240

Tatlisumak T, Takano K, Carana RA, Miller LP, Foster AC, Fisher M (1998) Delayed treatment with an adenosine kinase inhibitor, GP683, attenuates infarct size in rats with temporary middle cerebral artery occlusion. Stroke 29:1952–1958

Thompson SM, Haas HL, Gahwiler BH (1992) Comparison of the actions of adenosine at pre- and postsynaptic receptors in the rat hippocampus in vitro. J Physiol (Lond) 451:347–363

Thorn JA, Jarvis SM (1996) Adenosine Transporters. Gen Pharmac 27:613–620

Turski WA, Cavalheiro EA, Ikonomidou C, Mello LE, Bortolotto ZA, Turski L (1985) Effects of aminophylline and 2-chloroadenosine on seizures produced by pilocarpine in rats: morphological and electroencephalographic correlates. Brain Res 361:309–323

Von Lubitz DK (1999) Adenosine and cerebral ischemia: therapeutic future or death of a brave concept? Eur J Pharmacol 365:9–25

Von Lubitz DK, Beenhakker M, Lin RC, Carter MF, Paul IA, Bischofberger N, Jacobson KA (1996a) Reduction of postischemic brain damage and memory deficits following treatment with the selective adenosine A1 receptor agonist. Eur J Pharmacol 302:43–48

Von Lubitz DK, Lin RC, Paul IA, Beenhakker M, Boyd M, Bischofberger N, Jacobson KA (1996b) Postischemic administration of adenosine amine congener (ADAC): analysis of recovery in gerbils. Eur J Pharmacol 316:171–179

Von Lubitz DK, Lin RC, Jacobson KA (1995) Cerebral ischemia in gerbils: effects of acute and chronic treatment with adenosine A2A receptor agonist and antagonist. Eur J Pharmacol 287:295–302

Von Lubitz DK, Lin RC, Malman N, Ji XD, Carter MF, Jacobson KA (1994) Chronic administration of selective adenosine A1 receptor agonist or antagonist in cerebral ischemia. Eur J Pharmacol 256:161–167

Von Lubitz DK, Lin RC, Popik P, Carter MF, Jacobson KA (1994) Adenosine A3 receptor stimulation and cerebral ischemia. Eur J Pharmacol 263:59–67

Von Lubitz DK, Marangos PJ (1990) Cerebral ischemia in gerbils: postischemic administration of cyclohexyl adenosine and 8-sulfophenyl-theophylline. J Mol Neurosci 2:53–59

Wang J, Schaner ME, Thomassen S, Su S-F, Piquette-Miller M, Giacomini KM (1997) Functional and molecular characteristics of Na+-dependent nucleoside transporters. Pharmaceutical Research 14:1524–1532

Wiesner JB, Ugarkar BG, Castellino AJ, Barankiewicz J, Dumas DP, Gruber HE, Foster AC, Erion MD (1999) Adenosine kinase inhibitors as a novel approach to anticonvulsant therapy. J Pharmacol Exp Ther 289:1669–1677

Wiesner JB, Mullane KM, Foster AC (1997) A selective adenosine kinase inhibitor, GP3269, as a novel approach to anticonvulsant therapy. In: Okada Y (ed) The Role of Adenosine in the Nervous System, p 277, Elsevier, Amsterdam

Wiesner JB, Zimring ST (1994) Inhibition of maximal electroshock seizures (MES) by A1 adenosine receptor agonists. Abs Soc Neurosci 20:668–669

Wu LG, Saggau P (1994) Adenosine inhibits evoked synaptic transmission primarily by reducing presynaptic calcium influx in area CA1 of hippocampus. Neuron 12:1139–1148

Yang SN, Dasgupta S, Lledo PM, Vincent JD, Fuxe K (1995) Reduction of dopamine D2 receptor transduction by activation of adenosine A2a receptors in stably A2a/D2 (long-form) receptor co-transfected mouse fibroblast cell lines: studies on intracellular calcium levels. Neuroscience 68:729–736

Zhang G, Franklin PH, Murray TF (1993) Manipulation of endogenous adenosine in the rat prepiriform cortex modulates seizure susceptibility. J Pharmacol Exp Ther 264:1415–1424

Zeng YC, Domenici MR, Frank C, Sagratella S, Scotti de Carolis A (1992) Effects of adenosinergic drugs on hypoxia-induced electrophysiological changes in rat hippocampal slices. Life Sci 51:1073–1082

Zhou JG, Meno JR, Hsu SS, Winn HR (1994) Effects of theophylline and cyclohexyladenosine on brain injury following normo- and hyperglycemic ischemia: a histopathologic study in the rat. J Cereb Blood Flow Metab 14:166–173

CHAPTER 8
Sodium and Calcium Channel Blockers

C.P. Taylor

A. Introduction

The term "neuroprotection" means the pharmacologic prevention of the rapid (hours) or slower neuronal cell death caused by impaired blood flow or head trauma. Typically, neuroprotection is studied by acute drug treatment shortly after the insult in animal models of stroke, brain ischemia, or head trauma. The study of neuroprotection with organic cation channel blockers probably begins with nimodipine. This drug is a 1,4-dihydropyridine compound closely related to nitrendipine and other drugs used to treat hypertension by relaxing arterial smooth muscles. Subsequently, other compounds with varying specificity for blocking Ca^{2+} and Na^+ channels have proven to be neuroprotective in various animal models. Several of these have been studied in placebo-controlled clinical trials. Unfortunately, to date, none of these compounds has unequivocally reduced human brain damage from stroke, cardiac arrest, or head trauma.

Neuroprotection also occurs in animals treated with the unique polypeptide molecule, ziconotide (also known as SNX-111). Ziconotide is fundamentally different from the others mentioned above, in that it is a relatively large polypeptide that is highly specific for a single sub-class of Ca^{2+} channels and is highly potent (IC_{50} approximately 1.0 nM). Ziconotide blocks only the so-called N-type channels found exclusively on neurons and endocrine cells (see section on channel types). Although very selective, ziconotide suffers from the drawback that it is a relatively large and highly charged molecule (MW = 2.8 kDa), with much lower permeability to membranes (including the blood–brain barrier) than traditional synthetic drug molecules (MW = 250–400 Da). N-type channels are responsible for a significant fraction of rapid neurotransmitter release in brain and peripheral ganglia. Therefore, ziconotide and related peptides represent a novel approach to neuroprotection, and also provide a unique tool for the study of N-type Ca^{2+} channels.

Although none of the clinical studies with Na^+ or Ca^{2+} blockers have yet shown clear-cut reductions in brain damage with drug treatment, the rationale for treatment with this class of agents (based on results from animal models) remains strong. This review is an attempt to briefly summarize the recent literature with regard to the structure and function of voltage-gated Na$^+$ and Ca^{2+} channels in the central nervous system, and also to describe blockade by synthetic compounds that may be of therapeutic benefit for the treatment of ischemia. A recent review of the chemical patent literature (Cox and Denyer 1998) covers similar material from the point of view of medicinal chemistry.

B. Structure and Function of Voltage-Gated Cation Channels

I. Molecular Biology and Protein Structure

1. Basic Structure of Channel Proteins

Sodium and calcium ion channels are both members of a large family of voltage-gated cation channels that all have related molecular architecture and shared functional attributes (Armstrong and Hille 1998; Hille 1992). The structure and function of sodium (Catterall 1992; Taylor and Narasimhan 1997) and calcium channels (Nooney et al. 1997; Walker and De Waard 1998) are the subject of several recent reviews. In addition, there are selective Cl$^-$ channels and nonselective ion channels. Several genetic diseases are caused by mutations that alter the function of ion channels (Ackerman and Clapham 1997). All of the voltage-gated cation channels have a similar two-dimensional structural plan (Fig. 1). Furthermore, all are assumed to have a three-dimensional structure that forms a plasma membrane-spanning "doughnut" shape comprising an ion-conducting path from the extracellular space into the cell cytosol (Fig. 2). Recently, the three-dimensional structures of a small bacterial K$^+$ channel and one fragment (the intracellular domain) of a mammalian voltage-gated K$^+$ channel have been determined by X-ray crystallography (Sansom 1998). These studies show that each of the four separate subunits of the K$^+$ channel is arranged in a rosette around a central ion-conducting pore. It is assumed that Na$^+$ and Ca^{2+} channels have a similar rosette arrangement, with each of the four homologous domains (I–IV, Figs. 1, 2) surrounding a central pore.

Na$^+$ and Ca^{2+} channels are relatively large glycosylated and membrane-bound protein monomers (1600–1800 amino acids), each coded by a single gene. There are several different genes that code for subtypes of Na$^+$ and Ca^{2+} channels expressed in brain (Tables 1, 2). In sodium channels the ion-conducting protein is called the α subunit and in calcium channels the analogous protein is called α-1. In both Na$^+$ and Ca^{2+} channels, the four domains each consist of six membrane-spanning α helices of about 25 amino acids (Fig. 1). The amino acids lining the pore are highly charged, allowing a ready interface with water and ions (Fig. 2). There also are lipophilic amino acids within the helices that anchor to the plasma membrane. All the known channel

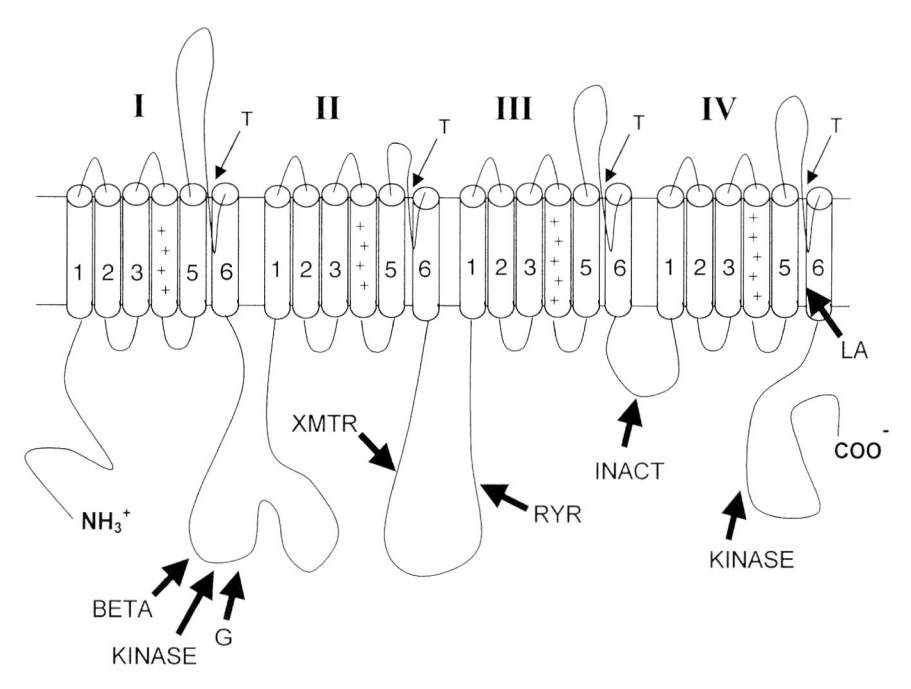

Fig. 1. Structural features of Na⁺ and Ca²⁺ channels include four homologous subunits (I–IV), each made up of six transmembrane domains (1–6). Intervening polypeptide loops extend extracellularly (*up*) or intracellularly (*down*). Domain 4 of each subunit has (+) charges that move to cause gating. Several extracellular loops are extensively glycosylated (not shown). Tetrodotoxin sensitivity is dependent on glutamate residues in each subunit at the extracellular mouth of the Na⁺ channel (*T*). Blockade by local anesthetics (*LA*), anticonvulsants and verapamil depends upon several residues near the cytosolic side of the ion-conducting pore. Inactivation of Na⁺ channels is dependent on folding of an intracellular loop (*INACT*) across the ion-conducting pathway; this loop is unimportant to Ca²⁺ channel inactivation. Protein kinase C or cyclic A-dependent protein kinases (*KINASE*) modulate channel function when they phosphorylate defined residues. N-type Ca²⁺ channels interact with proteins needed for rapid neurotransmitter release (*XMTR*) (syntaxin, Synaptotagmin, SNAP-25). L-type Ca²⁺ channels of striate muscle interact with the ryanodine receptor protein (*RYR*), which releases Ca²⁺ from intracellular stores. The Ca²⁺ channel β subunit (*BETA*) interacts with residues at the I–II cytosolic loop. G-protein β-γ subunits (*G*) bind to modify Ca²⁺ and Na⁺ channel function in response to G-protein receptor activation. The β-γ interaction in Ca²⁺ channels is partly competitive with binding of the β subunit

blockers and toxins interact directly with the ion conducting (α or α-1) subunits. When expressed artificially in cells using recombinant techniques, the isolated ion-conducting subunits provide functional channels that selectively conduct only a single ion species (however, see Sect. I.2. on auxiliary protein subunits).

Among the brain Ca²⁺ channels, neurotransmitter receptors such as GABA$_B$, adenosine, muscarinic acetylcholine, and metabotropic glutamate

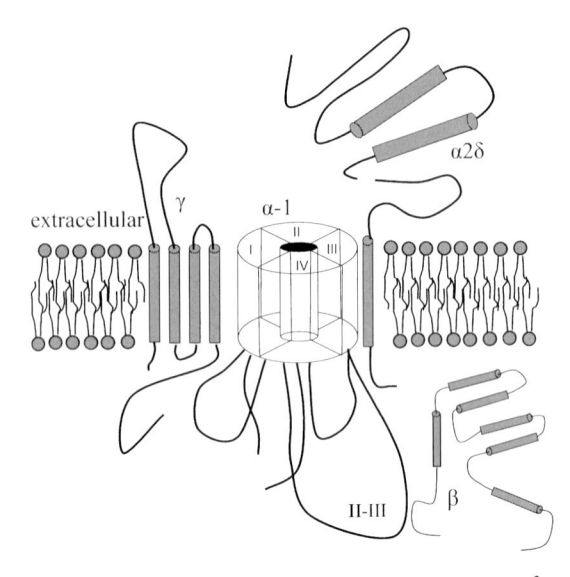

Fig. 2. Schematic three-dimensional diagram of voltage-gated Ca^{2+} channel showing auxiliary protein subunits (α-2-δ, β and γ), and the three protein subunits of the α-1 protein (*I–IV*) configured around the ion-conducting pore

receptors modulate particularly the N-type and Q-type channels (WHEELER et al. 1994). Furthermore, the N and P/Q channel types are largely responsible for neurotransmitter release at a variety of synapses in brain, and this has been studied particularly at glutamate synapses in hippocampus and neocortex. In order to mediate the modulatory effects of neurotransmitters on Ca^{2+} channels, intracellular messenger proteins interact with the ion-conducting subunit. Many of the amino acid residues that are needed for these interactions have been mapped (WALKER and DE WAARD 1998) (see Fig. 1).

Many neurotransmitter modulations are mediated by cytosolic G-proteins (WICKMAN and CLAPHAM 1995). G-protein coupled receptors reduce the current through Ca^{2+} channels through a membrane-delimited signaling pathway, and this interaction can reduce current by as much as 80%–90%. Many Ca^{2+} channels temporarily loose their G-protein mediated inhibition after the application of a brief voltage-clamp pulse to +50mV, which temporarily dissociates G-protein subunits. In the resting state, β-γ protein dimers associate with a G-α protein subunit near the neurotransmitter receptor site. Upon receptor activation, the dissociated G-β-γ dimer diffuses toward a Ca^{2+} channel and associates with the α-1 subunit, causing inhibition. The residues of the Ca^{2+} channel α-1 subunit that mediate this interaction are on the I–II cytosolic loop of class A, B, and E Ca^{2+} channels (Fig. 2). This pathway explains the modulation of N-type Ca^{2+} channels by somatostatin and by muscarinic acetylcholine agonists. Na$^+$ channels also are modulated directly by G-protein β-γ subunits (MA et al. 1994). One particular β-γ combination causes a large

Table 1. Na$^+$ channel characteristics

Sub-family	Major name	Other names	Human gene	Location of expression	TTX IC$_{50}$	Major functions
Nav 1.1	Type I	Brain I	SCN1A	Brain, DRG, brainstem (dendrites)	Low nM	Integration of synaptic potentials in cell bodies and dendrites
Nav 1.2	Type II	Brain II	SCN2A	Brain (axons)	Low nM	Includes splice variant, IIA, which is the most common type in brain
Nav 1.3	Type III	Brain III	SCN3A	Fetal brain, adult DRG (after axotomy)	Low nM	Expressed significantly only in fetal brain
Nav 1.4	SkM1		SCN4A	Skeletal muscle	Low nM	Skeletal muscle excitation
Nav 1.5	SkM2	rH1, hH1	SCN5A	Heart, denervated muscle	10–30 μM	Heart muscle excitation
Nav 1.6	Type VI	Scn8a, Cer III	SCN8A	Brain, cerebellar Purkinje cells	Low nM	Cerebellar function
Nav 1.7	PN1		SCN10A	DRG cells	Low nM	Sensory neuron function
Nav 1.8	PN3	SNS	SCN9 A	Pain-sensitive DRG cells only	20–50 μM	Pain-sensitive sensory neuronal function (upregulated with inflammation, neuropathic damage)
Nav 2.1	Nav 2.1		SCN6A	Brain, uterus		Not known
Nav 2.2			SCN7A	Glial cells, others?		
Nav 3.1	NaN	SNS2	SCN11A	Pain-sensitive DRG cells only	0.5–3 μM	Pain-sensitive sensory neuronal function (upregulated with axotomy)

Table 2. Ca^{2+} channel characteristics

α1 Gene	Pharmacol. name	Selective antagonists	Location of expression	Major functions
$α1_A$	P/Q-type	Ω-Aga IVA (selective for P) Ω-MVIIC (not highly selective)	P: cerebellar Purkinje cells Q: pyramidal neurons (splice variants)	Neurotransmitter release
$α1_B$	N-type	Ω-MVIIA (ziconotide) Ω-GVIA	Neurons (splice variants): α1B-d in brain α1B-b in sensory neurons	Neurotransmitter release
$α1_C$	L-type	1,4-Dihydropyridines	Heart, lung, vascular smooth muscle	Muscle contraction, change gene expression
$α1_D$	L-type (neuronal)	1,4-Dihydropyridines	Brain, endocrine	Regulate slow changes in intracellular Ca^{2+} concentration
$α1_E$	R-type	SNX-482	Brain	Not known
$α1_G, α1_H$	T-type	Mebafradil (not highly selective)	Brain, other tissues	Transient activation; neuronal firing, modulation of cardiac pacemaking

increase in the noninactivating component of Na⁺ current (MA et al. 1997), in contrast to the action of G-proteins on Ca^{2+} channels, which are mostly inhibitory.

A second type of ion channel modulation comes from phosphorylation of channel proteins by cytosolic protein kinases. Neither G-protein mediated modulation, nor protein kinase-mediated modulation has been the target of drug therapy, probably because these systems interact with a large number of additional proteins besides ion channels.

2. Auxiliary Ion Channel Subunit Proteins

Both Na⁺ and Ca^{2+} channel ion-conducting subunits associate with glycosylated auxiliary proteins. These auxiliary proteins may aid the proper folding and insertion of ion-conducting proteins into the membrane after synthesis in the cytosol. Therefore, co-expression of auxiliary subunits in recombinant systems increases the peak current observed in voltage-clamp experiments with single cells, and usually increases the rate of activation and inactivation. The structure and function of ion channel auxiliary subunits has been reviewed (ISOM et al. 1994; WALKER and DE WAARD 1998). Particularly with Ca^{2+} channels, the auxiliary subunits are important to normal function, and only very small Ca^{2+} currents can be recorded in recombinant systems unless the β and α-2 subunits also are expressed. In contrast, the β-1 and β-2 subunits of Na⁺ channels are less vital; almost normal Na⁺ currents are seen when only α subunits are expressed in mammalian cells. In addition, both β (PRAGNELL et al. 1994; STEA et al. 1993) and α-2-δ (SHIROKOV et al. 1998) subunits of Ca^{2+} channels appear to be needed for normal rapid inactivation in non-L-type channels. At present, there are four known subtypes of Ca^{2+} channel β subunits (SCOTT et al., 1996), and also three or four different α-2 subunits, one of whose expression is restricted to brain (KLUGBAUER et al. 1999). Recent studies describe a γ auxiliary protein subunit in brain that also contributes to Ca^{2+} channel function by increasing Ca^{2+} current and shifting the voltage dependence of inactivation (LETTS et al. 1998). Recent results indicate that Na⁺ channel β subunits serve cell adhesion functions and also anchor ion channels to cytoskeletal proteins (MALHOTRA et al. 2000). It is possible that other auxiliary proteins serve similar functions.

To date, the only therapeutically useful drug known to associate with auxiliary subunits is gabapentin, an anticonvulsant that binds to the α-2-δ subunit of Ca^{2+} channels (GEE et al. 1996). However, the functional correlate of gabapentin binding for Ca^{2+} channel function (if any) has not been described. In contrast, a recent in vitro study describes a synthetic compound that interferes specifically with the interaction between Ca^{2+} channel α-1 and β subunits (YOUNG et al. 1998), thereby reducing Ca^{2+} channel function. This novel kind of drug interaction might be utilized in the future to reduce Ca^{2+} channel function without a direct blockade of ion conduction.

II. Activation and Inactivation

1. Activation Gating

The ion-conducting subunits include several small but highly charged regions (S4 segments, Fig. 1) that sense depolarization by physically moving toward the extracellular space. These voltage sensors move backwards or forwards across the membrane within microseconds of a large change in membrane voltage. The movement of S4 segments within Na$^+$ channels has been studied with molecular techniques, and results confirm the idea that S4 movement is central to ion channel gating (YANG et al. 1996). Movement of S4 segments triggers a conformational change (called activation) opening the ion-conducting pathway and allowing the transit of either Na$^+$ or Ca^{2+} ions. These ions follow their concentration gradient inward across the plasma membrane. Recent studies using site-directed mutagenesis demonstrate that amino acid residues within the S4 region of domain IV are crucial to the action of α-scorpion toxins and sea anemone toxin (ROGERS et al. 1996). Furthermore, the S4 region of domain II is crucial for β-scorpion toxin (CESTELE et al. 1998). These toxins hold Na$^+$ channels in the open conformation, and prevent normal inactivation. Physiological inactivation (see Sect. II.3.) seems to more specifically depend upon movement of the S4 regions in domains III and IV (CHA et al. 1999). Therefore, gating and also toxin function are highly dependent upon movement of S4 regions. It is assumed that Ca^{2+} channels activate in a very similar manner.

2. Inactivation Processes

Within 1 or 2ms of activation (in the case of Na$^+$ channels) or tens to hundreds of milliseconds (in the case of most Ca^{2+} channels), a second time-dependent process called "inactivation" closes the ion pathway. In the case of Na$^+$ channels, inactivation is well understood: it occurs because of the movement of a "loop" of about 45 amino acids across the cytosolic mouth of the channel, binding at a specific spot that blocks ion conduction almost completely (e.g., CATTERALL 1994; KELLENBERGER et al. 1996). In contrast, Ca^{2+} channel inactivation is less complete, meaning that significant ionic current continues to flow during sustained depolarization. Ca^{2+} channel inactivation does not involve the folding of a cytosolic peptide loop, but instead may require the interaction of cytosolic β subunits with the α-1 subunit (see Sect. I.2. on auxiliary subunits). There are specific regions of the Ca^{2+} channel α-1 subunit that determine the rate of inactivation (ZHANG et al. 1994). These are located in the domain I, S6 region, and in the cytoplasmic and extracellular domains immediately adjacent to this region. Inactivation in L-type Ca^{2+} channels has been investigated in detail. There are specific amino acids in the IIIS6 region that alter the rate of inactivation when changed: conversion of two amino acids to alanine in this region almost completely eliminate L-type Ca^{2+} channel inactivation (HERING et al. 1997).

With both Na⁺ channels (HILLE 1992) and Ca²⁺ channels (ZHENG et al. 1992), the blockade of current by molecules that are not peptides or toxins is voltage-dependent, meaning that blockade occurs with lower drug concentrations if the cell membrane is depolarized. Recent studies have proposed that both inactivation and drug block (KUO and BEAN 1994a; KUO and BEAN 1994b) (BEAN 1984; ZHENG et al. 1992) depend upon the movement of S4 voltage sensors. A recent study shows that movement of S4 subunits in domains III and IV are particularly important for Na⁺ channel inactivation, and presumably also for drug binding (CHA et al. 1999). All of these analyses suggest that conformational changes needed for Ca²⁺ channel inactivation and drug-induced block by dihydropyridines or phenylalkylamines are interrelated (HERING et al. 1998; HOCKERMAN et al. 1997b).

III. Binding Sites for Blockers and Modulators

1. Tetrodotoxin and Conopeptides

The only parts of the channel molecule available for the binding of drugs directly from the extracellular space surround the extracellular entrance to the pore. These "extracellular loops" have been shown to bind tetrodotoxin in the case of Na⁺ channels and ω-conopeptides in the case of Ca²⁺ channels. Channel proteins with one or a few artificially-induced mutations to the native structure allow the identification of regions that bind blockers. High-affinity tetrodotoxin binding to Na⁺ channels is abolished by the mutation of two positively-charged glutamate residues at the mouth of the channel (NODA et al. 1989; TERLAU et al. 1991). ω-Conopeptide blockade of Ca²⁺ channels is reduced by mutation of at least four spatially distinct charged sites that surround the mouth of Ca²⁺ channels (ELLINOR et al. 1994). Binding at each site contributes "cooperatively" to the high affinity of the entire molecule. The multiple and somewhat spatially distant sites needed to account for conopeptide binding suggest that smaller nonpeptide molecules are unlikely to share the high affinity and selectivity of conopeptides. This is an important point when considering nonpeptide Ca²⁺ channel blockers (See Sect. II.3.).

2. Voltage-Dependent Blockers

Dihydropyridine compounds (and also phenylalkylamines like verapamil, and benzothiazepines like diltiazem) are relatively selective for L-type Ca²⁺ channels, although they also block Na⁺ and other cation channels at higher concentrations. The dihydropyridines bind to a modulatory site that is located on the extracellular face of L-type channels (HOCKERMAN et al. 1997b; STRIESSNIG et al. 1991). However, dihydropyridine function is not completely shared with other types of Ca²⁺ and Na⁺ channel blockers, since dihydropyridines can bind with high affinity and yet have different degrees of block or even increase channel opening (YU et al. 1988). Compounds such as the anticonvulsants phenytoin and carbamazepine and the local anesthetic lidocaine block the α subunit of sodium channels directly, as shown by voltage-clamp experiments with recombinant sodium channels (RAGSDALE et al. 1991). The anticonvulsants phenytoin, lamotrigine and topiramate and the neuroprotective compound riluzole block Na^+ currents in cultured or acutely-isolated neurons (KUO et al. 1997; KUO and BEAN 1994b; KUO and LU 1997; STEFANI et al. 1997; WHITE 1995; ZONA et al. 1997; ZONA and AVOLI 1997). However, many of these compounds block voltage-gated Ca^{2+} channels at similar or slightly higher concentrations (LEYBAERT and DE HEMPTINNE 1993; MCNAUGHTON et al. 1997; O'NEILL et al. 1997).

Various results implicate a cytosolic site for the binding of Ca^{2+} channel blockers related to verapamil and sodium blockers such as phenytoin or local anesthetics. Several Na^+ channel blockers interact with amino acids at the cytosolic side of the ion-conducting pathway. When six amino acids in this region are individually mutated, Na^+ channel physiology is relatively unchanged, but blockade of channels by local anesthetics such as etidocaine (RAGSDALE et al. 1994), or with anticonvulsants and antiarrhythmics (RAGSDALE et al. 1996) is markedly altered. These experiments confirm that drug molecules must first cross the plasma membrane, and then plug the ion-conducting pathway at a single set of closely related sites near the cytosolic mouth of the channel. Therefore, the voltage-dependent Na^+ channel blockers all require chemical properties that allow them to readily permeate cell membranes. Furthermore, these drugs interact with charged amino acids of the ion-conducting pathway (generally through a amine in the drug molecule), and also with lipophilic regions (generally by a hydrophobic aromatic ring or rings). See the structures of various blocking drugs in Table 3. A broad array of different structures can interact with cation channels.

Similar techniques have been used to map drug interactions with L-type Ca^{2+} channels, and the results show that phenylalkylamine blockers like verapamil interact with very specific parts of the α-1 subunit (HOCKERMAN et al. 1997a). Mutations in Ca^{2+} channels within domain III-S6 (very close to those that alter inactivation) change phenylalkylamine blockade without much change in inactivation (HOCKERMAN et al. 1997a). This result implies that inactivation and phenylalkylamine block involve neighboring parts of the α-1 subunit. The amino acids in this region are somewhat similar between Na^+ channels and Ca^{2+} channels. Therefore, it is understandable that many voltage-dependent synthetic blockers interact with all subtypes of Na^+ and Ca^{2+} channels with only modest selectivity. For example, the cytosolically-active Na^+ channel blocker and lidocaine derivative, QX-314, also blocks Ca^{2+} channels (TALBOT and SAYER 1996). It is one of the great (but mostly unfulfilled) hopes of ion channel pharmacology to discover synthetic compounds that have marked selectivity for single subtypes of voltage-gated neuronal Na^+ or Ca^{2+} channels.

Table 3. Neuroprotective actions of Na^+ and mixed Na^+/Ca^{2+} channel blockers

Drug	In vitro model	In vivo model
Relatively selective Na^+ channel blockers		
Tetrodotoxin	Hippocampal slice (Boening et al. 1989; Kass et al. 1992; Weber and Taylor 1994) Optic nerve (Stys et al. 1992) Neuronal culture (Lynch et al. 1995; Tasker et al. 1992) Spinal explant (Rothstein and Kuncl 1995)	Gerbil, rat forebrain ischemia (Lysko et al. 1995; Yamasaki et al. 1991) Isolated rat head (Xie et al. 1994) Rat cardiac arrest (Prenen et al. 1988) Rat spinal cord injury (Teng and Wrathall 1997)
Lidocaine (antiarrhythmic)	Hippocampal slice (Lucas et al. 1989; Weber and Taylor 1994) Optic nerve (Stys et al. 1992; Stys 1995)	Rabbit forebrain ischemia (Rasool et al. 1990)
Mexilitine (antiarrhythmic)	Optic nerve (Stys and Lesiuk 1996)	
Phenytoin (anticonvulsant)	Hippocampal slice (Stanton and Moskal 1991; Weber and Taylor 1994) Isolated optic nerve (Fern et al. 1993)	Rat MCAO (Boxer et al. 1990; Murakami and Furui 1994; Rataud et al. 1994) Gerbil forebrain ischemia (Taft et al. 1989) Levine rat pup (Hayakawa et al. 1994a)

Table 3. Continued

Drug	In vitro model	In vivo model
Fosphenytoin (anticonvulsant) [note: parenteral prodrug of phenytoin]		Clinical study: lack of neuroprotection in randomized, placebo-controlled study with acute posttreatment (1–4h after onset)
Carbamazepine (anticonvulsant)		Rat MCAO (MURAKAMI and FURUI 1994; RATAUD et al. 1994)
Lamotrigine (anticonvulsant)		Rat MCAO (MURAKAMI and FURUI 1994; SMITH and MELDRUM 1995) Gerbil forebrain ischemia (WIARD et al. 1995)

Rat MCAO (MELDRUM et al. 1992; TORP et al. 1993)
Rat MCAO w/ reperfusion (GASPARY et al. 1994)
Rat forebrain ischemia (LEKIEFFRE and MELDRUM 1993)
Levine rat pup (GILLAND et al. 1994)
Rat head trauma (OKIYAMA et al. 1995)

Rat MCAO (CHEN et al. 1995; LEACH et al. 1993; SMITH et al. 1993)
Rat forebrain ischemia (SMITH et al. 1993)
Rat head trauma (SUN and FADEN 1995)
Clinical study: (MUIR et al. 1995) Nonsignificant trend towards improved outcome with acute posttreatment for stroke

Levine rat pup (HAYAKAWA et al. 1994b)

Gerbil forebrain ischemia (O'NEILL et al. 1997; PRATT et al. 1992)
Rat MCAO (PRATT et al. 1992; WAHL et al. 1993)
Transgenic ALS mice (GURNEY et al. 1996)
MPTP monkeys (BENAZZOUZ et al. 1995)
Clinical study: Approved for clinical use as neuroprotection from slow cell death in amyotrophic lateral sclerosis

Spinal explant culture (ROTHSTEIN and KUNCL 1995)
Cerebellar granule cells in culture (DESSI et al. 1993)
Hippocampal slice (MALGOURIS et al. 1994)

BW1003C87

BW619C89

Zonisamide (anticonvulsant)

Riluzole (neuroprotective)

Table 3. *Continued*

Drug	In vitro model	In vivo model
Felbamate (anticonvulsant)	Hippocampal slice (WALLIS et al. 1992; WALLIS and PANIZZON 1995)	Levine rat pup (WASTERLAIN et al. 1992) Neocortical cell culture (KANTHASAMY et al. 1995)
PNU 151774E		Kainate-induced cell death (MAJ et al. 1998)
Nonselective Na⁺/Ca²⁺ blockers		
Flunarizine (anticonvulsant)	Neocortex cell cultures (PAUWELS et al. 1990) Hippocampal slice (HARA et al. 1990) Hippocampal cell culture (HARA et al. 1993b)	Rat forebrain ischemia (LU et al. 1990; VAN REEMPTS et al. 1986; WAUQUIER et al. 1989) Rat cardiac arrest (LU et al. 1990; WAUQUIER et al. 1989; XIE et al. 1995) Levine rat pup (SILVERSTEIN et al. 1986)

Lidoflazine	Neocortex cell cultures (PAUWELS et al. 1990)	Not active dog global forebrain ischemia (KUMAR et al. 1988) Gerbil forebrain ischemia (ALPS et al. 1988) Not active rabbit focal ischemia (LYDEN et al. 1988) Clinical study: (BRAIN RESUSCITATION CLINICAL TRIAL II STUDY GROUP 1991)
Lifarizine	Neocortex cell cultures (MAY et al. 1995)	Rat forebrain ischemia (ALPS et al. 1995; MCBEAN et al. 1995) Cat MCAO (KUCHARCZYK et al. 1993) Mouse MCAO (BROWN et al. 1995; CRAMER and TOOROP 1998) Rat cardiac arrest (XIE et al. 1995)
Lomerizine	Hippocampal slice (HARA et al. 1990) Hippocampal cell culture (HARA et al. 1993b)	Gerbil forebrain (YOSHIDOMI et al. 1989) Rat MCAO (HARA et al. 1993a)

Table 3. *Continued*

Drug	In vitro model	In vivo model
Cinnarizine	Neocortex cell cultures (Pauwels et al. 1990)	
Lubeluzole	Hippocampal slice (Ashton et al. 1997) Hippocampal cell cultures (Culmsee et al. 1998)	Rat focal ischemia (Buchkremer-Ratzmann and Witte 1995; Culmsee et al. 1998; de-Ryck et al. 1996) Clinical studies: No significant neuroprotection in randomized placebo-controlled studies with acute posttreatment within 6h of stroke onset (Diener et al. 1996; Diener 1998; Grotta 1997)
CNS1237		Rat MCAO (Goldin 1995) Gerbil forebrain ischemia (O'Neill et al. 1997)
Relatively selective Ca²⁺ blockers		
Ziconotide (SNX-111)	Hippocampal slice culture (Pringle et al. 1996) Not active hippocampal slice (Small et al. 1997)	Rat MCAO (Perez-Pinzon et al. 1997; Takizawa et al. 1995; Yenari et al. 1996) Rat transient MCAO (Bowersox et al. 1997) Rat forebrain ischemia (Buchan et al. 1994; Valentino et al. 1993; Zhao et al. 1994)

Sodium and Calcium Channel Blockers

Nimodipine

Not active hippocampal slice (SMALL et al. 1997)
Not active optic nerve (RANSOM et al. 1990)

Rat traumatic brain injury (HOVDA et al. 1994; XIONG et al. 1998)
Clinical studies: Not significantly different than placebo for acute posttreatment of head trauma (unpublished data)

Rat MCAO (BIELENBERG et al. 1990; BIELENBERG and BECK 1991; JACEWICZ et al. 1990; MARINOV et al. 1991; RODA et al. 1995)
Not active MCAO (ANDERSEN et al. 1991)
Not active rat focal ischemia (SNAPE et al. 1993)
Not active mouse MCAO (CRAMER and TOOROP 1998)
Gerbil forebrain ischemia (MOSSAKOWSKI and GADAMSKI 1990; NAKAMURA et al. 1993)
Rat forebrain ischemia (NUGLISCH et al. 1990; RAMI and KRIEGLSTEIN 1994; WELSCH et al. 1990)
Not active rat forebrain ischemia (CALLE et al. 1993; WAUQUIER et al. 1989)
Not active cat forebrain ischemia (TATEISHI et al. 1991)
Not active rat spinal cord injury (ROSS and TATOR 1991)
Clinical studies: (1990; 1992; MARTINEZ et al. 1990; WADWORTH and McTAVISH 1992) No significant neuroprotection in randomized, placebo-controlled studies with treatment within 12h of stroke onset

Isradipine

Rat MCAO (SAUTER and RUDIN 1990)

Table 3. *Continued*

Drug	In vitro model	In vivo model
Verapamil		Rat forebrain ischemia (Wauquier et al. 1989) Not active mouse MCAO (Cramer and Toorop 1998)
S-Emopamil		Rat forebrain ischemia (Block et al. 1990; Lin et al. 1990) Not active cat forebrain ischemia (Fleischer et al. 1992) Rat MCAO (Elger et al. 1994; Morikawa et al. 1991; Seega and Elger 1993) Mouse MCAO (Cramer and Toorop 1998) Not active cat MCAO (Gomi et al. 1995) Rat traumatic brain injury (Okiyama et al. 1992)
SB-201823	Not active hippocampal slice cultures (Pringle et al. 1996)	Not active gerbil forebrain ischemia (O'Neill et al. 1997)
U92032		Rat transient MCAO (Goto et al. 1994)

NNC-09-0026

Gerbil forebrain ischemia (BARONE et al. 1994; O'NEILL et al. 1997)

NS-649

Not active gerbil forebrain ischemia (O'NEILL et al. 1997)

Models are: (1) rat hippocampal slices deprived of oxygen and glucose in vitro; (2) optic nerve, rat optic nerve in vitro deprived of oxygen; (3) spinal explant culture: rat spinal organotypic cultures in glutamate uptake inhibitor; (4) gerbil or rat forebrain ischemia, global forebrain ischemia; (5) MCAO, unilateral middle cerebral artery occlusion (often with combined vertebral artery occlusion); (6) Levine rat pup, neonatal rat unilateral carotid occlusion with hypoxia; (7) rat cardiac arrest; (8) rat isolated head, arrested perfusion; (9) rat head trauma, brain damage from trauma; (10) ALS mice, transgenic mice with mutant superoxide dismutase gene (symptoms similar to ALS); (11) MPTP monkeys, monkeys treated with MPTP, producing symptoms similar to Parkinson's disease.

C. Peptides as Specific Ca²⁺ Channel Probes

Conopeptides were originally isolated from the venom of poisonous cone snails from the oceans of the South Pacific (OLIVERA et al. 1994). Cone snails hunt fish and other small organisms for food, and they paralyze their prey by injecting venom containing a wide variety of related polypeptides that each have about 25 amino acids linked with four or more disulfide bonds (NEWCOMB et al. 1995). This structure is highly compact and spatially defined (BASUS et al. 1995) (Fig. 3). The rigid structure allows a very specific and high-affinity interaction of peptide molecules with pharmacological targets. In particular, several peptides from the snail *Conus geographicus* (e.g., ω-conopeptide GVIA) and from *Conus magus* (e.g., ω-conopeptide MVIIA) are Ca²⁺ channel blockers, and they have significant activity only at N-type channels (MILJANICH and RAMACHANDRAN 1995). The synthetic version of ω-conopeptide MVIIA is called ziconotide or SNX-111 (2-dimensional structure in Table 3; 3-dimensional structure in Fig. 3). The Ca²⁺ channel conopeptides are uniquely potent blockers of Ca²⁺ channels at concentrations around 1nM, with no significant actions at Na⁺ or K⁺ channels at much higher concentrations. In addition, ω-conopeptide MVIIC (also from *Conus magus*), blocks N-type, P-type, and Q-type Ca²⁺ channels in central neurons. Conopeptides were first isolated by high-performance liquid chromatography, but methods were soon devised for the solid-state synthesis of significant quantities of conopeptides, allowing large-scale studies in animal models and clinical trials (NADASDI et al. 1995).

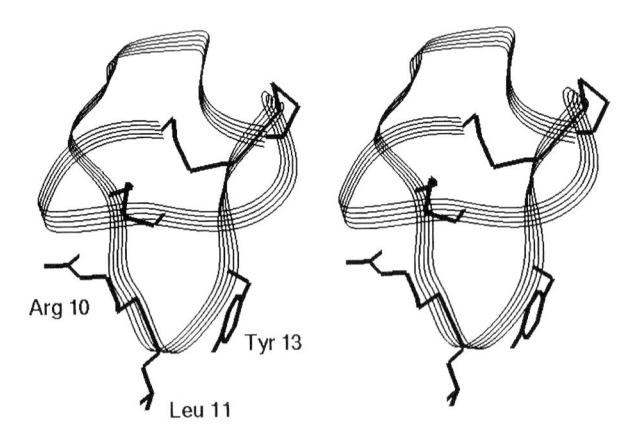

Fig. 3. Three-dimensional structure of ziconotide (*SNX-111*). Relaxed stereo view of the folding of ziconotide as determined from its solution structure (BASUS et al. 1995). The molecule is oriented with the loop that is critical for binding pointing down. A five-stranded ribbon represents the backbone. *Heavy black lines* depict the sidechain heavy atoms of six cysteines involved in 3-dimensional folding. The outlines of three functionally important residues within the loop region are shown in *dark black*

Peptide toxins have been used to define the contributions of various Ca^{2+} channels to the current recorded in voltage-clamp experiments of isolated neuronal cell bodies (McDonough et al. 1996; Randall and Tsien 1995). The same peptides also block components of Ca^{2+}-dependent neurotransmitter release (Newcomb and Palma 1994), and reduce synaptic potentials or synaptic currents in brain tissues (Randall and Tsien 1995; Regehr and Tank 1992) and block Ca^{2+} influx in synaptic endings of neuronal tissues (Regehr and Mintz 1994; Wu and Sagratella 1994).

The 48-amino acid polypeptide ω-Agatoxin IVA was derived from the venom of the spider *Agelenopsis aperta*. ω-Agatoxin IVA potently inhibits voltage-clamped Ca^{2+} currents [particularly the "P-type" currents in cerebellar Purkinje neurons (Adams et al. 1993; Mintz and Bean 1993)]. AgaIVA also reduces neurotransmitter release at many synapses. However, when given systemically to animals, ω-Agatoxin IVA is toxic and causes death by paralysis of skeletal muscles, probably because it blocks P-type Ca^{2+} channels and neurotransmitter release at neuromuscular junctions. Both ω-Agatoxin IVA and ω-Conotoxin MVIIC block neuromuscular transmission, while ω-Conotoxin MVIIA does not (Bowersox et al. 1995). Finally, the spider venom peptide (SNX-482) is a relatively selective blocker of neuronal R-type Ca^{2+} currents (E-class channels), and it is being studied as a selective pharmacological tool (Newcomb et al. 1999).

D. Mechanisms of Ischemic Neurotransmitter (Glutamate) Release

I. Ca^{2+}-Dependent Release

The textbook view of neurotransmitter release includes neurotransmitter molecules packaged into membrane-bound cytosolic vesicles that are transported to specialized locations near synapses. Vesicles are then released into the extracellular space in rapid response to elevated Ca^{2+}. Physiological neurotransmitter release is therefore directly linked to Ca^{2+} influx from voltage-gated Ca^{2+} channels, particularly N-type, P-type, and Q-type channels. Recently, it has been established that N-type and P/Q-type Ca^{2+} channel proteins associate directly with vesicle fusion proteins located near the membrane (Kim and Catterall 1997; Rettig et al. 1996; Rettig et al. 1997; Sheng et al. 1997). This close physical association between channels and vesicle release machinery assures that synaptic Ca^{2+} entry occurs in just the right place to rapidly trigger neurotransmitter release.

In accordance with this view, conopeptides and spider toxins that are specific Ca^{2+} channel blockers reduce neurotransmitter release and also block synaptic function (measured electrophysiologically) in a wide variety of experimental preparations. These experiments establish that several different Ca^{2+} channels are present in the membranes of each neuron. In particular, the so-called N-type but also P-type, and Q-type channels contribute to both glutamate release (see section on conopeptides) (WHEELER et al. 1994) and monoamine neurotransmitter release (OLIVERA et al. 1994). L-type Ca^{2+} channels (blocked selectively by low concentrations of dihydropyridines) play a negligible role in physiological neurotransmitter release in brain.

II. Ca^{2+}-Independent Release

During ischemia, cytosolic adenosine triphosphate concentrations decline within a few seconds. Vesicular neurotransmitter release virtually stops when ATP declines by more than 50% (LIPTON and WHITTINGHAM 1982). Therefore, the massive neurotransmitter release that occurs during ischemia proceeds in part by nonphysiological processes. Under conditions in vitro similar to those of ischemia, blockade of Ca^{2+} influx fails to reduce neurotransmitter release (TAYLOR et al. 1995). Additional results suggest that much of the ischemic neurotransmitter release is caused not by exocytosis of vesicles, but instead by reversal of plasma membrane neurotransmitter transporters (SZATKOWSKI and ATTWELL 1994; TAYLOR et al. 1992). Membrane-bound transporters for glutamate, GABA, and monoamines depend on the plasma membrane Na^+ gradient and also on membrane voltage. When ion gradients are disrupted and membranes depolarize (as in ischemia) neurotransmitter transporters spew glutamate and other transmitter substances into the extracellular space not only at synapses, but at other parts of neurons.

Although neurotransmitter release mediated by transporters is relatively slow, substantial amounts of neurotransmitter accumulate in the extracellular space over a minute or two. Furthermore, the loss of ion homeostasis during ischemia prevents the reuptake of neurotransmitters from the extracellular space, and so high neurotransmitter concentrations are sustained for relatively long periods of time. Ca^{2+} channel block (by elimination of extracellular Ca^{2+}) does not prevent acute "ischemic" neurotransmitter release in vitro. In contrast, Na^+ channel blockers such as tetrodotoxin delay the release of glutamate under ischemic conditions (TAYLOR et al. 1995). Reduced cellular Na^+ loading and an overall delay in the loss of ion homeostasis probably cause neuroprotection with tetrodotoxin (TAYLOR 1997).

With whole-animal models, it is likely that Ca^{2+} dependent release predominates in regions relatively distant from the energy-poor "core" of ischemia, whereas Ca^{2+} independent release occurs in areas with severe depletion of ATP. The accumulation of extracellular glutamate in brain tissue activates several types of ligand-gated Ca^{2+} and Na^+ channels, causing large inward fluxes of cations that are sustained for many minutes. This ionic imbalance is widely agreed to contribute to ischemic pathology. In addition, cytosolic Na^+ accumulation may secondarily cause Ca^{2+} influx by the reversal of plasma membrane Na^+/Ca^{2+} cotransport (LEHNING et al. 1996; ZHANG and LIPTON 1995) or mitochondrial Na^+/Ca^{2+} exchange into the cytosol (CROMPTON et al. 1978; SIMPSON and RUSSELL 1998; ZHANG and LIPTON 1999).

E. Neuroprotection with Na⁺ Channel Blockers

Although it is not useful for clinical therapy because of its obvious toxicity, the puffer fish toxin, tetrodotoxin (TTX), prevents or reduces functional and cellular damage in a variety of animal models of brain ischemia (Table 3). Since TTX is a very specific blocker of Na⁺ channels, with virtually no effect at other ion channels or receptors, this is strong evidence that neuroprotection can arise from selective blockade of Na⁺ channels. In contrast to TTX, a variety of synthetic compounds are less selective blockers of Na⁺ channels, and several of these compounds have neuroprotective actions in animal models (Table 3). It should be pointed out that most of these compounds also block Ca²⁺ channels and some have additional actions that are unrelated to ion channels. Therefore, the conclusion that compounds other than TTX are neuroprotective because of Na⁺ channel block is tentative. However, it is noteworthy that a rather broad range of different structural types bind to Na⁺ channels and are neuroprotective. The only shared structural motifs among the Na⁺ blockers seem to be at least one tertiary amine group and one or more spatially separate lipophilic regions, of which one is usually aromatic.

Clinically significant Na⁺ blockers include riluzole, which is approved for clinical treatment of the slow neurodegenerative disease amyotrophic lateral sclerosis. Lubeluzole has neuroprotective actions in animal models and has been studied in several clinical trials for acute stroke. Fosphenytoin is a water-soluble prodrug of the common anticonvulsant phenytoin that can be given parenterally. Fosphenytoin has been studied in a large placebo-controlled trial for acute treatment of stroke (less than 4h after onset). However, patients treated with fosphenytoin were not significantly improved in comparison to placebo.

F. Neuroprotection with Small Molecule Ca²⁺ Channel Blockers

Nimodipine is a 1,4-dihydropyridine compound (WADWORTH and MCTAVISH 1992) selective for L-type Ca²⁺ channels. It penetrates readily into the central nervous system (FANELLI et al. 1994). Dihydropyridines and phenylalkylamine drugs block L-type channels in vascular smooth muscle cells and in central neurons. Positive neuroprotection results with dihydropyridines at high concentrations in vitro (WEISS et al. 1990) may result at least in part from blockade of Na⁺ channels or non-L-type Ca²⁺ channels. Whole-animal models have only occasionally shown neuroprotection with selective L-type Ca²⁺ blockers, and several published studies have failed to see neuroprotection in vivo (see Table 3). This may be because such drugs lower systemic blood pressure, which can reduce perfusion of ischemic brain tissue.

Nimodipine has been extensively studied clinically, and it is approved for the treatment of vasospasm caused by subarachnoid hemorrhage in humans

(WHITFIELD and PICKARD 1994). In addition, nimodipine was studied clinically as acute posttreatment for stroke, without clear evidence of neuroprotection (TRUST STUDY GROUP 1990; AMERICAN NIMODIPINE STUDY GROUP 1992; BOGOUSSLAVSKY et al. 1990; GELMERS and HENNERICI 1990). In stroke clinical studies, nimodipine was given between 3 and 12h after stroke onset, and outcome was evaluated at several time points, weeks after stroke onset. Most reviewers believe that nimodipine was given too long after stroke onset to demonstrate a neuroprotective effect. In addition, treatment of acute stroke patients was complicated by direct vascular hypotensive actions of nimodipine, which also are seen in experimental animals. Since nimodipine is relatively selective for L-type channels, it is a potent dilator of vascular smooth muscle, and it has actions on the vasculature at dosages similar to those effective on neurons.

S-emopamil is a drug that is closely related to the phenylalkylamine vascular L-type Ca^{2+} blocker, verapamil. It penetrates well into the central nervous system, and it has neuroprotective actions in several animal models (CRAMER and TOOROP 1998) (Table 3). Like nimodipine, S-emopamil was studied clinically for treatment of acute stroke, without a clear-cut demonstration of positive effect.

G. Neuroprotection with Conopeptides

Although ω-conopeptides are selective blockers of Ca^{2+} channels, they can only be given intravenously (oral administration leads to degradation in the stomach) and they are only sparingly permeable to the blood–brain barrier. Typically, much less than 1% of conopeptide concentrations available in blood plasma are found in the brain extracellular space. Nevertheless, ziconotide reduces brain damage in both models of focal (BUCHAN et al. 1994; TAKIZAWA et al. 1995; ZHAO et al. 1994) and global ischemia (BUCHAN et al. 1994; VALENTINO et al. 1993) (Table 3). Neuroprotection with ziconotide is even more remarkable in models of global forebrain ischemia, where neuroprotection is seen with treatment that is delayed up to 24h after the onset of ischemia (BUCHAN et al. 1993 SOC. NEUROSCI. ABSTR.). Furthermore, ziconotide treatment reduces brain damage from head trauma in rats (HOVDA et al. 1994; VERWEIJ et al. 1997).

These and other results with ziconotide in animal models have lead to placebo-controlled clinical studies for acute posttreatment of severe head trauma. However, the systemic administration of ziconotide to humans, with dosages at or below those projected to result in neuroprotection, profoundly decreased systemic blood pressure (MCGUIRE et al. 1997). Based on results with animals, hypotension may be caused by blockade of norepinephrine release from sympathetic ganglia by specific blockade of N-type Ca^{2+} channels outside the blood–brain barrier. In animals, sympatholysis with ziconotide occurs at dosages lower than those needed for prevention of brain damage from ischemia (BOWERSOX et al. 1992). This may be exacerbated because ziconotide only penetrates the blood–brain barrier poorly, and is present at higher concentrations in the periphery. The clinical studies with ziconotide necessitated a clinical protocol with the use of pressor agents prior to intravenous drug infusion. Despite the apparent success of pressor treatements in maintaining proper blood pressure, these studies were terminated because an interim statistical analysis of the data showed no indication of efficacy in the drug-treated group compared to the placebo group. Thus, as with other types of neuroprotective treatments, conopeptides have failed in at least one clinical study of neuroprotection.

H. Mixed Na⁺ and Ca²⁺ Channel Blockers

Flunarizine blocks both Na⁺ (BROWN et al. 1995; KISKIN et al. 1993) and various types of Ca²⁺ channels (COHAN et al. 1993) Flunarizine was originally studied for treatment of epilepsy (LEVY et al. 1995) and migraine, but slightly later, it also was found to prevent ischemic damage in animal models of both global and focal ischemia (PAUWELS et al. 1991). It was presumed that the neuroprotection observed with flunarizine was caused by actions directly on nerve cells, since it is neuroprotective in vitro with neuronal systems devoid of vasculature (ASHTON et al. 1990; PAUWELS et al. 1989; PAUWELS et al. 1990). In the mid-1980s and 1990s, additional animal studies were done with other nonpeptide synthetic molecules that block both Ca²⁺ channels and Na⁺ channels, including several congeners of flunarizine, riluzole, BW619C89, lifarizine, and others. The pharmacology of these compounds in models of neuroprotection are outlined in Table 3.

I. Conclusions

The rationale for neuroprotection with selective Na⁺ channel blockers, and with selective N-type Ca²⁺ blockers, or alternatively with mixed blockers of Na⁺ and Ca²⁺ channels remains sound, with support from a large number of studies in laboratory animals and in vitro models. However, drug-induced depression of ion channels in the cardiovascular system must be avoided to provide a successful clinical treatment. Results with in vitro and animal experiments indicate that blockade of voltage-activated cation channels prevents or reduces neuronal damage from focal ischemia, global forebrain ischemia and head trauma. Despite these results, several full-blown clinical studies have failed to show significant neuroprotection after treatment with cation blockers. It is likely that many of the clinical studies failed because therapy was initiated too long after the onset of ischemia (e.g., more than 3–4h) or because the severity of ischemia studied varied widely between patients, making statistical comparison with a placebo group problematic. In order to show unambiguous clinical efficacy, Na⁺ or Ca²⁺ channel blockers may have to be given prior to ischemia, so that effective plasma concentrations are present at very early times after ischemic onset. This may be practical only in a small subset of ischemic conditions. Additional clinical studies will be required to test this idea.

Acknowledgements. I would like to thank several colleagues for allowing me to quote unpublished clinical data with ziconotide, and also colleagues at Elan Pharmaceuticals (formerly Neurex Corporation) for an interesting drug discovery collaboration over the past several years. Thanks also to Frank Marcoux and Michael Poole for reading an earlier version of the manuscript and to Michael D. Reily for the structure of ziconotide (Fig. 3).

References

Ackerman MJ, Clapham DE (1997) Ion Channels – Basic science and clinical disease. New Engl J Med 336:1575–1586
Adams ME, Mintz IM, Reily MD, Thanabal V, Bean BP (1993) Structure and properties of ω-agatoxin IVB, a new antagonist of P-type calcium channels. Mol Pharmacol 44:681–688
Alps BJ, Calder C, Hass WK, Wilson AD (1988) Comparative protective effects of nicardipine, flunarizine, lidoflazine and nimodipine against ischaemic injury in the hippocampus of the Mongolian gerbil. Br J Pharmacol 93:877–883
Alps BJ, Calder C, Wilson AD, Mcbean DE and Armstrong JM (1995) Reduction by lifarizine of the neuronal damage induced by cerebral ischaemia in rodents. Br J Pharmacol 115:1439–1446
American Nimodipine Study Group (1992) Clinical trial of nimodipine in acute ischemic stroke. [published erratum appears in Stroke 1992 Apr;23(4):615]. Stroke 23:3–8
Andersen CS, Andersen AB, Finger S (1991) Neurological correlates of unilateral and bilateral "strokes" of the middle cerebral artery in the rat. Physiol Behav 50: 263–269
Armstrong CM, Hille B (1998) Voltage-gated ion channels and electrical excitability. Neuron 20:371–380
Ashton D, Willems R, Marrannes R, Janssen PA (1990) Extracellular ions during veratridine-induced neurotoxicity in hippocampal slices: neuroprotective effects of flunarizine and tetrodotoxin. Brain Res 528:212–222
Ashton D, Willems R, Wynants J, van-Reempts J, Marrannes R, Clincke G (1997) Altered Na(+)-channel function as an in vitro model of the ischemic penumbra: action of lubeluzole and other neuroprotective drugs. Brain Res 745:210–221
Barone FC, Price WJ, Jakobsen P, Sheardown MJ, Feuerstein G (1994) Pharmacological profile of a novel neuronal calcium channel blocker includes reduced cerebral damage and neurological deficits in rat focal ischemia. Pharmacol Biochem Behav 48:77–85
Basus VJ, Nadasdi L, Ramachandran J, Miljanich GP (1995) Solution structure of omega-conotoxin MVIIA using 2D NMR spectroscopy. FEBS Lett 370:163–169
Bean BP (1984) Nitrendipine block of cardiac calcium channels: high-affinity binding to the inactivated state. Proc Natl Acad Sci USA 81:6388–6392
Benazzouz A, Borad T, Dubedat P, Boireau A, Stutzmann J-M, Gross C (1995) Riluzole prevents MPTP-induced parkinsonism i the rhesus monkey: a pilot study. Eur J Pharmacol 284:299–307
Bielenberg GW, Beck T (1991) The effects of dizocilpine (MK-801), phencyclidine, and nimodipine on infarct size 48h after middle cerebral artery occlusion in the rat. Brain Res 552:338–342

Bielenberg GW, Burniol M, Rosen R, Klaus W (1990) Effects of nimodipine on infarct size and cerebral acidosis after middle cerebral artery occlusion in the rat. Stroke 21:IV90–IV92

Block F, Jaspers RM, Heim C, Sontag KH (1990) S-emopamil ameliorates ischemic brain damage in rats: histological and behavioural approaches. Life Sci 47:1511–1518

Boening JA, Kass IS, Cottrell JE, Chambers G (1989) The effect of blocking sodium influx on anoxic damage in the rat hippocampal slice. Neurosci 33:263–268

Bogousslavsky J, Regli F, Zumstein V, Kobberling W (1990) Double-blind study of nimodipine in non-severe stroke. Eur Neurol 30:23–26

Bowersox SS, Miljanich GP, Sugiura Y, Li C, Nadasdi L, Hoffman BB, Ramachandran J, Ko CP (1995) Differential blockade of voltage-sensitive calcium channels at the mouse neuromuscular junction by novel omega-conopeptides and omega-agatoxin-IVA. J Pharmacol Exp Ther 273:248–256

Bowersox SS, Singh T, Luther RR (1997) Selective blockade of N-type voltage-sensitive calcium channels protects against brain injury after transient focal cerebral ischemia in rats. Brain Res 747:343–347

Bowersox SS, Singh T, Nadasdi L, Zukowska GZ, Valentino K, Hoffman BB (1992) Cardiovascular effects of omega-conopeptides in conscious rats: mechanisms of action. J Cardiovasc Pharmacol 20:756–764

Boxer PA, Cordon JJ, Mann ME, Rodolosi LC, Vartanian MG, Rock DM, Taylor CP, Marcoux FW (1990) Comparison of phenytoin with noncompetitive N-methyl-D-aspartate antagonists in a model of focal brain ischemia in rat. Stroke 21:47–51

Brain Resuscitation Clinical Trial II Study Group (1991) A randomized clinical trial of calcium entry blocker administration to comatose survivors of cardiac arrest. Design, methods, and patient characteristics [see comments]. Control Clin Trials 12:525–545

Brown CM, Calder C, Linton C, Small C, Kenny BA, Spedding M, Patmore L (1995) Neuroprotective properties of lifarizine compared with those of other agents in a mouse model of focal cerebral ischaemia. Br J Pharmacol 115:1425–1432

Buchan AM, Gertler SZ, Li H, Xue D, Huang ZG, Chaundy KE, Barnes K, Lesiuk HJ (1994) A selective N-type Ca(2+)-channel blocker prevents CA1 injury 24h following severe forebrain ischemia and reduces infarction following focal ischemia. J Cereb Blood Flow Metab 14:903–910

Buchkremer-Ratzmann I, Witte OW (1995) Periinfarct and transhemispheric diaschisis caused by photothrombotic infarction in rat neocortex is reduced by lubeluzole but not MK-801. J Cereb Blood Flow Metab 15:S381

Calle PA, Paridaens K, De-Ridder LI, Buylaert WA (1993) Failure of nimodipine to prevent brain damage in a global brain ischemia model in the rat [see comments]. Resuscitation 25:59–71

Catterall WA (1992) Cellular and molecular biology of voltage-gated sodium channels. Physiol Rev 72:S15–S48

Catterall WA (1994) Molecular mechanisms of inactivation and modulation of sodium channels. Renal Physiol Biochem 17:121–125

Cestele S, Qu Y, Rogers JC, Rochat H, Scheuer T, Catterall WA (1998) Voltage sensor-trapping: enhanced activation of sodium channels by beta-scorpion toxin bound to the S3-S4 loop in domain II. Neuron 21:919–931

Cha A, Ruben PC, George AL, Fujimoto E, Bezanilla F (1999) Voltage sensors in domains III and IV, but not I and II, are immobilized by Na$^+$ channel fast inactivation. Neuron 22:73–87

Chen J, Graham SH, Simon RP (1995) A comparison of the effects of a sodium channel blocker and an NMDA antagonist upon extracellular glutamate in rat focal cerebral ischemia. Brain Res 699:121–124

Cohan SL, Redmond DJ, Chen M, Wilson D, Cyr P (1993) Flunarizine blocks elevation of free cytosolic calcium in synaptosomes following sustained depolarization. J Cereb Blood Flow Metab 13:947–954

Cox B, Denyer JC (1998) N-type calcium channel blockers in pain and stroke. Exp Opin Ther Patents 8:1237–1250

Cramer WC, Toorop GP (1998) Focal cerebral ischemia in the mouse: hypothermia and rapid screening of drugs. Gen Pharmacol 30:195–200

Crompton M, moser R, Ludi H (1978) The interrelations between the transport of sodium and calcium in mitochondria of various mammalian tissues. Eur J Biochem 82:25–31

Culmsee C, Junker V, Wolz P, Semkova I, Krieglstein J (1998) Lubeluzole protects hippocampal neurons from excitotoxicity in vitro and reduces brain damage caused by ischemia. Eur J Pharmacol 342:193–201

de-Ryck M, Keersmaekers R, Duytschaever H, Claes C, Clincke G, Janssen M, Van-Reet G (1996) Lubeluzole protects sensorimotor function and reduces infarct size in a photochemical stroke model in rats. J Pharmacol Exp Ther 279:748–758

Dessi F, Ben Ari Y, Charriaut Marlangue C (1993) Riluzole prevents anoxic injury in cultured cerebellar granule neurons. Eur J Pharmacol 250:325–328

Diener HC (1998) Multinational randomised controlled trial of lubeluzole in acute ischaemic stroke. European and Australian Lubeluzole Ischaemic Stroke Study Group. Cerebrovasc Dis 8:172–181

Diener HC, Hacke W, Hennerici M, Radberg J, Hantson L, De Keyser J (1996) Lubeluzole in acute ischemic stroke. A double-blind, placebo-controlled phase II trial. Stroke 27:76–81

Elger B, Seega J, Raschack M (1994) Oedema reduction by levemopamil in focal cerebral ischaemia of spontaneously hypertensive rats studied by magnetic resonance imaging. Eur J Pharmacol 254:65–71

Ellinor PT, Zhang JF, Horne WA, Tsien RW (1994) Structural determinants of the blockade of N-type calcium channels by a peptide neurotoxin. Nature 372:272–275

Fanelli RJ, McCarthy RT, Chisholm J (1994) Neuropharmacology of nimodipine: from single channels to behavior. Ann N Y Acad Sci 747:336–350

Fern R, Ransom BR, Stys P, Waxman SG (1993) Pharmacological protection of CNS white matter during anoxia: actions of phenytoin, carbamazepine and diazepam. J Pharmacol Exp Ther 266:1549–1555

Fleischer JE, Nakakimura K, Drummond JC, Scheller MS, Zornow MH, Grafe MR, Shapiro HM (1992) Effects of levemopamil on neurologic and histologic outcome after cardiac arrest in cats. Crit Care Med 20:126–134

Gaspary HL, Simon RP, Graham SH (1994) BW1003C87 and NBQX but not CGS19755 reduce glutamate release and cerebral ischemic necrosis. Eur J Pharmacol 262:197–203

Gee NS, Brown JP, Dissanayake VU, Offord J, Thurlow R, Woodruff GN (1996) The novel anticonvulsant drug, gabapentin (Neurontin), binds to the alpha2-delta subunit of a calcium channel. J Biol Chem 271:5768–5776

Gelmers HJ, Hennerici M (1990) Effect of nimodipine on acute ischemic stroke. Pooled results from five randomized trials. Stroke 21:IV81–IV84

Gilland E, Puka Sundvall M, Andine P, Bona E, Hagberg H (1994) Hypoxic-ischemic injury in the neonatal rat brain: effects of pre- and post-treatment with the glutamate release inhibitor BW1003C87. Brain Res Dev Brain Res 83:79–84

Goldin SM (1995) Neuroprotective use-dependent blockers of Na^+ and Ca^{2+} channels controlling presynaptic release of glutamate. Ann N Y Acad Sci 765:210–229

Gomi S, Greenberg JH, Croul S, Reivich M (1995) Failure of levemopamil to improve histological outcome following temporary occlusion of the middle cerebral artery in cats. J Neurol Sci 130:128–133

Goto Y, Kassell NF, Hiramatsu K, Hong SC, Soleau SW, Lee KS (1994) Effects of two dual-function compounds, U92798 and U92032, on transient focal ischemia in rats. Neurosurg 34:332–337

Grotta J (1997) Lubeluzole treatment of acute ischemic stroke. The US and Canadian Lubeluzole Ischemic Stroke Study Group. Stroke 28:2338–2346

Gurney ME, Cutting FB, Zhai P, Doble A, Taylor CP, Andrus PK, Hall ED (1996) Benefit of vitamin E, riluzole, and gabapentin in a transgenic model of familial amyotrophic lateral sclerosis. Ann Neurol 39:147–157

Hara H, Harada K, Sukamoto T (1993a) Chronological atrophy after transient middle cerebral artery occlusion in rats. Brain Res 618:251–260

Hara H, Ozaki A, Yoshidomi M, Sukamoto T (1990) Protective effect of KB-2796, a new calcium antagonist, in cerebral hypoxia and ischemia. Arch Int Pharmacodyn Ther 304:206–218

Hara H, Yokota K, Shimazawa M, Sukamoto T (1993b) Effect of KB-2796, a new diphenylpiperizine Ca^{2+} antagonist, on glutamate-induced neurotoxicity in rat hippocampal primary cell cultures. Jpn J Pharmacol 61:361–365

Hayakawa T, Hamada Y, Maihara T, Hattori H, Mikawa H (1994a) Phenytoin reduces neonatal hypoxic-ischemic brain damage in rats. Life Sci 54:387–392

Hayakawa T, Higuchi Y, Nigami H, Hattori H (1994b) Zonisamide reduces hypoxic-ischemic brain damage in neonatal rats irrespective of its anticonvulsive effect. Eur J Pharmacol 257:131–136

Hering S, Aczel S, Kraus RL, Berjukow S, Striessnig J, Timin EN (1997) Molecular mechanism of use-dependent calcium channel block by phenylalkylamines: role of inactivation. Proc Natl Acad Sci USA 94:13323–13328

Hering S, Berjukow S, Aczel S, Timin EN (1998) Ca^{2+} channel block and inactivation: common molecular determinants. Trends in Pharmacological Sciences 19:439–443

Hille B (1992) Ionic Channels of Excitable Membranes. Sinauer Assoc., Inc., Sunderland, MA, USA, 607pp

Hockerman GH, Johnson BD, Abbott MR, Scheuer T, Catterall WA (1997a) Molecular determinants of high affinity phenylalkylamine block of L-type calcium channels in transmembrane segment IIIS6 and the pore region of the alpha1 subunit. J Biol Chem 272:18759–18765

Hockerman GH, Peterson BZ, Johnson BD, Catterall WA (1997b) Molecular determinants of drug binding and action on L-type calcium channels. Annu Rev Pharmacol Toxicol 37:361–396

Hovda DA, Fu K, Badie H, Samii A, Pinanong P, Becker DP (1994) Administration of an omega-conopeptide one hour following traumatic brain injury reduces 45calcium accumulation. Acta Neurochir Suppl Wien 60:521–523

Isom LL, DeJongh KS, Catterall WA (1994) Auxiliary subunits of voltage-gated ion channels. Neuron 12:1183–1194

Jacewicz M, Brint S, Tanabe J, Pulsinelli WA (1990) Continuous nimodipine treatment attenuates cortical infarction in rats subjected to 24 hours of focal cerebral ischemia. J Cereb Blood Flow Metab 10:89–96

Kanthasamy AG, Matsumoto RR, Gunasekar PG, Trunong DD (1995) Excitoprotective effect of felbamate in cultured cortical neurons. Brain Res 705:97–104

Kass IS, Abramowicz AE, Cottrell JE, Chambers G (1992) The barbiturate thiopental reduces ATP levels during anoxia but improves electrophysiological recovery and ionic homeostasis in the rat hippocampal slice. Neurosci 49:537–543

Kellenberger S, Scheuer T, Catterall WA (1996) Movement of the Na+ channel inactivation gate during inactivation. J Biol Chem 271:30971–30979

Kim DK, Catterall WA (1997) Ca2+-dependent and -independent interactions of the isoforms of the alpha1 A subunit of brain Ca2+ channels with presynaptic SNARE proteins. Proc Natl Acad Sci USA 94:14782–14786

Kiskin NI, Chizhmakov IV, Tsyndrenko AY, Krishtal OA, Tegtmeier F (1993) R56865 and flunarizine as Na^{+}-channel blockers in isolated Purkinje neurons of rat cerebellum. Neurosci 54:575–585

Klugbauer N, Lacinova L, Marais E, Hofmann F (1999) Molecular diversity of the dalcium channel alpha-$_2$-delta subunit. J Neurosci 19:684–691

Kucharczyk J, Mintorovitch J, Moseley ME, Asgari HS, Sevick RJ, Derugin N, Norman D, Leach MJ, Swan JH, Eisenthal D, Dopson M, Nobbs M (1993) Ischemic brain damage: reduction by sodium-calcium ion channel modulator RS- 87476. Stroke 24:1063–1067

Kumar K, White BC, Krause GS, Indrieri RJ, Evans AT, Hoehner TJ, Garritano AM, Koestner A (1988) A quantitative morphological assessment of the effect of lidoflazine and deferoxamine therapy on global brain ischaemia. Neurol Res 10: 136–140

Kuo CC, Bean BP (1994a) Na$^+$ channels must deactivate to recover from inactivation. Neuron 12:819–829

Kuo CC, Bean BP (1994b) Slow binding of phenytoin to inactivated sodium channels in rat hippocampal neurons. J Pharmacol Exp Ther 46:716–725

Kuo CC, Chen R-S, Lu L, Chen R-C (1997) Carbamazepine inhibition of neuronal Na$^+$ currents: quantitative distinction from phenytoin and possible therapeutic implications. Mol Pharmacol 51:1077–1083

Kuo CC, Lu L (1997) Characterization of lamotrigine inhibition of Na+ channels in rat hippocampal neurons. Br J Pharmacol 121:1231–1238

Leach MJ, Swan JH, Eisenthal D, Dopson M, Nobbs M (1993) BW619C89, a glutamate release inhibitor, protects against focal cerebral ischemic damage [see comments]. Stroke 24:1063–1067

Lehning EJ, Doshi R, Isaksson N, Stys PK, LoPachin RM, Jr (1996) Mechanisms of injury-induced calcium entry into peripheral nerve myelinated axons: role of reverse sodium-calcium exchange. J Neurochem 66:493–500

Lekieffre D, Meldrum BS (1993) The pyrimidine derivative, BW 1003C87, protects CA1 and striatal neurons following transient severe forebrain ischaemia in rats. A microdialysis and histological study. Neurosci 56:93–99

Letts VA, Felix R, Biddlecome GH, Arikkath J, Mahaffey CL, Valenzuela A, Bartlett FS, Mori Y, Campbell KP, Frankel WN (1998) The mouse stargazer gene encodes a neuronal CA^{2+} channel gamma subunit. Nature Genet 19:340–347

Levy RH, Mattson RH, Meldrum BS (1995) Antiepileptic Drugs. Fourth, 1–1120. New York, Raven Press

Leybaert L, de Hemptinne A (1993) A voltage-clamp study of calcium currents in neurons freshly isolated from the dorsal root ganglion of adult rats. Arch Int Physiol Biochim Biophys 101:315–323

Lin BW, Dietrich WD, Busto R, Ginsberg MD (1990) (S)-emopamil protects against global ischemic brain injury in rats. Stroke 21:1734–1739

Lipton P, Whittingham TS (1982) Reduced ATP concentrations a basis for synaptic transmission failure during hypoxia in the in vitro guinea-pig hippocampus. J Physiol (Lond) 325:51–65

Lu HR, Van Reempts J, Haseldonckx M, Borgers M, Janssen PAJ (1990) Cerebroprotective effects of flunarizine in an experimental rat model of cardiac arrest. Am J Emerg Med 8:1–6

Lucas LF, West CA, Rigor BM, Schurr A (1989) Protection against cerebral hypoxia by local anesthetics: a study using brain slices. J Neurosci Meth 28:47–50

Lyden PD, Zivin JA, Kochhar A, Mazzarella V (1988) Effects of calcium channel blockers on neurologic outcome after focal ischemia in rabbits. Stroke 19:1020–1026

Lynch JJ, Yu SP, Canzoniero LM, Sensi SL, Choi DW (1995) Sodium channel blockers reduce oxygen-glucose deprivation-induced cortical neuronal injury when combined with glutamate receptor antagonists. J Pharmacol Exp Ther 273:554–560

Lysko PG, Webb CL, Yue T-Y, Gu J-LG, Feuerstein MD (1995) Neuroprotective effects of tetrodotoxin as a Na$^+$ channel modulator and glutamate release inhibitor in cultured rat cerebellar neurons and in gerbil global brain ischemia. Stroke 25:2476–2482

Ma JY, Catterall WA, Scheuer T (1997) Persistent sodium currents through brain sodium channels induced by G protein betagamma subunits. Neuron 19:443–452

Ma JY, Li M, Catterall WA, Scheuer T (1994) Modulation of brain Na$^+$ channels by a G-protein-coupled pathway. Proc Natl Acad Sci USA 91:12351–12355

Maj R, Fariello RG, Ukmar G, Varasi M, Pevarello P, McArthur RA, Salvati P (1998) PNU-151774E protects against kainate-induced status epilepticus and hippocampal lesions in the rat. Eur J Pharmacol 359:27–32

Malgouris C, Daniel M, Doble A (1994) Neuroprotective effects of riluzole on N-methyl-D-aspartate- or veratridine-induced neurotoxicity in rat hippocampal slices. Neurosci Lett 177:95–99

Malhotra JD, Kaxen-Gillespie K, Hortsch M, Isom LL (2000) Sodium channel beta subunits mediate homophilic cell adhesion and recruit ankyrin to points of cell-to-cell contact. J Biol Chem 275:11383–11388

Marinov M, Wassmann H, Natschev S (1991) Effect of nimodipine in treatment of experimental focal cerebral ischaemia. Neurol Res 13:77–83

Martinez VE, Guillen F, Villanueva JA, Matias GJ, Bigorra J, Gil P, Carbonell A, Martinez LJ (1990) Placebo-controlled trial of nimodipine in the treatment of acute ischemic cerebral infarction. Stroke 21:1023–1028

Matyja E, Kida E (1991) Protective effect of nimodipine against quinolinic acid-induced damage of rat hippocampus in vitro. Neuropatol Pol 29:69–77

May GR, Rowand WS, McCormack JG, Sheridan RD (1995) Neuroprotective profile of lifarizine (RS-87476) in rat cerebrocortical neurones in culture. Br J Pharmacol 114:1365–1370

Mcbean DE, Winters V, Wilson AD, Oswald CB, Alps BJ, Armstrong JM (1995) Neuroprotective efficacy of lifarizine (RS-87476) in a simplified rat survival model of 2 vessel occlusion. Br J Pharmacol 116:3093–3098

McDonough SI, Swartz KJ, Mintz IM, Boland LM, Bean BP (1996) Inhibition of calcium channels in rat central and peripheral neurons by omega-conotoxin MVIIC. J Neurosci 16:2612–2623

McGuire D, Bowersox S, Fellmann JD, Luther RR (1997) Sympatholysis after neuron-specific, N-type, voltage-sensitive calcium channel blockade: first demonstration of N-channel function in humans. J Cardiovasc Pharmacol 30:400–403

McNaughton NCL, Leach MJ, Hainsworth AH, Randall AD (1997) Inhibition of human N-type voltage-gated Ca^{2+} channels by the neuroprotective agent BW 619C89. Neuropharmacol 36:1795–1798

Meldrum BS, Swan JH, Leach MJ, Millan MH, Gwinn R, Kadota K, Graham SH, Chen J, Simon RP (1992) Reduction of glutatmate release and protection against ischemic brain damage by BW1003C87. Brain Res 593:1–6

Miljanich GP, Ramachandran J (1995) Antagonists of neuronal calcium channels: structure, function, and therapeutic implications. Annu Rev Pharmacol Toxicol 35:707–734

Mintz IM, Bean BP (1993) Block of calcium channels in rat neurons by synthetic omega-Aga-IVA. Neuropharmacol 32:1161–1169

Morikawa E, Ginsberg MD, Dietrich WD, Duncan RC, Busto R (1991) Postischemic (S)-emopamil therapy ameliorates focal ischemic brain injury in rats. Stroke 22:355–360

Mossakowski MJ, Gadamski R (1990) Nimodipine prevents delayed neuronal death of sector CA1 pyramidal cells in short-term forebrain ischemia in Mongolian gerbils. Stroke 21:IV120–IV122

Muir KW, Lees KR, Hamilton SJ, George CF, Hobbiger SF, Lunnon MW (1995) A randomized, double-blind, placebo-controlled ascending dose tolerance study of 619C89 in acute stroke. Ann NY Acad Sci 765:328–329

Murakami A, Furui T (1994) Effects of the conventional anticonvulsants, phenytoin, carbamazepine and valproic acid on sodium-potassium-adenosine triphosphatase in acute ischemic brain. Neurosurg 34:1047–1051

Nadasdi L, Yamashiro D, Chung D, Tarczy HK, Adriaenssens P, Ramachandran J (1995) Structure-activity analysis of a Conus peptide blocker of N-type neuronal calcium channels. Biochem 34:8076–8081

Nakamura K, Hatakeyama T, Furuta S, Sakaki S (1993) The role of early Ca2+ influx in the pathogenesis of delayed neuronal death after brief forebrain ischemia in gerbils. Brain Res 613:181–192

Newcomb R, Gaur S, Bell JR, Cruz L (1995) Structural and biosynthetic properties of peptides in cone snail venoms. Peptides 16:1007–1017

Newcomb R, Palma A (1994) Effects of diverse omega-conopeptides on the in vivo release of glutamic and gamma-aminobutyric acids. Brain Res 638:95–102

Newcomb R, Szoke B, Palma A, Wang G, Chen X-H, Hopkins W, Cong R, Miller J, Urge L, Tarczy-Hornoch K, Loo JA, Dooley DJ, Nadasdi L, Tsien RW, Lemos J, Miljanich G (1999) A selective peptide antagonist of the class E calcium channel from the venom of the tarantula, *Hysterocrates gigas*. J Neurosci

Noda M, Suziki S, Numa S, Stuhmer W (1989) A single point mutation confers tetrodotoxin and saxitoxin insensitivity on the sodium channel II. FEBS Lett 259:213–216

Nooney JM, Lambert RC, Feltz A (1997) Identifying neuronal non-L Ca^{2+} channels – more than stamp collecting. Trends Pharmacol Sci 18:363–371

Nuglisch J, Karkoutly C, Mennel HD, Rossberg C, Krieglstein J (1990) Protective effect of nimodipine against ischemic neuronal damage in rat hippocampus without changing postischemic cerebral blood flow. J Cereb Blood Flow Metab 10:654–659

O'Neill MJ, Bath CP, Dell CP, Hicks CA, Gilmore J, Ambler SJ, Ward MA, Bleakman D (1997) Effects of Ca^{2+} and Na^+ channel inhibitors in vitro and in global cerebral ischaemia in vivo. Eur J Pharmacol 332:121–131

Okiyama K, Smith DH, Gennarelli TA, Simon RP, Leach M, McIntosh TK (1995) The sodium channel blocker and glutamate release inhibitor BW1003C87 and magnesium attenuate regional cerebral edema following experimental brain injury in the rat. J Neurochem 64:802–809

Okiyama K, Smith DH, Thomas MJ, McIntosh TK (1992) Evaluation of a novel calcium channel blocker, (S)-emopamil, on regional cerebral edema and neurobehavioral function after experimental brain injury. J Neurosurg 77:607–615

Olivera BM, Miljanich GP, Ramachandran J, Adams ME (1994) Calcium channel diversity and neurotransmitter release: the omega-conotoxins and omega-agatoxins. Annu Rev Biochem 63:823–867

Pauwels PJ, Leysen JE, Janssen PA (1991) Ca++ and Na+ channels involved in neuronal cell death. Protection by flunarizine. Life Sci 48:1881–1893

Pauwels PJ, Van-Assouw HP, Peeters L, Leysen JE (1990) Neurotoxic action of veratridine in rat brain neuronal cultures: mechanism of neuroprotection by Ca++ antagonists nonselective for slow Ca++ channels. J Pharmacol Exp Ther 255:1117–1122

Pauwels PJ, Van Assouw HP, Leysen JE, Janssen PA (1989) Ca2+-mediated neuronal death in rat brain neuronal cultures by veratridine: protection by flunarizine. Mol Pharmacol 36:525–531

Perez-Pinzon M, Yenari MA, Sun GH, Kunis DM, Steinberg GK (1997) SNX-111, a novel, presynaptic N-type calcium channel antagonist, is neuroprotective against focal cerebral ischemia in rabbits. J Neurol Sci 153:25–31

Pragnell M, De-Waard M, Mori Y, Tanabe T, Snutch TP, Campbell KP (1994) Calcium channel beta-subunit binds to a conserved motif in the I-II cytoplasmic linker of the alpha 1-subunit [see comments]. Nature 368:67–70

Pratt J, Rataud J, Bardot F, Roux M, Blanchard JC, Laduron PM, Stutzmann JM (1992) Neuroprotective actions of riluzole in rodent models of global and focal cerebral ischemia. Neurosci Lett 140:225–230

Prenen GHM, Gwan GK, Postema F, Zuiderveen F, Korf J (1988) Cerebral cation shifts in hypoxic-ischemic brain damage are prevented by the sodium channel blocker tetrodotoxin. Exp Neurol 99:118–132

Pringle AK, Benham CD, Sim L, Kennedy J, Iannotti F, Sundstrom LE (1996) Selective N-type calcium channel antagonist omega conotoxin MVIIA is neuroprotective against hypoxic neurodegeneration in organotypic hippocampal-slice cultures. Stroke 27:2124–2130

Ragsdale DS, McPhee JC, Scheuer T, Catterall WA (1994) Molecular determinants of state-dependent block of Na^+ channels by local anesthetics. Science 265:1724–1728

Ragsdale DS, McPhee JC, Scheuer T, Catterall WA (1996) Common molecular determinants of local anesthetic, antiarrhythmic, and anticonvulsant block of voltage-gated Na+ channels. Proc Natl Acad Sci USA 93:9270–9275

Ragsdale DS, Scheuer T, Catterall WA (1991) Frequency and voltage-dependent inhibition of type IIA Na$^+$ channels, expressed in a mammalian cell line, by local anesthetic, antiarrhythmic, and anticonvulsant drugs. Mol Pharmacol 40:756–765

Rami A, Krieglstein J (1994) Neuronal protective effects of calcium antagonists in cerebral ischemia. Life Sci 55:2105–2113

Randall A, Tsien RW (1995) Pharmacological dissection of multiple types of Ca2+ channel currents in rat cerebellar granule neurons. J Neurosci 15:2995–3012

Ransom BR, Stys PK, Waxman SG (1990) The pathophysiology of anoxic injury in central nervous system white matter. Stroke 21:III52–III57

Rasool N, Faroqui M, Rubinstein EH (1990) Lidocaine accelerates neuroelectric recovery after incomplete global ischemia in rabbits. Stroke 21:929–935

Rataud J, Debanot F, Mary V, Pratt J, Stutzmann JM, Silverstein FS, Buchanan K, Hudson C, Johnston MV (1994) Comparative study of voltage-sensitive sodium channel blockers in focal ischaemia and electric convulsions in rodents. Neurosci Lett 172:19–23

Regehr WG, Mintz IM (1994) Participation of multiple calcium channel types in transmission at single climbing fiber to Purkinje cell synapses. Neuron 12:605–613

Regehr WG, Tank DW (1992) Calcium concentration dynamics produced by synaptic activation of CA1 hippocampal pyramidal cells. J Neurosci 12:4202–4223

Rettig J, Heinemann C, Ashery U, Sheng ZH, Yokoyama CT, Catterall WA, Neher E (1997) Alteration of Ca2+ dependence of neurotransmitter release by disruption of Ca2+ channel/syntaxin interaction. J Neurosci 17:6647–6656

Rettig J, Sheng ZH, Kim DK, Hodson CD, Snutch TP, Catterall WA (1996) Isoform-specific interaction of the alpha1 A subunits of brain Ca2+ channels with the presynaptic proteins syntaxin and SNAP-25. Proc Natl Acad Sci USA 93:7363–7368

Roda JM, Carceller F, Diez TE, Avendano C (1995) Reduction of infarct size by intra-arterial nimodipine administered at reperfusion in a rat model of partially reversible brain focal ischemia. Stroke 26:1888–1892

Rogers JC, Qu Y, Tanada TN, Scheuer T, Catterall WA (1996) Molecular determinants of high affinity binding of alpha- scorpion toxin and sea anemone toxin in the S3-S4 extracellular loop in domain IV of the Na+ channel alpha subunit. J Biol Chem 271:15950–15962

Ross IB, Tator CH (1991) Further studies of nimodipine in experimental spinal cord injury in the rat. J Neurotrauma 8:229–238

Rothstein JD, Kuncl RW (1995) Neuroprotective strategies in a model of chronic glutamate-mediated motor neuron toxicity. J Neurochem 65:643–651

Sansom MSP (1998) Ion channels: a first view of K+ channels in atomic glory. Current Biology 8:R450–R452

Sauter A, Rudin M (1990) Calcium antagonists for reduction of brain damage in stroke. J Cardiovasc Pharmacol 15 Suppl 1:S43–S47

Seega J, Elger B (1993) Diffusion- and T2-weighted imaging: evaluation of oedema reduction in focal cerebral ischaemia by the calcium and serotonin antagonist levemopamil. Magn Reson Imaging 11:401–409

Sheng ZH, Yokoyama CT, Catterall WA (1997) Interaction of the synprint site of N-type Ca2+ channels with the C2B domain of synaptotagmin I. Proc Natl Acad Sci USA 94:5405–5410

Shirokov R, Ferreira G, Yi J, Rios E (1998) Inactivation of gating currents of L-type calcium channels. Specific role of the alpha 2 delta subunit. J Gen Physiol 111:807–823

Silverstein FS, Buchanan K, Hudson C, Johnston MV (1986) Flunarizine limits hypoxia-ischemia induced morphologic injury in immature rat brain. Stroke 17:477–482

Simpson PB, Russell JT (1998) Role of mitochondrial Ca^{2+} regulation in neuronal and glial cell signalling. Brain Res Rev 26:72–81

Small DL, Monette R, Buchan AM, Morley P (1997) Identification of calcium channels involved in neuronal injury in rat hippocampal slices subjected to oxygen and glucose deprivation. Brain Res 753:209–218

Smith SE, Lekieffre D, Sowinski P, Meldrum BS (1993) Cerebroprotective effect of BW619C89 after focal or global cerebral ischaemia in the rat. Neuroreport 4:1339–1342

Smith SE, Meldrum BS (1995) Cerebroprotective effect of lamotrigine after focal ischemia in rats. Stroke 26:117–121

Snape MF, Baldwin HA, Cross AJ, Green AR (1993) The effects of chlormethiazole and nimodipine on cortical infarct area after focal cerebral ischaemia in the rat. Neurosci 53:837–844

Stanton PK, Moskal JR (1991) Diphenylhydantoin protects against hypoxia-induced impairment of hippocampal synaptic transmission. Brain Res 546:351–354

Stea A, Dubel SJ, Pragnell M, Leonard JP, Campbell KP, Snutch TP (1993) A beta-subunit normalizes the electrophysiological properties of a cloned N-type Ca2+ channel alpha 1-subunit. Neuropharmacol 32:1103–1116

Stefani A, Spadoni F, Bernardi G (1997) Differential inhibition by riluzole, lamotrigine and phenytoin of sodium and calcium currents in cortical neurons: Implications for neuroprotective strategies. Exp Neurol 147:115–122

Striessnig J, Murphy BJ, Catterall WA (1991) Dihydropyridine receptor of L-type Ca^{2+} channels: identification of binding domains for $[^3H](+)$-PN200–110 and $[^3H]$azidopine within the α1 subunit. Proc Natl Acad Sci USA 88:10769–10773

Stys P (1995) Protective effects of antiarrhythmic agents against anoxic injury in CNS white matter. J Cereb Blood Flow Metab 15:425–432

Stys P, Waxman SG, Ransom BR (1992) Ionic mechanisms of anoxic injury in mammalian CNS white matter: role of Na^+ channels and Na^+-Ca^{2+} exchanger. J Neurosci 12:430–439

Stys PK, Lesiuk H (1996) Correlation between electrophysiological effects of mexiletine and ischemic protection in central nervous system white matter. Neurosci 71:27–36

Sun FY, Faden AI (1995) Neuroprotective effects of BW619C89, a use-dependent sodium channel blocker, in rat traumatic brain injury. Brain Res 637:133–140

Szatkowski M, Attwell D (1994) Triggering and execution of neuronal death in brain ischemia: two phases of glutamate release by different mechanisms. Trends Neurosci 17:359–365

Taft WC, Clifton GL, Blair RE, DeLorenzo RJ (1989) Phenytoin protects against ischemia-produced neuronal cell death. Brain Res 483:143–148

Takizawa S, Matsushima K, Fujita H, Nanri K, Ogawa S, Shinohara Y (1995) A selective N-type calcium channel antagonist reduces extracellular glutamate release and infarct volume in focal cerebral ischemia. J Cereb Blood Flow Metab 15: 611–618

Talbot MJ, Sayer RJ (1996) Intracellular QX-314 inhibits calcium currents in hippocampal CA1 pyramidal neurons. J Neurophysiol 76:2120–2124

Tasker RC, Coyle JT, Vornov JJ (1992) The regional vulnerability to hypoglycemia-induced neurotoxicity in organotypic hippocampal culture: protection by early tetrodotoxin or delayed MK-801. J Neurosci 12:4298–4308

Tateishi A, Scheller MS, Drummond JC, Zornow MH, Grafe MR, Fleischer JE, Shapiro HM (1991) Failure of nimodipine to improve neurologic outcome after eighteen minutes of cardiac arrest in the cat. Resuscitation 21:191–206

Taylor CP (1997) Sodium channels and therapy of central nervous system diseases. Adv Pharmacol 39:47–98

Taylor CP, Burke SP, Weber ML (1995) Hippocampal slices: glutamate release and cellular damage from ischemia are reduced by sodium channel blockade. J Neurosci Meth 59:121–128

Taylor CP, Geer JJ, Burke SP (1992) Endogenous extracellular glutamate accumulation in rat neocortical cultures by reversal of the transmembrane sodium gradient. Neurosci Lett 145:197–200

Taylor CP, Narasimhan LS (1997) Sodium channels and therapy of central nervous system diseases. Adv Pharmacol 39:47–98

Teng YD, Wrathall JR (1997) Local blockade of sodium channels by tetrodotoxin ameliorates tissue loss and long-term functional deficits resulting from experimental spinal cord injury. J Neurosci 17:4359–4366

Terlau H, Heinemann SH, Stuhmer W, Pusch M, Conti F, Imoto K, Numa S (1991) Mapping the site of block by tetrodotoxin and saxitoxin of sodium channel II. FEBS Lett 293:93–96

Torp R, Arvin B, Le Peillet E, Chapman AG, Ottersen OP, Meldrum BS (1993) Effect of ischaemia and reperfusion on the extra- and intracellular distribution of glutamate, glutamine, aspartate and GABA in the rat hippocampus, with a note on the effect of the sodium channel blocker BW1003C87. Exp Brain Res 96:365–376

Trust Study Group (1990) Randomised, double-blind, placebo-controlled trial of nimodipine in acute stroke. Lancet 336:1205–1209

Valentino K, Newcomb R, Gadbois T, Singh T, Bowersox S, Bitner S, Justice A, Yamashiro D, Hoffman BB, Ciaranello R et al. (1993) A selective N-type calcium channel antagonist protects against neuronal loss after global cerebral ischemia. Proc Natl Acad Sci USA 90:7894–7897

Van Reempts J, Haseldonckx M, Van Deuren B, Wouters L, Borgers M (1986) Structural damage of the ischemic brain: involvement of calcium and effects of postischemic treatment with calcium entry blockers. Drug Dev Res 8:387–395

Verweij BH, Muizelaar JP, Vinas FC, Peterson PL, Xiong Y, Lee CP (1997) Mitochondrial dysfunction after experimental and human brain injury and its possible reversal with a selective N-type calcium channel antagonist (SNX-111). Neurol Res 19:334–339

Wadworth AN, McTavish D (1992) Nimodipine. A review of its pharmacological properties, and therapeutic efficacy in cerebral disorders. Drugs Aging 2:262–286

Wahl F, Allix M, Plotkine M, Boulu RG (1993) Effect of riluzole on focal cerebral ischemia in rats. Eur J Pharmacol 230:209–214

Walker D, De Waard M (1998) Subunit interaction sites in voltage-dependent Ca2+ channels: role in channel function. Trends in Neurosceince 21:148–154

Wallis RA, Panizzon KL (1995) Felbamate neuroprotection against CA1 traumatic neuronal injury. Eur J Pharmacol 294:475–482

Wallis RA, Panizzon KL, Fairchild MD, Wasterlain CG (1992) Protective effects of felbamate against hypoxia in the rat hippocampal slices. Stroke 23:547–551

Wasterlain CG, Adams LM, Hattori H, Schwartz PH (1992) Felbamate reduces hypoxic-ischemic brain damage in vivo. Eur J Pharmacol 212:275–278

Wauquier A, Melis W, Janssen PAJ (1989) Long-term neurological assessment of the post-resuscitative effects of flunarizine, verapamil and nimodipine in a new model of global complete ischeamia. Neuropharmacol 28:837–846

Weber ML, Taylor CP (1994) Damage from oxygen and glucose deprivation in hippocampal slices is prevented by tetrodotoxin, lidocaine and phenytoin without blockade of action potentials. Brain Res 664:167–177

Weiss JH, Hartley DM, Koh J, Choi DW (1990) The calcium channel blocker nifedipine attenuates slow excitatory amino acid neurotoxicity. Science 247:1474–1477

Welsch M, Nuglisch J, Krieglstein J (1990) Neuroprotective effect of nimodipine is not mediated by increased cerebral blood flow after transient forebrain ischemia in rats. Stroke 21:IV105–IV107

Wheeler DB, Randall A, Tsien RW (1994) Roles of N-type and Q-type Ca2+ channels in supporting hippocampal synaptic transmission [see comments]. Science 264:107–111

White, G. The broad spectrum anticonvulsant ADCI blocks TTX sensitiive voltage-activated Na channel currents in NE!-115 mouse neuroblastoma cells. Society for Neuroscience Abstracts 20, 1239. 1995

Whitfield PC, Pickard JD (1994) Nimodipine. Br J Hosp Med 52:539–540

Wiard RS, Dickerson MC, Beek O, Norton R, Cooper BR (1995) Neuroprotective properties of the novel antiepileptic lamotrigine in a gerbil model of global cerebral ischemia. Stroke 26:466–472

Wickman K, Clapham DE (1995) Ion channel regulation by G-proteins. Physiol Rev 75:865–885

Wu L-G, Sagratella S (1994) Pharmacological identification of two types of presynaptic voltage-dependent calcium channels at CA3-CA1 synapses of the hippocampus. J Neurosci 14:5613–5622

Xie Y, Dengler K, Zacharias E, Wilffert B, Tegtmeier F (1994) Effects of the sodium channel blocker tetrodotoxin (TTX) on cellular ion homeostasis in rat brain subjected to complete ischemia. Brain Res 652:216–224

Xie Y, Zacharias E, Hoff P, Tegtmeier F (1995) Ion channel involvement in anoxic depolarization induced by cardiac arrest in rat brain. J Cereb Blood Flow Metab 15: 587–594

Xiong Y, Peterson PL, Verweij BH, Vinas FC, Muizelaar JP, Lee CP (1998) Mitochondrial dysfunction after experimental traumatic brain injury: combined efficacy of SNX-111 and U-101033 E. J Neurotrauma 15:531–544

Yamasaki Y, Kogure K, Hara H, Ban H, Akaike N (1991) The possible involvement of tetrodotoxin-sensitive ion channels in ischemic neuronal damage in the rat hippocampus. Neurosci Lett 121:251–254

Yang N, George ALJr, Horn R (1996) Molecular basis of charge movement in voltage-gated sodium channels. Neuron 16:113–122

Yenari MA, Palmer JT, Sun GH, de-Crespigny A, Mosely ME, Steinberg GK (1996) Time-course and treatment response with SNX-111, an N-type calcium channel blocker, in a rodent model of focal cerebral ischemia using diffusion-weighted MRI. Brain Res 739:36–45

Yoshidomi M, Hayashi T, Abe K, Kogure K (1989) Effects of a new calcium channel blocker, KB-2796, on protein synthesis of the CA1 pyramidal cell and delayed neuronal death following transient forebrain ischemia. J Neurochem 53:1589–1594

Young K, Lin S, Sun L, Lee E, Modi M, Hellings S, Husbands M, Ozenberger B, Franco R (1998) Identification of a calcium channel modulator using a high throughput yeast two-hybrid screen. Nature Biotechnology 16:946–950

Yu C, Jia M, Litzinger M, Nelson PG (1988) Calcium agonist (BayK 8644) augments voltage-sensitive calcium currents but not synaptic transmission in cultured mouse spinal cord neurons. Exp Brain Res 71:467–474

Zhang JF, Ellinor PT, Aldrich RW, Tsien RW (1994) Molecular determinants of voltage-dependent inactivation in calcium channels. Nature 372:97–100

Zhang Y-L, Lipton P (1995) Mitochondria and endoplasmic reticulum are major sources of increased cytosolic calcium during ischemia in the rat hippocampus. Society for Neuroscience Abstracts 21:217

Zhang Y, Lipton P (1999) Cytosolic Ca2+ changes during in vitro ischemia in rat hippocampal slices: major roles for glutamate and Na+-dependent Ca2+ release from mitochondria. J Neurosci 19:3307–3315

Zhao Q, Smith ML, Siesjo BK (1994) The omega-conopeptide SNX-111, an N-type calcium channel blocker, dramatically ameliorates brain damage due to transient focal ischaemia. Acta Physiol Scand 150:459–461

Zheng W, Stoltefuss J, Goldmann S, Triggle DJ (1992) Pharmacologic and radioligand binding studies of 1,4-dihydropyridines in rat cardiac and vascular preparations: stereoselectivity and voltage dependence of antagonist and activator interactions. Mol Pharmacol 41:535–541

Zona C, Avoli M (1997) Lamotrigine reduces voltage-gated sodium currents in rat central neurons in culture. Epilepsia 38:522–525

Zona C, Ciotti MT, Avoli M (1997) Topiramate attenuates voltage-gated sodium currents in rat cerebellar granule cells. Neurosci Lett 231:123–126

CHAPTER 9
Neuroprotection by Free Radical Scavengers and Other Antioxidants

J.W. Phillis

A. Introduction

Cells which depend on oxygen for survival are continuously subjected to oxidative stress due to the fact that during the respiratory process, in which molecular oxygen is reduced to water, a small fraction (2%–5%) of the oxygen is converted to superoxide radical ($O_2^{\bullet}$) by a one-electron reduction mechanism. Superoxide is further converted to hydrogen peroxide and related metabolites. Such semi-reduced species of oxygen (reactive oxygen species) are highly reactive and capable of initiating a series of oxidative reactions, which collectively constitute *oxidative stress*. Under normal physiological conditions, cells are protected from oxidative injury by the various endogenous antioxidant defenses which have evolved to reduce oxidative stress. Tissue injury can result from any disruption of normal cellular function to the extent that oxygen radical production is substantially increased and these cellular defenses are overwhelmed. The central nervous system is particularly sensitive to oxygen radical damage because of the high levels of polyunsaturated lipids present that, when peroxidized, compromise the integrity of neuronal cell membranes together with the specialized functions of those membranes.

Oxidative or free radical injury appears to be a fundamental mechanism of human disease (McCord 1993; Cross 1987). Increasing evidence suggests that such injury is important in the pathogenesis of a diverse group of neurological disorders, including cerebrovascular accidents (stroke, cardiac arrest), neurotrauma, effects of aging, epilepsy, and certain neurodegenerative diseases, including Huntington's disease, Parkinson's disease, Alzheimer's disease, and amyotrophic lateral sclerosis (Jenner 1994; Beal 1995; Williams 1995; Delanty and Dichter 1998; Beckman and Ames 1998). Other fundamental processes involved in the pathogenesis of neuronal injury appear to involve free radicals. For instance, free radicals may be involved in the induction or mediation of programmed cell death or apoptosis (Greenlund et al. 1995) and in signal transduction and gene expression (Palmer and Paulson 1997).

Endogenous antioxidant mechanisms which serve to protect against the oxidative stress associated with normal metabolism include enzyme systems such as superoxide dismutase, catalase and glutathione peroxidase, endogenous antioxidant compounds which can terminate free radical reactions such as ascorbic acid and α-tocopherol, and specific transport proteins and binding molecules for transition metals such as iron (ferritin and transferrin) and copper. The relatively low level of the antioxidant defenses in central nervous tissues (SAVOLAINEN 1978; USHIJIMA et al. 1986; COHEN 1988) may be one reason for the susceptibility of the brain and spinal cord to free radical-induced injury.

B. Free Radical Formation in the CNS

I. Reactive Oxygen Species

1. Mitochondrial Activity

Electron transport by mitochondrial membranes is the major source of free radicals in the brain. During ischemia/reperfusion or after trauma it is likely that the mitochondrial antioxidative defense mechanisms become inadequate as a result of an overproduction of oxygen radicals, inactivation of detoxification systems, and the consumption of antioxidants, with resultant increases in ROS levels. Microsomal and nuclear membranes also contain electron transport systems which may add to free radical production.

2. Excitotoxic Amino Acids

Excitotoxic amino acids including glutamate and aspartate, induce oxygen free radical formation in both in vitro and in vivo brain tissue (LAFON-CAZAL et al. 1993; LANCELOT et al. 1998). Cerebral ischemia and trauma are accompanied by massive increases in the extracellular concentrations of these amino acids (BENVENISTE et al. 1984; HAGBERG et al. 1985; SIMPSON et al. 1992; NILSSON et al. 1990; KATAYAMA et al. 1990). Glutamate, aspartate, and their analogs in excessive concentrations are neurotoxic and may cause cell death by generating ROS (DYKENS et al. 1987; CIANI et al. 1996). The interrelationship between excitatory amino acid release into the interstitial space and free radical formation has been emphasized by studies showing that degradation of free radicals by superoxide dismutase and catalase reduces glutamate release during ischemia (Fig. 1) (PELLEGRINI-GIAMPIETRO et al. 1990; O'REGAN et al. 1997). Conversely, incubation with enzymes and substrates known to cause the formation of free radicals such as xanthine oxidase and xanthine or prostaglandin synthase plus arachidonic acid increase ischemia-evoked excitotoxic amino acid release (Figs. 2, 3) from the rat cerebral cortex (O'REGAN et al. 1997). Free radical formation and glutamate release appear, therefore, to be mutually related and cooperate in precipitating neuronal death.

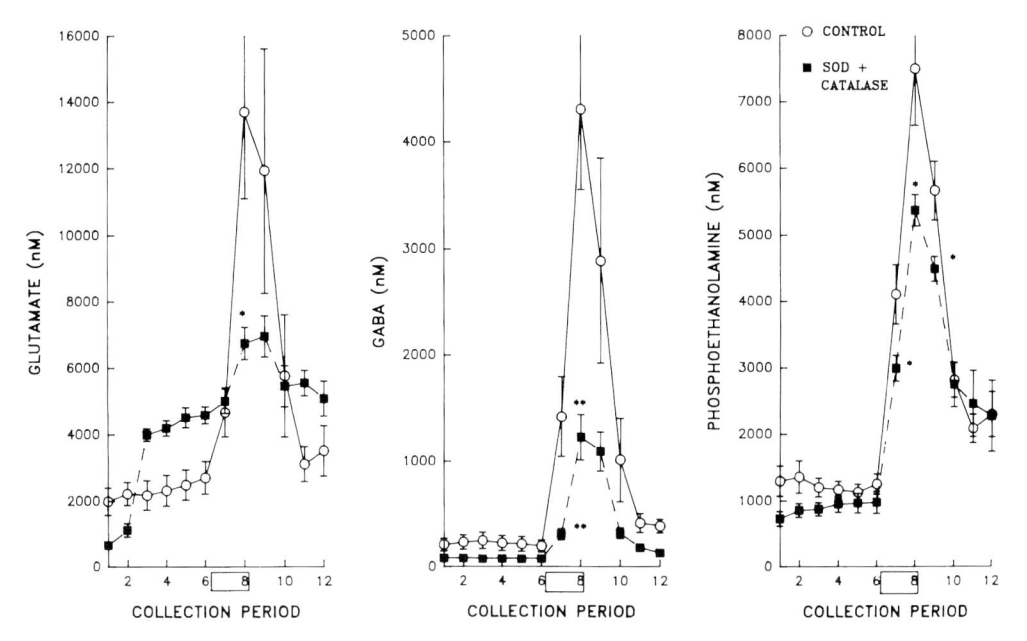

Fig. 1. Effects of superoxide dismutase (90 U/ml) plus catalase (3000 U/ml) on the ischemia-evoked release of glutamate, GABA, and phosphoethanolamine into rat cerebral cortical superfusates. *Line plots* show the time-course of amino acid efflux into 10-min superfusate collections before, during, and following a 20-min period of four-vessel occlusion cerebral ischemia. Enzyme application was initiated following collection period 2 and continued for the remainder of the experiment. Results are plotted in nM/l (mean ± SEM). The *open box* designates the period of ischemia (collections 7 and 8). The increase in basal glutamate levels following enzyme application was due to the presence of this amino acid in the catalase preparation. Ischemia-evoked release of all three amino acids was significantly reduced. (Reproduced with permission from O'REGAN et al. 1997)

3. Arachidonic Acid Metabolism

Arachidonic acid metabolism may be a significant source of ROS formation. Calcium entry into neurons during seizures, ischemia, or trauma activates phospholipases, including PLA_2, which cleaves a fatty acyl chain from the Sn-2 position of phospholipids to produce arachidonic acid, the level of which rises rapidly (ABE et al. 1987; REDDY and BAZAN 1987; SHOHAMI et al. 1989; UMEMURA et al. 1992; SALUJA et al. 1999). Arachidonic acid can then be metabolized by either cyclooxygenases or lipoxygenases to produce a variety of eicosanoids together with reactive oxygen species.

4. Purine Catabolism

Purine catabolism is another source of ROS formation in the seizing, ischemic, or traumatized brain. During these conditions ATP is utilized and its metabo-

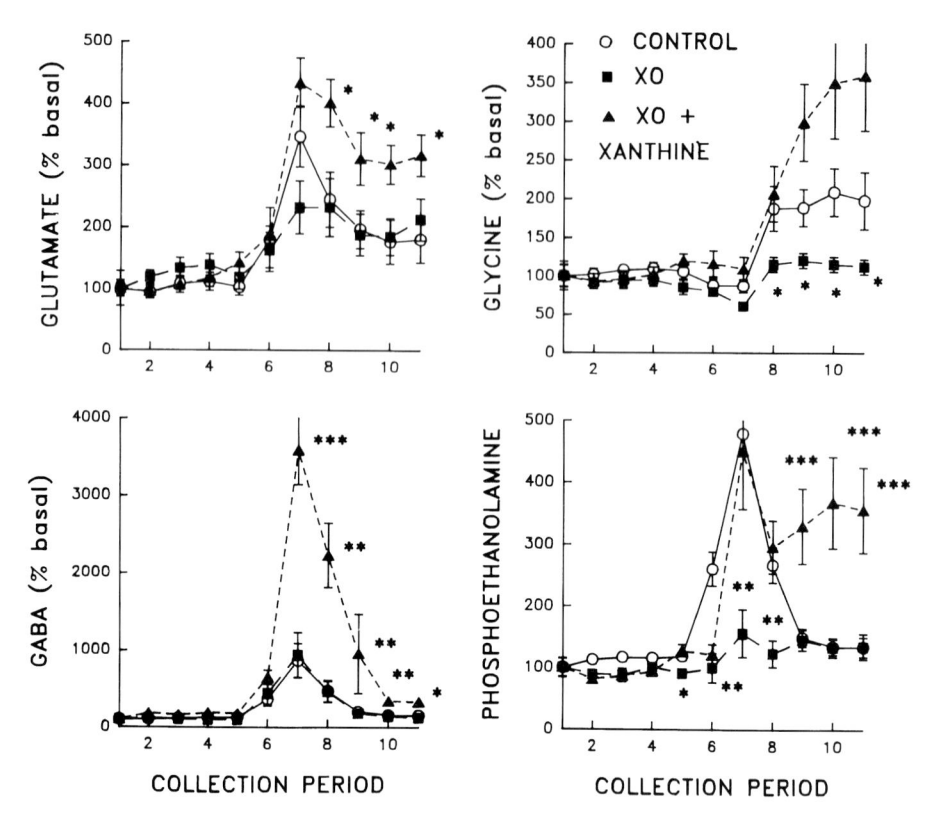

Fig. 2. Effects of xanthine oxidase (*filled square*, 20 mU/ml) and xanthine oxidase plus 0.5 mM xanthine (*filled triangle*) on the ischemia-evoked release of glutamate, glycine, GABA, and phosphoethanolamine into rat cerebral cortical superfusates. *Line plots* show the time-course of amino acid efflux (percent of preischemic basal levels, mean ± SEM) before, during, and following a 20-min period of four-vessel (*4VO*) occlusion (collection periods 6 and 7). Drug application was initiated prior to collection 2 and continued for the remainder of the experiment. Addition of xanthine plus xanthine oxidase significantly enhanced the ischemia/reperfusion-evoked release of all four amino acids. (Reproduced with permission from O'Regan et al. 1997)

lites adenosine, inosine, hypoxanthine, and xanthine accumulate (During and Spencer 1992; Simpson et al. 1992; Ikeda and Long 1990). In the normoxic brain, hypoxanthine and xanthine are metabolized by xanthine dehydrogenase in a reaction which does not produce ROS. During injury, the rise in intracellular calcium causes a conversion of xanthine dehydrogenase to xanthine oxidase which uses molecular oxygen rather than the nucleotide radical of NAD as its electron acceptor and produces $O_2^{\bullet}$ and H_2O_2.

5. Neutrophils

Neutrophils are considered to be significant contributors to ROS levels during injury to the brain. Neutrophils accumulate at injury sites and can release $O_2^{\bullet}$

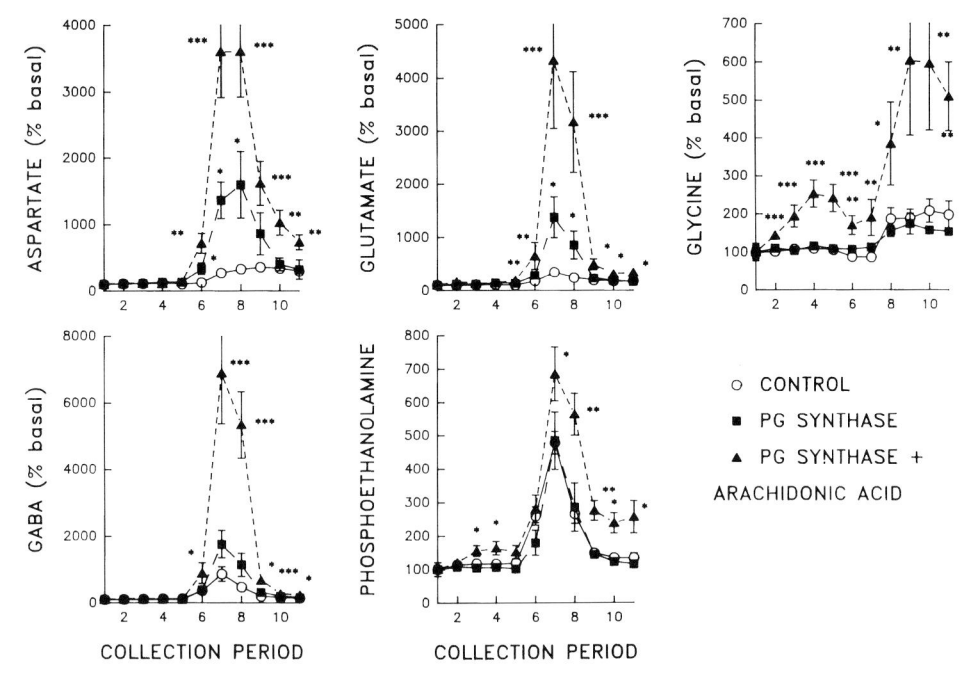

Fig. 3. Effects of prostaglandin synthase (*filled square*, 250 U/ml) and prostaglandin synthase plus 50 μM arachidonic acid (*filled triangle*) on ischemia-evoked release of aspartate, glutamate, glycine, GABA, and phosphoethanolamine into rat cerebral cortical superfusates. For further details see legend to Fig. 2. (Reproduced with permission from O'REGAN et al. 1997)

synthesized by NADP oxidase during the respiratory burst. They also secrete the enzyme myeloperoxidase, which catalyzes the formation of hypochlorous acid, a nonradical oxidizing agent, from H_2O_2 and chloride ions.

6. Autooxidation of Catecholamines

Autooxidation of catecholamines may contribute to free radical generation in the brain. Catecholamines are normally inactivated by pathways involving the enzymes monoamine oxidase and catechol-*O*-methyltransferase. When the enzymatic mechanisms are saturated or impaired, catecholamines can autooxidize with the generation of ROS.

II. Reactive Nitrogen Species

Nitric oxide, the endothelially-derived relaxing factor, is itself a free radical with both neuroprotective and neurodegenerative effects. NO•-mediated neurotoxicity is engendered, at least in part, by its reaction with superoxide anion, leading to the formation of peroxynitrite ($ONOO^-$) (LIPTON et al. 1993).

Inhibition of nitric oxide synthase by N^G-nitro-L-arginine methyl ester inhibits hydroxyl radical formation during glutamate infusion into the striatum (LANCELOT et al. 1998), providing evidence that NOS is involved in the generation of •OH in excitotoxic conditions (LANCELOT et al. 1998). The authors suggest that NO• might have combined with $O_2^{•-}$ to produce $ONOO^-$, which could spontaneously decompose to yield •OH (BECKMAN et al. 1990). Alternatively NOS may be able to directly generate ROS in highly excitotoxic conditions (POU et al. 1992).

C. Oxidative Stress in the CNS

Oxidative stress refers to a condition in which there is a serious imbalance between the production and elimination of free radicals. It could be a consequence of a diminished level of endogenous antioxidants as a result of a reduced intake of dietary antioxidants or other essential dietary constituents and possibly to mutations affecting antioxidant defense enzymes or to toxins that deplete such defenses. Increased production of ROS/RNS may involve excessive activation of natural ROS/RNS-producing systems such as occurs during epilepsy, ischemia/reperfusion, and trauma; or be a result of exposure to toxins which are either themselves reactive species (e.g., nitrogen dioxide, $NO_2^{•}$) or which are metabolized to generate free radicals, or as a consequence of exposure to ionizing radiation.

D. Evidence of Free Radical Generation in the CNS

Evidence of the involvement of oxygen-derived free radicals in brain injury following ischemia/reperfusion, trauma, and epilepsy has been slowly accumulating. Direct detection of free radicals in cerebral tissues has been hindered by their short half-lives and trace amounts. Until recently, most of the studies on free radical participation in such injuries have employed an indirect approach, using free radical scavengers or inhibition of radical synthesizing enzymes to infer a role of ROS/RNS in tissue damage.

Techniques used to monitor free radical production in brain have included nitro blue tetrazolium reduction (NELSON et al. 1992), salicylate trapping of hydroxyl radicals with detection of 2,3- and 2,5-dihydroxybenzoic acids (CAO et al. 1988; LANCELOT et al. 1998), malondialdehyde and 4-hydroxyalkenal production (NISHIO et al. 1997), and in vivo chemiluminescence (PETERS et al. 1998).

Direct evidence for free radical formation has been obtained by the application of spin trapping and electron paramagnetic resonance spectrometry techniques. Short-lived free radicals are trapped with nitrone, nitroso, or pyrroline N-oxide spin-trapping agents, upon which the resulting adduct gives rise to a stable free radical, which can then be measured with high specificity and sensitivity by EPR. Using this technique, it has been possible to demonstrate the presence of •OH and related free radicals in brain perfusates of rats subjected to global ischemia/reperfusion (SEN and PHILLIS 1993), focal ischemia/reperfusion (DUGAN et al. 1995), and cerebral trauma (SEN et al. 1994). EPR-computed tomographic brain imaging has been used to observe nitroxide radical images in the rat striatum following iron-induced epileptogenic seizures after intraperitoneal administration of carbamoyl-PROXYL (HIRAMATSU et al. 1996).

Free radical generation in the rat cerebral cortex was detected by measuring the levels of a hydroxyl radical adduct of the spin trapping agent pyridyl-N-oxide-*tert*-butylnitrone (POBN) in cortical superfusates prior to and during global ischemia and following reperfusion (SEN and PHILLIS 1993). Preischemic samples did not elicit an EPR signal. During and following a 30-min period of cerebral ischemia •OH radical adduct signals (characterized by a 6-line triplet of doublets) were detected in the superfusate samples, the most intense signals being evident during the initial reperfusion period (Fig. 4). DUGAN et al. (1995) used 5,5-dimethyl-1-pyrroline N-oxide to examine free radical production in the cerebral cortex of a rat focal ischemia preparation. As in the global ischemia model, spin adduct signals were observed during the 90-min period of middle cerebral artery occlusion and increased during the first phase of reperfusion.

POBN was also used to observe free radical release from the rat cerebral cortex following a moderate closed head concussive injury (SEN et al. 1994). EPR analysis of cortical superfusate samples revealed the 6-line spectra characteristic of POBN-OH radical adducts, the intensity of which peaked 40 min posttrauma (Fig. 5). The signal then declined and could no longer be detected after 120 min.

MALINSKI et al. (1993), using a porphyrinic microsensor to measure the time course of nitric oxide formation during focal ischemia/reperfusion, observed increases in the levels of this radical. Since NO is paramagnetic, it can also be detected by spin-trapping and electron paramagnetic resonance. Cerebral cortical NO production was potentiated after 5 min of focal ischemia and its levels continued to rise during a further 55 min of ischemia. The levels had declined significantly following 30 min of reperfusion (SATO et al. 1994).

The relationship between free radical production (EPR spectroscopy) and lipid peroxidation (measured as malondialdehyde production) in ischemic/reperfused rat brains was examined by SAKAMOTO et al. (1991). Ischemia/reperfusion initiated an abrupt burst of free radical formation, which was accompanied by lipid peroxidation, a result consistent with the hypothesis that lipid peroxidation may be a direct consequence of the action of free radicals.

E. Antioxidant Defenses

One definition of antioxidants (HALLIWELL and GUTTERIDGE 1989) is that they are "substances that, when present at low concentrations compared to those

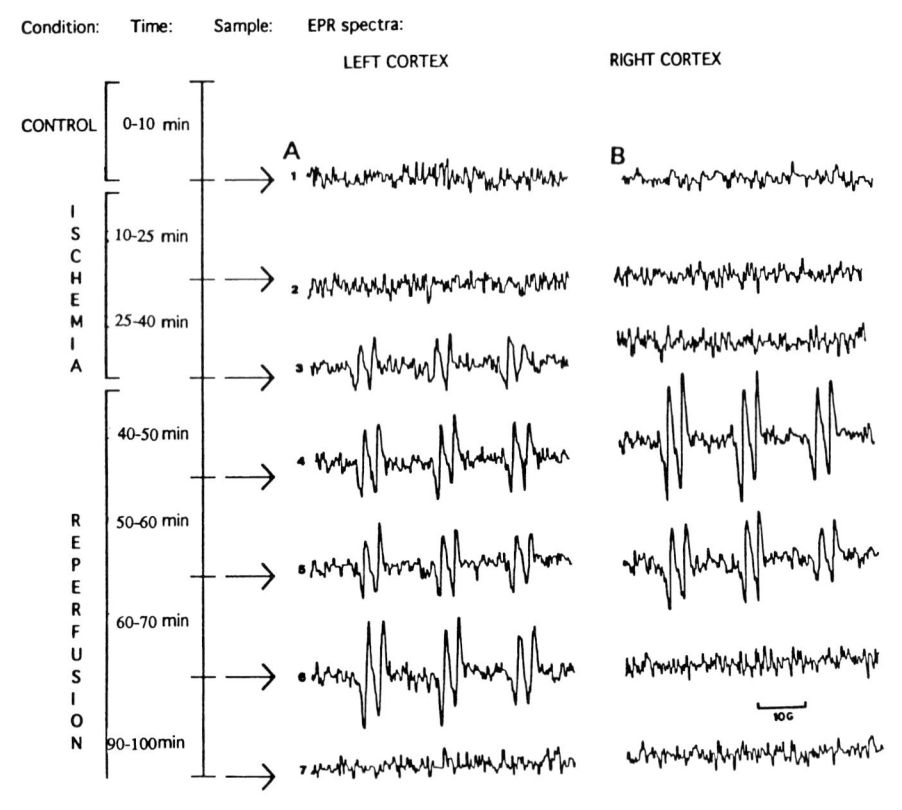

Fig. 4A,B. EPR spectra obtained from cerebral cortical superfusates of a rat at different time points of brain 4VO ischemia/reperfusion using POBN as a spin-trapping agent. Superfusate samples were obtained from the left (**A**) and right (**B**) cortical hemispheres respectively. The *left* panel indicates the experimental conditions and timings of the collection of superfusate samples. (Reproduced with permission from SEN and PHILLIS 1993)

of an oxidisable substrate, significantly delay or prevent oxidation of that substrate." A practical approach is to classify antioxidants as agents capable of; (1) scavenging ROS/RNS, (2) preventing the formation of free radicals, (3) upregulating endogenous defense systems, and (4) promoting repair of the damage (HALLIWELL 1997). A primary focus of this chapter will be on the scavenging of free radicals by a variety of naturally occurring or synthetic compounds.

Examples of the prevention of free radical formation include the use of xanthine oxidase inhibitors, cyclooxygenase/lipoxygenase inhibitors and of synthetic metal ion chelators (to prevent the reduction of H_2O_2 to ˙OH by the Fenton reaction). Upregulation of endogenous defense systems can be achieved by appropriate dietary adjustments or by enhancing the levels of endogenous antioxidant enzymes. For instance, controlled exposure to oxida-

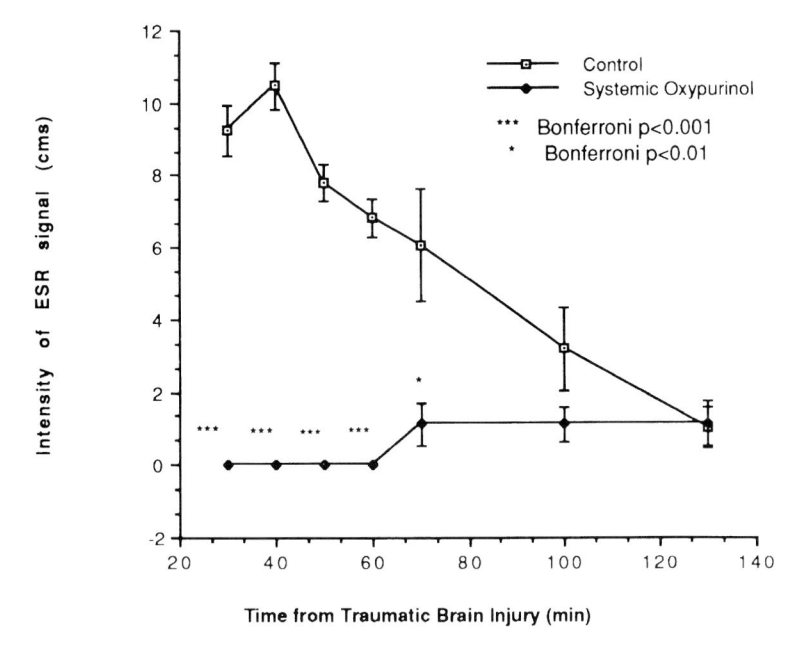

Fig. 5. Variation of EPR signal intensity with time in rats subjected to traumatic concussive brain injury (*TBI*). The EPR signal intensity was measured peak to peak and expressed in cm. The two groups of animals represented include POBN administered control TBI rats (*squares*) and rats injected with POBN plus oxypurinol (40mg/kg, i.p.) 10min prior to surgery. Oxypurinol attenuated free radical generation in TBI rats. The 30-min interval between brain injury and the onset of superfusate collection was required for placement of the cortical cup. (Reproduced with permission from SEN et al. 1993)

tive stresses can upregulate endogenous antioxidant systems. Examples of the latter have been demonstrated in the heart, where a brief exposure of the myocardium to oxygen radicals results in a preconditioning response whereby cardiac myocytes become resistant against the lethal injury of prolonged ischemia (ZHOU et al. 1996). The protection was attributed to increased Mn-SOD activity in the myocytes. Pre-exposure of rabbit hearts to oxygen radicals generated by xanthine and xanthine oxidase resulted in a significant reduction in infarct size in comparison with control ischemic hearts following a 30-min period of coronary artery ligation followed by 2.5h of reperfusion (TRITTO et al. 1995). The beneficial effect of oxygen radicals in this instance was attributed to the activation of protein kinase C.

Promoting repair of oxidative injury remains an elusive goal but may be achievable with the administration of appropriate neurotrophic and other growth factors (GORIO and DIGIULIO 1996). Treatment of neurons with nerve growth factor is able to confer resistance to further oxidative damage, apparently by induction of antioxidant enzymes (JACKSON et al. 1990).

F. Cerebroprotective Effects of Administered Antioxidants

In addition to the endogenous antioxidant systems, mentioned above, many other chemicals possess an ability to either prevent the formation of or to scavenge ROS, and a growing number of these have been used to reduce or prevent free radical-induced injuries in the central nervous system.

I. Free Radical Scavengers

1. Spin-Trapping Agents

Reports of significant protection against myocardial ischemia/reperfusion injury with the spin-trap agents α-phenyl-*tert*-butylnitrone (PBN) and DMPO (HEARSE and TOSAKI 1987; BOLLI et al. 1988) opened up potential new avenues for the treatment of cerebral ischemia/reperfusion injury. The rationale for these initial studies was that such agents would react with free radicals to form relatively stable adducts, thus interrupting the cascade of reactions that ultimately result in membrane lipid peroxidation and cellular injury. PBN was selected for use in these experiments because of its lipophilicity and ability to cross cell membranes and scavenge free radicals (LAI et al. 1986). A related spin-trapping agent, POBN, which is considerably less lipophilic than PBN, was also less potent in protecting hearts from adriamycin-induced, free radical-mediated, cardiotoxicity (MONTI et al. 1991).

In a preliminary account lacking technical details, FLOYD (1990) reported that PBN administration prevented lethality in gerbils following a 10-min global forebrain ischemia. Administration of PBN (32 mg/kg, i.p. twice daily) for 14 days rendered aged gerbils essentially resistant to 10 min of brain ischemia when lethality was assessed at day 7 (FLOYD and CARNEY 1996). The PBN-mediated protection was still evident when ischemia was introduced 3 days after the last PBN injection and some protective activity was noted even after a 5 day interval. Such long-lasting effects suggested that the change in susceptibility to ischemia may have been a consequence of the reduction in the levels of oxidized proteins that these authors observed in the brains of aged gerbils treated with PBN in comparison with those in control aged gerbils. Other investigators have, however, had difficulties in reproducing these effects of PBN on protein oxidation (CAO and CUTLER 1995a and b; DUBEY et al. 1995). Administration of PBN from age 20 weeks throughout the life of a short-lived strain of mouse increased its life span from 42 to 56 weeks (EDAMATSU et al. 1995). Equally impressive was the performance of PBN-treated old gerbils in a test of short term memory. Whereas old gerbils were twice as likely as young gerbils to make errors, they performed as well as the young animals following PBN administration (CARNEY et al. 1991).

Pretreatment of animals with PBN prior to the onset of ischemia resulted in a marked reduction of the previously described OH-POBN adduct signal in rat cerebral cortical superfusates (Fig. 6), together with the appearance of

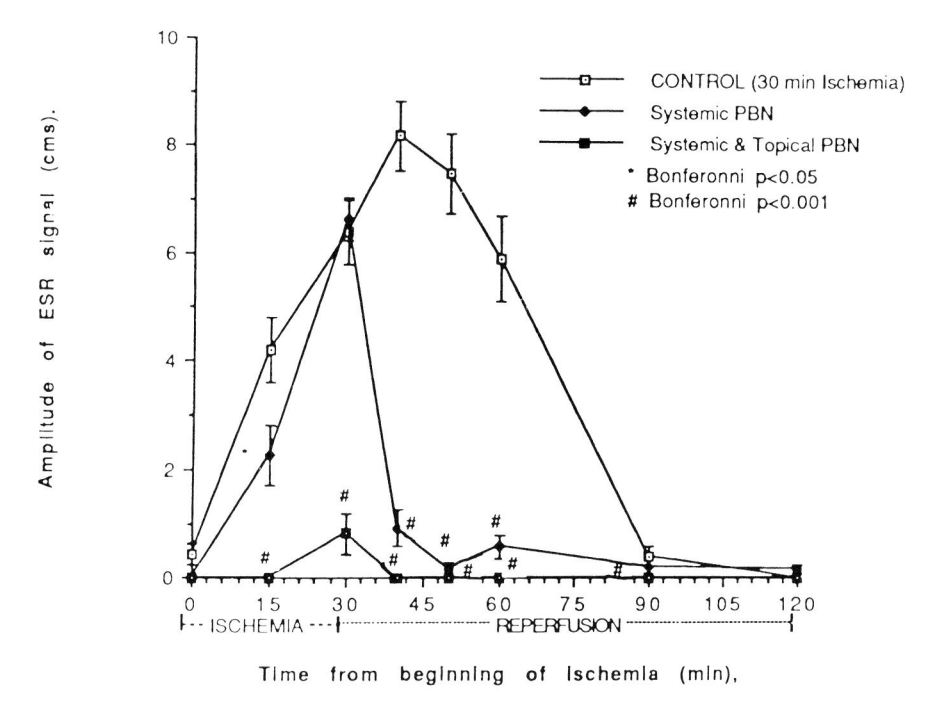

Free Radical Scavengers and Other Antioxidants 255

Fig. 6. Variation of EPR signal intensity with time in POBN administered rats subjected to 30 min 4VO cerebral ischemia followed by 90 min of reperfusion. The EPR signal was measured peak to peak and expressed in cm. Three groups of animals include (a) control ischemia; (b) treated prior to ischemia with PBN (100 mg/kg, i.p.) and (c) treated prior to ischemia with PBN (100 mg/kg, i.p.) and topically (100 mM in aCSF). Free radical formation was significantly attenuated by both pretreatments, and especially by the combination of systemic and topical PBN administration. (Reproduced with permission from SEN and PHILLIS 1993)

an ascorbyl radical adduct signal (Fig. 7) (SEN and PHILLIS 1993). The disappearance of the POBN-OH adduct signals from cortical superfusates in the presence of PBN may be a consequence of the different degrees of lipophilicity of the two spin trap agents referred to above. PBN is more lipophilic and has a different intracellular distribution, including access to the nucleus and mitochondria (CHENG et al. 1993; COVA et al. 1992). Thus, PBN, but not POBN, may be able to achieve adequate concentrations in mitochondrial and nuclear membranes to interrupt the free radical cascade at its site of initiation by forming a stable adduct with superoxide anion radicals, thus preventing the formation of H_2O_2. H_2O_2, which is able to diffuse freely across cell membranes, could otherwise access the extracellular space, where, after conversion to ·OH radicals, it could form the OH adduct of POBN recorded in cerebral cortical superfusates. The appearance of ascorbyl radical adducts in PBN containing superfusate may be a consequence of the scavenging by ascorbate of the active oxygen species that were initially trapped by PBN.

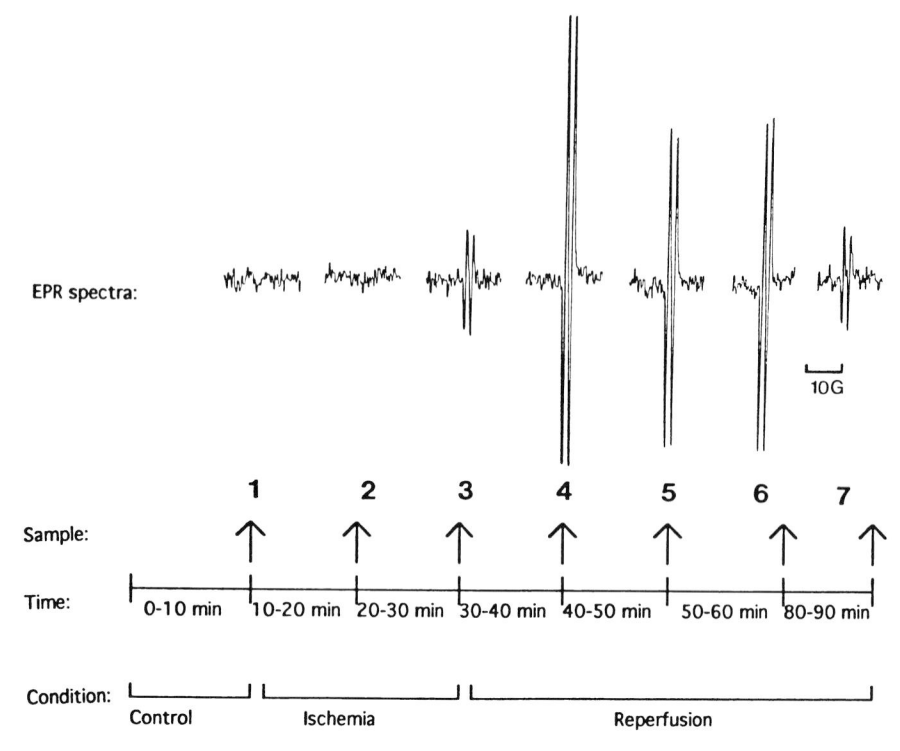

Fig. 7. EPR spectra characteristic of ascorbyl radical recorded from cortical superfusate samples taken at different time intervals from a PBN-treated rat subjected to 4VO cerebral ischemia/reperfusion. (Reproduced with permission from SEN and PHILLIS 1993)

EPR and spin-trapping were used to detect brain •OH generation in acute traumatic brain injury using POBN as the spin-trap. When the rats were pretreated with PBN the •OH signal was no longer detectable, again suggesting that PBN had trapped free radicals at their putative site of origin in nuclear and mitochondrial membranes (SEN et al. 1994). Direct detection by EPR spectroscopy of ascorbyl radicals in cerebral cortical microdialysates following trauma and in the absence of spin traps has been demonstrated by KIHARA et al. (1995). In this instance, when the rats were challenged with cold-induced brain injury, ascorbyl radical levels in the microdialysates were increased. Intravenous injections of superoxide dismutase or catalase, which did not alter the basal levels of ascorbyl radicals, resulted in a significant attenuation of the cold-induced increase in extracellular ascorbyl radicals.

Reference has already been made to the ability of excitotoxic amino acids to generate free radicals in the brain (LAFON-CAZAL et al. 1993). NOS inhibition with L-NAME (LANCELOT et al. 1998) or N^G-nitro-L-arginine (HAMMER et al. 1993) significantly reduced glutamate or NMDA-evoked free radical formation in the rat striatum, providing evidence that NOS is involved in the generation of OH• in excitotoxic conditions in vivo. NO can combine with $O_2^{•-}$ to produce ONOO⁻, which could then spontaneously decompose to yield •OH. Alternatively, it is possible that NOS may be able to directly generate $O_2^{•-}$ (Pou et al. 1992; Xia et al. 1996).

PBN (100mg/kg) administered either prior to, or 0.5h after a 5-min period of forebrain ischemia had cerebroprotective activity in gerbils (Clough-Helfman and Phillis 1991), reducing both the increase in locomotor activity and the extent of CA1 hippocampal pyramidal cell loss. These effects were no longer apparent when PBN was administered 2h postischemia. Yue et al. (1992) reported that PBN attenuated both forebrain edema and hippocampal CA1 neuronal loss in ischemia/reperfused gerbils and that it protected rat cerebellar neurons in primary culture from glutamate induced neurotoxicity. A PBN derivative *N-tert*-butyl-α(2-sulfophenyl)-nitrone (S-PBN), significantly attenuated in vivo striatal excitotoxicity produced by NMDA, kainic acid, and α-amino-3-hydroxy-5-methyl-isoxazole-4-propionic acid (Schulz et al. 1995).

The effects of PBN and the more water-soluble PBN analogue S-PBN were evaluated in a rat forebrain ischemia/reperfusion model (15min bilateral carotid artery occlusion + hypotension). PBN (100mg/kg) reduced neuronal necrosis in the neocortex when given 30min postischemia but not when administered either before or 6h after ischemia, and it failed to reduce damage in the CA1 sector of the hippocampus or caudatoputamen (Pahlmark and Siesjo 1996). The sulphonyl derivative of PBN failed to reduce damage in any area, presumably due to its failure to traverse the blood–brain barrier. The discrepancy between the results obtained in gerbil and rat forebrain models may be due to the different durations of the ischemias (5min in gerbil, 15min in rat).

PBN markedly attenuated both the degree of neurological disability and the magnitude of focal cerebral cortical infarcts following permanent middle cerebral artery occlusion in rats, even when administered 12h after the induction of ischemia (Fig. 8) (Cao and Phillis 1994). Furthermore, PBN dramatically reduced infarct size when given 3h after the onset of recirculation following a 2-h period of MCA occlusion in rats (Zhao et al. 1994). PBN improved the recovery of brain energy and lactate status in rats after a 2-h period of MCA occlusion, even when administered 1h after the onset of recirculation, preventing a secondary energy failure (Folbergrova et al. 1995). It is likely that this effect was a consequence of PBN's ability to ameliorate the secondary mitochondrial dysfunction evident in control animals after 4h of recirculation (Kuroda et al. 1996). MDL 101002, a cyclic variant of PBN, has also proven to be an effective neuroprotective agent in both permanent and transient models of rat focal cerebral ischemia (Johnson et al. 1998).

These dramatic cerebroprotective effects of PBN, which are consistent with its established ability to quench •OH radical adduct formation, are fully

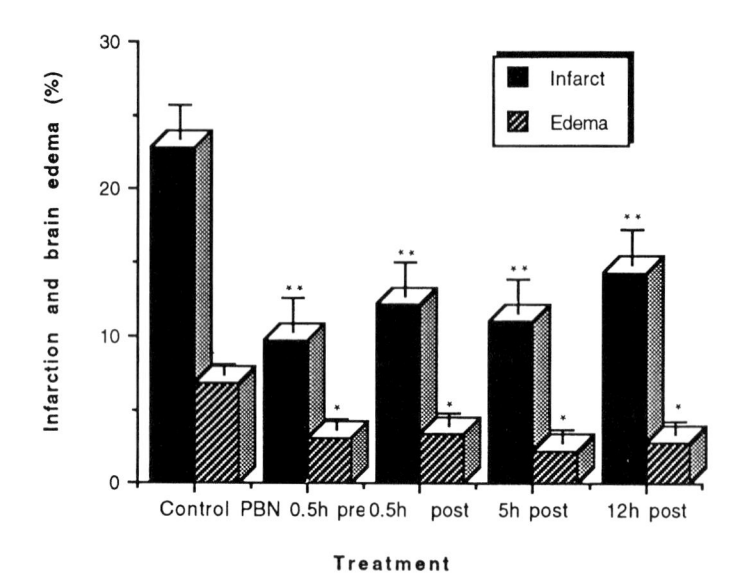

Fig. 8. The effects of PBN administered either prior to, or 0.5, 5, and 12 h postischemia in a rat permanent middle cerebral artery occlusion model. The data are expressed as % change in infarct volume and hemispheric edema from that in untreated controls. PBN administration at all time points significantly reduced both infarct volume and edema. Neurological behavior scores were also significantly reduced (not shown). (Data from CAO and PHILLIS 1994)

consistent with the concept that free radicals play an important role in the pathogenesis of cerebral ischemia/reperfusion lesions.

2. Mannitol, Dimethylthiourea and Dimethylsulfoxide

Mannitol, a hydrophilic free radical scavenger (MORI et al. 1996) may have cerebroprotective actions in the brain in addition to its routine use in traumatic brain injury as a hyperosmolar agent to produce an osmotic pressure gradient between the blood and brain, reducing brain volume and intracranial pressure, and thus increasing blood flow. Mannitol pretreatment prevented free radical formation, detected by chemiluminescence, during reoxygenation of homogenates from hypoxic rat brains (SUZUKI et al. 1985) and reduced ischemia-evoked aspartate and glutamate release from rat hippocampal slices (PELLIGRINI-GIAMPIETRO et al. 1990). It prevented iron-induced lipid peroxidation in spinal cord (ANDERSON and MEANS 1985).

Dimethylthiourea (DMTU) is a lipophilic free radical scavenger that readily crosses the blood–brain barrier to exert an antioxidative effect on central neurons. DMTU significantly reduced infarct volume and brain edema 24 h after middle cerebral artery occlusion in rats (MARTZ et al. 1989). DMTU was also effective in protecting vulnerable neurons in the hippocampus, cerebral cortex, and caudatoputamen against forebrain ischemia after 7 days of recovery (PAHLMARK et al. 1993).

Dimethylsulfoxide, a relatively effective hydroxyl radical scavenger (PANGANAMALA et al. 1976), is of interest because of its widespread use as a solubilizer of lipophilic agents undergoing testing for cerebroprotective activity. DMSO itself has significant cerebroprotective action, protecting gerbil CA1 pyramidal neurons from forebrain ischemic injury (SUGITA et al. 1993; PHILLIS et al. 1998).

3. α-Lipoic Acid

α-Lipoic acid (thioctic acid), a naturally occurring compound, is a cofactor for mitochondrial α-ketoacid dehydrogenases and is covalently bound as a lipoamide to some enzymes. Together with its reduced form, dihydrolipoic acid, it also participates in reactions that reduce the oxidized forms of α-tocopherol and glutathione (PRUIJN et al. 1991). α-Lipoic acid also possesses intrinsic free radical scavenging properties (SUZUKI et al. 1991; SCOTT et al. 1994).

Prolonged treatment of cultured chick telencephalic neurons with α-lipoic acid protected against hypoxic, glutamate, or iron induced injury (MULLER and KRIEGLSTEIN 1995). When administered acutely in vivo neither α-lipoic acid nor dihydrolipoic acid were able to reduce CA1 pyramidal neuron injury in rats subjected to global ischemia/reperfusion, although dihydrolipoic acid was moderately protective in mouse and rat middle cerebral artery models of focal ischemia (PREHN et al. 1992). After a 10-day period of administration, both α-lipoic and dihydrolipoic acids protected rat striatal neurons against excitotoxic death mediated by NMDA receptor antagonists or malonic acid (GREENAMYRE et al. 1994).

α-Lipoic acid, administered twice per diem for 7 days, protected CA1 hippocampal pyramidal cells in stroked gerbils (CAO and PHILLIS 1995). Ischemic injury levels were also assessed by monitoring the increases in locomotor activity following a 5-min bilateral occlusion of the carotid arteries. By both criteria, the treated animals were significantly less injured than vehicle-injected controls. Dietary supplementation with α-lipoic acid, which readily enters the brain (PACKER 1996), may be a useful prophylactic therapy for individuals at risk for cerebrovascular accidents.

α-Lipoic acid administration has also been shown to improve the performance of aged mice in an open-field memory test (STOLL et al. 1993). Treatment did not enhance performance in young mice, suggesting that, as was the case with PBN, α-lipoic acid may have been able to reduce the levels of oxidized protein in the brain. α-Lipoic acid administration may therefore be useful for reducing age-related memory losses and for the treatment of neurodegenerative diseases.

4. α-Tocopherol

Administration of α-tocopherol (vitamin E) prior to the onset of ischemia attenuates lipid peroxidation during reperfusion (YAMAMOTO et al. 1983). Conversely, rats raised on diets deficient in α-tocopherol have accentuated cerebral lipid peroxidation in response to ischemia/reperfusion in comparison with rats raised on a diet with excess α-tocopherol (YOSHIDA et al. 1984). Dogs given α-tocopherol in combination with mannitol and phenytoin prior to ischemia showed a significant recovery in brain electrical activity and decreased edema in comparison with dogs receiving phenytoin alone (SUZUKI et al. 1987). α-Tocopherol had a protective effect on hippocampal neuronal damage in stroked gerbils (HARA et al. 1990) and protected primary cultured rat cerebellar neurons exposed to glutamate by more than 50%. Synergistic effects were observed between α-tocopherol and ascorbic acid (CIANI et al. 1996).

EPC-K1, a phosphate diester of α-tocopherol and ascorbic acid, reduced functional deficits after global ischemia in rats, measured as spatial learning in a Morris water maze (BLOCK et al. 1995). CA1 hippocampal cell injury was nonsignificantly reduced in these animals.

5. Superoxide Dismutase and Catalase

Superoxide dismutase (SOD) catalyses the conversion of superoxide anion radical into H_2O_2, which is then removed either by glutathione peroxidase (which catalyses its conversion to H_2O while converting reduced glutathione (GSH) into oxidized glutathione (GSSG)), or by catalase, which decomposes H_2O_2 to oxygen and water. Catalase is largely confined to peroxisomes (COTGREAVE et al. 1988), which may explain why the brain, which lacks peroxisomes (BOVERIS and CHANCE 1973), contains low levels of catalase.

Initial studies on the effects of exogenously supplied SOD on ischemic or traumatic brain injury provided mixed results. Postcardiac arrest administration of SOD and deferoxamine, an iron chelator, ameliorated changes in cerebral blood flow and enhanced neuronal recovery (CERCHIARI et al. 1987). Administration of CuZn-SOD (SOD1) did not attenuate cellular edema induced by arachidonic acid in brain slices (CHAN and FISHMAN 1980). Free CuZn-SOD has a brief (6 min) half-life in plasma and is rapidly cleared from the body. The blood–brain barrier results in a virtually complete exclusion of SOD entry into the brain and even if it did penetrate the barrier, the enzyme would not be taken up by neurons and astrocytes and would, therefore, fail to reach the sites of $O_2^{\bullet}$ generation.

Conjugation of SOD with polyethylene glycol (PEG-SOD) has been used to extend the half-life of the enzyme to 38 h (BECKMAN et al. 1988). PEG-SOD and PEG-catalase, when administered intravenously prior to ischemia, reduced infarct volume in a rat model of focal ischemia by 24% (LIU et al. 1989). Administration of PEG-SOD (10000 U/kg) significantly improved the outcome of patients with severe head injury in a Phase II trial (MUIZELAAR et al. 1993). However, no significant differences in neurologic outcome or mortality were observed in a subsequent trial (YOUNG et al. 1996). When liposome-encapsulated SOD, which should enhance its entry into the brain, was administered prior to injury, it reduced the infarct volume in a rat model of focal cerebral ischemia (IMAIZUMI et al. 1990).

To circumvent the limitations imposed by exogenously-administered antioxidant enzymes, several laboratories have begun to study brain injury in transgenic animals with genetic modifications that result in elevated levels of these antioxidant enzymes. Transgenic mice heterozygous for the human SOD-1 transgene had significant reductions in infarct size, water content, and permeability to Evan's Blue 24h after the application of a cold probe in comparison with control mice (CHAN et al. 1991). Similar results were observed following cold-injury in transgenic mice expressing the human form of extracellular SOD (OURY et al. 1993). Acute and chronic damage following traumatic concussive injury was significantly attenuated in transgenic mice expressing CuZn-SOD (MIKAWA et al. 1996). Mice heterozygous for the SOD-1 transgene had significantly smaller cortical infarcts following a transient focal cerebral ischemia (YANG et al. 1994), but were not protected when the mice were subjected to a permanent focal ischemia (CHAN et al. 1993). Overexpression of SOD-1 in transgenic rats protected vulnerable neurons in several brain areas from global ischemia/reperfusion injury (CHAN et al. 1998).

Studies in EC-SOD transgenic mice suggest that increased levels of SOD may protect tissues by decreasing the formation of $ONOO^-$. The enzyme may be colocalized with NOS and play a beneficial role by protecting NO from conversion into the toxic peroxynitrite, a highly reactive oxidizing agent (OKABE et al. 1998). This conclusion is supported by observations that L-NAME administration reduced the level of cerebral edema in nontransgenic, but not in transgenic mice overexpressing SOD (OURY et al. 1993).

II. Inhibition of ROS/RNS Formation

1. Xanthine Oxidase Inhibitors

Xanthine oxidase in the brain is primarily localized in the cerebral capillary endothelial cells (BETZ 1985; TERADA et al. 1991), which are able to generate •OH radicals during anoxia/reoxygenation (GRAMMAS et al. 1993; BEETSCH et al. 1998). Moreover, the enzyme is able to leak from ischemically-injured cells in the liver, and perhaps small intestine, into the plasma where it is completely converted into the XO form (BATTELLI et al. 1992) and could injure the vascular endothelium at a distance from the original injury (YOKOYAMA et al. 1990; TAN et al. 1993).

Cerebroprotective actions of the xanthine oxidase inhibitor, allopurinol, in stroked rats were first reported more than a decade ago (ITOH et al. 1986; TAYLOR et al. 1984) and this has been confirmed in subsequent studies (MARTZ et al. 1989; PALMER et al. 1990). Allopurinol blocks the metabolism of hypoxanthine and xanthine by XO by virtue of its being a competing substrate for the enzyme, whereas its metabolite, oxypurinol, is a true dead-end inhibitor of the enzyme (SPECTOR 1988). Moreover, as the conversion of allopurinol to oxypurinol is accompanied by superoxide radical formation, it is clearly evident that oxypurinol should be superior to allopurinol for protecting tissue against ischemia/reperfusion injury. EPR studies demonstrated that oxypurinol greatly attenuates the formation of hydroxyl radicals during ischemia/reperfusion and concussive trauma (Fig. 5) of the rat cerebral cortex confirming a role for XO in ROS formation (PHILLIS and SEN 1993; SEN et al. 1993). Oxypurinol significantly reduced the extent of hippocampal CA1 pyramidal cell loss in ischemia/reperfused gerbils, also, decreasing the extent of the ischemia/reperfusion evoked increase in locomotor activity (PHILLIS 1989; PHILLIS and CLOUGH-HELFMAN 1990) and decreased both the degree of neurological deficit and infarct volume in middle cerebral artery lesioned rats (LIN and PHILLIS 1992). In a microwave fixation study, oxypurinol was observed to result in a significant increase in the cortical energy charge and ATP levels during ischemia and in the initial stages of reperfusion (PHILLIS et al. 1995). These beneficial effects were likely a consequence of its purine sparing (salvage) effects and of its ability to protect mitochondria by inhibiting free radical formation.

It has been suggested that the protective effects of allopurinol and oxypurinol are due to their ability to scavenge free radicals rather than inhibition of xanthine oxidase (MOORHOUSE et al. 1987; DAS et al. 1987). However, allopurinol did not enhance the antioxidant properties of extracellular fluid, indicating that scavenging of free radicals is not its primary mode of action (ZIMMERMAN et al. 1988). Confirmation of the role of XO in ischemia/reperfusion-evoked ROS formation was obtained when it was observed that amflutizole, an unrelated XO inhibitor, caused a marked attenuation of POBN-OH adduct levels in rat cerebral cortical superfusates as measured by EPR spectroscopy (O'REGAN et al. 1994).

2. Inhibition of Phospholipases and Arachidonic Acid Metabolism

The breakdown of phospholipids in the outer cell membrane by Ca^{2+}-activated phospholipases A_2 and C commences within a minute of the onset of ischemia, as demonstrated by the rapid appearance of arachidonic and stearic acids (ABE et al. 1987; KATSURA et al. 1993). Similar events occur following brain and spinal cord trauma (IKEDA and LONG 1990; DEMEDIUK et al. 1987). The stage is then set for the release of ROS during the metabolism of arachidonic acid by lipoxygenases and cyclooxygenases. Phospholipase A_2 appears to play a predominant role in the arachidonic acid formation because it cleaves a fatty acyl chain from the β-position of phospholipids and arachidonic acid in mammalian cells is esterified almost exclusively at the β-position. Knockout mice lacking cytoplasmic PLA_2 had smaller infarcts, less edema, and fewer neurological deficits following transient middle cerebral artery occlusion (BONVENTRE et al. 1997).

Inhibitors of cyclooxygenase (including indomethacin and ibuprofen) have been demonstrated to have beneficial actions in cerebral ischemia resulting from stroke or cardiac arrest (DEMPSEY et al. 1985; SASAKI et al. 1988; KUHN et al. 1986) and on neurological recovery of head-injured mice (HALL 1985). Recent evidence suggests that cyclooxygenase-2 (COX-2), rather than COX-1, activity is of particular significance in promoting neuronal death following cerebral ischemia (NOGAWA et al. 1997; NAKAYAMA et al. 1998). Lipoxygenase inhibitors, including nordihydroguaiaretic acid, substantially reduced neuronal cell death produced by the exposure of cultured rat hippocampal neurons to N-methyl-D-aspartic acid (ROTHMAN et al. 1993), but were not effective in reducing hippocampal injury in stroked gerbils (NAKAGOMI et al. 1989). The phospholipase A_2 inhibitor mepacrine had a cerebroprotective action in global (PHILLIS 1996) and focal (ESTEVEZ and PHILLIS 1997) ischemia/reperfusion experiments on gerbils and rats respectively. Mepacrine, several other phospholipase A_2 and C inhibitors, and indomethacin significantly reduced the ischemia-evoked release of glutamate and aspartate into cerebral cortical superfusates (O'REGAN et al. 1995; PHILLIS and O'REGAN 1996; PHILLIS et al. 1994). Reductions in the release of these potentially injurious amino acids may be a contributing factor in the cerebroprotective actions of mepacrine and indomethacin.

3. Inhibitors of Nitric Oxide Synthases

Nitric oxide, the recently discovered endothelial relaxing factor and putative neurotransmitter, is a free radical and appears to have both neuroprotective and neurotoxic actions. Its neuroprotective actions may result from a down-regulation of NMDA receptor activity by a reaction with thiol groups of the receptor's redox modulating site (LIPTON et al. 1993). Nitric oxide is also a potent vasodilator with a demonstrable role in the regulation of the cerebral circulation, where it couples cerebral metabolic activity to cerebral blood flow (FARACI and BRIAN 1994). Anoxic injury of rat brain endothelial cells in vitro increases the production of both nitric oxide and •OH radicals (KUMAR et al. 1996). PBN blocked •OH formation but not that of nitric oxide. The cerebral microcirculation may contribute to ROS/RNS production during ischemia/reperfusion or trauma. An ability to increase cerebral blood flow during ischemia/reperfusion may also be a significant factor in its neuroprotective actions. $NO_2^{•}$-mediated neurotoxicity is engendered, at least in part, by its reaction with superoxide anion, leading to the formation of peroxynitrite (LIPTON et al. 1993).

The enzyme responsible for nitric oxide formation, NO synthase, is present in both constitutive [endothelial (e) and neuronal (n)] and inducible (i) forms. NO synthase catalyzes NO formation from L-arginine by a Ca^{2+}/calmodulin-dependent mechanism. Thus, the increases in intracellular Ca^{2+} during neuronal injury can activate the enzyme. In the absence of adequate levels of L-arginine (a condition which may occur during ischemic and other forms of injury) brain NOS can generate superoxide (Pou et al. 1992). As glutamate neurotoxicity is primarily mediated by NMDA receptor activation with Ca^{2+} entry into cells, a possible participation of NO in excitotoxic injury has been hypothesized. NO synthase inhibitors prevent neurotoxicity elicited by NMDA and related amino acids in rat primary cortical cultures (Dawson et al. 1991). Protection against cerebral ischemic injury in vivo following the administration of NO synthase inhibitors has been reported by several groups of investigators (Buisson et al. 1993; Nagafugi et al. 1992; Caldwell et al. 1994; Yoshida et al. 1994) although this has not been a universal finding (Buchan et al. 1994; Sancessario et al. 1994; Weissman et al. 1992). Cerebroprotective effects of NO precursors such as L-arginine and nitroprusside have also been observed (Morikawa et al. 1994; Zhang and Iadecola 1993). It appears likely that the conflicting results obtained by different investigators were a consequence of the specific balance between the cerebroprotective and cerebrotoxic effects of NO formation (described in this paragraph and in Sect. B.II.) in each experimental situation.

III. Chelation of Metal Ions

The free radical hypothesis suggests that some groups of neurons may be selectively vulnerable to injury as a consequence of a high endogenous content of iron and/or low glutathione peroxidase activity (Zaleska and Floyd 1985; Ushijima et al. 1986). Studies on the histochemical localization of lipid peroxidation products following ischemia/reperfusion have demonstrated that the reaction product is localized to neurons in such selectively vulnerable zones, and that the lipid peroxidation products appear to be concentrated in the Golgi apparatus, an important site for membrane recycling (White et al. 1993). A key mechanism of tissue injury following ischemia or trauma may involve the release of iron from intracellular stores, and in some instances extravasation of blood with hemolysis and the deposition of iron extracellularly in the tissues, with an acceleration of •OH formation via the Fenton reaction.

Deferoxamine, a chelator of ferric iron, has been used to inhibit iron-catalyzed tissue damage. Deferoxamine, in combination with superoxide dismutase, facilitated the recovery of somatosensory evoked potentials and normalized cerebral blood flow profiles in dogs subjected to cardiac arrest followed by resuscitation (Cerchiari et al. 1987). There was a decreased formation of lipid peroxides and decreased brain edema during reperfusion of dogs pretreated with deferoxamine prior to ischemia induced by cardiac arrest (Komara et al. 1986). In some studies deferoxamine pretreatment improved overall survival rate of animals exposed to cardiac arrest with resuscitation (Kompala et al. 1986), but failed to improve the neurological status of survivors (Fleischer et al. 1987). Conjugation of deferoxamine with hydroxyethyl starch to increase plasma half-life resulted in a significant improvement in mortality rates and neurological outcome in rats subjected to cardiac arrest and resuscitation (ROSENTHAL et al. 1992). Deferoxamine conjugated to dextran improved neurologic outcome following concussive head injury in mice (PANTER et al. 1992). Modified forms of deferoxamine may therefore yet prove to be of value for the treatment of traumatic and ischemic/reperfusion injuries.

G. Oxidative Injury and Age-Related Neurodegeneration

The free radical theory of aging was initially proposed by HARMAN (1956). More recent evidence that oxidative stress can affect aging comes from the report of ORR and SOHAL (1994), who observed a significant extension of life-span in the fly, *Drosophila melanogaster*, following overexpression of the enzymes superoxide dismutase and catalase. Overexpression of SOD by itself had no effect on life span, demonstrating a requirement for catalase to detoxify the H_2O_2 produced by superoxide dismutase. Life-span of frogs was increased following induction of SOD, glutathione reductase, glutathione, and ascorbate (LOPES-TORRES et al. 1993). An inherited enhancement of hydroxyl radical generation and lipid peroxidation contributed to premature aging in a particular S strain of rats (SALGANIK et al. 1994). Dietary or caloric restriction has been implicated as a technique for extending life-span. In rats caloric restriction increases life-span in comparison with that of litter mates given unrestricted access to food (MASORO et al. 1982). DNA damage due to oxidative stress was significantly reduced in rats fed on a calorie-restricted diet (DJURIC et al. 1992).

Decreased levels of GSH and mitochondrial cytochromes, as well as mitochondrial dysfunction, have been observed in the brains of aging rodents (BOSSI et al. 1993; BENZI et al. 1992). The activity of antioxidant enzymes is reduced and the levels of oxidized proteins may be increased with decreased protein turnover in aged brains (BARJA DE QUIROGA et al. 1990; CAND and VERDETTI 1989; CARNEY et al. 1991; LEBEL and BONDY 1992; MATSUO et al. 1992; ZHANG et al. 1993). FUNAHASHI et al. (1994), utilizing magnetic resonance to directly monitor pH and high-energy phosphates in ischemia/reperfused gerbil brains, observed that recovery of pH during reperfusion was markedly delayed in older gerbils. Reoxygenation caused a rapid restoration of high energy phosphate levels in young, but not older gerbils, in which normal levels were not restored even after 1h. These results implicate deficiencies in key mitochondrial functions and suggest that PBN-induced protection against ischemia-induced lethality in aging gerbils (FLOYD and CARNEY 1996) involves an improvement in mitochondrial function which may be related to free radical scavenging and reductions in protein oxidation. Increased brain levels of iron, which may enhance ROS production, are associated with aging in rats (ZALESKA and FLOYD 1985). The characteristic accumulation of lipofuscin in the aging brain is an indication of increased lipid peroxidation (PORTA 1991). α-Tocopherol deficiency in both experimental animals and man leads to an excessive accumulation of lipofuscin (SOKOL 1989) and levels are reduced when animals are treated with α-tocopherol supplements (MONJI et al. 1994).

In sum, these observations suggest that oxidative stress can contribute to senescence in the aging individual or animal. The gradual accumulation of degraded oxidized proteins, representing oxidative events integrated over time, may be especially significant for long-lived, nondividing cells such as neurons.

H. Oxidative Stress and Neurogenerative Diseases
I. Amyotrophic Lateral Sclerosis

ALS is a motor disease involving a progressive degeneration of upper motor neurons in the cerebral cortex and lower motor neurons of the brain stem and spinal cord (INCE et al. 1998). The discovery that some familial forms of ALS are caused by point mutations in the gene encoding for CuZn superoxide dismutase (ROSEN et al. 1993) generated enormous interest in the role of oxidative stress in the pathogenesis of this disease. The mutations are associated with reduced SOD activity (BOWLING et al. 1993). The mechanism by which the mutation leads to neuronal death could be a result of either a loss of ability to catalyze the conversion of $O_2^{\bullet}$ to H_2O_2 or to the "gain of a novel function." The potential effect of a loss of function is uncertain as SOD knockout mice fail to develop as an animal model of the disease (GURNEY 1997) although chronic inhibition of the enzyme can cause neurodegeneration (ROTHSTEIN et al. 1994). Consistent with a "gain of function" is the result of studies on mice overexpressing the mutant SOD gene causing human familial ALS, which develop a progressive degeneration of motor neurons resembling ALS (GURNEY et al. 1994). Although the SOD mutation is not found in sporadic ALS, there is evidence that markers of protein and lipid peroxidation are elevated in these ALS patients (BOWLING et al. 1993). Antioxidant trials in ALS patients with N-acetylcysteine (to enhance glutathione levels) or the MAO-B inhibitor L-deprenyl (selegiline) have all been negative (LOUVEL et al. 1997).

II. Alzheimer's Disease

AD is a neurodegenerative disease characterized by gradual loss of memory, reasoning, orientation, and judgement (KATZMAN 1986; BEHL 1999). The formation of amyloid deposits, a characteristic feature of the disease, is associated with aberrant processing and folding of a protein called amyloid β-precursor protein to form amyloid plaques (SELKOE 1994). Although a causative role for oxidative damage in the pathogenesis of AD is still being debated (MARKESBERRY 1997), there is evidence that the in vitro formation of synthetic amyloid plaques from the precursor protein is accelerated by the presence of oxygen and that the amyloid protein, which is cytotoxic to neurons in vitro, appears to stimulate the generation of oxidants (SMITH et al. 1995; HENSLEY et al. 1996). Increased lipid peroxidation, protein and DNA oxidation, and formation of advanced glycosylation end products have been reported in patients with AD (SMITH et al. 1995).

Administration of α-tocopherol (2000 IU/diem) for 2 years reduced the incidence of death, severe dementia, loss of daily activities, and institutionalization in patients with moderate AD (SANO et al. 1997). In a controlled trial, Ginkgo biloba extracts, which protect neurons in vitro from H_2O_2 induced oxidative stress (OYAMA et al. 1996), were observed to be of benefit to AD patients (LEBARS et al. 1997).

III. Parkinson's Disease

The possibility that oxidative damage is involved in the pathogenesis of PD has attracted considerable attention. This condition is characterized by the loss of dopaminergic neurons in the substantia nigra pars compacta and a concomitant loss of dopaminergic input to the striatum. The loss of this circuitry results in tremor, bradykinesia, and rigidity and can also be associated with dementia (DUVOISIN 1992). Evidence for an involvement of oxidative stress in PD comes from the finding of increases in malondialdehyde and lipid peroxides in the PD substantia nigra (DEXTER et al. 1994). There is a selective increase in iron levels in the pars compacta (GRIFFITHS and CROSSMAN 1993) where it can bind to melanin in the dopaminergic neurons. Hydrogen peroxide is a product of dopamine metabolism by the catabolic enzyme monoamine oxidase B; in the presence of iron it can readily be converted to toxic $^\bullet$OH radicals, which could then initiate a destructive lipid peroxidation cascade, peroxidize proteins and DNA, fatally injuring the dopaminergic neurons (ADAMS and ODUNZE 1991; OLANOW 1993; SHOULSON 1992). Reduced glutathione, which is necessary for the detoxification of H_2O_2 by glutathione peroxidase, is significantly *lowered* in the substantia nigra of postmortem PD brains (SIAN et al. 1994).

PD-like symptoms are induced in humans and animals by the neurotoxin, N-methyl-4-phenyl-1,2,3,6-tetrahydropyridine and its monoamine oxidase-B metabolite, N-methyl-4-phenylpyridinium. The MPTP-induced form of PD is also associated with increased oxidative stress (WEI et al. 1996; OBATA and CHIUEH 1992). MPTP neurotoxicity is reduced in transgenic mice overexpressing SOD-1 activity (PRZEDBORSKI et al. 1992). α-Tocopherol (2000 IU/diem) for 14 months failed to benefit PD patients in a controlled trial (PARKINSON'S DISEASE STUDY GROUP 1993). It remains possible, however, that more potent antioxidants will provide effective therapy for this disease. L-Deprenyl, an MAO-B inhibitor, has proven to be effective in delaying the development of the disease (LEWITT 1994), but whether this was a consequence of its ability to preserve brain dopamine or to block of the formation of H_2O_2, the toxic product of dopamine metabolism (or both), is unresolved.

IV. Huntington's Disease

HD is an autosomal dominant disorder characterized by emotional disturbances, choreic movements, seizures, and dementia. Oxidative stress has been considered as a potential contributor to the striatal pathology of this disease (BROWNE et al. 1997). Increased levels of 8-hydroxydeoxyguanosine, indicative of oxidation, were found in caudate nuclear DNA in postmortem HD brains in comparison with normal controls, although no differences in the levels of SOD activity were evident. Administration of α-tocopherol (3000 IU/diem) for 1 year failed to ameliorate neurological or neuropsychiatric symptoms in the active treatment group in comparison with the control group (PEYSER et al. 1995). However, both groups were also given additional antioxidant treatment with ascorbic acid and vitamin A, and this may have compromised the likelihood of observing benefits from the α-tocopherol.

I. Conclusions

The data presented in this review of neural injury resulting from ischemia, trauma, epilepsy, aging, or neurodegenerative diseases provide substantial evidence for a role of oxidative stress and free radicals in the pathogenesis of these conditions. During the past decade, a number of potent inhibitors of free radical production have been identified for use in experimentally induced CNS trauma or ischemia. These include enzymes such as superoxide dismutase and catalase, and a series of nonenzymic inhibitors of free radical production, including agents which inhibit the enzymes generating the radicals and others that neutralize or inactivate free radicals. The xanthine oxidase inhibitors, allopurinol and oxypurinol, reduce cerebral injury in both global and focal models of ischemia and facilitate the recovery of energy-rich phosphates during reperfusion. Oyxpurinol also inhibits free radical formation in traumatic brain injured rats, suggesting that it will be useful in the treatment of such injuries. Inhibitors of nitric oxide synthase have been cerebroprotective in some, but not all, experimental studies. Modified forms of the iron-chelating agent, deferoxamine, have been beneficial in the treatment of cerebral ischemia and traumatic injuries to experimental animals.

Cerebroprotective free radical scavenging compounds discussed in this chapter include α-tocopherol, ascorbic acid, α-lipoic acid, dimethylthiourea, mannitol, and spin-trapping agents, including α-phenyl-*tert*-butylnitrone (PBN). The latter compound, which inactivates toxic free radicals by forming stable nitroxide adducts, has been dramatically effective in the treatment of experimental focal and global ischemia's and amelioration of aging-induced neurodegeneration. Spin-trapping agents, in particular, may provide a series of promising new candidates for the prevention and treatment of oxidative stress-induced injuries to the brain and spinal cord.

Acknowledgements. Thanks are due to Drs. A.Y. Estevez, L.L. Guyot and M.H. O'Regan for critiquing the manuscript and to Ms. L. McCraw for her expert assistance in the preparation of the typescript.

Abbreviations

AD	Alzheimer's disease
ALS	amyotrophic lateral sclerosis
AMPA	α-amino-3-hydroxy-5-methyl-isoxazole-4-propionic acid
ATP	adenosine triphosphate
CNS	central nervous system
DMPO	5,5-dimethyl-1-pyrroline N-oxide
DMSO	dimethylsulfoxide
DNA	deoxyribonucleic acid
EPR	electron paramagnetic resonance
GSH	reduced glutathione
GSSG	oxidized glutathione
HD	Huntington's disease
MPTP	N-methyl-4-phenyl-1,2,3,6-tetrahydropyridine
NAD	nicotinamide adenine dinucleotide
NADP	nicotinamide adenine dinucleotide phosphate
L-NAME	N^G-nitro-L-arginine methyl ester
NO	nitric oxide
NOS	nitric oxide synthase
PBN	α-phenyl-*tert*-butylnitrone
PD	Parkinson's disease
POBN	pyridyl-N-oxide-*tert*-butylnitrone
RNS	reactive nitrogen species
ROS	reactive oxygen species
SOD	superoxide dismutase
SPBN	N-*tert*-butyl-α(2-sulfophenyl) nitrone
XO	xanthine oxidase

References

Abe K, Kogure K, Yamamoto H, Imazawa M, Miyamoto K (1987) Mechanism of arachidonic acid liberation during ischemia in gerbil cortex. J Neurochem 48: 503–509

Adams JD, Odunze IN (1991) Oxygen free radicals and Parkinson's disease. Free Rad Biol Med 10:161–169

Anderson DK, Means ED (1985) Iron-induced lipid peroxidation in spinal cord: protection with mannitol and methylprednisolone. J Free Rad Biol Med 1:59–64

Barja de Quiroga G, Perez Campo R, Lopez Torres M (1990) Anti-oxidant defences and peroxidation in liver and brain of aged rats. Biochem J 272:247–250

Battelli MG, Abbondanza A, Stirpe F (1992) Effects of hypoxia and ethanol on the xanthine oxidase of isolated hepatocytes: conversion of D to O form and leakage from cells. Chem-Biol Interactions 83:73–84

Beal MF (1995) Aging, energy, and oxidative stress in neurodegenerative diseases. Ann Neurol 38:357–366

Beckman JS, Beckman TW, Chen J, Marshall PA, Freeman BA (1990) Apparent hydroxyl radical production by peroxynitrite: implications for endothelial injury from nitric oxide and superoxide. Proc Natl Acad Sci USA 87:1620–1624

Beckman JS, Minor RL, White CW, Repine JE, Rosen GM, Freeman BA (1998) Superoxide dismutase and catalase conjugated to polyethylene glycol increases endothelial enzyme activity and oxidant resistance. J Biol Chem 263:6884–6892

Beckman KB, Ames BN (1998) The free radical therapy of aging matures. Physiol Rev 78:547–581

Beetsch JW, Park TS, Dugan LL, Shah AR, Gidday JM (1998) Xanthine oxidase-derived superoxide causes reoxygenation injury of ischemic cerebral endothelial cells. Brain Res 786:89–95

Behl C (1999) Alzheimer's disease and oxidative stress: implications for novel therapeutic approaches. Progr Neurobiol 57:301–323

Benzi G, Pastoris O, Marzatico F, Villa RF, Dagani F, Curti D (1992) The mitochondrial electron transfer alteration as a factor involved in the brain aging. Neurobiol Aging 13:361–368

Benveniste H, Drejer J, Schousboe A, Diemer NH (1984) Elevation of the extracellular concentration of glutamate and aspartate in rat hippocampus during transient cerebral ischemia monitored by intracerebral microdialysis. J Neurochem 43:1369–1374

Betz AL (1985) Identification of hypoxanthine transport and xanthine oxidase activity in brain capillaries. J Neurochem 44:574–579

Block F, Kunkel M, Sontag K-H (1995) Posttreatment with EPC-K1, an inhibitor of lipid peroxidation and of phospholipase A_2 activity, reduces functional deficits after global ischemia in rats. Brain Res Bull 36:257–260

Bolli R, Jeroudi MO, Patel BS, Aruoma OI, Halliwell B, Lai EK, McCay PB (1989) Marked reduction of free radical generation and contractile dysfunction by antioxidant therapy begun at the time of reperfusion: Evidence that myocardial "stunning" is a manifestation of reperfusion injury. Circ Res 65:607–622

Bonventre JV, Huang Z, Taheri MR, O'Leary E, Li E, Moskowitz MA, Sapirstein A (1997) Reduced fertility and postischaemic brain injury in mice deficient in cytosolic phospholipase A_2. Nature 390:622–625

Bossi SR, Simpson JR, Isacson O (1993) Age dependence of striatal neuronal death caused by mitochondrial dysfunction. Neuroreport 4:73–76

Boveris A, Chance B (1973) The mitochondrial generation of hydrogen peroxide. Biochem J 134:707–716

Bowling AC, Schulz JB, Brown RH, Beal MF (1993) Superoxide dismutase activity, oxidative damage, and mitochondrial energy metabolism in familial and sporadic amyotrophic lateral sclerosis. J Neurochem 61:2322–2325

Browne SE, Bowling AC, MacGarvey U, Baik MJ, Berger SC, Muquit MM, Bird ED, Beal MF (1997) Oxidative damage and metabolic dysfunction in Huntington's disease: selective vulnerability of the basal ganglia. Ann Neurol 41:646–653

Buchan AM, Gertler SZ, Huang Z-G, Li H, Chaundy KE, Xue D (1994) Failure to prevent selective CA1 neuronal death and reduce cortical infarction following cerebral ischemia with inhibition of nitric oxide synthase. Neuroscience 61:1–11

Buisson A, Margaill I, Callebert J, Plotkine M, Boulu RG (1993) Mechanisms involved in the neuroprotective activity of a nitric oxide synthase inhibitor during focal cerebral ischemia. J Neurochem 61:690–695

Caldwell M, O'Neill M, Earley B, Leonard B (1994) N^G-Nitro-L-arginine protects against ischemia-induced increases in nitric oxide and hippocampal neurodegeneration in the gerbil. Eur J Pharm 260:191–200

Cand F, Verdetti J (1989) Superoxide dismutase, glutathione peroxidase, catalase, and lipid peroxidation in the major organs of the aging rats. Free Rad Biol Med 7:59–63

Cao G, Cutler RG (1995a) Protein oxidation and aging, I, Difficulties in measuring reactive protein carbonyls in tissues using 2,4-dinitrophenylhydrazine. Arch Biochem Biophys 320:106–114

Cao G, Cutler RG (1995b) Protein oxidation and aging, II, Difficulties in measuring alkaline protease activity in tissues using the fluorescamine procedure. Arch Biochem Biophys 320:195–201

Cao W, Carney J-M, Duchon A, Floyd RA, Chevion M (1988) Oxygen free radical involvement in ischemia and reperfusion injury to brain. Neurosci Lett 88:233–238

Cao X, Phillis JW (1994) α-Phenyl-*tert*-butyl-nitrone reduces cortical infarct and edema in rats subjected to focal ischemia. Brain Res 644:267–272

Cao X, Phillis JW (1995) The free radical scavenger, α-lipoic acid, protects against cerebral ischemia-reperfusion injury in gerbils. Free Radical Res 23:365–370

Carney JM, Starke-Reed PE, Oliver CN, Landum RW, Cheng MS, Wu JF, Floyd RA (1991) Reversal of age-related increases in brain protein oxidation, decrease in enzyme activity, and loss in temporal and spatial memory by chronic administration of the spin-trapping compound N-tert-butyl-α-phenylnitrone. Proc Natl Acad Sci USA 88:3633–3636

Cerchiari EL, Hoel TM, Safar P, Sclabassi RJ (1987) Protective effects of combined superoxide dismutase and desferrioxamine on recovery of cerebral blood flow and function after cardiac arrest in dogs. Stroke 18:869–878

Chan PH, Fishman RA (1980) Transient formation of superoxide radicals in polyunsaturated fatty acid-induced brain swelling. J Neurochem 35:1004–1007

Chan PH, Chu L, Chen SF, Carlson E, Epstein CJ (1991) Post-traumatic brain injury and edema are reduced in transgenic mice overexpressing CuZn-superoxide dismutase. Ann Neurol 21:482–486

Chan PH, Kamii H, Yang G, Gafni J, Epstein CJ, Carlson E, Reola L (1993) Brain infarction is not reduced in SOD-1 transgenic mice after a permanent focal cerebral ischemia. Neuroreport 5:293–296

Chan PK, Kawase M, Murakami K, Chen SF, Li Y, Calagui B, Reola L, Carlson E, Epstein CJ (1998) Overexpression of SOD-1 in transgenic rats protects vulnerable neurons against ischemic damage after global cerebral ischemia and reperfusion. J Neurosci 18:8292–8299

Cheng HY, Liu T, Feuerstein G, Barone FC (1993) Distribution of spin trapping compounds in rat blood and brain – in vivo microdialysis determination. Free Rad Biol Med 14:243–250

Ciani E, Grøneng L, Voltattorni M, Rolseth V, Contestabile A, Paulsen RE (1996) Inhibition of free radical production or free radical scavenging protects from the excitotoxic cell death mediated by glutamate in cultures of cerebellar granule neurons. Brain Res 728:1–6

Clough-Helfman C, Phillis JW (1991) The free radical trapping agent N-*tert*-butyl-α-phenylnitrone (PBN) attenuates cerebral ischemic injury in gerbils. Free Rad Res Comm 15:177–186

Cohen G (1988) Oxygen radicals and Parkinson's disease. In: Halliwell B (ed) Oxygen radicals and tissue injury. FASEB, Bethesda MD, pp 130–135

Cotgreave IA, Moldevs P, Orrenius S (1988) Host biochemical defense mechanisms against prooxidants. Ann Rev Pharmacol Toxicol 28:189–212

Cova D, Deangelis L, Monti E, Piccinini F (1992) Subcellular distribution of two spin-trapping agents in rat heart – possible explanation for their different protective effects against doxorubicin-induced cardiotoxicity. Free Rad Res Commun 15:353–360

Cross CE (1981) Moderator: Oxygen radicals and human disease. Ann Intern Med 107:526–545

Das DK, Engelman RM, Clement R, Otani H, Prasad MR, Rao PS (1987) Role of xanthine oxidase inhibitor as free radical scavenger: a novel mechanism of action of allopurinol and oxypurinol in myocardial salvage. Biochem Biophys Res Commun 148:314–319

Dawson VL, Dawson TM, London ED, Bredt DS, Snyder SH (1991) Nitric oxide mediates glutamate neurotoxicity in primary cortical cultures. PNAS 88:6368–6371

Delanty N, Dichter MA (1998) Oxidative injury in the nervous system. Acta Neurol Scand 98:145–153

Demediuk P, Saunders RD, Anderson DK, Means ED, Horrocks LA (1987) Early membrane lipid changes in laminectomized and traumatized cat spinal cord. Neurochem Pathol 7:79–89

Dempsey RJ, Roy MW, Meyer KL, Donaldson DL (1985) Indomethacin-mediated improvement following middle cerebral artery occlusion in cats: effects of anesthesia. J Neurosurg 62:874–881

Dexter DT, Holley AE, Flitter WD, Slater TF, Wells FR, Daniel SE, Lees AJ, Jenner P, Marsden CD (1994) Increased levels of lipid hydroperoxides in the Parkinsonian substantia nigra: an HPLC and ESR study. Mov Disord 9:92–97

Djuric Z, Lu MH, Lewis SM, Luongo DA, Shen XW, Heilbrun LK, Reading BA, Duffy PH, Hart RW (1992) Oxidative DNA damage levels in rats fed low-fat, high-fat or calorie-restricted diets. Toxicol Appl Pharmacol 115:156–160

Dubey A, Forster MJ, Sohal RS (1995) Effect of the spin-trapping compound N-*tert*-butyl-alpha-phenylnitrone on protein oxidation and life span. Arch Biochem Biophys 324:249–254

Dugan L, Sensi S, Canzoniero L, Handran S, Rothman S, Lin T-S, Goldberg M, Choi D (1995) Mitochondrial production of reactive oxygen species in cortical neurons following exposure to N-methyl-D-aspartate. J Neurosci 15:6377–6388

During MJ, Spencer DD (1992) Adenosine: a potential mediator of seizure arrest and postictal refractoriness. Ann Neurol 32:618–624

Duvoisin RC (1992) Overview of Parkinson's disease, Ann NY Acad Sci 648:187–193

Dykens JA, Stern A, Trenker E (1987) Mechanism of kainate toxicity to cerebellar neurons in vitro is analogous to reperfusion tissue injury. J Neurochem 49:1222–1228

Edamatsu R, Mori A, Packer L (1995) The spin-trap N-*tert*-alpha-phenyl-butylnitrone prolongs the life span of the senescence accelerated mouse. Biochem Biophys Res Commun 211:847–849

Estevez AY, Phillis JW (1997) The phospholipase inhibitor, quinacrine, reduces infarct size in rats after transient middle cerebral artery occlusion. Brain Res 752:203–208

Faraci FM, O'Brian JE (1994) Nitric oxide and the cerebral circulation. Stroke 25:692–703

Fleischer JE, Lanier WL, Milde JH, Michenfelder JD (1987) Failure of deferoxamine, an iron chelator, to improve neurologic outcome following complete cerebral ischemia in dogs. Stroke 18:124–127

Floyd RA (1990) Role of oxygen free radicals in carcinogenesis and brain ischemia. FASEB J 4:2587–2597

Floyd RA, Carney JM (1996) Nitrone radical traps protect in experimental neurodegenerative diseases. In: Olanow CW, Jenner P, Youdim M (eds) Neurodegeneration and neuroprotection in Parkinson's Disease. Academic Press, San Diego, pp 70–90

Folbergrova J, Zhao Q, Katsura K, Siesjo BK (1995) N-tert-butyl-α-phenylnitrone improves recovery of brain energy state in rats following transient focal ischemia. Proc Natl Acad Sci USA 92:5057–5061

Funahashi T, Floyd RA, Carney JM (1994) Age effect on brain pH during ischemia/reperfusion and pH influence on peroxidation. Neurobiol Aging 15:161–167

Gorio A, Di Giulio AM (1996) Neuroregeneration as a repair mechanism in brain injuries: Pharmacological role of growth promoting factors. In: Peterson PL, Phillis JW (eds) Novel therapies for CNS injuries: Rationales and results. CRC Press, Boca Raton, pp 273–290

Grammas P, Liu G-J, Wood K, Floyd RA (1993) Anoxia/reoxygenation induces hydroxyl free radical formation in brain microvessels. Free Rad Biol Med 14:553–557

Greenamyre JT, Garcia-Osuna M, Greene JG (1994) The endogenous cofactors, thioctic acid and dihydrolipoic acid, are neuroprotective against NMDA and malonic acid lesions of striatum. Neurosci Lett 171:17–20

Greenlund LJS, Deckwerth TL, Johnson EM (1995) Superoxide dismutase delays neuronal apoptosis: a role of reactive oxygen species in programmed neuronal death. Neuron 14:303–315

Griffiths PD, Crossman AR (1993) Distribution of iron in the basal ganglia and neocortex in postmortem tissue in Parkinson's disease and Alzheimer's disease. Dementia 4:61–65

Gurney M (1997) The use of transgenic mouse models of amyotrophic lateral sclerosis in preclinical drug studies. J Neurol Sci 152 [Suppl 1]:S67–S73

Gurney ME, Pu H, Chiu AY, Dal Canto MC, Polchow CY, Alexander DD, Caliendo J, Hentati A, Kwon YW, Deng H-X, Chen W, Zhai P, Sufit RL, Siddique T (1994) Motor neuron degeneration in mice that express a human Cu,Zn superoxide dismutase mutation. Science 264:1772–1775

Hagberg H, Lehmann A, Sandberg M, Nystrom B, Jacobson I, Hamberger A (1985) Ischemia-induced shift of inhibitory and excitatory amino acids from intra- to extracellular compartments. J Cereb Blood Flow Metab 5:413–419

Hall E (1985) Beneficial effects of acute intravenous ibuprofen on neurologic recovery of head-injured mice: comparison of cyclooxygenase inhibition with inhibition of thromboxane A_2 synthetase or 5-lipoxygenase. J Neurotrauma 2:75–83

Halliwell B (1997) Introduction: Free radicals and human disease – trick or treat? In: Thomas CE, Kalyanaraman B (eds) Oxygen radicals and the disease process. Harwood Academic Publishers, Amsterdam, pp 1–14

Halliwell B, Gutteridge JMC (1989) Free radicals in biology and medicine, 2nd edn. Clarendon Press, Oxford

Hammer B, Parker WD, Bennett JP (1993) NMDA receptors increase OH radicals in vivo by using nitric oxide synthase and protein kinase C. Neuroreport 5:72–74

Harman D (1956) Aging: a theory based on free radical and radiation chemistry. J Gerontol 11:289–300

Hara H, Kato H, Kogure K (1990) Protective effect of alpha-tocopherol on ischemic neuronal damage in the gerbil hippocampus. Brain Res 510:335–338

Hearse DJ, Tosaki A (1987) Free radicals and reperfusion-induced arrythmias: protection by spin trap agent PBN in the rat heart. Circ Res 60:375–383

Hensley K, Butterfield DA, Hall N, Cole P, Subramanian R, Mark P, Mattson MP, Markesbery WR, Harris ME, Aksenov M, Aksenova M, Wu JF, Carney JM (1996) Reactive oxygen species as causal agents in the neurotoxicity of Alzheimer's disease-associated amyloid beta peptide. Ann NY Acad Sci 786:120–134

Hiramatsu M, Nakajima M, Komatsu M, Oikawa K, Ueda Y, Nakai O, Willmore LJ (1996) Electron spin resonance – computed tomography: Brain imaging and epilepsy. In: Packer L, Hiramatsu M, Yoshikawa T (eds) Free radicals in brain physiology and disorders. Academic Press, San Diego, pp 185–195

Ikeda Y, Long DM (1990) The molecular basis of brain injury and brain edema: the role of oxygen free radicals. Neurosurgery 27:1–11

Imaizumi S, Woolworth V, Kinouchi H, Chen SF, Fishman RA, Chan PH (1990) Liposome-entrapped superoxide dismutase ameliorates infarct volume in focal cerebral ischemia. Acta Neurochirurgica 51 [Suppl]:236–238

Ince PG, Lowe J, Shaw PJ (1998) Amyotrophic lateral sclerosis: current issues in classification, pathogenesis and molecular pathology. Neuropathol Appl Neurobiol 24:104–117

Itoh T, Kawakami M, Yamauchi Y, Shimizu S, Nakamura M (1986) Effect of allopurinol on ischemia and reperfusion-induced cerebral injury in spontaneously hypertensive rats. Stoke 17:1284–1287

Jackson GR, Apffel L, Werbach-Perez K, Perez-Polo JR (1990) Role of nerve growth factor in oxidant-antioxidant balance and neuronal injury. I. Stimulation of hydrogen peroxide resistance. J Neurosci Res 25:360–368

Jenner P (1994) Oxidative damage in neurodegenerative disease. Lancet 344:796–798

Johnson MP, McCarty DR, Velayo NL, Markgraf CG, Chmielewski PA, Ficorilli JV, Cheng HC, Thomas CE (1998) MDL 101002, a free radical spin trap, is efficacious in permanent and transient focal ischemia models. Life Sci 63:241–253

Katayama Y, Becker DP, Tamura T, Hovda DA (1990) Massive increases in extracellular potassium and indiscriminate release of glutamate following concussive brain injury. J Neurosurg 73:889–900

Katsura K-I, Rodriguez De Turco EB, Folbergrova J, Bazan NG, Siesjo BK (1993) Coupling among energy failure, loss of ion homeostasis, and phospholipase A_2 and C activation during ischemia. J Neurochem 61:1677–1684

Katzman R (1986) Alzheimer's disease. N Engl J Med 314:964–973

Kihara T, Sakata S, Ikeda M (1995) Direct detection of ascorbyl radical in experimental brain injury: microdialysis and an electron spin resonance spectroscopic study. J Neurochem 65:282–286

Komara JS, Nayini NR, Bialick HA, Indrieri RJ, Evans AT, Garritano AM, Hoehner TJ, Jacobs WA, Huang RR, Krause GS, White BC, Aust SD (1986) Brain iron delocalization and lipid peroxidation following cardiac arrest. Ann Emerg Med 15:384–389

Kompala S-D, Babbs CF, Blako KE (1986) Effect of deferoxamine on late deaths following CPR in rats. Ann Emerg Med 15:405–407

Kuhn JE, Steimle CN, Zelenock GB, D'Alecy LG (1986) Ibuprofen improves survival and neurologic outcome after resuscitation from cardiac arrest. Resuscitation 14:199–212

Kumar M, Liu G-J, Floyd RA, Grammas P (1996) Anoxic injury of endothelial cells increases production of nitric oxide and hydroxyl radicals. Biochem Biophys Res Comm 219:497–501

Kuroda S, Katsura K-I, Tsuchidate R, Siesjo BK (1996) Secondary bioenergetic failure after transient focal ischaemia is due to mitochondrial injury. Acta Physiol Scand 156:149–150

Lafon-Cazal M, Pietri S, Culcasi M, Bockaert J (1993) NMDA-dependent superoxide production and neurotoxicity. Nature 364:535–537

Lai EK, Crossley C, Sridhar R, Misra HP, Janzen EG, McCay PB (1986) In vivo spin-trapping of free radicals generated in brain, spleen and liver during gamma radiation of mice. Arch Biochim Biophys 244:156–160

Lancelot E, Revaud ML, Boulu RG, Plotkine M, Callebert J (1998) A microdialysis study investigating the mechanisms of hydroxyl radical formation in rat striatum exposed to glutamate. Brain Res 809:294–296

Le Bars PL, Katz MM, Berman N, Itil TM, Freedman AM, Schatzberg AF (1997) A placebo-controlled, double-blind, randomized trial of an extract of Ginkgo biloba for dementia. JAMA 278:1327–1332

Lebel CP, Bondy SC (1992) Oxidative damage and cerebral aging. Progr Neurobiol 38:601–609

Le Witt PA (1994) Clinical trials of neuroprotection in Parkinson's disease: long term selegiline and alpha-tocopherol treatment. J Neural Transm Suppl 43:171–181

Lin Y, Phillis JW (1992) Deoxycoformycin and oxypurinol: protection against focal ischemic injury in the rat. Brain Res 571:272–280

Lipton SA, Choi Y-B, Pan Z-H, Lei SZ, Chen, Z-HV, Sucher NJ, Loscalzo J, Singel DJ, Stamler JS (1993) A redox-based mechanism for the neuroprotective and neurodestructive effects of nitric oxide and related nitroso-compounds. Nature 364:626–632

Liu TH, Beckman JS, Freeman BA, Hogan EL, Hsu CY (1989) Polyethylene glycol-conjugated superoxide dismutase and catalase reduce ischemic brain damage. Am J Physiol 256:H589–H593

Lopez-Torres M, Perez-Campo R, Rojas C, Cadenas S, Barja G (1993) Simultaneous induction of SOD, glutathione reductase, GSH, and ascorbate in liver and kidney correlates with survival during aging. Free Rad Biol Med 15:133–142

Louvel E, Hugon J, Doble A (1997) Therapeutic advances in amyotrophic lateral sclerosis. TIPS 18:196–203

Malinski T, Bailey F, Zhang ZG, Chopp M (1993) Nitric oxide measured by a porphyrinic microsensor in rat brain after transient middle cerebral artery occlusion. J Cereb Blood Flow Metab 13:355–358

Markesbery WR (1997) Oxidative stress hypothesis in Alzheimer's disease. Free Rad Biol Med 23:134–147

Martz D, Rayos G, Schielke GP, Betz AL (1989) Allopurinol and dimethylthiourea reduce brain infarction following middle cerebral artery occlusion in rats. Stroke 20:488–494

Masoro EJ, Yu BP, Bertrand HA (1982) Action of food restriction in delaying the aging process. Proc Natl Acad Sci USA 79:4239–4241

Matsuo M, Gomi F, Dooley MM (1992) Age-related alterations in antioxidant capacity and lipid peroxidation in brain, liver and lung homogenates of normal and vitamin-E deficient rats. Mech Aging Dev 64:273–292

McCord JM (1993) Human disease, free radicals, and the oxidant/antioxidant balance. Clin Biochem 26:351–357

Mikawa S, Kinouchi H, Kamii H, Gobbel GT, Chen SF, Carlson E, Epstein CJ, Chan PH (1996) Attenuation of acute and chronic damage following traumatic brain injury in copper, zinc-superoxide dismutase transgenic mice. J Neurosurg 85:885–891

Monji A, Morimoto N, Okuyama I, Yamashita N, Tashiro N (1994) Effect of dietary vitamin E on lipofuscin accumulation with age in the rat brain. Brain Res 634:62–68

Monti E, Paracchini L, Perletti G, Piccinini F (1991) Protective effects of spin-trapping agents on adriamycin-induced cardiotoxicity in isolated rat atria. Free Rad Res Comms 14:41–45

Moorhouse PC, Grootveld M, Halliwell B, Quinlan JG, Gutteridge JMC (1987) Allopurinol and oxypurinol are hydroxyl radical scavengers. FEBS Lett 213:23–28

Mori Y, Takashima H, Seo H, Satoh K, Ohkawa M, Tanabe M (1996) N,N'-Propylenedinicotinamide: scavengers of free radicals and inhibitor of free radical promoted damage to benzoate, deoxyribose, and amino acids. In: Packer L, Hiramatsu M, Yoshikawa T (eds) Free radicals in brain physiology and disorders. Academic Press, San Diego, pp 233–241

Morikawa E, Moskowitz MA, Huang Z, Yoshida T, Irikura K, Dalkara T (1994) L-Arginine infusion promotes nitric oxide-dependent vasodilation, increases in regional blood flow, and reduces infarction volume in the rat. Stroke 25:429–435

Muizelaar JP, Marmarou A, Young HF, Choi SC, Wolf A, Schneider RL, Kontos HA (1993) Improving the outcome of severe head injury with the oxygen radical scavenger polyethylene glycol-conjugated superoxide dismutase: a phase II trial. J Neurosurg 78:375–382

Muller U, Krieglstein J (1995) Prolonged treatment with α-lipoic acid protects cultured neurons against hypoxic, glutamate-, or iron-induced injury. J Cereb Blood Flow Metab 15:624–630

Nagafuji T, Matsui T, Koide T, Asano T (1992) Blockade of nitric oxide formation by N^ω-nitro-L-arginine mitigates ischemic brain edema and subsequent cerebral infarction in rats. Neurosci Lett 147:159–162

Nakagomi T, Sasaki T, Kirino T, Tamura A, Noguchi M, Saito I, Takakura K (1989) Effect of cyclooxygenase and lipoxygenase inhibitors on delayed neuronal death in the gerbil hippocampus. Stroke 20:925–929

Nakayama M, Uchimura K, Zhu RL, Nagayama T, Rose ME, Stetler RA, Isakson PC, Chen J, Graham SH (1998) Cyclooxygenase-2 inhibition prevents death of CA1 hippocampal neurons following global ischemia. Proc Natl Acad Sci USA 95:10954–10959

Nelson CW, Wei EP, Povlishock JT, Kontos HA, Moskowitz MA (1992) Oxygen radicals in cerebral ischemia. Am J Physiol 263:H1356–H1362

Nilsson P, Hillered L, Ponten U, Ungerstedt U (1990) Changes in cortical extracellular levels of energy-related metabolites and amino acids following concussive brain injury in rats. J Cereb Blood Flow Metab 10:631–637

Nishio S, Yunoki M, Noguchi Y, Kawauchi M, Asari S, Ohmoto T (1997) Detection of lipid peroxidation and hydroxyl radicals in brain contusion in rats. Acta Neurochir 70 [Suppl]:84–86

Nogawa S, Zhang F, Ross ME, Iadecola C (1997) Cyclo-oxygenase-2 gene expression in neurons contributes to ischemic brain injury. J Neurosci 17:2746–2755

Obata T, Chieuh CC (1992) In vivo trapping of hydroxyl free radicals in the striatum utilizing intracranial microdialysis perfusion of salicylate: Effects of MPTP, MPDP[+] and MPP[+]. J Neural Trans 89:139–145

Okabe M, Saito S, Saito T, Ito K, Kimura S, Niioka T, Kurasaki M (1998) Histochemical localization of superoxide dismutase activity in rat brain. Free Rad Biol Med 24:1470–1476

Olanow CW (1993) A radical hypothesis for neurodegeneration. TINS 16:439–444

O'Regan MH, Smith-Barbour M, Perkins LM, Cao X, Phillis JW (1994) The effect of amflutizole, a xanthine oxidase inhibitor, on ischemia-evoked purine release and free radical formation in the rat cerebral cortex. Neuropharmacol 33:1197–1201

O'Regan MH, Smith-Barbour M, Perkins LM, Phillis JW (1995) A possible role for phospholipases in the release of neurotransmitter amino acids from ischemic rat cerebral cortex. Neurosci Lett 185:191–194

O'Regan MH, Song D, Vanderheide SJ, Phillis JW (1997) Free radicals and the ischemia-evoked extracellular accumulation of amino acids in rat cerebral cortex. Neurochem Res 22:273–280

Orr WC, Sohal RS (1994) Extension of life span by overexpression of superoxide dismutase and catalase in Drosophila melanogaster. Science 263:1128–1130

Oury TD, Piantadosi CA, Crapo JD (1993) Cold-induced brain edema in mice: involvement of extracellular superoxide dismutase and nitric oxide. J Biol Chem 268: 15394–15398

Oyama Y, Chikahisa L, Ueha T, Kanemaru K, Noda K (1996) Ginkgo biloba extract protects brain neurons against oxidative stress induced by hydrogen peroxide. Brain Res 712:349–352

Packer L (1996) Prevention of free radical damage in the brain: Protection by α-lipoic acid. In: Packer L, Hiramatsu M, Yoshikawa T (eds) Free radicals in brain physiology and disorders. Academic Press, San Diego, pp 19–34

Pahlmark K, Folbergrova J, Smith M-L, Siesjo BK (1993) Effects of dimethylthiourea on selective neuronal vulnerability in forebrain ischemia in rats. Stroke 24:731–737

Pahlmark K, Siesjo BK (1996) Effects of the spin trap α-phenyl-N-tert-butyl nitrone (PBN) in transient forebrain ischaemia in the rat. Acta Physiol Scand 157:41–51

Palmer C, Vannucci RC, Towfighi J (1990) Reduction of perinatal hypoxic-ischemic brain damage with allopurinol. Pediatr Res 27:332–336

Palmer HJ, Paulson KE (1997) Reactive oxygen species and antioxidants in signal transmission and gene expression. Nutr Rev 55:353–361

Panganamala RV, Sharma HM, Heikkila RE, Green JC, Cornwell DG (1976) Role of hydroxyl radical scavengers, dimethyl sulfoxide, alcohols, and methional in inhibition of prostaglandin synthesis. Prostaglandins 11:599–607

Panter SS, Braughler JM, Hall ED (1992) Dextran-coupled deferoxamine improves outcome in a murine model of head injury. J Neurotrauma 9:47–53

Pellegrini-Giampietro DE, Cherici G, Alesiani M, Carla V, Moroni F (1990) Excitatory amino acid release and free radical formation may cooperate in the genesis of ischemia-induced neuronal damage. J Neurosci 10:1035–1041

Peters O, Back T, Lindauer U, Busch C, Megow D, Dreier J, Dirnagl U (1998) Increased formation of reactive oxygen species after permanent and reversible middle cerebral artery occlusion in the rat. J Cereb Blood Flow Metab 18:196–205

Peyser CE, Folstein M, Chase GA, Starkstein S, Brandt J, Cockrell JR, Bylsma F, Coyle JT, McHugh PR, Folstein SE (1995) Trial of D-α-tocopherol in Huntington's disease. Am J Psychiat 152:1771–1775

Phillis JW (1989) Oxypurinol attenuates ischemia-induced hippocampal damage in the gerbil. Brain Res Bull 23:467–470

Phillis JW, Clough-Helfman C (1990) Oxypurinol, but not deoxycoformycin, administered post-ischemia, protects against CA1 hippocampal damage in the gerbil. Int J Purine Pyrimid Res 1:31–35

Phillis JW, Sen S (1993) Oxypurinol attenuates hydroxyl radical production during ischemia/reperfusion injury of the rat cerebral cortex: an ESR study. Brain Res 628:309–312

Phillis JW, Smith-Barbour M, Perkins LM, O'Regan MH (1994) Indomethacin modulates ischemia-evoked release of glutamate and adenosine from the rat cerebral cortex. Brain Res 652:353–356

Phillis JW, Sen S, Cao X (1994) Amflutizole, a xanthine oxidase inhibitor, inhibits free radical generation in the ischemic/reperfused rat cerebral cortex. Neurosci Lett 169:188–190

Phillis JW, Perkins LM, Smith-Barbour M, O'Regan MH (1995) Oxypurinol-enhanced post-ischemic recovery of the rat brain involves preservation of adenine nucleotides. J Neurochem 64:2177–2184

Phillis JW (1996) Cerebroprotective action of the phospholipase inhibitor quinacrine in the ischemia/reperfused gerbil hippocampus. Life Sci 58:PL97–PL101

Phillis JW, O'Regan MH (1996) Mechanisms of glutamate and aspartate release in the ischemic rat cerebral cortex. Brain Res 730:150–164

Phillis JW, Estevez AY, O'Regan MH (1998) Protective effect of the free radical scavengers, dimethyl sulfoxide and ethanol, in cerebral ischemia in gerbils. Neurosci Lett 244:109–111

Porta EA (1991) Advances in age pigment research. Arch Gerontol Geriatr 12:303–320

Pou S, Pou WS, Bredt DS, Snyder SH, Rosen GM (1992) Generation of superoxide by purified brain nitric oxide synthase. J Biol Chem 267:24173–24176

Prehn J-HM, Karkoutly C, Nuglisch J, Peruche B, Krieglstein J (1992) Dihydrolipoate reduces neuronal injury after cerebral ischemia. J Cereb Blood Flow Metab 12:78–87

Pruijn FB, Haenen GRMM, Bast A (1991) Interplay between vitamin E, glutathione and dihydrolipoic acid in protection against lipid peroxidation. Fat Sci Technol 93:216–221

Przedborski S, Kostic V, Jackson-Lewis V, Naini AB, Simonetti S, Fahn S, Carlson E, Epstein CT, Cadet JL (1992) Transgenic mice with increased Cu/Zn-superoxide dismutase activity are resistant to N-methyl-4-phenyl-1,2,3,6-tetrahydropyridine-induced neurotoxicity. J Neurosci 12:1658–1667

Reddy TS, Bazan NG (1987) Arachidonic acid, stearic acid and diacylglycerol accumulation correlates with the loss of phosphatidylinositol 4,5-bisphosphate in cerebrum 2 seconds after electroconvulsive shock. Complete reversion of changes 5 minutes after stimulation. J Neurosci Res 18:449–455

Rosen DR, Siddique T, Patterson D, Figlewicz DA, Sapp P, Hentati A, Donaldson D, Goto J, O'Regan JP et al. (1993) Mutations in Cu/Zn superoxide dismutase gene are associated with familial amyotrophic lateral sclerosis. Nature 362:59–62

Rosenthal RE, Chanderbhan R, Marshall G, Fiskum G (1992) Prevention of post-ischemic brain lipid conjugated diene production and neurological injury by hydroxyethyl starch-conjugated deferoxamine. Free Rad Biol Med 12:29–33

Rothman SM, Yamada KA, Lancaster N (1993) Nordihydroguaiaretic acid attenuates NMDA neurotoxicity-action beyond the receptor. Neuropharmacol 32:1279–1288

Rothstein JD, Bristol LA, Hosler B, Brown RH, Kuncl RW (1994) Chronic inhibition of superoxide dismutase produces apoptotic death of spinal neurons. Proc Natl Acad Sci USA 91:4155–4159

Sakamoto A, Ohnishi ST, Ohnishi T, Ogawa R (1991) Protective effect of a new antioxidant on the rat brain exposed to ischemia-reperfusion injury: inhibition of free radical formation and lipid peroxidation. Free Rad Biol Med 11:385–391

Salganik RI, Solovyova NA, Dikalov SI, Grishaeva ON, Semenova LA, Popovsky AV (1994) Inherited enhancement of hydroxyl radical generation and lipid peroxidation in the S strain rats results in DNA rearrangements, degenerative diseases, and premature aging. Biochem Biophys Res Commun 199:726–733

Saluja I, O'Regan MH, Song D, Phillis JW (1999)Activation of cPLA$_2$, PKC and ERKs in the rat cerebral cortex during ischemia/reperfusion. Neurochem Res 24:669–677

Sancesario G, Iannone M, Morello M, Nistico G, Bernardi G (1994) Nitric oxide inhibition aggravates ischemic damage of hippocampal but not of NADPH neurons in gerbils. Stroke 25:436–444

Sano M, Ernesto C, Thomas RG, Klauber MR, Schafer K, Grundman M, Woodbury P, Growdon J, Cotman C, Pfeiffer W, Schneider LS, Thal LJ (1997) A controlled trial of selegiline, alpha-tocopherol, or both as treatment for Alzheimer's disease. N Engl J Med 336:1216–1222

Sasaki T, Nakagomi T, Kirino T, Tamura A, Noguchi M, Saito I, Takakura K (1988) Indomethacin ameliorates ischemic neuronal damage in the gerbil hippocampal CA1 sector. Stroke 19:1399–1403

Sato S, Tominaga T, Ohnishi T, Ohnishi ST (1994) Electron paramagnetic resonance study on nitric oxide production during brain focal ischemia and reperfusion in the rat. Brain Res 647:91–96

Savolainen H (1978) Superoxide dismutase and glutathione peroxidase activities in rat brain. Res Comm Chem Pathol Pharmacol 21:173–175

Schulz JB, Henshaw DR, Siwek D, Jenkins BG, Ferrante RJ, Cipolloni PB, Kowall NW, Rosen BR, Beal MF (1995) Involvement of free radicals in excitotoxicity in vivo. J Neurochem 64:2239–2247

Scott BC, Aruoma OI, Evans PJ, O'Neill C, Vadervliet A, Cross CE, Tritschler H, Halliwell B (1994) Lipoic and dihydrolipoic acids as antioxidants. A critical evaluation. Free Rad Res 20:119–133

Selkoe DJ (1994) Cell biology of the amyloid beta-protein precursor and the mechanism of Alzheimer's disease. Ann Rev Cell Biol 10:373–403

Sen S, Goldman H, Morehead M, Murphy S, Phillis JW (1993) Oxypurinol inhibits free radical release from the cerebral cortex of closed head injured rats. Neurosci Lett 162:117–120

Sen S, Phillis JW (1993) Alpha-phenyl-*tert*-butyl-nitrone (PBN) attenuates hydroxyl radical production during ischemia/reperfusion injury of rat brain: an EPR study. Free Rad Res Comm 19:255–265

Sen S, Goldman H, Morehead M, Murphy S, Phillis JW (1994) α-Phenyl-tert-butyl nitrone inhibits free radical release in brain concussion. Free Rad Biol Med 16:685–691

Shohami E, Shapira Y, Yadid G, Reisfeld N, Yedgar S (1989) Brain phospholipase A$_2$ is activated after experimental closed head injury in the rat. J Neurochem 53:1541–1546

Shoulson I (1992) Antioxidant therapeutic strategies for Parkinson's disease. Ann NY Acad Sci 648:37–41

Sian J, Dexter DT, Lees AJ, Daniel S, Agid Y, Javoy-Agid F, Jenner P, Marsden CD (1994) Alterations in glutathione levels in Parkinson's disease and other neurodegenerative disorders affecting basal ganglia. Ann Neurol 36:348–355

Simonian NA, Coyle JT (1996) Oxidative stress in neurodegenerative diseases. Ann Rev Pharmacol Toxicol 36:83–106

Simpson RE, O'Regan MH, Perkins LM, Phillis JW (1992) Excitatory transmitter amino acid release from the ischemic rat cerebral cortex: effects of adenosine receptor agonists and antagonists. J Neurochem 58:1683–1690

Smith MA, Sayre LM, Monnier VM, Perry G (1995) Radical AGEing in Alzheimer's disease. TINS 18:172–176

Sokol RJ (1989) Vitamin E and neurologic function in man. Free Rad Biol Med 6:189–207

Spector T (1988) Oxypurinol as an inhibitor of xanthine oxidase-catalyzed production of superoxide radical. Biochem Pharmacol 37:349–352

Stoll S, Hartmann H, Cohen SA, Muller WE (1993) The potent free radical scavenger α-lipoic acid improves memory in aged mice: putative relationship to NMDA receptor deficits. Pharmacol Biochem Behav 46:799–805

Sugita Y, Kondo T, Kanazawa A, Itou T, Mizuno Y (1993) Protective effect of FPF1070 (cerebrolysin) on delayed neuronal death in the gerbil-detection of hydroxyl radicals with salicylic acid (Japanese). No-to-Shinkei-Brain and Nerve 45:325–331

Suzuki J, Imaizumi S, Kayama T, Yoshimoto T (1985) Chemiluminescence in hypoxic brain – the second report, cerebral protective effect of mannitol, vitamin E and glucocorticoid. Stroke 16:695–700

Suzuki J, Mizoi AK, Oba M, Yoshimoto T (1987) Protective effect of phenytoin and its enhanced action by combined administration with mannitol and vitamin E in cerebral ischemia. Acta Neurochir 88:56–64

Tan S, Yokoyama Y, Dickens E, Cash TG, Freeman BA, Parks DA (1993) Xanthine oxidase activity in the circulation of rats following hemorrhagic shock. Free Rad Biol Med 15:407–414

Taylor MD, Mellert TK, Parmentier JL, Eddy LJ (1984) Pharmacological protection of reoxygenation damage to in vitro brain slice tissue. Brain Res 347:268–273

Terada LS, Willingham IR, Rosandich ME, Leff JA, Kindt GW, Repine JE (1991) Generation of superoxide anion by brain endothelial cell xanthine oxidase. J Cell Physiol 148:191–196

The Parkinson's Disease Study Group (1993) Effects of tocopherol and deprenyl on the progression of disability in early Parkinson's disease. N Engl J Med 328: 176–183

Tritto I, Scognamiglio A, D'Andrea D, Eramo N, Elia PP, Simone CD, Exposito A, Violante A, Chiariello M, Ambrosio G (1995) Role of protein kinase C activation in oxygen radical induced preconditioning. Circulation 92:I-716–717

Ushijima K, Miyazaki H, Morioka T (1986) Immunohistochemical localization of glutathione peroxidase in the brain of the rat. Resuscitation 13:97–105

Wei Q, Yeung M, Jurma OP, Anderson JK (1996) Genetic elevations of monoamine oxidase levels in dopaminergic PC12 cells results in increased free radical damage and sensitivity to MPTP. J Neurosci Res 46:666–673

Weissman BA, Kadar T, Brandeis R, Shapira S (1992) N^G-Nitro-L-arginine enhances neuronal death following transient forebrain ischemia in gerbils. Neurosci Lett 146:139–142

White BC, Rafols JA, DeGracia DJ, Skjaerlund JM, Krause GS (1993) Fluorescent histochemical localization of lipid peroxidation during brain reperfusion. Acta Neuropathol (Berlin) 86:1–9

Williams LR (1995) Oxidative stress, age-related neurodegeneration, and the potential for neurotrophic treatment. Cerebrovasc Brain Metabol Rev 7:55–73

Xia Y, Dawson VL, Dawson TM, Snyder SH, Zweier JH (1996) Nitric oxide synthase generates superoxide and nitric oxide in arginine-depleted cells leading to peroxynitrite-mediated cellular injury. Proc Natl Acad Sci USA 93:6670–6674

Yamamoto M, Shina T, Vozumi T, Sogabe T, Yamada K, Kawasaki T (1983) A possible role of lipid peroxidation in cellular damage caused by cerebral ischemia and the protective effect of α-tocopherol administration. Stroke 14:977–982

Yang GY, Chan PH, Chen J, Carlson E, Chen SF, Weinstein P, Epstein CJ, Kamii H (1994) Human copper-zinc superoxide dismutase transgenic mice are highly resistant to reperfusion injury after focal cerebral ischemia. Stroke 25:165–170

Yokoyama Y, Beckman JS, Beckman TK, Wheat JK, Cash TG, Freeman BA, Parks DA (1990) Circulating xanthine oxidase: potential mediator of ischemic injury. Am J Physiol 258:G564–G570

Yoshida S, Busto R, Santiso M, Ginsberg M (1984) Brain lipid peroxidation induced by postischemic reoxygenation in vitro: effect of vitamin E. J Cereb Blood Flow Metab 4:466–469

Yoshida T, Limmroth V, Irikura K, Moskowitz MA (1994) The NOS inhibitor, 7-nitro-indazole, decreases focal infarct volume but not the response to topical acetylcholine in pial vessels. J Cereb Blood Flow Metab 14:924–929

Young B, Runge JW, Waxman KS, Harrington T, Wilberger J, Muizelaar JP, Boddy A, Kupiec JW (1996) Effects of pegorgotein on neurologic outcome of patients with severe head injury. A multicenter, randomized controlled trial. JAMA 276:538–543

Yue T-L, Gu J-L, Lysko PG, Cheng H-Y, Barone FC, Feuerstein G (1992) Neuroprotective effects of phenyl-t-butyl-nitrone in gerbil global brain ischemia and in cultured rat cerebellar neurons. Brain Res 574:193–197

Zaleska MM, Floyd RA (1985) Regional lipid peroxidation in rat brain in vitro. Possible role of endogenous iron. Neurochem Res 10:397–410

Zhang F, Iadecola C (1993) Nitroprusside improves blood flow and reduces brain damage after focal ischemia. Neuroreport 4:559–562

Zhang J-R, Andrus PK, Hall ED (1993) Age-related changes in hydroxyl radical stress and antioxidants in gerbil brain. J Neurochem 61:1640–1647

Zhao Q, Pahlmark K, Smith ML, Siesjo BK (1994) Delayed treatment with the spin trap α-phenyl-N-tert-butyl nitrone (PBN) reduces infarct size following transient middle cerebral artery occlusion in rats. Acta Physiol Scand 152:349–350

Zhou X, Zhai X, Ashraf M (1996) Direct evidence that initial oxidative stress triggered by preconditioning contributes to second window of protection by endogenous antioxidant enzyme in myocytes. Circulation 93:1177–1184

Zimmerman BJ, Parks DA, Grisham MB, Granger DN (1988) Allopurinol does not enhance antioxidant properties of extracellular fluid. Am J Physiol 255: H202–H206

CHAPTER 10

Chemokines and Chemokine Receptors in the Central Nervous System: New Opportunities for Novel Therapeutics in Brain Ischemia and Trauma

G.Z. FEUERSTEIN and X. WANG

A. Introduction

I. The Chemokine Family of Polypeptides and Their Receptors

The superfamily of chemokines (chemoattractant cytokines) consists of a broad array of polypeptides of diverse biological actions and structures (Fig. 1). The current number of chemokines exceeds 50 related proteins. These proteins range in size from 68–120 amino acids (in the mature form) and can be segregated into at least four structural branches: C, C-C, C-X-C, and CXXXC according to variations in a shared cysteine motif. The largest branch, i.e., the C-C or β-chemokines, has nearly twenty members in humans while the smallest branch, the C class, has only one (OPPENHEIM et al. 1991; ROLLINS 1997; MANTOVANI 1999; KELNER et al. 1994; PAN et al. 1997). The C-X-C, or the β-chemokine branch can be further subdivided by structure and function into proteins containing the amino acid motif E-L-R-C-X-C (the majority) and those few that do not have the E-L-R motif adjunct to C-X-C. Structural distinctions of the different branches of the superfamily of chemokines have been shown to parallel general (though not absolute) distinctions in their biological activities (Fig. 2). For example, most C-X-C chemokines (those with E-L-R) are chemoattractants for neutrophils but not for monocytes, whereas C-C chemokines generally attract monocytes and lymphocytes, but not neutrophils. Basophils and eosinophils are also affected by C-C chemokines. The C chemokine appears thus far to be lymphocyte-specific. The expanding array of chemokine ligands is rapidly matched by a growing number of chemokine receptors.

Four separate classes of chemokine receptors have been defined: "specific," "shared," "promiscuous," and "viral" (Fig. 1), (WELLS et al. 1998; MURPHY 1994; HOLMES et al. 1991; LEE et al. 1992; NEOTE et al. 1993). The class of specific receptors, which bind only one known ligand of the chemokine family, contains only two examples: the CXCR1, formally known as the interleukin-8 (IL-8) receptor A (IL8RA), and the CXCR4, Fusin or SDF-1 receptor. Shared receptors are those that will bind more than one chemokine within either the C-X-C or the C-C branches (but not across the branches); this

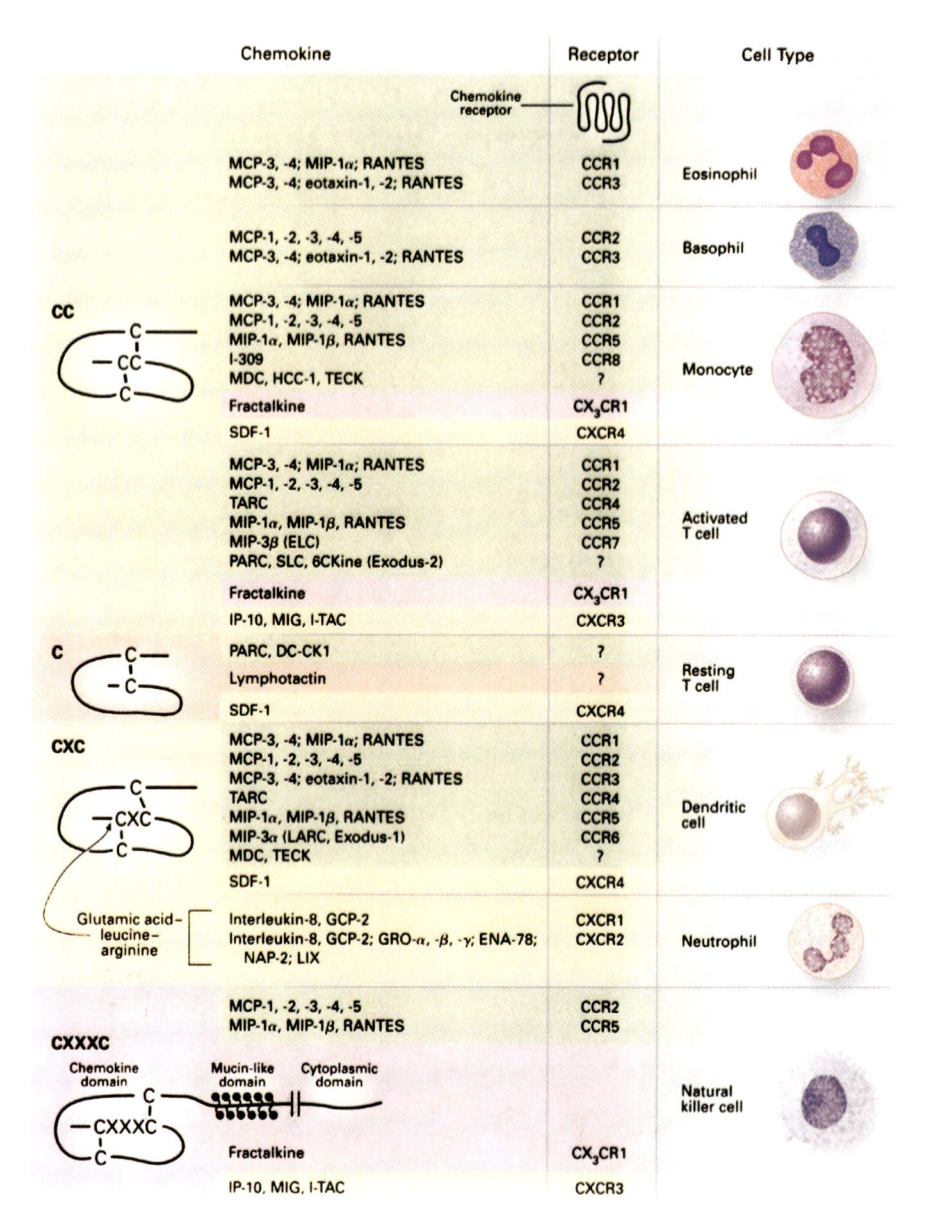

Chemokine	Receptor	Cell Type
MCP-3, -4; MIP-1α; RANTES MCP-3, -4; eotaxin-1, -2; RANTES	CCR1 CCR3	Eosinophil
MCP-1, -2, -3, -4, -5 MCP-3, -4; eotaxin-1, -2; RANTES	CCR2 CCR3	Basophil
MCP-3, -4; MIP-1α; RANTES MCP-1, -2, -3, -4, -5 MIP-1α, MIP-1β, RANTES I-309 MDC, HCC-1, TECK	CCR1 CCR2 CCR5 CCR8 ?	Monocyte
Fractalkine	CX_3CR1	
SDF-1	CXCR4	
MCP-3, -4; MIP-1α; RANTES MCP-1, -2, -3, -4, -5 TARC MIP-1α, MIP-1β, RANTES MIP-3β (ELC) PARC, SLC, 6CKine (Exodus-2)	CCR1 CCR2 CCR4 CCR5 CCR7 ?	Activated T cell
Fractalkine	CX_3CR1	
IP-10, MIG, I-TAC	CXCR3	
PARC, DC-CK1	?	Resting T cell
Lymphotactin	?	
SDF-1	CXCR4	
MCP-3, -4; MIP-1α; RANTES MCP-1, -2, -3, -4, -5 MCP-3, -4; eotaxin-1, -2; RANTES TARC MIP-1α, MIP-1β, RANTES MIP-3α (LARC, Exodus-1) MDC, TECK	CCR1 CCR2 CCR3 CCR4 CCR5 CCR6 ?	Dendritic cell
SDF-1	CXCR4	
Interleukin-8, GCP-2 Interleukin-8, GCP-2; GRO-α, -β, -γ; ENA-78; NAP-2; LIX	CXCR1 CXCR2	Neutrophil
MCP-1, -2, -3, -4, -5 MIP-1α, MIP-1β, RANTES	CCR2 CCR5	Natural killer cell
Fractalkine	CX_3CR1	
IP-10, MIG, I-TAC	CXCR3	

Fig. 1. Chemokines and their receptors. Four subfamilies of chemokines (i.e., C, CC, CXC and CXXXC chemokines) have been described based upon the number and position of conserved cysteine residues. Chemokine receptors are G-protein-coupled proteins with seven transmembrane domains. Ligands that bind to the individual receptor(s) on specific cell types are depicted. As described in the text, four classes of chemokine receptors have been identified, i.e., *specific*, *shared*, *promiscuous*, and *viral* according to the nature of their interactions with matched chemokine ligands. (Adapted from LUSTER 1998 by permission of *NEJM*)

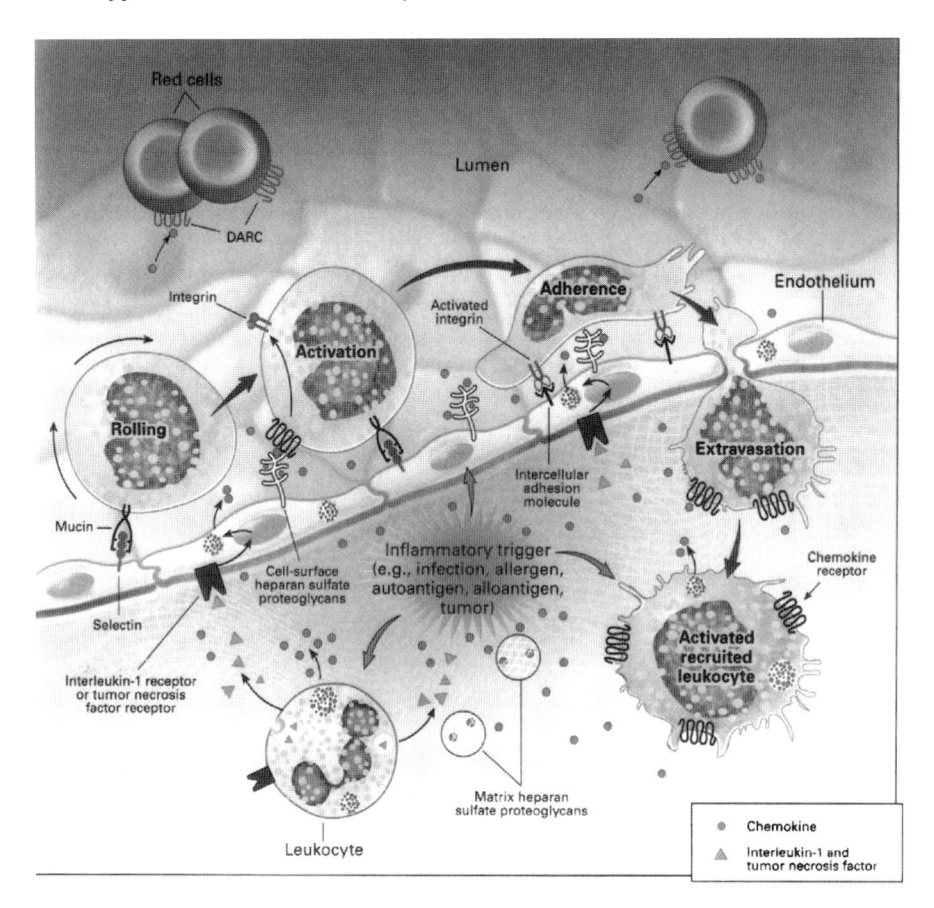

Fig. 2. Chemokine regulation of leukocyte movement. Chemokines are secreted at sites of inflammation and infection by the resident cells and infiltrated leukocytes. Chemokines are locally retained on the matrix and cell surface heparan sulfate proteoglycans, establishing a chemokine concentration gradient surrounding the inflammatory stimulus, as well as on the surface of the overlying endothelium. Leukocytes rolling on the endothelium (in a selectin-mediated process) and chemokine–chemokine-receptor interaction leads to firm adhesion and infiltration of leukocytes into the inflammatory lesions. The recruited leukocytes are activated by local inflammatory cytokines and may become desensitized to further chemokine signaling. The Duffy antigen receptor for chemokines (*DARC*) functions as a sink to maintain a tissue–bloodstream chemokine gradient. (Adapted from LUSTER1998 by permission of *NEJM*)

category has the largest number of members to date. One shared C-X-C receptor, the CXCR2 or IL-8 receptor B (IL8RB), binds to all E-L-R C-X-C chemokines (IL-8, Gro$\alpha/\beta/\gamma$, NAP-2, LIX, GCP-2, ENA-78); another, CXCR3, appears to bind to nonE-L-R C-X-C chemokines (IP-10, MIG, I-TAC; Fig. 1). The majority of shared receptors cloned to date bind the C-C chemokines; these are the CCR1 through CCR5 receptors, which are depicted in Fig. 1, with their complement of ligands. A promiscuous receptor is commonly regarded as a receptor that will bind many chemokine ligands of either C-X-C or C-C branches. There is only one known promiscuous receptor, the Duffy Blood Group antigen (Duffy Ag), originally identified as an erythrocyte chemokine receptor (NEOTE et al. 1993). Duffy Ag is widely divergent from the rest of the receptor family, while still belonging to the G-protein-coupled receptor GPCR class. Last, virally encoded chemokine receptors represent shared receptors that have been transduced into the viral genomes. There are currently two reports of the receptors in γ herpes viruses; one encoded by cytomegalovirus (CMV US28) (GAO and MURPHY 1994) and one from Kaposi's sarcoma-associated herpes virus (KSHV), (ARVANITAKIS et al. 1997), although others have been recently detected.

The assignment of receptors to particular classes should be considered provisional. For example, it may be the case that as all receptor-ligand pairings are revealed, the class of specific receptors will no longer exist. Nevertheless, it may be instructive to note that the first chemokine receptor cloned, the CXCR1, has been studied extensively, yet no ligand beyond IL-8 was identified. Similarly, CXCR4 has been probed with all known chemokines and was only recently shown to interact with SDF-1 (LUSTER 1998).

Chemokine receptors are structurally related. Although the overall degree of amino acid identity between different types of chemokine receptors can be as low as ~30% (for example, between CCR1 and CXCR1), the CXCR1 and CXCR2 receptors are approximately 65% identical. The molecular basis of chemokine binding and selectivity is not fully understood, but mutagenesis studies have provided some clues. Domain swapping experiments show that the N-terminal portion of chemokine receptors is key to determining ligand-binding specificity. Other mapping studies point to conserved Cys, Arg, and Asp residues in the extracellular loops as important for high-affinity binding of IL-8 to the specific receptor CXCR1 (MURPHY 1994). The shared C-C receptors CCR1 through CCR5 are remarkably homologous through large segments, particularly in the transmembrane regions and the second and third intracellular loops. This supports the notion that the conserved regions are critically involved in maintaining structural relationships within the mature protein and in coupling to intracellular signaling molecules. By contrast, the N- and C-terminal portions and extracellular loops are highly variable and likely to be involved in ligand binding and receptor-specific interactions with signaling components. The viral chemokine-like receptor, CMV US28, accomplishes a broader spectrum of C-C chemokine interactions and signaling than any of the individual human CCR with only about 30% identity. Certain chemokine receptor classes are also genetically clustered. The shared C-C receptors are localized to human chromosome 3, whereas CXCR1, CXCR2, and CXCR4 are clustered on human chromosome 2 (ROLLINS 1997).

It is still unclear how the chemokines achieve their diverse functions. The relationship of ligand and receptor in vivo is likely to involve a dynamic relationship between at least four components: (1) a "Presentation" molecule (such as glycosaminoglycan determinant on a proteoglycan) (2) the nominal chemokine ligand, (3) the receptor on the target cell, or (4) the receptor-linked cytoplasmic G-protein complex. It is likely that the presence, absence, or perturbation of any of these components affects the finding interaction and the signaling properties of the complex. Thus, it is plausible to suggest that chemokine specificity is achieved only "in context." A given chemokine, bound to a specific presentation molecule, may preferentially bind to a receptor linked to a specific G-protein. Thus, a complex interaction in vivo could determine a specificity not detected in in vitro assays and could explain in part the apparent paucity of cloned receptors of a specific class.

Many other questions in chemokine-receptor biology remain unanswered. Virtually nothing is known of the nature of the C class (lymphotactin) receptor, or whether chemokines may mediate some of their actions through non-GPCR. Even the distribution of chemokine receptors is not yet fully defined. Polymerase chain reaction (PCR) and blotting analyses suggest that most of the shared receptors are broadly distributed among leukocyte classes, but clear definition of tissue distribution of the receptors await wider availability of specific reagents such as monoclonal antibodies and quantitative PCR methodology. Targeted gene deletion analysis has been performed for only a few chemokine receptors thus far. For example, deletion of the CXCR2 in mice results in lymphadenopathy and splenomegaly (CACALANO 1994) with an attendant increase in circulating B cells and neutrophils; deletion of the CCR2 (MCP-1 receptor) conveys marked resistance to atherosclerosis (BORING et al. 1998). Because of the genetic linkage of chemokine receptors described above, it may prove to be a challenge to obtain targeted multiple deletions of receptors by the cross-breeding of different chemokine receptor knockout mice.

B. Chemokine and the Central Nervous System

Leukocyte traffic in brain tissue is known for its limited and restricted capacity as compared to peripheral organs. This phenomenon may not be the result of lack of adhesion mechanisms since adhesion molecules are readily induced in central nervous system (CNS) microvessels by intraparenchymal injection of lipopolysaccharide (LPS), indicating that failure of neutrophil infiltration could not be accounted for by absence of the required adhesive molecules. In fact, it has been suggested that neutrophils might fail to enter the CNS after LSP injection (and, by extension, other pathological states) because the required neutrophil chemoattractants were not elicited by the stimuli (CASTANO et al. 1996).

Studies conducted by PERRY et al. used recombinant adenoviruses to express a neutrophil-specific chemokine (MIP-2) in the nervous system of mice, which ultimately developed massive neutrophil inflammation in their CNS, associated with blood-brain barrier disruption at sites of extravasation.

Mice that expressed a related α-chemokine, GRO-α, governed by the CNS-specific myelin basic protein (MBP) promoter, showed an identical phenotype at the time of peak transgene expression (TANI et al. 1996). These results strongly supported the concept that the CNS fails to recruit neutrophils because of insufficient chemokines expression. One highexpression line of MBP-GRO-α transgenic mice began to exhibit neurological symptoms of ataxia with pronounced postural instability at about six weeks of age (TANI et al. 1996). Neuropathological examination of affected transgenic animals revealed extensive microgliosis and astrogliosis without signs of neutrophil activation or CNS myelination. Thus, one may suggest that the CNS has limited capacity to express α-chemokines because their high amounts of synthesis might compromise neuronal function.

Studies of CNS overexpression of a β-chemokine, MCP-1, yielded different results; mice that expressed MCP-1 under the control of the MBP promoter had no consistent perivascular inflammation; however, after intraperitoneal LPS injections to MCP-1 overexpressing (but not normal) mice, marked monocyte infiltration to the brain was noted (FUENTES et al. 1995). The data suggests that overexpression of the β-chemokine MCP-1 primes the CNS for leukocyte infiltration upon systemic inflammatory conditions.

C. Chemokine Expression in Brain Trauma

Evidence that chemokines selectively recruit specific leukocytes focused attention on the possibility that restricted chemokine expression could account, in part, for the unusual monocyte-rich inflammation that occurs after CNS trauma. That chemokines might mediate selective recruitment of leukocytes to the CNS was supported by the unexpected finding that parenchymal neural cells were competent to express chemokines under physiological circumstances. It was found that CNS chemokine response to distinct insults was characterized primarily by expression of the monocyte chemoattractant proteins, e.g., MCP-1, and closely related products. This selective CNS production of MCPs corresponded to selective monocyte accumulation. MCP-1 was initially monitored at the mRNA and protein levels in models of mechanical brain injury that used various stimuli for gliosis (nitrocellulose membrane stab) in either adult or neonatal animals. Increased expression of MCP-1 mRNA and protein was restricted to injury models characterized by astroglial reaction and macrophage response. MCP-1 mRNA levels were strikingly elevated three h after injury and well before appearance of inflammatory leukocytes, implying expression by in situ, parenchymal cells (GLABINSKI et al. 1996).

In situ hybridization analysis of MCP-1 expression in coronal sections of injured brain disclosed hybridization-positive cells surrounding the wound as early as six h after trauma. MCP-1-expressing cells were shown by immunohistochemical colocalization studies with GFAP to be astrocytes. It was concluded that astrocytes selectively produce the chemokine MCP-1 during inflammation following certain forms of mechanical injuries, and that chemokine synthesis constitutes one element of the reactive astroglial program of gene expression.

A second series of experiments was conducted in mice subjected to cortical cryoprobe injury. This lesion elicits a larger and much more destructive lesion followed by inflammatory reaction than the lesion produced by a cerebral stab with a nitrocellulose membrane. In this model, steady-state levels of MCP-1 mRNA were markedly elevated six h after lesion placement, with a return to baseline by 48h. In the contralateral cortex, MCP-1 expression was significantly elevated to approximately one-third the extent as observed ipsilaterally, with a superimposed time course. This result suggested that damage to axon termini from contralateral projection fields elicited chemokine expression in proximity to the neuronal cell bodies from which the axons originated.

D. Chemokines and Cerebral Ischemia

I. Cytokine-Induced Neutrophil Chemoattractant

Cytokine-induced neutrophil chemoattractant (CINC) was originally identified as a secreted product of IL-1-activated epithelial cells. Subsequently, it was shown to be the homologue of the human GRO and the mouse KC gene. KC, human GRO, and CINC are α-chemokines that exhibit significant chemotactic properties for neutrophils. CINC mRNA has been found to be highly upregulated in focal ischemic brain tissue of rats (LIU et al. 1993). The expression of CINC/KC gene, as reflected by its mRNA levels, increased rapidly within a few hours after ischemia, peaked at 24h, and downregulated within 2–5 days. This time course of mRNA upregulation fits to the "inflammatory Phase III" described in Fig. 3. Interestingly, the expression of this neutrophil chemotactic agent precedes the robust leukocyte accumulation that follows focal ischemia. While the stimuli/mechanism of KC/CINC upregulation in brain ischemia is unknown, it is plausible to suggest that cytokine such as TNF-α and IL-1β might play a role (at least in part) in upregulation of this α-chemokine. The cellular origin of KC/CINC in ischemic brain has not been elucidated as yet; macrophages have been shown to be capable of producing this chemokine and their influx into the stroke zone may contribute to KC/CINC production at the late time points (1 day and beyond). It is also possible that microglia may be responsible for the earlier production of this chemokine, but this possibility awaits further investigation.

II. Monocyte Chemoattractant Protein 1

Monocyte chemoattractant protein 1 (MCP-1) was originally discovered as a rapidly induced gene in PDGF-stimulated murine fibroblasts (ROLLINS et al. 1998). MCP-1 functions as a chemoattractant for monocytes and basophils, but

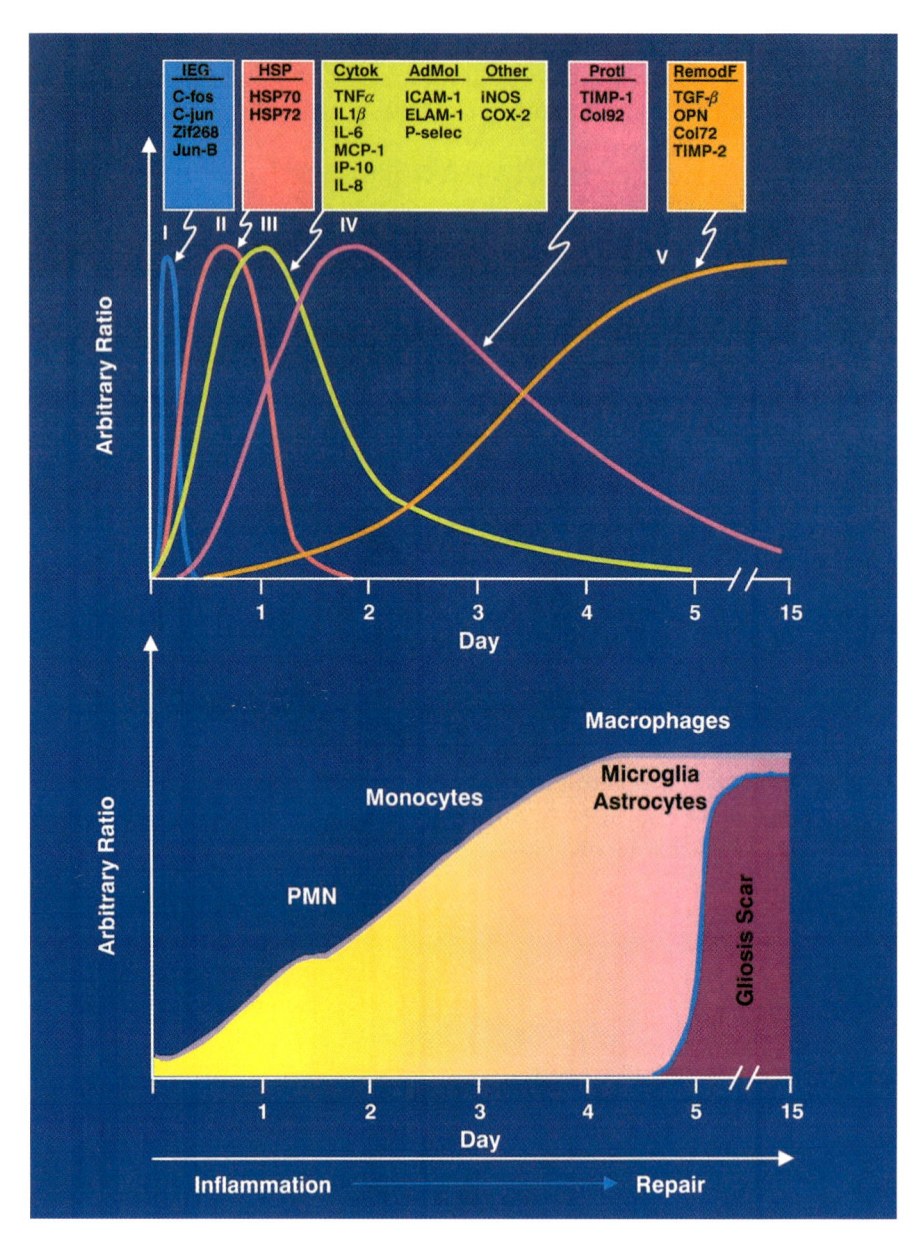

Fig. 3. Temporal expression of genes after focal brain ischemia in the rat. Five waves of gene expression are induced after ischemic injury, which include a broad range of upregulated transcription factors (phase 1), heat shock proteins (phase 2), proinflammatory mediators (phase 3), proteases and protease inhibitors (phase 4), and delayed tissue remodeling proteins (phase 5) (*upper panel*). The corresponding temporal profile of leukocytes infiltration/accumulation in the ischemic cortex, as well as the astrocyte activation/gliosis, is depicted in the *lower panel*

not for neutrophils (BISCHOFF et al. 1992). MCP-1 is induced in a variety of nonimmune cells, including endothelium and smooth muscle cells, in response to a wide array of cytokines. MCP-1 expression is absent (or at extremely low levels) in normal brain tissue (WANG et al. 1995a). However, in response to an ischemic condition, rapid upregulation of MCP-1 mRNA can be detected by Northern blotting; the time course of MCP-1 mRNA upregulation parallels that of KC/CINC and other cytokines as depicted in Fig. 3. MCP-1 mRNA is upregulated even faster in conditions of ischemia and reperfusion (WANG et al. 1995a). MCP-1 was expressed initially by astrocytes and later by both astrocytes and infiltrating macrophages following cerebral ischemia (GOURMALA et al. 1997).

III. Monocyte Chemottractant Protein 3

Monocyte chemotactic protein 3 (MCP-3) is a member of the CC or α-chemokine subfamily. Using the suppression subtractive hybridization (SSH) strategy for differential gene cloning, the rat homologue to murine and human MCP-3 was cloned from rat ischemic cortex (WANG et al. in press b). MCP-3 demonstrates chemotactic preference for monocytes, lymphocytes, eosinophils, and basophils (ROLLINS 1997). MCP-3 interacts with two primary CC chemokine receptors, i.e., CCR2 and CCR3 (RANSOHOFF and TANI 1998). The temporal expression of MCP-3 mRNA was examined in brain tissue rendered ischemic by permanent or temporary occlusion of the middle cerebral artery (MCAO). A marked increase in MCP-3 mRNA was observed 12 h postischemia, with 49-fold over control in the permanent MCAO. Significant induction of MCP-3 in ischemic cortex was sustained up to 5 days after ischemic injury (Fig. 4). The profile of MCP-3 mRNA induction paralleled leukocyte infiltration and accumulation that occur after focal stroke, suggesting a role for MCP-3 in recruiting these inflammatory cells into the ischemic tissue. The cellular sources of MCP-3 after stroke remains to be investigated although previous in vitro studies (MINTY et al. 1993) suggested that the infiltrated leukocytes in the ischemic lesion may contribute to MCP-3 production.

IV. Interferon-γ-Inducible Protein 10

Interferon-γ-inducible protein 10 (IP-10) is a C-X-C chemokine that is chemoattractant for monocytes and activated T-lymphocytes (TAUB et al. 1993). IP-10 is also a potent chemotactic and mitogenic factor for vascular smooth muscle cells (WANG et al. 1996c), and possibly chemotactic for astrocytes (WANG et al. 1998d). The temporal expression of IP-10 mRNA after MCAO in rat was examined by Northern analysis and by in situ hybridization. Different from other chemokines, IP-10 mRNA expression following focal stroke demonstrated a unique biphasic induction profile (Fig. 5). IP-10 mRNA was markedly increased early at 3 h (4.9-fold over control) and peaked at 6 h

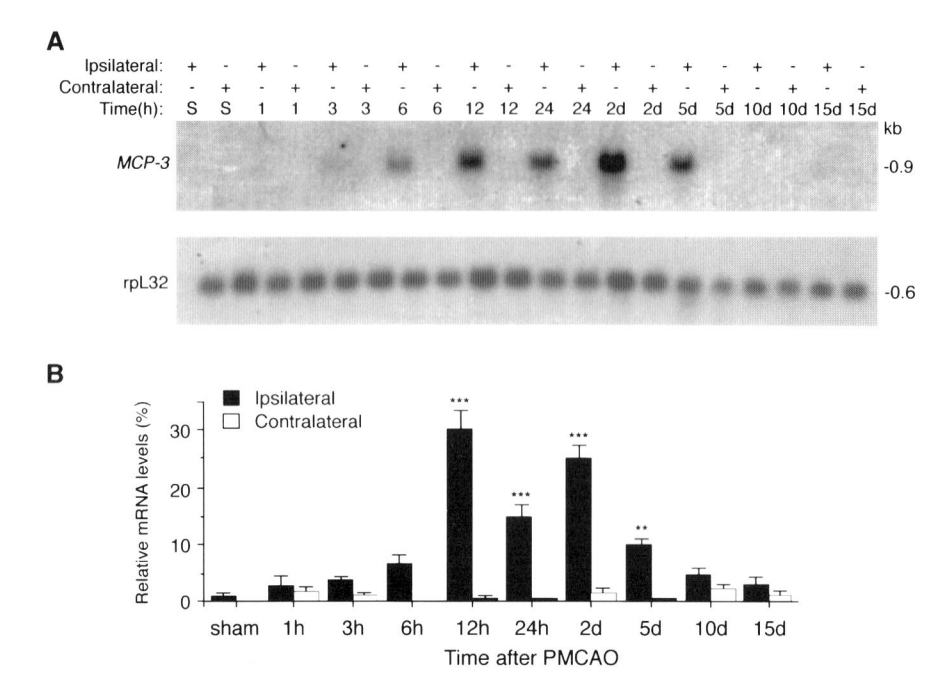

Fig. 4A,B. Time-course study of MCP-3 mRNA induction in rat ischemic cortex following permanent MCAO. **A** A representative Northern blot for MCP-3 and rpL32 cDNA probes to the samples from spontaneously hypertensive rats (*SHR*) following permanent MCAO. Total cellular RNA (40µg/lane) was resolved by electrophoresis, transferred to a nylon membrane, and hybridized to the indicted cDNA probe. Ipsilateral and contralateral cortical samples (denoted by +) from individual rats of sham surgery (*S*; 12h) or following 1, 3, 6, 12, and 24h, and 2, 5, 10, and 15 days of permanent MCAO and depicted. **B** Quantitative Northern blot data (*n* = 4) for MCP-3 mRNA expression after focal stroke. The data were generated using PhosphorImager analysis and displayed graphically after normalizing with rpL32 mRNA signals. $**P < 0.01$, $***P < 0.001$, compared to sham-operated animals

(14.5-fold) after MCAO; a second wave induction was observed 10–15 days postischemia (7.2- and 9.3-fold increase for 10 and 15 days, respectively). In situ hybridization confirmed the induced expression of IP-10 mRNA and revealed its spatial distribution after focal stroke (Fig. 6). Immunohistochemical studies demonstrated the expression of IP-10 peptide in neurons (3–12h) and astroglial cells (6h–15 days) in the ischemic zone (Wang et al. 1998d). The temporal and spatial expression of IP-10 after focal stroke, along with the biochemical features of this chemokine, suggests that ischemia-induced IP-10 may play a pleiotropic role in prolonged leukocyte recruitment and astrocyte migration/activation after focal stroke.

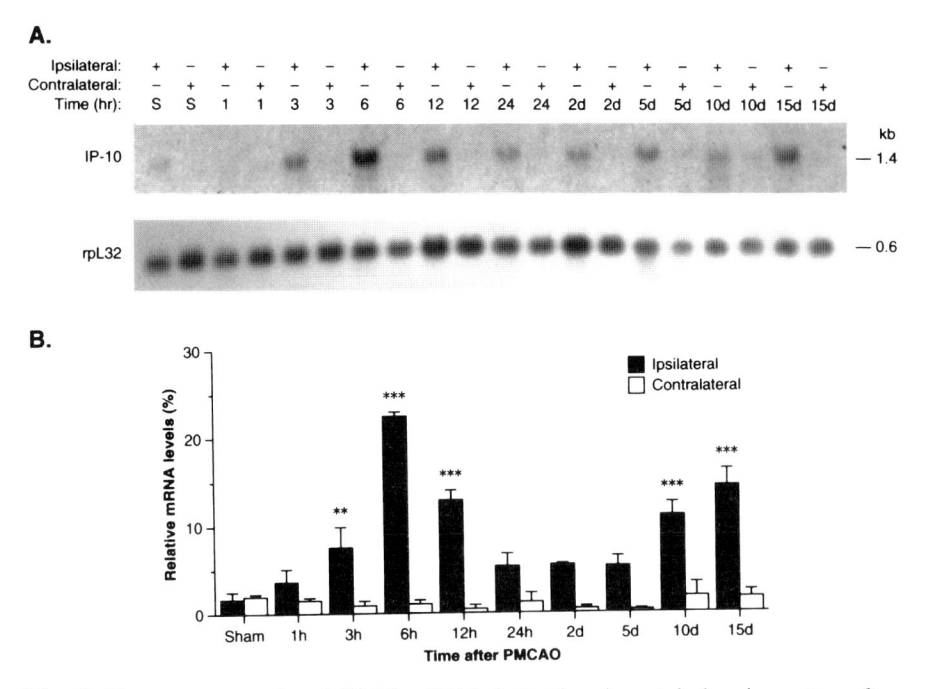

Fig. 5. Time-course study of IP-10 mRNA induction in rat ischemic cortex after MCAO. A representative Northern blot for IP-10 and rpL32 probes to the samples isolated at various time points and conditions from rats subjected to MCAO. The figure is illustrated as described in the legend of Fig. 4

E. Interleukin-8 in Human Brain Injury

IL-8 is a CXC chemokine that acts via CXCR1 and CXCR2 receptors (see Fig. 1). Increased level of IL-8 was reported in serum of patients with brain injury (OTT et al. 1994). In an elaborated study by (KOSSMAN et al. 1997) conducted in patients with severe traumatic brain injury, IL-8 was identified in the cerebrospinal fluid (CSF) and at higher concentrations than in the serum. The maximal levels of IL-8 correlated with the severity of the blood brain barrier dysfunction. An interesting correlation between levels of IL-8 and nerve growth factors (NGF) was also noted. The possible relationships between IL-8 and NGF was further suggested in studies where CSF containing high IL-8 levels stimulated NGF production in cultured astrocytes more than CSF samples that contained lower IL-8 levels; furthermore, antibodies against IL-8 reduced NGF release. IL-8 has been implicated in a variety of human CNS diseases such as bacterial meningitis, moningoencephalitis (HANDA 1992; MASTROIANNI et al. 1994), and malignant brain tumors (TADA et al. 1993; MORITA et al. 1993). These reports provide strong evidence, by association, that IL-8 may be involved in diverse brain pathologies, yet the functional

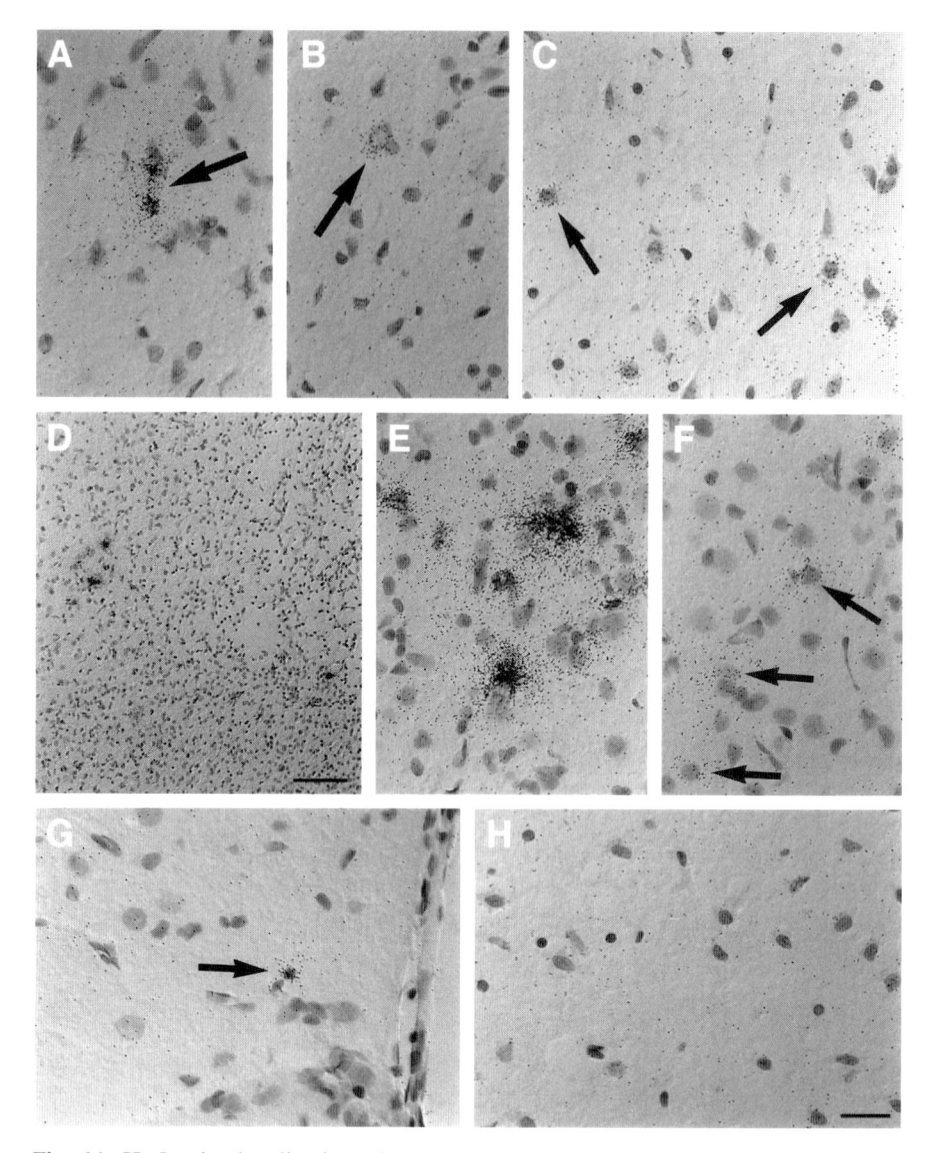

Fig. 6A–H. In situ localization of IP-10 mRNA expression in the ischemic cortex. **A, B** IP-10 mRNA positive cells (*arrows*) in the infarct border at 6h post-MCAO. **C** Relative high numbers of IP-10 mRNA positive cells (*arrows*) in the lesion at 6h post-MCAO. **D–F** (low power) The IP-10 mRNA-positive cells around lesion rim at 15 days post-MCAO. Note that some cells expressed high levels of IP-10 mRNA (**E**) while some relatively low levels (**F**, indicated with *arrows*), and that the distribution of IP-10 mRNA-positive cells appears to be lined up along the lesion. **G** Very few IP-10 mRNA positive cells (*arrow*) in the ventral site of sham-operated animals (2 days). **H** Sense riboprobe control in the core infarct of 6h post-MCAO. **A–C, E–H** ×400, *scale bar* = 20 µm; **D** ×100, *scale bar* = 100 µm

significance of IL-8 production in these pathological states awaits further evaluation. At this time, no direct evidence exists that IL-8 per se perturbs BBB, glial, or neuronal cell function; the evidence to date may suggest a role for IL-8 as a chemotactic agent for early neutrophil recruitment and hence an indirect role in neutrophil-mediated injury.

F. Chemokines and Neuro-Acquired Immune Deficiency Syndrome

Recent evidence suggests that certain chemokines and their receptors may have a role in neuro-acquired immune deficiency syndrome (AIDS). Neuro-AIDS is believed to be largely the result of viral reservoir and progression via brain macrophage/microglia, not neurons. The chemokine receptors CXCR4, CCR5, and CCR3 bind to HIV-coat protein gp120 and facilitate infection of targeted cells. Microglia express both CCR3 and CCR5, by which HIV infection of these cells is facilitated (VALLAT et al.1998; HE et al. 1997). In addition to the chemokine receptors, various chemokines are increased in HIV with neuro-AIDS (e.g., MCP-1, RANTES, MIP-1α, β) thereby promoting further recruitment of HIV-infected monocytes and lymphocytes. In addition to promoting HIV infection in the brain, some chemokines may also mediate neurotoxicity; the chemokine SDF-1, which binds to CXCR4, has been shown to induce apoptosis in hNT (neuronal) cells (KURIHARA et al. 1997). However, certain chemokines, e.g., SDF-1, may actually suppress HIV gp120-induced neuronal toxicity by downregulation of the facilitatory chemokine CXCR4 (MEUCCI et al. 1998). These in vitro findings may as well reflect the in vivo pathology since neuro-AIDS is associated with neuronal loss by apoptosis (GELBARD et al. 1995; ADE-BIASSETTE et al. 1995).

G. Chemokines in Neurodegenerative Disorders

The histopathology of Alzheimer's disease (AD) is now acknowledged as having an inflammatory component in addition to the known hallmarks such as the neurofibrillary tangles, senile plaques, and neurodegeneration. AD is recognized for the marked activation of astrocytes and microglia observed in the pathologic area (ROGERS et al. 1996; MRAK et al. 1995). The inflammatory process in AD does not involve leukocyte infiltration or BBB disruption. There is, however, strong upregulation and expression of the CXCR2 receptors localized to dystrophic neurons in the senile plaque (HORUK et al. 1997; XIA et al. 1997) which correlate to APP levels (XIA et al. 1997). In vitro studies further implicate chemokines to AD pathology; thus, Aβ was shown to stimulate the production of IL-8 in human astrocytoma cells as well as monocytes (GITTER et al. 1995; MEDA et al. 1996). Furthermore, Aβ is also capable of stimulating the production of MCP-1 and MIP-1 α/β from human monocytes; MCP-1 was also found in the human AD plaques and reactive microglia (FIALA et al. 1998; ISHIZUKA et al. 1997). Taken together, emerging evidence implicates certain chemokines and their receptors in the key pathological features of AD. Clearly, the casual role of these (or other) chemokines remains to be elucidated.

H. Summary and Conclusion

The chemokines are a diverse group of polypeptides that possess broad biological actions mediated by a large number of GPCRs. While originally described as chemotactic agents for various leukocytic migratory functions, the gamut of the chemokine functions includes cell activation, viral infections, apoptosis, and host defense. The brain has been no exception to the peripheral systems; it is capable of in situ production of various chemokines and their receptors associated with numerous forms of brain pathological reactions ranging from infections to stroke to neurodegeneration. However, from the data gathered so far, several important conclusions can be drawn:

1. The normal brain has little if any expression of chemokines or their receptors; therefore, one may assume (with the caveat of technical detection levels by current methodology) that chemokines may not bear on normal CNS functions.
2. In situ upregulation of chemokine synthesis and release, along with their respective receptors, has been repeatedly reported in diverse forms of neuroinjury. Therefore, it is plausible to anticipate that chemokines may have roles in broad neuropathological reactions in the brain.
3. Evidence generated so far on the regulation and effects of chemokines in brain injury is largely circumstantial and implicated chemokine to brain pathology by association. Thus, the robust upregulation of chemokines in ischemic brain injury (experimental stroke models) that precedes neutrophils/monocyte migration into the brain implicates chemokines such as IL-8, MCP-1, and IP-10 in the leukocyte response to injury by their "reputation" and temporal relationships to the cellular response. Similarly, in AD, the presence of chemokines in the pathology of AD may be a "smoking gun," yet no proof of a causative role of any chemokines in AD has been provided.
4. Since no potent and specific antagonists for any of the chemokines are available as yet to explore the specific contributions of chemokines to human pathology, caution must be exercised in drawing conclusions on their biological or pathological role in humans. Experiments with mice which have been genetically manipulated to carry disruption of chemokine or chemokine receptor genes may be useful to discern such roles. For example, MCP-1 knockout (null mutation) of MCP-1 or its receptor CCR2 results in significant tolerance to develop atherosclerosis in response to high

fat diets. It would be of interest to explore whether CNS injury is also modified in such models.
5. Although it is common to assign proinflammatory and pathological roles for chemokines in various immune and inflammatory conditions based on their chemotactic and stimulatory effect on leukocytes, one needs to keep in mind that inflammatory reactions in response to tissue injury serve to clear the insult and tissue debris as well as laying grounds for healing and repair. In this context, interfering with certain chemokine actions that mediate elements of the inflammatory reaction may ultimately result in poor healing and restoration of organ function.

In summary, CNS injuries, whether of infectious, toxic, ischemic, or degenerative in nature, have been shown to be associated with upregulation of chemokines and their receptors in the brain. It is plausible that chemokines indeed serve an important role in brain cell response to injury, yet the precise cellular origin of each chemokine and their receptors, and the temporal and spatial actions and their significance is largely unknown. However, major efforts by pharmaceutical organizations are aimed at discovering and developing highly potent and selective antagonists which would be critical in establishing the role of chemokines in CNS diseases.

References

Adle-Biassette H, Levy Y, Colombel M, Poron F, Natchev S, Keohane C, Gray F (1995) Neuronal apoptosis in HIV infection in adults. Neuropathol Appl Neurobiol 21: 218–227

Arvanitakis L, Geras-Raaka E, Varma A, Gershongom MC, Cesarman E (1997) Human herpesvirus KSHV encodes a constitutively active G-protein coupled receptor linked to cell proliferation. Nature 385:347–350

Bischoff SC, Krieger M, Brunner T, Dahinden CA (1992) Monocyte chemotactic protein-1 is a potent activator of human basophils. J Exp Med 175:1271–1275

Boring L, Gosling J, Cleary M, Charo IF (1998) Decreased lesion formation in CCR2(–/–) mice reveals a role for chemokines in the initiation of atherosclerosis. Nature 394:894–897

Cacalano G, Lee J, Kikly K, Ryan AM, Pitts-Meek S, Hultgren B, Wood WI, Moore MW (1994) Neutrophil and B cell expansion in mice that lack the murine IL-8 receptor homolog. Science 265:682–684

Castano A, Bell MD, Perry VH (1996) Unusual aspects of inflammation in the nervous system: Wallerian degeneration. Neurobiol Aging 17:745–751

Fiala M, Zhang L, Gan X, Sherry B, Taub D, Graves MC, Hama S, Way D, Weinand M, Witte M, Lorton D, Kuo YM, Roher AE (1998) Amyloid-β induces chemokine secretion and monocyte migration across a human blood-brain barrier model. Mol Med 4:480–489

Fuentes ME, Durham SK, Swerdel MR, Lewin AC, Barton DS, Megill JR, Bravo R, Lira SA (1995) Controlled recruitment of monocytes and macrophages to specific organs through transgenic expression of monocyte chemoattractant protein-1. J Immunol 155:5769–5796

Gao JL, Murphy PM (1994) Human cytomegalovirus open reading frame US 28 encodes a functional β-chemokine receptor. J Biol Chem 269:28539–28542

Gelbard HA, James HJ, Sharer LR, Perry SW, Saito Y, Kazee AM, Blumberg BM, Epstein LG (1995) Apoptotic neurons in brains from paediatric patients with HIV-1 encephalitis and progressive encephalopathy. Neuropathol Appl Neurobiol 21:208–217

Gitter BD, Cox LM, Rydel RE, May PC (1995) Amyloid β peptide potentiates cytokine secretion by interleukin-1 β-activated human astrocytoma cells. Proc Natl Acad Sci USA 92:10738–10741

Glabinski AR, Balasingam V, Tani M, Kunkel SL, Strieter RM, Yong VW, Ransohoff RM (1996) Chemokine monocyte chemoattractant protein-1 is expressed by astrocytes after mechanical injury to the brain. J Immunol 156:4363–4368

Gourmala NG, Buttini M, Limonta S, Sauter A, Boddeke HWGM (1997) Differential and time-dependent expression of monocyte chemoattractant protein-1 mRNA by astrocytes and macrophages in rat brain effects of ischemia and peripheral lipopolysaccharide administration. J Neuroimmunol 74:35–44

Handa S (1992) Concentration of interleukin-1β, interleudin-6, interleudin-8 and TNFα in cerebrospinal fluid from children with septic meningitis. Kurume Med J 39: 257–265

He J, Chen Y, Farzan M, Choe H, Ohagen A, Gartner S, Busciglio J, Yang X, Hofmann W, Newman W, Mackay CR, Sodroski J, Gabuzda D (1997) CCR3 and CCR5 are co-receptors for HIV-1 infection of microglia. Nature 385:645–649

Holmes WE, Lee J, Kuang WJ, Rice GC, Wood WI (1991) Structure and functional expression of human interleukin-8 receptor. Science 253:1278–1280

Horuk R, Martin AW, Wang Z, Schweitzer L, Gerassimides A, Guo H, Lu Z, Hesselgesser J, Perez HD, Kim J, Parker J, Hadley TJ, Peiper SC (1997) Expression of chemokine receptors by subsets of neurons in the central nervous system. J Immunol 158:2882–2890

Ishizuka K, Kimura T, Igata-yi R, Katsuragi S, Takamatsu J, Miyakawa T (1997) Identification of monocyte chemoattractant protein-1 in senile plaques and reactive microglia of Alzheimer's disease. Psychiatry Clin Neurosci 51:135–138

Kelner GS, Kennedy J, Bacon KB, Kleyensteuber S, Largaespada DA, Jenkins NA, Coepland NG, Bazan JF, Moore KW, Schall TJ, Zlotnik A (1994) Lymphotactin: A cytokine that represents a new class of chemokines. Science 266–1395

Kossmann T, Stahel PF, Lenzlinger PM, Redl H, Dubs RW, Trentz O, Schlag G, Morganti-Kossmann MC (1997) Interleukin-8 released into the cerebrospinal fluid after brain injury is associated with blood-brain barrier dysfunction and nerve growth factor production. J Cereb Blood Flow Metab 17:280–289

Kurihara T, Warr G, Loy J, Bravo R (1997) Defects in macrophage recruitment and host defense in mice lacking the CCR2 chemokine receptor. J Exp Med 186: 1757–1762

Lee J, Horuk R, Rice GC, Bennett GL, Camerato T, Wood WI (1992) Characterization of two high affinity human interleukin-8 receptors. J Biol Chem 267:6283–6287

Liu T, Young PR, McDonnel PC, While RF, Barone FC, Feuerstein GZ (1993) Cytokine induced neutrophil chemoattractant mRNA expressed in cerebral ischemia. Neurosci Lett 164:125–128

Luster AD (1998) Chemokines-chemotactic cytokines that mediate inflammation. The New England Journal of Med 338:436–445

Mantovani A (1999) The chemokine system: redundancy for robust outputs. Immunology Today 20:254–257

Mastroianni CM, Paoletti F, Rivosecchi RM et al. (1994) Cerebrospinal fluid interleukin-8 bacterial and tuberculous meningitis. Pediat Infect Dis J 13:1008–1010

Meda L, Bernasconi S, Bonaiuto C, Sozzani S, Zhou D, Otvos L Jr, Mantovani A, Rossi F, Cassatella MA (1996) β-amyloid (25–35) peptide and IFN-γ synergistically induce the production of the chemotactic cytokine MCP-1/JE in monocytes and microglial cells. J Immunol 157:1213–1218

Meucci O, Fatatis A, Simen AA, Bushell TJ, Gray PW, Miller RJ (1998) Chemokines regulate hippocampal neuronal signaling and gp120 neurotoxicity. Proc Natl Acad Sci USA 95:14500–14505

Minty A, Chalon P, Guillemot JC, Kadhad M, Liauzun KP, Magazin M, Milous B, Minty BC, Ramond P, Vita N, Lupker J, Shire D, Ferrara P, Caput D (1993) Molecular cloning of the MCP-3 chemokine gene and regulation of its expression. Cur Cytokine Netw 4:99–110

Morita M, Kasahara T, Mubouda N, Matsushima K, Nagashima T, Nishizawa M, Yoshida M (1993) Induction and regulation of interleukin-8 and MCAF production in human brain tumor cell lines and brain tumor tissue. Eur Cytokine Network 4:351–358

Mrak RE, Sheng JG, Griffin WS (1995) Glial cytokines in Alzheimer's disease: review and pathogenic implications. Hum Pathol 26:816–823

Murphy PM (1994) The molecular biology of leukocytes chemotactic receptors. Annu Rev Immunol 12:593–633

Neote K, Darbonne W, Ogez J, Horuk R, Scahll TJ (1993) Identification of a promiscuous inflammatory peptide receptor on the surface of red blood cells. J Biol Chem 268:12247–••

Oppenheim JJ, Zachariac COC, Mukaida N, Matsushima K (1991) Properties of a novel pro-inflammatory supergene "intercrine" family. Ann Rev Immunol 9:617–648

Ott L, McClouin CJ, Gillespie M, Young B (1994) Cytokines and metabolic dysfunction after severe head injury. J Neurotrauma 11:447–472

Pan Y, Lloyd C, Zhou H, Dolich S, Deeds J, Gonzalo JA, Vath J, Gosselin M, Ma J, Dussault B, Wolf E, Alperin G, Culpepper J, Gutierrez-Ramos JC, Gearing D (1997) Neurotactin, a membrane-anchored chemokine upregulated in brain inflammation. Nature 387–611

Ransohoff RM, Tani M (1998) Do chemokines mediate leukocyte recruitment in post-traumatic CNS inflammation. TINS 21:154–159

Rogers J, Webster S, Lue LF, Brachova L, Civin WH, Emmerling M, Shivers B, Walker D, McGeer P (1996) Inflammation and Alzheimer's disease pathogenesis. Neurobiol Aging 17:681–686

Rollins B, Morrison E, Stiles C (1998) Cloning and expression of JE, a gene inducible by platelet-derived growth factor and whose product has cytokine-like properties. Pro Natl Acad Sci USA 85:3738–3742

Rollins BJ (1997) Chemokines. Blood 90:909–928

Tada M, Susuki K, Yamakawa Y, Sawamura Y, Saduma S, Abe H, Van Mier E, Tribblet N (1993) Human givblastoma cells produce 77 amino acid interleukin-8. J Neurooncol 16:25–34

Tani M, Fuentes ME, Peterson JW, Trapp BD, Durham SK, Loy JK, Bravo R, Ransohoff RM, Lira SA (1996) Neutrophil infiltration, glial reaction, and neurological disease in transgenic mice expressing the chemokine N51/KC in oligodendrocytes. J Clin Invest 98:529–539

Taub DD, Lloyd RA, Conlon K, Wang JM, Ortaldo JR, Harada A, Matsushima K, Kelvin JD, Oppenheim JJ (1993) Recombinant human interferon-inducible protein 10 is a chemoattractant for human monocytes and T lymphocytes and promotes T cell adhesion to endothelial cells. J Exp Med 177:1809–1814

Vallat AV, De Girolami U, He J, Mhashilkar A, Marasco W, Shi B, Gray F, Bell J, Keohane C, Smith TW, Gabuzda D (1998) Localization of HIV-1 co-receptors CCR5 and CXCR4 in the brain of children with AIDS. Am J Pathol 152:167–178

Wang XK, Yue T-L, Barone FC, Feuerstein GZ (1995) Monocyte chemoattractant protein-1 messenger RNA expression in rat ischemic cortex. Stroke 26:661–666

Wang XK, Li X, Yaish-Ohad S, Sarau HM, Barone FC, Feuerstein GZ (in press) Molecular cloning and expression of the rate monocyte chemotactic protein-3: a possible role in stroke. Mol Brain Res

Wang XK, Yue TL, Ohlstein EH, Sung CP, Feuerstein GZ (1996) Interferon-inducible protein-10 involves vascular smooth muscle cell migration, proliferation, and inflammatory response. J Biol Chem 271:24286–24293

Wang XK, Ellison JA, Siren AL, Lysko PG, Yue TL, Barone FC, Shatzman A, Feuerstein GZ (1998) Prolonged expression of interferon-inducible protein-10 in

ischemic cortex after permanent occlusion of the middle cerebral artery in rat. J Neurochem 71:1194–1204
Wells TNC, Power CA, Proudfoot AEI (1998) Definition, function and pathophysiological significance of chemokine receptors. TIPS 19:376–380
Xia M, Qin S, McNamara M, Mackay C, Hyman BT (1997) Interleukin-8 receptor B immunoreactivity in brain and neuritic plaques of Alzheimer's disease. Am J Pathol 150:1267–1274

Section III
CNS Delivery of Neuroprotective Therapies

CHAPTER 11
Ex Vivo Gene Therapy in the Central Nervous System

A. BLESCH and M.H. TUSZYNSKI

A. Introduction

The targeted delivery of genes into the adult central nervous system (CNS) has received considerable interest in recent years with the development of improved viral vector systems and suitable strategies for therapeutic intervention. Experimental gene therapy in animal models has been studied to prevent or slow the progression of chronic neurodegenerative diseases, to improve recovery after traumatic CNS injury and to kill malignant brain tumors. Genes that have been investigated in these various models include those that code for neurotrophic factors, neurotransmitter synthesis enzymes, modulators of neuronal and glial function, and inducers of cell death. Generally, two different types of gene therapy have been distinguished: in vivo and ex vivo gene therapy. The direct injection of genes into the CNS using viral vectors or DNA-liposome suspensions is termed in vivo gene therapy. Ex vivo gene therapy is based on genetic modification of cells in vitro followed by the grafting of these cells into the CNS. Ex vivo approaches to gene therapy will be the focus of this review.

Gene therapy was classically viewed as a tool for the replacement of missing or defective genes. However, the delivery of pharmacological amounts of therapeutic agents has also become a major focus of gene therapy since this approach provides a means of locally delivering therapeutic molecules to precise in vivo targets. This is especially important in the CNS, where many diseases are restricted to specific, relatively small groups of cells. In these cases of localized cellular dysfunction, gene therapy has clear advantages over the systemic delivery of pharmaceuticals (see Sect. B.). Like every new form of therapeutic intervention, ex vivo gene delivery needs to be thoroughly evaluated for each specific application with regard to potential benefits and risk factors before starting clinical trials. To date, ex vivo gene therapy has been more extensively studied for therapeutic intervention and as a potential tool for elucidating CNS function than in vivo gene transfer. The base of knowledge regarding ex vivo approaches is large, and efficacy in a number of CNS disease models has been established. However, direct in vivo gene delivery has some advantages over ex vivo gene transfer including a lower degree of invasiveness, a smaller risk of tumor formation, and a potentially reduced expense of treatment per patient. Viral vectors for in vivo gene delivery like adenovirus, adeno-associated virus (AAV), herpes virus (HSV) and lentivirus have been substantially improved in recent years and are discussed in more detail in the following chapter.

B. Gene Therapy Versus Conventional Drug Delivery

One major obstacle for the delivery of bioactive drugs into the CNS is the inability of large molecules, including most proteins, to cross the blood–brain barrier (BBB). Various approaches have been taken to circumvent this problem including temporal breakdown of the BBB, modification of proteins to make them more lipophilic, or chemical linkage to proteins that are actively transported across the BBB (e.g., transferrin-receptor antibodies; for review see PARDRIDGE 1994). Generally, doses of systemically administered molecules need to be very high to reach sufficient concentrations at the relevant therapeutic CNS target. As a result, the number and extent of side effects increase proportionately to sometimes intolerable levels, which prevent further treatment. Alternatively, direct intracerebral drug delivery, including intraventricular or intraparenchymal infusion into the CNS, has been applied. Diffusion through the ventricular wall into the brain parenchyma remains a major problem especially in the primate brain, where the target area and ventricles can be several centimeters apart. As treatments in chronic degenerative diseases are necessary over long periods of time, pumps would need to be refilled or exchanged. They can shift their position or may be obstructed, leading to repeated invasive surgeries. Furthermore, the instability of proteins can result in the loss of bioactivity over extended time periods, and exposing broad areas of the brain to high amounts of bioactive molecules can result in severe adverse effects (see Sect. D.I.1.).

Gene therapy provides a rational alternative to these delivery approaches. The targeted delivery of genes to a specific region restricts the bioavailability of a protein to the site of interest, minimizing undesirable side effects. Cell-based delivery systems can serve as biological "mini-pumps" over extended time periods, delivering bioactive substances in a regionally specific, accurately targeted, chronic, and well-tolerated fashion (BLESCH et al. 1998; BLESCH and TUSZYNSKI 1995; GAGE et al. 1987; TUSZYNSKI et al. 1994b).

C. Practical Considerations

I. Cell Types Suitable for Gene Transfer and Grafting to the CNS

A wide range of cells has been used for genetic modification in vitro and subsequent transplantation into the CNS. Each of these cell types has some advantages and some disadvantages. The target area for transplantation, the desired function, and the in vivo and in vitro properties of cells influence the choice of the cell type in transplantation paradigms. Reactions of the host immune system, in vivo survival, and risk factors like tumorigenesis of grafted cells also need to be taken into account.

In initial studies, immortalized cell lines were often used for CNS transplantation (HORELLOU et al. 1990; ROSENBERG et al. 1988). Although these studies showed that the transplantation of genetically modified cells into the brain is a means of eliciting significant biological responses, tumor formation and immunological rejection limited their usefulness to short term studies in animals (WIDNER and BRUNDIN 1988). Primary cells obtained from either the host itself (autologous) or from a heterologous host [allogeneic (same species) or xenogenic (different species)] transplants have been employed as an alternative strategy. As expected, the host immune response is one of the main problems with allografts or xenografts. One way to circumvent some of these problems is the isolation of grafted cells from the host immune response. Cells can be encapsulated in semipermeable polymers allowing the flow of nutrients like glucose and oxygen into the capsule and the flow of therapeutic substances secreted by the grafted cells from the capsule into the surrounding host parenchyma (AEBISCHER et al. 1994, 1996a, 1991; EMERICH et al. 1994, 1996, 1997). Alternatively, autologous cells can be isolated, propagated, and genetically engineered in vitro before grafting them back into the same host. Thereby, graft rejection can be prevented and survival over long time periods postgrafting can be achieved (TUSZYNSKI et al. 1994b).

Several primary neural and nonneural cell types have been previously used as vehicles for gene transfer into the CNS: astrocytes (LIN et al. 1997; LUNDBERG et al. 1996; YOSHIMOTO et al. 1995), Schwann cells (MENEI et al. 1998; TUSZYNSKI et al. 1998), fibroblasts parentheses (FISHER et al. 1991; KAWAJA et al. 1992; ROSENBERG et al. 1988; TUSZYNSKI et al. 1994a, 1996a; WINKLER et al. 1995), myoblasts (DÉGLON et al. 1996), neural progenitor cells (MARTINEZ-SERRANO and BJÖRKLUND 1996; MARTINEZ-SERRANO et al. 1995) and others.

Fibroblasts are probably one of the most commonly used autologous cell types for ex vivo gene transfer. These cells can be easily obtained from a skin biopsy and cultured under standard cell culture conditions. Fibroblasts survive for long times in vivo without invasive growth or tumorigenesis. On the other hand, nonneural cells do not distribute in the host parenchyma after grafting, but rather stay directly at the injection site as an isolated cell mass.

Schwann cells have also been used as targets for ex vivo gene therapy and grafting to the CNS. Schwann cells naturally possess properties that support regeneration, including the secretion of trophic molecules, permissive extracellular matrix substances, and the ability to remyelinate injured axons. They also appear to survive for extended time periods after grafting to the CNS (TUSZYNSKI et al. 1998). Because of these properties, they are being more thoroughly investigated as potential vehicles of ex vivo gene therapy in models of CNS degeneration, particularly models characterized by extensive axonal degeneration.

The main advantage of neural progenitor cell lines as candidates for ex vivo gene therapy lies in their ability to migrate, their integration into the CNS, and their potential to repopulate injured regions of the CNS. The ability of neural progenitor cells to migrate from the site of injection permits the potential distribution of genetically introduced therapeutic molecules over widely dispersed regions of the diseased brain. This property would be primarily of benefit in neural repair strategies that aim to correct inborn errors of metabolism, such as phenylketonuria. Additionally, neural progenitor cells have the potential to differentiate into neuronal and glial cell types, thereby enabling the replacement of lost neurons; it may even be possible to direct their differentiation patterns toward neurons or glia (reviewed in FISHER 1997; MARTINEZ-SERRANO and SNYDER 1999; SNYDER 1994). However, neural stem cells are more difficult to obtain and to culture in vitro than cells such as fibroblasts; in addition, they are often derived from allogeneic hosts, resulting in immunological rejection. Further developments in this relatively young field will continue to reveal its feasibility for neural repair.

In summary, an ideal cell for ex vivo gene transfer into the CNS would be easy to obtain and culture in vitro, integrate well into the CNS parenchyma, would be neither tumorigenic nor induce any immune responses and survive for long time period after transplantation. Such a cell type is purely hypothetical and depending on target and disease state some compromise is necessary.

II. Methods of Ex Vivo Gene Transfer

The modification of cells in vitro can be achieved by a number of methods. The major goal of genetic modification is the generation of cells that have high and possibly regulatable expression of the gene to be transferred. Sustained, long-term gene expression after in vivo grafting of these cells is clearly important for the treatment of slowly progressing neurodegenerative diseases.

A simple way to introduce genetic material into cells is the direct delivery of a gene therapy "vector" containing the gene of interest into a cell that is the target for genetic modification (i.e., "transfection of plasmid DNA"). The major disadvantages of this approach is that the injected genetic material is expressed by the cell with low efficiency, especially in primary and neural cells, and gene expression is lost over short time periods in the vast majority of cells. A more effective and frequently used method is to introduce genes into cells using recombinant replication-defective retroviruses, mostly based on Moloney murine leukemia virus (Mo-MLV) (MILLER et al. 1986, 1993). These retroviral vectors integrate into the genome of the host cell and the new gene ("transgene") is permanently anchored into the host genetic material. This is clearly an advantage in terms of transgene stability and long-term expression, but also contains risks. Due to the random integration of retroviruses into the host genome, insertional mutagenesis can result in activation of oncogenes or inactivation of tumor suppressor genes. In addition, with this approach target cells must be proliferating in order to efficiently integrate the retrovirally based transgene. Nonetheless, retroviral vectors are now the most widely used ex vivo gene therapy vectors, primarily based upon their ability to sustain long-term transgene expression.

Retroviral vectors are RNA viruses that are constructed using producer cell lines that contain a partially deleted retroviral genome with all genes intact that are necessary to package and replicate retroviral RNAs (the *gag-pol-env* genes). However, this genome cannot be packaged into infectious viral particles, since the so called psi-sequence necessary for packaging, and additional elements, are missing. Production of wild-type, replication-competent retroviruses as a result of natural recombination is prevented by several methods: (1) Deletion of the 5' and 3' long terminal repeats (LTRs), replacing them with other promoters and polyadenylation sequences, respectively; and (2) Separating the retroviral genome into different transcription units. Therefore, multiple recombination events would be necessary to reassemble all elements into a functional wild-type retrovirus, an event that is very highly unlikely to occur. The most useful packaging cell lines are amphotropic producer cells that can infect a wide range of mammalian cells (for a review of available packaging cell lines see MILLER 1990).

To construct a recombinant retrovirus, genes are introduced into a retroviral transfer vector. An example of a prototypical retroviral vector is shown in Fig. 1. Only three parts of a wild-type retrovirus are maintained in this configuration: the 5' LTR, which can also serve as a strong promoter/enhancer, the 3' LTR, containing a polyadenylation sequence, and the psi-sequence, necessary for packaging of the retroviral RNA in viral coats. After introducing the transfer vector into a packaging cell line, structural genes are provided in trans by the packaging cell line mentioned above, and infectious retroviral particles are generated. By including a dominant selectable marker for drug resistance into the viral vector, e.g., a neomycin-phosphotransferase gene, untransfected cells can be eliminated in vitro, providing a pool of cells that all express the gene of interest. Cells are usually characterized in vitro for trans-

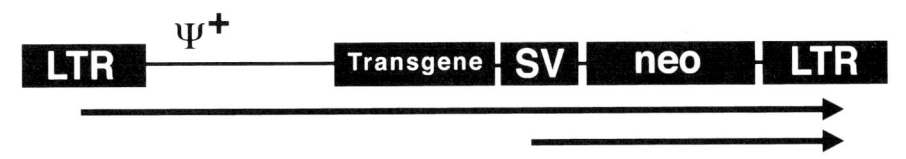

Fig. 1. Prototype of a retroviral vector. A recombinant retrovirus consists of the 5'-long terminal repeat (*5'LTR*) serving as a promoter/enhancer to drive the expression of a gene of interest (transgene). An internal promoter, such as SV40 (*SV*), is necessary to express the neomycin resistance gene (*neo*), allowing selection of transfected cells in vitro. The polyadenylation sequence located in the *3'LTR* is used by both transcripts. The Ψ-2 sequence is needed for the packaging of the retroviral RNA into viral particles. The two transcripts from the 5' and the internal SV40 promotor are indicated by *arrows*

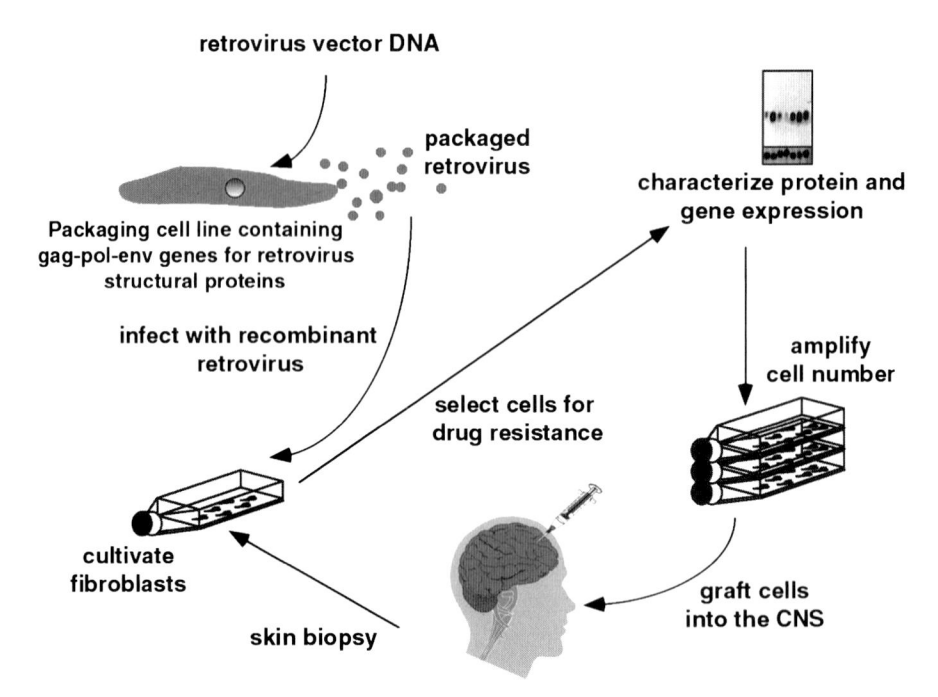

Fig. 2. Schematic outline of ex vivo gene therapy. The retroviral vector (outlined in Fig. 1) is transfected into a producer cell line, which provides the genes necessary for packaging the recombinant retrovirus. Autologous fibroblasts obtained from a skin biopsy are cultivated in vitro and infected with the recombinant retrovirus. After selection for drug resistance, cells are characterized for gene expression, expanded, and grafted back into the CNS of the host

gene expression, and to ensure that wild-type virus is not present, before they are grafted into the CNS (Fig. 2).

The maximum size of a retroviral vector that can be efficiently packaged in the MLV vector is approximately 8 kb; however, the inclusion of internal promoters and selectable markers significantly reduces the available space for inserting foreign genes to about 3–4 kb. Inserts larger than 3–4 kb in the MLV vector result in low retroviral titers and a diminished capacity to transfect target cells.

The choice of promoters for use in ex vivo gene therapy vectors largely depends on the targeted cell type. Several strategies have been employed to drive gene expression at high and sustained levels: genes can be linked to the 5′LTR to provide gene expression, internal promoters can be included, and alternative splicing or internal ribosome entry sites can be used to express multiple proteins from one or more transcripts. In fibroblasts, the 5′LTR has been shown to give high gene expression in vitro and sustained, long-term expression in vivo (GRILL et al. 1997a; TUSZYNSKI et al. 1996b). The choice of promoters will clearly influence the amount of gene product and the duration of in vivo gene expression, as some promoters will be shut off in vivo earlier than others.

III. Transplantation of Genetically Modified Cells to In Vivo Models of CNS Disease

As cellular grafting constitutes an invasive procedure that can disrupt normal circuitry, several issues need to be considered in performing in vivo studies with grafts of genetically modified cells: the cell number to be grafted, the location, the number of graft sites, the cell type, protection from the immune system if nonautologous cells are used, and survival of cells in vivo. Some of these points have already been discussed above (see Sect. C.I.).

It is known that intraparenchymally grafted cells are exposed to unspecific immune reactions due to the disruption of the normal host parenchyma. These host reactions together with neovascularization and glial responses may compromise the survival of grafted cells. It is therefore of importance to determine how many cells can be and need to be grafted to minimize damage to the host tissue while still providing sufficient amounts of transgene product. The number of grafted cells, in addition to the promoter and the number of gene copies, will influence the amount of gene product that is delivered. Further, the location of the grafted cells has an important influence on the availability of the secreted products. Especially in the primate brain where the target can be spread over considerable distances, several grafting sites might be needed. Studies discussed in detail below (Sect. D.) have used stereotaxic graft placement and magnetic resonance imaging to accurately target exact locations for cellular grafts in the primate brain (BANKIEWICZ et al. 1997; TUSZYNSKI et al. 1996a). MRI can also be used to identify disruptions of the BBB and changes in the grafts over time. For clinical applications, the reversibility of a grafting procedure might be an important issue. This could be accomplished by introduction of suicide genes into the grafted cells as a means of destroying the graft in the event that adverse effects occurred. Current efforts are focused on developing these strategies.

Depending on the specific application, cells can either be directly injected into the CNS as cell suspensions, embedded into extracellular matrix prior to grafting, or encapsulated prior to transplantation. As summarized in several examples below, cell type and specific disease application will often dictate some of the above issues.

D. Experimental Therapies in Animal Models of Neurodegenerative Diseases

Two major strategies have evolved in recent years for the treatment of neurodegenerative diseases using ex vivo gene therapy. The first strategy directly targets the group of neurons undergoing degeneration to slow their degener-

Table 1. Neurotrophic factors used in ex vivo gene therapy animal models of neurodegenerative diseases

Neurodegenerative diseases	Neurotrophic factors
Alzheimer's disease (AD)	NGF, BDNF, FGF-2
Amyotrophic lateral sclerosis (ALS)	CNTF, BDNF, GDNF
Huntington's disease (HD)	NGF, CNTF
Parkinson's disease (PD)	BDNF, GDNF

NGF, nerve growth factor; BDNF, brain-derived neurotrophic factor; GDNF, glial cell line-derived neurotrophic factor; CNTF, ciliary neurotrophic factor; FGF-2, fibroblast growth factor-2 (or basic FGF).

ation (i.e., a *neuroprotective* strategy) and, concurrently, to augment their function. Neurotrophic factors constitute the majority of molecules involved in these studies. Table 1 summarizes some of the neurotrophic factors that have been used in rodent and primate models of neurodegenerative diseases using ex vivo gene therapy.

Neurotrophic factors act through specific receptors with restricted expression in the CNS. However, no receptor is expressed in only a single neuronal population. Systemic deliveries of growth factors have therefore resulted in significant numbers of side effects. Indeed, clinical trials of neurotrophic factors to date have been disappointing, either because trophic factors fail to reach degenerating neurons of the CNS in sufficiently high concentrations after peripheral administration, or because central infusions elicit deleterious growth responses from nontargeted neuronal systems. This has been extensively documented for nerve growth factor (NGF) in the treatment of Alzheimer's Disease. Gene therapy offers a means of potentially circumventing problems associated with growth factor delivery to the CNS.

A second gene therapy strategy in neurodegenerative disorders offers a means of *compensating* for neuronal loss, and the concomitant loss of neurotransmitter production. Cells modified to produce certain neurotransmitters can be grafted into the target of lost neuronal connections to biochemically restore neurotransmitter function. Models of Parkinson's disease and Alzheimer's disease are the main targets of this approach.

I. Alzheimer's Disease

Cellular and biochemical changes in the brain of Alzheimer's disease (AD) patients have been extensively described, but the exact etiology is still unresolved. Pathological hallmarks of AD include the formation of neurofibrillary tangles, amyloid deposition, and loss of neurons and synapses (PRICE 1986; SELKOE 1996; TERRY and KATZMAN 1983; YANKNER 1996). These changes are not restricted to a single system, but are rather widespread, with some areas of the brain more extensively involved than others. One of the systems that is severely affected during AD progression is the cholinergic system of the basal forebrain [medial septum (MS) and vertical limb of the diagonal band (VDB)], and the nucleus basalis of Meynert (NBM). Loss of cholinergic function is closely correlated with the decline of cognitive function, extent of clinical dementia, and pathological severity (BARTUS et al. 1982; COYLE et al. 1983; MASLIAH et al. 1991; PERRY et al. 1978; THAL 1994). Currently, the only partially effective treatment for AD targets the cholinergic system (using anticholinesterase compounds). For these reasons, a substantial part of research has focused on augmenting cholinergic function either by directly modulating cholinergic neurons in the basal forebrain and NBM, or by ameliorating acetylcholine levels in their target area.

1. Nerve Growth Factor

In the early 1980s it was found that nerve growth factor (NGF), a member of the neurotrophin family, is expressed in the target areas of cholinergic basal forebrain neurons (CBF) (CRUTCHER and COLLINS 1982; KORSCHING et al. 1985; SHELTON and REICHARDT 1986). Further investigations revealed that these neurons possess receptors for NGF (SOBREVIELA et al. 1994) and that NGF is retrogradely transported from the target area to cholinergic cell somata (SCHWAB et al. 1979; SEILER and SCHWAB 1984). These discoveries suggested that NGF might be a molecule that can modulate cholinergic function, a hypothesis that has now been convincingly demonstrated.

In a set of experiments it was shown that intracerebroventricular (ICV) infusions of NGF can entirely prevent the loss of the cholinergic phenotype in basal forebrain cholinergic (BFC) neurons after fimbria–fornix transection, a model that axotomizes the projection of cholinergic and other neurons to the hippocampus. (GAGE et al. 1988; HEFTI 1986; KROMER 1987). Although responses to fimbria–fornix transections are a result of injury, compared to the slowly progressing neuronal degeneration seen in AD, these results raised the possibility that NGF might ameliorate cholinergic dysfunction in AD. Further support for this hypothesis came from subsequent studies in aged rodents which showed that NGF can reverse age-related declines in cholinergic neuronal morphology and prevent age-associated deficits in memory function (FISCHER et al. 1991; FISCHER et al. 1987). These beneficial effects of NGF are also generally observed in the nucleus basalis of Meynert and vertical limb of the diagonal band.

Experiments conducted in nonhuman primates led to similar results: intracerebroventricular infusions of NGF, as well as polymer-encapsulated NGF-producing cells, efficiently prevented neuronal degeneration in BFC neurons. (EMERICH et al. 1994; KOLIATSOS et al. 1990; KORDOWER et al. 1994; TUSZYNSKI et al. 1990, 1991).

If indeed NGF is so potent a molecule for preventing neuronal degeneration in the rodent and primate brain, why has it not yet been rigorously tested in clinical trials? Indeed, three patients in Sweden have received ICV infusions of murine NGF, but treatments were discontinued due to development of a pain syndrome (OLSON et al. 1992). Adverse effects of ICV infusions of NGF have also been documented in animal studies (LEVI-MONTALCINI 1987; SAFFRAN et al. 1989; WILLIAMS 1991; WINKLER et al. 1997). Side effects resulting from "flooding" of the brain with NGF include weight loss, proliferation and migration of Schwann cells surrounding the brain stem and spinal cord, and the sprouting of sympathetic and sensory axons. These studies led to the conclusion that neurotrophic factors need to be delivered in a well-targeted, spatially restricted and intraparenchymal manner in the CNS to ensure that they reach only the target neurons of the basal forebrain.

As mentioned above, grafting of genetically modified cells is one way of accomplishing intraparenchymal, long-term, targeted delivery of molecules into the CNS. To test if NGF gene therapy can prevent the degeneration of basal forebrain cholinergic neurons, fibroblasts genetically modified to express and secrete NGF were grafted to the medial septum of adult rats after a fimbria–fornix transection. Cellularly delivered NGF was as efficient in preventing BFC degeneration as previously shown with ICV NGF infusions (KAWAJA et al. 1992; ROSENBERG et al. 1988), and could also ameliorate behavioral deficits in aged memory-impaired rats (CHEN and GAGE 1995). These results were later replicated using NGF-secreting neural progenitor cells in the fimbria–fornix transection model (MARTINEZ-SERRANO et al. 1995) and in aged rodents (MARTINEZ-SERRANO et al. 1995, 1996).

As a step toward clinical applications, similar studies needed to be conducted in nonhuman primates. Thus, human and monkey fibroblasts were infected with an NGF-expressing retrovirus (TUSZYNSKI et al. 1994b). Cells showed continuous NGF expression in vitro over successive days and increasing passage number. To test survival and biological activity of NGF-secreting fibroblasts in vivo, monkeys received grafts of autologous or allogeneic NGF-producing fibroblasts into several regions of the primate brain (TUSZYNSKI et al. 1994b). Cells showed survival for at least 6 months in vivo. Cholinergic axons from the host brain penetrated NGF-secreting grafts for at least 6 months postgrafting, but showed little penetration in control grafts lacking the NGF gene. As expected, the survival of autografts was superior compared to allografts; the latter were infiltrated by inflammatory cells. In all primates that have been grafted to date (totaling over 100 grafts), tumor formation has never been observed. As a further, logical step towards human application, it had to be determined whether intraparenchymal transplants of primary autologous cells genetically modified to produce NGF would prevent injury-induced degeneration of cholinergic neurons. Adult monkeys underwent unilateral fimbria-fornix transections and received grafts of either NGF-producing or control fibroblasts into the septal BFC nucleus (TUSZYNSKI et al. 1996a). One month later NGF-grafted animals showed a significantly higher number of p75- and ChAT-labeled neurons on the lesioned side compared to control-grafted animals (Fig. 3). Rescue of the cholinergic phenotype correlated with the size and location of the graft. In one animal with the most accu-

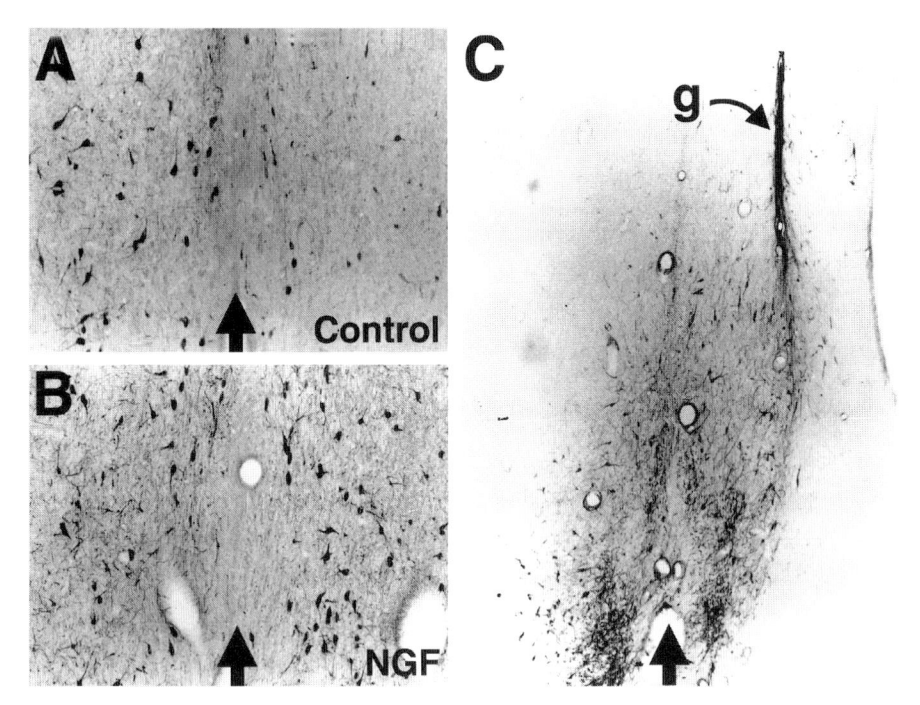

Fig. 3A–C. NGF-producing fibroblasts prevent the degeneration of basal forebrain cholinergic neurons after unilateral fimbria–fornix lesions in adult primates. **A** NGF-producing grafts rescue the degenerating cholinergic neurons on the right (lesioned) side of the brain, compared to (**B**) control-grafted animals that show a loss of ChAT-labeled neuronal profiles after lesions. **C** Low magnification of the medial septal region of an NGF-grafted animal after fimbria–fornix lesion demonstrates that cholinergic neurons (labeled for the low-affinity nerve growth factor receptor, p75) extend axons toward the graft (*g*). These axons densely penetrate the graft. *Straight arrows* in **A–C** indicate the midline; *g*, graft

rately located graft, 92% of cholinergic degeneration was prevented using NGF amounts that were at least 40 times lower than previous experiments that had used NGF infusions (Tuszynski et al. 1991).

To determine if cellularly delivered NGF can also prevent morphological declines of cholinergic systems in the aged primate, labeling for the low affinity p75 receptor was compared between normal young rhesus monkeys and aged rhesus monkeys, and between aged monkeys that received grafts of either NGF or control fibroblasts (Smith et al. 1998). Using unbiased stereology, it was determined that normal aged monkeys have a 44% reduction in the number of p75 labeled neurons in the NBM compared to a young control group. Cellular delivery of NGF was able to completely reverse these changes, whereas control grafted animals showed no difference compared to unoperated aged controls. These results are the first indication that age-related loss of neurotrophin receptor expression occurs in the primate brain, and that

NGF gene transfer is able to ameliorate changes in the cholinergic system resulting from age-related degeneration. Influences of NGF on functional declines in the aged nonhuman primate are currently under investigation, potentially establishing the rationale for clinical trials of NGF gene therapy in AD.

2. Biochemical Modulation of Acetylcholine Levels

A second experimental approach for the treatment of AD is based on a strategy of increasing acetylcholine levels by ex vivo gene therapy. Fibroblasts were modified to express the acetylcholine synthesis-promoting enzyme choline acetyl-transferase (ChAT). Perhaps somewhat surprisingly, ChAT-expressing fibroblasts can produce and release acetylcholine in vitro and in vivo, and the levels of acetylcholine produced can be regulated by the amount of choline available (FISHER et al. 1993). When these cells were implanted into the frontal and parietal cortex of rats after lesioning the nucleus basalis of Meynert (NBM), increased neocortical acetylcholine levels were able to ameliorate learning deficits in a behavioral test (WINKLER et al. 1995). Previous experiments using fetal cholinergic transplantation into the neocortex had suggested that reconstruction of cholinergic circuits in rats with NBM lesions can restore memory function (DUNNETT et al. 1985; FINE et al. 1985; HODGES et al. 1991). Fetal grafts appear to reinnervate the host cortex; however, acetylcholine provided by genetically modified fibroblasts has effects similar to neuronal grafts on behavioral function, suggesting that diffuse cholinergic activation is sufficient to induce behavioral recovery after lesions of the NBM, and that regulated synaptic release of acetylcholine may not be necessary to mediate functional recovery. Thus, genetically modified cells can provide a local supply of active neurotransmitters to promote functional recovery and repair. Currently, research is directed toward regulating the ChAT activity in these cells using tetracycline regulatable gene expression systems (PIZZO et al. 1998).

In summary, ex vivo gene therapy approaches for AD have aimed at the augmentation of cholinergic function. Both strategies described above have shown that morphological recovery is achievable in primates and that both morphological and functional recovery are achievable in rodents. It remains to be determined if the aged human brain, and particularly the brain affected by AD, has the same responsiveness to NGF as the CNS of rodents and nonhuman primates, a question that can only be answered by clinical trials.

II. Parkinson's Disease

Parkinson's disease is characterized by the progressive loss of dopaminergic neurons in the substantia nigra (SN) and concurrent diminishment of dopamine levels in nigral target areas including the striatum. Oral administration of L-DOPA, a precursor of dopamine, is still the most widely used therapy. This symptomatic approach has several drawbacks, however: (1) L-DOPA delivery is not site-specific, resulting in unwanted side effects, (2) due to fluctuation in drug levels in the CNS, "on–off" effects occur throughout the day, and (3) dopaminergic cells continue to degenerate. Two different therapeutic approaches using ex vivo gene therapy in experimental models of PD have been taken to counteract or compensate for the loss of midbrain dopaminergic neurons: (1) grafting of cells genetically modified to produce survival factors for dopaminergic neurons and (2) grafting of cells genetically modified to produce L-DOPA. Due to the focal degeneration and cellular dysfunction of a specific, localized population of neurons in PD, this disease is one of the most attractive targets for gene therapy.

1. Neuroprotective Strategies

A large number of neurotrophic factors has been shown to promote survival of dopaminergic neurons in vitro, including the neurotrophins BDNF (brain-derived neurotrophic factor) (HYMAN et al. 1991), NT-3 (neurotrophin-3), NT-4/5 (neurotrophin-4/5) (HYMAN et al. 1994), and members of the TGF-β superfamily including GDNF (glia cell line-derived neurotrophic factor) (LIN et al. 1993), PSP (persephin) (MILBRANDT et al. 1998), NTN (neurturin) (HORGER et al. 1998), and TGF-β (transforming growth factor-β) (KRIEGLSTEIN et al. 1995). Only two of these factors that have been used in ex vivo gene transfer experiments will be discussed here in detail: BDNF and GDNF.

Although initial studies using intraparenchymal infusions of BDNF failed to show a protective effect on dopaminergic neurons after axotomy-induced degeneration (KNÜSEL et al. 1992; LAPCHAK et al. 1993), subsequent experiments using modified fibroblasts showed promising effects. Grafts of immortalized fibroblast cell lines genetically modified to produce BDNF were able to protect dopaminergic neurons from MPTP-induced neurotoxicity (FRIM et al. 1994; GALPERN et al. 1996). In these studies, cells were implanted dorsal to the substantia nigra 1 week prior to intrastriatal infusions of MPP$^+$. Similarly, intrastriatal grafts of primary BDNF-secreting fibroblasts have been shown to rescue dopaminergic neurons after intrastriatal 6-hydroxydopamine (6-OHDA) lesions (LEVIVIER et al. 1995).

The more recently cloned member of the TGF-β superfamily, GDNF, seems to be one of the most potent trophic molecules for dopaminergic neurons (LIN et al. 1993). Injections of GDNF are able to protect dopaminergic neurons in the substantia nigra (SN) from axotomy and from MPTP-induced lesions in rodents (BECK et al. 1995; TOMAC et al. 1995) and in nonhuman primates (GASH et al. 1996). To date, only two studies have investigated the effects of GDNF delivered by ex vivo gene therapy in models of Parkinson's disease (EMERICH et al. 1996; TSENG et al. 1997). GDNF provided by genetically modified encapsulated cells is able to rescue dopaminergic neurons and to ameliorate amphetamine-induced rotational abnormalities after medial forebrain bundle axotomy (TSENG et al. 1997). As the improvement in rotational behavior was not accompanied by dopaminergic striatal reinnervation, the mechanism leading to the behavioral recovery is unclear and might involve neuronal plasticity in the substantia nigra.

Promising results in models of PD using GDNF delivery by recombinant adenoviral (BILANG-BLEUEL et al. 1997; CHOI-LUNDBERG et al. 1997) and adeno-associated viral vectors (DURING and LEONE 1997) further support the usefulness of GDNF gene therapy in the treatment of PD.

2. Biochemical Modulation of Dopamine Levels

Since the loss of striatal dopamine as a result of dopaminergic neuronal degeneration in the SN is thought to be the cause of PD, a second gene therapy approach has focused on elevating the levels of dopamine by genetically modified cells to express tyrosine hydroxylase (TH), the rate-limiting step in dopamine synthesis. Cells modified to express TH will release L-DOPA; this dopamine precursor then must be converted to dopamine by L-aromatic amino acid decarboxylase (L-AACD). The localized delivery of L-DOPA to the striatum could diminish side effects currently observed with systemic L-DOPA treatment.

Initial short term studies using tumorigenic cell lines expressing TH demonstrated that fibroblasts can survive transplantation into the striatum, express TH, and secrete dopamine in vivo (HORELLOU et al. 1990; WOLFF et al. 1989). Apomorphine-induced rotations after 6-OHDA lesions were significantly ameliorated. These results were replicated at time points of up to 10 weeks postgrafting using primary fibroblasts expressing TH (FISHER et al. 1991). Additionally, L-DOPA-producing astrocytes have been shown to ameliorate apomorphine-induced rotational asymmetry in a rat model of PD (LUNDBERG et al. 1996). More recent studies reported in vivo and ex vivo gene therapy approaches for the delivery of TH to the CNS of nonhuman primates (BANKIEWICZ et al. 1997).

III. Huntington's Disease

Huntington's disease (HD) is an autosomal dominant disorder, leading to progressive motor, cognitive and psychiatric disturbances. The genetic defect causing HD is located at the IT-15 Huntington gene, and is a result of an expansion of CAG trinucleotide repeats (THE HUNTINGTON'S DISEASE COLLABORATIVE RESEARCH GROUP 1993). Pathologically, the most severely affected cell types in HD are the medium spiny neurons in the striatum. These neurons receive afferent input from the cortex and the substantia nigra, and extend fibers into the globus pallidus and substantia nigra pars reticulata, thus comprising an important part of the basal ganglia.

Although the genetic abnormality underlying Huntington's disease is now known, the exact mechanism leading to neuronal cell death remains unknown.

To date, a cure has not yet been defined. Four general mechanisms of cell death have been implicated in HD: excitotoxicity, metabolic stress, free radical formation and apoptosis (SHARP and ROSS 1996).

Even before the genetic abnormality in HD was identified, different models for the replication of neuronal death observed in HD were developed using glutamate receptor agonists (reviewed in BEAL 1994; DUNNETT and SVENDSEN 1993). Initially, excitotoxic lesions of the striatum with kainic acid were widely used as a model system to replicate changes seen in HD (COYLE and SCHWARCZ 1976; MCGEER and MCGEER 1976). More recently it was shown that quinolinic acid lesions more accurately reflect the pathological changes seen in HD because of the relative sparing of somatostatin and neuropeptide Y levels, while depleting GABA and substance P levels (BEAL et al. 1986, 1991). Other lesion models use the mitochondrial toxin 3-nitropropionic acid, reflecting the impaired energy metabolism that may precede or be responsible for the excitotoxic damage in HD.

Based on the fact that neurotrophic factors have been proposed to influence energy metabolism and neuronal survival, and might protect neurons against excitotoxicity, a number of trophic molecules were tested for their ability to protect striatal neurons from excitotoxic lesions in animal models. The following neurotrophic factors were tested: NGF (ANDERSON et al. 1996; DAVIES and BEARDSALL 1992; EMERICH et al. 1994; FRIM et al. 1993, 1993a; KORDOWER et al. 1996, 1997; MARTINEZ-SERRANO and BJÖRKLUND 1996; VENERO et al. 1994), BDNF (FRIM et al. 1993; MARTINEZ-SERRANO and BJÖRKLUND 1996; VOLPE et al. 1998), NT-3 (ANDERSON et al. 1996), NT-4/5 (ALEXI et al. 1997), GDNF (ARAUJO and HILT 1997; PÉREZ-NAVARRO et al. 1996), TGF-β (ALEXI et al. 1997), and the neuropoetic cytokine CNTF (ANDERSON et al. 1996; EMERICH et al. 1996; EMERICH et al. 1997). This strategy was based on the assumption that neurons can be protected before they are irreversibly lesioned and destined to die. Indeed, individuals with a high risk for HD can be identified by genetic testing prior to the onset of the disease. Thus, neuroprotective strategies could be started in an early phase of the disease and might slow the progression of HD.

The most intensively and successfully studied trophic molecules in rodent and primate models of HD are NGF and CNTF. Initial studies used intrastriatal injections or infusions of NGF prior to excitotoxic lesions in rats. The results demonstrated that the survival of cholinergic interneurons was augmented by NGF, but the overall size of the lesion was not reduced (ANDERSON et al. 1996; DAVIES and BEARDSALL 1992; VENERO et al. 1994). These data are consistent with the expression of the high affinity NGF receptor trkA in cholinergic striatal neurons, but not in other neurons of the striatum. The sparing of cholinergic neurons is, however, of relatively little importance to HD, as these neurons are only partially affected in the disease. Surprisingly, consequent studies were able to show that grafts of NGF-secreting tumorigenic fibroblasts can reduce the overall lesion size after QA injection by up to 80% in rats (FRIM et al. 1993, 1993a). Results from these experiments, which were restricted to short time points due to tumor formation by grafted cells, were recently confirmed at longer time points using either NGF-producing progenitor cells (KORDOWER et al. 1997; MARTINEZ-SERRANO and BJÖRKLUND 1996) or encapsulated NGF-producing cells (EMERICH et al. 1994; KORDOWER et al. 1996). These studies not only showed that the size of QA-induced striatal lesions was significantly reduced by NGF in Nissl-stained sections, but that medium-sized spiny neurons were also protected, and that astroglial and microglial responses are reduced. The neuronal protection was also reflected in changes in apomorphine-induced rotations. Rats grafted with encapsulated NGF-producing cells showed a significant decrease in the extent of rotational abnormality (EMERICH et al. 1994).

Thus, NGF seems to be capable of protecting certain cells from excitotoxic lesions, even if these cells do not possess receptors for NGF. The exact biological mechanism leading to this neuroprotective effect is unclear. It has been speculated that NGF may have a general "detoxifying" effect by promoting free radical scavenging, or might influence specific neuron–glia interactions to support neuronal survival. Support for the first hypothesis comes from studies with the mitochondrial neurotoxin 3-NPA. NGF-secreting fibroblasts were able to reduce the size of 3-NPA-induced lesions in the striatum (FRIM et al. 1993b). This might be in part due to the inhibition of peroxynitrite formation, a highly reactive oxidant that is a reaction product of nitric oxide and superoxide (GALPERN et al. 1996).

The second trophic factor implicated in neuroprotective strategies of HD is CNTF (ANDERSON et al. 1996; EMERICH et al. 1996, 1997). Intrastriatal infusions of CNTF were shown to protect striatal neurons in Nissl-stained sections lesioned by QA injections (ANDERSON et al. 1996). Furthermore, CNTF-producing encapsulated cells placed into the ventricles of rats reduced QA-induced lesion size, protected medium-sized spiny neurons, and reduced apomorphine-induced rotational abnormalities (EMERICH et al. 1996). Based on these studies, experiments were conducted in a primate model of HD (EMERICH et al. 1997). Monkeys received capsules of CNTF-producing cells or untransfected cells (as controls) into the striatum, followed by QA lesions. Four weeks later it was reported that CNTF prevented the loss of GABAergic, cholinergic, and diaphorase-positive neurons in the striatum (Fig. 4). Furthermore, innervation of striatal targets was protected and atrophy of layer V motor neurons that project to the striatum was reduced (EMERICH et al. 1996). Similar to the protective effects of NGF, it is unclear if the CNTF-induced protection is mediated through receptor-signaling mechanisms (EMERICH et al. 1996). As mentioned above, a pre-treatment of the striatum generally for several days with NGF or CNTF is necessary before injecting the excitotoxins to result in neuroprotective effects. This might also explain the observed differences between NGF infusions and cellularly delivered NGF. Further studies are needed to determine the exact biological mechanisms involved in this protection, and to investigate if NGF and CNTF act via common pathways.

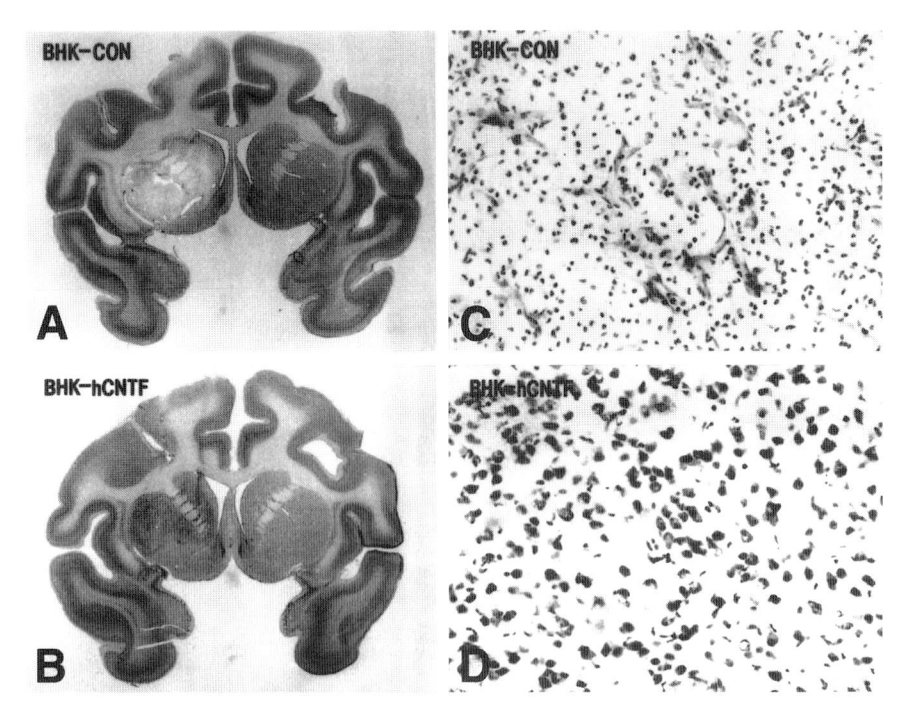

Fig. 4A–D. Low-power photomicrographs of Nissl-stained sections through the forebrain of monkeys receiving unilateral intracaudate and intraputamenal lesions with quinolic acid (180 nmoles per site) and encapsulated (**A**) BHK-Con (baby hamster kidney-control) or (**B**) BHK-CNTF-producing implants. Note in panel **A** the extensive lesion area on the left side of the brain (*lightly stained region*) which encompasses much of the striatum. In contrast, lesion size is markedly reduced in **B**, where monkeys received grafts of BHK cells genetically modified to secrete hCNTF. **C, D** Medium-power photomicrographs of Nissl-stained sections through the striatum of (**C**) BHK-con and (**D**) BHK-hCNTF-grafted monkeys. **C** A paucity of neurons is observed within the striatum of control-grafted monkeys. **D** In contrast, numerous healthy appearing neurons are observed close to the quinolinic acid injection site in BHK-hCNTF-implanted animals. (Figure provided by Dr. Jeffrey Kordower, Dept. of Neurosciences, Rush-Presbyterian Medical Center)

IV. Amyotrophic Lateral Sclerosis

Amyotrophic lateral sclerosis (ALS) is characterized by the progressive degeneration of upper motor neurons in the cerebral cortex and lower motor neurons in the brain stem and spinal cord. Two forms of ALS have been identified: the more common sporadic form, and the rare familial form of ALS. Only the recent discovery that a subset of familial ALS is caused by a gain of function mutation of the Cu/Zn superoxide dismutase (SOD-1) gene has given some insight in the etiology of the disease, and has resulted in an improved animal model of this devastating neurological disease. Previously, injury-induced motor neuron degeneration and hereditary murine mutants of ALS such as the progressive motor neuronopathy (*pmn*) mouse, *wobbler* mouse, and *wasted* mutant were the only available models for ALS, but their exact relationship to the pathogenesis of ALS remains to be determined. Other transgenic models overexpressing heavy or light chain neurofilaments have been shown to disrupt the morphological and functional integrity of motor neurons, and neurofilament mutations have been detected in some ALS patients. Besides the discovered gene mutations in familial ALS, the cause of ALS remains unknown and hypotheses for the underlying cause of ALS include glutamate excitotoxicity, autoimmune disorders, and a lack of neurotrophic support.

Neurotrophic factors have been shown to promote motor neuron survival in vitro and in vivo during development, adulthood, and after injury, thus they may be useful for the treatment of ALS. Based on in vitro data and on studies in different animal models (IKEDA et al. 1995; MITSUMOTO et al. 1994; SENDTNER et al. 1992), clinical trials using subcutaneous administration of the neurotrophic factor CNTF have been conducted [ALS CNTF TREATMENT STUDY GROUP 1996; THE ALS CNTF TREATMENT STUDY (ACTS) PHASE I-II STUDY GROUP 1995; MILLER et al. 1996]. Similar trials have been conducted with BDNF and IGF-1 (BORASIO et al. 1998; LANGE et al. 1996). To date, these clinical trials have essentially failed, an outcome, which is at least partially attributable to the mode of neurotrophic factor delivery and the instability of the administered molecules. As discussed in detail above, for trophic molecules to be effective, localized and sustained delivery is necessary.

To date, few studies have evaluated the delivery of neurotrophic factors in models of ALS by ex vivo gene therapy, either into the CNS or into muscle. One approach has used subcutaneous implants of CNTF-producing encapsulated cells in *pmn* mice. Similar to the above-mentioned studies, disease progression was delayed and motor function improved (SAGOT et al. 1995). A phase I safety and toxicity clinical study was therefore initiated in which CNTF-producing encapsulated cells were implanted intrathecally (AEBISCHER et al. 1996a,b). Results from this study indicated that the implants produced CNTF for at least 17 weeks in vivo, and adverse side effects that had been observed after peripheral systemic delivery did not occur (AEBISCHER et al. 1996a).

In the future it will be advisable to study other motor neuron trophic factors, such as GDNF, in models of ALS. Combinations of neurotrophic factors have been shown in some in vivo studies to be more efficient in promoting motor neuronal survival than each factor administered independently (MITSUMOTO et al. 1994). The potential of ex vivo gene therapy has not been fully explored in models of ALS, but with the availability of new models of ALS, such as the SOD mutant mice, the data on the possible impact of trophic factors in ALS treatment should be available in the near future.

E. Experimental Therapies in Animal Models of Spinal Cord Injury

In contrast to therapies of neurodegenerative diseases that aim to prevent neuronal degeneration, experimental treatments of spinal cord injury (SCI) are aimed primarily at augmenting axonal growth and regeneration, rather than simply preventing cell death. Indeed, the classical view that regeneration and functional recovery cannot occur in the adult mammalian spinal cord has recently been challenged by a number of studies (BLESCH et al. 1999; BREGMAN et al. 1993, CHENG 1996; BREGMAN et al. 1995; GRILL et al. 1997a; LI et al. 1997; Z'GRAGGEN et al. 1998).

The ability of CNS axons to *spontaneously* regenerate after injury is limited for several reasons: insufficient neurotrophic factor support, CNS myelin- and white matter-based growth inhibitors, unfavorable growth substrates, the upregulation of putatively inhibitory extracellular matrix molecules such as proteoglycans of the chondroitin-sulfate family (DAVIES et al. 1997), and deleterious inflammatory responses (KREUTZBERG 1996; RAPALINO et al. 1998; ZEEV-BRANN et al. 1998). To overcome some of these limitations, different strategies to promote axonal growth have been employed: transplantation of Schwann cells (LI and RAISMAN 1994; TUSZYNSKI et al. 1998; XU et al. 1994, 1995, 1997), ensheathing cells (LI et al. 1997; RAMON-CUETO and NIETO-SAMPEDRO 1994), fetal tissue (BREGMAN et al. 1993; MORI et al. 1997; STOKES and REIER 1992; TESSLER 1991; TESSLER et al. 1997), or peripheral nerves (CHENG et al. 1996; DAVID and AGUAYO 1981; RICHARDSON et al. 1980). In addition, some axonal growth-promoting strategies have aimed to neutralize myelin-associated inhibitors (BREGMAN et al. 1995; SCHNELL and SCHWAB 1990; SCHNELL and SCHWAB 1993; Z'GRAGGEN et al. 1998), or to provide trophic factors to injured CNS axons (BLESCH et al. 1999; GRILL et al. 1997a,b; SCHNELL et al. 1994; TUSZYNSKI et al. 1994a, 1996b, 1997a, 1998; YE and HOULE 1997).

One way of providing suitable axonal growth substrates to the injured spinal cord environment together with growth-promoting molecules such as neurotrophic factors is ex vivo gene therapy. Cells are genetically modified to produce neurotrophic factors and are grafted to the injured spinal cord, thus providing: (1) potential "bridges" for growing axons, (2) a suitable substrate for axonal regeneration that is free of myelin- or white-matter-associated inhibitors, and (3) augmented amounts of trophic molecules. Two cell types have been used to date for genetic engineering and grafting to the spinal cord: primary Schwann cells or fibroblasts; these have been genetically modified to produce either NGF, BDNF, NT-3, GDNF, or leukemia inhibitory factor (LIF).

In initial studies, it was shown that fibroblasts genetically modified to produce NGF and injected into the intact rat spinal cord induced growth of sensory and supraspinal noradrenergic axons into the cellular graft, whereas control grafts showed only sparse penetration by these axons (TUSZYNSKI et al. 1994a). In a subsequent study the same cells were used in a dorsal hemisection model of SCI. In this lesion model, the dorsal half of the rat spinal cord is transected, leaving the white and gray matter of the ventral half of the spinal cord intact. Growing axons are thereby provided a choice of potential axon growth milieus: they can either extend into the cellular graft, into the remaining host gray matter ventral to the lesion site, or into the remaining host white matter ventral to the lesion site. Cells embedded in vitro into a collagen matrix were grafted as gelatinous blocks into the dorsal hemisection lesion cavity, thereby locally providing NGF as a neurotrophic factor for injured axons, together with collagen and fibronectin naturally produced by fibroblasts as growth substrate. As in the intact spinal cord, sensory axons identified by immunolabeling for calcitonin gene related peptide, and coeruleospinal axons labeled for tyrosine hydroxylase (observed to originate from the host coeruleospinal projection), robustly penetrated NGF-secreting grafts; few axons mostly of sensory origin penetrated control-grafted animals (TUSZYNSKI et al. 1996b). Additionally, cholinergic axons were found in some NGF-producing grafts. The axonal growth persisted for at least one year, the longest time point examined, and Schwann cells migrated into the lesion site. These Schwann cells partially remyelinated and ensheathed the growing axons. This extensive growth of spinal sensory and supraspinal coeruleospinal axons was also elicited in chronic stages after spinal cord injury by ex vivo NGF gene delivery, recapitulating previous results from acute lesion models (GRILL et al. 1997b; TUSZYNSKI et al. 1996b). Thus, neurotrophins produced by genetically modified fibroblasts can augment the growth of specific subpopulations of axons responsive to neurotrophic factors, resulting in long-lasting, persistent morphological changes.

Another growth factor with influences on axonal growth and spinal cord development is neurotrophin-3 (NT-3). To determine if continuous delivery of NT-3 by genetically modified fibroblasts was able to augment axonal growth and ameliorate functional recovery, rats underwent dorsal hemisection lesions, as described above, and received either NT-3, NGF, or control β-galactosidase-producing fibroblast grafts (GRILL et al. 1997a). Four weeks and 3 months later the animals were trained to walk across a grid, and the number of failures to grasp a rung were measured. This grid task was used as a test for sensorimotor integration. Animals with NT-3-secreting grafts showed significantly fewer footfalls than control-grafted animals at both time points tested. This recovery was partial, as NT-3 grafted animals still exhibited significantly greater numbers of footfalls than unlesioned animals. No differences were found between animals that received NGF- and control β-galactosidase producing cells. Histological analysis revealed significant increases in corticospinal axon growth up to 8mm distal to the injury site in NT-3 grafted animals compared to control-grafted animals (GRILL et al. 1997a). Interestingly, axons did not extend through the graft, but rather used the remaining host gray matter as substrate. It remains to be determined if the substrate of the graft is not suited for corticospinal axon growth, or if this response is due to an inhibitory influence at the host/graft interface or to a preference for gray matter as a growth substrate.

In a recent study, fibroblasts genetically modified to secrete the cytokine growth factor leukemia inhibitory factor (LIF) were also found to enhance the growth of corticospinal axons after spinal cord injury (BLESCH et al. 1999). In this experiment, LIF-secreting fibroblast implants placed in the lesioned mid-thoracic spinal cord significantly enhanced the growth of corticospinal axons 2 weeks after injury when compared to lesioned animals with control grafts expressing the reporter gene green fluorescent protein (GFP). To investigate whether these effects of LIF-secreting cells on axon growth were directly or indirectly mediated, expression levels of other growth factors at the lesion site were compared in LIF-grafted or GFP-grafted animals. Of note, LIF-grafted animals showed a significant increase in NT-3 expression in the vicinity of the lesion, but no significant change in the expression of other neurotrophins including NGF, BDNF, GDNF and CNTF. These findings establish that LIF-secreting cell grafts can influence CNS responses to injury, augment levels of other growth factors, and directly or indirectly enhance corticospinal axon growth.

Initial studies on regenerative responses to GDNF after spinal cord injury have also been reported (BLESCH et al. 1998). GDNF-secreting fibroblasts grafted to the hemisected spinal cord are densely penetrated by motor and sensory axons up to 3 months postgrafting, whereas few sensory and motor axons are found in control grafts (Fig. 5). Responses of supraspinal axons are currently under investigation in this model.

Cell types other than fibroblasts are also attractive candidates for gene therapy approaches to promote axonal growth in the injured CNS. For example, Schwann cells possess several natural properties that promote axonal regeneration in the peripheral nervous system, including: (1) natural expression of trophic molecules, (2) an ability to support and guide axonal growth by secreting extracellular matrix and cell adhesion molecules, and (3) the ability to remyelinate axons. When grafted to the CNS, Schwann cells have been reported to remyelinate central axons. Schwann cells can be readily obtained from peripheral nerve biopsies and cultured in vitro. The introduction of growth factor expressing genes into Schwann cells could potentially enhance their regeneration-supporting abilities in the injured adult CNS. To test this hypothesis, primary Schwann cells were genetically modified to express enhanced amounts of NGF and were grafted as cell suspensions to the unlesioned spinal cord. Compared to control nonmodified Schwann cell grafts, NGF-transduced Schwann cells induced robust growth of sensory and tyrosine-hydroxylase labeled coeruleospinal axons, similar to NGF-producing fibroblasts, and remyelinated host axons penetrating the grafts (TUSZYNSKI et al. 1998). In addition, after grafting to the lesioned spinal cord, genetically modified NGF-secreting Schwann cells spontaneously oriented into linear arrays in vivo and guided growing CNS axons (WEIDNER and TUSZYNSKI, unpublished observations). In another study, Schwann cells genetically

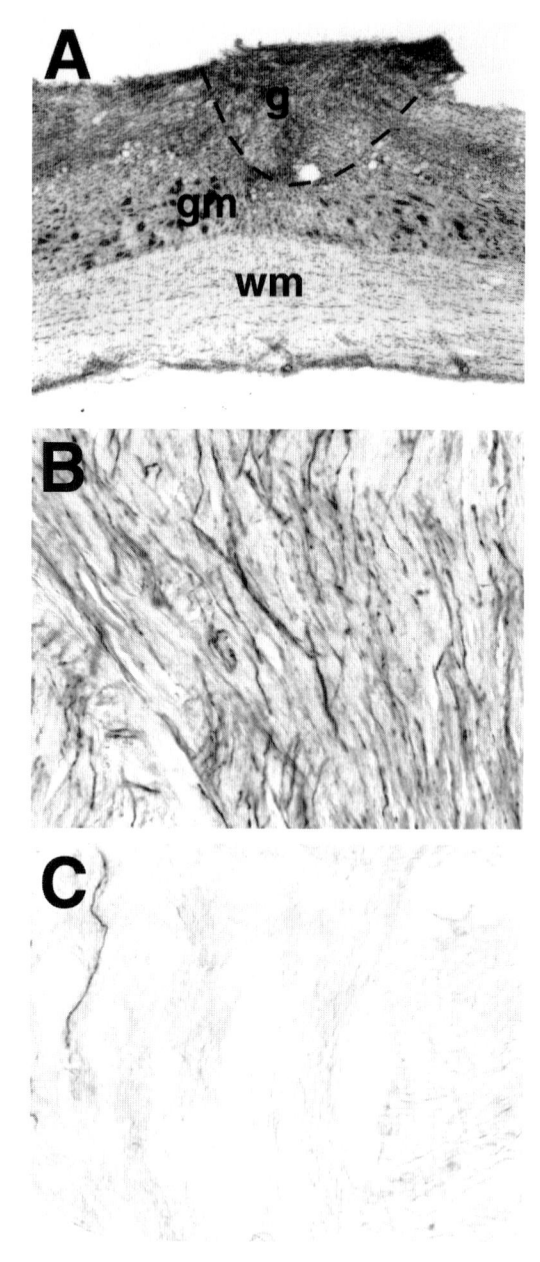

Fig. 5A–C. Genetically modified fibroblasts grafted to the dorsally hemisected spinal cord induce axonal growth. A Sagittal Nissl-stained section of a control-grafted animal 3 months postinjury. Grafted fibroblasts (g) fill the lesion cavity in the dorsal half of the spinal cord providing a substrate for axonal elongation. B Cholinergic motor axons densely penetrate GDNF-producing grafts whereas few cholinergic axons are found in (C) control grafts. g, graft; gm, gray matter; wm, white matter modified to secrete BDNF augmented growth of sensory axons and increase the number of retrogradely labeled brainstem neurons that had grown through a Schwann cell transplant to the injured spinal cord (MENEI et al. 1998). Results from these investigations indicate that transduced Schwann cells merit further study in CNS injury paradigms.

Grafts of genetically modified cells may also influence the nature of the host cellular responses after spinal cord injury. A recent study reported that grafts of NT-3- and BDNF-producing fibroblasts in the contused spinal cord contained significantly increased numbers of myelin basic protein-immunolabeled profiles, together with an increased number of oligodendrocyte precursor cells compared to control-grafted animals (MCTIGUE et al. 1998). Genetic delivery of neurotrophins might therefore not only enhance axonal growth, but induce myelination by expanding endogenous oligodendrocyte progenitor cell populations.

In summary, ex vivo neurotrophic factor delivery can substantially enhance axonal regeneration and sprouting after spinal cord injury, resulting in some cases in functional recovery. Future studies must address combinatory approaches to maximize axonal growth, integrating different successful strategies for therapeutic intervention.

F. Experimental Therapies of Brain Tumors

The prognosis for patients with malignant gliomas is extremely poor. The time from diagnosis to death is usually only several months, despite treatment with surgery, radiation therapy, and chemotherapy. Ex vivo gene therapy approaches to treat brain cancer focus on introducing cell death-inducing genes into rapidly dividing tumor cells.

Malignant cells undergo rapid cell division and are therefore unique in the normally quiescent CNS. This unique feature of tumor cells in the brain can be utilized to selectively target malignant cells with retroviruses that infect only dividing cells efficiently (see Sect. II.). Normal nondividing glia and neurons are not affected. Vectors bearing a suicide gene have been successfully used to treat brain tumors in experimental animals (BARBA et al. 1993; CULVER et al. 1992; EZZEDDINE et al. 1991; TAKAMIYA et al. 1993). The protocol used is a mixture of ex vivo and in vivo gene therapy: retrovirus-producing cell lines are first constructed in vitro and then transplanted into the brain to infect the tumor cells. The reason for grafting producer cells instead of injecting the retroviruses directly into the brain lies in the instability of retroviruses, and their low efficiency in infecting large numbers of cells in vivo when directly injected. The gene transduced by the retroviral vectors is the herpes simplex virus thymidine kinase gene (HSV-TK), which is capable of killing transduced cells after peripheral ganciclovir administration. HSV-TK phosphorylates ganciclovir, blocking DNA synthesis, resulting in cell death. In addition to the direct effect by ganciclovir, the so-called bystander effect also occurs in which toxic products produced by cells possessing the TK gene are passed to neighboring cells, killing them also. Systemic and local immune responses caused by the grafted retrovirus-producing mouse cells further contribute to the tumor cell death in this protocol.

The HSV-TK-mediated killing of brain tumor cells is currently the only phase III experimental clinical gene therapy protocol under study in the United States. Several hundred people are enrolled in the protocol and results are expected within the next few years.

G. Future Perspectives

Two important development issues for the clinical use of gene therapy in the future are: (1) exogenous control of gene expression and (2) the need to eliminate grafted cells if deleterious effects occur. Both of these issues can be addressed. For example, several systems for potentially regulating gene expression have been recently described: the method that has been most widely described utilizes a tetracycline-regulatable promoter that can either be turned "on" or "off" by the addition of tetracycline (GOSSEN and BUJARD 1992; GOSSEN et al. 1995). The incorporation of inducible suicide genes, or the inclusion of the previously mentioned HSV-TK gene, are potential means of eliminating grafted cells should the need arise.

New cell types for ex vivo gene therapy might also result in further improvements. Grafting of neural precursor cells that are genetically engineered holds potential promise as a future treatment strategy.

As outlined in this review, the potential of ex vivo gene therapy has been clearly demonstrated in many rodent and primate models of neurodegenerative diseases and CNS trauma. Some protocols are close to or in early phases of clinical trials. The outcome of these trials and further developments in vector design, animal models, and treatment strategies will determine the usefulness of gene therapy for the treatment of human disease in the foreseeable future.

Acknowledgements. This work was supported by grants from National Institutes of Health (NS37083, AG0435), the Hollfelder Foundation, the Paralysis Project, and Veterans Affairs. Special thanks to Jeffrey Kordower for providing Fig. 4.

References

Aebischer P, Goddard M, Signore AP, Timpson RL (1994) Functional recovery in hemiparkinsonian primates transplanted with polymer-encapsulated PC12 cells. Exp Neurol 126:151–158

Aebischer P, Pochon NA, Heyd B, Deglon N, Joseph JM, Zurn AD, Baetge EE, Hammang JP, Goddard M, Lysaght M, Kaplan F, Kato AC, Schluep M, Hirt L, Regli F, Porchet F, De Tribolet N (1996b) Gene therapy for amyotrophic lateral sclerosis (ALS) using a polymer encapsulated xenogenic cell line engineered to secrete hCNTF. Hum Gene Ther 7:851–860

Aebischer P, Schluep M, Déglon N, Joseph JM, Hirt L, Heyd B, Goddard M, Hammang JP, Zurn AD, Kato AC, Regli F, Baetge EE (1996a) Intrathecal delivery of CNTF using encapsulated genetically modified xenogeneic cells in amyotrophic lateral sclerosis patients. Nature Med 2:696–699

Aebischer P, Tresco PA, Winn SR, Greene LA, Jaeger CB (1991) Long-term cross-species brain transplantation of a polymer-encapsulated dopamine-secreting cell line. Exp Neurol 111:269–275

Alexi T, Venero JL, Hefti F (1997) Protective effects of neurotrophin-4/5 and transforming growth factor-alpha on striatal neuronal phenotypic degeneration after excitotoxic lesioning with quinolinic acid. Neurosci 78:73–86

Anderson KD, Panayotatos N, Corcoran TL, Lindsay RM, Wiegand SJ (1996) Ciliary neurotrophic factor protects striatal output neurons in an animal model of Huntington disease. Proc Natl Acad Sci USA 93:7346–7351

Araujo DM, Hilt DC (1997) Glial cell line-derived neurotrophic factor attenuates the excitotoxin-induced behavioral and neurochemical deficits in a rodent model of Huntington's disease. Neurosci 81:1099–1110

Bankiewicz KS, Leff SE, Nagy D, Jungles S, Rokovich J, Spratt K, Cohen L, Libonati M, Snyder RO, Mandel RJ (1997) Practical aspects of the development of ex vivo and in vivo gene therapy for Parkinson's disease. Exp Neurol 144:147–156

Barba D, Hardin J, Ray J, Gage FH (1993) Thymidine kinase-mediated killing of rat brain tumors. J Neurosurg 79:729–735

Bartus RT, Dean RLD, Beer B, Lippa AS (1982) The cholinergic hypothesis of geriatric memory dysfunction. Science 217:408–414

Beal MF (1994) Neurochemistry and toxin models in Huntington's disease. Current Opinion Neurol 7:542–547

Beal MF, Ferrante RJ, Swartz KJ, Kowall NW (1991) Chronic quinolinic acid lesions in rats closely resemble Huntington's disease. J Neurosci 11:1649–1659

Beal MF, Kowall NW, Ellison DW, Mazurek MF, Swartz KJ, Martin JB (1986) Replication of the neurochemical characteristics of Huntington's disease by quinolinic acid. Nature 321:168–171

Beck KD, Valverde J, Alexi T, Poulsen K, Moffat B, Vandlen RA, Rosenthal A, Hefti F (1995) Mesencephalic dopaminergic neurons protected by GDNF from axotomy-induced degeneration in the adult brain. Nature 373:339–341

Bilang-Bleuel A, Revah F, Colin P, Locquet I, Robert JJ, Mallet J, Horellou P (1997) Intrastriatal injection of an adenoviral vector expressing glial-cell-line-derived neurotrophic factor prevents dopaminergic neuron degeneration and behavioral impairment in a rat model of Parkinson disease. Proc Natl Acad Sci USA 94: 8818–8823

Blesch A, Diergardt N, Tuszynski MH (1998) Cellularly delivered GDNF induces robust growth of motor and sensory axons in the injured adult spinal cord. Soc for Neuroscience Abstracts 24:555

Blesch A, Grill RJ, Tuszynski MH (1998) Neurotrophin gene therapy in CNS models of trauma and degeneration. Progr Brain Res 117:473–484

Blesch A, Tuszynski MH (1995) Ex vivo gene therapy for Alzheimer's disease and spinal cord injury. Clinical Neurosci 3:268–274

Blesch A, Uy HS, Grill RJ, Cheng JG, Patterson PH, Tuszynski MH (1999) LIF augments corticospinal axon growth and neurotrophin expression after adult CNS injury. J Neurosci 19:3556–3566

Borasio GD, Robberecht W, Leigh PN, Emile J, Guiloff RJ, Jerusalem F, Silani V, Vos PE, Wokke JH, Dobbins T (1998) A placebo-controlled trial of insulin-like growth factor-I in amyotrophic lateral sclerosis. European ALS/IGF-I Study Group. Neurol 51:583–586

Bregman BS, Kunkel-Bagden E, Reier PJ, Dai HN, McAtee M, Gao D (1993) Recovery of function after spinal cord injury: mechanisms underlying transplant-mediated recovery of function differ after spinal cord injury in newborn and adult rats. Exp Neurol 123:3–16

Bregman BS, Kunkel-Bagden E, Schnell L, Dai HN, Gao D, Schwab ME (1995) Recovery from spinal cord injury mediated by antibodies to neurite growth inhibitors. Nature 378:498–501

Chen KS, Gage FH (1995) Somatic gene transfer of NGF to the aged brain: behavioral and morphological amelioration. J Neurosci 15:2819–2825

Cheng H, Cao Y, Olson L (1996) Spinal cord repair in adult paraplegic rats: partial restoration of hind limb function. Science 273:510–513

Choi-Lundberg DL, Lin Q, Chang YN, Chiang YL, Hay CM, Mohajeri H, Davidson BL, Bohn MC (1997) Dopaminergic neurons protected from degeneration by GDNF gene therapy. Science 275:838–841

Coyle JT, Price DL, DeLong MR (1983) Alzheimer's disease: a disorder of cortical cholinergic innervation. Science 219:1184–1190

Coyle JT, Schwarcz R (1976) Lesion of striatal neurones with kainic acid provides a model for Huntington's chorea. Nature 263:244–246

Crutcher KA, Collins F (1982) In vitro Evidence for two distinct hippocampal growth factors: basis of neuronal plasticity? Science, 67–68

Culver KW, Ram Z, Wallbridge S, Ishii H, Oldfield EH, Blaese RM (1992) In vivo gene transfer with retroviral vector-producer cells for treatment of experimental brain tumors. Science 256:1550–1552

David S, Aguayo AJ (1981) Axonal elongation into peripheral nervous system "bridges" after central nervous system injury in adult rats. Science 214:931–933

Davies SJ, Fitch MT, Memberg SP, Hall AK, Raisman G, Silver J (1997) Regeneration of adult axons in white matter tracts of the central nervous system. Nature 390:680–683

Davies SW, Beardsall K (1992) Nerve growth factor selectively prevents excitotoxin induced degeneration of striatal cholinergic neurones. Neurosci Letters 140:161–164

Déglon N, Heyd B, Tan SA, Joseph JM, Zurn AD, Aebischer P (1996) Central nervous system delivery of recombinant ciliary neurotrophic factor by polymer encapsulated differentiated C2C12 myoblasts. Hum Gene Ther 7:2135–2146

Dunnett SB, Svendsen CN (1993) Huntington's disease: animal models and transplantation repair. Current Opinion Neurobiol 3:790–796

Dunnett SB, Toniolo G, Fine A, Ryan CN, Björklund A, Iversen SD (1985) Transplantation of embryonic ventral forebrain neurons to the neocortex of rats with lesions of nucleus basalis magnocellularis–II. Sensorimotor and learning impairments. Neurosci 16:787–797

During MJ, Leone P (1997) Targets for gene therapy of Parkinson's disease: growth factors, signal transduction, and promoters. Exp Neurol 144:74–81

Emerich DF, Hammang JP, Baetge EE, Winn SR (1994) Implantation of polymer-encapsulated human nerve growth factor-secreting fibroblasts attenuates the behavioral and neuropathological consequences of quinolinic acid injections into rodent striatum. Exp Neurol 130:141–150

Emerich DF, Lindner MD, Winn SR, Chen EY, Frydel BR, Kordower JH (1996) Implants of encapsulated human CNTF-producing fibroblasts prevent behavioral deficits and striatal degeneration in a rodent model of Huntington's disease. J Neurosci 16:5168–5181

Emerich DF, Plone M, Francis J, Frydel BR, Winn SR, Lindner MD (1996) Alleviation of behavioral deficits in aged rodents following implantation of encapsulated GDNF-producing fibroblasts. Brain Res 736:99–110

Emerich DF, Winn SR, Hantraye PM, Peschanski M, Chen EY, Chu Y, McDermott P, Baetge EE, Kordower JH (1997) Protective effect of encapsulated cells producing neurotrophic factor CNTF in a monkey model of Huntington's disease. Nature 386:395–399

Emerich DF, Winn SR, Harper J, Hammang JP, Baetge EE, Kordower JH (1994) Implants of polymer-encapsulated human NGF-secreting cells in the nonhuman primate: rescue and sprouting of degenerating cholinergic basal forebrain neurons. J Comp Neurol 349:148–164

Ezzeddine ZD, Martuza RL, Platika D, Short MP, Malick A, Choi B, Breakefield XO (1991) Selective killing of glioma cells in culture and in vivo by retrovirus transfer of the herpes simplex virus thymidine kinase gene. New Biologist 3:608–614

Fine A, Dunnett SB, Björklund A, Iversen SD (1985) Cholinergic ventral forebrain grafts into the neocortex improve passive avoidance memory in a rat model of Alzheimer disease. Proc Natl Acad Sci USA 82:5227–5230

Fischer W, Björklund A, Chen K, Gage FH (1991) NGF improves spatial memory in aged rodents as a function of age. J Neurosci 11:1889–1906

Fischer W, Wictorin K, Bjorklund A, Williams LR, Varon S, Gage FH (1987) Amelioration of cholinergic neuron atrophy and spatial memory impairment in aged rats by nerve growth factor. Nature 329:65–68

Fisher LJ (1997) Neural precursor cells: applications for the study and repair of the central nervous system. Neurobiol Dis 4:1–22

Fisher LJ, Jinnah HA, Kale LC, Higgins GA, Gage FH (1991) Survival and function of intrastriatally grafted primary fibroblasts genetically modified to produce L-dopa. Neuron 6:371–380

Fisher LJ, Raymon HK, Gage FH (1993) Cells engineered to produce acetylcholine: therapeutic potential for Alzheimer's disease. Ann N Y Acad Sci 695:278–284

Fisher LJ, Schinstine M, Salvaterra P, Dekker AJ, Thal L, Gage FH (1993) In vivo production and release of acetylcholine from primary fibroblasts genetically modified to express choline acetyltransferase. J Neurochem 61:1323–1332

Frim DM, Short MP, Rosenberg WS, Simpson J, Breakefield XO, Isacson O (1993a) Local protective effects of nerve growth factor-secreting fibroblasts against excitotoxic lesions in the rat striatum. J Neurosurg 78:267–273

Frim DM, Simpson J, Uhler TA, Short MP, Bossi SR, Breakefield XO, Isacson O (1993b) Striatal degeneration induced by mitochondrial blockade is prevented by biologically delivered NGF. J Neurosci Res 35:452–458

Frim DM, Uhler TA, Galpern WR, Beal MF, Breakefield XO, Isacson O (1994) Implanted fibroblasts genetically engineered to produce brain-derived neurotrophic factor prevent 1-methyl-4-phenylpyridinium toxicity to dopaminergic neurons in the rat. Proc Natl Acad Sci USA 91:5104–5108

Frim DM, Uhler TA, Short MP, Ezzedine ZD, Klagsbrun M, Breakefield XO, Isacson O (1993) Effects of biologically delivered NGF, BDNF and bFGF on striatal excitotoxic lesions. Neuroreport 4:367–370

Gage FH, Armstrong DM, Williams LR, Varon S (1988) Morphological response of axotomized septal neurons to nerve growth factor. J Comp Neurol 269:147–155

Gage FH, Wolff JA, Rosenberg MB, Xu L, Yee JK, Shults C, Friedmann T (1987) Grafting genetically modified cells to the brain: possibilities for the future. Neurosci 23:795–807

Galpern WR, Frim DM, Tatter SB, Altar CA, Beal MF, Isacson O (1996) Cell-mediated delivery of brain-derived neurotrophic factor enhances dopamine levels in an MPP+ rat model of substantia nigra degeneration. Cell Transplantat 5:225–232

Galpern WR, Matthews RT, Beal MF, Isacson O (1996) NGF attenuates 3-nitrotyrosine formation in a 3-NP model of Huntington's disease. Neuroreport 7:2639–2642

Gash DM, Zhang Z, Ovadia A, Cass WA, Yi A, Simmerman L, Russell D, Martin D, Lapchak PA, Collins F, Hoffer BJ, Gerhardt GA (1996) Functional recovery in parkinsonian monkeys treated with GDNF. Nature 380:252–255

Gossen M, Bujard H (1992) Tight control of gene expression in mammalian cells by tetracycline-responsive promoters. Proc Natl Acad Sci USA 89:5547–5551

Gossen M, Freundlieb S, Bender G, Muller G, Hillen W, Bujard H (1995) Transcriptional activation by tetracyclines in mammalian cells. Science 268:1766–1769

Grill R, Murai K, Blesch A, Gage FH, Tuszynski MH (1997a) Cellular delivery of neurotrophin-3 promotes corticospinal axonal growth and partial functional recovery after spinal cord injury. J Neurosci 17:5560–5572

Grill RJ, Blesch A, Tuszynski MH (1997b) Robust growth of chronically injured spinal cord axons induced by grafts of genetically modified NGF-secreting cells. Exp Neurol 148:444–452

ALS CNTF Treatment Study Group (1996) A double-blind placebo-controlled clinical trial of subcutaneous recombinant human ciliary neurotrophic factor (rHCNTF) in amyotrophic lateral sclerosis. Neurology 46:1244–1249

The ALS CNTF Treatment Study (ACTS) Phase I-II Study Group (1995) A phase I study of recombinant human ciliary neurotrophic factor (rHCNTF) in patients with amyotrophic lateral sclerosis. Clin Neuropharmacol 18:515–532

Hefti F (1986) Nerve growth factor promotes survival of septal cholinergic neurons after fimbrial transections. J Neurosci 6:2155–2162

Hodges H, Allen Y, Sinden J, Lantos PL, Gray JA (1991) Effects of cholinergic-rich neural grafts on radial maze performance of rats after excitotoxic lesions of the forebrain cholinergic projection system–II. Cholinergic drugs as probes to investigate lesion-induced deficits and transplant-induced functional recovery. Neurosci 45:609–623

Horellou P, Brundin P, Kalen P, Mallet J, Bjorklund A (1990) In vivo release of dopa and dopamine from genetically engineered cells grafted to the denervated rat striatum. Neuron 5:393–402

Horger BA, Nishimura MC, Armanini MP, Wang LC, Poulsen KT, Rosenblad C, Kirik D, Moffat B, Simmons L, Johnson E Jr, Milbrandt J, Rosenthal A, Bjorklund A, Vandlen RA, Hynes MA, Phillips HS (1998) Neurturin exerts potent actions on survival and function of midbrain dopaminergic neurons. J Neurosci 18:4929–4937

Hyman C, Hofer M, Barde YA, Juhasz M, Yancopoulos GD, Squinto SP, Lindsay RM (1991) BDNF is a neurotrophic factor for dopaminergic neurons of the substantia nigra. Nature 350:230–232

Hyman C, Juhasz M, Jackson C, Wright P, Ip NY, Lindsay RM (1994) Overlapping and distinct actions of the neurotrophins BDNF, NT-3, and NT-4/5 on cultured dopaminergic and GABAergic neurons of the ventral mesencephalon. J Neurosci 14:335–347

Ikeda K, Wong V, Holmlund TH, Greene T, Cedarbaum JM, Lindsay RM, Mitsumoto H (1995) Histometric effects of ciliary neurotrophic factor in wobbler mouse motor neuron disease. Ann Neurol 37:47–54

Kawaja MD, Rosenberg MB, Yoshida K, Gage FH (1992) Somatic gene transfer of nerve growth factor promotes the survival of axotomized septal neurons and the regeneration of their axons in adult rats. J Neurosci 12:2849–2864

Knüsel B, Beck KD, Winslow JW, Rosenthal A, Burton LE, Widmer HR, Nikolics K, Hefti F (1992) Brain-derived neurotrophic factor administration protects basal forebrain cholinergic but not nigral dopaminergic neurons from degenerative changes after axotomy in the adult rat brain. J Neurosci 12:4391–4402

Koliatsos VE, Nauta HJ, Clatterbuck RE, Holtzman DM, Mobley WC, Price DL (1990) Mouse nerve growth factor prevents degeneration of axotomized basal forebrain cholinergic neurons in the monkey. J Neurosci 10:3801–3813

Kordower JH, Chen EY, Mufson EJ, Winn SR, Emerich, DF (1996) Intrastriatal implants of polymer encapsulated cells genetically modified to secrete human nerve growth factor: trophic effects upon cholinergic and noncholinergic striatal neurons. Neurosci 72:63–77

Kordower JH, Chen EY, Winkler C, Fricker R, Charles V, Messing A, Mufson EJ, Wong SC, Rosenstein JM, Björklund A, Emerich DF, Hammang J, Carpenter MK (1997) Grafts of EGF-responsive neural stem cells derived from GFAP-hNGF transgenic mice: trophic and tropic effects in a rodent model of Huntington's disease. J Comp Neurol 387:96–113

Kordower JH, Winn SR, Liu YT, Mufson EJ, Sladek JR Jr, Hammang JP, Baetge EE, Emerich DF (1994) The aged monkey basal forebrain: rescue and sprouting of axotomized basal forebrain neurons after grafts of encapsulated cells secreting human nerve growth factor. Proc Natl Acad Sci USA 91:10898–10902

Korsching S, Auburger G, Heumann R, Scott J, Thoenen H (1985) Levels of nerve growth factor and its mRNA in the central nervous system of the rat correlate with cholinergic innervation. Embo J 1389–1393

Kreutzberg GW (1996) Microglia: a sensor for pathological events in the CNS. Trends Neurosci 19:312–318

Krieglstein K, Suter-Crazzolara C, Fischer WH, Unsicker K (1995) TGF-beta superfamily members promote survival of midbrain dopaminergic neurons and protect them against MPP+ toxicity. Embo J 14:736–742

Kromer LF (1987) Nerve growth factor treatment after brain injury prevents neuronal death. Science 235:214–216

Lange DJ, Felice KJ, Festoff BW, Gawel MJ, Gelinas DF, Kratz R, Lai EC, Murphy MF, Natter HM, Norris FH, Rudnicki S (1996) Recombinant human insulin-like growth factor-I in ALS: description of a double-blind, placebo-controlled study. North American ALS/IGF-I Study Group. Neurology 47:S93–S95

Lapchak PA, Beck KD, Araujo DM, Irwin I, Langston JW, Hefti F (1993) Chronic intranigral administration of brain-derived neurotrophic factor produces striatal dopaminergic hypofunction in unlesioned adult rats and fails to attenuate the decline of striatal dopaminergic function following medial forebrain bundle transection. Neurosci 53:639–650

Levi-Montalcini R (1987) The nerve growth factor 35 years later. Science 237:1154–1162

Levivier M, Przedborski S, Bencsics C, Kang UJ (1995) Intrastriatal implantation of fibroblasts genetically engineered to produce brain-derived neurotrophic factor prevents degeneration of dopaminergic neurons in a rat model of Parkinson's disease. J Neurosci 15:7810–7820

Li Y, Field M, Raisman G (1997) Repair of adult rat corticospinal tract by transplants of olfactory ensheathing cells. Science 277:2000–2002

Li Y, Raisman G (1994) Schwann cells induce sprouting in motor and sensory axons in the adult rat spinal cord. J Neurosci 4050–4063

Lin LF, Doherty DH, Lile JD, Bektesh S, Collins F (1993) GDNF: a glial cell line-derived neurotrophic factor for midbrain dopaminergic neurons. Science 260:1130–1132

Lin Q, Cunningham LA, Epstein LG, Pechan PA, Short MP, Fleet C, Bohn MC (1997) Human fetal astrocytes as an ex vivo gene therapy vehicle for delivering biologically active nerve growth factor. Hum Gene Ther 8:331–339

Lundberg C, Horellou P, Mallet J, Björklund A (1996) Generation of DOPA-producing astrocytes by retroviral transduction of the human tyrosine hydroxylase gene: in vitro characterization and in vivo effects in the rat Parkinson model. Exp Neurol 139:39–53

Martinez-Serrano A, Björklund A (1996) Protection of the neostriatum against excitotoxic damage by neurotrophin-producing, genetically modified neural stem cells. J Neurosci 16:4604–4616

Martinez-Serrano A, Fischer W, Bjorklund A (1995) Reversal of age-dependent cognitive impairments and cholinergic neuron atrophy by NGF-secreting neural progenitors grafted to the basal forebrain. Neuron 473–484

Martinez-Serrano A, Fischer W, Soderstrom S, Ebendal T, Bjorklund A (1996) Long-term functional recovery from age-induced spatial memory impairments by nerve growth factor gene transfer to the rat basal forebrain. Proc Natl Acad Sci USA 6355–60

Martinez-Serrano A, Lundberg C, Horellou P, Fischer W, Bentlage C, Campbell K, McKay RD, Mallet J, Bjorklund A (1995) CNS-derived neural progenitor cells for gene transfer of nerve growth factor to the adult rat brain: complete rescue of axotomized cholinergic neurons after transplantation into the septum. J Neurosci 5668–5680

Martinez-Serrano A, Snyder EY (1999) Neural stem cell lines for CNS repair. In: Tuszynski MH, Kordower JH (eds) CNS Regeneration, (San Diego: Academic Press), pp 203–250

Masliah E, Terry RD, Alford M, DeTeresa R, Hansen LA (1991) Cortical and subcortical patterns of synaptophysinlike immunoreactivity in Alzheimer's disease. Am J Pathol 138:235–246

McGeer EG, McGeer PL (1976) Duplication of biochemical changes of Huntington's chorea by intrastriatal injections of glutamic and kainic acids. Nature 263:517–519

McTigue DM, Horner PJ, Stokes BT, Gage FH (1998) Neurotrophin-3 and brain-derived neurotrophic factor induce oligodendrocyte proliferation and myelination of regenerating axons in the contused adult rat spinal cord. J Neurosci 18: 5354–5365

Menei P, Montero-Menei C, Whittemore SR, Bunge RP, Bunge MB (1998) Schwann cells genetically modified to secrete human BDNF promote enhanced axonal regrowth across transected adult rat spinal cord. Eur J Neurosci 10:607–621

Milbrandt J, de SFJ, Fahrner TJ, Baloh RH, Leitner ML, Tansey MG, Lampe PA, Heuckeroth RO, Kotzbauer PT, Simburger KS, Golden JP, Davies JA, Vejsada R, Kato AC, Hynes M, Sherman D, Nishimura M, Wang LC, Vandlen R, Moffat B, Klein RD, Poulsen K, Gray C, Garces A, Johnson EM Jr et al. (1998) Persephin, a novel neurotrophic factor related to GDNF and neurturin. Neuron 20:245–253

Miller AD (1990) Retrovirus packaging cells. Hum Gene Ther 1:5–14

Miller AD, Miller DG, Garcia JV, Lynch CM (1993) Use of retroviral vectors for gene transfer and expression. Methods Enzymol 217:581–599

Miller AD, Palmer TD, Hock RA (1986) Transfer of genes into human somatic cells using retrovirus vectors. Cold Spring Harb Symp Quant Biol 51 Pt 2:1013–1019

Miller RG, Petajan JH, Bryan WW, Armon C, Barohn RJ, Goodpasture JC, Hoagland RJ, Parry GJ, Ross MA, Stromatt SC (1996) A placebo-controlled trial of recombinant human ciliary neurotrophic (rhCNTF) factor in amyotrophic lateral sclerosis. rhCNTF ALS Study Group. Ann Neurol 39:256–260

Mitsumoto H, Ikeda K, Holmlund T, Greene T, Cedarbaum JM, Wong V, Lindsay RM (1994) The effects of ciliary neurotrophic factor on motor dysfunction in wobbler mouse motor neuron disease. Ann Neurol 36:142–148

Mitsumoto H, Ikeda K, Klinkosz B, Cedarbaum JM, Wong V, Lindsay RM (1994) Arrest of motor neuron disease in wobbler mice cotreated with CNTF and BDNF. Science 265:1107–1110

Mori F, Himes BT, Kowada M, Murray M, Tessler A (1997) Fetal spinal cord transplants rescue some axotomized rubrospinal neurons from retrograde cell death in adult rats. Exp Neurol 143:45–60

Olson L, Nordberg A, von HH, Backman L, Ebendal T, Alafuzoff I, Amberla K, Hartvig P, Herlitz A, Lilja A et al. (1992) Nerve growth factor affects 11C-nicotine binding, blood flow, EEG, and verbal episodic memory in an Alzheimer patient (case report). J Neural Transm Park Dis Dement Sect 4:79–95

Pardridge WM (1994) New approaches to drug delivery through the blood-brain barrier. Trends Biotechnol 12:239–245

Pérez-Navarro E, Arenas E, Reiriz J, Calvo N, Alberch J (1996) Glial cell line-derived neurotrophic factor protects striatal calbindin-immunoreactive neurons from excitotoxic damage. Neurosci 75:345–352

Perry EK, Tomlinson BE, Blessed G, Bergmann K, Gibson PH, Perry RH (1978) Correlation of cholinergic abnormalities with senile plaques and mental test scores in senile dementia. Br Med J 2:1457–1459

Pizzo DL, Paban V, Winkler J, Gage FH, Thal LJ (1998) Characterization of a tetracycline regulatable ChAT fibroblast line. Soc Neurosci Abstracts 24:1055

Price DL (1986) New perspectives on Alzheimer's disease. Annu Rev Neurosci 9:489–512

Ramon-Cueto A, Nieto-Sampedro M (1994) Regeneration into the spinal cord of transected dorsal root axons is promoted by ensheathing glia transplants. Exp Neurol 127:232–244

Rapalino O, Lazarov-Spiegler O, Agranov E, Velan GJ, Yoles E, Fraidakis M, Solomon A, Gepstein R, Katz A, Belkin M, Hadani M, Schwartz M (1998) Implantation of stimulated homologous macrophages resulkts in partial recovery of paraplegic rats. Nature Med 4:814–821

Richardson PM, McGuinness UM, Aguayo AJ (1980) Axons from CNS neurons regenerate into PNS grafts. Nature 284:264–265

Rosenberg MB, Friedmann T, Robertson RC, Tuszynski M, Wolff JA, Breakefield XO, Gage FH (1988) Grafting genetically modified cells to the damaged brain: restorative effects of NGF expression. Science 242:1575–1578

Saffran BN, Woo JE, Mobley WC, Crutcher KA (1989) Intraventricular NGF infusion in the mature rat brain enhances sympathetic innervation of cerebrovascular targets but fails to elicit sympathetic ingrowth. Brain Res 492:245–254

Sagot Y, Tan SA, Baetge E, Schmalbruch H, Kato AC, Aebischer P (1995) Polymer encapsulated cell lines genetically engineered to release ciliary neurotrophic factor can slow down progressive motor neuronopathy in the mouse. Eur J Neurosci 7:1313–1322

Schnell L, Schneider R, Kolbeck R, Barde YA, Schwab ME (1994) Neurotrophin-3 enhances sprouting of corticospinal tract during development and after adult spinal cord lesion. Nature 367:170–173

Schnell L, Schwab ME (1990) Axonal regeneration in the rat spinal cord produced by an antibody against myelin-associated neurite growth inhibitors. Nature 343:269–272

Schnell L, Schwab ME (1993) Sprouting and regeneration of lesioned corticospinal tract fibres in the adult rat spinal cord. Eur J Neurosci 5:1156–1171

Schwab ME, Otten U, Agid Y, Thoenen H (1979) Nerve growth factor (NGF) in the rat CNS: absence of specific retrograde axonal transport and tyrosine hydroxylase induction in locus coeruleus and substantia nigra. Brain Res 168:473–483

Seiler M, Schwab ME (1984) Specific retrograde transport of nerve growth factor (NGF) from neocortex to nucleus basalis in the rat. Brain Res 300:33–39

Selkoe DJ (1996) Cell biology of the beta-amyloid precursor protein and the genetics of Alzheimer's disease. Cold Spring Harb Symp Quant Biol 61:587–596

Sendtner M, Schmalbruch H, Stöckli KA, Carroll P, Kreutzberg GW, Thoenen H (1992) Ciliary neurotrophic factor prevents degeneration of motor neurons in mouse mutant progressive motor neuronopathy. Nature 358:502–504

Sharp AH, Ross CA (1996) Neurobiology of Huntington's disease. Neurobiol Dis 3:3–15

Shelton DL, Reichardt LF (1986) Studies on the expression of the beta nerve growth factor (NGF) gene in the central nervous system: level and regional distribution of NGF mRNA suggest that NGF functions as a trophic factor for several distinct populations of neurons. Proc Natl Acad Sci USA 83:2714–2718

Smith DE, McCay HL, Gage FH, Roberts JA, Tuszynski MH (1998) Intraparenchymal delivery of NGF by ex vivo gene transfer reverses age-related loss of expression of p75-NTR in basal forebrain cholinergic neurons. Soc Neurosci Abstr 24(1):541

Snyder EY (1994) Grafting immortalized neurons to the CNS. Current Opin Neurobiol 4:742–751

Sobreviela T, Clary DO, Reichardt LF, Brandabur MM, Kordower JH, Mufson EJ (1994) TrkA-immunoreactive profiles in the central nervous system: colocalization with neurons containing p75 nerve growth factor receptor, choline acetyltransferase, and serotonin. J Comp Neurol 350:587–611

Stokes BT, Reier PJ (1992) Fetal grafts alter chronic behavioral outcome after contusion damage to the adult rat spinal cord. Exp Neurol 116:1–12

Takamiya Y, Short MP, Moolten FL, Fleet C, Mineta T, Breakefield XO, Martuza RL (1993) An experimental model of retrovirus gene therapy for malignant brain tumors. J Neurosurg 79:104–110

Terry RD, Katzman R (1983) Senile dementia of the Alzheimer type. Ann Neurol 14:497–506

Tessler A (1991) Intraspinal transplants. Ann Neurol 29:115–123

Tessler A, Fischer I, Giszter S, Himes BT, Miya D, Mori F, Murray M (1997) Embryonic spinal cord transplants enhance locomotor performance in spinalized newborn rats. Adv Neurol 72:291–303

Thal L (1994) Clinical trials in Alzheimer disease. In: Terry R, Katzman R, Bick K (eds) Alzheimer disease, (New York: Raven Press), pp 431–444

The-Huntington's-Disease-Collaborative-Research-Group (1993) A novel gene containing a trinucleotide repeat that is expanded and unstable on Huntington's disease chromosomes. The Huntington's Disease Collaborative Research Group. Cell 72:971–983

Tomac A, Lindqvist E, Lin LF, Ogren SO, Young D, Hoffer BJ, Olson L (1995) Protection and repair of the nigrostriatal dopaminergic system by GDNF in vivo. Nature 373:335–339

Tseng JL, Baetge EE, Zurn AD, Aebischer P (1997) GDNF reduces drug-induced rotational behavior after medial forebrain bundle transection by a mechanism not involving striatal dopamine. J Neurosci 17:325–333

Tuszynski MH, Gabriel K, Gage FH, Suhr S, Meyer S, Rosetti A (1996b) Nerve growth factor delivery by gene transfer induces differential outgrowth of sensory, motor, and noradrenergic neurites after adult spinal cord injury. Exp Neurol 137:157–173

Tuszynski MH, Murai K, Blesch A, Grill R, Miller I (1997a) Functional characterization of NGF-secreting cell grafts to the acutely injured spinal cord. Cell Transplant 6:361–368

Tuszynski MH, Peterson DA, Ray J, Baird A, Nakahara Y, Gage FH (1994a) Fibroblasts genetically modified to produce nerve growth factor induce robust neuritic ingrowth after grafting to the spinal cord. Exp Neurol 126:1–14

Tuszynski MH, Roberts J, Senut MC, U HS, Gage FH (1996a) Gene therapy in the adult primate brain: intraparenchymal grafts of cells genetically modified to produce nerve growth factor prevent cholinergic neuronal degeneration. Gene Ther 3:305–314

Tuszynski MH, Sang H, Yoshida K, Gage FH (1991) Recombinant human nerve growth factor infusions prevent cholinergic neuronal degeneration in the adult primate brain. Ann Neurol 30:625–636

Tuszynski MH, Senut MC, Ray J, Roberts J (1994b) Somatic gene transfer to the adult primate central nervous system: In vitro and in vivo characterization of cells genetically modified to secrete nerve growth factor. Neurobiol Dis 1:67–78

Tuszynski MH, U HS, Amaral DG, Gage FH (1990) Nerve growth factor infusion in the primate brain reduces lesion-induced cholinergic neuronal degeneration. J Neurosci 10:3604–3614

Tuszynski MH, Weidner N, McCormack M, Miller I, Powell H, Conner J (1998) Grafts of genetically modified Schwann cells to the spinal cord: survival, axon growth, and myelination. Cell Transplant 7:187–196

Venero JL, Beck KD, Hefti F (1994) Intrastriatal infusion of nerve growth factor after quinolinic acid prevents reduction of cellular expression of choline acetyltransferase messenger RNA and trkA messenger RNA, but not glutamate decarboxylase messenger RNA. Neurosci 61:257–268

Volpe BT, Wildmann J, Altar CA (1998) Brain-derived neurotrophic factor prevents the loss of nigral neurons induced by excitotoxic striatal-pallidal lesions. Neurosci 83:741–748

Widner H, Brundin P (1988) Immunological aspects of grafting in the mammalian central nervous system. A review and speculative synthesis. Brain Res 472:287–324

Williams LR (1991) Hypophagia is induced by intracerebroventricular administration of nerve growth factor. Exp Neurol 113:31–37

Winkler J, Ramirez GA, Kuhn HG, Peterson DA, Day-Lollini PA, Stewart GR, Tuszynski MH, Gage FH, Thal LJ (1997) Reversible Schwann cell hyperplasia and sprouting of sensory and sympathetic neurites after intraventricular administration of nerve growth factor. Ann Neurol 41:82–93

Winkler J, Suhr ST, Gage FH, Thal LJ, Fisher LJ (1995) Essential role of neocortical acetylcholine in spatial memory. Nature 375:484–487

Wolff JA, Fisher LJ, Xu L, Jinnah HA, Langlais PJ, Iuvone PM, O'Malley KL, Rosenberg MB, Shimohama S, Friedmann T et al. (1989) Grafting fibroblasts genetically modified to produce L-dopa in a rat model of Parkinson disease. Proc Natl Acad Sci USA 86:9011–9014

Xu XM, Chen A, Guenard V, Kleitman N, Bunge MB (1997) Bridging Schwann cell transplants promote axonal regeneration from both the rostral and caudal stumps of transected adult rat spinal cord. J Neurocytol 26:1–16

Xu XM, Guenard V, Kleitman N, Aebischer P, Bunge MB (1995) A combination of BDNF and NT-3 promotes supraspinal axonal regeneration into Schwann cell grafts in adult rat thoracic spinal cord. Exp Neurol 134:261–272

Xu XM, Guenard V, Kleitman N, Bunge MB (1994) Axonal regeneration into Schwann cell-seeded guidance channels grafted into transected adult rat spinal cord. J Comp Neurol 351:145–160

Yankner BA (1996) Mechanisms of neuronal degeneration in Alzheimer's disease. Neuron 16:921–932

Ye JH, Houle JD (1997) Treatment of the chronically injured spinal cord with neurotrophic factors can promote axonal regeneration from supraspinal neurons. Exp Neurol 143:70–81

Yoshimoto Y, Lin Q, Collier TJ, Frim DM, Breakefield XO, Bohn MC (1995) Astrocytes retrovirally transduced with BDNF elicit behavioral improvement in a rat model of Parkinson's disease. Brain Res 691:25–36

Z'Graggen WJ, Metz GA, Kartje GL, Thallmair M, Schwab ME (1998) Functional recovery and enhanced corticofugal plasticity after unilateral pyramidal tract lesion and blockade of myelin-associated neurite growth inhibitors in adult rats. J Neurosci 18:4744–4757

Zeev-Brann AB, Lazarov-Spiegler O, Brenner T, Schwartz M (1998) Differential effects of central and peripheral nerves on macrophages and microglia. Glia 23:181–190

CHAPTER 12
Developing Gene-Based Neuroprotection Strategies Using Herpes Amplicon Vectors

M.W. Halterman and H.J. Federoff

A. Introduction

Ischemia in the central nervous system produces a heterogeneous pattern of neuronal loss, which is dependent on the extent, duration, and distribution of vessel occlusion. Profound hypoperfusion produces neuronal necrosis in a matter of hour, resulting in plasma membrane rupture and subsequent release of synaptic and somal contents to the interstitial space. An appreciation for the role of elevated extracellular glutamate and intracellular calcium in this pathologic sequence has prompted the development of glutamate and calcium receptor antagonists. Unfortunately, these strategies have not produced in humans the degree of protection observed in animal models of acute necrotic stroke (Zivin and Miller 1999).

Alternatively, ischemia can trigger delayed neuronal death in the post-ischemic period through the activation of an ordered sequence of genetic and biochemical events (Pulsinelli et al. 1982; Du et al. 1996). This pathologic progression involves initiation, commitment, and execution steps, which require *de novo* gene expression, the participation of neuronal mitochondria, and the activation of cellular proteases, respectively. Elucidation of the mechanisms that govern delayed death offers the means to develop novel gene or small molecule-based neuroprotective therapies.

As an alternative to traditional gene disruption modalities, the genetic manipulation of the mammalian central nervous system using virus vector-mediated gene delivery has deepened our understanding for the role gene expression plays in delayed forms of neuronal death. Most recently, experiments from our laboratory have suggested that interventions which target the initiation phase of delayed neuronal death are capable of interrupting this pathologic sequence, and may provide a rational alternative to the widely investigated anti-commitment- and anti-execution-based therapies. In this chapter, we provide an overview of the mechanisms that underlie ischemia-induced pathology, paying specific attention to the role of *de novo* gene expression. Examples of how our group and others have used viral vector-based neuroprotective approaches to characterize the activity of putative neuroprotective genes will also be provided.

B. Central Nervous System Ischemia Induces Distinct Patterns of Neuronal Loss

Acute reductions in cerebrocortical blood flow activate whole organ as well as subcellular compensatory mechanisms in an attempt to normalize the metabolic imbalance between supply and demand. In addition to increasing regional blood flow through vasodilatation, these mechanisms include alterations in metabolic enzyme activity as well as an overall reduction in level of protein synthesis to match the level of blood flow to the affected region (ASTRUP et al. 1981). Distal to the occlusion, a reduction in normal flow from 100ml/100g per minute to 50–80ml/100g per minute induces a modest suppression in protein translation, while reduction below 40–50ml/100g per minute results in the complete cessation of protein synthesis (JACEWICZ et al. 1986; XIE et al. 1989). Once the cerebral vascular system becomes maximally dilated, neurons enhance glucose absorption and rely on alternative energy sources for ATP production. Dual analyses of blood flow and glucose utilization have shown that a minimum perfusion of 20ml/100g per minute is required to maintain neuronal homeostasis (PASCHEN et al. 1992). When ATP supplies fall below a critical level, the activity of energy-dependent ion channels decline, resulting in the suppression of electro encephalographic activity. This period of silence is reversible providing reperfusion occurs within a short window period. However, once ATP reserves are depleted and ion pumps fail, inability to regulate glutamate release or the rapid influx of extracellular ions, results in neuronal lysis. This necrotic form of cell death matures within 24h after the onset of ischemia. While inhibition of gene expression does not affect the outcome of such an event, reduction in ion flux with the use of channel-specific inhibitors does provide protection in the postischemic period (TASKER et al. 1992; MA et al. 1998). Although neurons in the ischemic core can transiently exhibit an apoptotic morphology (CHARRIAUT-MARLANGUE et al. 1996), antiapoptotic drugs are for the most part ineffective in preventing necrosis.

Transient global ischemia induces the delayed and asynchronous loss of cortical, hippocampal, striatal and cerebellar neurons beginning 48–72h after the onset of ischemia (KIRINO et al. 1984; PETITO et al. 1987). This form of delayed neuronal death, also termed the "maturation phenomenon" (ITO et al. 1975), is morphologically and mechanistically distinct, whereby neurons in the CA1 region of the hippocampus die more slowly and lack the cytoplasmic vacuolization commonly associated with acute necrosis. Although the middle cerebral artery occlusion (MCAO) model of transient focal ischemia has classically been used to investigate mechanisms of neuronal necrosis, substantial neuronal loss occurring in the penumbra (the region surrounding the necrotic core) has been described. DU and colleagues found that while 30-min occlusion produces no observable damage 24h later, infarct volumes when measured 14 days later equaled that produced in the permanent MCAO model (DU et al. 1996). Importantly, systemic administration of the translation inhibitor cycloheximide prevented the evolution of this delayed lesion. Factors which include gradients of hypoxia, glutamate, and growth factors, the effects of cortical spreading depression as well as increasing concentrations of inflammatory molecules secreted by resident glia and infiltrating peripheral mononuclear cells are all potential mediators of neuron loss in the ischemic brain.

Delayed neuronal death is capable of producing damage equivalent to that seen after extensive occlusion. Furthermore, since the penumbral lesion matures over the course of days to weeks, a window of opportunity for intervention is feasible. Lastly, the persistence of residual circulation in the penumbra provides a means to deliver neuroprotective compounds to this ischemic population of cells (PULSINELLI 1992). Taken together, delayed neuronal death in the ischemic penumbra represents a clinically relevant therapeutic target. It is likely that studies of antinecrotic drugs which used time points earlier than 48h to assess infarct volume underestimated the contributions to delayed death which would manifest days later. This is one potential explanation for why neuroprotective strategies based on the interruption of glutamate and calcium-mediated activities have failed to demonstrate efficacy in clinical trials. Further study is required to assess the impact of such therapies on mature ischemic lesions.

C. A Cell-Autonomous Program Activates the Delayed Form of Neuronal Death

Programmed cell death is a ubiquitous process occurring in many differentiated and progenitor cell populations (YOSHIDA et al. 1998). Delayed death plays a significant role in the developing and diseased nervous system and can be modeled using dissociated neuronal preparations and hippocampal slice cultures (GWAG et al. 1995; WALLIS and PANIZZON 1995). Studies in these systems recapitulate the timing of delayed neuronal death as well as the dependence on gene expression. However, by modeling ischemia *in vitro*, the characteristics of delayed death can be studied without interference from immune cell infiltrates, compensatory cardiovascular responses, or fluctuating hormone and catecholamine levels. Further investigation using purified cortical or cerebellar neuronal preparations has demonstrated that delayed neuronal death cascades can be initiated in a cell-autonomous fashion.

Mutational analyses in the nematode *C. elegans* and in the mouse have identified individual factors and signaling pathways which orchestrate the commitment and execution phases of the cell death program (YUAN 1996; MILLER et al. 1997). While these phases of apoptosis exhibit cell-type dependent specialization, a shared feature among many prodeath pathways, is the involvement of the mitochondria in this process (Fig. 1). Ironically, some of the early morphologic studies performed on nonneuronal, apoptotic cells detected no change in mitochondrial structure (SEARLE et al. 1975). However,

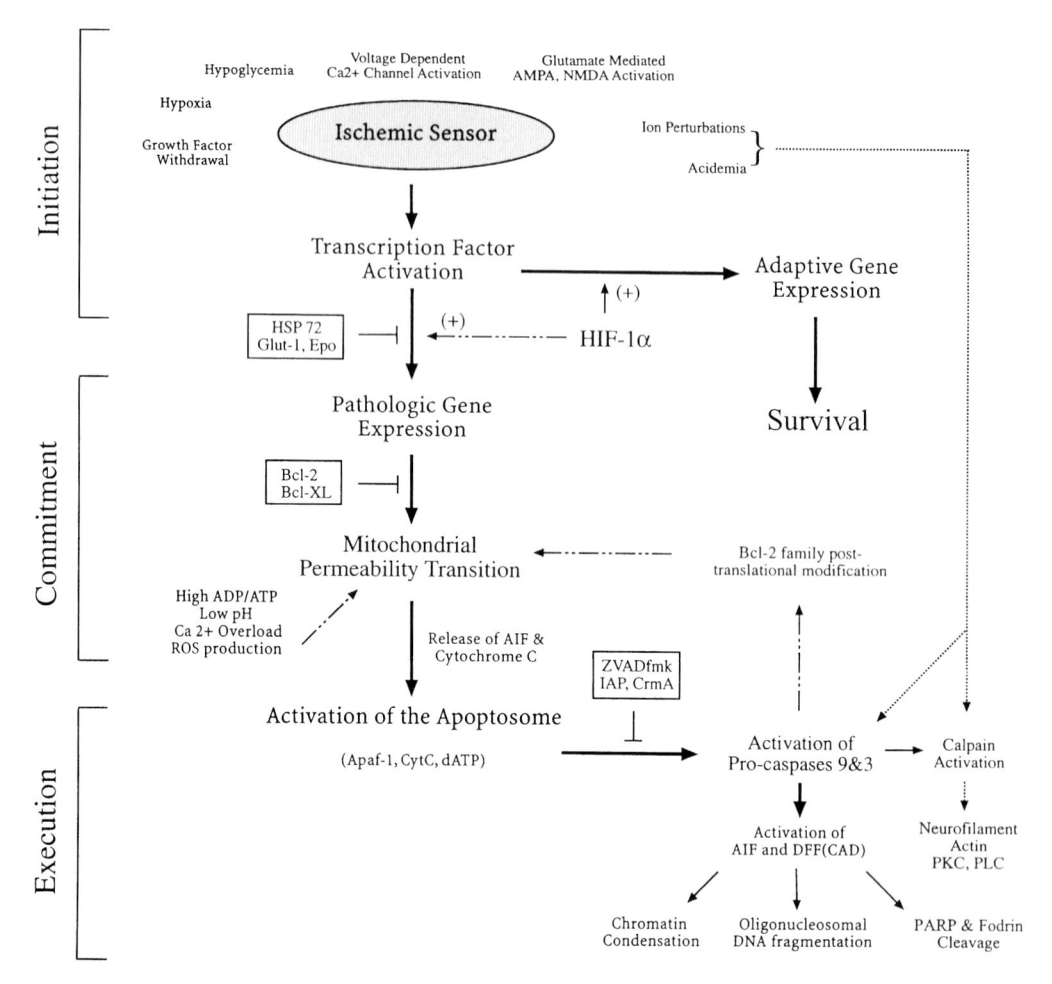

Fig. 1. Ischemia promotes adaptive and pathologic responses in the neuronal compartment. Extrinsic and intrinsic perturbations in the ischemic brain activate a neuronal ischemic sensor, which in turn promotes adaptive and pathologic gene expression (*solid lines*). In addition, select stimuli can activate neuronal death independent of *de novo* gene expression (*dashed line*). Specific gene products, which have demonstrated the ability to block the initiation, commitment or execution phases of the programmed death pathway, are also included (*boxed items*)

subsequent analysis has revealed that in the ischemic neuron, mitochondria undergo dramatic electrochemical, biochemical and macromolecular changes (DESHMUKH and JOHNSON 1998; FUJIMURA et al. 1998).

I. Mitochondria Can Sense and Activate Delayed Death

Mitochondria possess the necessary structural and biochemical complexity to regulate oxidative phosphorylation and the release of an armament of apoptotic molecules. The mitochondrial inner membrane, which surrounds an internal colloidal matrix, is the primary site of cellular ATP production. Inserted into the inner membrane are respiratory multiprotein complexes I, III, and IV, which along with the matrix associated complex II, donate electrons to molecular oxygen through a series of oxidation reduction reactions, while concomitantly shuttling hydrogen ions from the mitochondrial matrix to the intramembrane space. This net accumulation of H^+ lowers the pH of the mitochondrial intermembrane space and generates a -150 to $-200\,mV$ potential across the inner membrane promoting the production of ATP by the proton-translocating ATP synthase. The efficiency of ATP production is enhanced by elevating mitochondrial matrix calcium, which directly influences the activity of complex II (MCCORMACK and DENTON 1990), a phenomenon which has been shown to provide protection to ischemic neurons (OHTA et al. 1996).

The outer mitochondrial membrane contains nonspecific pores that permit diffusion of molecules of up to 10 kDa, yet sequesters the proapoptotic proteins cytochrome c, AIF (apoptosis-inducing factor) as well as caspases 2, 3, and 9 within the intermembrane space thereby prohibiting their association with cytoplasmic targets (SUSIN et al. 1999). In the ischemic brain, mitochondrial release of these factors, which can be induced by transmembrane (extrinsic) and intracellular (intrinsic) signaling events, trigger the structural changes which give apoptotic cells their characteristic morphology. Once released, cytochrome c binds with dATP and Apaf-1 to form the apoptosome, which consequently activates procaspase 9. It should be noted, however, that intracellular injection of cytochrome c is not sufficient to induce caspase activation (DESHMUKH and JOHNSON 1998). It has been proposed that additional factors or costimulatory signals are required to activate the apoptosome – a process which has been termed "competence to die." How cytochrome c, AIF, and the other mitochondrial death factors are liberated from the mitochondrial intermembrane space is unclear although mechanisms involving channel-mediated translocation or passive leak through the ruptured outer mitochondrial membranes have been proposed. The precise temporal relationship between mitochondrial permeability and the release of these apoptogens is also currently under debate. In several models, however, permeability transition occurs prior to the evolution of DNA fragmentation, placing this event early in the apoptotic cascade (VAYSSIERE et al. 1994; ZAMZAMI et al. 1995; MIGNOTTE and VAYSSIERE 1998).

The electrochemical and structural integrity of the mitochondria is regulated in part by the multiprotein permeability transition pore, which bridges the mitochondrial matrix and intermembrane space with the cytosol. Although not completely characterized, the voltage dependent anion channel (VDAC) and the adenine nucleotide transporter (ANT) present in the outer and inner membranes respectively, comprise the major structural elements of this pore. Pore opening is favored by conditions common to ischemic neurons, which include elevated matrix calcium, high ADP/ATP ratios, elevated cytosolic reactive oxygen species, and low pH conditions. By effectively short-circuiting the proton motive force and discharging the inner mitochondrial membrane potential, pore opening can promote homeostasis in metabolically overloaded mitochondria. The appreciation for the role that mitochondria play in the activation and execution of ischemia-induced neuronal apoptosis has defined a rational basis for the development of neuroprotective strategies that target the mitochondrial metabolic and homeostatic machinery.

D. Herpes Virus-Mediated Gene Delivery

Altering the function of genes involved in the ischemic process can be accomplished either by pharmacological inhibition, the transient delivery of plasmid-based vectors, or through permanent manipulation of an organism's genome. However, sustained alterations in the expression patterns of a cellular gene can confound the interpretation of a given disease phenotype in the adult. For example, if the gene in question subserves essential housekeeping or developmental functions, global inactivation often results in embryonic lethality (COPP 1995). Commonly, molecules that contribute to disease susceptibility in the mature nervous system also influence survival during development. In this way, engineered germline mutations could alter developmentally determined neuronal properties, secondarily altering their response to an ischemic insult.

As an alternative to transgenic and knockout models, a number of viral vector strategies (herpes virus, adenovirus, adeno-associated virus, retrovirus, lentivirus) have been used. Virus-mediated gene delivery permits rapid and focal modification of target cells, thus providing the means to test gene function in the absence of a potentially confounding global modification of the genome. Shared among these viral strategies is the ability to control both the timing and location of vector delivery. The size of the genetic construct, tropism of the virus, and the mitotic activity of the target cell are all germane to selecting a suitable vector platform. For greater detail regarding these and other considerations, see (GESCHWIND et al. 1994; HERMENS and VERHAAGEN 1998).

I. The Herpes Amplicon Vector System

Herpes virus vector systems are unique in that they exploit the highly evolved neuronal tropism exhibited by this class of viruses. The amplicon system uses a minimal plasmid, which in addition to the transcription unit of interest, contains the necessary *cis* elements allowing for helper virus-directed replication and packaging of the amplicon DNA into mature virions. One of the amplicon's greatest attributes is its flexibility for accepting large transcription units of variable configuration. Constructs are limited to the theoretical size of the herpes genome (150 kb); however, smaller transcription units are more easily produced using standard cloning techniques. Amplicon constructs can include transcription units which exhibit small molecule regulated expression, cell type-restricted expression, and the independent and constitutive expression of multiple gene products. Because herpes viruses package viral DNA in serial repeated units or concatamers, amplicon viruses are capable of delivering multiple transcriptional units per infection event. This can result in higher levels of accumulated transgene product as compared to other viral vectors encoding identical transcription units (H.J. FEDEROFF, unpublished observations).

Amplicon virus stocks, in addition to containing the cloned amplicon construct, also contain the original helper virus deleted in a single, essential immediate early (IE) gene. This packaging strategy restricts viral proliferation to the packaging cell line, which expresses the complimentary immediate early gene. Single-deletion helper viruses continue to express remaining immediate early gene products, some of which produce cytotoxicity. Once concentrated, amplicon stocks routinely contain $1–3 \times 10^8$ infectious amplicon particles/ml with helper titers ranging between $1–5 \times 10^8$ plaque forming units/ml. Alternative strategies in which the wild-type herpes genome, represented either by a set of overlapping cosmids or a single bacterial artificial chromosome (BAC), resulting in the production of infectious, pure amplicon vector stocks have been developed (FRAEFEL et al. 1996; STAVROPOULOS and STRATHDEE 1998). Detailed protocols describing the components and production of amplicon virus stocks have been recently described.

II. Application of Amplicon Vectors in Neuroscience Research

The HSV amplicon virus system has been used as a vehicle for gene delivery to address a vast array of questions regarding developmental and disease mechanisms in the central nervous system. To investigate the amplicon virus' expression characteristics, initial studies were performed using the reporter construct HSVLac which expresses the β-galactosidase (*LacZ*) gene under the control of the herpes IE 4/5 promoter. The IE 4/5 promoter directs robust reporter gene expression early after infection with the number of β-galactosidase positive cells having reached 30% of maximum levels just 4h after infection. This is likely due to the association of the strong viral transactivator VP16 with the IE 4/5 promoter at the time of infection. Furthermore, HSV amplicon vectors demonstrate highly efficient construct delivery to neurons as well as to other mammalian cell types (H.J. FEDEROFF, unpublished results).

By combining HSV-mediated gene delivery with organotypic hippocampal slice cultures, it becomes possible to investigate a given gene's function in the context of intact, mature neuronal circuits while continuously providing access to the site of gene expression. Initial attempts at delivering virus directly to the culture media produced diffuse infection of glia, occurring mostly at the slice border. We subsequently developed a protocol, which used a microinjection system, mounted on a vibration-free table, to deliver small volumes of virus vector (25–500nl) to subregions of the hippocampal slice. Twenty-four hours after focal delivery of virus at a flow rate of 100nl/min, infected cultures were fixed and assayed for β-galactosidase by X-gal histochemistry. Delivery of 250nl of HSVLac to CA3 stratum pyramidale (at a flow rate of 100nl/min) resulted in intense neuronal and glial staining with efficiencies averaging 83 ± 18%. Similar efficiencies were seen after microapplication of virus into CA1 stratum pyramidale or stratum granulosum. Importantly, the application of 125 HSVLac infectious transducing particles to the CA3 stratum pyramidale did not increase or dampen the excitability of the CA3 region as judged by assessment of evoked field responses in CA3 neurons following depolarization of mossy fibers (CASACCIA-BONNEFIL et al. 1993).

Viral vector delivery across the blood-brain barrier is a highly inefficient process. Although HSV and adenoviral vectors can be delivered to brain regions from peripheral or distant sites in the CNS by retrograde transport (JIN et al. 1996; MAIDMENT et al. 1996; HERMENS et al. 1997), access is most often achieved through stereotactic injection at the intended target site. To transduce specific regions of the cortex and hippocampus using viral vectors, we have applied a similar technology as described above. With the use of a microprocessor-controlled injector system, we have improved consistency in the volumetric distribution of viral vector delivery and the reproducibility of gene product expression between replicate infection events (BROOKS et al. 1998). This is probably due to increased transduction efficiency, a reduction in pressure-related toxicity, and a reduction in pressure-related virus inactivation.

E. Developing Neuroprotective Viral Vector Strategies

Generic milestones in the cell death pathway provide a useful framework to view the increasingly complicated prodeath circuitry. During the initiation phase, signals originate from the activation of surface receptors or intracellular sensors either to convey pro-death messages to the mitochondrion or directly to the latent apoptosome (Fig. 1). It is believed that commitment to a program of cell death occurs at the level of the mitochondrial permeability transition and is thought to proceed in an all-or-none fashion. In neurons, it is also likely that permeability transition precedes caspase activation, although caspases have been shown to promote mitochondrial transition (MARZO et al. 1998). Determining the stage at which the ischemic neuron reaches "the point of no return" is of crucial importance. Provided in Sect. F are several examples of how the manipulation of genes involved in the ischemic pathologic cascade have helped to define the mechanisms which underlie delayed neuronal death.

I. Targeting the Late Stage of Delayed Death: IL-1β and Caspase Activity

Following CNS ischemia, the elevated expression of cytokines in the brain is a known contributor to neuronal death in the postischemic period. The inflammatory cytokine IL-1β, expressed mostly by resident microglia and infiltrating peripheral leukocytes (BHAT et al. 1996), is cleaved to an active form by the interleukin-converting enzyme (FRIEDLANDER et al. 1996). Delivery of either recombinant IL-1 receptor (nonmembrane bound) or the IL-1 receptor antagonist have both been shown to reduce postinfarct lesions by 40% as well as decrease edema and leukocyte infiltration in the affected cortex (STROEMER and ROTHWELL 1997). By delivering a viral construct expressing a secreted factor, diffusion of the therapeutic protein provides protection to neurons distant from the injection site.

In addition to their role in the conversion of the proenzymes like IL-1β, gene disruption studies have established that caspases 3 and 9 play an essential role in commuting neuronal death in developmental and disease paradigms (HAKEM et al. 1998; KUIDA et al. 1998). Caspases may facilitate mitochondrial permeability transition by acting directly on mitochondria or indirectly by cleaving substrates like Bcl-2, Bax, and Bid, thereby enhancing their apoptotic potential through the exposure of potent BH3 domains (CHENG et al. 1997; LUO et al. 1998). Viral gene products and the cellular homologs which potently inhibit activated caspases have been identified (Table 1). Overexpression of the cowpox virus protein CrmA and the mammalian IAP family of proteins can block caspase activity and apoptosis following diverse stimuli (LISTON et al. 1996; SIMONS et al. 1999). Small, cell permeant peptide inhibitors of caspase activity modeled after these proteins also prevent the manifestation of apoptotic morphologies in *in vitro* and in

Table 1. Classification of endogenous neuroprotective genes

Transcript class	Mechanism of action	cDNAs
Bcl-2 superfamily	Regulate mitochondrial permeability transition	Bcl-2, Bcl-X_L
Caspase inhibitors	Viral or endogenous cellular genes which bind to the active site of mature caspases	XIAP, NAIP, CrmA, c-IAP 1&2, survivin, ICH-1_s
Inhibitors of ROS accumulation	Catalyze ROS degradation directly by enzymatic reaction or indirectly by accepting electrons from donating species	SOD-1, Catalase, HO-1
Heat shock proteins	Stress proteins which aid in refolding denatured proteins	HSP72
Glucose handling	Increase the extraction of glucose from the circulation and enhance ATP production	Glucose transporter, ALD-A, LDH-A
Oxygen delivery	Enhance the delivery of oxygen to hypoxic tissue regions	VEGF, Epo
Neurotrophins	Promote survival through receptor-mediated signaling	GDNF, BDNF, NT-3, NGF
Calcium buffering agents	Enhance ability to manage intracellular calcium transients	Calbindin, Calmodulin

vivo ischemic paradigms and are being investigated for their potential use in ischemic neurologic disease (GOTTRON et al. 1997). Treatment with the caspase inhibitor zDEVD has been shown to block caspase-mediated damage protecting neurons for perhaps longer than 3 days when injected up to 9h after mild 30-min transient ischemia (FINK et al. 1998). However, since ischemia induces global perturbations upstream of mitochondrial commitment and caspase activation including the suppression of protein synthesis, it is unclear whether these strategies can interrupt the initiation of delayed death signaling, or restore function to neurons endangered by this process. In fact, CA1 hippocampal neurons previously exposed to transient ischemia appear morphologically normal prior to the onset of delayed death (DESHPANDE et al. 1992).

II. Targeting the Intermediate Stage of Delayed Death: *bcl-2*

The Bcl-2 superfamily of proteins has been demonstrated to modulate mitochondrial function in a vast array of cell types and cell death models. The anti-apoptotic gene *bcl-2* is overexpressed in neurons surviving focal and global ischemia (SHIMAZAKI et al. 1994; CHEN et al. 1997). Moreover, analysis of the regional expression of the Bcl-2 homolog Bax within susceptible brain regions further supports their role in ischemic pathology (CHEN et al. 1996). Many of the Bcl-2 homology domain containing proteins impart their biologic function through interactions with the mitochondria and associated membrane-bound proteins. Although the details remain incomplete, Bax and Bcl-X_L are believed to influence the permeability transition by interacting with the adenine nucleotide transporter and the voltage-dependent anion channel, respectively (MARZO et al. 1998; HEIDEN et al. 1999). Bcl-2 proteins may also control activation of cell death programs by regulating ion flux across organelle membranes, sequestering and inactivating other proapoptotic factors from the cytosol, as well as regulating the levels of cytoplasmic calcium and cytochrome c (MERRY and KORSMEYER 1997; REED 1997).

The activity of the Bcl-2 superfamily is regulated at the transcriptional and post-translational level. One such regulatory mechanism already described involves the caspase-mediated cleavage of targets, including Bid as well as Bcl-2, which activates or enhances prodeath signaling by exposing potent BH3 death domains. Bax is expressed within neurons under basal conditions, and although its increased expression correlates with cell death, the participation of posttranslational processes apart from proteolytic cleavage are required to cause cell death (GOPING et al. 1998). In the absence of pro-death stimuli, the enforced homodimerization of Bax is sufficient to induce translocation of the complex to mitochondria (GROSS et al. 1998). Neurotrophic factor-induced tyrosine kinase receptor signaling also regulates *bcl-2* proteins. For example, withdrawal of NGF from either sympathetic or cerebellar neurons diminishes Akt-mediated phosphorylation of Bad and promotes the ability of Bad to free Bax from Bcl-X_L/Bax heterodimer complexes, thus stimulating cell death

(YANG et al. 1995; DUDEK et al. 1997). This is likely to be of significance in the ischemic brain given the ability of the related neurotrophins GDNF or BDNF to confer neuroprotection in pre- and post-delivery models of focal ischemia (BECK et al. 1994; WANG et al. 1997).

1. Viral Delivery of Bcl-2 Protects Ischemic Neurons

Neuron-specific expression of Bcl-2 reduces infarct size in transgenic mice following permanent focal ischemia (MARTINOU et al. 1994). Since stroke volume measurements were made 7 days after the onset of ischemia, it is possible Bcl-2 protected cortical tissue from necrotic as well as delayed death. However, longterm expression of Bcl-2 is also known to induce catalase activity and superoxide dismutase activity, as well as increase total GSH levels while inhibiting nuclear calcium concentrations. It was therefore difficult to know whether transient expression of Bcl-2 protein in the peri-ischemic period would recapitulate the protection observed in the transgenic model (ELLERBY et al. 1996). To test this, we constructed an amplicon vector expressing the human *bcl-2* gene and tested its ability to protect cortical neurons when delivered prior to ischemia. Two microliters of virus (1×10^5 *bcl-2* transducing particles) was delivered to two cortical injection sites, each selected to reside 1 mm from the medial edge of the infarction as determined in control, stroked animals. Twenty-four hours after virus delivery, rats were subjected to permanent tandem occlusion of the right middle cerebral artery and right common carotid artery. The effect of virus injection on ischemic damage was subsequently evaluated by whole brain staining with 2% TTC (2,3,5-triphenyl-tetrazolium chloride) in PBS (phosphate buffered saline), followed by the determination of the distance from the midline of the brain to the medial edge of the infarct at each injection point. Using HSVLac for comparison, significantly more viable tissue was present at both anterior and posterior injection sites in HSVbcl-2 (LINNIK et al. 1995). These results suggest that expression of Bcl-2 prior to the onset of ischemia is neuroprotective. In agreement with the MCAO model, pre-treatment with HSVbcl-2 has also been shown to protect CA1 neurons from transient global ischemia (ANTONAWICH et al. 1999). Alternatively, others have shown that in a post-treatment paradigm, Bcl-2 delivery to the striatum 30 min following 1h of focal occlusion prevented significant neuronal loss when assessed 48h later (LAWRENCE et al. 1996). However, delivery of Bcl-2 4h after the beginning of the reperfusion phase was no longer protective. In aggregate these results demonstrate that acute infection with *bcl-2* vectors in both pre- and post-insult paradigms can protect neurons in the ischemic penumbra. Although not directly tested, the most parsimonious explanation for this protective activity favors Bcl-2 activity at the mitochondrion rather than through the mechanisms documented in longterm expression paradigms.

We further clarified Bcl-2's neuroprotective activity using an *in vitro* model of hypoxia-induced delayed neuronal death (GWAG et al. 1995). In this system, mixed cortical cultures are made anoxic and hypoglycemic in the presence of the AMPA and glutamate receptor antagonists CNQX and MK801. Blockade of glutamate excitotoxicity during oxygen glucose deprivation (OGD) blocks the initial wave of neuronal necrosis and promotes the maturation of the delayed form of neuronal death. This process, which matures 48h post-insult, can be blocked with caspase and protein synthesis inhibitors. In our experience nearly 35–40% of neurons from mixed cultures and 50–60% of neurons from purified cortical cultures die following ischemic exposure using this model (Fig. 2). Delivery of HSVLac amplicon virus at an MOI of 1.25 does not effect culture viability when assessed 24h after vector delivery, and produces similar profiles of cell death when compared to uninfected, OGD-treated cultures. Importantly, we have found that addition of HSVbcl-2 can provide protection in the delayed paradigm of hypoxia-induced death while at the same time providing no apparent protection to neurons under-

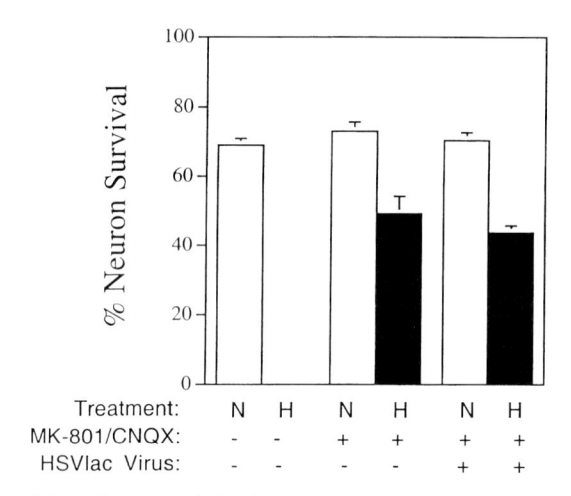

Fig. 2. Modeling delayed neuronal death *in vitro*. To test the activity of putative amplicon containing viral stocks, dissociated cortical cultures were infected with HSV amplicon virus, exposed to oxygen glucose deprivation and assesed for neuron viability. Mixed cortical cultures were harvested according to protocols described in GOTTRON et al. (1996). Select wells from 24-well cultures were infected with HSVlac amplicon virus (*MOI 1.25*) 24h prior to hypoxic insult, which results in 70% of neurons expressing β-galactosidase. Maintenance media was replaced with EBSS containing 0mM glucose, select wells were also exposed to 10μM of MK-801 and 100μM CNQX, and cultures were either incubated under normoxic (*N*) or hypoxic (*H*; 0.5% O_2) conditions at 37°C /6%CO_2. After 3-h exposure, media was replaced with EBSS containing 25mM glucose and cultures were stained 48h later with 0.4% trypan blue, and neuron survival was calculated from an average of four fields taken from quadruplicate wells. Untreated mixed cultures undergo a rapid necrotic cell death (*first pair of bars*). Addition of the glutamate antagonists MK-801 and CNQX blocks necrosis and allows a delayed neuron death to develop 48h later (*second pair of bars*). Finally, pretreatment with HSVlac amplicon virus which expresses the inert β-galactosidase, neither enhances nor inhibits neuron survival (*last pair of bars*)

going hypoxia-induced necrosis. These results suggest that acute delivery of a Bcl-2-expressing amplicon virus protects cortical neurons from undergoing delayed, but not necrotic cell death.

While transient Bcl-2 expression is capable of protecting neurons from ischemia-induced death in the short term, it is less clear whether this protection will promote long-term survival. For example, while the overexpression of Bcl-2 in the mutant superoxide dismutase (SOD-1) transgenic model of familial amyotrophic lateral sclerosis (ALS) delayed the onset of disease, prolonging the life span of mutant SOD mice, it did not shorten the disease duration (Kostic et al. 1997). Similar genetic studies performed in the *pmn* mouse model of progressive neurodegeneration cast similar doubts on Bcl-2's therapeutic utility. Although expression of Bcl-2 in these experiments promoted the survival of neuronal cell bodies in select populations, it did not prevent the degeneration of myelinated axons or increase the life span of the animals. The description of a *bcl-2*-independent prodeath signaling pathway in Lurcher mutant mice crossed with *bcl-2* transgenics offers a potential explanation as to why Bcl-2 expression fails to provide complete protection in the above experimental paradigms (Sagot et al. 1995; Zanjani et al. 1998).

F. Using Viral Vectors to Regulate Early Ischemic Transcriptional Responses

When exposed to a sublethal hypoxic stimulus, neurons become refractory to subsequent severe ischemic stimuli. This phenomenon, known as ischemic tolerance, occurs rapidly at the biochemical level as well as on a more delayed timescale through the elaboration of protective gene products (Barone et al. 1998). For example, 30 min of ischemia, delivered in either continuous or fractioned doses, can protect up to 30% of neurons from subsequent death as well as providing protection to glia and blood vessel endothelium (Chen et al. 1996). Tolerance in these experiments required 48 h to mature and lasted up to 7 days and required *de novo* protein synthesis. Chemically or electrically induced neuronal depolarization can also engender ischemic tolerance, suggesting that this phenomenon may apply to disease processes other than ischemia (Kobayashi et al. 1995).

Preconditioning studies have identified a subset of ischemic tolerance transcripts involved in the inactivation of reactive oxygen species (ROS), growth factor signaling, DNA repair, and the regulation of cytoplasmic calcium levels. The neuroprotective properties of many of these gene products have been confirmed using transgenic, knockout as well as virus-mediated gene delivery paradigms (Table 1). Neuronal protein synthesis is directly related to the degree of blood flow to a given region. Ultrastructural and biochemical studies have described the disaggregation of polyribosomes and a global reduction in the incorporation of radioactive amino acids in ischemic neurons (Petito and Pulsinelli 1984; Araki et al. 1990; Neumar et al. 1998). It is remarkable that despite this generalized phenomenon, ischemia potently upregulates the levels of a select subset of proteins.

I. HIF-1α and p53 Regulate Ischemic Gene Expression

The process by which neurons change their profiles of gene expression to favor death over survival, and which stimuli induce this change, remain poorly understood. Cell culture experiments have established that hypoxia, hypoglycemia, and acidemia potently modulate mRNA abundance at the transcriptional and post-transcriptional level. However, since hypoxia is the leading physiologic perturbation in the model of transient global ischemia, we hypothesized that in addition to controlling the expression of many adaptive gene species, hypoxia is also responsible for inducing pro-death gene expression. Vertebrates have evolved elaborate mechanisms to regulate the rapid and high level expression of hypoxia-inducible proteins. Members of the PER-ARNT-SIM (PAS) family of basic-helix-loop-helix transcription factors have been shown to mediate hypoxia inducible transcriptional events. The necessity for these specific PAS family members to engender the adaptive response has been established through gene disruption. Loss of either HIF-1α or ARNT (HIF-1β) in embryonic stem cells renders them largely incapable of upregulating the expression of hypoxia- and hypoglycemia-responsive target genes. Unfortunately, embryonic lethality has prevented the use of these genetic mutants in models of CNS ischemia (MALTEPE et al. 1997; IYER et al. 1998).

While the nuclear factor HIF-1β is present under normoxic conditions, its heterodimerization partner HIF-1α is highly labile and is subject to ubiquitin-mediated proteolysis, suggesting that HIF-1α is the rate-limiting factor in the formation of the HIF-1 complex (SALCEDA and CARO 1997). At the molecular level, a critical iron-containing moiety in the carboxy-terminus of HIF-1α as well as the participation of reactive oxygen species produced by the mitochondrial complex III have been shown to regulate the stabilization of the HIF-1α protein (CHANDEL et al. 1998). Under hypoxic conditions, the stabilized HIF-1α protein dimerizes with HIF-1β forming the HIF-1 complex that translocates to the nucleus and binds to hypoxia-responsive elements (HRE) (FIRTH et al. 1995; MELILLO et al. 1995). HREs have been identified upstream of a number of "adaptive" transcriptional targets which include the glucose transporter and enzymes involved in glycolysis, as well as the growth factor erythropoietin. Hypoxic neuronal cell lines and glial cultures also exhibit HIF-1 binding activity and regulate some of the HIF-1 transcriptional targets described above (RUSCHER et al. 1998). Our laboratory has also shown that HIF-1α mRNA and HIF-1 transcriptional activity are present in normoxic purified cortical neurons and hypoxic cortical cultures, respectively. These data suggest that HIF-1 activation is conserved throughout mammalian tissues, including cell types resident in the CNS.

The tumor suppressor protein p53 is among the best-studied prodeath transcriptional activators. Under conditions of extreme hypoxia, p53 promotes growth arrest of dividing cells and apoptosis through the transactivation of genes such as p21$^{waf1/cip1}$ and *bax*. Human tumors harboring mutated p53 allele(s) exhibit reduced apoptosis under hypoxic conditions (GRAEBER et al. 1996; LEVINE 1997), and restoration of p53 function in these lines restrains cell growth and often reestablishes apoptotic potential (CHEN et al. 1996). The role of p53 in producing pathology in neurons is supported by the reduction of infarct volumes measured in mice deficient in p53 (CRUMRINE et al. 1994), and by increased neuronal p53 expression, which temporally precedes neuronal death in the ischemic brain (LI et al. 1994, 1997).

The recent discovery that HIF-1α stabilizes p53 in tumor cell lines suggests a model in which cellular hypoxia can lead to either adaptive or pathologic responses depending on the presence and degree of p53 stabilization. Exposure to hypoxia, or hypoxia-mimetic agents, stabilizes p53 and HIF-1α protein levels resulting in enhanced target gene transcription (AN et al. 1998; BLAGOSKLONNY et al. 1998). Also, removal of HIF-1α from embryonic stem cells blocks the induction of p53 levels following extreme hypoxic exposure and attenuates stress-induced cell death (CARMELIET et al. 1998). This relationship between p53 and HIF-1α highlights a novel therapeutic target for regulating the balance of adaptive and pathologic mRNA species.

II. Disruption of HIF-1α Signaling Protects Against Ischemia-Induced Delayed Neuronal Death

We have recently investigated whether the transcriptional regulators HIF-1α and p53 are involved in delayed neuronal death. To probe for activation of the HIF-1 complex in neurons, we constructed a series of amplicon reporter constructs containing HREs (BLANCHARD et al. 1992). When transiently transfected into Hep3B cells, HRE constructs exhibited increasing induction profiles under hypoxic conditions commensurate with the number of HIF-1α binding sites present (Fig. 3). To test for the activation of the HIF-1 complex in the ischemic penumbra, the single HRE containing reporter plasmid was packaged into HSV virus particles and used in an in vivo paradigm of cerebral ischemia. HSVLac and HREprLac (1μl each) were injected into the cortices of mice in a region determined in preliminary stroke experiments to fall along the medial stroke border. Mice were subjected to permanent focal ischemia 24h later and sections were harvested and analyzed for β-galactosidase activity, correcting for recovery of the injection site using semi-quantitative PCR directed against sequences in the amplicon backbone. Results showed that reporter activity was increased 24h after the onset of ischemia only in ischemic hemispheres injected with HREprLac (HALTERMAN et al. 1996). Conversely, expression from the HSVLac-infected cortices was reduced. Companion experiments performed in mixed cortical cultures

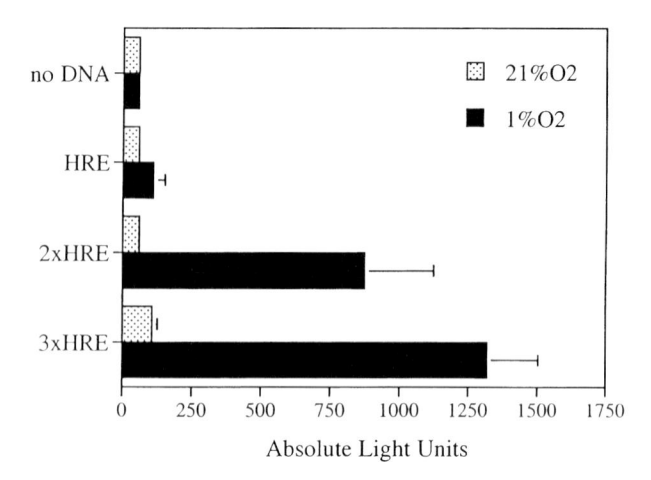

Fig. 3. HRE duplication enhances hypoxia-responsive transcription in transfected Hep3B cells. To produce a hypoxia-responsive amplicon virus vector, hypoxia-responsive elements were cloned into the amplicon reporter plasmid HSVori^sLac. Reporter plasmids containing either 1, 2, or 3 copies of the HRE were transfected into the human hepatoma line Hep3B and subsequently exposed to either normoxic (21%O_2) or hypoxic (1%O_2) conditions in normal growth media (DMEM-HG/10% FCS/1xP/S) for 24h. Monolayers were harvested and analyzed for β-galactosidase activity using the Galactolite kit (Tropix, Bedford, Mass., USA). Data represent the average (±SD) absolute luminescence from four wells

suggest the reduction seen in HSVLac expression is likely due to an overall reduction in the activity of the translational machinery in the ischemic cortex.

Having identified that HIF-1 activity was increased in cortical regions containing at risk neuronal populations, we determined whether dominant negative disruption of the HIF-1 transcriptional signaling pathway could protect ischemic neurons from undergoing OGD-induced delayed neuronal death. The neuroprotective potential of the HIF-1α dominant negative (HIFdn) amplicon construct, which lacks both basic DNA binding and transcriptional activation domains, was tested using in vitro models of oxygen glucose deprivation-induced necrosis and delayed neuronal death. While HIFdn provided no protection against necrosis, neuron death in mixed cortical cultures subjected to the delayed paradigm was significantly reduced, presumably through the disruption of HIF-1-mediated signaling (HALTERMAN and FEDEROFF 1999). These findings were recapitulated using a glia-free neuronal preparation, suggesting that oxygen glucose deprivation triggers a cell-autonomous, HIF-1α-mediated cell death pathway in neurons. Given the association between HIF-1α and p53, we performed a similar series of protection experiments using cortical cultures made from p53 deficient mice. In contrast to its ability to protect wild-type neurons, virus-mediated delivery of the HIFdn construct did not promote survival in p53 null neurons exposed to lethal levels of hypoxia. In fact, p53 null neurons infected with the HIFdn construct exhibited reduced survival, which may in part reflect an inhibition of adaptive transcripts in the peri-ischemic period. These data support the thesis that hypoxic activation of HIF-1 signaling depends on p53 for its apoptotic readout.

Thus the simplest model posits that under basal conditions where p53 levels are low, hypoxia induces HIF-1 signaling through heterodimeric PAS member associations and leads to upregulation of adaptive genes. Under extreme conditions, HIF-1α or other HIF homologues associate with p53 and direct the expression of pro-death transcripts. These data define a new and important node of regulation of ischemic cell death. Implicit in this model is the testable notion of an integrative hypoxia sensing mechanism that not only determines which fate, adaptive or pathologic, the cell will follow, but also provides a rational regulatory node at which to direct neuron-sparing therapeutics.

G. Conclusions

In the post-ischemic period, paracrine as well as intracellular signaling events play a major role in determining neuronal survival. It has been recognized that ischemia promotes both adaptive and maladaptive programs of gene expression depending on the duration and severity of the stimulus, and consequently on the nature of the transcriptional response. Identifying the relevant physiologic stimuli, the identity of the ischemic sensor, and the means by which these components interact to activate programmed cell death cascade will greatly accelerate the development of neuroprotective compounds active against delayed death. If the concept of a commitment point in ischemic death signaling is valid, therapies designed to target events which transpire during the execution phase (i.e., caspase inhibitor strategies) may provide only transient protection. Alternatively, interventions which target the precommitment transcriptional or biochemical events are likely to provide broader, longer-lasting protection against delayed neuronal death.

The studies described herein illustrate how virus-mediated overexpression of HIF-1α dominant negative can inhibit delayed neuronal death likely by limiting the expression of death-inducing hypoxic transcripts during the early stages of neuronal ischemia. These findings highlight a novel regulatory node in ischemia-induced cell death signaling toward which neuron-sparing therapeutics can be applied. Further definition of the factor(s) which control the ischemic threshold and prodeath gene transcription will provide the opportunity to develop pharmacological agents that inactivate pathologic gene expression, while sparing the adaptive transcriptional response.

References

An WG, Kanekal M, Simon MC, Maltepe E, Blagosklonny MV, Neckers LM (1998) Stabilization of wild-type p53 by hypoxia-inducible factor 1alpha. Nature 392 (6674):405–408

Antonawich FJ, Federoff HJ, Davis JN (1999) bcl-2 transduction, using a herpes simplex virus amplicon, protects hippocampal neurons from transient global ischemia. Exp Neurol 156(1):130–137

Araki T, Kato H, Inoue T, Kogure K (1990) Regional impairment of protein synthesis following brief cerebral ischemia in the gerbil. Acta Neuropathol 79(5):501–505

Astrup J, Siesjo B, Symon L (1981) Thresholds in cerebral ischemia – the ischemic penumbra. Stroke 12(6):723–725

Barone FC, White RF, Spera PA, Ellison J, Currie RW, Wang X, Feuerstein GZ (1998) Ischemic preconditioning and brain tolerance: temporal histological and functional outcomes, protein synthesis requirement, and interleukin-1 receptor antagonist and early gene expression. Stroke 29(9):1937–1950; discussion 1950–1951

Beck T, Lindholm D, Castren E, Wree A (1994) Brain-derived neurotrophic factor protects against ischemic cell damage in rat hippocampus. J Cereb Blood Flow Metab 14(4):689–692

Bhat RV, DiRocco R, Marcy VR, Flood DG, Zhu Y, Dobrzanski P, Siman R, Scott R, Contreras PC, Miller M (1996) Increased expression of IL-1β converting enzyme in hippocampus after ischemia: Selective localization in microglia. J Neurosci 16(13):4146–4154

Blagosklonny MV, An WG, Romanova LY, Trepel J, Fojo T, Neckers L (1998) p53 inhibits hypoxia-inducible factor-stimulated transcription. J Biol Chem 273(20): 11995–11998

Blanchard KL, Acquaviva AM, Galson DL, Bunn HF (1992) Hypoxic induction of the human erythropoietin gene: cooperation between the promoter and enhancer, each of which contains steroid receptor response elements. Mol Cell Biol 12(12): 5373–5385

Brooks AI, Halterman MW, Chadwick CA, Davidson BL, Haak-Frendscho M, Radel CA, Porter C, Federoff HJ (1998) Reproducible and efficient murine CNS gene delivery using a microprocessor-controlled injector. Journal of Neuroscience Methods 80(2):137–147

Carmeliet P, Dor Y, Herbert JM, Fukumura D, Brusselmans K, Dewerchin M, Neeman M, Bono F, Abramovitch R, Maxwell P, Koch CJ, Ratcliffe P, Moons L, Jain RK, Collen D, Keshet E (1998) Role of HIF-1alpha in hypoxia-mediated apoptosis, cell proliferation and tumour angiogenesis. Nature 394(6692):485–490

Casaccia-Bonnefil P, Benedikz E, Shen H, Stelzer A, Edelstein D, Geschwind MD, Brownlee M, Federoff HJ, Bergold PJ (1993) Localized gene transfer into hippocampal slice cultures and acute hippocampal slices. J Neurosci Meth 50:341–351

Chandel NS, Maltepe E, Goldwasser E, Mathieu CE, Simon MC, Schumacker PT (1998) Mitochondrial reactive oxygen species trigger hypoxia-induced transcription. Proc Natl Acad Sci USA 95(20):11715–11720

Charriaut-Marlangue C, Margaill I, Represa A, Popovici T, Plotkine M, Ben-Ari Y (1996) Apoptosis and necrosis after reversible focal ischemia: an in situ DNA fragmentation analysis. J Cereb Blood Flow Metab 16(2):186–194

Chen J, Graham S, Zhu R, Simon R (1996) Stress proteins and tolerance to focal cerebral ischemia. J Cereb Blood Flow Metab 16(4):566–577

Chen J, Graham SH, Nakayama M, Zhu RL, Jin K, Stetler RA, Simon RP (1997) Apoptosis repressor genes bcl-2 and bcl-x-long are expressed in the rat brain following global ischemia. Journal of Cerebral Blood Flow and Metabolism 17:2–10

Chen J, Zhu RL, Nakayama M, Kawaguchi K, Jin K, Stetler RA, Simon RP, Graham SH (1996) Expression of the apoptosis-effector gene, Bax, is up-regulated in vulnerable hippocampal CA1 neurons following global ischemia. J Neurochem 67(1): 64–71

Chen X, Ko LJ, Jayaraman L, Prives C (1996) p53 levels, functional domains, and DNA damage determine the extent of the apoptotic response of tumor cells. Genes Dev 10(19):2438–2451

Cheng EH, Kirsch DG, Clem RJ, Ravi R, Kastan MB, Bedi A, Ueno K, Hardwick JM (1997) Conversion of Bcl-2 to a Bax-like death effector by caspases. Science 278(5345):1966–1968

Copp AJ (1995) Death before birth: Clues from gene knockouts and mutations. TIG 11(3):87–93

Crumrine RC, Thomas AL, Morgan PF (1994) Attenuation of p53 expression protects against focal ischemic damage in transgenic mice. J Cereb Blood Flow Metab 14(6):887–891

Deshmukh M, Johnson EM Jr (1998) Evidence of a novel event during neuronal death: development of competence-to-die in response to cytoplasmic cytochrome c. Neuron 21(4):695–705

Deshpande J, Bergstedt K, Linden T, Kalimo H, Wieloch T (1992) Ultrastructural changes in the hippocampal CA1 region following transient cerebral ischemia: evidence against programmed cell death. Exp Brain Res 88(1):91–105

Du C, Hu R, Csernansky CA, Hsu CY, Choi DW (1996) Very delayed infarction after mild focal cerebral ischemia: a role for apoptosis? J Cereb Blood Flow Metab 16(2):195–201

Dudek H, Datta SR, Franke TF, Birnbaum MJ, Yao R, Cooper GM, Segal RA, Kaplan DR, Greenberg ME (1997) Regulation of neuronal survival by the serine-threonine protein kinase Akt. Science 275(5300):661–665

Ellerby LM, Ellerby HM, Park SM, Holleran AL, Murphy AN, Fiskum G, Kane DJ, Testa MP, Kayalar C, Bredesen DE (1996) Shift of the cellular oxidation-reduction potential in neural cells expressing *bcl-2*. J Neurochem 67(3):1259–1267

Fink K, Zhu J, Namura S, Shimizu-Sasamata M, Endres M, Ma J, Dalkara T, Yuan J, Moskowitz MA (1998) Prolonged therapeutic window for ischemic brain damage caused by delayed caspase activation. J Cereb Blood Flow Metab 18(10):1071–1076

Firth JD, Ebert BL, Ratcliffe PJ (1995) Hypoxic regulation of lactate dehydrogenase A. Interaction between hypoxia-inducible factor 1 and cAMP response elements. J Biol Chem 270(36):21021–21027

Fraefel C, Song S, Lim F, Lang P, Yu L, Wang Y, Wild P, Geller AI (1996) Helper virus-free transfer of herpes simplex virus type 1 plasmid vectors into neural cells. J Virol 70(10):7190–7197

Friedlander RM, Gagliardini V, Rotello RJ, Yuan J (1996) Functional role of interleukin 1β (IL-1β) in IL-1β-converting enzyme-mediated apoptosis. J Exp Med 184:717–724

Fujimura M, Morita-Fujimura Y, Murakami K, Kawase M, Chan PH (1998) Cytosolic redistribution of cytochrome c after transient focal cerebral ischemia in rats. J Cereb Blood Flow Metab 18(11):1239–1247

Geschwind MD, Lu B, Federoff HJ (1994) Expression of neurotrophic genes from herpes simplex virus type 1 vectors: Modifying neuronal phenotype. Methods Neurosci 21:462–482

Goping IS, Gross A, Lavoie JN, Nguyen M, Jemmerson R, Roth K, Korsmeyer SJ, Shore GC (1998) Regulated targeting of BAX to mitochondria. J Cell Biol 143(1):207–215

Gottron FJ, Ying HS, Choi DW (1997) Caspase inhibition selectively reduces the apoptotic component of oxygen-glucose deprivation-induced cortical neuronal cell death. Mol Cell Neurosci 9(3):159–169

Graeber TG, Osmanian C, Jacks T, Housman DE, Koch CJ, Lowe SW, Giaccia AJ (1996) Hypoxia-mediated selection of cells with diminished apoptotic potential in solid tumours. Nature 379(6560):88–91

Gross A, Jockel J, Wei MC, Korsmeyer SJ (1998) Enforced dimerization of BAX results in its translocation, mitochondrial dysfunction and apoptosis. Embo J 17(14):3878–3885

Gwag BJ, Lobner D, Koh JY, Wie MB, Choi DW (1995) Blockade of glutamate receptors unmasks neuronal apoptosis after oxygen- glucose deprivation in vitro. Neuroscience 68(3):615–619

Hakem R, Hakem A, Duncan GS, Henderson JT, Woo M, Soengas MS, Elia A, de la Pompa JL, Kagi D, Khoo W, Potter J, Yoshida R, Kaufman SA, Lowe SW,

Penninger JM, Mak TW (1998) Differential requirement for caspase 9 in apoptotic pathways in vivo. Cell 94(3):339–352

Halterman M, Federoff H (1999) HIF-1α Cooperates with p53 to Promote Hypoxia-induced Neuronal Death. Keystone Symposia – Apoptosis and Programmed Cell Death, Breckenridge, CO

Halterman M, Panahian N, Federoff H (1996) Inclusion of hypoxia responsive elements confers regulated gene expression in the HSV amplicon system. Society for Neuroscience, Washington, D.C.

Heiden MG, Chandel NS, Schumacker PT, Thompson CB (1999) Bcl-xL prevents cell death following growth factor withdrawal by facilitating mitochondrial ATP/ADP exchange. Mol Cell 3(2):159–167

Hermens WT, Giger RJ, Holtmaat AJ, Dijkhuizen PA, Houweling DA, Verhaagen J (1997) Transient gene transfer to neurons and glia: analysis of adenoviral vector performance in the CNS and PNS. J Neurosci Methods 71(1):85–98

Hermens WT, Verhaagen J (1998) Viral vectors, tools for gene transfer in the nervous system. Prog Neurobiol 55(4):399–432

Ito U, Spatz M, Walker J, Katzo I (1975) Experimental cerebral ischemia in mongolian gerbils. I. Light microscopic observations. Acta Neuropath 32(3):209–223

Iyer NV, Kotch LE, Agani F, Leung SW, Laughner E, Wenger RH, Gassmann M, Gearhart JD, Lawler AM, Yu AY, Semenza GL (1998) Cellular and developmental control of O2 homeostasis by hypoxia- inducible factor 1 alpha. Genes Dev 12(2): 149–162

Jacewicz M, Kiessling M, Pulsinelli WA (1986) Selective gene expression in focal cerebral ischemia. J Cereb Blood Flow Metab 6(3):263–272

Jin BK, Belloni M, Conti B, Federoff HJ, Starr R, Son JH, Baker H, Joh TH (1996) Prolonged in vivo gene expression driven by a tyrosine hydroxylase promoter in a defective herpes simplex virus amplicon vector. Hum Gene Ther 7(16):2015–2024

Kirino T, Tamura A, Sano K (1984) Delayed neuronal death in the rat hippocampus following transient forebrain ischemia. Acta Neuropathol 64:139–147

Kobayashi S, Harris VA, Welsh FA (1995) Spreading depression induces tolerance of cortical neurons to ischemia in rat brain. J Cereb Blood Flow Metab 15(5):721–727

Kostic V, Jackson-Lewis V, de Bilbao F, Dubois-Dauphin M, Przedborski S (1997) Bcl-2: prolonging life in a transgenic mouse model of familial amyotrophic lateral sclerosis. Science 277(5325):559–562

Kuida K, Haydar TF, Kuan CY, Gu Y, Taya C, Karasuyama H, Su MS, Rakic P, Flavell RA (1998) Reduced apoptosis and cytochrome c-mediated caspase activation in mice lacking caspase 9. Cell 94(3):325–337

Lawrence MS, Sun GH, Kunis DM, Saydam TC, Dash R, Ho DY, Sapolsky RM, Steinberg GK (1996) Overexpression of the glucose transporter gene with a herpes simplex viral vector protects striatal neurons against stroke. J Cereb Blood Flow Metab 16(2):181–185

Levine AJ (1997) p53, the cellular gatekeeper for growth and division. Cell 88(3): 323–331

Li Y, Chopp M, Powers C (1997) Granule cell apoptosis and protein expression in hippocampal dentate gyrus after forebrain ischemia in the rat. J Neurol Sci 150(2): 93–102

Li Y, Chopp M, Zhang ZG, Zaloga C, Niewenhuis L, Gautam S (1994) p53-immunoreactive protein and p53 mRNA expression after transient middle cerebral artery occlusion in rats. Stroke 25(4):849–855; discussion 855–856

Linnik MD, Zahos P, Geschwind MD, Federoff HJ (1995) Expression of bcl-2 from a defective herpes simplex virus-1 vector limits neuronal death in focal cerebral ischemia. Stroke 26(9):1670–1674; discussion 1675

Liston P, Roy N, Tamai K, Lefebvre C, Baird S, Cherton-Horvat S, Farahani R, McLean M, Ikeda JE, MacKenzie A, Korneluk RG (1996) Suppression of apoptosis in mammalian cells by NAIP and a related family of IAP genes. Nature 379(6563): 349–353

Luo X, Budihardjo I, Zou H, Slaughter C, Wang X (1998) Bid, a Bcl-2 interacting protein, mediates cytochrome c release from mitochondria in response to activation of cell surface death receptors. Cell 94(4):481–490

Ma J, Endres M, Moskowitz MA (1998) Synergistic effects of caspase inhibitors and MK-801 in brain injury after transient focal cerebral ischaemia in mice. Br J Pharmacol 124(4):756–762

Maidment NT, Tan AM, Bloom DC, Anton B, Feldman LT, Stevens JG (1996) Expression of the lacZ reporter gene in the rat basal forebrain, hippocampus, and nigrostriatal pathway using a nonreplicating herpes simplex vector. Exp Neurol 139(1):107–114

Maltepe E, Schmidt JV, Baunoch D, Bradfield CA, Simon MC (1997) Abnormal angiogenesis and responses to glucose and oxygen deprivation in mice lacking the protein ARNT. Nature 386(6623):403–407

Martinou JC, Dubois-Dauphin M, Staple JK, Rodriguez I, Frankowski H, Missotten M, Albertini P, Talabot D, Catsicas S, Pietra C, et al (1994) Overexpression of *bcl-2* in transgenic mice protects neurons from naturally occurring cell death and experimental ischemia. Neuron 13(4):1017–1030

Marzo I, Brenner C, Zamzami N, Jurgensmeier JM, Susin SA, Vieira HL, Prevost MC, Xie Z, Matsuyama S, Reed JC, Kroemer G (1998) Bax and adenine nucleotide translocator cooperate in the mitochondrial control of apoptosis. Science 281(5385):2027–2031

Marzo I, Brenner C, Zamzami N, Susin SA, Beutner G, Brdiczka D, Remy R, Xie ZH, Reed JC, Kroemer G (1998) The permeability transition pore complex: a target for apoptosis regulation by caspases and Bcl-2-related proteins. J Exp Med 187(8):1261–1271

McCormack JG, Denton RM (1990) The role of mitochondrial Ca2+ transport and matrix Ca2+ in signal transduction in mammalian tissues. Biochim Biophys Acta 1018(2–3):287–291

Melillo G, Musso T, Sica A, Taylor TS, Cox GW, Varesio L (1995) A hypoxia-responsive element mediates a novel pathway of activation of the inducible nitric oxide synthase promoter. J Exp Med 182(6):1683–1693

Merry DE, Korsmeyer SJ (1997) *bcl-2* gene family in the nervous system. Annu Rev Neurosci 20:245–267

Mignotte B, Vayssiere JL (1998) Mitochondria and apoptosis. Eur J Biochem 252(1):1–15

Miller TM, Moulder KL, Knudson CM, Creedon DJ, Deshmukh M, Korsmeyer SJ, Johnson EM Jr (1997) Bax deletion further orders the cell death pathway in cerebellar granule cells and suggests a caspase-independent pathway to cell death. J Cell Biol 139(1):205–217

Neumar RW, DeGracia DJ, Konkoly LL, Khoury JI, White BC, Krause GS (1998) Calpain mediates eukaryotic initiation factor 4G degradation during global brain ischemia. J Cereb Blood Flow Metab 18(8):876–881

Ohta S, Furuta S, Matsubara I, Kohno K, Kumon Y, Sakaki S (1996) Calcium movement in ischemia-tolerant hippocampal CA1 neurons after transient forebrain ischemia in gerbils. J Cereb Blood Flow Metab 16(5):915–922

Paschen W, Mies G, Hossmann KA (1992) Threshold relationship between cerebral blood flow, glucose utilization, and energy metabolites during development of stroke in gerbils. Exp Neurol 117(3):325–333

Petito C, Feldman E, Pulsinelli W, Plum F (1987) Delayed hippocampal damage in humans following cardiorespiratory arrest. Neurology 37(8):1281–1286

Petito CK, Pulsinelli WA (1984) Delayed neuronal recovery and neuronal death in rat hippocampus following severe cerebral ischemia: possible relationship to abnormalities in neuronal processes. J Cereb Blood Flow Metab 4(2):194–205

Pulsinelli W (1992) Pathophysiology of acute ischaemic stroke. Lancet 339(8792):533–536

Pulsinelli WA, Brierley JB, Plum F (1982) Temporal profile of neuronal damage in a model of transient forebrain ischemia. Ann Neurol 11:491–498

Reed JC (1997) Double identity for proteins of the Bcl-2 family. Nature 387(6635): 773–776

Ruscher K, Isaev N, Trendelenburg G, Weih M, Iurato L, Meisel A, Dirnagl U (1998) Induction of hypoxia inducible factor 1 by oxygen glucose deprivation is attenuated by hypoxic preconditioning in rat cultured neurons. Neurosci Lett 254(2): 117–120

Sagot Y, Dubois-Dauphin M, Tan SA, de Bilbao F, Aebischer P, Martinou JC, Kato AC (1995) bcl-2 overexpression prevents motoneuron cell body loss but not axonal degeneration in a mouse model of a neurodegenerative disease. J Neurosci 15(11): 7727–7733

Salceda S, Caro J (1997) Hypoxia-inducible factor 1alpha (HIF-1alpha) protein is rapidly degraded by the ubiquitin-proteasome system under normoxic conditions. Its stabilization by hypoxia depends on redox-induced changes. J Biol Chem 272(36):22642–22647

Searle J, Lawson TA, Abbott PJ, Harmon B, Kerr JF (1975) An electron-microscope study of the mode of cell death induced by cancer-chemotherapeutic agents in populations of proliferating normal and neoplastic cells. J Pathol 116(3):129–138

Shimazaki K, Ishida A, Kawai N (1994) Increase in Bcl-2 oncoprotein and the tolerance to ischemia-induced neuronal death in the gerbil hippocampus. Neurosci Res 20(1):95–99

Simons M, Beinroth S, Gleichmann M, Liston P, Korneluk RG, MacKenzie AE, Bahr M, Klockgether T, Robertson GS, Weller M, Schulz JB (1999) Adenovirus-mediated gene transfer of inhibitors of apoptosis protein delays apoptosis in cerebellar granule neurons. J Neurochem 72(1):292–301

Stavropoulos TA, Strathdee CA (1998) An enhanced packaging system for helper-dependent herpes simplex virus vectors. J Virol 72(9):7137–7143

Stroemer RP, Rothwell NJ (1997) Cortical protection by localized striatal injection of IL-1ra following cerebral ischemia in the rat. J Cereb Blood Flow Metab 17(6): 597–604

Susin SA, Lorenzo HK, Zamzami N, Marzo I, Snow BE, Brothers GM, Mangion J, Jacotot E, Costantini P, Loeffler M, Larochette N, Goodlett DR, Aebersold R, Siderovski DP, Penninger JM, Kroemer G (1999) Molecular characterization of mitochondrial apoptosis-inducing factor [see comments]. Nature 397(6718): 441–446

Tasker RC, Coyle JT, Vornov JJ (1992) The regional vulnerability to hypoglycemia-induced neurotoxicity in organotypic hippocampal culture: Protection by early tetrodotoxin or delayed MK-801. J Neurosci Lett 12(11):4298–4308

Vayssiere JL, Petit PX, Risler Y, Mignotte B (1994) Commitment to apoptosis is associated with changes in mitochondrial biogenesis and activity in cell lines conditionally immortalized with simian virus 40. Proc Natl Acad Sci USA 91(24): 11752–11756

Wallis RA, Panizzon KL (1995) Delayed neuronal injury induced by sub-lethal NMDA exposure in the hippocampal slice. Brain Res 674(1):75–81

Wang Y, Lin SZ, Chiou AL, Williams LR, Hoffer BJ (1997) Glial cell line-derived neurotrophic factor protects against ischemia-induced injury in the cerebral cortex. J Neurosci 17(11):4341–4348

Xie Y, Mies G, Hossmann KA (1989) Ischemic threshold of brain protein synthesis after unilateral carotid artery occlusion in gerbils. Stroke 20(5):620–626

Yang E, Zha J, Jockel J, Boise LH, Thompson CB, Korsmeyer SJ (1995) Bad, a heterodimeric partner for Bcl-XLL and Bcl-2, displaces Bax and promotes cell death. Cell 80(2):285–291

Yoshida H, Kong YY, Yoshida R, Elia AJ, Hakem A, Hakem R, Penninger JM, Mak TW (1998) Apaf1 is required for mitochondrial pathways of apoptosis and brain development. Cell 94(6):739–750

Yuan J (1996) Evolutionary conservation of a genetic pathway of programmed cell death. J Cell Biochem 60(1):4–11

Zamzami N, Marchetti P, Castedo M, Zanin C, Vayssiere JL, Petit PX, Kroemer G (1995) Reduction in mitochondrial potential constitutes an early irreversible step of programmed lymphocyte death in vivo. J Exp Med 181(5):1661–1672

Zanjani HS, Vogel MW, Martinou JC, Delhaye-Bouchaud N, Mariani J (1998) Postnatal expression of Hu-*bcl-2* gene in Lurcher mutant mice fails to rescue Purkinje cells but protects inferior olivary neurons from target- related cell death. J Neurosci 18(1):319–327

Zivin J, Miller L (1999) Stroke: Present and Future Therapy. Stroke Therapy: Basic, Preclinical and Clinical Directions. L. Miller, Wiley-Liss, Inc.: 401–421

Section IV
Disease Targeting of Neuroprotective Therapies

CHAPTER 13
Stroke

M.P. Goldberg

A. Introduction

The last two decades have witnessed impressive progress in the science of brain ischemia. As other chapters of this monograph document, there is now considerable understanding from animal models of complex pathways and cascades leading to brain injury after cerebral ischemia. Dozens of pharmacological agents have been proven effective in reducing brain injury in animal models. There has also been progress in clinical stroke management. Clinical stroke trials have greatly expanded in patient numbers, and have improved in trial design and experimental endpoints. Several large, well-designed, and clinically meaningful trials have provided advances in primary and secondary stroke prevention. In addition, there is now an effective treatment for patients with acute stroke, tissue plasminogen activator, which improves outcome if given early following stroke onset (NINDS rtPA Stroke Study Group 1995). However, despite substantial effort and expense, no neuroprotective agent has yet been shown to improve clinical outcome in large, randomized multi-center trials. This review examines the data on neuroprotection in clinical stroke, with a focus on future directions for animal and human study.

I. Cerebrovascular Disorders Defined

Advances in basic and preclinical stroke research require a clear understanding of the clinical disease under study. Stroke is defined as an abrupt neurological deficit related to disease of the cerebral vasculature. This broad term actually encompasses three distinct disease processes which are commonly grouped for statistical and epidemiological analysis: ischemic stroke, intracerebral hemorrhage, and subarachnoid hemorrhage. The present review focuses on ischemic stroke, which makes up approximately 80% of the total incidence.

Approximately 20% of strokes are hemorrhagic, following rupture of cerebral vessels at two locations. Bleeding directly into the brain tissue is termed intracerebral or intraparenchymal hematoma. The most common cause is chronic hypertension; less common etiologies include malformations of cerebral blood vessels, brain tumors, trauma, and coagulation disorders. Bleeding from the vessels surrounding the brain is subarachnoid hemorrhage. This is usually due to rupture of cerebral aneurysms, congenital defects in the arteries at the circle of Willis. Hemorrhagic stroke may be followed by ischemia as a secondary complication due to compression of surrounding brain tissue and vessels (intracerebral hemorrhage) or to delayed vasospasm of cerebral arteries (subarachnoid hemorrhage). No neuroprotective agent has been tested in large-scale trials of intracerebral hemorrhage. The L-type calcium channel blocker, nimodipine, reduces deficits related to delayed vasospasm after subarachnoid hemorrhage (FEIGIN et al. 2000). It is not known whether this effect is mediated at the level of cerebral vessels or brain parenchyma.

Ischemic stroke is initiated by abrupt occlusion of arterial flow to any portion of the brain. There is considerable diversity in ischemic stroke etiology. Arterial occlusion may originate within the brain, at the level of the small penetrating arterioles of the brain (lacunar stroke) or within the intracerebral arteries. Alternatively, impairment of cerebral circulation may follow embolism from the extracranial arteries (carotid, vertebral or basilar), heart (typically because of atrial fibrillation, myocardial infarction, or valvular heart disease), or aortic arch. These are the most common causes of stroke in older adults. Many other conditions can lead to stroke (particularly in children and young adults); these include blood abnormalities such as sickle cell anemia or coagulation disorders, illicit drug use, or congenital malformations of the cerebral circulation. One might ask whether the variety of stroke types and causes can be viewed as a single condition for development of acute therapies. It is reasonable to conclude that ischemic stokes may share similar cellular injury mechanisms despite considerable diversity in vascular pathology.

Regardless of the source of occlusion, the severity and reversibility of neurological deficit depend on the region of perfusion loss, collateral circulation, and duration of occlusion. Many arterial occlusions resolve spontaneously. If blood flow reduction is brief or incomplete, there may be no permanent injury. Transient ischemic attack (TIA) has been defined as any neurological deficit which resolves spontaneously within 24h of onset. However, the vast majority of TIAs last less than 20min (LEVY 2000), and it is now recognized that ischemic episodes lasting more than hour are likely to be associated with permanent lesions, even if this is recognized only by CT or MRI scan (BOGOUSSLAVSKY et al. 1985). For this reason, current stroke management and clinical trials are initiated as soon as possible, and ideally within the first few hours after symptom onset. If blood flow reduction is severe and sustained for more than a few minutes (typically more than 1h), then there is permanent loss of brain tissue at the center of the vascular territory (ischemic core). Over a period of several days, the pathology becomes that of pannecrosis, in which all brain cells – neurons, glia, vascular cells – degenerate and are replaced by a cystic cavity. The existence and extent of a surrounding rim of partially or reversibly injured tissue (variously defined as ischemic penumbra) is an extremely important and unresolved question for human stroke. Experimental stroke models suggest that neuroprotective agents can reduce injury only in this area.

Stroke is not the only clinical disease of brain ischemia. Transient, global ischemia occurs when there is a loss of blood flow to the entire brain, such as cardiac arrest. It is important to recognize that the pathology of the ischemic core in focal ischemia is entirely distinct from that observed after transient but global ischemia. Global ischemic injury is characterized by delayed, selective neuronal loss within defined areas of the hippocampus, cerebral cortex, cerebellum, and striatum. For this reason, results from animal models of global ischemia should be used with great caution to draw conclusions about focal ischemia, or stroke. Cerebral ischemia also contributes to neurological dysfunction in such disparate clinical settings as perinatal asphyxia (leading to periventricular leukomalacia, a major cause of cerebral palsy), cardiac surgery, vascular dementia and traumatic brain injury. Like global ischemia, these conditions share limited aspects of pathophysiology with ischemic stroke and must be viewed separately for preclinical and clinical drug development.

II. Scope of the Problem

Stroke is an enormously significant target for therapeutic intervention. It is the most common serious neurologic disease in Western countries. Stroke is the third leading cause of death (after heart attack and cancer) and leading cause of adult neurological disability. In the United States, current estimates for annual stroke incidence range from 600000 to 720000. Approximately 150000 deaths are attributed to stroke annually (AMERICAN HEART ASSOCIATION 2001). Stroke prevalence is strongly related to age, sex, and ethnic or racial background. Stroke is relatively uncommon below age 55, and increases dramatically in each subsequent decade. The age-adjusted stroke incidence rates (per 1000 person-years) are 1.78 for white men, 4.44 for black men, 1.24 for white women and 3.10 for black women. The aftermath of stroke represents a major source of long-term care medical expenses. In 1997, $3.8 billion ($5955 per discharge) was paid to Medicare beneficiaries discharged from short-stay hospitals for stroke (AHA 2001). Clearly, even a small increment in effective stroke therapy could have substantial economic and public health impact.

The most effective management of stroke is by prevention (GORELICK et al. 1999). Stroke incidence is dramatically reduced by effective treatment for hypertension. Other stroke risk factors (such as diabetes, smoking, obesity) are also well documented, but the impact of treatment is less significant. Stroke risk can be significantly reduced in specific patient groups with recognized risks: by management of elevated cholesterol (in patients with previous myocardial infarction), oral anticoagulation (in patients with atrial fibrillation), carotid endarterectomy (in patients with high-grade carotid stenosis), or antiplatelet agents (in patients with previous TIA or stroke) (WOLF et al. 1999; GORELICK et al. 1999; NASCET COLLABORATORS 1991).

III. Rationale for Development of Neuroprotective Agents

In contrast to many other diseases considered in this monograph, the etiology of ischemic stroke is well understood and potentially preventable. Optimal management should consist of primary and secondary prevention to reduce stroke incidence. If this fails, stroke patients should receive rapid intervention to restore blood flow. In other words, it seems that stroke therapy should be directed to treatment of the underlying cerebrovascular disorder. Is there still a need for development of potential neuroprotective therapy? The answer is clearly positive.

Stroke incidence remains very high despite extensive knowledge of risk factors. Reductions in stroke incidence from treatment of hypertension and other effective therapies have been limited by poor patient compliance and insufficient physician recognition. Age-adjusted stroke mortality has been declining over the last few decades, but stroke incidence and stroke hospitalizations have increased over the last decade (FANG and ALDERMAN 2001). Therefore, the requirement for acute therapeutic intervention in stroke has not diminished, and is expected to continue to increase substantially as the overall population ages.

Acute stroke management by restoration of blood flow has thus far failed to yield significant public health impact. Thrombolytic and antithrombotic agents, effective tools in management of myocardial infarction, have only recently emerged as proven therapies in ischemic stroke. The pivotal trial of intravenous tissue plasminogen activator (tPA, Alteplase) was published in 1995 (NINDS 1995). This NIH-supported study was a randomized, double-blind, multicenter trial which compared intravenous tPA to placebo in 624 patients who received the treatment within 3h of the onset of ischemic stroke. Three months later, there was a statistically significant, 30% increase in the number of patients who had little or no disability by an array of assessment scales. The treatment was not without risk. Tissue plasminogen activator treatment was complicated by a tenfold increase in the number of patients with early, symptomatic intracranial hemorrhage (placebo 0.6%, tPA 6.4%), but the overall mortality was similar in both groups. Significant improvement was sustained for at least one year after randomization (KWIATKOWSKI et al. 1999). The incidence of thrombolytic-induced brain hemorrhage increases rapidly after stroke onset, and several large trials have failed to demonstrate a benefit of tPA given beyond 3h (HACKE et al. 1998; CLARK et al. 1999b).

The net impact of this therapy has been disappointing. In the NINDS trial, the absolute proportion of patients with good outcome was increased by 11%–13%. For example, 28% of the placebo group and 41% of the tPA group had a favorable outcome on the Rankin scale 1 year later (KWIATKOWSKI 1999). This represents a large clinical impact, but it shows that even among those who receive tPA, as many as 60% will be left with death or significant disability at one year. Furthermore, five years after approval of tPA, no more than 1%–4% of all patients with ischemic stroke receive the drug (JOHNSTON et al. 2001). This is due to poor recognition of the importance of rapid diagnosis and treatment, within the general public as well as emergency medical systems and hospitals. Currently well fewer than 1 in 200 current victims of acute ischemic stroke (12% of the <4% who receive drug) are likely to derive benefit from tPA. Other thrombolytics drugs and drug delivery methods are still in early clinical testing. Therefore, despite availability of tPA, new therapeutic directions are urgently required.

B. Clinical Trials of Neuroprotective Agents

I. Progress in Clinical Trials and Trial Designs

The explosion in stroke basic science has been accompanied by more gradual advances in clinical stroke trial design, numbers, and size. Acute ischemic stroke poses significant obstacles for clinical trials; these issues are discussed in detail below in Sect. III. Large sample numbers are needed because of variability in presentation, outcome, and pathophysiology. Unfortunately, patient enrolment in acute stroke trials is costly and time consuming. Since strokes occur at all hours of the day, rapid trial entry requires 24-h stroke trial teams which include physicians, study coordinators, emergency room personnel, and often laboratory and radiology personnel. Even in large centers, only a few patients each month meet entry criteria for acute therapy trials. Most acute trials published prior to 1990 enrolled fewer than 100 subjects, a sample size now recognized to provide little chance of demonstrating statistically significant differences (SAMSA and MATCHAR 2001). Moreover, almost all trials published before 1990 initiated treatment no earlier than 1–2 days (sometimes up to 1–2 weeks) after stroke onset, a time window which is now recognized to have little hope for effective intervention. Only in the last decade have there been sufficient numbers of large, well-designed trials to allow an initial assessment of certain neuroprotective strategies.

Tables 1 and 2 show representative phase III trials of potential neuroprotective agents. This partial list is limited to trials for which there is sufficient published information and in which the tested intervention was initiated within 24h of stroke onset. This information is derived from published sources and from a website edited by the author, the Stroke Trials Directory (GOLDBERG 2001). Several approaches thought to work primarily by increasing cerebral perfusion or oxygen delivery are not included.

II. Glutamate Antagonists

Excitotoxic damage due to excessive activation of neuronal glutamate receptors plays a pivotal role in the pathogenesis of ischemic damage in animal and

Table 1. Receptor antagonists and agonists in multicenter stroke clinical trials

Drug category	Chemical name	Generic, trade names	n	Window (h)	Active?
Glutamate receptor antagonists					
NMDA antagonists					
Competitive antagonist	CGS 19755	Selfotel	567	6	No
Glycine site antagonists	GV150526	Gavestinel	1367	6	No
			1804	6	
Polyamine site antagonists	SL 82–0715	Eliprodil		8	No
	CP-101, 606		Ongoing	8	Yes
Uncompetitive antagonists	CNS 1102	Aptiganel, Cerestat	628	6	No
(channel blockers)	Magnesium		Ongoing	12	Yes
AMPA/Kainate receptor antagonists	YM872		Ongoing		Yes
GABA agonist		Chlomethiazole	1360	12	No
		Zendra	1200	12	
		Diazepam, Valium	Ongoing	12	Yes
Opioid agonist	ORF 11676	Nalmefene, Cervene	368	6	No
Serotonin agonist	Bay x 3702	Repinotan	Ongoing	6	Yes

Representative neuroprotective agents reaching phase III randomized, multicenter trials in published and unpublished research are shown. Studies enrolling patients more than 24 h after stroke onset are not included. *n*, Total number of patients enrolled (placebo + treatment); window, maximum time in hours from onset of stroke symptoms to initiation of treatment; active, compound is in ongoing clinical development for acute stroke. Information from published sources given in text, and GOLDBERG 2001.

cell culture models. The substantial evidence demonstrating efficacy of glutamate receptor blockade in preclinical models has been reviewed (ALBERS et al. 1989), and these agents have been in clinical trial development since the late 1980s.

Glutamate receptors are classified according to pharmacological specificity and signal transduction mechanisms. All have been cloned and sequenced at the subunit level. The two major categories of postsynaptic excitatory amino acid receptors are metabotropic and ionotropic. Metabotropic receptors (mGluRs) do not gate an ion channel directly but are coupled to activation of intracellular second messenger systems; these receptors have diverse physiological functions and receptor antagonism has not been shown conclusively to limit ischemic injury. The ionotropic receptors include a cation channel (permeable to Na, K, and sometimes Ca) and are named according to selective agonists which activate them: N-methyl-D-asparate (NMDA), alpha-amino-3 hydroxy-5 methyl-4 isoxazole proprionic acid (AMPA), and kainate. AMPA and kainate receptors share pharmacological properties and are often

Table 2. Other neuroprotective agents in multicenter stroke clinical trials

Drug category	Chemical name	Generic, trade names	n	Window (h)	Active?
Ion channel agents					
Sodium channel blockers	ACC-9653	Fosphenytoin, Cerebryx	462	4	No
		Lubeluzole	721	6	No
		Prosynap	725	6	
			1786	8	
Calcium channel blockers		Nimodipine, Nimotop	454	6	No
		Flunarizine, Sibelium	331	24	No
Potassium channel activator	BMS-204352	Maxi-Post	1978	6	No
Antioxidants	U74006F	Tirilazad, Freedox	556	6	No
			126	4	
			1023		
	PZ 51	Ebselen, Harmokisane	Ongoing	24	Yes
	NXY-059		Ongoing		Yes
Inhibitors of inflammation or leukocyte activation	Ab R 6.5	Anti-ICAM antibody, Enlimomab	625	6	No
	Hu23F2G	Leukarrest	310	12	No
	UK-279,276	Neutrophil inhibitory factor, Corleukin	Ongoing	6	Yes
Phosphatidylcholine precursor		Citicoline, CerAxon	259	24	
			394	24	
			899	24	
Mechanism unknown		GM-1 ganglioside	502	12	No
		Sysgen	792	5	
		Piracetam	927	12	
			Ongoing	7	Yes

Representative neuroprotective agents reaching phase III randomized, multicenter trials in published and unpublished research are shown. Studies enrolling patients more than 24h after stroke onset are not included. *n*, Total number of patients enrolled (placebo + treatment); window, maximum time in hours from onset of stroke symptoms to initiation of treatment; active, compound is in ongoing clinical development for acute stroke. Information from published sources given in text, and GOLDBERG 2001.

referred to as AMPA/kainate (AMPA/KA) receptors. Cloned receptors subunits in mouse are termed NR1, NR2 (NMDA), GluR1–4 (AMPA), GluR5–7 and KA1–2 (kainate), and there are many subunit and splice variations. The NMDA receptor can be blocked by competitive interaction with its glutamate binding domain (competitive antagonists), and also by blockade of a required glycine modulatory site (glycine site antagonists), a polyamine binding site (also known as NR2B-selective antagonists), or at the ion channel itself (noncompetitive antagonists).

1. NMDA Receptor Antagonists

Development of glutamate agents for cerebral ischemia initially focused on NMDA receptors for several reasons. The NMDA receptor has the unique property of gating a large calcium conductance, which is thought to contribute to rapid neurotoxic death by intraneuronal calcium overload. In contrast, AMPA and kainate receptors are most often calcium-impermeant, and tend to exhibit rapid desensitization which would tend to limit toxic ion fluxes. Early studies of cultured hippocampal and cortical neurons showed that NMDA receptor activation plays by far the major role in neurotoxicity induced by glutamate exposure or by deprivation of oxygen and glucose (ROTHMAN et al. 1987; GOLDBERG et al. 1987). Finally, problems of solubility and blood-brain barrier penetration have greatly limited development of selective AMPA/KA antagonists as suitable candidates for drug development.

At least 30 NMDA blockers have reached clinical trial stages, encompassing agents with many distinct chemical structures and receptor mechanisms. In addition to stroke, such agents have been considered in a wide range of indications including cardiac surgery, traumatic brain injury, epilepsy, pain, neurodegenerative diseases, and drug dependence. Initial development was limited by adverse effects observed in animals or humans. NMDA antagonists at high concentrations produce notable behavioral actions including confusion, agitation, and hallucinations, typified by the drug of abuse, phencyclidine (PCP, angel dust) and related general anesthetic agent, ketamine. Similar dose-limiting CNS effects have been reported with other NMDA antagonists, including the potent and selective ion channel blocker, dizocilpine (MK-801), lower affinity antagonist dextrorphan (ALBERS et al. 1995) and competitive antagonist selfotel (CGS 19755; GROTTA et al. 1995). Another concern with acute administration is effects on blood pressure, including both hypertension and hypotension. Drug-related reduction in blood pressure is particularly worrisome because of the theoretical possibility that brain perfusion might be selectively reduced in partially ischemic areas. Drugs known to be in early phase clinical trials but no longer under active development for stroke include CPP-ene, dizocilpine, dextrorphan, dextromethorphan, licostinel, and NPS 1506 (GOLDBERG 2001).

Early NMDA antagonist trials were also delayed by the discovery by OLNEY and colleagues (1989) of neuronal vacuolization in the retrosplenial cortex and cingulated gyrus of rats treated acutely with MK-801 and PCP. Similar changes have been observed with lower affinity noncompetitive antagonists such as dextrorphan (ORTIZ et al. 1999) and competitive antagonists (HARGREAVES et al. 1993), but not with other agents including glycine-site blockers (HAWKINSON et al. 1997; BORDI et al. 1999). The significance of these findings remains unknown. In some cases the changes resolve spontaneously, and neuronal death has been reported only after the highest doses. Although little has been published in higher species, similar changes have not been reported in nonhuman primates (AUER et al. 1996) or in humans who have received ketamine or drugs of abuse.

Despite some positive results from phase II trials, NMDA antagonists have not been proven effective in large multicenter phase III trials (Table 1). Completed studies involve at least one agent from each NMDA drug class. The polyamine-site blocker eliprodil was among the first antagonists tested in large-scale trials; although the results remain unpublished it is assumed that the trial did not demonstrate a therapeutic benefit. The competitive antagonist, Selfotel (CGS 1975), was studied in two large randomized trials which compared placebo to a single intravenous 1.5mg/kg dose of Selfotel, administered within 6h of stroke onset. Trials were halted after enrolment of 567 patients because of a significant excess in early mortality detected in the treatment group (DAVIS et al. 1997, 2000). Parallel studies in traumatic brain injury were halted after enrolment of 693 patients; Selfotel did not alter outcome but no difference in mortality was noted (MORRIS et al. 1999). The competitive antagonist Cerestat (CNS 1102, aptiganel) met a similar fate: phase III stroke trials (unpublished) were discontinued after enrolment of 620 patients, because interim analysis showed little possibility of a positive drug impact. Traumatic brain injury trials were also abandoned.

The only published, completed large-scale trials to date involve the glycine-site NMDA antagonist, gavistinel (GV150526). In two large multicenter trials, 3171 patients were randomized to receive gavistinel (intravenous loading dose of 800 mg followed by 200 mg every 12h for five doses) or matching placebo, within 6h of stroke onset (LEES et al. 2000; SACCO et al. 2001). There were no statistically significant differences in 3-month mortality or functional status. No significant adverse effects were reported.

At least two additional NMDA antagonists remain in ongoing large-scale acute stroke trials. Magnesium acts as an NMDA channel blocker, in addition to its other effects such as blockade of voltage-gated calcium channels. IMAGES, a UK MRC funded study, is a multicenter, randomized trial of patients who will receive placebo or IV $MgSO_4$ (given as 16 mmol over 15 min followed by 65 mmol over 24h) within 12h of stroke onset. Planned enrolment is 2700 by 2002. CP-101606 is a selective antagonist of NMDA receptors containing the NR2B subunit (polyamine site blocker). It was reported to be well tolerated in preliminary stroke and traumatic brain injury trials, and is now in multicenter stroke trials.

2. AMPA Receptor Antagonists

AMPA/KA receptor blockade is neuroprotective in preclinical global and focal ischemia models (GILL 1994). Available agents have been constrained by difficulties with solubility and blood-brain barrier penetration. In addition, since AMPA/KA receptors account for most rapid synaptic excitatory neurotransmission, complete receptor blockade at high doses can be expected to cause significant sedation and possible respiratory depression. No trials involving AMPA antagonists in stroke patients are yet published. YM90k and ZK20075 are among agents which have been abandoned after early clinical experience. The water-soluble, selective AMPA receptor antagonist YM872 is currently in ongoing multicenter stroke trials, and clinical development is proceeding for at least one other agent, SPD 502 (NS 1209).

III. Other Receptor Agents

Activation of inhibitory neurotransmitters may reduce glutamate-mediated excitotoxicity or reduce synaptic spread of injury. The $GABA_A$ agonist, clomethiazole (Zendra), reduced infarct volume in rodent models and was tested in two large clinical trials. The CLASS trial randomized 1,360 patients presenting within 12h of stroke onset to receive placebo or clomethiazole (75 mg/kg as an intravenous infusion over a 24-h period). The two groups did not differ significantly overall, but there was suggestion of benefit in a predefined subgroup of patients with total anterior circulation (middle cerebral artery) syndrome (WAHLGREN et al. 1999). A second trial involving 1,200 patients with this stroke subtype was recently concluded with negative results (GOLDBERG 2001). Not unexpectedly, the major adverse effect of clomethiazole was sedation, requiring discontinuation of study drug in more than 10%. There is currently a European study to examine effects of the benzodiazepine, diazepam, in acute stroke with a 12-h time window.

Evidence from animal studies of stroke, trauma, and spinal cord injury suggests that endogenous opioids might contribute to tissue injury, perhaps by enhancing glutamate release. Nalmefene (Cervene) is an opioid antagonist with relative selectivity for κ-opiate receptors. Studies with this agent in rodent stroke models have not been published. Despite suggestive results from subgroup analysis of a phase IIb trial, no benefit was seen in a randomized trial of 368 patients who received treatment within 6h of stroke onset (CLARK et al. 2000). BAY x 3702 (Repinotan) is a serotonin ($5HT_{1A}$) agonist which has in vivo and in vitro neuroprotective actions, possibly related to inhibition of glutamate release or neuronal apoptosis. Multicenter trial results are not yet published.

IV. Ion Channel Blockers

Neuronal voltage-gated sodium and calcium channels contribute to ischemic brain injury by at least two mechanisms. Sodium channels are required for action potential propagation. Some subtypes of voltage-gated calcium channels (for example, N-type channels but not L-type channels) localized at presynaptic terminals allow calcium entry required for synaptic vesicle release.

Inhibition of either of these channels inhibits synaptically released glutamate and may reduce propagation of excitotoxic damage in stroke. However, it should be noted that calcium-dependent vesicular release is only one source of extracellular glutamate. Increasing evidence suggests that a more important source of toxic glutamate release in ischemia may be calcium-independent reversal of glutamate transporters in neurons and astrocytes. Independent of effects on excitotoxicity, activation of ion channels during energy failure may contribute to injury through disruption of cellular sodium and calcium homeostasis.

Multicenter trials have failed to demonstrate protective actions of sodium channel blockade (Table 2). The anticonvulsant pro-drug, fosphenytoin (Cerebryx), was evaluated in a trial with a 4-h treatment window. The trial was halted after enrolment of 452 patients showed little chance of a therapeutic benefit (GOLDBERG 2001). Another sodium channel blocker, BW619C89, has completed phase II development (MUIR et al. 1998).

Lubeluzole (Prosynap) inhibits sodium channels, reduces glutamate release, and interferes with nitric oxide production. This drug was tested in three large multicenter trials. A trial 721 patients in the United States and Canada showed a significantly greater improvement at week 12 in neurological recovery, functional status, and overall disability (GROTTA 1997). In contrast, a European–Australian phase III with 725 patients and similar design found no effect of lubeluzole on mortality or clinical outcome (DIENER 1998). A third international trial of 1,786 patients (800 patients at 0–6 h, 986 patients at 6–8 h) did not demonstrate therapeutic efficacy of lubeluzole in acute stroke (DIENER 2000).

A metaanalysis of 46 trials including more than 7,000 patients failed to show benefit for L-type calcium antagonists in acute stroke (HORN and LIMBURG 2000). However, these trials were limited by variable outcome measures and long treatment windows (typically 24–72 h). There was a suggestion of adverse effects at the highest doses. The dihydropyridine antagonist of L-type voltage-gated calcium channels, nimodopine, is effective in reducing mortality after subarachnoid hemorrhage, perhaps by attenuating vasospasm-induced ischemia. A recent trial of 454 patients treated within 6h of acute stroke onset found no difference between placebo and nimodipine (30 mg orally, four times daily for 10 days). N-type calcium channels are involved in synaptic release. SNX 111 is a selective N-type antagonist which is in clinical trials for pain, but the status of stroke trials is unknown.

Two novel approaches to reducing toxic calcium accumulation have been evaluated. DP-b99 is a membrane-activated calcium chelator currently in phase II acute stroke trials. If intracellular calcium accumulation contributes to neurotoxicity, then cell-permeant calcium chelators may be neuroprotective (TYMIANSKI et al. 1994). BMS-204352 (Maxi-Post), activates a class of calcium-activated potassium channels. The net effect of potassium channel activation is predicted to be membrane hyperpolarization, which would tend to inactivate voltage-dependent cation channels (and also NMDA channels). BMS-

204352 reduced infarct volume in rats even when given 2h after focal ischemia onset (GRIBKOFF et al. 2001). A trial of 1978 patients treated within 6h of acute stroke onset showed no benefit of BMS-204352 (GOLDBERG 2001).

V. Antioxidants and Lipid Peroxidation Inhibitors

Substantial evidence supports the role of toxic free radicals and lipid injury in ischemic neuronal injury. Tirilazad mesylate is a nonglucocorticoid, 21-aminosteroid that inhibits lipid peroxidation which has been tested in large trials of ischemic stroke, subarachnoid hemorrhage, and traumatic brain injury. A meta-analysis of six ischemic stroke trials involving 1,757 patients showed no benefit of tirilazad, with a possible increase in mortality and disability (TIRILAZAD INTERNATIONAL STEERING COMMITTEE 2000; RANTTAS INVESTIGATORS 1996; HALEY 1998). Two antioxidant compounds have shown promising results in early clinical development and are now in multicenter trials: the seleno-organic antiinflammatory agent, ebselen (YAMAGUCHI et al. 1998) and the nitrone spin-trapping agent, NXY-059 (LEES et al. 2001).

VI. Inhibitors of Inflammation or Leukocyte Activation

The inflammatory response may exacerbate injury after cerebral ischemia (DEL ZOPPO et al. 2001). An important question is whether inflammatory cellular or chemical responses occur before, rather than after, the development of irreversible injury. One possible target is the CNS migration of circulating neutrophils, which must first attach to the endothelial intercellular adhesion molecule (ICAM-1) through the leukocyte integrin, CD11/CD18. Enlimomab is a monoclonal antibody against the intercellular adhesion molecule ICAM-1. A trial of 625 patients enrolled within 6h of acute stroke onset found that enlimomab increased morbidity compared to placebo. A later trial involved Hu23F2G (LeukArrest), an antibody to leukocyte integrin CD11/CD18. A humanized antibody may be less likely to develop adverse effects observed with enlimomab, including fever. This trial was stopped after enrolment of 310 patients. A humanized antibody to β-2 integrin, LDP-01, was reported to be in phase II development and neutrophil inhibitory factor (Corleukin) is in phase III trials. Other approaches to reducing inflammation remain in preclinical and early clinical development.

VII. Agents with Other Mechanisms of Action

The number and diversity of compounds shown to reduce ischemic stroke damage in animals models is bewildering. Other representative agents which have reached phase III trials are shown in Table 2. These include a precursor of neuronal membrane phospholipids, CDP-choline (Citicoline; CLARK et al. 1997, 1999), a CNS growth factor, basic epidermal growth factor (bFGF, trafermin, Fiblast), GM-1 ganglioside (LENZI et al. 1994), and piracetam (Nootropil;

RICCI et al. 2000). None of these agents has demonstrated significant protection in large trials and only piracetam remains in clinical development.

C. Issues Regarding Preclinical Stroke Research

Why have neuroprotective agents failed to demonstrate benefit in randomized clinical stroke trials? Results of trials to date have been unequivocally disappointing. Despite sometimes extensive animal data, none of the agents in trials published or presented so far has shown a statistically significant benefit for stroke outcome or mortality. These negative results have occasioned significant consideration and debate (DORMAN and SANDERCOCK 1996; LEE et al. 1999; STAIR 1999; DEGRABA and PETTIGREW 2000; SAMSA 2001; STAIR II 2001). The explanations may be divided as follows: (a) animal experiments do not properly model the human condition of ischemic stroke, (b) clinical stroke trials have not duplicated the conditions of animal experiments.

Animal research has obvious utility for understanding mechanisms of disease, identifying approaches to therapy, and screening potential neuroprotective agents for subsequent development. However, the value of animal stroke models in preclinical drug development can be debated (e.g., WIEBERS et al. 1990; ZIVIN and GROTTA 1990; GINSBERG 1996; DEGRABA and PETTIGREW 2000). Rodent models are clearly predictive for many nervous system diseases, such as epilepsy. In contrast, no model has proven predictive value for neuroprotective efficacy in human stroke. This follows from the fact that no drug of this type has yet proven to improve outcomes in large-scale randomized clinical trials. The optimal parameters for animal models in neuroprotective drug development can only be surmised. The assumed goal is that important features of the preclinical model should match as closely as possible the clinical situation.

The STROKE THERAPY ACADEMIC INDUSTRY ROUNDTABLE (STAIR 1999) has described consensus aspects of animal models for stroke drug development (Table 3). The most important recommendations are that drug dosing and delivery should match interventions which can be delivered in human practice. In particular, animal models should assess the longest time after ischemia onset at which efficacy can be demonstrated. It also seems important to consider later time points for analysis (several days to weeks, rather than 24h, which is most commonly used), and behavioral as well as histological outcomes. Still to be determined is the optimal duration of ischemia (transient vs. permanent?) and the best species for preclinical studies (is it sufficient to use rats alone?). Clearly a promising drug candidate is one which shows robust protection in a variety of models, species, and laboratories.

An important question is whether species differences are critical for predicting neuroprotective efficacy in cerebral ischemia. Within mammals there is considerable sequence homology for most known CNS genes. Nonetheless, rodents and humans sometimes differ in aspects of receptor pharmacology,

Table 3. Summary of STAIR recommendations for preclinical drug development [modified from STROKE THERAPY ACADEMIC INDUSTRY ROUNDTABLE (STAIR) 1999]

Topic	Major recommendations
Drug dosage	Animal studies should determine minimal effective and maximum tolerated doses Evidence should be available to indicate that a neuroprotective target concentration can be achieved in humans
Therapeutic window	Time window studies should demonstrate benefit when therapy is initiated at delayed time periods after stroke onset
Animal models	Permanent middle cerebral artery occlusion models should be studied first Initial studies should be completed in rats A second larger species (cats, nonhuman primates) should be strongly considered for preclinical assessment Neuroprotective efficacy should be demonstrated in at least two laboratories, one of which is independent of sponsor Animal studies should be randomized and blinded
Physiological monitoring	Blood pressure, blood gases, hemoglobin, glucose, and cerebral blood flow should be monitored Brain temperature should be monitored and controlled Adverse effects and drug interactions should be noted
Outcome measures	Animal studies should include measures of infarct volume and functional responses Outcome measures should be determined for both short – (1–3 days) and long – (7–30 days) term effects
Publication	Preclinical study data should be published or submitted for review in peer-reviewed journals

and they differ substantially in drug metabolism and blood-brain barrier penetration. Human stroke is a highly heterogeneous condition, involving diverse brain areas, severities, and pathophysiologies; rodent models aim for uniform and reproducible infarcts. It is difficult to assess drug effects on functional outcome in rodents because even large infarcts produce relatively few lasting behavioral changes. Most animal models use healthy adult animals whereas human strokes occur in elderly individuals with long-standing risk factors and diseases.

Species vary substantially in brain structure. In addition the massive increase in neuronal number and brain size, higher primates differ from rodents in their proportion of gray and white matter. The cerebral cortex of lissencephalic species such as rats and mice is comprised of only 10%–15% white matter, whereas human brains are evenly divided between white and gray. There is growing awareness of the importance of protecting white matter in stroke (STYS 1998; DEWAR et al. 1999). White matter is vulnerable to ischemia, but its mechanisms of injury are only partially overlapping with those of gray matter. For example, NMDA receptors are localized to neuronal cell bodies and dendrites, and are not found in axons or glial cells of white matter. Therefore, there is little chance that NMDA antagonists can directly reduce white matter ischemic injury. In contrast, oligodendrocytes express AMPA receptors, and AMPA receptor blockade reduces hypoxic injury in cultured oligodendrocytes (MCDONALD et al. 1998) and white matter brain slices (TEKKOK et al. 2001). Some agents, such as antioxidants, may reduce damage in white or gray matter (IMAI et al. 2001). Routine preclinical studies of neuroprotective agents do not identify effects on white matter, since this comprises such a small proportion of the rat or mouse brain. The significance of these observations for clinical trials is that neuroprotective agents which protect only gray matter may not be effective in human stroke. These problems could be resolved by development of reproducible stroke models in gyrencephalic animals, such as cats, dogs, sheep, or nonhuman primates.

D. Issues Regarding Clinical Stroke Research

The most obvious explanation for the discrepancy between animal experiments and clinical trial results is that no human study has fully duplicated experimental conditions in which drugs were found effective in rodents. Many of these issues are considered in detail in reports of the STROKE THERAPY ACADEMIC INDUSTRY ROUNDTABLE (STAIR) (1999, 2001). Recommendations of this group are summarized in Table 4. The most compelling problems are in drug dosing and therapeutic window.

I. Drug Dosing

For most neuroprotective agents, there is insufficient science to relate rodent to human dosing levels. Therapeutic doses in animal models are generally selected for the maximum efficacy which does not increase mortality. It is possible in experimental models to determine brain or CSF concentrations, which may be compared with concentrations known to block the target receptor or molecule. In contrast, investigational drug dosing in human studies is generally determined by the highest dose which causes no adverse effects. Since neuroprotective agents by definition target the brain, the absence of CNS effects for certain compounds strongly suggests that they are not present in sufficient concentration. For example, it is difficult to imagine that effective blockade of AMPA/KA receptors, or activation of GABA receptors, can be accomplished without significant sedation. Likewise, effective blockade of NMDA receptors is expected to cause CNS behavioral changes including agitation, confusion, and hallucinations (these side effects may possibly be minimized with agents which bind the receptor with low affinity, or with NR2B subunit-selective antagonists).

This is a difficult problem to solve. Certainly it is reasonable to expect that agents to be tested in clinical trials should achieve plasma concentrations

Table 4. Summary of STAIR II recommendations for phase III clinical trial evaluation of acute stroke therapies [modified from STROKE THERAPY ACADEMIC INDUSTRY ROUNDTABLE (STAIR) II 2001]

Topic	Major recommendations
Drug dosage	Dosing should be selected based on preclinical models and phase I and II trial data Dosing should achieve plasma and CNS levels within predicted levels of efficacy
Therapeutic window	Time window for patient enrolment should reflect data from preclinical models and phase II trials Investigators should be encouraged to enroll patients as quickly as possible after stroke onset Stratification by enrolment time should be considered
Patient selection	Include patients with moderate deficits at enrolment (e.g., NIH Stroke Scale 7–22) Inclusion of stroke subtypes and localization should take into consideration putative mechanism of study drug Trials of drugs which do not protect white matter should exclude subcortical strokes confined to white matter Trials of thrombolytic drugs should exclude patients for whom reperfusion has already been established Hemorrhagic stroke should be excluded unless there is reason to believe intervention is effective
Outcome measures	Impairment, disability, and handicap measures all have advantages and disadvantages Consider using a global statistic combining several measures 90 days after stroke onset is typically used as the primary end point Covariate analysis should be prespecified
Trial logistics	Appropriate relationships should be fostered between sponsors, academicians, and investigators Data quality, trial performance, end-point evaluation, and patient safety should be monitored by appropriate committees Agreement should be reached between sponsors and participants that trial results will be presented and submitted for publication, even if negative

similar to their effective levels in animal studies (STAIR II 2001). However, it is widely recognized that circulating drug levels are at best loosely related to CNS levels, and that pharmacokinetics and blood brain barrier penetration differ markedly between species. Drug CNS penetration can be estimated by measurement of cerebrospinal fluid levels, although these may vary dramatically from brain concentrations (STEINBERG et al. 1996). Actual brain levels can be measured by obtaining samples from neurosurgical patients undergoing intracranial surgery (STEINBERG et al. 1996), but the practical implementation of this method is necessarily limited. A promising approach is use of positron emission tomography (PET) or other imaging technology to determine brain delivery of radiolabeled pharmaceutical compounds, or to measure effects on CNS function noninvasively (Fischman et al. 1997). Clinical trial development would be greatly enhanced by more thorough consideration of these issues.

It is also worth noting that some drug effects may have adverse consequences without necessarily being dose-limiting in an acute stroke setting. For example, transient sedation would be a problem for a stroke prevention drug but is easily managed in an acute stroke setting. The approved acute stoke agent, tPA, is associated with lethal intracerebral hemorrhage in 6% of patients, but even this catastrophic adverse effect is offset by observed therapeutic benefit.

II. Time of Drug Delivery

Clinical trials diverge most dramatically from preclinical models in the time of treatment after stroke onset. Indeed, the vast majority of in vitro and in vivo experimental studies have described effects of neuroprotective agents given only before or during ischemia. This has practical clinical relevance only for situations in which CNS ischemic injury may be immediately anticipated, such as cardiac, vascular or neurological surgery, or in which rapid hospital transport is typical, as in cardiac arrest or neurotrauma.

Neuroprotective agents in experimental stroke models have demonstrated efficacy at most two to 3h after initiation of focal ischemia. In contrast, patients with ischemic stroke typically present many hours or days after onset. Explanations for this delay include the frequently painless presentation of stroke, its resemblance to more benign conditions such as transient peripheral nerve compression, poor public recognition of stroke symptoms, insufficient training of paramedical and medical personnel, and inadequate organization of acute stroke services (Wester et al. 1999; Morris et al. 2000; Barber et al. 2001). Large clinical stroke trials in the 1970s and 1980s typically enrolled stroke patients no earlier than 24h after onset and sometimes as long as two to seven days later. Initial attempts failed to reduce this time by constraining treatment intervals, as demonstrated by the experience of one center which enrolled a single patient of 192 screened (LaRue 1988). However, it is clearly possible to treat a substantially larger proportion of patients within an early time window. Intensive public and professional education in the course of recruitment for initial thrombolytic clinical trials can result in measurable improvement in patient arrival times (Alberts et al. 1992; Barsan et al. 1994), and this gain has continued with major national efforts to optimize prehospital stroke care.

There remain significant limitations to widespread patient recruitment in early neuroprotective drug trials. For the foreseeable future, many stroke patients will continue to delay seeking medical attention. For example, the exact time of stroke onset cannot be determined for a sizable proportion of patients, including the 25% or so who awaken with stroke (Lago et al. 1998;

WROE et al. 1992). Some therapies cannot be initiated before completion of detailed clinical testing, laboratory studies, or brain imaging. This may consume an hour or more in the emergency room. Therefore there is substantial utility in selecting neuroprotective approaches and study designs which can start without such testing, ideally in the ambulance. Finally, the existence of an approved therapy for acute stroke prevents enrolment in placebo-controlled studies for tPA-eligible patients who choose this treatment (in countries where tPA is approved for stroke). This is unlikely to represent a significant problem for neuroprotective drug development. Neuroprotective drug mechanisms are not expected to interact with thrombolytic agents, and indeed should be enhanced by the possibility of drug delivery after reperfusion. Therefore, clinical trials should be stratified to include patients who receive tPA in both placebo and neuroprotective treatment arms (e.g., LYDEN et al. 2001).

Advanced imaging studies in stroke patients may help to determine the longest duration after stroke onset in which ischemic tissue remains viable. Beyond this time, there is clearly no hope of neuroprotection. Since there is currently no way to relate treatment window in animals to that in humans, clinical trials should aim to treat patients as soon as possible, and to test therapies with the greatest potential for delayed effectiveness.

III. Clinical Trial Design and Outcomes

Effective randomized trials must include a sufficient number of patients to convincingly exclude a therapeutic benefit, if one is present. Many stroke trials have been far too small to demonstrate or exclude a treatment effect. There are reasons to be concerned that even larger trials may be statistically underpowered if experimental variation have been underestimated or treatment effects overestimated (SAMSA 2001). It should be noted that a substantial proportion of the current neuroprotective trials have been halted prior to accumulating the planned sample size, because of interim analyses suggesting that further enrolment would be futile.

Several important considerations relate to study design. Unlike most experimental models, clinical stroke is characterized by extreme diversity in patient characteristics, vascular territory, stroke mechanisms, disease severity and expected outcomes. Selection of broad entry criteria provides larger patient numbers and greater generalizability of study results. For example, the INTERNATIONAL STROKE TRIAL examined effects of aspirin and subcutaneous heparin in a simple study design which allowed enrolment of 19435 patients within 48h of symptom onset (IST COLLABORATIVE GROUP 1997). On the other hand, more restrictive criteria which define a more uniform patient selection could increase statistical power by reducing expected variability in study endpoints or treatment effects. For some therapeutic approaches, it is necessary to enroll only patients with stroke types predicted to respond to the therapy. For example, patients with intracerebral hemorrhage are excluded from trials of thrombolytic or antithrombotic drugs because of the likelihood that treatment will be detrimental. Likewise, the mechanism of vascular occlusion (embolic, atherosclerotic, small vessel disease) could influence a choice of reperfusion strategies. However, it is not clear that these considerations must apply to neuroprotective drugs. Assuming adequate tissue delivery, most such agents may reduce tissue damage regardless of the cause of vascular occlusion.

As noted above, there is growing recognition that white and gray matter may have distinct cellular mechanisms of injury. Some authors have proposed that patients with small, deep white matter infarctions (lacunar stroke) might be excluded from studies of drugs targeted only at gray matter (STAIR II 2001). This reasonable suggestion neglects the fact that virtually *all* ischemic strokes in humans include both white and gray matter. Arterial supply of the cerebral cortex is shared between white and gray, which are occupy approximately equal volumes in humans. It an important goal for preclinical trials to evaluate the efficacy of drugs on both brain areas.

Substantial questions persist regarding the optimal outcome measure for randomized clinical trials (Tables 4, 5). Most preclinical models assess therapeutic efficacy by a reduction of the volume of infarcted tissue. Infarct volume is readily measured in patients by CT or MR scans but this is not routinely used as a primary outcome for stroke trials. Stroke volume is only modestly correlated with clinical endpoints such as mortality and measures of neurologic disability (SAVER et al. 1999). Imaging modalities may have limited value as surrogate markers, and are more likely to be useful in patient selection or stratification (for example, to determine tissue at risk). Early trials

Table 5. Outcome measures for stroke clinical trials (from US AHCPR 1995; KELLY-HAYES et al. 1998; most scales are reproduced in EDWARDS 2001)

Outcome type	Examples
Mortality	
General neurologic impairment scores	NIH Stroke Scale (NIHSS)
	Canadian Stroke Scale
	Scandinavian Stroke Scale
	Orgogozo Stroke Scale
Specific neurological impairment scores	Folstein Mini-Mental State Examination
	Fugl-Myer Scale (motor)
	Porch Index of Communicative Ability
Global disability	Modified Rankin Scale
Activities of daily living (ADLs)	Functional Independence Measure (FIM)
	Barthel Index
Global impairment and disability	AHA stroke outcome classification
Radiological endpoints	Infarct volume (CT or MRI)
	Perfusion-diffusion mismatch (MRI)

emphasized clinical scales of neurologic deficits, such as the NIH Stroke Scale, Scandinavian Stroke Scale, and others. These standardized exam measures have reasonable reliability and reproducibility but do not measure functional abilities which lead to independence and quality of life. A variety of tools have been developed to assess these outcomes (Table 4), but the optimal choice has not been determined. Some recent trials, including the pivotal tPA study (NINDS 1995), have utilized a composite score which balances a variety of impairment and disability measures. There is also substantial disagreement regarding the stratification of outcome measures. Should a neuroprotective agent be evaluated according to the number of patients who return to normal or nearly normal function? Or alternatively, by the number of patients who are spared severe disability or death?

IV. Publication of Trial Results

A serious problem limiting stroke drug development has been delayed communication of major clinical trial results. Results of early trials (phase I or II) trials are often unpublished. While some major trials have been published quickly, several others remain unpublished up to ten years after completion. One consequence is that there is no way to fully evaluate the fate of therapeutic approaches which appeared to work in animals but not in people.

There are two solutions to this situation. Before enrolling patients, clinical investigators should insist that phase III trial results will be published and presented, regardless of outcome. There is an ethical obligation to the patients who volunteer to receive experimental treatment with the sole reward of advancing medical knowledge. Second, investigators should list studies in clinical trial registries. These are databases which catalog trials, and sometimes include protocols, results, and publications. The International Cochrane Collaboration provides systematic reviews of therapeutic trials; Cochrane abstracts are indexed in Medline and full reviews are available by subscription. The Cochrane Stroke Group maintains a registry of stroke trial protocols which are used to prepare reviews. This registry itself is not publicly available. The US National Institutes of Health has developed a major on-line registry, ClinicalTrials.gov, to include all trials of life-threatening conditions. This resource includes several stroke trials, but presently it is limited to studies funded by the NIH; industry-sponsored trials will be included in the future. There are no plans to include trials outside of the United States. The Stroke Trials Directory (www.stroketrials.org) was developed to address these problems (GOLDBERG 2001). This is a continuously updated, on-line database of ongoing and completed major trials for stroke prevention, acute therapy, and recovery. It is a component of the Internet Stroke Center stroke education project at Washington University with support from the American Heart Association and National Institute of Neurologic Disorders and Stroke. Readers with current stroke trial information are urged to contact the author with updates and corrections.

D. Conclusions

It is premature to draw conclusions about the efficacy of neuroprotective agents on outcome in human stroke. For a limited number of drugs randomized clinical trial have demonstrated no improvement in outcome when treatment is initiated 6–8 h after acute stroke onset. However, such trials have not yet evaluated any agent within a time window close to that shown to be protective in animals. There are additional concerns regarding trial design. It is not clear whether the completed studies had adequate dosing to reproduce effects found in preclinical models. Recent clinical trial designs and sample sizes appear impressive, but it remains possible that a real therapeutic effect may have been masked by large outcome variability and insufficient statistical power, particularly as many trials were terminated prematurely. Therefore, it is possible that one or more drugs which has been abandoned after clinical trial may actually have neuroprotective properties, although this is unlikely to be tested.

Many promising agents remain to be examined in large-scale trials. Some drug candidates were discontinued because of problems with pharmacodynamics or adverse effects rather than tested lack of efficacy. While NMDA antagonists have received extensive study (none earlier than 6 h after stroke onset), AMPA antagonists and potent antioxidants are only now entering phase III trials. As yet, few drugs have been tested in combination with therapies to improve perfusion.

As reviewed in this monograph, the wealth of basic research in neuroprotection offers many potential targets for subsequent clinical development. Drugs should be selected from animal models based on stringent criteria, including extended post-ischemic therapeutic window, robust dose-response pharmacology, improvement in functional outcomes as well as reduction of infarct size, and efficacy in a range of experimental conditions and species.

The approved treatment for acute stroke, intravenous tissue plasminogen activator, provides proof of the concept that stroke outcome can be improved. The history of this intervention is also instructive: thrombolysis for stroke was proven effective only after publication of negative results from more than a dozen trials involving several thousand patients – almost all with time windows longer than 3 h. Therefore, the disappointing clinical experience of neuroprotection for acute stroke should not discourage future efforts. Acute ischemic stroke is a problem with enormous public health impact. Effective therapy is both possible and urgently needed.

References

Albers GW, Goldberg MP, Choi DW (1989) N-methyl-D-aspartate antagonists: ready for clinical trial in brain ischemia? Ann Neurol 25:398–403

Albers GW, Atkinson RP, Kelley RE, Rosenbaum DM (1995) Safety, tolerability, and pharmacokinetics of the N-methyl-D-aspartate antagonist dextrorphan in patients with acute stroke. Dextrorphan Study Group. Stroke 26:254–258

Alberts MJ, Perry A, Dawson DV, Bertels C (1992) Effects of public and professional education on reducing the delay in presentation and referral of stroke patients. Stroke 23:352–356

American Heart Association (2001) 2001 Heart and Stroke Statistical Update. Dallas, Texas: American Heart Association

Auer RN, Coupland SG, Jason GW, Archer DP, Payne J, Belzberg AJ, Ohtaki M, Tranmer BI (1996) Postischemic therapy with MK-801 (dizocilpine) in a primate model of transient focal brain ischemia. Mol Chem Neuropathol 29:193–210

Barber PA, Zhang J, Demchuk AM, Hill MD, Buchan AM (2001) Why are stroke patients excluded from TPA therapy? An analysis of patient eligibility. Neurology 56:1015–1020

Barsan WG, Brott TG, Broderick JP, Haley EC Jr, Levy DE, Marler JR (1994) Urgent therapy for acute stroke. Effects of a stroke trial on untreated patients. Stroke 25:2132–2137

Bogousslavsky J, Regli F (1985) Cerebral infarct in apparent transient ischemic attack. Neurology 35:1501–1503

Bordi F, Terron A, Reggiani A (1999) The neuroprotective glycine receptor antagonist GV150526 does not produce neuronal vacuolization or cognitive deficits in rats. Eur J Pharmacol 378:153–160

Candelise L, Ciccone A (2000) Gangliosides for acute ischaemic stroke. Cochrane Database Syst Rev CD000094

Clark WM, Warach SJ, Pettigrew LC, Gammans RE, Sabounjian LA (1997) A randomized dose-response trial of citicoline in acute ischemic stroke patients. Citicoline Stroke Study Group. Neurology 49:671–678

Clark WM, Williams BJ, Selzer KA, Zweifler RM, Sabounjian LA, Gammans RE (1999a) A randomized efficacy trial of citicoline in patients with acute ischemic stroke. Stroke 30:2592–2597

Clark WM, Wissman S, Albers GW, Jhamandas JH, Madden KP, Hamilton S (1999b) Recombinant tissue-type plasminogen activator (Alteplase) for ischemic stroke 3–5 h after symptom onset. The ATLANTIS Study: a randomized controlled trial. Alteplase Thrombolysis for Acute Noninterventional Therapy in Ischemic Stroke. JAMA 282:2019–2026

Clark WM, Raps EC, Tong DC, Kelly RE (2000) Cervene (Nalmefene) in acute ischemic stroke: final results of a phase III efficacy study. The Cervene Stroke Study Investigators. Stroke 31:1234–1239

Cornu C, Boutitie F, Candelise L, Boissel JP, Donnan GA, Hommel M, Jaillard A, Lees KR (2000) Streptokinase in acute ischemic stroke: an individual patient data meta-analysis: The Thrombolysis in Acute Stroke Pooling Project. Stroke 31:1555–1560

Davis SM, Albers GW, Diener HC, Lees KR, Norris J (1997) Termination of Acute Stroke Studies Involving Selfotel Treatment. ASSIST Steering Committed. Lancet 349:32

Davis SM, Lees KR, Albers GW, Diener HC, Markabi S, Karlsson G, Norris J (2000) Selfotel in acute ischemic stroke: possible neurotoxic effects of an NMDA antagonist. Stroke 31:347–354

DeGraba TJ, Pettigrew LC (2000) Why do neuroprotective drugs work in animals but not humans? Neurol Clin 18:475–493

del Zoppo GJ, Becker KJ, Hallenbeck JM (2001) Inflammation after stroke: is it harmful? Arch Neurol 58:669–672

Dewar D, Yam P, McCulloch J (1999) Drug development for stroke: importance of protecting cerebral white matter. Eur J Pharmacol 375:41–50

Diener HC (1998) Multinational randomised controlled trial of lubeluzole in acute ischaemic stroke. European and Australian Lubeluzole Ischaemic Stroke Study Group. Cerebrovasc Dis 8:172–181

Diener HC, Cortens M, Ford G, Grotta J, Hacke W, Kaste M, Koudstaal PJ, Wessel T (2000) Lubeluzole in acute ischemic stroke treatment: A double-blind study with an 8-h inclusion window comparing a 10-mg daily dose of lubeluzole with placebo. Stroke 31:2543–2551

Dorman PJ, Sandercock PA (1996) Considerations in the design of clinical trials of neuroprotective therapy in acute stroke. Stroke 27:1507–1515

Dorman PJ, Counsell C, Sandercock P (1999) Reports of randomized trials in acute stroke, 1955 to 1995. What proportions were commercially sponsored? Stroke 30: 1995–1998

Edwards DF (2001) Stroke scales and neurological assessment tools. In: Goldberg MP (ed) Internet Stroke Center website: www.strokecenter.org/trials/scales

Fang J, Alderman MH (2001) Trend of stroke hospitalization, united states, 1988–1997. Stroke 32:2221–2226

Feigin VL, Rinkel GJ, Algra A, Vermeulen M, van Gijn J (2000) Calcium antagonists for aneurysmal subarachnoid haemorrhage. Cochrane. Database. Syst Rev CD000277

Fischman AJ, Alpert NM, Babich JW, Rubin RH (1997) The role of positron emission tomography in pharmacokinetic analysis. Drug Metab Rev 29:923–956

Franke CL, Palm R, Dalby M, Schoonderwaldt HC, Hantson L, Eriksson B, Lang-Jenssen L, Smakman J (1996) Flunarizine in stroke treatment (FIST): a double-blind, placebo-controlled trial in Scandinavia and the Netherlands. Acta Neurol Scand 93:56–60

Gill R (1994) The pharmacology of alpha-amino-3-hydroxy-5-methyl-4-isoxazole propionate (AMPA)/kainate antagonists and their role in cerebral ischaemia. Cerebrovasc Brain Metab Rev 6:225–256

Ginsberg MD (1996) The validity of rodent brain-ischemia models is self-evident. Arch Neurol 53:1065–1067

Goldberg MP (ed) (2001) Stroke Trials Directory. Internet Stroke Center website: www.stroketrials.org

Goldberg MP, Weiss JH, Pham PC, Choi DW (1987) N-methyl-D-aspartate receptors mediate hypoxic neuronal injury in cortical culture. J Pharmacol Exp Ther 243: 784–791

Gorelick PB, Sacco RL, Smith DB, Alberts M, Mustone-Alexander L, Rader D, Ross JL, Raps E, Ozer MN, Brass LM, Malone ME, Goldberg S, Booss J, Hanley DF, Toole JF, Greengold NL, Rhew DC (1999) Prevention of a first stroke: a review of guidelines and a multidisciplinary consensus statement from the National Stroke Association. JAMA 281:1112–1120

Gribkoff VK, Starrett JE Jr, Dworetzky SI, Hewawasam P, Boissard CG, Cook DA, Frantz SW, Heman K, Hibbard JR, Huston K, Johnson G, Krishnan BS, Kinney GG, Lombardo LA, Meanwell NA, Molinoff PB, Myers RA, Moon SL, Ortiz A, Pajor L, Pieschl RL, Post-Munson DJ, Signor LJ, Srinivas N, Taber MT, Thalody G, Trojnacki JT, Wiener H, Yeleswaram K, Yeola SW (2001) Targeting acute ischemic stroke with a calcium-sensitive opener of maxi-K potassium channels. Nat Med 7:471–477

Grotta J, Clark W, Coull B, Pettigrew LC, Mackay B, Goldstein LB, Meissner I, Murphy D, LaRue L (1995) Safety and tolerability of the glutamate antagonist CGS 19755 (Selfotel) in patients with acute ischemic stroke. Results of a phase IIa randomized trial. Stroke 26:602–605

Grotta J (1997) Lubeluzole treatment of acute ischemic stroke. The US and Canadian Lubeluzole Ischemic Stroke Study Group. Stroke 28:2338–2346

Hacke W, Kaste M, Fieschi C, von Kummer R, Davalos A, Meier D, Larrue V, Bluhmki E, Davis S, Donnan G, Schneider D, Diez-Tejedor E, Trouillas P (1998) Randomised double-blind placebo-controlled trial of thrombolytic therapy with intravenous alteplase in acute ischaemic stroke (ECASS II). Second European-Australasian Acute Stroke Study Investigators. Lancet 352:1245–1251

Haley EC Jr (1998) High-dose tirilazad for acute stroke (RANTTAS II). RANTTAS II Investigators. Stroke 29:1256–1257

Hargreaves RJ, Rigby M, Smith D, Hill RG, Iversen LL (1993) Competitive as well as uncompetitive N-methyl-D-aspartate receptor antagonists affect cortical neuronal morphology and cerebral glucose metabolism. Neurochem Res 18: 1263–1269

Hawkinson JE, Huber KR, Sahota PS, Han HH, Weber E, Whitehouse MJ (1997) The N-methyl-D-aspartate (NMDA) receptor glycine site antagonist ACEA 1021 does not produce pathological changes in rat brain. Brain Res 744:227–234

Horn J, Limburg M (2000) Calcium antagonists for acute ischemic stroke. Cochrane Database Syst Rev CD001928

Imai H, Masayasu H, Dewar D, Graham DI, Macrae IM (2001) Ebselen protects both gray and white matter in a rodent model of focal cerebral ischemia. Stroke 32:2149–2154

International Stroke Trial (IST) Collaborative Group (1997) The International Stroke Trial (IST): a randomised trial of aspirin, subcutaneous heparin, both, or neither among 19435 patients with acute ischaemic stroke. Lancet 349(9065):1569–1581

Johnston SC, Fung LH, Gillum LA, Smith WS, Brass LM, Lichtman JH, Brown AN (2001) Utilization of intravenous tissue-type plasminogen activator for ischemic stroke at academic medical centers: the influence of ethnicity. Stroke 32:1061–1068

Koudstaal PJ, van Gijn J, Lodder J, Frenken WG, Vermeulen M, Franke CL, Hijdra A, Bulens C (1991) Transient ischemic attacks with and without a relevant infarct on computed tomographic scans cannot be distinguished clinically. Dutch Transient Ischemic Attack Study Group. Arch Neurol 48:916–920

Kwiatkowski TG, Libman RB, Frankel M, Tilley BC, Morgenstern LB, Lu M, Broderick JP, Lewandowski CA, Marler JR, Levine SR, Brott T (1999) Effects of tissue plasminogen activator for acute ischemic stroke at one year. National Institute of Neurological Disorders and Stroke Recombinant Tissue Plasminogen Activator Stroke Study Group. N Engl J Med 340:1781–1787

Lago A, Geffner D, Tembl J, Landete L, Valero C, Baquero M (1998) Circadian variation in acute ischemic stroke: a hospital-based study. Stroke 29:1873–1875

LaRue LJ, Alter M, Traven ND, Sterman AB, Sobel E, Kleiner J (1988) Acute stroke therapy trials: problems in patient accrual. Stroke 19:950–954

Lee JM, Zipfel GJ, Choi DW (1999) The changing landscape of ischaemic brain injury mechanisms. Nature 399 [Suppl 6738]:A7–A14

Lees KR, Asplund K, Carolei A, Davis SM, Diener HC, Kaste M, Orgogozo JM, Whitehead J (2000) Glycine antagonist (gavestinel) in neuroprotection (GAIN International) in patients with acute stroke: a randomised controlled trial. GAIN International Investigators. Lancet 355:1949–1954

Lees KR, Sharma AK, Barer D, Ford GA, Kostulas V, Cheng YF, Odergren T (2001) Tolerability and pharmacokinetics of the nitrone NXY-059 in patients with acute stroke. Stroke 32:675–680

Lenzi GL, Grigoletto F, Gent M, Roberts RS, Walker MD, Easton JD, Carolei A, Dorsey FC, Rocca WA, Bruno R (1994) Early treatment of stroke with monosialoganglioside GM-1. Efficacy and safety results of the Early Stroke Trial. Stroke 25:1552–1558

Levy DE (1988) How transient are transient ischemic attacks? Neurology 38:674–677

Lyden P, Jacoby M, Schim J, Albers G, Mazzeo P, Ashwood T, Nordlund A, Odergren T (2001) The Clomethiazole Acute Stroke Study in tissue-type plasminogen activator-treated stroke (CLASS-T): Final results. Neurology 57:1199–1205

McDonald JW, Althomsons SP, Hyrc KL, Choi DW, Goldberg MP (1998) Oligodendrocytes from forebrain are highly vulnerable to AMPA/kainate receptor-mediated excitotoxicity. Nat Med 4:291–297

Morris DL, Rosamond W, Madden K, Schultz C, Hamilton S (2000) Prehospital and emergency department delays after acute stroke: the Genentech Stroke Presentation Survey. Stroke 31:2585–2590

Morris GF, Bullock R, Marshall SB, Marmarou A, Maas A, Marshall LF (1999) Failure of the competitive N-methyl-D-aspartate antagonist Selfotel (CGS 19755) in the treatment of severe head injury: results of two phase III clinical trials. The Selfotel Investigators. J Neurosurg 91:737–743

Muir KW, Hamilton SJ, Lunnon MW, Hobbiger S, Lees KR (1998) Safety and tolerability of 619C89 after acute stroke. Cerebrovasc Dis 8:31–37

National Institute of Neurological Disorders and Stroke (NINDS) rt-PA Stroke Study Group (1995) Tissue plasminogen activator for acute ischemic stroke. N Engl J Med 333:1581–1587

North American Symptomatic Carotid Endarterectomy Trial (NASCET) Collaborators (1991) Beneficial effect of carotid endarterectomy in symptomatic patients with high-grade carotid stenosis. N Engl J Med 325:445–453

Ogawa A, Yoshimoto T, Kikuchi H, Sano K, Saito I, Yamaguchi T, Yasuhara H (1999) Ebselen in acute middle cerebral artery occlusion: a placebo-controlled, double-blind clinical trial. Cerebrovasc Dis 9:112–118

Olney JW, Labruyere J, Price MT (1989) Pathological changes induced in cerebrocortical neurons by phencyclidine and related drugs. Science 244:1360–1362

Ortiz GG, Guerrero JM, Reiter RJ, Poeggeler BH, Bitzer-Quintero OK, Feria-Velasco A (1999) Neurotoxicity of dextrorphan. Arch Med Res 30:125–127

RANTTAS Investigators (1996) A randomized trial of tirilazad mesylate in patients with acute stroke (RANTTAS). Stroke 27:1453–1458

Ricci S, Celani MG, Cantisani TA, Righetti E (2000) Piracetam in acute stroke: a systematic review. J Neurol 247:263–266

Rothman SM, Thurston JH, Hauhart RE, Clark GD, Solomon JS (1987) Ketamine protects hippocampal neurons from anoxia in vitro. Neuroscience 21:673–678

Sacco RL, DeRosa JT, Haley EC Jr, Levin B, Ordronneau P, Phillips SJ, Rundek T, Snipes RG, Thompson JL (2001) Glycine antagonist in neuroprotection for patients with acute stroke: GAIN Americas: a randomized controlled trial. JAMA 285:1719–1728

Saver JL, Johnston KC, Homer D, Wityk R, Koroshetz W, Truskowski LL, Haley EC (1999) Infarct volume as a surrogate or auxiliary outcome measure in ischemic stroke clinical trials. The RANTTAS Investigators. Stroke 30:293–298

Samsa GP, Matchar DB (2001) Have randomized controlled trials of neuroprotective drugs been underpowered? An illustration of three statistical principles. Stroke 32:669–674

Steinberg GK, Bell TE, Yenari MA (1996) Dose escalation safety and tolerance study of the N-methyl-D-aspartate antagonist dextromethorphan in neurosurgery patients. J Neurosurg 84:860–866

Stroke Therapy Academic Industry Roundtable (STAIR) (1999) Recommendations for standards regarding preclinical neuroprotective and restorative drug development. Stroke 30:2752–2758

Stroke Therapy Academic Industry Roundtable (STAIR) II (2001) Recommendations for clinical trial evaluation of acute stroke therapies. Stroke 32:1598–1606

Stys PK (1998) Anoxic and ischemic injury of myelinated axons in CNS white matter: from mechanistic concepts to therapeutics. J Cereb Blood Flow Metab 18:2–25

Tekkok SB, Goldberg MP (2001) AMPA/kainate receptor activation mediates hypoxic oligodendrocyte death and axonal injury in cerebral white matter. J Neurosci 21:4237–4248

Tirilazad International Steering Committee (2000) Tirilazad mesylate in acute ischemic stroke: A systematic review. Stroke 31:2257–2265

Tymianski M, Spigelman I, Zhang L, Carlen PL, Tator CH, Charlton MP, Wallace MC (1994) Mechanism of action and persistence of neuroprotection by cell-permeant Ca^{2+} chelators. J Cereb Blood Flow Metab 14:911–923

Wahlgren NG, Ranasinha KW, Rosolacci T, Franke CL, van Erven PM, Ashwood T, Claesson L (1999) Clomethiazole acute stroke study (CLASS): results of a randomized, controlled trial of clomethiazole versus placebo in 1360 acute stroke patients. Stroke 30:21–28

Wester P, Radberg J, Lundgren B, Peltonen M (1999) Factors associated with delayed admission to hospital and in-hospital delays in acute stroke and TIA: a prospective, multicenter study. Seek- Medical-Attention-in-Time Study Group. Stroke 30:40–48

Wiebers DO, Adams HP Jr, Whisnant JP (1990) Animal models of stroke: are they relevant to human disease? Stroke 21:1–3

Wolf PA, Clagett GP, Easton JD, Goldstein LB, Gorelick PB, Kelly-Hayes M, Sacco RL, Whisnant JP (1999) Preventing ischemic stroke in patients with prior stroke and transient ischemic attack: a statement for healthcare professionals from the Stroke Council of the American Heart Association. Stroke 30:1991–1994

Wroe SJ, Sandercock P, Bamford J, Dennis M, Slattery J, Warlow C (1992) Diurnal variation in incidence of stroke: Oxfordshire community stroke project. BMJ 304:155–157

Yamaguchi T, Sano K, Takakura K, Saito I, Shinohara Y, Asano T, Yasuhara H (1998) Ebselen in acute ischemic stroke: a placebo-controlled, double-blind clinical trial. Ebselen Study Group. Stroke 29:12–17

Zivin JA, Grotta JC (1990) Animal stroke models. They are relevant to human disease. Stroke 21:981–983

CHAPTER 14
Traumatic Central Nervous System Injury

R.M. POOLE

A. Introduction

Traumatic injury to the central nervous system is an important public health problem. Severe head and spinal cord injuries are a major cause of disability and death, particularly among young adults. The medical sequelae of central nervous system injury are substantial, resulting in large demands upon the patient's family, the health care system, and society in general. The National Head and Spinal Cord Injury Survey estimates the rate of occurrence of severe head injury at 200 per 100000 (KRAUS et al. 1996). Motor vehicle accidents are the primary cause of both brain and spinal cord injury, and young healthy males are the primary victims.

Despite years of productivity and exciting developments in preclinical neuroprotection research, few of these advances have been translated into approved treatments in the clinic. Although nimodipine has recently been approved in Europe for posttraumatic subarachnoid hemorrhage, and two large trials have shown that administration of methylprednisolone improves outcome in patients with spinal cord injury, there are no other pharmacologic therapies available to treat patients with traumatic injuries of the central nervous system. One explanation for this lack of success may be that, to date, the compounds that have been in clinical trials simply are not effective in humans. Alternatively, the translation from animal experiments to clinical trials may have been inadequate or the trials themselves may have been flawed. The purpose of this chapter is to review the clinical experience with trials in traumatic central nervous system injury and to outline factors that may improve the chances for moving from successful neuroprotective experiments in laboratory animals to convincing clinical studies in humans.

B. Clinical Trial Experience in Traumatic Brain Injury

Excellent reviews of the experience with clinical trials in traumatic brain injury (TBI) over the past two decades have already been published (ALVES and EISENBERG 1996; DOPPENBERG and BULLOCK 1997). Many different pharmacologic approaches have been attempted in the clinic both in small phase II studies and in adequately powered phase III trials. These trials have examined drugs with such widely varying pharmacologic properties as scopolamine, high-dose corticosteroids, barbiturates, tromethamine buffer, L and N-type calcium channel antagonists, free radical scavengers (tirilazad and superoxide dismutase), subtype-selective NMDA antagonists, bradykinin antagonists, and cannabinoids. DOPPENBERG and BULLOCK (1997) tabulated the results of many of the trials of neuroprotective agents that have been performed to date. Based on the experience gleaned from these studies, they proposed five theoretical requirements for advancing a neuroprotective agent into pivotal clinical trials: (1) the mechanism on which the drug acts can be shown in animal models of TBI, (2) the mechanism is blocked by the drug under study in the same models, (3) the same mechanism is present in human TBI, (4) brain penetration of the drug is adequate to block the mechanism, and (5) the drug is safe and tolerable in TBI patients. They argue that in most of the negative clinical trials performed to date, one or more of these requirements was missing from the knowledge base about the compound prior to beginning the trial.

Characterizing a compound with respect to these requirements appears to represent the minimum necessary knowledge base prior to embarking on definitive pivotal efficacy trials for a neuroprotective drug for TBI. There are, however, a number of additional factors that may have affected the results of previous clinical trials and that need to be considered in the design and conduct of studies of new drugs for TBI. These include the need for more comprehensive testing of compounds in preclinical models; stricter adherence in clinical trials to the conditions present in preclinical experiments; better characterization of compounds in early clinical development in humans; and special challenges related to careful control of patient characteristics at entry, the choice of outcome measures, and methods of analysis.

I. Adequate Characterization of Compounds in Preclinical Experiments

Over the past 5 years, considerable advances have been made in understanding the pathophysiology of traumatic brain injury (TEASDALE and GRAHAM 1998b) and pharmacologic approaches to its treatment (MCINTOSH et al. 1998). Head injury is a complex pathophysiologic state that cannot be adequately modeled by any one preclinical experiment. Because of the lack of success in the clinic, no particular preclinical model can be said to be predictive of efficacy in human trials. Preclinical investigation of new compounds should therefore involve brain injury models in several species as well as both focal and global ischemia models. Multiple different endpoints such as histologic appearance, infarction and edema volume, and neuropsychometric measures should be examined in animals. Robust efficacy in a number of models provides confidence that the mechanism of action of the compound is relevant to primary brain injury and to the secondary cellular processes that complicate head injury. The window of efficacy in animal ischemia and brain injury models needs to be carefully characterized with respect to the maximum allowable delay of administration after injury and the optimal duration of treatment. With some exceptions (TOULMOND 1993), detailed information about the therapeutic window has not been provided for many compounds, leaving clinicians to guess about the optimal regimen and window to use in clinical trials.

Some compounds have not been successful in the clinic despite exhaustive preclinical testing. This disparity between preclinical and clinical results raises the fundamental question as to whether any current animal models (or combinations of different models) are useful predictors of clinical outcomes with new therapies for TBI. Alternatively, given the complex nature of TBI pathophysiology, the principle difficulty may stem from overly simplistic translation of animal models to human TBI. It is tempting to use these reservations about animal models and their interpretation to justify employing treatment paradigms in human trials that are different from those observed to be effective in animals. Unfortunately, clinical trial design based on the best available animal data is the only feasible approach to a rational exploration of strategies in this area. Deviations from the time window and treatment conditions identified in animal models should be based not on generic reference to the complexities of the situation but rather on specific supportive experimental human or animal data.

The results of efforts to find an effective neuroprotective agent for stroke offer some insight in this regard. With the exception of tissue plasminogen activator (t-PA) administered within 3h of symptom onset (NINDS RT-PA STROKE STUDY GROUP 1995), no neuroprotective agent has shown clear efficacy in stroke clinical trials despite numerous attempts. One possible explanation for these results is that, except for t-PA, no agent has been studied in human trials with a window of administration identical to the effective time window in animals (Table 1). ZIVIN has argued that although animal ischemia data fairly closely mirror clinical experience with patients and tend to support a 6-h window of administration for clinical trials in stroke, new drugs should probably be administered earlier to give the best chance for success (ZIVIN 1998). Given the experience with t-PA, it can be argued that initial clinical trials of neuroprotective agents for stroke and TBI should be performed with a window that falls within the effective window determined in animal models.

Although time windows as short as 90min have been successfully employed in stroke clinical trials (BROTT et al. 1992), most researchers agree that a 3-h window for study drug administration is about the shortest time feasible for a large pivotal trial. Fortunately, the majority of patients with TBI are rapidly transported to medical facilities and short time windows in TBI studies are somewhat less a problem than for stroke clinical trials. Even so, enrolling patients within a 3-h time window is an enormously challenging undertaking.

Table 1. Comparison of maximal preclinical and clinical windows of administration for neuroprotective agents that have failed in stroke trials

Compound	Preclinical efficacy	Maximum preclinical poststroke interval	Maximum clinical poststroke interval
Nimodipine Ca^{++} Channel antagonist	MCAO, MCAOR, global ischemia	6 h	24 h
Tirilazad lipid peroxidation inhibitor	MCAO, MCAOR, global ischemia	1 h	6 h
Cerestat, Selfotel NMDA antagonist	MCAO, MCAOR	1 h/10 min	6 h/24 h
Eliprodil SS-NMDA antagonist	MCAO	45 min	8 h
BW619C89 Na^+ Channel antagonist	MCAO, global ischemia	?	12 h
Fosphenytoin Na^+ Channel antagonist	MCAO, neonatal hypoxia	30 min	4 h

Neuroprotective drugs with long time windows of administration in animal experiments, if they exist, stand the best chance for ultimate success in the clinic.

II. Characterization of Compounds in Early Clinical Development

Prior to beginning exploratory or confirmatory efficacy trials, as much as possible should be known about the basic pharmacokinetics, pharmacodynamics, metabolism, side effects, and rationale for dose selection and regimen of a neuroprotective agent. Phase I studies in volunteers are the main arena in which this information is generated although valuable data can be collected in small phase I trials in TBI patients. This is particularly true when a drug has pharmacodynamic effects that may complicate patient management in the intensive care unit or when certain aspects of TBI management might influence drug delivery or metabolism.

In addition to standard pharmacokinetic assessments, studies should be performed to define gender differences in metabolism and to detect drug/drug interactions with medications that are likely to be coadministered with the compound of interest (e.g., vasopressors or sedatives). These studies should be completed as early as possible in the development process. Although specific protocols can be designed and implemented to evaluate pharmacokinetic data early on in large studies, correcting dosing regimens for variations in drug metabolism in a particular subgroup is difficult and expensive once a large trial is underway.

Lessons about the importance of adequate pharmacokinetic characterization come from the clinical development of the lipid peroxidation inhibitor tirilazad. The failure to account for gender differences in metabolism, for example, may have been a crucial factor in the unsuccessful first international trial of tirilazad for neuroprotection in the setting of subarachnoid hemorrhage (KASSELL et al. 1996). Similarly, phenytoin, which accelerates tirilazad metabolism and which was coadministered to many patients in the North American subarachnoid hemorrhage study, may have played a role in the negative outcome of that trial by reducing tirilazad plasma concentrations (HALEY et al. 1997; FLEISHAKER et al. 1998).

When a neuroprotective drug affects hemodynamic stability, its human pharmacology must be understood in as much detail as possible. Since systemic hypotension carries such a significant risk for patients with TBI (CHESNUT et al. 1993; CHESNUT 1997), the ability to successfully manage any blood pressure-lowering effect of a compound must be demonstrated in volunteers and in small numbers of patients with TBI prior to beginning larger efficacy studies. Careful documentation of the effect of the drug on hemodynamics, intracranial pressure, cerebral perfusion pressure, and jugular vein oxygen saturation, as well as other physiologic parameters are necessary to define a safe treatment regimen for such a drug. BARDT described a monitoring system used in a neurointensive care unit setting that allows continuous monitoring of complex patterns of brain function and metabolism (BARDT et al. 1998). This kind of multimodality cerebral monitoring can be carried out in early phase I studies in TBI patients and would be an essential component of the early clinical evaluation of a drug with a risky hemodynamic profile prior to beginning larger efficacy studies.

It is essential to have information about the brain penetration of a compound and to know whether the purported mechanism of action is operative at achievable brain concentrations. This data is difficult to obtain in large numbers of patients and would be most valuable early in development. Using intracerebral microdialysis, BULLOCK demonstrated that the excitatory amino acids (EAA) glutamate and aspartate rose to very high levels in 30% of 80 consecutive patients with severe brain injury and that these increases persisted for up to 4 days in many of the patients studied (BULLOCK 1998). For a drug whose mechanism of action involves the presynaptic interruption of EAA release, it might be possible to correlate dialysate concentrations of EAA with parenchymal concentrations of the active compound and to rationally select dosage regimens that result in effective brain concentrations. Further, it might be possible to make a rational decision to stop developing a drug if brain concentrations adequate to affect the relevant mechanism could not be achieved. This approach may not be relevant for every compound but where possible, such information would be very valuable. Theoretically this data could be gathered early on in special centers participating in a small phase I study or in an open label run-in period prior to starting a larger placebo-controlled exploratory efficacy trial.

The value of knowing as much as possible about a neuroprotective compound prior to embarking on large pivotal efficacy trials seems obvious. There are, however, potent incentives at work in the pharmaceutical industry to shorten timelines at every stage of development in the clinical arena. Neuroprotective drug development in particular is a high-risk venture with a low probability of success that commands large manpower and financial resources. There is often significant pressure from corporate managers to generate pivotal efficacy data as quickly as possible so that critical drug development decisions can be made rapidly. Because of the continuous erosion of compound patent life, the need for speed in drug development is particularly important. In some respects time is more important than overall development costs since the longer a drug is on the market without generic competition the more opportunity there is to recoup costs and to allow a company to generate profit. Although the profit motive may seem distasteful to some, pharmaceutical companies may be the only entities in our society that can commit the massive resources necessary to support neuroprotective drug development and companies cannot survive long in the marketplace without providing returns to their investors.

The importance of the time element in the pharmaceutical industry has resulted in compression of clinical development programs and the performance of trials that are a hybrid between classical phase II and phase III studies. These "phase II/III" studies combine dose ranging and collection of exploratory safety and efficacy endpoints typical of phase II studies but are powered like a phase III study to enroll enough patients to draw scientifically valid conclusions. Such a trial is considered an "exploratory" efficacy trial, the purpose of which is to determine efficacy and safety in patients as well as the dose and regimen for a second "confirmatory" efficacy trial. The confirmatory trial evaluates efficacy, safety, and sometimes the dose-response relationship in a large group of patients and together with the exploratory trial forms the basis for a regulatory submission (SANDAGE 1998).

Although this approach may work well for drugs with relatively robust efficacy measures and a strong track record of approval (e.g., cholesterol lowering agents, hypertension drugs, H_2 antagonists, or NSAIDs) there is probably no clear advantage of this style of development for neuroprotective drugs. For this scheme to work in TBI, exploratory efficacy trials would necessarily be complex, expensive and highly risky. The trial would have numerous critical objectives and would have to be very carefully designed to avoid overloading investigative sites with data collection requirements. One important objective would be clear identification of the optimal dose and regimen since carrying multiple dose arms into a second trial would be complicated and costly. Careful pharmacokinetic characterization of the patient population would be critical, as would an exploration of the operational characteristics of the primary and secondary efficacy measures. Since definitive outcome measures in TBI are typically collected 6 months after randomization, a significant time lag would occur before flaws in the experimental design could be discovered and corrected.

As a trial becomes more complicated, there is a higher probability of failing to recognize study design problems early on and of failing to reach critical scientific objectives. Smaller trials in TBI patients with clear objectives such as those outlined above, performed early in development, would provide a better foundation for the large complex efficacy trials that are ultimately necessary. A traditional development approach is more logical if we are to advance the science of neuroprotective therapy for TBI.

III. Special Challenges of Trial Design and Conduct for TBI

Clinical trials in TBI are complex and difficult to conduct. There are special challenges related to the short time window for treatment, characterization of the patient population, the need for treatment uniformity among investigative centers, and a need for better outcome measures.

1. Short Enrollment Time Window

The importance of early treatment of traumatic brain injury has already been stressed. Although patients with TBI typically arrive in the hospital quickly, enrollment in an experimental drug study requires obtaining legal informed consent. Patients with TBI are usually not capable of giving informed consent themselves and very often it is difficult to find relatives or next-of-kin who can give permission for a patient to participate in a clinical research study. Because the time window for administration of many treatments is short, the ability to conduct trials in TBI can be significantly compromised by the inability to obtain informed consent.

The background of US regulations regarding the participation of human subjects in clinical trials involving investigational drugs for emergency conditions has been reviewed (WICHMAN 1997). In several early drug studies in TBI, researchers interpreted an existing Food and Drug Administration (FDA) regulation that allows the emergency use of investigational drugs in single patients to apply to subjects in randomized trials. This led to the use of "deferral" of informed consent in two TBI trials, which was approved by the institutional review boards (IRB) in some centers as long as written informed consent could be obtained from a relative within 24h after study initiation (YOUNG et al. 1996; MARSHALL et al. 1998). Because of controversy over the use of deferral of informed consent and concern over inconsistencies in FDA and Department of Health and Human Services regulations on informed consent for emergency research, the FDA wrote new regulations that went into effect in November 1996. The new FDA regulation allows waiver of informed consent requirements in emergency research as long as certain criteria are met (WICHMAN 1997). Much of the regulation is directly applicable to TBI but meeting some of the criteria will pose a considerable challenge to clinical investigators.

For example, the regulation requires IRBs to seek consultation with representatives of the community in which the trial will be conducted and to publicly disclose the plans for the research along with its risks and possible benefits. While such public disclosure is reasonable, there is no guidance in the regulation about what would constitute sufficient community consultation and disclosure. The regulation also requires that if a legally authorized representative is not available, the investigator will attempt to contact a family member who is not a legally authorized representative to ask whether there is an objection to the patient's participation in the study. The IRB is also responsible for approving the procedures for contacting patient representatives as well as the consent forms and documents (such as telephone scripts) used as part of the procedure. Although well intentioned, it is not clear that these requirements will truly protect human subjects. The legal liability of those conducting the research or approving waiver of informed consent under these new regulations is also unclear. Nevertheless, the law now provides a mechanism for allowing waiver of informed consent in trials involving life-threatening conditions.

Clinical investigators and sponsors need to work together diligently and commit the resources necessary to incorporate waiver of informed consent into future clinical trials so that neuroprotective drugs can be tested with the shortest possible time window. One potential benefit of the new waiver of informed consent regulation, depending on how the community notification requirement is met, may be that there will be much greater awareness of the problem of TBI and of studies being performed to treat it. This may help to decrease community resistance to participation in trials of new medications.

2. Characterization and Stratification of the Study Population

Careful characterization and adequate randomization of the patient population is a basic but vital part of good study design in TBI. Fortunately, a great deal is known about factors that influence prognosis in TBI (CHOI and BARNES 1996). Information from the Traumatic Coma Data Bank (MARSHALL et al. 1983) allows investigators to accurately predict the outcome of placebo patients with particular clinical characteristics at admission in TBI trials. This information is critical to the calculation of appropriate sample sizes and to the interpretation of outcome data generated in small phase II trials.

The predictability of outcome in severe traumatic brain injury is one of the advantages of studying severely injured patients in clinical drug trials; moderately and mildly injured patients have a much more variable clinical course. CHOI studied prognostic variables for outcome prediction in severe head injury and devised charts that could be used at the time of presentation to hospital and at the fourth day after admission (CHOI et al. 1983). Their study showed that Glasgow Coma Scale (GCS) score, oculocephalic responses, and age reliably predicted outcome on the Glasgow Outcome Scale (GOS) at 6 months. Later, these observations were refined utilizing data from 523 patients admitted to the neurosurgical service at Medical College of Virginia over 10 years from 1976 to 1986 (CHOI et al. 1988). In this study, a step-wise discriminant analysis method was used to empirically determine the best prognostic model for predicting GOS outcome based on 21 clinical variables noted at admission. While many factors showed a strong correlation with GOS outcome at 6 months, the combination of age, GCS motor score, and pupillary response provided a simple and highly predictive model.

Using this model, it is possible to accurately predict the outcome of patients receiving placebo in a clinical trial and to properly balance treatment arms for certain prognostic factors. Because TBI is such a complex disorder, it is possible to conduct a large randomized trial and have a significant imbalance in treatment arms with respect to either underlying pathology or one or more prognostic variables. This problem occurred in the European trial of tirilazad for moderate and severe TBI (MARSHALL et al. 1998). In this study of 1120 patients, there were significant imbalances favoring the placebo group in the percentages of patients with pretreatment hypotension and hypoxia, factors that have a strong influence on mortality after TBI. In addition, there were nearly twice as many epidural hematomas in the placebo arm compared to the tirilazad arm. Because surgical drainage of an epidural hematoma can rapidly and dramatically improve a patient's neurologic condition, the imbalance favoring the placebo group may have contributed to the overall negative outcome of the trial.

Patients with many different types of intracranial pathology are enrolled into TBI clinical trials. Some researchers have argued for the exclusion of certain types of injuries in TBI trials in order to homogenize the study population. For example, patients that have suffered isolated diffuse axonal shearing injury have a different prognosis for recovery than patients with focal intracranial mass lesions and severe edema (GENNARELLI et al. 1982). Patients with posttraumatic subarachnoid hemorrhage have higher rates of poor outcome (GAETANI et al. 1995), and might be excluded or analyzed separately from those with severe injury without subarachnoid hemorrhage.

Using a computerized central randomization system, it is possible to stratify patients at the time of randomization into prognostically important subsets. Investigators provide information on clinical variables such as age, GCS motor score, and pupillary response to the computer using a telephonic interactive voice response (IVR) system. The randomization system ensures that prognostic factors are balanced between treatment groups and provides a treatment assignment. These systems can also provide an accurate inventory of drug supply and a system for emergency code-break. Although in a large study the need to correct for treatment-by-center interactions may effectively limit the number of strata that can reasonably be employed (CHOI et al. 1995), theoretically a large number of prognostic variables can be programmed into an IVR system. IVR systems are ideal for this kind of randomization scheme since most important prognostic variables can be accurately captured by a brief series of yes/no, multiple choice or numeric entry questions.

3. Standardization of Care

Pivotal efficacy trials in TBI are necessarily large affairs that involve numerous centers spread sometimes across many geographical regions. There is often wide variation in neurosurgical and intensive care unit practice at these centers resulting in different thresholds for operating on intracranial mass lesions and differences in the methods employed to treat patients with brain edema and critically elevated intracranial pressure (GHAJAR et al. 1995). Because variability in patient management is a significant confounding variable, there is a critical need for standardization of therapy across centers and for real-time surveillance of patient management in TBI trials.

Two groups have published guidelines for managing patients with severe brain injury. A North American version, published by the Brain Trauma Foundation in cooperation with the Joint Section on Neurotrauma and Critical Care of the American Association of Neurologic Surgeons and the Congress of Neurologic Surgeons (BULLOCK et al. 1996), used evidence-based methods to establish management recommendations. Recommendations were rank ordered as standards, guidelines, or options depending on a literature-based assessment of high, moderate, or unclear degrees of clinical certainty, respectively. The European Brain Injury Consortium published its own recommendations for severe head injury management based on initial guidelines drafted by an expert working group of Consortium members with subsequent feedback and refinement by participating centers (MAAS et al. 1997).

Despite the different methods used to develop these guidelines, the recommendations are similar and provide clear direction for standardized management of severe brain injury in clinical trials. Both groups stopped short of strongly recommending routine monitoring of intracranial pressure (ICP) for all severely injured patients because of the lack evidence from prospective, randomized controlled trials supporting monitoring and because of legitimate differences in expert opinion on the necessity for monitoring in certain patient groups. Because management of raised ICP is such a critical part of the care of severely brain-injured patients and because some drugs may exert their neuroprotective effect through lowering ICP, monitoring should be required for patients enrolled in pivotal efficacy trials in severe TBI.

The need for standardization of patient management across centers and for real-time feedback on center performance in clinical trials is being addressed by two groups of academic neurosurgeons interested in head injury management and clinical trials. The American Brain Injury Consortium (MARMAROU 1996) and the European Brain Injury Consortium (TEASDALE et al. 1997) were formed separately by groups of clinical scientists who were grappling with the challenges posed by conducting trials with private sector funding. Their major efforts focus on assuring proper scientific trial design, certifying the capabilities of investigative centers and providing surveillance during the conduct of a trial to ensure that centers care for their patients according to standard management guidelines. The consortia also exert control over which compounds will be selected to advance into efficacy trials in their member centers, based on what they believe to be the scientific merit of the mechanism and the clinical approach. They are also keenly (and appropriately) interested in publishing the data generated in clinical trials.

Although industry sponsors bristle at the level of control demanded by the consortia and point to the lack of success in trials with significant consortium input, these groups provide an important function for sponsors interested in advancing compounds in TBI. The management of severe head injury patients is enormously complex, even for clinicians whose sole professional activity is neurocritical care. There are few pharmaceutical sponsors that possess the internal expertise and resources to adequately monitor center performance and interpret data generated during the course of a complicated trial in severe brain injury. The consortia can help to provide this guidance.

One particularly useful function provided by the American Brain Injury Consortium is a critical review by an expert clinician of ICP management during the first several days of treatment. For each patient enrolled in a trial, they provide an assessment of adherence to standard treatment guidelines and give appropriate feedback to the center regarding the need for changes in patient management. Given the complexity of ICP management, it is difficult for a sponsor physician without neurocritical care experience to know when to question the medical practice of an investigator in a trial; in this respect the oversight provided by the consortium is very valuable. The consortium can also be instrumental in helping to sort out safety questions that arise during a study both with respect to individual serious adverse events and to outcomes in a particular treatment arm.

4. Outcome Measures

In 1991, a meeting of TBI experts was held in Houston to develop recommendations for outcome measures in traumatic brain injury clinical trials (CLIFTON et al. 1992). Functional outcome measures such as the Glasgow Outcome Scale (GOS) (JENNETT and BOND 1975; JENNETT et al. 1981; TEASDALE et al. 1998a) or the Disability Rating Scale (DRS) (RAPPAPORT 1982) were recommended as primary outcome measures for trials in severe or moderate brain injury. The Neurobehavioural Rating Scale (LEVIN et al. 1987) was recommended for measuring behavioral changes and a brief battery of neuropsychometric measures was recommended as a supplemental secondary outcome parameter. Most of the large pivotal efficacy trials in TBI performed since that meeting have adhered to these recommendations.

In most trials in severe TBI, the primary outcome measure has been the GOS administered at 6 months after injury. The GOS classifies patient outcome into one of five categories: good outcome, moderate disability, severe disability, vegetative state, and dead. In many trials the GOS was collapsed into a dichotomized scale that includes "favorable" (good recovery and moderate disability) and "unfavorable" (severe disability, vegetative, and dead)

Table 2. Functional status of two hypothetical patients 6 months after severe traumatic brain injury with appropriate GOS and DRS ratings

	Patient 1	Patient 2
Eye opening	Spontaneous	Opens eyes only when spoken to
Orientation	Knows name and place but not date	Unable to communicate
Motor function	Obeys commands	Obeys simple commands
Feeding	Initiates and eats independently	No awareness of feeding needs
Toileting	Has occasional bladder accidents accidents	No awareness of how or when to use toilet
Grooming	Initiates and completes grooming independently	No awareness of grooming needs
Independence	Lives at home, takes care of self partially but needs another person there	Lives in a nursing home, needs help with all major activities at all times
Employability	Unemployable	Unemployable
GOS Rating	Severely disabled	Severely disabled
DRS rating	8	21

outcomes. Some of the advantages of the GOS are its long history of use in TBI, its relative simplicity, and validity. However, because no study has shown a statistically significant benefit of treatment, concern has been raised about the sensitivity and reliability of outcome assignments on the GOS.

The GOS is not a particularly sensitive instrument in that in some of its categories it fails to differentiate patients with very widely different degrees of recovery. Table 2 shows an example of the functional characteristics of two different patients at 6 months after severe TBI, both of whom would be correctly classified as "severely disabled" by the GOS. While it is true that both of the patients in this example are severely disabled in that both are dependent on others for their care, there is no question that there is a marked difference in their functional levels and that is highly clinically significant. No one would argue that if a neuroprotective therapy were able, on average, to produce a difference in recovery of this magnitude that it would be meaningful to patients and to society. This difference would never be detected using the five-point GOS.

When the GOS is dichotomized, there is even poorer discrimination of outcomes. Consider a study in severe TBI in which the GOS at 6 months is the primary outcome measure. The GOS ratings are dichotomized into favorable (good recovery and moderate disability) and unfavorable (severe disability, vegetative survival, and dead) outcomes and the percentage of patients achieving a favorable outcome is the primary criterion for evaluation of efficacy. In this study, a demonstration of 10% improvement in the percentage of patients with a favorable outcome is considered a clinically important intervention. Dead and vegetative survival are combined since it is difficult to decide which of these two outcomes is worse. Assume that with placebo treatment, the distribution of GOS scores in the top graph of Fig. 1 is expected (that is, 60% favorable and 40% unfavorable outcomes on the dichotomized GOS).

Suppose that a neuroprotective treatment is so effective that half of the patients in a category shift their GOS outcome one grade upward. That is, half of the patients expected to die or survive in a vegetative state will survive as severely disabled patients, half of those expected to have severe disability will have moderate disability, and so on. This 50% shift is depicted in the middle graph. The bottom graph displays the result of the 50% shift. In this hypothetical case, despite a very powerful therapeutic effect, the rate of favorable outcomes under the dichotomized model has only increased by 6%. No drug that has been in clinical trials to date has had such a potent therapeutic effect yet many of these trials used the dichotomized GOS as the primary efficacy parameter. There was virtually no chance for these drugs to demonstrate efficacy under this design.

An expanded form of the GOS has been described that seeks to provide better discrimination of outcomes at the severe disability, moderate disability and good recovery levels. In the expanded GOS, each of these outcome categories is expanded into upper and lower divisions. Recently, WILSON described the use of a structured questionnaire-based interview aimed at substantially improving the reliability of assignment to the standard GOS and expanded GOS categories (WILSON et al. 1998). These refinements of the GOS should substantially improve the performance of the scale in TBI trials.

The Disability Rating Scale (DRS) has also been used in TBI clinical trials but generally is collected as a secondary outcome measure. The DRS was developed by physicians in brain injury rehabilitation centers who were frustrated by the inability of available scales to reflect the improvements they observed in their patients. The scale takes into account the wide variation in recovery that occurs after TBI and tracks patients from "coma to community" (RAPPAPORT 1982). The DRS utilizes a 30-point scale that includes ratings based on clinical examination, performance of activities of daily living and employment. Its advantages are that it has been widely validated and is reliable and can be completed fairly quickly. Although it is easy to learn, accurate rating takes practice and errors in assignment are possible if the scale is not used regularly. Using the case examples in Table 2 again, patient 1 would have a DRS score of 8 and patient 2 would have a DRS score of 21. Because of the larger number of categories for capturing specific improvements in function, on the surface it seems that the DRS would be a more sensitive outcome measure for TBI clinical trials.

To date however, the DRS has not performed as well as expected in large TBI efficacy trials. In a study of the operating characteristics of the GOS and DRS in two trials of neuroprotective agents for TBI, despite its larger range,

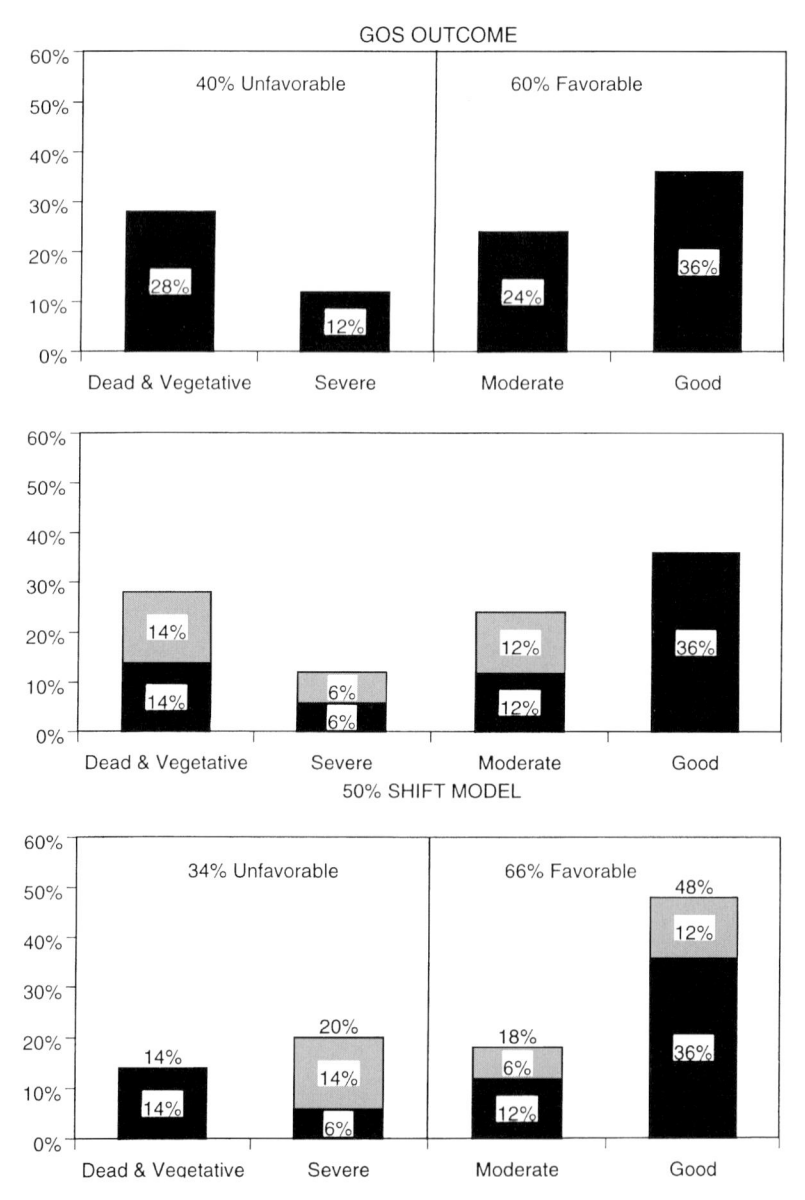

Fig. 1. Hypothetical Glasgow Outcome Scale (*GOS*) distributions for a clinical trial in traumatic brain injury. The percentages of "favorable and "unfavorable" outcomes reflect the standard method for dichotomizing GOS outcomes. See text for details

the DRS was shown to be no more sensitive than the GOS (CHOI et al. 1998). In this study there was a very surprising overlap of the DRS scores for patients that were rated as severely disabled or vegetative on the GOS. Although the authors state that in both studies almost all of the DRS and GOS ratings were performed by neuropsychologists or nurses who were well trained in the outcome measures, significant errors must have been made in one or both of the DRS or GOS ratings in many cases.

This study points to some of the difficulties inherent in using multi-item rating scales in large clinical trials. At the very least, clinical outcome raters must be carefully trained in performing each outcome measure at the start of the study. It is helpful to have a training session for outcome raters along with a certifying examination at the study investigator's meeting to help ensure this. Ongoing training during the course of the trial can help to ensure that rating skills have not deteriorated, particularly at sites where enrollment has been slow. For example, the American Brain Injury Consortium provides an Internet web page with hypothetical patient vignettes on which investigative sites can practice GOS ratings. The site provides a discussion for each case and gives instant feedback on performance.

Another approach may be to establish a central committee of trained outcome raters to perform the outcome measures on every patient in a clinical trial. This committee's function would be similar to that of an endpoint adjudicating body, such as is used in cardiovascular or stroke studies or a central CT scan reading site, a method used commonly in TBI trials to ensure the reliability of CT interpretation. Each member of the committee would be responsible for traveling to a proportion of the sites and would report back to the committee the results of ratings for the patients enrolled at those sites. The committee would review ratings and agree on the scores assigned to each patient. Although this system would be expensive and logistically challenging, it has several advantages. First, it would eliminate the need to train outcome raters at every center. This would especially helpful for more difficult rating scales such as the DRS or scales like the Functional Independence Measure (FIM) (HAMILTON et al. 1987) that require raters to undergo a formal certification process. It would also eliminate concerns about the quality of ratings performed at sites with low enrollment and therefore less experience with using rating instruments. For drugs with pharmacologic characteristics that could potentially unblind a study coordinator, a central rating committee would help to ensure blinded outcome ratings. Lastly, a central rating committee could help to reduce the inter-rater variability in outcome measures. Having the committee utilize structured interview instruments for each rating scale, if they are available, could improve this even further.

Neither the GOS nor the DRS directly capture the dimensions of neuropsychological dysfunction that results from traumatic brain injury. Specific neuropsychometric testing batteries are necessary to characterize the effects of TBI on higher cortical function. A recent study provides a striking observation of the degree to which cognitive dysfunction affects patients with the best outcomes in TBI. Functional and neuropsychological outcomes were measured in 139 severely brain-injured patients in a double blind, randomized, placebo controlled trial of the bradykinin antagonist, deltibant (MARMAROU 1999). A neuropsychometric testing battery consisting of the Rey Complex Figure, Rey Recall, Controlled Oral Word Association Test, Symbol Digit, and Grooved Pegboard was administered at 3 and 6 months following injury. Neuropsychometric testing scores were converted to percentiles and adjusted for age and level of education and an average composite neuropsychological score was calculated. At the 3-month assessment, the average composite neuropsychological score attained by patients in the placebo arm reached only 19.9% of normal. At 6 months, the composite score improved from 19.9% to 31.5% of normal. Most importantly, patients whose GOS rating was "good recovery," on average had composite neuropsychological scores only 40% of normal.

These data point to the need for better measures of therapeutic efficacy, perhaps combining the gross functional outcomes provided by scales like the GOS and DRS with formal neuropsychometric testing. This kind of combined analysis may be limited to patients on the better end of the recovery scale however, since collecting neuropsychometric data in severely brain injured patients is challenging. A recent report from the National Acute Brain Injury Study: Hypothermia showed that although follow-up rates at the 6-month assessment were excellent, up to one-third of patients were unable to participate in standard neuropsychometric testing because of cognitive or behavioral deficits (LEVIN et al. 1998).

Neuropsychological deficits are particularly important sequelae of moderate brain injury, and combined functional and neuropsychometric outcomes may be most suitable for this group of TBI patients. There is very little controlled clinical trial experience with neuroprotective agents in this patient group, in part because of difficulties with predicting prognosis, the wide variation in outcomes, and because of the dire unmet medical need in severe injury. Much work would need to be done to identify appropriate patient populations for study and to identify crucial prognostic variables in order to successfully study less severely injured patients. Nevertheless, many of the same cellular perturbations and biochemical mechanisms at work in severe TBI are present in mild and moderate brain injury (HAYES and DIXON 1994) and there are undoubtedly opportunities for demonstrating neuroprotective efficacy in this patient group. It may be possible to demonstrate significant treatment effects by concentrating more on neuropsychometric endpoints in less severely injured patients, thereby avoiding some of the difficulties associated with the effect of high rates of mortality and vegetative outcomes on functional outcome measures. The clinical significance of an effect on a neuropsychometric measure may be less compelling than a change in a functional outcome category but the effect could be correlated with global ratings completed by patients, caregivers, and clinicians.

C. Spinal Cord Injury Clinical Trials

The experience with randomized, controlled clinical trials for spinal cord injury has been somewhat more promising and illustrates some of the issues already discussed with respect to TBI. Several important lessons can be gleaned from these trials with respect to the significance of animal models, timing of drug administration, regimen, and outcome measures.

In the first National Acute Spinal Cord Injury Study (NASCIS I), which ran over 5 years from 1977 to 1982, 330 patients were randomized within 24h of injury to receive either high dose (1000mg bolus followed by daily infusion) or conventional dose (100mg bolus followed by daily infusion) methylprednisolone (BRACKEN et al. 1984). Treatment was administered for 10 days after injury. No statistically significant difference in motor or sensory recovery was observed between the two treatment arms in this trial. Although the study suggested that corticosteroids were not effective in acute spinal cord injury, the results were questioned because of the low dose employed compared to the effective dose in animal models, the long delay in treatment, and the lack of a placebo control group.

The second National Acute Spinal Cord Injury Study (NASCIS II) was designed to address some of these concerns (BRACKEN et al. 1990). In this trial, 487 patients with acute spinal cord injury were randomized within 12h of injury to receive high-dose methylprednisolone (30mg/kg followed by continuous infusion for 23h), naloxone, or placebo. Patients who received methylprednisolone within 8h of injury had significant improvement in motor and sensory function at the 6-month outcome visit. Patients who received methylprednisolone more than 8h after injury or naloxone had no better recovery than those in the placebo group.

In the third National Acute Spinal Cord Injury Study (NASCIS III) trial, the optimal duration of methylprednisolone treatment was examined (BRACKEN et al. 1997). In this trial, 499 patients who presented within 8h of injury received a bolus dose of methylprednisolone and were then randomized to receive a 24-h infusion of methylprednisolone, a 48-h infusion of methylprednisolone, or a 48-h infusion of tirilazad mesylate. Overall, patients who received methylprednisolone for 48h showed improved recovery at 6 weeks and 6 months but the differences were not statistically significant. Significant improvement was observed in patients in the 48-h methylprednisolone treatment arm if treatment was begun 3–8h after injury. The rate of recovery was identical for all three regimens among patients who began treatment within 3h.

One important conclusion from these studies is that the timing of drug administration and the total duration of treatment had important influences on outcome and were appropriately modeled by the data from animal experiments. The importance of the short time window for treatment, and the very high 30mg/kg initial dose of methylprednisolone were accurately predicted by experimental spinal cord injury in the cat (HALL 1992).

The NASCIS trials utilized outcome measures that are relatively more sensitive to treatment effect than the scales used in TBI. In all three trials, motor recovery was assessed using a scale that summed the scores of motor function in each of 14 muscles such that the possible scores ranged from 0 (no contraction in any muscle) to 70 (all normal responses). Similarly, sensory function was measured by assessing pinprick and light touch sensation in each dermatome from C-2 through S-5, and summing the scores obtained so that the scale ranged from 6 (all responses abnormal) to 18 (all responses normal). The magnitude of the treatment effect observed on these scales in NASCIS II and NASCIS III was small, but in some subgroups there was also a significant effect on the FIM, which more globally assesses functionality.

Overall, the NASCIS experience provides a good example in the neuroprotective arena of the importance of designing trials based on information from animal models, sensitive and relevant outcome measures, and logical sequencing of human experiments based on data.

D. Conclusions

Overall, the experience to date with clinical trials for neuroprotective agents in traumatic central nervous system injury has been disappointing, but many important lessons have been learned that can be applied to future efforts. Keys to success may lie in the careful application of animal models to the design of clinical trials, thorough clinical testing as early as possible during development, avoidance of overrapid drug development programs, and the use of more sensitive outcome measures.

References

Alves WM, Eisenberg HM (1996) Head Injury Trials-Past and Present. In: Narayan RK, Wilberger JE Jr, Povlishock JT (eds) Neurotrauma. McGraw-Hill, New York, pp 946–967

Bardt TF, Unterberg AW, Kiening KL, Schneider GH, Lanksch WR (1998) Multimodal cerebral monitoring in comatose head-injured patients. Acta Neurochir (Wien) 140:357–365

Bracken MB, Collins WF, Freeman DF, Shepard MJ, Wagner FW, Silten RM, Hellenbrand KG, Ransohoff J, Hunt WE, Perot PL Jr et al. (1984) Efficacy of methylprednisolone in acute spinal cord injury. JAMA 251:45–52

Bracken MB, Shepard MJ, Collins WF, Holford TR, Young W, Baskin DS, Eisenberg HM, Flamm E, Leo Summers L, Maroon J et al. (1990) A randomized, controlled trial of methylprednisolone or naloxone in the treatment of acute spinal-cord injury. Results of the Second National Acute Spinal Cord Injury Study. N Engl J Med 322:1405–1411

Bracken MB, Shepard MJ, Holford TR, Leo Summers L, Aldrich EF, Fazl M, Fehlings M, Herr DL, Hitchon PW, Marshall LF, Nockels RP, Pascale V, Perot PL Jr, Piepmeier J, Sonntag VK, Wagner F, Wilberger JE, Winn HR, Young W (1997) Administration of methylprednisolone for 24 or 48 h or tirilazad mesylate for 48 h in the treatment of acute spinal cord injury. Results of the Third National Acute Spinal Cord Injury Randomized Controlled Trial. National Acute Spinal Cord Injury Study. JAMA 277:1597–1604

Brott TG, Haley EC Jr, Levy DE, Barsan W, Broderick J, Sheppard GL, Spilker J, Kongable GL, Massey S, Reed R et al. (1992) Urgent therapy for stroke. Part I. Pilot study of tissue plasminogen activator administered within 90 min. Stroke 23:632–640

Bullock R, Chesnut RM, Clifton G, Ghajar J, Marion DW, Narayan RK, Newell DW, Pitts LH, Rosner MJ, Wilberger JW (1996) Guidelines for the management of severe head injury. Brain Trauma Foundation. Eur J Emerg Med 3:109–127

Bullock R, Zauner A, Woodward JJ, Myseros J, Choi SC, Ward JD, Marmarou A, Young HF (1998) Factors affecting excitatory amino acid release following severe human head injury. J Neurosurg 89:507–518

Chesnut RM, Marshall SB, Piek J, Blunt BA, Klauber MR, Marshall LF (1993) Early and late systemic hypotension as a frequent and fundamental source of cerebral ischemia following severe brain injury in the Traumatic Coma Data Bank. Acta Neurochir Suppl (Wien) 59:121–125

Chesnut RM (1997) Avoidance of hypotension: conditio sine qua non of successful severe head-injury management. J Trauma 42:S4–S9

Choi SC, Ward JD, Becker DP (1983) Chart for outcome prediction in severe head injury. J Neurosurg 59:294–297

Choi SC, Narayan RK, Anderson RL, Ward JD (1988) Enhanced specificity of prognosis in severe head injury. J Neurosurg 69:381–385

Choi SC, Germanson TP, Barnes TY (1995) A simple measure in defining optimal strata in clinical trials. Control Clin Trials 16:164–171

Choi SC, Barnes TY (1996) Predicting Outcome in the Head Injured Patient. In: Narayan RK, Wilberger JE Jr, Povlishock JT (eds) Neurotrauma. McGraw-Hill, New York, pp 779–792

Choi SC, Marmarou A, Bullock R, Nichols J, Wei X, Pitts L, The American Brain Injury Consortium Study Group (1998) Primary endpoints in phase III clinical trials of severe head trauma: DRS versus GOS. J Neurotrauma 15:771–776

Clifton GL, Hayes RL, Levin HS, Michel ME, Choi SC (1992) Outcome measures for clinical trials involving traumatically brain-injured patients: report of a conference. Neurosurgery 31:975–978

Doppenberg EM, Bullock R (1997) Clinical neuro-protection trials in severe traumatic brain injury: lessons from previous studies. J Neurotrauma 14:71–80

Fleishaker JC, Pearson LK, Peters GR (1998) Induction of tirilazad clearance by phenytoin. Biopharm Drug Dispos 19:91–96

Gaetani P, Tancioni F, Tartara F, Carnevale L, Brambilla G, Mille T, Rodriguez y Baena R (1995) Prognostic value of the amount of posttraumatic subarachnoid haemorrhage in a six month follow up period. J Neurol Neurosurg Psychiatry 59:635–637

Gennarelli TA, Spielman GM, Langfitt TW, Gildenberg PL, Harrington T, Jane JA, Marshall LF, Miller JD, Pitts LH (1982) Influence of the type of intracranial lesion on outcome from severe head injury. J Neurosurg 56:26–32

Ghajar J, Hariri RJ, Narayan RK, Iacono LA, Firlik K, Patterson RH (1995) Survey of critical care management of comatose, head-injured patients in the United States. Crit Care Med 23:560–567

Haley EC Jr, Kassell NF, Apperson Hansen C, Maile MH, Alves WM (1997) A randomized, double-blind, vehicle-controlled trial of tirilazad mesylate in patients with aneurysmal subarachnoid hemorrhage: a cooperative study in North America. J Neurosurg 86:467–474

Hall ED (1992) The neuroprotective pharmacology of methylprednisolone. J Neurosurg 76:13–22

Hamilton BB, Granger CV, Sherwin FS (1987) A uniform national data system for medical rehabilitation. In: Fuhrer MJ (ed) Rehabilitation Outcomes: Analysis and Measurement. Brooks, Baltimore, pp 137–147

Hayes RL, Dixon CE (1994) Neurochemical changes in mild head injury. Semin Neurol 14:25–31

Jennett B, Bond M (1975) Assessment of outcome after severe brain damage. Lancet 1:480–484

Jennett B, Snoek J, Bond MR, Brooks N (1981) Disability after severe head injury: observations on the use of the Glasgow Outcome Scale. J Neurol Neurosurg Psychiatry 44:285–293

Kassell NF, Haley EC Jr, Apperson Hansen C, Alves WM (1996) Randomized, double-blind, vehicle-controlled trial of tirilazad mesylate in patients with aneurysmal subarachnoid hemorrhage: a cooperative study in Europe, Australia, and New Zealand. J Neurosurg 84:221–228

Kraus JF, McArthur DL, Silverman TA, Jayaraman M (1996) Epidemiology of Brain Injury. In: Narayan RK, Wilberger JE Jr, Povlishock JT (eds) Neurotrauma, vol 1. McGraw-Hill, New York, pp 13–30

Levin HS, High WM, Goethe KE, Sisson RA, Overall JE, Rhoades HM, Eisenberg HM, Kalisky Z, Gary HE (1987) The neurobehavioural rating scale: assessment of the behavioural sequelae of head injury by the clinician. J Neurol Neurosurg Psychiatry 50:183–193

Maas AI, Dearden M, Teasdale GM, Braakman R, Cohadon F, Iannotti F, Karimi A, Lapierre F, Murray G, Ohman J, Persson L, Servadei F, Stocchetti N, Unterberg A (1997) EBIC-guidelines for management of severe head injury in adults. European Brain Injury Consortium. Acta Neurochir (Wien) 139:286–294

Marmarou A (1996) Conduct of head injury trials in the United States: the American Brain Injury Consortium (ABIC). Acta Neurochir Suppl (Wien) 66:118–121

Marmarou A, Nichols J, Burgess J, Newell D, Troha J, Burnham D, Pitts L, The American Brain Injury Consortium Study Group (1999) Effects of the bradykinin antagonist Bradycor (Deltibant, CP-1027) in severe brain injury: results of a multi-center, randomized, placebo-controlled trial. In Press

Marshall LF, Becker DP, Bowers SA, Cayard C, Eisenberg H, Gross CR, Grossman RG, Jane JA, Kunitz SC, Rimel R, Tabaddor K, Warren J (1983) The National Traumatic Coma Data Bank. Part 1: Design, purpose, goals, and results. J Neurosurg 59:276–284

Marshall LF, Maas AI, Marshall SB, Bricolo A, Fearnside M, Iannotti F, Klauber MR, Lagarrigue J, Lobato R, Persson L, Pickard JD, Piek J, Servadei F, Wellis GN, Morris GF, Means ED, Musch B (1998) A multicenter trial on the efficacy of using tirilazad mesylate in cases of head injury. J Neurosurg 89:519–525

McIntosh TK, Juhler M, Wieloch T (1998) Novel pharmacologic strategies in the treatment of experimental traumatic brain injury: 1998. J Neurotrauma 15:731–769

NINDS rt-PA Stroke Study Group (1995) Tissue plasminogen activator for acute ischemic stroke. N Engl J Med 333:1581–1587

Rappaport M, Hall KM, Hopkins K, Belleza T, Cope DN (1982) Disability rating scale for severe head trauma: coma to community. Arch Phys Med Rehabil 63:118–123

Sandage BW (1998) Balancing phase II and III efficacy trials. Drug Inf J 32:977–980

Scheibel RS, Levin HS, Clifton GL (1998) Completion rates and feasibility of outcome measures: experience in a multicenter trial of systemic hypothermia for severe head injury. J Neurotrauma 15:685–692

Teasdale GM, Braakman R, Cohadon F, Dearden M, Iannotti F, Karimi A, Lapierre F, Maas A, Murray G, Ohman J, Persson L, Servadei F, Stocchetti N, Trojanowski T, Unterberg A (1997) The European Brain Injury Consortium. Nemo solus satis sapit: nobody knows enough alone. Acta Neurochir (Wien) 139:797–803

Teasdale GM, Pettigrew L, Wilson JT, Murray G, Jennett B (1998) Analyzing outcome of treatment of severe head injury: a review and update on advancing the use of the Glasgow Outcome Scale. J Neurotrauma 15:587–597

Teasdale GM, Graham DI (1998) Craniocerebral trauma: protection and retrieval of the neuronal population after injury. Neurosurgery 43:723–737

Toulmond S, Serrano A, Benavides J, Scatton B (1993) Prevention by eliprodil (SL 82.0715) of traumatic brain damage in the rat. Existence of a large (18h) therapeutic window. Brain Res 620:32–41

Wichman A, Sandler AL (1997) Research involving critically ill subjects in emergency circumstances: new regulations, new challenges. Neurology 48:1151–1155

Wilson JT, Pettigrew L, Teasdale GM (1998) Structured interviews for the Glasgow Outcome Scale and the extended Glasgow Outcome Scale: guidelines for their use. J Neurotrauma 15:573–585

Young B, Runge JW, Waxman KS, Harrington T, Wilberger J, Muizelaar JP, Boddy A, Kupiec JW (1996) Effects of pegorgotein on neurologic outcome of patients with severe head injury. A multicenter, randomized controlled trial. JAMA 276:538–543

Zivin JA (1998) Factors determining the therapeutic window for stroke. Neurology 50:599–603

CHAPTER 15

Neurohormonal Signalling Pathways and the Regulation of Alzheimer β-Amyloid Metabolism
Integrating Basic and Clinical Observations for Developing Successful Strategies to Delay, Retard or Prevent Cerebral Amyloidosis in Alzheimer's Disease

S. GANDY

Alzheimer's disease (AD) is characterized by the intracranial accumulation of the 4-kDa amyloid-β peptide (Aβ), following proteolysis of a ~700-amino acid, integral membrane precursor, the Alzheimer amyloid precursor protein (APP). The best evidence causally linking APP to Alzheimer's disease has been provided by the discovery of mutations within the APP coding sequence that segregate with disease phenotypes in autosomal dominant forms of familial Alzheimer's disease (FAD). Though FAD is rare (<10% of all AD), the hallmark features – amyloid plaques, neurofibrillary tangles, synaptic and neuronal loss, neurotransmitter deficits, dementia – are indistinguishable when FAD is compared with typical, common, "non-familial," or sporadic, AD (SAD). Studies of some clinically relevant mutant APP molecules from FAD families have yielded evidence that APP mutations can lead to enhanced generation or aggregability of Aβ, consistent with a pathogenic role in AD. Other genetic loci for FAD have been discovered which are distinct from the immediate regulatory and coding regions of the APP gene, indicating that defects in molecules other than APP can also specify cerebral amyloidogenesis and FAD. To date, all APP and non-APP FAD mutations can be demonstrated to have the common feature of promoting amyloidogenesis of Aβ.

Epidemiological studies indicate that postmenopausal women on estrogen replacement therapy (ERT) have their relative risk of developing SAD diminished by about 1/3–1/2 as compared with age-matched women not receiving ERT (TANG et al. 1996). Because of the key role of cerebral Aβ accumulation in initiating AD pathology, it is most attractive that estradiol might modulate SAD risk or age-at-onset by inhibiting Aβ accumulation. A possible mechanistic basis for such a scenario is reviewed here.

A. Alzheimer's Disease Is Associated with an Intracranial Amyloidosis

Amyloid is a generic description applied to a heterogeneous class of tissue protein precipitates that have the common feature of β-pleated sheet secondary structure, a characteristic that confers affinity for the histochemical dye Congo red. Amyloids may be deposited in a general manner throughout the body (systemic amyloids) or confined to a particular organ (e.g., cerebral amyloid, renal amyloid). AD is characterized by clinical evidence of cognitive failure in association with cerebral amyloidosis, cerebral intraneuronal neurofibrillary pathology, neuronal and synaptic loss, and neurotransmitter deficits. The cerebral amyloid of AD is deposited around meningeal and cerebral vessels, as well as in gray matter. In gray matter, the deposits are multifocal, coalescing into miliary structures known as plaques. Parenchymal amyloid plaques are distributed in brain in a characteristic fashion, differentially affecting the various cerebral and cerebellar lobes and cortical laminae.

The main constituent of cerebrovascular amyloid was purified and sequenced by GLENNER and WONG in 1984. This 40–42 amino acid polypeptide, designated "β protein" (or, according to MASTERS et al., "A4;" now standardized as "Aβ" by THE HUSBY COMMISSION), is derived from a 695–770 amino acid precursor, termed the amyloid precursor protein (APP; Fig. 1), which was discovered by molecular cloning using oligonucleotide probes from the published amyloid peptide sequence. The primary citations for this molecular neuropathology review can be found elsewhere (GANDY and GREENGARD 1994).

B. Aβ Is a Catabolite of an Integral Precursor

The deduced amino acid sequence of APP predicts a protein with a single transmembrane domain (Fig. 1, upper panel). Isoform diversity is generated by alternative mRNA splicing, and isoforms of 751 and 770 amino acids include a protease inhibitor domain ("Kunitz-type protease inhibitor" domain, or KPI) in the extracellular region of the APP molecule. The ectodomains of the protease inhibitor-bearing isoforms of APP are identical to molecules previously identified according to their tight association with proteases, and thus were designated "protease nexin II" (PN-II). Identical molecules are also present in the platelet α-granules, where they were described under the name of "factor XI$_a$ inhibitor" (XI$_a$I). Upon degranulation of the platelet, factor XI$_a$I (PN-II/APP) exerts an antiproteolytic effect on activated factor XI$_a$ at late steps of the coagulation cascade. Some evidence suggests that KPI-lacking isoforms may also act as regulators of proteolysis. While many bioactivities for APP have been catalogued in the literature, Aβ might simply be a by-product of APP metabolism that is not necessarily intimately involved with APP function.

The proteolytic processing steps for APP have by now been definitively identified by purification and sequencing. The first to be identified involves cleavage within the Aβ domain. A large amino-terminal fragment of the APP extracellular domain (PN-II; or s-APPα or APPs$_\alpha$, for soluble APP$_\alpha$) is released into the medium of cultured cells and into the cerebrospinal fluid,

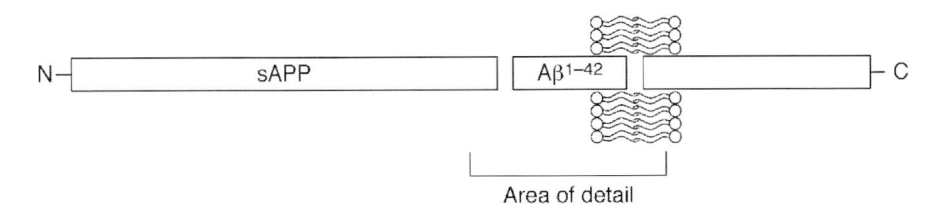

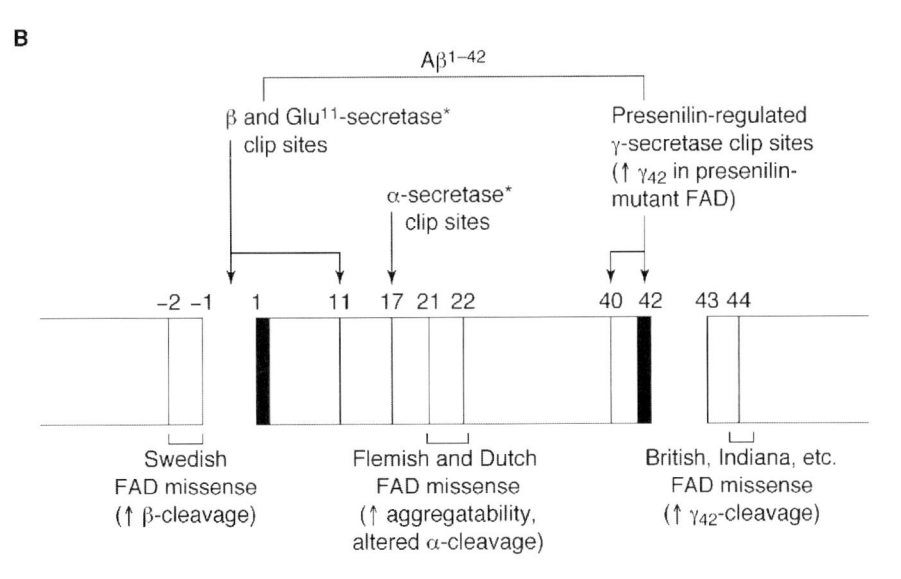

The Regulation of Alzheimer β-Amyloid Metabolism 411

Fig. 1A,B. Structure of the Alzheimer amyloid precursor protein (*APP*; **A**), including the fine structure around the amyloid-β peptide domain (**B**). The long intralumenal (or, under certain circumstances, extracellular) domain is directed leftward, whereas the short cytoplasmic domain is directed rightward. *FAD*, familial Alzheimer's disease. (From GANDY 1999)

leaving a small nonamyloidogenic carboxyl-terminal fragment associated with the cell. This pathway is designated the α-secretory cleavage/release processing pathway for APP, and the enzyme(s) which perform(s) this nonamyloidogenic cleavage/release has been designated α-secretase. Thus, one important processing event in the biology of APP acts to preclude amyloidogenesis by proteolyzing APP within the Aβ domain, between residues $A\beta^{16}$ and $A\beta^{17}$ (Fig. 1, lower panel).

Until mid-1992, the prevailing notion was that Aβ production might be restricted to the brain and perhaps even then only in association with aging and AD. This concept became obsolete with the discovery by several groups that a soluble Aβ species (presumably a forerunner of the aggregated fibrillar species which is deposited in senile plaque cores) is detectable in body fluids from various species and in the conditioned medium of cultured cells. Still, "soluble Aβ" (to distinguish it from deposited Aβ amyloid) is not readily detectable in the lysates of most cultured cells. Soluble Aβ is apparently generated in several cellular compartments, including the endoplasmic reticulum, the Golgi apparatus, the *trans*-Golgi network, and endosomes. Aβ production might begin by cleavage at the Aβ amino-terminus by another enzyme designated β-secretase (Fig. 1, lower panel), probably beginning in the constitutive secretory pathway, i.e., the *trans*-Golgi network (TGN). This model accounts especially well for providing a codistribution opportunity involving APP and β-secretase prior to APP's encounter with the Aβ-domain-destroying α-secretase which appears to act primarily at the plasma membrane. Data from one FAD mutant (the "Swedish" FAD mutant; Fig. 1, lower panel; discussed in detail below) support this model strongly. Other data indicate that at least some Aβ arises from APP molecules following their residence at the cell surface; presumably these APP molecules have escaped cell-surface α-secretase and have been internalized intact, encountering β-secretase in an endosome. An Aβ variant bearing Glu^{11} at its N-terminus has been recently noted as a prominent product of APP metabolism in cultured neurons (Gouras et al. 1998). The generation and pharmacologic regulation of [Glu^{11}] Aβ formation appears to parallel that observed for standard [Asp^1] Aβ. The cellular and biochemical origin of the alternative cleavage at Glu^{11} and its role in amyloidogenesis, if any, are all yet to be elucidated.

It is important to appreciate that one area of particular current interest is the elucidation of the molecular and cellular mechanisms by which the carboxyl-terminus of Aβ is generated, since this region of the APP molecule resides within an intramembranous domain (Fig. 1). The protease(s) responsible for that cleavage, designated the γ-secretase(s) (Fig. 1, lower panel), are particularly interesting not only because of the novelty of the active site-substrate reaction (i.e., within a membranous domain) but also owing to evidence that another class of FAD [known as presenilin-related FAD; (Sherrington et al. 1995)] appears to act via control of γ-secretase function (Scheuner et al. 1996; Borchelt et al. 1996). Along this line, the Aβ C-terminus is heterogeneous, being composed mostly of peptides terminating at residue Aβ40 (Fig. 1, lower panel); however, a small but important minority of peptides terminate at residue Aβ42 (Fig. 1), and it is these highly aggregatable peptides that are believed to initiate Aβ accumulation in all forms of the disease (Iwatsubo et al. 1994; Lemere et al. 1996).

The primary citations for the foregoing review of APP processing can be found elsewhere (Gandy and Greengard 1994).

C. Pathogenic APP Mutations Occur Within or Near the Aβ Domain, and Yield APP Molecules That Display Proamyloidogenic Properties In Vitro and Alzheimer-Like Phenotypes in Transgenic Mouse Models In Vivo

Certain mutations associated with familial cerebral amyloidoses have been identified within or near the Aβ region of the coding sequence of the APP gene. These mutations segregate with the clinical phenotypes of either hereditary cerebral hemorrhage with amyloidosis, Dutch type (HCHWAD, or FAD-Dutch; Fig. 1, lower panel) or more typical familial Alzheimer's disease ("British," "Indiana" FAD; Fig. 1, lower panel), and provide support for the notion that aberrant APP metabolism is a key feature of AD.

In FAD-Dutch, an uncharged glutamine residue is substituted for a charged glutamate residue at position 22 of Aβ (Fig. 1, lower panel). This mutated residue is located in the extracellular region of APP, within the Aβ domain, where it apparently exerts its proamyloidogenic effect by generating Aβ molecules which bear enhanced aggregation properties and perhaps also alter α-secretase cleavage. A "Flemish" form of FAD involves a mutation at position 21 of Aβ (Fig. 1, lower panel) and generates a phenotype which includes features of both the relatively pure vascular amyloidosis of "Dutch FAD" and the more typical FAD involving parenchymal and vascular amyloidosis.

Mutations in APP that are apparently pathogenic for more typical FAD have also been discovered. In the first discovered FAD mutation, an isoleucine residue is subsituted for a valine residue within the transmembrane domain, at a position two residues downstream from the carboxyl-terminus of the Aβ domain (Fig. 1, lower panel), a position equivalent to "Aβ44" (if such a position existed). Although a conservative substitution, the mutation segregates with FAD in pedigrees of American, European ("British"), and Asian origins, arguing against the possibility that the mutations represent irrelevant polymorphisms. Other pedigrees have been discovered in which affected members have either phenylalanyl ("Indiana") or glycyl residues at position "Aβ44." Neuropathological examination has verified the similarity of these individuals to typical SAD. These "Aβ44" mutant APPs are the most common of the FAD-causing APP mutations, and the mechanism by which the mutations exert their effects appears to be by enhancing generation of hyperaggregable, C-terminally extended Aβ peptides (especially Aβ42).

Another FAD pedigree has been established involving a large Swedish kindred. In this instance, tandem missense mutations occur just upstream of the amino terminus of the Aβ domain. Transfection of cultured cells with APP molecules containing the "Swedish" missense mutations results in the production of six- to eightfold excess soluble Aβ above that generated from wild-type APP. This is the first (and, to date, only) example of Alzheimer's disease apparently caused by excessive Aβ production. An important issue for clarification in sporadic Alzheimer's disease will be to establish whether hyperaggregation or hyperproduction of Aβ (or neither) is/are important predisposing factor(s) to this much more commonly encountered clinical entity. The primary citations for the foregoing review of the genetic-neuropathological correlations of APP mutant FAD can be found elsewhere (GANDY and GREENGARD 1994).

Some of the most exciting and newest evidence in support of the pathogenicity of these mutants has arisen from the recent demonstration that certain mutant APP transgenic mice exhibit not only cerebral amyloidosis but also neurofibrillary tangle epitopes and neuronal loss (CALHOUN et al. 1998). If these animals also meet key behavioral criteria, Koch's postulates will have been met for the re-creation of all key features of Alzheimer's disease in a species not otherwise susceptible to the illness.

D. Signal Transduction Regulates the Relative Utilization of APP Processing Pathways in Cultured Cell Lines, in Primary Neurons, and in Rodents In Vivo

Substantial progress has been made toward understanding the regulation of APP cleavage. For example, the relative utilization of the various alternative APP processing pathways appears to be at least partially cell-type determined, with transfected AtT20 cells secreting virtually all APP molecules, whereas glia release little or none. In neuronal-like cells, the state of differentiation plays a role in determining the relative utilization of the pathways, with the differentiated neuronal phenotype being associated with relatively diminished basal utilization of the nonamyloidogenic α-secretase cleavage/release pathway.

It is also well-accepted that certain signal transduction systems involving protein phosphorylation are important regulators of APP cleavage, acting in many cases by stimulating the relative activity of nonamyloidogenic cleavage by α-secretase. The role of protein kinase C (PKC) and PKC-linked first messengers has received the most attention. In many types of cultured cells, activation of PKC by phorbol esters dramatically stimulates APP proteolysis (BUXBAUM et al. 1990) and sAPP$_\alpha$ cleavage/release (CAPORASO et al. 1992; GILLESPIE et al. 1992; SINHA and LIEBERBURG 1992). PKC-stimulated α-secretory cleavage of APP may also be induced by the application of neurotransmitters and other first messenger compounds whose receptors are linked to PKC (BUXBAUM et al. 1992; NITSCH et al. 1992). Okadaic acid, an inhibitor of protein phosphatases 1 and 2A, also increases APP proteolysis and release via the α-secretase pathway (BUXBAUM et al. 1992). Indeed, it has recently been shown in primary neuronal cultures that protein phosphatase inhibition by okadaic acid is much more effective in lowering Aβ production than is PKC activation (GOURAS et al. 1998). Thus, either stimulation of PKC or inhibition of protein phosphatases 1 and 2A is sufficient to produce a dramatic acceleration of nonamyloidogenic APP degradation. PKC-activated APP processing can be demonstrated to diminish generation of Aβ in virtually every situation tested to date, including primary neuronal cultures and rodent brain in vivo (BUXBAUM et al. 1992; HUNG et al. 1993; SAVAGE et al. 1998). This phenomenon is sometimes referred to as "regulated cleavage of APP" or "reciprocal regulation of APP metabolism;" i.e., when certain signals are activated, α-secretase cleavage increases and β and Glu^{11}-secretase cleavages decrease (GOURAS et al. 1998) (Fig. 1), although these two phenomena can sometimes be dissociated.

Steroid hormones such as 17β-estradiol (JAFFE et al. 1994; XU et al. 1998; CHANG et al. 1998) and dihydroepiandrostenedione (DHEA) (DANENBOERG et al. 1996) are other signal transduction compounds which can apparently regulate APP metabolism in cultured cells. Several of these investigators (XU et al. 1998; CHANG et al. 1997) have demonstrated that estradiol diminishes Aβ generation, while others (JAFFE et al. 1994; DANENBOERG et al. 1996; XU et al. 1998) have documented accumulation of sAPP in the conditioned media of cultured cells treated with estradiol (JAFFE et al. 1994; XU et al. 1998) or DHEA (DANENBOERG et al. 1996). One attractive explanation for the accumulation of sAPP is that steroids might stimulate either the amount or the activity of α-secretase. This explanation is consistent with the finding that the cell-associated product of this activity (the 9-kDa nonamyloidogenic carboxyl-terminal fragment) is increased by estrogen treatment (JAFFE et al. 1994). For estradiol in particular, one means by which the activity of α-secretase might increase is by an increase in the level or activity of PKC, leading to increased "regulated APP cleavage" along the α-secretase pathway. Estrogen has been shown to regulate protein kinase C in both normal and neoplastic tissue (MAEDA and LLOYD 1993; MAIZELS et al. 1992; KAMEL and KUBAJAK 1988) and to potentiate the phorbol ester-stimulated release of luteinizing hormone from cultured pituitary cells (DROUVA et al. 1990). The effects of estrogen on protein kinase C might be mediated by polypeptide growth factors (DICKSON et al. 1986; BATES et al. 1988; LIPPMAN et al. 1988), many of which activate protein kinase C and a number of which have been demonstrated to enhance the accumulation of sAPP in the conditioned media of cultured cells (REFOLO et al. 1989; FUKUYAMA et al. 1993). The colocalization of estrogen receptors in the basal forebrain, cerebral cortex, and hippocampus with the receptors for nerve growth factor, as well as with nerve growth factor itself, is consistent with the idea that growth factors might play a role in mediating the effects of estrogen in the central nervous system (MIRANDA et al. 1993).

Extending this line of reasoning, a more complex relationship involving intercellular signals and APP metabolism is currently evolving: both steroid hormones and neurotrophins have been demonstrated to potentiate the ability of incoming neurotransmitters to activate the "regulated cleavage of APP," as measured by sAPP release (DANENBERG et al. 1995; FISHER et al. 1996; HARING et al. 1998; ROBNER et al. 1998). The "regulated cleavage" model makes it likely that when compounds from either or both of these classes of substances (i.e., steroid hormones, neurotrophins) are present at the time of neurotransmitter application, the fold-diminution of Aβ and [Glu^{11}]Aβ will be greatly enhanced. Assessment of this model is currently underway.

E. Insights into Mechanism(s) of Regulated APP Processing

In general terms, the possible mechanisms for activated processing of integral molecules can be conceptualized as involving either activation or redistribution of either the substrate (i.e., APP) or the enzyme (i.e., α-secretase). Based on APP cytoplasmic tail mutational analyses (DA CRUZ E SILVA et al. 1993; SAHASRABUDHE et al. 1992), the "substrate activation" model (GANDY et al. 1988) is inadequate to explain activated processing of APP, since abrogation of APP cytoplasmic tail phosphorylation fails to abolish PKC-regulated α-cleavage. In order to test the model of PKC-regulated APP redistribution, XU et al. demonstrated that the PKC-stimulated biogenesis of nascent transport vesicles from the *trans*-Golgi network includes vesicles which bear [^{35}SO$_4$]APP (XU et al. 1995). These vesicles would be predicted to fuse with the plasma membrane where codistribution of APP and α-secretase would take place, leading to generation of sAPP$_\alpha$ at the plasma membrane and release of that sAPP$_\alpha$ into the culture medium. Given the preponderant localization of APP within the TGN (CAPORASO et al. 1994), it is likely that PKC-regulated TGN vesicle biogenesis plays a role in regulating APP metabolism, although its importance cannot be completely assessed until the identification of all relevant PKC targets of this apparatus has been completed.

Along a related line of investigation, BOSENBERG et al. (1993) succeeded in reconstituting faithful activated processing of the TGF-α precursor in porated cells in the virtual absence of cytosol, and in the presence of N-ethylmaleimide or 2.5M NaCl. The preservation of activated processing under such conditions suggests that extensive vesicular trafficking is unlikely to be required for all forms of activated processing, and supports a model of direct or indirect enzyme activation which involves an intrinsic plasma membrane mechanism (e.g., direct phosphorylation of an α-secretase or its regulatory subunit). Cell biological and molecular cloning data indicate that APP α-secretases include, or are related to, the metalloproteinase family of adamalysins, and TGFα and protumor necrosis factor convertases. These enzymes appear to have overlapping specificities for these substrates. Several candidate secretases have recently been successfully cloned, and these data should facilitate resolution of the question of which enzymes are most important for cleaving which substrates (BLACK et al. 1997; PESCHON et al. 1998; BUXBAUM et al. 1998). Aspartyl proteinases are the most important α-secretases in yeast, and the potential roles of their mammalian homologues, if any, in mammalian APP metabolism, remain to be elucidated (KOMANO et al. 1998).

F. Therapeutic Manipulation of Aβ Generation Via Ligand or Hormonal Manipulation

From a therapeutic standpoint, then, the modulation of APP metabolism via signal transduction pharmacology might be beneficial in individuals with, or at risk for, AD (SINHA and LIEBERBURG 1992; GANDY and GREENGARD 1992, 1994), and it is along this line that 17β-estradiol – via its possible ability to elevate PKC activity – came to be identified as a potential modulator of Aβ metabolism. This phenomenon has been confirmed in living animals. Such investigations should begin to clarify whether estradiol can indeed regulate Aβ generation and/or accumulation in vivo.

Estradiol might also modify other factors contributing to Aβ deposition and fibril formation, including the processing of soluble Aβ into an aggregated form (BURDICK et al. 1992; PIKE et al. 1993) and/or the association of Aβ with other molecules, such as α-1-antichymotrypsin (ABRAHAM et al. 1988), heparan sulfate proteoglycan (SNOW et al. 1994), and apolipoprotein E (WISNIEWSKI and FRANGIONE 1992). In particular, it has recently been recognized that apolipoprotein E is an obligatory participant in Aβ accumulation (BALES et al. 1997), and that apolipoprotein E isoforms play important roles as differential determinants of risk for AD (SAUNDERS et al. 1993; STRITTMATTER et al. 1993). Human postmortem (REBECK et al. 1993; POLVIKOSKI et al. 1995) and cell culture (YANG et al. 1999) data are also beginning to indicate that apolipoprotein E isoforms exert at least some of their effects via controlling Aβ accumulation, perhaps at the stage of clearance of Aβ peptides. Apolipoprotein E-isoform-specific synaptic remodeling also appears to be a consistent and potentially relevant observation (ARENDT et al. 1997; SUN et al. 1998).

Events beyond Aβ deposition may also be crucial in determining the eventual toxicity of Aβ plaques. While aggregation of Aβ is important for in vitro models of neurotoxicity (PIKE et al. 1993), the relevance of these phenomena for the pathogenesis of AD is unclear, since Aβ deposits may occur in normal aging in the absence of any evident proximate neuronal injury (CRYSTAL et al. 1988; MASLIAH et al. 1990; BERG et al. 1993; DELAERE et al. 1993). This suggests that other events must distinguish "diffuse" cerebral amyloidosis with wispy deposits of Aβ alone (MASLIAH et al. 1990) from "full-blown" AD with neuritic plaques, and neuron and synapse loss. One intriguing possible contributing factor is the association of complement components with Aβ (ROGERS et al. 1992). In cerebellum, where Aβ deposits appear to cause no injury, plaques are apparently free of associated complement, while in the forebrain, complement associates with plaques, perhaps becoming activated and injuring the surrounding cells (LUE and ROGERS 1992). Other, as-yet undiscovered plaque-associated molecules may also play important roles.

It should also be possible to apply animal models of Aβ deposition and neurodegeneration (CALHOUN et al. 1998) to begin to elucidate whether the effect of estrogen to delay or prevent AD in postmenopausal women is due to one or a combination of its various documented activities relevant to the pathogenesis of AD, i.e., lowering Aβ load, as reviewed above, sustaining the basal forebrain cholinergic system (LUINE 1985), modulating interactions with growth factors and/or their receptors (MIRANDA et al. 1993; SOHRABJI et al. 1994), supporting neuritic plasticity (WOOLLEY 1998; STONE et al. 1998), or serving as an antioxidant (GRIDLEY et al. 1998). One cannot yet exclude the possibility that estradiol plays several of these roles in its apparent (YAFFE et al. 1998; MAYEUX and GANDY 1999) ability to delay and/or prevent AD.

Acknowledgements. This work was supported by the USPHS, the New York State Office of Mental Health, and the Research Foundation for Mental Hygiene. Dr. GANDY receives or has received extramural support from the Women's Health Research Institute of Wyeth-Ayerst Pharmaceuticals, a division of American Home Corporation, and honoraria for consulting for Pfizer, Pharmacia, Parke-Davis Pharmaceuticals, and Hoffman La Roche.

References

Abraham CR, Selkoe DJ, Potter H (1988) Immunochemical identification of the serine protease inhibitor α_1-antichymotrypsin in the brain amyloid deposits of Alzheimer's disease. Cell 52:487–501

Arendt T, Schindler C, Bruckner MK, Escherich K, Bigl V, Zedlick D, Marcova L (1997) Plastic neuronal remodeling is impaired in patients with Alzheimer's disease carrying apolipoprotein epsilon 4 allele. J Neurosci 17(2):516–529

Bales KR, Verina T, Dodel RC, Du Y, Altstiel L, Bender M, Hyslop P, Johnstone EM, Little SP, Cummins DJ, Piccardo P, Ghetti B, Paul SM (1997) Lack of apolipoprotein E dramatically reduces amyloid beta-peptide deposition. Nat Genet 17(3): 263–264

Bates SE, Davidson NE, Valverius EM, Freter CE, Dickson RB, Tam JP, Kudlow JE, Lippman ME, Salomon DS (1988) Expression of transforming growth factor alpha and its messenger ribonucleic acid in human breast cancer: its regulation by estrogen and its possible functional significance. Mol Endocrinol 2:543–555

Berg L, McKeel, DW, Miller JP, Baty J, Morris JC (1993) Neuropathological indexes of Alzheimer's disease in demented and nondemented persons aged 80 years and older. Arch Neurol 50:349–358

Black RA, Rauch CT, Kozlosky CJ, Peschon JJ, Slack JL, Wolfson MF, Castner BJ, Stocking KL, Reddy P, Srinivasan S, Nelson N, Boiani N, Schooley KA, Gerhart M, Davis R, Fitzner JN, Johnson RS, Paxton RJ, March CJ, Cerretti DP (1997) A metalloproteinase disintegrin that releases tumour-necrosis factor-alpha from cells. Nature 385(6618):729–733

Borchelt DR, Thinakaran G, Eckman CB, Lee MK, Davenport F, Ratovitsky T, Prada CM, Kim G, Seekins S, Yager D, Slunt HH, Wang R, Seeger M, Levey AI, Gandy SE, Copeland NG, Jenkins NA, Price DL, Younkin SG, Sisodia SS (1996) Familial Alzheimer's disease-linked presenilin 1 variants elevate Abeta1–42/1–40 ratio in vitro and in vivo. Neuron 17(5):1005–1013

Bosenberg MW, Pandiella A, Massague J (1993) Activated release of membrane-anchored TGF-α in the absence of cytosol. J Cell Biol 122:95–101

Burdick D, Soreghan B, Kwon M, Kosmoski J, Knauer M, Henschen A, Yates J, Cotman C, Glabe C (1992) Assembly and aggregation properties of synthetic Alzheimer's A4/β amyloid peptide analogs. J Biol Chem 267:546–554

Buxbaum JD, Oishi M, Chen HI, Pinkas-Kramarski R, Jaffe EA, Gandy SE, Greengard P (1992) Cholinergic agonists and interleukin I regulate processing and secretion of the Alzheimer β/A4 amyloid protein precursor. Proc Natl Acad Sci USA 89:10075–10078

Buxbaum JD, Liu KN, Luio Y, Slack JL, Stocking KL, Peschon JJ, Johnson RS, Castner BJ, Cerretti DP, Black RA (1998) Evidence that tumor necrosis factor alpha converting enzyme is involved in regulated alpha-secretase cleavage of the Alzheimer amyloid protein precursor. J Biol Chem 273(43):27765–27767

Buxbaum JD, Gandy SE, Cicchetti P, Ehrlich ME, Czernik AJ, Fracasso RP, Ramabhadran TV, Unterbeck AJ, Greengard P (1990) Processing of Alzheimer β/A4 amyloid precursor protein: Modulation by agents that regulate protein phosphorylation. Proc Natl Acad Sci USA 87:6003–6006

Buxbaum JD, Koo EH, Greengard P (1993) Protein phosphorylation inhibits production of Alzheimer amyloid β/A4 peptide. Proc Natl Acad Sci USA 90:9195–9198

Calhoun ME, Wiederhold KH, Abramowski D, Phinney AL, Probst A, Sturchler-Pierrat C, Staufenbiel M, Sommer B, Jucker M (1998) Neuron loss in APP transgenic mice. Nature 395:755–756

Caporaso G, Takei K, Gandy S, Matteoli M, Mundigl I, Greengard P, de Camilli P (1994) Morphologic and biochemical analysis of the intracellular trafficking of the Alzheimer β/A4 amyloid precursor protein. J Neurosci 14 (5 Pt 2):3122–3138

Caporaso GL, Gandy SE, Buxbaum JD, Ramabhadran TV, Greengard P (1992) Protein phosphorylation regulates secretion of Alzheimer β/A4 amyloid precursor protein. Proc Natl Acad Sci USA 89:3055–3059

Chang D, Kwan J, Timiras PS (1997) Estrogens influence growth, maturation and amyloid beta-peptide production in neuroblastoma cells and in a beta-APP transfected kidney 293 cell line. Adv Exp Med Biol 429:261–271

Crystal H, Dickson D, Fuld P, Masur D, Scott R, Mehler M, Masdeu J, Kawas C, Aronson M, Wolfson L (1988) Clinico-pathologic studies in dementia: Nondemented subjects with pathologically confirmed Alzheimer's disease. Neurology 38:1682–1687

da Cruz e Silva 0, Iverfeldt K, Oltersdorf T, Sinha S, Lieberburg I, Ramabhadran T, Suzuki T, Sisodia S, Gandy S, Greenard P (1993) Regulated cleavage of Alzheimer β-amyloid precursor protein in the absence of the cytoplasmic tail. Neuroscience 57:873–877

Danenberg HD, Haring R, Heldman E, Gurwitz D, Ben-Nathan D, Pittel Z, Zuckerman A, Fisher A (1995) Dehydroepiandrosterone augments M1-muscarinic receptor-stimulated amyloid precursor protein secretion in desensitized PC12M1 cells. Ann NY Acad Sci 774:300–303

Danenboerg HD, Haring R, Fisher A, Pittel Z, Gurwitz D, Heldman E (1996) Dehydroepiandrosterone (DHEA) increases production and release of Alzheimer's amyloid precursor protein. Life Sci 59(19):1651–1657

Delaere P, He Y, Fayet G, Duyckaerts C, Hauw JJ (1993) βA4 deposits are constant in the brain of the older old: An immunocytochemical study of 20 French centenarians. Neurobiol Aging 14:191–194

Dickson RB, Huff KK, Spencer EM, Lippman ME (1986) Induction of epidermal growth factor-related polypeptides by 17 beta-estradiol in MCF-7 human breast cancer cells. Endocrinology 118:138–142

Dickson RB, McManaway ME, Lippman ME (1986) Characterization of estrogen responsive transforming activity in human breast cancer cell lines. Cancer Res 46:1707–1713

Dickson RB, McManaway ME, Lippman ME (1986) Estrogen-induced factors of breast cancer cells partially replace estrogen to promote tumor growth. Science 232:1540–1543

Drouva SV, Geronne I, Laplante E, Rerat E, Enjalbert A, Kordon C (1990) Estradiol modulates protein kinase C activity in the rat pituitary in vivo and in vitro. Endocrinology 126:536–544

Fisher A, Heldman E, Gurwitz D, Haring R, Karton Y, Meshulam H, Pittel Z, Marciano D, Brandeis R, Sadot E, Barg Y, Pinkas-Kramarski R, Vogel Z, Ginzburg I, Treves TA, Verchovsky R, Klimowsky S, Korczyn AD (1996) M1 agonists for the treatment of Alzheimer's disease: Novel therapeutic properties and clinical update. Ann NY Acad Sci 777:189–196

Fukuyama R, Chandrasekaran K, Rapoport SI (1993) Nerve growth factor-induced neuronal differentiation is accompanied by differential induction and localization of the amyloid precursor protein (APP) in PC12 cells and variant PC12S cells. Brain Res Mol Brain Res 17:17–22

Gandy S (1999) Neurohormonal signaling pathways and the regulation of Alzheimer β-amyloid precursor metabolism (review). Trends Endocrinol Metab 10:273–279

Gandy S, Czernik AJ, Greengard P (1988) Phosphorylation of Alzheimer's disease amyloid precursor peptide by protein kinase C and Ca^{+2}/calmodulin-dependent protein kinase II. Proc Natl Acad Sci USA 85:6218–6221

Gandy S, Greengard P (1992) Amyloidogenesis in Alzheimer's disease: Some possible therapeutic opportunities. Trends Pharmacol Sci 13:108–113

Gandy S, Greengard P (1994) Processing of Aβ-amyloid precursor protein: Cell biology, regulation, and role in Alzheimer's disease. Intl Rev Neurobiol 36:29–50

Gillespie SL, Golde TE, Younkin SG (1992) Secretory processing of the Alzheimer amyloid β/A4 protein precursor is increased by protein phosphorylation. Biochem Biophys Res Commun 187:1285–1290

Gouras GK, Xu H, Jovanovic JN, Buxbaum JD, Wang R, Greengard P, Relkin NR, Gandy S (1998) Generation and regulation of beta-amyloid peptide variants by neurons. J Neurochem 71(5):1920–1925

Gridley KE, Green PS, Simpkins JW (1998) A novel, synergistic interaction between 17β-estradiol and glutathione in the protection of neurons against Aβ 25–35-induced toxicity In vitro. Mol Pharmacol 4(5):874–880

Haring R, Fisher A, Marciano D, Pittel Z, Kloog Y, Zuckerman A, Eshhar N, Heldman E (1998) Mitogen-activated protein kinase-dependent and protein kinase C-dependent pathways link the m1 muscarinic receptor to β-amyloid precursor protein secretion. J Neurochem 71:2094–2103

Hung AY, Haass C, Nitsch RM, Qiu WQ, Citron M, Wurtman RJ, Growdon JH, Selkoe DJ (1993) Activation of protein kinase C inhibits cellular production of the amyloid β-protein. J Biol Chem 268:22959–22962

Iwatsubo T, Odaka A, Suzuki N, Mizusawa H, Nukina N, Ihara Y (1994) Visualization of A beta 42(43) and A beta 40 in senile plaques with end-specific A beta monoclonals: evidence that an initially deposited species is A beta 42(43). Neuron 12(1):45–53

Jaffe AB, Toran-Allerand CD, Greengard P, Gandy SE (1994) Estrogen regulates metabolism of Alzheimer amyloid beta precursor protein. J Biol Chem 269(18): 13065–13068

Kamel F, Kubajak CL (1988) Gonadal steroid effects on LH response to arachidonic acid and protein kinase C. Am J Physiol 155:314–321

Komano HM, Seeger M, Gandy S, Wang GT, Krafft GA, Fuller RS (1998) Involvement of cell surface glycosyl-phosphatidylinositol-linked aspartyl proteases in alpha-secretase-type cleavage and ectodomain solubilization of human Alzheimer beta-amyloid precursor protein in yeast. J Biol Chem 273(48):31648–31651

Lemere CA, Blusztajn JK, Yamaguchi H, Wisniewski T, Saido TC, Selkoe DJ (1996) Sequence of deposition of heterogeneous amyloid beta-peptides and *APOE* in Down syndrome: implications for initial events in amyloid plaque formation. Neurobiol Dis 3(1):16–32

Lippman ME, Dickson RB, Gelmann EP, Rosen N, Knabbe C, Bates S, Bronzert D, Huff K, Kasid A (1988) Growth regulatory peptide production by human breast carcinoma cells. J Steroid Biochem 30:53–61

Lue L-F, Rogers J (1992) Full complement activation fails in diffuse plaques of the Alzheimer's disease cerebellum. Dementia 3:308–313

Luine VN (1985) Estradiol increases choline acetyltransferase activity in specific basal forebrain nuclei and projection areas of female rats. Exp Neurol 89(2):484–490

Maeda T, Lloyd RV (1993) Protein kinase C activity and messenger RNA modulation by estrogen in normal and neoplastic rat pituitary tissue. Lab Invest 68:472–480

Maizels ET, Miller JB, Cutler RJ, Jackiw V, Carney EM, Mizuno K, Ohno S, Hunzicker DM (1992) Estrogen modulates Ca(2+)-independent lipid-stimulated kinase in the rabbit corpus luteum of pseudopregnancy. Identification of luteal estrogen-modulated lipid-stimulated kinase as protein kinase C delta. J Biol Chem 267: 17061–17068

Masliah E, Terry RD, Mallory M, Alford M, Hansen LA (1990) Diffuse plaques do not accentuate synapse loss in Alzheimer's disease. Am J Pathol 137:1293–1297

Mayeux R, Gandy S (1999) Cognitive Impairment, Alzheimer's Disease and Other Dementias. In: Goldman MB, Hatch MC (eds) Women and Health. Academic Press, San Diego CA

Miranda RC, Sohrabji F, Toran-Allerand CD (1993) Neuronal colocalization of mRNAs for neurotrophins and their receptors in the developing central nervous system suggests a potential for autocrine interactions. Proc Natl Acad Sci USA 90:6439–6443

Miranda RC, Sohrabji F, Toran-Allerand CD (1993) Interactions of estrogen with the neurotrophins and their receptors during neural development. Mol Cell Neurosci 4:510–525

Nitsch RM, Slack BE, Wurtman RJ, Growdon JH (1992) Release of Alzheimer amyloid precursor derivatives stimulated by activation of muscarinic acetylcholine receptors. Science 258:304–307

Peschon JJ, Slack JL, Reddy P, Stocking KL, Sunnarborg SW, Lee DC, Russell WE, Castner BJ, Johnson RS, Fitzner JN, Boyce RW, Nelson N, Kozlosky CJ, Wolfson MD, Rauch CT, Cerretti DP, Paxton RJ, March CJ, Black RA (1998) An essential role for ectodomain shedding in mammalian development. Science 282(5392): 1281–1284

Pike CJ, Burdick D, Walencewicz AJ, Glabe CG, Cotman CW (1993) Neurodegeneration induced by β-amyloid peptides in vitro: The role of peptide assembly state. J Neurosci 13:1676–1687

Polvikoski T, Sulkava R, Haltia M, Kainulainen K, Vuorio A, Verkkoniemi A, Niinisto L, Halonen P, Kontula K (1995) Apolipoprotein E, dementia, and cortical deposition of beta-amyloid protein. N Engl J Med 333(19):1242–1247

Rebeck GW, Reiter JS, Strickland DK, Hyman BT (1993) Apolipoprotein E in sporadic Alzheimer's disease: allelic variation and receptor interactions. Neuron 11(4):575–580

Refolo LM, Salton SR, Anderson JP, Mehta P, Robakis NK (1989) Nerve and epidermal growth factors induce the release of the Alzheimer amyloid precursor from PC 12 cell cultures. Biochem. Biophys. Res Commun 164:664–670

Rossner S, Ueberham U, Schliebs R, Perez-Polo JR, Bigl V (1998) The regulation of amyloid precursor protein metabolism by cholinergic mechanisms and neurotrophic receptor signalling. Prog Neurobiol 56:541–569

Rossner S, Ueberham U, Schliebs R, Perez-Polo JR, Bigl V (1998) p75 and trkA receptor signalling independently regulate amyloid precursor protein mRNA expression, isoform composition and protein secretion in PC12 cells. J Neurochem 71:757–766

Rogers J et al. (1992) Complement activation by β-amyloid in Alzheimer's disease. Proc. Natl Acad USA 89:10016–10020

Sahasrabudhe SR, Spruyt MA, Muenkel HA, Blume AJ, Vitek MP, Jacobsen JS (1992) Release of amino-terminal fragments from amyloid precursor protein reporter and mutated derivatives in cultured cells. J Biol Chem 267:25062–25608

Saunders AM, Strittmatter WJ, Schmechel D, St George-Hyslop PH, Pericak-Vance MA, Joos SH, Rosi BL, Gusella JF, Crapper-MacLachlan DR, Alberts MJ, Hulette C, Crain B, Goldgaber D, Roses ADL (1993) Association of apolipoprotein E allele $\varepsilon 4$ with late-onset familial and sporadic Alzheimer's disease. Neurology 43:1467–1472

Savage MJ, Trusko SP, Howland DS, Pinsker LR, Mistretta S, Reaume AG, Greenberg ND, Siman R, Scott RW (1998) Turnover of amyloid beta-protein in mouse brain and acute reduction of its level by phorbol ester. J Neurosci 18(5):1743–1752

Scheuner D, Eckman C, Jensen M, Song X, Citron M, Suzuki N, Bird TD, Hardy J, Hutton M, Kukull W, Larson E, Levy-Lahad E, Viitanen M, Peskind E, Poorkaj P, Schellenberg G, Tanzi R, Wasco W, Lannfelt L, Selkoe D, Younkin S (1996) Secreted amyloid beta-protein similar to that in the senile plaques of Alzheimer's disease is increased in vivo by the presenilin 1 and 2 and APP mutations linked to familial Alzheimer's disease. Nat Med 2(8):864–870

Sherrington R, Rogaev EI, Liang Y, Rogaeva EA, Levesque G, Ikeda M, Chi H, Lin C, Li G, Holman K, Tsuda T, Mar L, Foncin L-F, Bruni AC, Montesi MP, Sorbi S, Rainero I, Pinessi L, Nee L, Chumakov I, Pollen D, Brookes A, Sanseau P, Polinsky RJJ, Wasco W, da Silva HAR, Haines JL, Pericak-Vance MA, Tanzi RE, Roses AD, Fraser PE, Rommens JM, St. George-Hyslop PH (1995) Cloning of a gene bearing missense mutations in early-onset familial Alzheimer's disease. Nature 375:754–760

Sinha S, Lieberburg I (1992) Normal metabolism of the amyloid precursor protein (APP). Neurodegeneration 1:169–175

Snow AD, Sekiguchi R, Nochlin D, Kimata K, Schreier WA, Morgan DG (1994) An important role of heparan sulfate proteoglycan (perlecan) in a model system for the deposition and persistence of fibrillar Aβ amyloid in rat brain. Neuron 12:219–234

Sohrabji F, Greene LA, Miranda RC, Toran-Allerand CD (1994) Reciprocal regulation of estrogen and NGF receptors by their ligands in PC12 cells. J Neurobiol 25(8):974–988

Stone DJ, Rozovsky I, Morgan TE, Anderson CP, Finch CE (1998) Increased synaptic sprouting in response to estrogen via an apolipoprotein E-dependent mechanism: Implications for Alzheimer's disease. J Neurosci 18(9):3180–3185

Strittmatter WJ, Saunders AM, Schmechel D, Pericak-Vance M, Enghild J, Saivesen GS, Roses AD (1993) Apolipoprotein E: High-avidity binding to β-amyloid and increased frequency of type 4 allele in late-onset familial Alzheimer's disease. Proc Natl Acad Sci USA 90:1977–1981

Sun Y, Wu S, Bu G, Onifade MK, Patel SN, LaDu MJ, Fagan AM, Holtzman DM (1998) Glial fibrillary acidic protein-apolipoprotein E (*APOE*) transgenic mice: astrocyte-specific expression and differing biological effects of astrocyte-secreted apoE3 and apoE4 lipoproteins. J Neurosci 18(9):3261–3272

Tang MX, Jacobs D, Stern Y, Marder K, Schofield P, Gurland B, Andrews H, Mayeux R (1996) Effect of estrogen during menopause on risk and age at onset of Alzheimer's disease. Lancet 348(9025):429–432

Wisniewski T, Frangione B (1992) Apolipoprotein E: A pathological chaperone in patients with cerebral and systemic amyloid. Neurosci Lett 135:235–238

Woolley CS (1998) Estrogen-mediated structural and functional synaptic plasticity in the female rat hippocampus. Horm Behav 34(2):140–148

Xu H, Gouras GK, Greenfield JP, Vincent B, Naslund J, Mazzarelli L, Fried G, Javanovic JN, Seeger M, Relkin NR, Liao F, Checler F, Buxbaum JD, Chait BT, Thinakaran G, Sisodia SS, Wang R, Greengard P, Gandy S (1998) Estrogen reduces neuronal generation of Alzheimer beta-amyloid peptides. Nat Med 4(4):447–451

Xu H, Greengard P, Gandy S (1995) Regulated formation of Golgi secretory vesicles containing Alzheimer beta-amyloid precursor protein. J Biol Chem 270(40):23243–23245

Yaffe K, Sawaya G, Lieberburg I, Grady D (1998) Estrogen therapy in postmenopausal women. JAMA 279:688–695

Yang DS, Small DH, Seydel U, Smith JD, Hallmayer J, Strickland D, Gandy SE, Martins RN Apolipoprotein E promotes the association of amyloid-β with Chinese Hamster Ovary cells in an isoform-specific manner. Neuroscience (in press)

CHAPTER 16
Amyotrophic Lateral Sclerosis

M. JACKSON and J.D. ROTHSTEIN

A. Introduction

Amyotrophic lateral sclerosis (ALS) is a progressive fatal neurological disorder that results from loss of motor neurons. Many hypotheses have been put forward, including the involvement of environmental factors, autoimmune phenomena, oxidative stress, excitotoxicity, cytoskeletal abnormalities and loss of trophic factor support, to explain the mechanism and processes that lead to neuronal cell death. In this chapter we will discuss the evidence that supports the various hypotheses and describe the preclinical and clinical trials that have taken place to test potential therapies suggested by the different etiological hypotheses. These trials have provided the first agents that reproducibly alter the course of amyotrophic lateral sclerosis.

ALS is the term applied to the most common form of adult motor neuron disease in which there is progressive degeneration of both the upper motor neurons in the cortex and the lower motor neurons in the brainstem and spinal cord (MARTIN and SWASH 1995). The progression of disease is rapid, leading to muscle atrophy, paralysis, and death, usually from respiratory failure, generally within 5 years of onset. The disease is almost entirely confined to the motor system, with sensory and cognitive functions not being affected. Some motor neurons are also spared, notably those supplying ocular muscles (HUGHES 1982) and those involved in voluntary control of bladder sphincters (MANNEN et al. 1977).

The majority of ALS cases are apparently sporadic (SALS), accounting for greater than 90% of reported cases (SIDDIQUE et al. 1996). The remainder have a family history of the disease and this subset of patients is classified as familial ALS (FALS). Much effort has gone into the quest for finding the genetic defect(s) responsible for the familial ALS cases in the hope that it might shed light on the pathophysiology of the more common sporadic form of the disease. A breakthrough came in 1991 when it was reported that a subset of autosomal dominant FALS cases was linked to chromosome 21q22.1 markers (SIDDIQUE et al. 1991). By 1993 it was shown that chromosome 21-linked FALS cases are associated with mutations in SOD-1 (ROSEN et al. 1993), the gene that encodes copper-zinc superoxide dismutase (CuZnSOD). Since then, more than 90 different mutations have been identified in the SOD-1 gene, which account for approximately 20% of all FALS cases and only a small number of SALS cases (JONES et al. 1994; ROBBERECHT et al. 1996; JACKSON et al. 1997). It must therefore be acknowledged that no single insult can explain all cases of ALS as only about 1% of all ALS patients carry a SOD-1 mutation. Instead, multiple mechanisms are likely to be responsible for motor neuron degeneration in ALS and all theories must ultimately explain the mechanism and processes that lead to neuronal cell death and the relative selectivity for motor neurons.

B. Pathological Mechanisms

I. Free Radicals and Oxidative Stress

The discovery of mutations in the gene for superoxide dismutase-1 (SOD-1), one function of which is the detoxification of oxygen free radicals to hydrogen peroxide and water, stimulated the research into the potential role of free radical damage in ALS. It was found that experimentally induced oxidative stress by the depletion of glutathione, an important antioxidant, could kill embryonic cortical neurons (RATAN et al. 1994). Moreover, chronic inhibition of SOD-1 by the administration of either antisense oligodeoxynucleotides or diethyldithiocarbamate to spinal cord organotypic cultures results in apoptotic degeneration of spinal neurons, suggesting that oxidative stress can induce cell death in neurons (ROTHSTEIN et al. 1994). However, most SOD-1 mutations identified in FALS show only a modest loss in the activity of the enzyme (BORCHELT et al. 1994), implying that SOD-1 mutants are deleterious to motor neurons, not via a decrease in the activity of the enzyme, but through a toxic gain-of-function. This was subsequently proven by the development of transgenic mice expressing point mutations in SOD-1, including Ala4Val, Gly93Ala (GURNEY et al. 1994), Gly37Arg (WONG et al. 1995) and Gly85Arg (TU et al. 1996). There are numerous possibilities as to how the SOD-1 mutations could result in the acquisition of a cytotoxic gain-of-function activity, outlined in Table 1, but the mechanism is as yet unknown. Certainly, understanding how the SOD-1 mutations are toxic to motor neurons may reveal other potential neurotoxic pathways.

II. Cytoskeletal Abnormalities

Neurofilaments are the most abundant cytoskeletal protein in large myelinated axons such as those of lower and upper motor neurons. They play a particularly important role in the development and maintenance of axonal integrity in cells that possess long processes, such as motor neurons. Clumping of phosphorylated neurofilament protein into spheroids in the cell body and the proximal axons is one of the histopathological hallmarks of ALS (CARPENTER et al. 1968). This raises the question of whether aberrant accu-

Table 1. Possible cytotoxic gain-of-function activities of mutant SOD-1

Gain-of-function	Evidence
Enhancement of apoptosis	Overexpression of A4V and G37R SOD-1 mutants enhanced apoptosis in neuralcells, overexpression of wild-type SOD-1 inhibited apoptosis (RABIZADEH et al. 1994)
Formation of novel protein interactions	Mutant G85R and G93A interacted with lysyl-tRNA synthetase and translocon-associated protein delta. May result in improper cellular localization or function (KUNST et al. 1997)
Unstable, precipitates and forms toxic aggregates	Immunopositive SOD-1 inclusions in FALS patients (SHIBATA et al. 1996; KATO et al. 1996) and SOD-1 transgenic mice (BRUIJN et al. 1997a and 1998; WATANABE, unpublished observations)
Ineffective in buffering Cu and Zn	Cu and Zn known to be neurotoxic (SCHEINBERG 1988; KOH and CHOI 1988). Cu chelators are neuroprotective against mutant SOD-1 (WIEDAU-PAZOS et al. 1996)
Altered substrate affinity	More efficient use of peroxynitrite (BECKMAN et al. 1993), leading to elevated nitrotyrosine levels (BRUIJN et al. 1997b), or increased peroxidase activity (WIEDAU-PAZOS et al. 1996; YIM et al. 1996), Possible cytotoxic gain-of-function activities of mutant SOD-1

mulations of neurofilaments are merely by-products of the pathogenic process or whether they contribute to the process of motor neuron degeneration. Transgenic mice have been used to try and resolve this issue.

Over expression of wild-type murine NF-L (XU et al. 1993) or wild-type human NF-H (COTE et al. 1993) protein leads to a similar pathology as that reported in both sporadic and familial ALS. There is, however, no significant motor neuron death which makes neither set of mice a satisfying model for ALS. A more compelling transgenic model of ALS is obtained from expression of a point mutation in the murine NF-L subunit (LEE et al. 1994). This model, unlike the first two, produces selective degeneration and death of motor neurons of the lumbar and cervical spinal cord. Interestingly, new studies suggest that these results might also be explained by alterations in RNA processing due to the transgenic construct titrating out ribonucleoprotein complexes (CANETE-SOLER et al. 1999). Despite the evidence in transgenic mice that a mutation in a neurofilament subunit may cause neuron degeneration, only two small deletions in the NF-H gene in 5 of 356 SALS patients, but not in over 300 normal individuals, were originally identified (FIGLEWICZ et al. 1994). A subsequent study identified two novel deletions within the NF-H gene in 2 of 196 patients (AL-CHALABI et al. 1997), but it is not clear if any of these rare mutations could actually account for the disease. Examination of almost all of the coding domains of all three neurofilament subunits in 100

FALS patients has failed to find mutations that are linked to the disease (VECHIO et al. 1996). Nevertheless, even if the neurofilamentous abnormalities are secondary, it is likely that blockage of axonal transport by neurofilament accumulation can be a central pathogenic factor leading to axonal swelling and degeneration. In fact, experiments that altered the expression of neurofilament genes in SOD-1 transgenic mice were found, in some cases, to prolong survival, by as yet unknown mechanisms (COUILLARD-DESPRES et al. 1998; WILLIAMSON et al. 1998).

III. Autoimmune Factors

A series of experimental and clinical studies have implicated an autoimmune reaction in the etiology of ALS by some, but not all investigators. The clinical evidence has been indirect and circumstantial but includes an increased incidence of some autoimmune diseases (APPEL et al. 1986) and paraproteinemias (SHY et al. 1986) in patients with ALS. The experimental evidence includes the presence of IgG in motor neurons as well as infiltrates of T lymphocytes and microglia in spinal gray matter and motor cortex of patients with ALS (TROOST et al. 1989). Recently one group has detected serum antibodies that bind to the L-type voltage-gated calcium channels in 75% of sporadic ALS patients but not in patients with FALS (SMITH et al. 1992). However, other groups have failed to detect such antibodies in ALS sera (Dr. DRACHMAN, Johns Hopkins and Dr. VINCENT, Oxford) and the antisera has also been shown not to be toxic to motor neurons in culture. Nevertheless, passive transfer of the immunoglobulin-containing fractions into mice produces changes at the neuromuscular junction that are very similar to changes observed in patients with SALS (APPEL et al. 1991). This ability of serum immunoglobulins to passively transfer the disease is a typical feature of an antibody-mediated autoimmune disorder, yet immunosuppressive therapy using cyclosporine (APPEL et al. 1988) and the more powerful method of total lymphoid irradiation (DRACHMAN et al. 1994) failed to ameliorate the course of the disease. This argues against autoimmune processes playing a major role in the disease process and the presence of such antibodies may be a consequence, rather than a cause, of motor neuron degeneration. It also remains possible, however, that the course of the disease cannot be altered even if the pathogenic antibodies are suppressed.

IV. Excitotoxicity

The evidence that glutamate excitotoxicity contributes to neurodegeneration in ALS has received a lot of support in recent years. The first evidence that glutamate was involved in the pathogenesis of ALS came from the discovery that there was an increase in glutamate levels in plasma, CSF, and postmortem tissue from ALS patients (PLAITAKIS et al. 1987; ROTHSTEIN et al. 1990). A recent study confirmed these observations, but the elevated CSF glutamate levels were only detected in a subset (49%) of ALS patients (SHAW et al. 1995), indicating the heterogeneity of the ALS population. Subsequently transport was found to be decreased in motor cortex and spinal cord from postmortem tissue of ALS patients (ROTHSTEIN et al. 1992). Moreover, the expression of the astroglial-specific EAAT2 glutamate transporter was shown, by immunoblot and immunohistochemical analysis, to be reduced in the same brain regions (ROTHSTEIN et al. 1995a). This loss of glutamate transporter protein is supported by an earlier study in which a reduction in density of [^{3}H]D-aspartate binding was found in ALS brain tissue compared to control tissue (SHAW et al. 1994).

The hypothesis that chronic loss of glutamate transport could lead to neurodegeneration was tested using organotypic spinal cord cultures. Chronic blockade of glutamate transport by the administration of drugs that nonselectively block all glutamate transporter subtypes led to increased extracellular glutamate and a slow selective loss of motor neurons (ROTHSTEIN et al. 1993). Moreover, molecular knockout studies by the intercerebroventricular administration of antisense oligonucleotides to rats revealed a predominant role for the astroglial transporters GLAST (EAAT1) and GLT-1 (EAAT2), with their loss producing progressive limb weakness and motor neuron degeneration (ROTHSTEIN et al. 1996). Mutational analysis at the genomic level has provided no evidence for disease-specific mutations in EAAT2 (AOKI et al. 1998; JACKSON et al. 1999), except for a single sporadic ALS patient that was found to have a mutation in exon 5 (AOKI et al. 1998). The mutation resulted in a substitution of serine for an asparagine that might be involved in N-linked glycosylation of the EAAT2 protein. However, abnormal mRNA species of EAAT2 have been identified in 65% of SALS patients (LIN et al. 1998). In vitro expression studies indicate that the proteins translated from these aberrant mRNA species may undergo rapid degradation and exert a dominant negative effect on the wild-type protein, resulting in loss of protein activity. These findings support the hypothesis that loss of tissue glutamate transport and increases in CSF glutamate in ALS patients are likely to be due to the selective loss of the EAAT2 protein.

The mechanism that leads to the generation of the aberrant EAAT2 mRNA species in ALS patients is as yet unknown but could involve defects in RNA processing. Interestingly, the survival motor neuron gene (*SMN*), one of the genes responsible for childhood-onset proximal spinal muscular atrophy (SMA) (LEFEBVRE et al. 1995), another disorder specific to motor neurons, encodes an RNA binding protein. The protein was found to interact with heterogeneous nuclear ribonucleoproteins (LIU and DREYFUSS 1996) and play an important role in the assembly of the spliceosomal complex (FISCHER and DREYFUSS 1997). However, the deletions of *SMN* that are detected in over 98% of SMA patients were not found in either familial or sporadic ALS patients (JACKSON et al. 1996) and conversely, the aberrant EAAT2 RNA species were not detected in SMA patients (C.L.G. LIN, unpublished observations).

Further evidence for the occurrence of an excitotoxic process in ALS has come from a study in which CSF from ALS patients was shown to kill rat cortical neurons cultured in vitro whereas CSF from controls was found to be innocuous (COURATIER et al. 1993). The toxic factor responsible was not identified, but its activity was found to be antagonized by the AMPA receptor antagonist 6-cyano-7-nitroquinoxaline-2,3-dione (CNQX).

C. Mechanism of Motor Neuron Degeneration

Whatever the underlying genetic or environmental factors may turn out to be in the pathogenesis of ALS, they must ultimately explain the selectivity of the disease process to motor neurons and why certain motor neuron populations are spared (oculomotor neurons and Onuf's nuclei). It is possible that motor neurons are more susceptible to glutamate toxicity than other spinal neurons. One contributing factor may be differences in the ability to buffer calcium, as motor neurons are unusual in that they do not contain the Ca^{2+}-buffering proteins parvalbumin and calbindin D-28K (ALEXIANU et al. 1994). Interestingly, the exceptions are those in the oculomotor tract and Onuf's nucleus, which are the motor neurons spared in ALS.

Motor neurons may also contain an unusual complement of glutamate receptors. In most neuronal types, toxicity to glutamate is mediated principally by the NMDA receptor subtype, whereas in motor neurons the AMPA/kainate receptor subtype appears to be more important. This was shown by the fact that non-NMDA antagonists could effectively prevent neuronal degeneration induced by inhibition of glutamate transport, whereas NMDA antagonists were not neuroprotective (ROTHSTEIN et al. 1993). Recently it was shown by in situ hybridization (WILLIAMS et al. 1997) and by immunohistochemistry (BAR-PELED et al. 1999) that the GluR2 AMPA/kainate receptor subunit is absent from spinal motor neurons. In AMPA receptors the GluR2 subunit is important in regulating the permeability of Ca^{2+} (HUME et al. 1991), and the absence of this subunit renders the receptors permeable to Ca^{2+}. The presence of Ca^{2+} permeable non-NMDA receptors on motor neurons that are affected in ALS is corroborated by earlier work using a cobalt-based stain on spinal cord organotypic cultures. Using this histochemical method, it was shown that the most sensitive population of neurons permeable to calcium via non-NMDA glutamate receptors was the ventral motor neurons, those affected in ALS (PRUSS et al. 1991). Thus, the localization of calcium-permeable non-NMDA glutamate receptors on motor neurons may make them selectively vulnerable to excitotoxicity.

Alternatively, because motor neurons possess very high levels of SOD-1 mRNA (TSUDA et al. 1994) they may become susceptible targets for adverse effects if free radical scavenging mechanisms are defective. They are very large, with long axons and so are dependent on axonal transport mechanisms for maintaining their integrity. It is these energy-requiring transport mechanisms that may be critically sensitive to oxidative damage. Evidence to support the theory that components of the axonal transport machinery are among the early targets of oxidative damage comes from work with the SOD-1 transgenic mice. It was found that axonal transport slowed in mice expressing the G93A (ZHANG et al. 1997), the G37R and the G85R SOD-1 mutation (WILLIAMSON and CLEVELAND 1998). This may explain why, in the human disease, it is the largest caliber, neurofilament rich motor axons that are lost and the smaller axons with lower neurofilament content that are spared.

It is therefore very likely that multiple different insults may intersect leading to the common phenotype of motor neuron death. For example, both glutamate excess and anticalcium antibodies could lead to increased intracellular calcium. Once an initial factor generates a rise in intracellular Ca^{2+}, calcium-dependent enzymes are activated and electron leakage from the respiratory chain is stimulated. Both events lead to enhanced production of reactive oxygen species. The free radicals generated may further damage the mitochondria, leading to membrane depolarization and increased influx of Ca^{2+} through voltage-gated calcium channels and NMDA receptors (BEAL 1995). In addition, the free radicals may damage cellular proteins such as neurofilaments, resulting in blockage of axonal transport, and/or glutamate transporters (VOLTERA et al. 1994; TROTTI et al. 1996), which would further augment the excitotoxic process. The relative importance of the different routes is likely to vary from individual to individual. For example, in the FALS cases, which possess a SOD-1 mutation, free radical damage may be the predominant factor, though the different routes are not likely to be independent, but mutually propagating.

Therefore, the slowly developing and spreading pattern of motor neuron degeneration in ALS may reflect a multistep process of: (1) an initiating event and (2) a propagating series of cyclic cascades. If the hypothesis of initiator and propagators is correct, then drugs may be designed for ALS that could interfere with either step. As diagnosis of ALS is often made well into the disease process, drugs that block the initiating steps may not be as effective at halting the disease progression as those directed toward altering the cyclical, propagating steps.

D. Clinical Trials

In the past few years there has been an unprecedented increase in the number of therapeutic trials in ALS. The main focus has been the use of drugs that have neurotrophic properties or drugs that may promote neuronal survival by reducing the action of either glutamate or oxidative species. However, as it seems that the etiology of ALS is multifactorial, it is unrealistic to expect a single drug treatment to stop disease progression completely. Combinations of treatments attacking different pathogenic pathways may be necessary, as has proved successful in cancer and AIDS therapy.

I. Neurotrophic Factors

The reported effects of neurotrophic factors on neuronal survival in experimental models (ARAKAWA et al. 1990; SENDTNER et al. 1990, 1992a) has led to their use in clinical trials. Several trials have been performed with recombinant human CNTF (rhCNTF) and although positive results were reported in a pilot study these were not confirmed in two larger Phase III trials (ALS CNTF TREATMENT STUDY GROUP 1996; MILLER et al. 1996a). Moreover, safety concerns about the systemic use of rhCNTF were also raised as rhCNTF caused significant side effects which included cough, nausea, anorexia, and weight loss (MILLER et al. 1996b). It was proposed that the failure of rhCNTF to alter the progression of ALS could be due to the antigenicity of the molecule as patients receiving rhCNTF developed antibodies which may have neutralized any therapeutic activity (ALS CNTF TREATMENT STUDY GROUP 1996). It has also been argued that the clinical trials failed because the peripheral administration of CNTF could not deliver sufficient trophic factor to the CNS. CNTF has a short plasma half-life and so injected CNTF is rapidly degraded, resulting in low yields reaching the motor neurons (DITTRICH et al. 1994). The short half-life and the associated side effects may limit the clinical usefulness of systemically administered CNTF in the treatment of ALS.

Other neurotrophic factors currently undergoing clinical trials in ALS are brain-derived neurotrophic factor (BDNF) and insulin growth factor (IGF-I). The first large trial of BDNF in the USA did not reveal clinical efficacy, but the two controlled studies that have been performed with recombinant human insulin growth factor (rhIGF-I) were more promising. The first study showed that rhIGF-I significantly slowed the deterioration of function (LANGE et al. 1996) and the second study showed a trend towards the same result, but there was no statistical significance (BORASIO et al. 1996). Further data will probably be required in order to draw a firm conclusion as to the efficacy of rhIGF-I.

II. Antioxidant Drugs

The identification of some familial ALS families possessing mutations in the free radical scavenging enzyme CuZn SOD has led to the use of antioxidant agents as therapies. If high levels of free radicals are involved, then it might be possible to treat ALS patients with compounds that can detoxify these radicals. Acetylcysteine is a potent free radical scavenging agent and in a placebo-controlled trial of N-acetylcysteine there was a trend towards slowing of deterioration of muscle strength and decreased mortality, but this did not reach statistical significance (LOUWERSE et al. 1995). Three trials were also carried out using the drug selegiline. Selegiline is a selective and irreversible inhibitor of type B monoamine oxidase and has been found to slow the progression of Parkinson's disease. Unfortunately all three studies failed to show any statistically significant effect in modifying the progression of ALS (JOSSAN et al. 1994; LANGE et al. 1998; MAZZANI et al. 1994). Many ALS patients are also taking a variety of over-the-counter antioxidants, including vitamin E and C, β-carotene, and selenium.

III. Antiexcitotoxic Agents

Interest in the role of excitotoxicity in the pathology of ALS has inspired a number of large clinical trials of antiexcitotoxic drugs over recent years. There are several possible molecular targets at which potential neuroprotective drugs could attenuate the excitotoxic process and these include the inhibition of glutamate release, the blockade of excitatory amino acid receptors, inhibition of voltage-sensitive calcium channels on nerve endings or cell bodies and the activation of intracellular calcium-buffering processes.

It was originally suggested that a partial defect of glutamate dehydrogenase (GDH), the enzyme that is responsible for metabolizing glutamate to α-ketoglutarate, was responsible for the increase in plasma glutamate levels in ALS (PLAITAKIS et al. 1987). The first clinical trials were therefore conducted with branched-chain amino acids (BCAAs), L-leucine, L-isoleucine, and L-valine, since these amino acids can activate GDH. The pilot study showed promising results with a significant reduction in the progression rate of disease (PLAITAKIS et al. 1988). These results were not confirmed in later larger studies which found BCAAs ineffective in slowing the progression of the disease (PLAITAKIS et al. 1992; TESTA et al. 1989; THE ITALIAN ALS STUDY GROUP 1993; STEINER 1994), and one study showed increased mortality (THE ITALIAN ALS STUDY GROUP 1993).

An alternative approach is to block the effect of glutamate at the receptor level. Five subtypes of the glutamate receptor have been identified, but selective antagonists suitable for human use currently exist only for the NMDA receptor. Of these, the one most suitable for clinical trials is dextromethorphan, a selective, noncompetitive inhibitor. Several studies have been performed with this drug and all have been negative (ASKMARK et al. 1993; APPELBAUM et al. 1991; BLIN et al. 1996; GREDAL et al. 1996). However, measures of plasma dextromethorphan levels suggest that the doses used (300 mg daily) may not have been high enough to produce significant levels of NMDA receptor block (HOLLANDER et al. 1994). It is also worth noting that glutamate neurotoxicity may be mediated by non-NMDA receptors as motor neuron loss in organotypic spinal cord cultures can be prevented by AMPA antagonists rather than NMDA antagonists (ROTHSTEIN et al. 1993). Furthermore, the toxic factor in ALS CSF was found to be antagonized by an AMPA receptor antagonist (COURATIER et al. 1993). There is an ongoing small phase II trial of the non-NMDA antagonist, LY300164 in ALS, but development of other non-NMDA receptor antagonists for therapeutic trials is desirable.

More promising results have been obtained with riluzole, which blocks glutamatergic neurotransmission in the central nervous system (CNS) and has been shown to reduce neuronal damage in a number of experimental models (COURATIER et al. 1994; ESTEVEZ et al. 1995). The mechanism of action of riluzole, however, seems to differ from that of a pure excitatory amino acid receptor antagonist. In addition, it appears to activate a G-protein-dependent process which leads to inhibition of glutamic acid release (MARTIN et al. 1993) and blockade of some of the postsynaptic events mediated by activation of NMDA receptors, including mobilization of calcium (HUBERT et al. 1994). The neuroprotective activity of riluzole may be due to these various mechanisms acting in synergy to block excitotoxicity. In two placebo-controlled studies there was a beneficial impact upon the progress of ALS in over 1,100 patients (MILLER et al. 1996c; LACOMBLEZ et al. 1996). Both studies demonstrated a 35% decrease in risk of death or tracheotomy over an 18-month period for patients given 100mg/day compared with placebo, but there was no effect on muscle strength. As a result of these studies, riluzole has been approved and marketed for the treatment of ALS in many countries.

Lamotrigine is a drug that inhibits glutamate release and is in current clinical use as a treatment for epilepsy. Disappointingly, in the one trial that has been conducted to date, no effect on disease progression was observed after 18 months (EISEN et al. 1993). However, the dose used was almost tenfold less than that used in epilepsy. Gabapentin, another antiepileptic drug, also acts to inhibit the glutamatergic system. The mechanism of gabapentin action is not well understood, but it may act as a presynaptic inhibitor of glutamate biosynthesis (TAYLOR 1994). Gabapentin inhibits the enzyme branched-chain amino acid transferase (BCAA-T) (GOLDLUST et al. 1995), which contributes to the synthesis of glutamate from α-ketoglutarate by catalyzing the transfer of an amino group from a branched-chain amino acid donor. In one placebo-controlled trial, gabapentin showed a trend towards a slowing of the rate of deterioration of function, although the effects did not achieve statistical significance (MILLER et al. 1996d). A recent trial in Italy, however, did show a statistically significant decrease in the decline of muscle strength (MAZZINI et al. 1998).

A controlled trial of nimodipine, a calcium channel blocker, has also been carried out in ALS patients to test the hypothesis that blocking the influx of calcium to motor neurons can halt the progression of the disease. Unfortunately, there was no significant difference in the rate of decline of pulmonary function or limb strength during treatment with drug or placebo (MILLER et al. 1996e). Verapamil, another calcium channel blocker, was also found to be ineffective in slowing the clinical progression in ALS patients (MILLER et al. 1996f).

E. Development of Preclinical Screening Models

Much effort has gone into the characterization of appropriate animal and cell culture models of ALS. These models can be used to try and unravel the sequence of events in motor neuronal degeneration and test potential therapies (Table 2). Until the recent development of transgenic FALS mice carrying different SOD-1 mutations, cell culture models and hereditary mouse models of motor neuron degeneration were used to identify potential therapeutic agents.

The cell culture models typically used cultures of pure motor neurons or dissociated spinal cord tissue. Organotypic cultures of spinal cord have two main advantages: first, they make it possible to study postnatal motor neurons with intact local synaptic connections, and second, they can be maintained for months at a time, which is essential for modeling the chronic insults hypothesized to occur in ALS. This model has been used to study the effects of chronic excitotoxicity (ROTHSTEIN et al. 1993) and oxidative stress (ROTHSTEIN et al. 1994) on motor neurons. From these studies, a number of agents were found to be neuroprotective, including riluzole, gabapentin, N-acetylcysteine, IGF-1 (ROTHSTEIN et al. 1993 and 1994), and the synthetic nonimmunosuppressive ligand of FKBP-12, GPI-1046 (STEINER et al. 1998). Studies from these models also suggest that non-NMDA glutamate receptor antagonists, such as GYKI-52466, CNQX, and NBQX, may be promising antiexcitotoxic drugs for the treatment of ALS. The outcomes of trials with such drugs are awaited with interest.

The facial nerve axotomy model and hereditary mouse models, including the *wobbler* mouse and the progressive motor neuronopathy (*pmn*) mouse have been used extensively to study the ability of trophic factors to protect against motor neuron degeneration. These studies have resulted in the identification of a number of agents that may be therapeutic for ALS, including CNTF (MITSUMOTO et al. 1994; IKEDA et al. 1995; SENDTNER et al. 1992b), BDNF (YAN et al. 1992), IGF-I (VAUGHT et al. 1996), and GDNF (SAGOT et al. 1996; YAN et al. 1995). Although CNTF was shown to be neuroprotective in all three models, it was completely ineffective in two separate large clinical trials of ALS (ALS CNTF TREATMENT STUDY GROUP 1996; MILLER et al. 1996a). However, low yields of CNTF actually reached the motor neurons and the injected doses shown to produce a therapeutic effect in animals (MITSUMOTO et al. 1994) were above the toxicity threshold in humans (CEDARBAUM et al. 1995; MILLER et al. 1996a). The clinical use of neurotrophic factors may therefore require the development of more suitable modes of delivery. Gene therapy and/or encapsulated cell delivery systems appear to be potential alternatives. The intramuscular administration of an adenoviral vector coding for neurotrophin-3 (NT-3) provided substantial therapeutic benefit to *pmn* mice, resulting in an increased life span (HASSE et al. 1997), and the effects of NT-3 could be potentiated by coadministering a CNTF adenoviral vector. Recently, the gene transfer of NT-3 was also shown to delay the onset of clinical disease in the G93A SOD-1 mice (AZZOUZ et al. 1998). The other promising technique involves the implantation of polymer-encapsulated cells which have been genetically modified to secrete neurotrophic factors. It was found that encapsulated cell lines secreting CNTF delayed the disease progression in *pmn* mice (SAGOT et al.

Table 2. Preclinical and clinical trials of neuroprotective agents for ALS

Action	Drug/Transgenic cross	Organotypic cell culture[a]	Facial axotomy model[b]	wobbler/pmn mice[c]	SOD-1 transgenic mice[d]	Clinical ALS trials[e]
Trophic	CNTF	−	+	+	−	−
	IGF-1	+	?	?	−	±
	BDNF	−	+	+	−	−
	GDNF	++	+	+	?	nt
	GPI-1046	+	nt	nt	+ (onset)	nt
Antiglutamate	Riluzole	+	nt	nt	+ (survival)	+
	Gabapentin	+	nt	nt	+ (survival)	+
	GYKI-52466	++	nt	nt	+ (survival)	nt
Antioxidant	Vitamin E	−	nt	nt	+ (onset)	−
	Phenyl-butyl nitrone	±	nt	nt	?	nt
	N-acetylcysteine	+	nt	nt	nt	−
	Carboxyfullerenes	nt	nt	nt	+ (onset)	nt
	Putricine-catalase	nt	nt	nt	+ (onset)	nt
Cu-chelator	d-penicillamine	nt	nt	nt	+ (onset)	±
Antiapoptotic	bcl-2 overexpression	nt	nt	nt	+ (onset)	nt
	Bax loss	nt	nt	nt	+ (onset and survival)	nt
	ICE inhibition	nt	nt	nt	+ (survival)	nt
Ca^{2+}-binding	NF-H overexpression	nt	nt	nt	+ (onset)	nt
Energy/mitochondria	Creatine	nt	nt	nt	+ (survival)	nt

nt, not tested.
[a] Steiner et al. 1998; Rothstein et al. 1993, 1994, 1995, and unpublished observations.
[b] Sendtner et al. 1990 and 1992a; Yan et al. 1992 and 1995.
[c] Mitsumoto et al. 1994; Sagot et al. 1996; Sendtner et al. 1992b.
[d] Couillarrd-Despres et al. 1998; Dugan et al. 1997; Friedlander et al. 1997; Gurney et al. 1996; Hottinger et al. 1997; Houseweart et al. 1998; Klivenyi et al. 1999; Kostic et al. 1997; Lindsay R, oral presentation at International Symposium on ALS/MND, Dublin, 1995; Nagano and Rothstein, unpublished observations; Putcha et al. 1998; Reinholz et al. 1999; Jackson and Steiner, unpublished observations.
[e] ALS CNTF Treatment Study Group, 1996; Borasio et al. 1996; Conradi et al. 1982; Jusic and Sostasko, 1977; Lacomblez et al. 1996; Lange et al. 1996; Louwerse et al. 1995; Mazzini et al. 1998; Miller et al. 1996a,c,d.

1995), rescued facial motor neurons following axotomy (TAN et al. 1996), and were able to continuously deliver CNTF within the CSF of humans without the associated side-effects observed with systemic delivery (AEBISCHER et al. 1996).

Nevertheless, the development of transgenic mice carrying different SOD-1 mutations found in FALS cases provide the best animal models for studying ALS as they all show a classic, progressive motor neuron degeneration characteristic of ALS. The first therapeutic success with the mutant SOD-1 transgenic model was obtained with vitamin E, riluzole, and gabapentin (GURNEY et al. 1996). Interestingly, vitamin E was shown to delay disease onset but not affect overall survival, whereas the antiglutamate agents, riluzole and gabapentin, were ineffective at altering disease onset but were able to prolong survival. Riluzole had about the same efficacy in the transgenic mice as seen in human ALS trials (MILLER et al. 1996c; LACOMBLEZ et al. 1996). These observations have led to the concept that in FALS cases an initiating event, possibly due to toxic free radicals, is then followed by a propagating series of cyclical cascades, such as glutamate excitotoxicity, reinforcing the need for combinatorial therapy.

Subsequent therapeutic screens in FALS transgenic mice have also shown specific effects on altering either disease onset or overall survival. The copper chelator d-penicillamine was shown to significantly delay the onset of disease (HOTTINGER et al. 1997), supporting the theory that copper may mediate a toxic gain-of-function of the mutant enzyme. Putrescine-modified catalase (REINHOLZ et al. 1999), an antioxidant enzyme, and carboxyfullerenes (DUGAN et al. 1997), free radical scavengers, were also shown to significantly delay disease onset, providing further evidence that free-radical-mediated oxidative damage is involved in the initiation of motor neurodegeneration in FALS mice. GPI-1046, the synthetic nonimmunosuppressive ligand of FKBP-12 was also recently found to delay the onset of disease in G93A FALS transgenic mice (Jackson and Steiner, unpublished observations), contrary to the recent finding that immunophilins actually enhance mutant SOD-1-mediated cell death (LEE et al. 1999)

Other potential therapeutic pathways have been investigated by crossbreeding the SOD-1 transgenic mice with other transgenic strains. Despite the prior finding that mice expressing high levels of human NF-H develop similar pathology to ALS patients (COTE et al. 1993), expression of human NF-H at lower levels reduced SOD-1 toxicity, increasing the mean lifespan of G37R mice by up to 65% (COUILLARD-DESPRES et al. 1998). The increased synthesis of NF-H trapped an increased proportion of neurofilaments in the perikarya of motor neurons with a corresponding reduction in axonal neurofilaments. Similarly by deleting NF-L, the onset and progression of disease were significantly slowed in G85R mice (WILLIAMSON et al. 1998). There was a concomitant reduction of NF-M and NF-H in the motor axons, but the levels of the two remaining neurofilament subunits were elevated in the cell bodies. It is possible that reducing axonal neurofilaments is neuroprotective, but since neurofilament proteins have multiple calcium-binding sites (LEFEBVRE and MUSHYNSKI 1987 and 1988) it is also conceivable that increasing neurofilament subunit protein levels in perikarya might confer protection against rises in intracellular calcium. Overexpression of bcl-2, a potent inhibitor of most types of apoptotic cell death, was also found to delay onset of motor neuron disease in G93A SOD-1 mice (KOSTIC et al. 1997), in mice possessing a point mutation in the NF-L subunit, but not in G37R SOD-1 mice (HOUSEWEART et al. 1998). Bcl-2 inhibits the activation of caspases, which are cysteine proteases that are important in the occurrence of apoptosis in mammalian cells. Interestingly, inhibition of interleukin-1β-converting enzyme (ICE), one such cysteine protease, was found to prolong survival in G93A mice, but not delay the onset of disease (FRIEDLANDER et al. 1997). Finally, the deletion of Bax, a pro-apoptotic protein, was found to delay onset and lengthen the duration of the symptomatic period in G93A mice (PUTCHA et al. 1998). However, a confounding variable in experiments that alter developmental motor neuron death (e.g., Bax and Bcl-2) is that animals may originally have a greater number of motor neurons compared to controls, making interpretation of outcomes problematic.

As shown in Table 2, success of drugs in preclinical screens has often correlated with effectiveness in clinical trials of ALS patients, underlying the importance of testing therapeutics in these disease models. However, because the transgenic mice only represent a defect found in 1%–2% of ALS patients, it is worth noting that a drug that is not effective at altering the onset of disease or survival in the FALS mice may still be effective in some ALS patients. Nevertheless, it would appear that the non-SOD1 mediated forms of the disease do indeed share common elements of pathogenesis with the SOD-1 mediated form of disease.

F. Future Therapies

The availability of riluzole is an important first step in our progress towards finding a treatment for ALS, even if its effects are small in ALS patients. It provides new directions for research since the positive results lend support to the hypothesis that glutamate excitotoxicity is an important mechanism in ALS. It is anticipated that in the future clinical trials using combinations of riluzole with neurotrophic factors and antioxidants, as well as other antiglutamate compounds, will be undertaken as, if motor neuron degeneration is a multistep process, a combination of treatments may have greater success (Table 3).

There are ongoing trials of neurotrophic factors, such as IGF-I, and perhaps others will be tested in the near future (e.g., GDNF, neurotrophin-3, and cardiotrophin). Synthetic agents that have trophic factor-like activity, such as GPI-1046 and SR57746A, are also being explored as they have shown notable neurotrophic activity in a variety of neurodegenerative models

Table 3. Future therapies for amyotrophic lateral sclerosis

Type of treatment	Drug/Strategy
Neurotrophic factor	GDNF NT-3 NT-4/5 Cardiotrophin-1 Synthetic agents (e.g., GPI-1046) Encapsulated cell delivery of trophic factors Gene therapy
Antioxidants/free radical scavengers	Lazeroids Free radical spin traps (e.g., α-phenyl-N-t-butyl-nitrone) Metal chelators (Cu/Zn) Nitric oxide inhibitors Fullerenes
Antiglutamate	Non-NMDA antagonists (e.g., GYKI-52466) NMDA antagonists Release inhibitors Synthesis inhibitors Increase glutamate transporter expression/activity
Antiapoptotic	Bcl-2 or bcl-2 homologs Caspase/cysteine protease inhibitors Bax inhibitors
Anticalcium	Increase Ca^{2+}-binding proteins (e.g., 1,25-dihydroxyvitamin D3 analogs) Antagonists of intracellular calcium release Calcium channel blockers
Combinatorial	Riluzole + neurotrophic factor Riluzole + antioxidant Riluzole + other antiglutamate

(STEINER et al. 1998; JACKSON and STEINER, unpublished observations; FOURNIER et al. 1993). There is ongoing development of antiglutamate and antioxidant agents, some of which will be clinically evaluated in the near future. However, since the biochemical nature of the cytotoxic gain-of-function of the mutant SOD-1 protein is not yet known, a simple "replacement of SOD-1" therapy is not appropriate. In fact, under certain conditions, addition of SOD-1 could actually exacerbate the toxicity associated with the mutations. Nevertheless, FALS patients may be the most appropriate to treat with potent new antioxidants. An ongoing area of interest focuses on the possibility that the cytotoxic gain-of-function of mutant SOD-1 involves the formation of protein aggregates, e.g., SOD/CCS interactions (ROTHSTEIN et al. 1998). If protein trafficking/folding is relevant, then future therapies may need to be directed towards this problem.

Novel therapeutic approaches may also focus on increasing glutamate transport. This could either be achieved by increasing the expression of EAAT2, the glutamate transporter found to be lost in ALS patients (GANEL et al. 1998), or by altering the post-translational regulation of glutamate transporters (e.g., phosphorylation or membrane trafficking). The identification and analysis of proteins that interact with, and regulate the glutamate transporter family of proteins (Jackson et al. 2001; Lin et al. 2001) will increase the understanding of the biology of glutamate transporters and how they normally maintain extracellular glutamate concentrations at levels low enough to prevent excitotoxicity. Hopefully, this may identify other molecular targets at which potential neuroprotective drugs could attenuate the excitotoxic process thought to occur in ALS. Understanding the biology that is responsible for the defects in RNA processing in ALS may also someday provide novel therapeutic opportunities.

If calcium homeostasis does play a role in neuronal injury, then the ability to increase calcium-binding proteins may have possible therapeutic value in ALS. It has been shown that 1,25-dihydroxyvitamin D3 increases the immunoreactivity for calcium-binding proteins in motor neurons in vitro and in vivo (Alexianu et al. 1998), suggesting that analogs of 1,25-dihydroxyvitamin D3 may have therapeutic benefits.

The mechanism of motor neuron cell death in ALS is still not known, but there has been great interest in the involvement of apoptosis in ALS (Rabizadeh et al. 1995; Weidau-Pazos et al. 1996) and other neurodegenerative diseases (Choi 1996). If apoptosis is involved, then antiapoptotic drugs may provide another therapeutic approach for the treatment of ALS. For example, the use of synthetic antagonists or inhibitors of cysteine proteases may have merit.

Finally, due to the serious safety concerns of systemic administration of neurotrophic factors, other modes of delivery need to be investigated for future clinical trials. The use of gene therapy with neurotrophic factors in vectors targeted to motor neurons or encapsulated cell delivery systems are two promising new therapeutic strategies.

G. Conclusions

In recent years, enormous progress has been made toward unraveling the pathogenesis of sporadic and familial ALS. This information is crucial for the development of new therapies. In fact, the identification of the possible role of oxidative stress and excitotoxicity has already produced small inroads into therapy. Riluzole, and more recently, gabapentin and IGF-1, have been shown to have small effects in ALS patients, demonstrating that progression of the disease can be altered. There are several new potential therapeutic agents that have been shown to alter the course of motor neuron degeneration in animal and cell culture models of ALS, and these must be seriously considered for clinical trials in ALS. However, complete preclinical evaluation, including chronic toxicity studies, are essential for all new ALS therapies. Finally, if ALS is attributable to multiple different causes, then some therapies may only be effective in a subset of patients. This may mean that certain drugs will fail to show efficacy, when in fact it could be beneficial to a subset of patients. For this reason, understanding the pathogenesis and the development of biochemical markers that distinguish such patients will be mandatory and hopefully, by interfering with the multiple pathways of motor neuron degeneration, our ambitions to halt this fatal disease may be realized.

References

Aebischer P, Schluep M, Deglon N, Joseph JM, Hirt L, Heyd B, Goddard M, Hammang JP, Zurn AD, Kato AC, Regli F, Baetge EE (1996) Intrathecal delivery of CNTF using encapsulated genetically modified xenogeneic cells in amyotrophic lateral sclerosis patients. Nature Medicine 2:696–699

Al-Chalabi A, Powell JF, Russ C, Leigh PN (1997) Novel mutations in a hypervariable site of the heavy neurofilament subunit in ALS. Neurology 48:A349

Alexianu ME, Ho B-K, Mohamed H, La Bella V, Smith RG, Appel SH (1994) The role of calcium-binding proteins in selective motoneuron vulnerability in amyotrophic lateral sclerosis. Ann Neurol 36:846–858

Alexianu ME, Robbins E, Carswell S, Appel SH (1998) 1alpha,25 dihydroxyvitamin D3-dependent up-regulation of calcium-binding proteins in motoneuron cells. J Neurosci Res 51:58–66

ALS CNTF Treatment Study Group (1996) A double-blind placebo-controlled clinical trial of subcutaneous recombinant human ciliary neurotrophic factor (rHCNTF) in amyotrophic lateral sclerosis. Neurology 46:1244–1249

Aoki M, Lin CLG, Rothstein JD, Geller BA, Hosler BA, Munsat TL, Horvitz HR, Brown RH (1998) Mutations in the glutamate transporter EAAT2 gene do not cause abnormal transcripts in ALS. Ann Neurol 43:645–653

Appel SH, Stockton-Appel V, Stewart SS, Kerman RH (1986) Amyotrophic lateral sclerosis: associated clinical disorders and immunological evaluations. Arch Neurol 43:234–238

Appel SH, Stewart SS, Appel V, Harati Y, Mietlowski W, Weiss W, Belendiuk GW (1988) A double blind study of the effectiveness of cyclosporine in amyotrophic lateral sclerosis. Arch Neurol 45:381–386

Appel SH, Engelhardt JI, Garcia J, Stefani E (1991) Immunoglobulins from animal models of motor neuron disease and from human amyotrophic lateral sclerosis patients passively transfer physiological abnormalities to the neuromuscular junction. Proc Natl Acad Sci USA 88:647–651

Appelbaum JS, Salazar-Grueso EF, Richman JG, Shanahan M, Ross RP (1991) Dextromethorphan in the treatment of ALS: A pilot study. Neurology 41 [Suppl 1]:393

Arakawa Y, Sendtner M, Thoenen H (1990) Survival effect of ciliary neurotrophic factor (CNTF) on chick embryonic motoneurons in culture: comparison with other neurotrophic factors and cytokines. J Neurosci 10:3507–3515

Askmark H, Aquilonius S-M, Gillberg P-G, Liedholm LJ, Stalberg E, Wuopio R (1993) A pilot trial of dextromethorphan in amyotrophic lateral sclerosis. J Neurol Neurosurg Psychiatry 56:197–200

Azzouz M, Paterna J-C, Hottinger AF, Zurn AD, Aebischer P, Bueler H (1998) Gene therapy in an animal model of ALS using an adeno-associated virus expressing neurotrophin-3. Society for Neuroscience 24:188.5

Bar-Peled O, O'Brien J, Morrison JH, Rothstein JD (1999) Cultured motor neurons possess calcium-permeable AMPA/kainate receptors. Neuroreport 10:855–859

Beal MF (1995) Aging, energy, and oxidative stress in neurodegenrative diseases. Ann Neurol 38:357–366

Beckman JS, Carson M, Smith CD, Koppenol WH (1993) ALS, SOD and peroxynitrite [letter]. Nature 364:584

Blin O, Azulay JP, Desnuelle C, Bille-Turc F, Braguer D, Besse D, Branger E, Crevat A, Serratrice G, Pouget JY (1996) A controlled one-year trial of dextromethorphan in amyotrophic lateral sclerosis. Clin Neuropharmacol 19:189–192

Borasio GD, De Jong JMBV, Emile J, Guiloff R, Jerusalem F, Leigh N, Murphy M, Robberecht W, Silani V, Wokke J, the Europena ALS/IGF-I study group (1996) Insulin-like growth factor I in the treatment of amyotrophic lateral sclerosis: results of the European multicenter, double-blind, placebo-controlled trial. J Neurol 243 [Suppl 2]:S26

Borchelt DR, Lee MK, Slunt HS, Guarnieri M, Xu Z-S, Wong PC, Brown RH, Price DL, Sisodia SS, Cleveland DW (1994) Superoxide dismutase 1 with mutations linked to familial amyotrophic lateral sclerosis possesses significant activity. Proc Natl Acad Sci USA 91:8292–8296

Bruijn LI, Becher MW, Lee MK, Anderson KL, Jenkins NA, Copeland NG, Sisodia SS, Rothstein JD, Borchelt DR, Price DL, Cleveland DW (1997a) ALS-linked SOD1 mutant G85R mediates damage to astrocytes and promotes rapidly progressive disease with SOD1-containing inclusions. Neuron 18:327–338

Bruijn LI, Beal MF, Becher MW, Schulz JB, Wong PC, Price DL, Cleveland DW (1997b) Elevated free nitrotyrosine levels, but not protein-bound nitrotyrosine or hydroxyl radicals, throughout amyotrophic lateral sclerosis (ALS)-like disease implicate tyrosine nitration as an aberrant in vivo property of one familial ALS-linked superoxide dismutase 1 mutant. Proc Natl Acad Sci USA 94:7606–7611

Bruijn LI, Houseweart M, Kato S, Anderson KL, Anderson SD, Ohama E, Reaume AG, Scott RW, Cleveland DW (1998) Aggregation and motor neuron toxicity of an ALS-linked SOD1 mutant independent from wild-type SOD1. Science 281:1851–1854

Canete-Soler R, Silberg DG, Gershon MD, Schlaepfer WW (1999) Mutation in neurofilament transgene implicates RNA processing in the pathogenesis of neurodegenerative disease. J Neurosci 19:1273–1283

Carpenter S (1968) Proximal axonal enlargement in motor neuron disease. Neurology 18:842–851

Cedarbaum JM (1995) A phase I study of recombinant human ciliary neurotrophic factor (rHCNTF) in patients with amyotrophic lateral sclerosis. Clin Neuropharmacol 18:515–532

Choi DW (1996) Ischemia-induced neuronal apoptosis. Curr Opin Neurobiol 6:667–672

Conradi S, Ronnevi LO, Nise G, Vesterberg O (1982) Long-time penicillamine-treatment in amyotrophic lateral sclerosis with parallel determination of lead in blood, plasma and urine. Acta Neurol Scand 65:203–211

Cote F, Collard JF, Julien JP (1993) Progressive neuronopathy in transgenic mice expressing the human neurofilament heavy gene: a mouse model of amyotrophic lateral sclerosis. Cell 73:35–46

Couillard-Despres S, Zhu Q, Wong PC, Price DL, Cleveland DW, Julien J-P (1998) Protective effect of neurofilament heavy gene overexpression in motor neuron disease induced by mutant superoxide dismutase. Proc Natl Acad Sci USA 95:9626–9630

Couratier P, Hugon J, Sindou P, Vallat JM, Dumas M (1993) Cell culture evidence for neuronal degeneration in amyotrophic lateral sclerosis being linked to glutamate AMPA/kainate receptors. Lancet 341:265–268

Couratier P, Sindou P, Esclaire F, Louvel E, Hugon J (1994) Neuroprotective effects of riluzole in ALS CSF toxicity. NeuroReport 5(8):1012–1014

Dittrich F, Thoenen H, Sendtner M (1994) Ciliary neurotrophic factor: Pharmokinetics and acute-phase response in rat. Ann Neurol 35:151–163

Drachman DB, Chaudhry V, Cornblath D, Kuncl RW, Pestronk A, Clawson L, Mellits ED, Quaskey S, Quinn T, Calkins A, Order S (1994) Trial of immunosuppression in amyotrophic lateral sclerosis using total lymphoid irradiation. Ann Neurol 35(2):142–150

Dugan LL, Turetsky DM, Cheng D, Lobner D, Wheeler M, Almli CR, Shen CK-F, Luh T-YL, Choi DW, Lin T-S (1997) Carboxyfullerenes as neuroprotective agents. Proc Natl Acad USA 94:9434

Eisen A, Stewart H, Schulzer M, Cameron D (1993) Antiglutamate therapy in amyotrophic lateral sclerosis: a trial using lamotrigine. Can J Neurol Sci 20:297–301

Estevez AG, Stutzmann JM, Barbeito L (1995) Protective effect of riluzole on excitatory amino acid-mediated neurotoxicity in motoneuron-enriched cultures. Eur J Pharmacol 280(1):47–53

Fischer U, Dreyfuss G (1997) The SMN-SIP1 complex has an essential role in spliceosomal snRNP biogenesis. Cell 90:1023–1029

Figlewicz DA, Krizus A, Martinoli MG, Meininger V, Dib M, Rouleau GA, Julien J-P (1994) Variants of the heavy neurofilament subunit are associated with the development of amyotrophic lateral sclerosis. Hum Mol Genet 3:1757–1761

Fournier J, Steinberg R, Gauthier T, Keane PE, Guzzi U, Coude FX, Bougault I, Maffrand JP, Soubrie P, Le Fur G (1993) Protective effects of SR 57746 A in central and peripheral models of neurodegenerative disorders in rodents and primates. Neuroscience 55:629–641

Friedlander RM, Brown RH, Gagliardini V, Wang J, Yuan J (1997) Inhibition of ICE slows ALS in mice. Nature 388:31

Ganel R, Ho T, Coccia C, Sakal C, Steiner J, Dykes-Hoberg M, Robinson MB, Rothstein JD (1998) Excitotoxicity and neurodegeneration–a novel therapeutic approach. Society for Neuroscience 24:825.19

Goldlust A, Su T-Z, Welty DF, Taylor CP, Oxender DL (1995) Effects of the anticonvulsant drug gabapentin on the enzymes in the metabolic pathways of glutamate and GABA. Epilepsy Res 22:1–11

Gredal O, Werdelin L, Bak S, Christensen PB, Boysen G, Kristensen MO, Jespersen JH, Regeur L, Hinge HH, Jensen TS (1996) A clinical trial of dextromethorphan in amyotrophic lateral sclerosis. Acta Neurol Scand 96:8–13

Gurney ME, Pu H, Chiu AY, Dal Canto MC, Polchow CY, Alexander DD, Caliendo J, Hentati A, Kwon YW, Deng H-X, Chen W, Zhai P, Sufit RL, Siddique T (1994) Motor neuron degeneration in mice that express a human Cu,Zn superoxide dismutase mutation. Science 264:1772–1775

Gurney ME, Cutting FB, Zhai P, Doble A, Taylor CP, Andrus PK, Hall ED (1996) Benefit of vitamin E, riluzole and gabapentin in a transgenic model of familial amyotrophic lateral sclerosis. Ann Neurol 39:147–157

Hasse G, Kennel P, Pettmann B, Vigne E, Akli S, Revah F, Schmalbruch H, Kahn A (1997) Gene therapy of murine motor neuron disease using adenoviral vectors for neurotrophic factors. Nature Medicine 3:429–436

Hollander D, Pradas J, Kaplan R, McLeod HL, Evans WE, Munsat TL (1994) High-dose dextromethorphan in amyotrophic lateral sclerosis: Phase I safety and pharmacokinetic studies. Ann Neurol 36:920–924

Hottinger AF, Fine EG, Gurnet ME, Zurn AD, Aebischer P (1997) The copper chelator d-penicillamine delays onset of disease and extends survival in a transgenic mouse model of familail amyotrophic lateral sclerosis. Eur J Neurosci 9:1548–1551

Houseweart MK, Martinou JC, Lee MK, Cleveland DW (1998) Effects of bcl-2 overexpression on neuronal death in two transgenic mouse models of motor neuron disease. Society for Neuroscience 24:184.1

Hubert JP, Delumeau JC, Glowinski J, Premont J, Doble A (1994) Antagonism by riluzole of entry of calcium evoked by NMDA and veratridine in rat cultured granule cells: evidence for a dual mechanism of action. Br J Pharmacol 113(1):261–267

Hughes JT (1982) Pathology of amyotrophic lateral sclerosis. Adv Neurol 36:61–73

Hume RI, Dingledine R, Heinemann SF (1991) Identification of a site in glutamate receptor subunits that controls calcium permeability. Science 253:1028–1031

Ikeda K, Wong V, Holmlund TH, Greene T, Cedarbaum JM, Lindsay RM, Mitsumoto H (1995) Histometric effects of ciliary neurotrophic factor in wobbler mouse motor neuron disease. Ann Neurol 37:47–54

Jackson M, Morrison KE, Al-Chalabi A, Bakker M, Leigh PN (1996) Analysis of chromosome 5q13 genes in amyotrophic lateral sclerosis: Homozygous NAIP deletion in a sporadic case. Ann Neurol 39:796–800

Jackson M, Al-Chalabi A, Enayat ZE, Chioza B, Leigh PN, Morrison KE (1997) Copper/Zinc superoxide dismutase 1 and sporadic amyotrophic lateral sclerosis: Analysis of 155 cases and identification of a novel insertion mutation. Ann Neurol 42:803–807

Jackson M, Steers G, Leigh PN, Morrison KE (1999) Polymorphisms in the glutamate transporter gene EAAT2 in European ALS patients. J Neurol 246:1140–1144

Jackson M, Song W, Liu M-Y, Jin L, Dykes-Hoberg M, Lin C-LG, Bowers WJ, Federoff HJ, Sternweis PC, Rothstein JD (2001) Modulation of the neuronal glutamate transporter EAAT4 by two interacting proteins. Nature 410:89–93

Jones CT, Shaw PJ, Chari G, Brock DJ (1994) Identification of a novel exon 4 SOD-1 mutation in a sporadic amyotrophic lateral sclerosis patient. Mol Cell Probes 8: 329–330

Jossan SS, Ekblom J, Gudjonsson O, Hagbarth KE, Aquilonius SM (1994) Double blind cross over trial with deprenyl in amyotrophic lateral sclerosis. J Neural Transm Suppl 41:237–241

Jusic A, Sostarko M (1977) Improvement of spinal amyotrophy by penicillamine therapy. Lancet 2:1034–1035

Kato S, Shimoda M, Watanabe Y, Nakashima K, Takahashi K, Ohama E (1996) Familial amyotrophic lateral sclerosis with a two base pair deletion in superoxide dismutase 1 gene: Multisystem degeneration with intracytoplasmic hyaline inclusions in astrocytes. J Neuropathol and Exper Neurol 55:1089–1101

Klivenyi P, Ferrante RJ, Matthews RT, Bogdanov MB, Klein AM, Andreassen OA, Mueller G, Wermer M, Kaddurah-Daouk R, Beal F (1999) neuroprotective effects of creatine in a transgenic animal model of amyotrophic lateral sclerosis. Nature Medicine 5(3):347–350

Koh JY, Choi DW (1988) Zinc alters excitatory amino acid neurotoxicity on central neurons. J Neurosci 8:2164–2171

Kostic V, Jackson-Lewis V, de Bilbao F, Dubois-Dauphin M, Przedborski S (1997) Bcl-2: Prolonging life in a transgenic mouse model of familial amyotrophic lateral sclerosis. Science 277:559–562

Kunst CB, Mezey E, Brownstein MJ, Patterson (1997) Mutations in SOD1 associated with amyotrophic lateral sclerosis cause novel protein interactions. Nature Genetics 15:91–94

Lacomblez L, Bensimon G, Leigh PN, Guillet P, Powe L, Durrleman S, Delumeau JC, Meininger V (1996) A confirmatory dose ranging study of riluzole in ALS. ALS/Riluzole Study Group II. Neurology 47 [6 Suppl 4]:S242–S250

Lange DJ, Felice KJ, Festoff BW, Gawel MJ, Gelinas DF, Kratz R, Lai EC, Murphy MF, Natter HM, Norris FH, Rudnicki S, the North American ALS/IGF-I Study Group (1996) Recombinant human insulin-like growth factor-I in ALS: Description of a double-blind, placebo-controlled study. Neurology 47 [Suppl 2]:S93–S95

Lange DJ, Murphy PL, Diamond B, Appel V, Lai EC, Younger DS, Appel SH (1998) Selegiline is ineffective in a collaborative double-blind, placebo-controlled trial for treatment of amyotrophic lateral sclerosis. Arch Neurol 55:93–96

Lee J-P, Palfrey HC, Bindokas VP, Ghadge GD, Ma L, Miller RJ, Roos RP (1999) The role of immunophilins in mutant superoxide dismutase-1-linked familial amyotrophic lateral sclerosis. Proc Natl Acad Sci USA 96:3251–3256

Lee MK, Marszalek JR, Cleveland DW (1994) A mutant neurofilament subunit causes massive, selective motor neuron death: implications for the pathogenesis of human motor neuron disease. Neuron 13:975–988

Lefebvre S, Mushynski WE (1987) Calcium binding to untreated and dephosphorylated porcine neurofilaments. Biochem Biophys Res Commun 145:1006–1011

Levebvre S, Mushynski WE (1988) Characterization of the cation-binding properties of porcine neurofilaments. Biochemistry 27:8503–8508

Lefebvre S, Burglen L, Reboullet S, Clermont O, Burlet P, Viollet L, Benichou B, Cruaud C, Millasseau P, Zeviani M, Le Paslier D, Frezal J, Cohen D, Weissenbach J, Munnich A, Melki J (1995) Identification and characterization of a spinal muscular atrophy-determining gene. Cell 80:155–165

Lin CLG, Bristol LA, Jin L, Dykes-Hoberg M, Crawford T, Clawson L, Rothstein JD (1998a) Aberrant RNA processing in a neurodegenerative disease: a common cause for loss of glutamate transport EAAT2 protein in sporadic amyotrophic lateral sclerosis. Neuron 20:589–602

Lin CLG, Orlov I, Ruggerio AM, Dykes-Hober M, Lee A, Jackson M, Rothstein JD (2001) modulation of the neuronal glutamate transporter EAAC1 by an interacting protein GTRAP3-18. Nature 410:84-88

Liu Q, Dreyfuss G (1996) A novel nuclear structure containing the survival of motor neurons protein. EMBO J 15:3555–3565

Louwerse ES, Weverling GJ, Bossuyt PM, Meyjes FE, de Jong JM (1995) Randomized, double-blind, controlled trial of acetylcysteine in amyotrophic lateral sclerosis. Arch Neurol 52:559–564

Mannen T, Iwata M, Toyokura Y, Nagashima K (1977) Preservation of a certain motoneurone group of the sacral cord in amyotrophic lateral sclerosis: its clinical significance. J Neurol Neurosurg Psychiatry 40:464–469

Martin D, Thompson MA, Nadler JV (1993) The neuroprotective agent riluzole inhibits release of glutamate and aspartate from slices of hippocampal area CA1. Eur J Pharmacol 250:473–476

Martin JE, Swash M (1995) The pathology of motor neuron disease. In: Leigh PN, Swash M (eds) Motor neuron disease: biology and management. Springer-Verlag, London, pp 93–118

Mazzani L, Testa D, Balzarini C, Mora G (1994) An open randomized clinical trial of selegiline in amyotrophic lateral sclerosis. J Neurol 241:223–227

Mazzini L, Mora G, Balzarini C, Brigatti M, Pirali I, Comazzi F, Pastore E (1998) The natural history and the effects of gabapentin in amyotrophic lateral sclerosis. J Neurol Sci 160 [Suppl 1]:S57–S63

Miller RG, Petajan JH, Bryan WW, Armon C, Barohn RJ, Goodpasture JC, Hoagland RJ, Parry GJ, Ross MA, Stromatt SC, the rhCNTF ALS Study Group (1996a) A placebo-controlled trial of recombinant human ciliary neurotrophic (rhCNTF) factor in amyotrophic lateral sclerosis. Ann Neurol 39:256–260

Miller RG, Bryan WW, Dietz MA, Munsat TL, Petajan JH, Smith SA, Goodpasture JC (1996b) Toxicity and tolerability of recombinant human ciliary neurotrophic factor in patients with amyotrophic lateral sclerosis. Neurology 47:1329–1331

Miller RG, Bouchard JP, Duquette P, Eisen A, Gelinas D, Harati Y, Munsat TL, Powe L, Rothstein J, Salzman P, Sufit RL, the ALS/riluzole Study Group-II (1996c) Clinical trials of riluzole in patients with ALS. Neurology 47 [Suppl 2]:S86–S92

Miller RG, Moore D, Young LA, Armon C, Barohn RJ, Bromberg MB, Bryan WW, Gelinas DF, Mendoza MC, Neville HE, Parry GJ, Petajan JH, Ravits JM, Ringel SP, Ross MA, the WALS Study Group (1996d) Placebo-controlled trial of gabapentin in patients with amyotrophic lateral sclerosis. Neurology 47:1383–1388

Miller RG, Shepherd R, Dao H, Khramstov A, Mendoza M, Graves J, Smith S (1996e) Controlled trial of nimodipine in amyotrophic lateral sclerosis. Neuromusc Disord 6:101–104

Miller RG, Smith SA, Murphy JR, Brinkmann JR, Graves J, Mendoza M, Sands ML, Ringel SP (1996f) A clinical trial of verapamil in amyotrophic lateral sclerosis. Muscle and Nerve 19:511–515

Mitsumoto H, Ikeda K, Klinkosz B, Cedarbaum JM, Wong V, Lindsay RM (1994) Arrest of motor neuron disease in wobbler mice cotreated with CNTF and BDNF. Science 265:1107–1110

Plaitakis A, Caroscio JT (1987) Abnormal glutamate metabolism in amyotrophic lateral sclerosis. Ann Neurol 22(5):575–579

Plaitakis A, Smith J, Mandeli J, Yahr MD (1988) Pilot trial of branched-chain aminoacids in amyotrophic lateral sclerosis. Lancet 1:1015–1018

Plaitakis A, Sivak M, Fesdjian CO, Mandeli J (1992) Treatment of amyotrophic lateral sclerosis with branched chain amino acids (BCAA): Results of a second study. Neurology 42 [Suppl 3]:454

Pruss RM, Akeson RL, Racke MM, Wilburn JL (1991) Agonist activated cobalt uptake identifies divalent cation-permeable kainate receptors on neurons and glial cells. Neuron 7:509–518

Putcha GV, Knudson CM, Johnson Jr EM, Korsmeyer SJ (1998) Bax deletion delays mortality in amurine model of familial ALS and neuronal cell death in SOD1-deficient mice. Society for Neuroscience 24:184.2

Rabizadeh S, Butler Gralla E, Borchelt DR, Gwinn R, Valentine JS, Sisodia S, Wong P, Lee M, Hahn H, Bredesen DE (1994) Mutations associated with amyotrophic lateral sclerosis convert superoxide dismutase from an antiapoptotic gene to a proapoptotic gene: Studies in yeast and neural cells. Proc Natl Acad Sci USA 92:3024–3028

Ratan RR, Murphy TH, Baraban M (1994) Oxidative stress induces apoptosis in embryonic cortical neurons. J Neurochem 62(1):376–379

Reinholz MM, Merkle CM, Poduslo JF (1999) Therapeutic benefits of putrescine-modified catalase in a transgenic mouse model of familial amyotrophic lateral sclerosis. Exp Neurol 159:204–216

Robberecht W, Aguirre T, Van Den Bosch L, Tilkin P, Cassiman JJ, Matthijs G (1996) D90 A heterozygosity in the SOD1 gene is associated with familial and apparently sporadic amyotrophic lateral sclerosis. Neurology 47:1336–1339

Rosen DR, Siddique T, Patterson D, Figlewicz DA, Sapp P, Hentati A, Donaldson D, Goto J, O'Regan JP, Deng H-X, Rahmani Z, Krizus A, McKenna-Yasek D, Cayabyab A, Gaston SM, Berger R, Tanzi RE, Halperin JJ, Herzfeldt Van den Bergh R, Hung W-Y, Bird T, Deng G, Mulder DW, Smyth C, Laing NG, Soriano E, Pericak-Vance MA, Haines J, Rouleau GA, Gusella JS, Horvitz HR, Brown RH (1993) Mutations in Cu/Zn superoxide dismutase gene are associated with familial amyotrophic lateral sclerosis. Nature 362:59–63

Rothstein JD, Tsai G, Kuncl RW, Clawson L, Cornblath DR, Drachman DB, Pestronk A, Stauch BL, Coyle JT (1990) Abnormal excitatory amino acid metabolism in amyotrophic lateral sclerosis. Ann Neurol 28:18–25

Rothstein JD, Martin LJ, Kuncl RW (1992) Decreased glutamate transport by the brain and spinal cord in amyotrophic lateral sclerosis. N Engl J Med 22:1464–1468

Rothstein JD, Jin L, Dykes-Hoberg M, Kuncl RW (1993) Chronic glutamate uptake inhibition produces a model of slow neurotoxicity. Proc Natl Acad Sci USA 90:6591–6595

Rothstein JD, Bristol LA, Hosler B, Brown RH, Kuncl RW (1994) Chronic inhibition of superoxide dismutase produces apoptotic death of spinal neurons. Proc Natl Acad Sci USA 91:4155–4159

Rothstein JD, Van Kammen M, Levey AI, Martin L, Kuncl RW (1995a) Selective loss of glial glutamate transporter GLT-1 in amyotrophic lateral sclerosis. Ann Neurol 38:73–84

Rothstein JD, Kuncl RW (1995b) Neuroprotective strategies in a model of chronic glutamate-mediated motor neuron toxicity. J Neurochem 65:643–651

Rothstein JD, Dykes-Hoberg M, Pardo CA, Bristol LA, Jin L, Kuncl RW, Kanai Y, Hediger M, Wang Y, Schielke J, Welty DF (1996) Knockout of glutamate transporters reveals a major role for astroglial transport in excitotoxicity and clearance of glutamate. Neuron 16:675–686

Rothstein JD, Becker M, Dykes-Hoberg M, Culotta V, Corson L, Cleveland D, Price D, Wong P (1998) CCS- A new copper chaperone protein that may contribute to familial ALS. Society for Neuroscience 24:184.15

Sagot Y, Tan SA, Baetge E, Schmalbruch H, Kato AC, Aebischer P (1995) Polymer encapsulated cell lines genetically engineered to release ciliary neurotrophic factor can slow down progressive motor neuronopathy in the mouse. Eur J Neurosci 7:1313–1322

Sagot Y, Tan SA, Hammang JP, Aebischer P, Kato AC (1996) GDNF slows loss of motoneurons but not axonal degeneration or premature death of pmn/pmn mice. J Neurosci 16:2335–2341

Scheinberg H (1988) The neurotoxicity of copper. In: Bondy SC, Prasad KN (eds) Metal Neurotoxicity. CRC Press, Boca Raton Florida, pp 55–60

Sendtner M, Kreutzberg GW, Thoenen H (1990) Ciliary neurotrophic factor prevents the degeneration of motor neurons after axotomy. Nature 345:440–441

Sendtner M, Holtmann B, Kolbeck R, Thoenen H, Barde YA (1992a) Brain derived neurotrophic factor prevents the death of motoneurons in newborn rats after nerve transection. Nature 360:757–759

Sendtner M, Schmalbruch H, Stockli KA, Carroll P, Kreutzberg GW, Thoenen H (1992b) Ciliary neurotrophic factor prevents degeneration of motor neurons in mouse mutant progressive motor neuronopathy. Nature 358:502–504

Shaw PJ, Chinnery RM, Ince PG (1994) [3H]D-aspartate binding sites in the normal human spinal cord and changes in motor neuron disease: a quantitative autoradiographic study. Brain Res 655:195–201

Shaw PJ, Forrest V, Ince PG, Richardson JP, Wastell HJ (1995) CSF and plasma amino acid levels in motor neuron disease: Elevation of CSF glutamate in a subset of patients. Neurodegeneration 4(2):209–216

Shibata N, Hirano A, Kobayashi M, Siddique T, Deng H-X, Hung W-Y, Kato T, Asayama (1996) Intense superoxide dismutase-1 immunoreactivity in intracytoplasmic hyaline inclusions of familial amyotrophic lateral sclerosis with posterior column involvement. J Neuropathol and Exper Neurol 55:481–490

Shy ME, Rowland LP, Smith T, Trojaborg W, Latov N, Sherman W, Pesce MA, Lovelace RE, Osserman EF (1986) Motor neuron disease and plasma cell dyscrasia. Neurology 36:1429–1436

Siddique T, Figlewicz DA, Pericak-Vance MA, Haines JL, Rouleau G, Jeffers AJ, Sapp P, Hung W-Y, Bebout J, McKenna-Yasek D, Deng G, Horvitz HR, Gusella JF, Brown RH, Roses AD, collaborators (1991) Linkage of a gene causing familial amyotrophic lateral sclerosis to chromosome 21 and evidence of genetic-locus heterogeneity. N Engl J Med 324:1381–1384

Siddique T, Nijhawan D, Hentati A (1996) Molecular genetic basis of familial ALS. Neurology 47 [Suppl 2]:S27–S35

Smith RG, Hamilton S, Hofmann F, Schneider T, Nastainczyk W, Birnbaumer L, Stefani E, Appel SH (1992) Serum antibodies to L-type calcium channels in patients with amyotrophic lateral sclerosis. N Engl J Med 327:1721–1728

Steiner JP, Ho TW, Lai M, Coccia C, Griffin JW, Snyder S, Rothstein JD (1998) The neuroimmunophilin GPI-1046 protects motor neurons from chronic excitotoxicity. Society for Neuroscience 24:121.1

Steiner TJ, for SPECIALS (1994) Multinational trial of branched-chain amino acids in amyotrophic lateral sclerosis. Muscle Nerve [Suppl 1]:S166

Tan SA, Deglon N, Zurn AD, Baetge EE, Bamber B, Kato AC, Aebischer P (1996) Rescue of motoneurons from axotomy-induced cell death by polymer encapsulated cells genetically engineered to release CNTF. Cell Transplantation 5:577–587

Taylor CP (1994) Emerging perspectives on the mechanism of action of gabapentin. Neurology 44 [Suppl 5]:S10–S16

Testa D, Caraceni T, Feloni V (1989) Branched-chain amino acids in the treatment of amyotrophic lateral sclerosis. J Neurol 236:445–447

The Italian ALS Study Group (1993) Branched-chain amino acids and amyotrophic lateral sclerosis: A treatment failure? Neurology 43:2466–2470

Troost D, Van den Oord JJ, de Jong JMVB, Swaab DF (1989) Lymphocytic infiltration in the spinal cord of patients with amyotrophic lateral sclerosis. Clin Neuropathol 8:289–294

Trotti D, Rossi D, Gjesdal O, Levy LM, Racagni G, Danbolt NC, Volterra A (1996) Peroxynitrite inhibits glutamate transporter subtypes. J Biol Chem 271:5976–5979

Tsuda T, Munthasser S, Fraser PE, Percy ME, Rainero I, Vaula G, Pinessi L, Bergamini L, Vignocchi G, McLachlan DR (1994) Analysis of the functional effects of a mutation in SOD-1 associated with familial amyotrophic lateral sclerosis. Neuron 13:727–736

Tu P-H, Raju P, Robinson KA, Gurney ME, Trojanowski JQ, Lee VM-Y (1996) Transgenic mice carrying a human mutant superoxide dismutase transgene develop neuronal cytoskeletal pathology resembling human amyotrophic lateral sclerosis lesions. Proc Natl Acad Sci USA 93:3153–3160

Vaught JL, Contreras PC, Glicksman MA, Neff NT (1996) Potential utility of rhIGF-1 in neuromuscular and/or degenerative disease. Ciba Foundation Symposium 196: 18–27

Vechio JD, Bruijn LI, Xu Z, Brown RH, Cleveland DW (1996) Sequence variants in human neurofilament proteins: absence of linkage to familial amyotrophic lateral sclerosis. Ann Neurol 40(4):603–610

Volterra A, Trotti D, Tromba C, Floridi S, Racagni G (1994) Glutamate uptake inhibition by oxygen free radicals in rat cortical astrocytes. J Neurosci 14:2924–2932

Wiedau-Pazos M, Goto JJ, Rabizadeh S, Gralla EB, Roe JA, Lee MK, Valentine JS, Bredesen DE (1996) Altered reactivity of superoxide dismutase in familial amyotrophic lateral sclerosis. Science 271:515–518

Williams TL, Day NC, Ince PG, Kamboj RK, Shaw PJ (1997) Calcium-permeable α-amino-3-hydroxy-5-methyl-4-isoxazole propionic acid receptors: a molecular determinant of selective vulnerability in amyotrophic lateral sclerosis. Ann Neurol 42:200–207

Williamson TL, Cleveland DW (1998) Slowing of axonal transport in mice expressing two different familial ALS-linked SOD1 mutations. Society for Neuroscience 24: 184.9

Williamson TL, Bruijn LI, Zhu Q, Anderson KL, Anderson SD, Julien J-P, Cleveland DW (1998) Absence of neurofilaments reduces the selective vulnerability of motor neurons and slows disease caused by a familial amyotrophic lateral sclerosis-linked superoxide dismutase 1 mutant. Proc Natl Acad Sci USA 95:9631–9636

Wong PC, Pardo CA, Borchelt DR, Lee MK, Copeland NG, Jenkins NA, Sisodia SS, Cleveland DW, Price DL (1995) An adverse property of a familial ALS-linked SOD1 mutation causes motor neuron disease characterized by vacuolar degeneration of mitochondria. Neuron 14:1105–1116

Xu Z, Cork LC, Griffin JW, Cleveland DW (1993) Increased expression of neurofilament subunit NF-L produces morphological alterations that resemble the pathology of human motor neuron disease. Cell 73:23–33

Yan Q, Elliot J, Snider WD (1992) Brain-derived neurotrophic factor rescues spinal motor neurons from axotomy-induced cell death. Nature 360:753–755

Yan Q, Matheson C, Lopez OT (1995) In vivo neurotrophic effects of GDNF on neonatal and adult facial motor neurons. Nature 373:341–344

Yim MB, Kang J-H, Yim H-S, Kwak H-S, Chock PB, Stadtman ER (1996) A gain of function of an amyotrophic lateral sclerosis-associated Cu,Zn-superoxide dismutase mutant: An enhancement of free radical formation due to a decrease in K_m for hydrogen peroxide. Proc Natl Acad Sci USA 93:5709–5714

Zhang B, Tu P, Abtahian F, Trojanowski JQ, Lee VM (1997) Neurofilaments and orthograde transport are reduced in ventral root axons of transgenic mice that express human SOD1 with a G93A mutation. J Cell Biol 139:1307–1315

CHAPTER 17
Therapeutics in Huntington's Disease

H.D. Rosas and W. Koroshetz

A. Background

Huntington's disease (HD) is an autosomal dominantly inherited, progressive neurodegenerative disorder which usually manifests in midlife. It affects psychiatric, cognitive, and motor systems. Severe disability results from progressive cognitive deficits and loss of voluntary motor control, which worsen over approximately 15–20 years until the affected individual is left unable to ambulate, communicate, swallow, or perform any of the activities of daily living. Death in this late stage of the illness is commonly due to aspiration pneumonia. Exciting advances in the understanding of the neurobiology of HD have occurred in the past decade as a consequence of the discovery and cloning of the gene and the development of genetically determined animal models of the disease.

Chorea, the Greek word for "dance," is the name given to the involuntary, jerky movements which are so conspicuous in this illness. These movements cease only in sleep. Chorea occurs most prominently in midlife, and most physicians await its presence to establish clinical onset. However, some evidence of mood disorders such as depression, emotional dyscontrol, mania, or cognitive disorders of attention and memory are often present months to years before the motor symptoms occur. With progression of the disease, chorea blends into dystonia, with rigidity occurring upon activation of any muscle groups. Younger patients often present without chorea but instead have bradykinesia, usually with abnormally slow saccadic eye movements. In fact, eye movement abnormalities are common in most persons with HD and include slowing of saccades, delay in generating saccades, and jerky pursuit eye movements. Motor control of fine movements of the fingers and tongue is impaired early; gait imbalance with frequent falls occurs in the mid to later stages of the illness.

The genetic defect has been identified as the expansion of an unstable trinucleotide repeat (CAG) within the coding region of the IT15 ("interesting transcript 15") gene on chromosome 4, which is translated into repeated glutamine units near the N-terminus of the protein huntingtin (HUNTINGTON'S STUDY GROUP 1993). In HD, this expansion ranges from 6–32 in normal individuals, and 40–180 in HD-affected individuals. CAG repeats between 32 and 40 are associated with low penetrance (RUBINSZTEIN 1996; SATHASIVAM et al. 1997; MYERS 1993). The length of the repeat correlates inversely with the age of onset of disease. This effect on age of onset is prominent in the range of CAG repeats above 60 but exerts minimal effect in the more common range of 40–50. The repeat length is unstable during spermatogenesis, which can lead to inheritance of a longer CAG repeat length from affected males (MACDONALD et al. 1993).

Inheritance of expanded polyglutamines in a variety of genes causes several neurodegenerative diseases, including dentatorubral-pallydolysian atrophy, spinobulbar muscular atrophy, and five spinocerebellar ataxias (REDDY and HOUSMAN 1997); each however, displays a characteristic set of neurological symptoms and unique neuropathology. Intranuclear inclusions containing aggregates of polyglutamine may be a common feature in all of these illnesses though it is not at all clear that the inclusions themselves are responsible for the cell dysfunction (see Sect. C.) (DAVIES 1997; SCHERZINGER et al. 1997; Ross 1997).

Despite the widespread expression of huntingtin in both neurons and extraneuronal tissues, the neuropathologic hallmark of Huntington's disease is the selective and progressive degeneration of the basal ganglia. Some studies have suggested that abnormalities in red cell fragility and glucose control occur in HD. Patients with HD frequently have enormous caloric requirements to maintain body weight. Higher sedentary energy expenditure in patients with Huntington's disease has been reported (PRATLEY 2000). Preliminary studies also suggest that muscle metabolism may be disordered (KOROSHETZ et al. 1997). However, no dysfunction of organs other than brain has been definitely proven to occur in HD patients.

Prior to the discovery of the genetic abnormality, several theories were proposed which might explain the selective neurodegeneration; among the most compelling are those related to excitotoxicity, abnormalities in energy metabolism and alterations in the expression of specific receptors. Recent advances in the cell biology of the mutated huntingtin, as well as the mutated proteins underlying the other CAG repeat disorders, has led to a host of new pathophysiologic hypotheses. We review the postulated pathophysiologic mechanisms underlying the disease process and discuss potential targeted therapeutic interventions, some of which have been used successfully in animal models of HD.

B. The Pathology of Huntington's Disease

Though the striatum, a component of the basal ganglia, is most heavily affected in HD, at autopsy the entire brain is usually markedly atrophic (Fig. 1). The clinical symptoms of HD reflect the progressive neurodegeneration of basal ganglia-thalamo-cortical circuits, which are briefly reviewed below. The

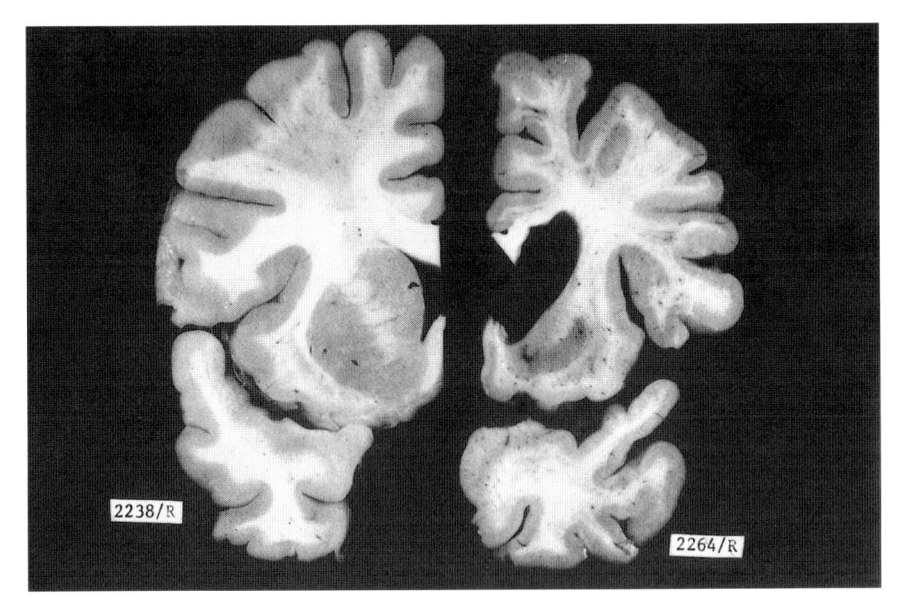

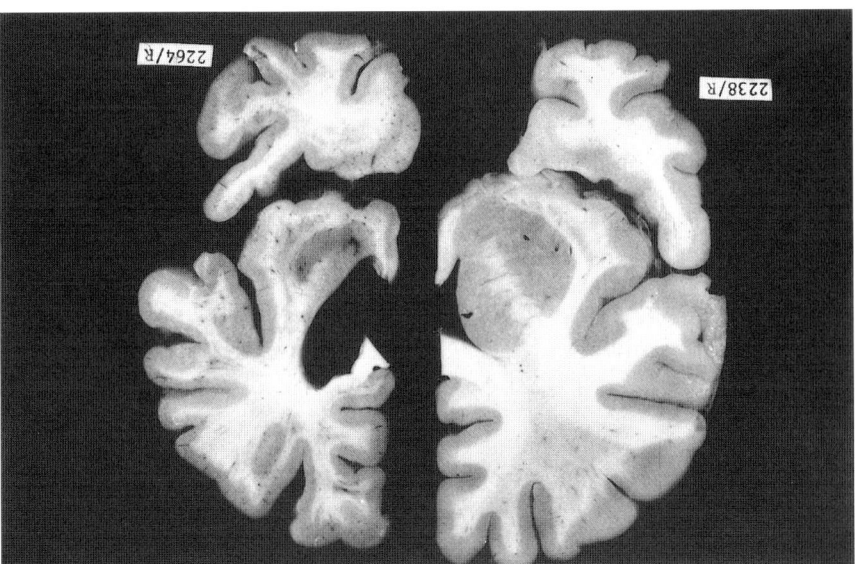

Fig. 1. The sections on the *left* are those of the normal left hemisphere of a 34-year-old man who committed suicide (brain weight 1680 g); those on the *right* are those of a 48-year-old patient with HD (brain weight 1100 g), which demonstrates diffuse atrophy and a moderately to severely atrophied striatum

functional anatomy of the basal ganglia can be conceptualized as two distinct compartments: one for input, consisting of the caudate nucleus and putamen, and one for output, consisting of the globus pallidus, subthalamic nucleus, and substantial nigra. Afferents to the caudate and putamen are glutamatergic, thus excitatory, and come from most cortical brain areas. The putamen has the densest input from motor system cortex.

The striatum receives excitatory glutamatergic input from the entire neocortex. Cortical inputs project to the globus pallidus and substantia nigra pars reticulata by two different routes, the direct and indirect pathways. The direct loop starts with striatal cells, which project inhibitory axons containing γ-aminobutryric acid (GABA) and substance P to the internal segment of the globus pallidus and the substantia nigra pars reticulata. The substantia nigra pars reticulata project their inhibitory efferents to the globus pallidus and subthalamic nucleus. Internal pallidal cells in turn send inhibitory efferents to thalamic cells (with one direct projection to superior colliculus) whose efferents complete the circuit by projecting back to cortex. Activation of this pathway results in disinhibition of thalamic activity.

An indirect loop in this circuit starts with striatal cells which project inhibitory axons containing GABA and eukephlin P to the external pallidum. The external pallidal cells send inhibitory fibers containing GABA to the subthalamic nucleus. Glutamatergic cells in the subthalamic nucleus project to the internal pallidum and substantia nigra pars reticulata which send inhibitory efferents to thalamic targets. It has been proposed that a relative imbalance in these two pathways results in the symptoms of HD; indeed, it has been proposed that there is preferential disturbance in this indirect pathway early in the course of disease (REINER et al. 1988; ALBIN et al. 1989).

Neuronal loss first involves the medial caudate nucleus adjacent to the lateral ventricle, the dorsal putamen, and the tail of the caudate nucleus, and proceeds in a posterior to anterior, dorsal to ventral, and medial to lateral manner (VONSATTEL et al. 1985). Furthermore, there is a division of the neurotransmitter makeup into compartments termed the "striosome" and "matrix." The striosome is acetylcholinesterase poor, low in NADPH diaphorase, and calbindin staining. These regions have a poorly staining core surrounded by dense staining for substance P, enkephalin, TGF-α, and cholecystokinin. The surrounding matrix is rich in somatostatin, NPY, NADPH diaphorase, calbindin, and acetylcholinesterase.

Neuronal loss and gliosis occur in both compartments; however, these first appear in the striosomes, indicating that neurons in striosomes may be more vulnerable in the early stages of disease (KREMER et al. 1994). Indeed, autopsy cases from presymptomatic or early affected persons with HD pathology confined to isolated islands of neuronal loss and astrocytosis within the striosome compartment of the neostriatum have been reported (HEDREEN and FOLSTEIN 1995). Ultimately, however, the matrix compartment most heavily undergoes degeneration, especially matrix neurons projecting to the external pallidum. Extrastriate degeneration occurs in HD, including degeneration of the subthalamic nucleus, thalamus, and cortex, most prominently in layers IV, V, and VI (CUDKOWICZ and KOWALL 1990; SOTREL et al. 1991), but the cortical neurodegeneration does not correlate with the severity of the disease and is dwarfed by the striatal changes (GREENAMYRE and SHOULSON 1994).

The cellular anatomy of the basal ganglia consists of two major categories of neurons: spiny neurons, which are projection neurons, and aspiny neurons, which have local connections. Both the spiny and aspiny neurons are represented by small-, medium-, and large-sized neurons (GRAVELAND et al. 1985). All spiny neurons contain the inhibitory neurotransmitter GABA and are the principal input and output neurons of the striatum. They represent either striatopallidal efferent neurons, and connect the striatum with the pallidum, or striatonigral neurons. Subsets of medium spiny neurons contain enkephalin, substance P, dynorphin, or calbindin. The medium to large aspiny interneurons contain acetylcholine, nicotinamide adenine dinucleotide phosphate diaphorase (NADPH-d), nitric oxide synthase, somatostatin, or neuropeptide Y (GRAYBIEL and RAGSDALE 1983; PULSINELLI 1985; FERRANTE et al. 1987).

Early in the course of HD, there is selective loss of striatal enkephalin staining neurons projecting to the external segment of the globus pallidus, a component of the indirect pathway. This leads to increased inhibition of subthalamic, glutamatergic neurons. The resultant, early loss of inhibitory input to pallidoreceptive thalamocortical neurons and therefore abnormal activation of excitatory glutamatergic output to cortex has been proposed to cause chorea. Loss of substance P containing striatal neurons projecting to the internal segment of the globus pallidus and the substantia nigra pars reticulata occurs next. Interestingly, this early loss of enkephlin staining has been noted in genetic animal models of HD as well (MENALLED et al. 2000). Aspiny interneurons and larger cholinergic interneurons are relatively spared and degenerate only in later stages of disease (FERRANTE et al. 1985; FERRANTE et al. 1987). This implicated a special vulnerability of a subset of GABAergic striatopallidal neurons (REINER et al. 1988; ALBIN et al. 1989; RICHFIELD et al. 1995).

In addition to decreased concentrations of glutamate receptors and decreases in NMDA-containing neurons, decreases in receptors for dopamine, GABA and muscarinic-cholinergic containing neurons, as well as increases in receptors for GABA have been described (PENNEY and YOUNG 1982; BEAL et al. 1986; DURE et al. 1991).

C. Molecular Basis of Huntington's Disease

I. Wild-Type Huntingtin

The discovery of the gene was a major breakthrough and led to an effort to understand the function of the normal huntingtin protein and how this function is altered in HD (HUNTINGTON'S DISEASE COLLABORATIVE RESEARCH GROUP 1993). The normal huntingtin protein is expressed ubiquitously in somatic tissues, in contrast to the restricted and regional pathology of the disease. The highest level of expression is in the brain, particularly layers II and V of the cortex and in the cerebellum (TROTTIER et al. 1995a). It is a cytoplasmic protein purported to have a nuclear localization sequence (NLS) that would mediate translocation to the nucleus (BESSERT et al. 1995).

Several hypotheses about the role of the normal huntingtin gene product have been proposed. Deletion of both huntingtin alleles in mice is lethal at embryonic day 8 (GUTEKUNST et al. 1995; WHITE et al. 1997; DRAGATSIS et al. 1998). The huntingtin protein is essential for gastrulation and neurogenesis (DUYA et al. 1997). The mutant gene is able to rescue the knock-out phenotype. These data are consistent with the evidence suggesting that the mutant huntingtin is not completely nonfunctional. Its sequence contains a possible leucine zipper, which has led to the hypothesis that huntingtin might be a nuclear transcription factor.

Wild-type huntingtin has been shown to upregulate transcription of brain-derived neurotrophic factor (BDNF), an important growth and survival factor made in cortex and released onto striatal neurons (ZUCCATO et al. 2001). BDNF levels have recently been reported to be decreased in HD brain (FERRER et al. 2000) and this may be significant as BDNF is a potent trophic factor for striatal cells.

Subcellular fractionation studies have suggested that huntingtin may associate with synaptic vesicles, microtubules, and the cytoskeleton (DIFIGLIA et al. 1995). These and other data suggest that huntingtin may play a role in trafficking and transport of vesicles and/or cytoskeletal anchorin (KEGEL et al. 2000).

How the N-terminal expansion of polyglutamine repeats translates into progressive, selective neurodegeneration is an active area of investigation. The mutant protein has the same neuronal distribution and intracellular location as normal huntingtin; therefore, it is heterogeneously expressed throughout the brain and not limited to the neurons which preferentially degenerate in HD (ARONIN et al. 1995; GOURFINKEL-AN et al. 1997).

II. Huntingtin with Expanded Polyglutamine Repeat

Normal huntingtin has been demonstrated to interact with a variety of proteins; an altered function of these associated proteins resulting from association with mutant huntingtin might then be related to cell dysfunction. Indeed, several proteins interact more strongly with mutant huntingtin (Fig. 2). One protein, termed huntingtin-associated protein-1 (HAP-1), is confined to the brain and shows an expression similar to that of neuronal nitric oxide synthase (LI et al. 1995). Huntingtin binds to calmodulin and forms a large calcium-dependent complex; the interaction is strong and becomes calcium-independent with increasing polyglutamine length. Huntingtin has also been shown to interact with glyceraldehyde-3-phosphate dehydrogenase (GADH), a critical glycolytic enzyme which catalyzes the oxidative phosphorylation of

Huntingtin in HD

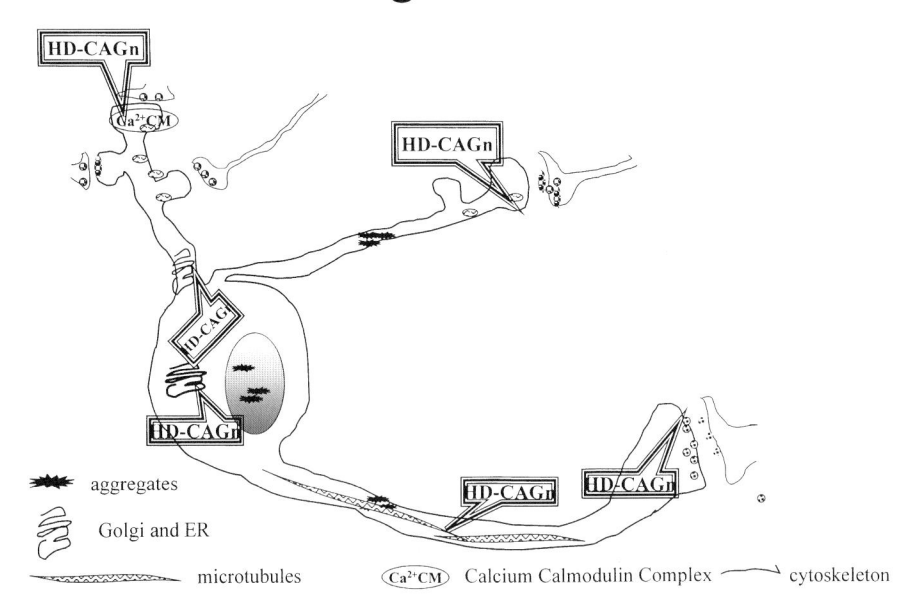

Fig. 2. Mutant huntingtin binds to a wide array of proteins and cellular structures, including the calcium calmodulin complex, the cytoskeleton, microtubules, golgi, and endoplasmic reticulum (*ER*). Aggregates which are ubiquinated and insoluble are found in the cytoplasm and nucleus in both transgenic mouse models and human HD

glyceraldehyde-3-phosphate to 1,3 biphosphylglycerate (BURKE et al. 1996; COOPER et al. 1997). This association suggested that the mutated huntingtin might play a role in altering cellular energetics. However, neither normal nor mutant huntingtin have been found in mitochondria, arguing against a direct association.

The mutant huntingtin with an expanded polyglutamine repeat may also associate with the wild-type allele and disturb the latter's normal function (ZUCCATO et al. 2001). The expression of mutant Huntingtin has also been reported to negatively effect the expression of the normal huntingtin (ONA et al. 1999).

These protein–protein interactions and their implications in the progressive neurodegeneration in HD remain an area of active investigation; clues to the function of the mutated form have come from transfection experiments and genetically determined animal models as discussed below.

Alternate theories regarding the chemical pathology of mutant huntingtin, and the other polyglutamine disorders have been proposed. Most focus on the role of an expanded polyglutamine stretch in protein–protein interaction. Long stretches of polyglutamines could form hydrogen bonds to other similar stretches in an antiparallel fashion, producing polar zippers. This process could lead to protein aggregation and precipitation, and perturb the function of multiple proteins in the cell containing amino acid stretches that might tightly adhere to the exaggerated polyglutamine in huntingtin. (PERUTZ et al. 1994). In vitro studies have shown that exon 1 of the huntingtin protein can spontaneously aggregate in a time, concentration, and repeat dependent manner (SCHERZINGER et al. 1997), forming insoluble aggregates. Indeed, abnormal ubiquitin-positive aggregates have been found in both human HD and transgenic mouse models (see Sect. C.II.2.). Alternatively, expanded polyglutamine stretches could be better substrates for transglutaminase, a calcium-dependent transamidating enzyme which has been implicated in cellular differentiation, than for wild-type huntingtin. Expanded polyglutamines could preferentially become cross-linked with polypeptides containing lysyl groups to form covalently bonded aggregates (KAHLEM et al. 1996).

1. Huntingtin and Altered Gene Transcription

Gene transcription has been reported to be disordered in a specific fashion in mice models of HD and other polyglutamine disorders. Complexation of mutant huntingtin with transcription factors may be at fault. The expanded huntingtin was found to aggregate with multiple transcription factors including p53, CREB-binding protein (CBP) and mSin3a, and TATA-binding protein TBP in cell culture, genetic animal models of HD or human HD brain (STEFFAN et al. 2000; HUANG et al. 1999; BOUTELL et al. 1999). Expression of the mutant huntingtin alters transcription of several genes including brain-derived growth factor (BDNF) (ZUCCATO 2000) and specific neurotransmitter receptors (CHA et al. 2000).

Specifically, mutant huntingtin has been found to repress transcription of genes regulated by the tumor suppressor/proapoptotic transcription factor, p53. This effect of mutant huntingtin is dependent upon the presence of the huntingtin's proline-rich region and the length of the polyglutamine stretch (STEFFAN et al. 2000). Interestingly, mutant huntingtin itself was also found to repress transcription of p53-dependent genes even in the absence of p53. This raises the possibility that mutant huntingtin can mimic the action of p53s itself. A unifying pathological mechanism falls out as P53 leads to apoptosis in part due to its ability to repress gene transcription; cells transfected with mutant huntingtin undergo apoptosis, and apoptosis is the likely cause of cell death in human Huntington's disease.

Polyglutamine-dependent interference by huntingtin with the CREB-binding protein (CBP), a cofactor for CREB-mediated gene transcription, has been proposed as an important pathogenic mechanism in HD and other polyglutamine disorders (STEFAN et al. 2000; NUCIFORA et al. 2001). In vitro assays demonstrated that transfection with the huntingtin gene containing an expanded CAG, but not with normal repeat length, inhibited CBP-mediated gene transcription. In transgenic mice with the expanded HD gene the cellular localization of CBP was abnormal and levels of CBP reduced. The in vitro toxicity of huntingtin and ataxin-1 containing expanded polyglutamine stretches was ameliorated by overexpression of CBP. Postmortem HD brain was also reported to have decreased levels of CBP and CBP was found to localize with nuclear aggregates of huntingtin (Nucifora et al. 2001). CBP contains an acetyltransferase domain and huntingtin binds to this domain more avidly when the polyglutamine repeat expands from the normal range (Steffan et al. 2001). Histone acetylation mediates gene transcription. Levels of acetylated histones are decreased in cells transfected with mutant huntingtin. Remarkably, agents which blocked deacetylation rescued neurodegeneration in a transgenic drosophila model containing the expanded HD exon 1. The authors hypothesize that treatments aimed at raising the body's global level of acetylation may be protective in Huntington's disease (Steffan et al. 2001). N-butyric acid inhibits histone deacetylation and it, and its derivatives, are nontoxic compounds under development for leukemia treatment (Santini et al. 2001).

Since demonstrated gene transcription is mediated by CBP, impaired CBP-mediated gene transcription could also account for the decreased production of cortical BDNF found in human HD and transgenic animals models. In fact, overexpression of normal huntingtin has been found to increase expression of BDNF. In contrast, mutant huntingtin with an expanded polyglutamine repeat suppressed BDNF expression in a neuron specific manner (Zuccato et al. 2001). In a mouse model of HD overexpressing mutant full-length huntingtin BDNF was found to be decreased in cortex and hippocampus and there was a 50% reduction in the cortically derived BDNF levels in the striatum. Thus depletion of this important growth factor, made in the cortex and released in the striatum, could lead to apoptosis of striatal neurons.

2. Pathologic Intracellular Inclusions of Huntingtin

Several hypotheses have been proposed about the role intracellular inclusions formed by aggregates of huntingtin and associated proteins may have in the disease process. Some have suggested that the inclusions play no direct role in the pathogenesis of the disease but occur secondarily (Sapp et al. 1999). Others have maintained that inclusions may be directly toxic (Davies et al. 1997; DiFiglia et al. 1997). With this in mind, techniques have been developed to clear huntingtin aggregates from the cell using intracellular antibodies (Lecerf et al. 2001). Protein aggregates forming intracellular inclusions resist degradation, may prevent ubiquitin recycling, and/or disrupt the proteosome (ubiquination tags proteins for proteolysis by proteosomes) (Hochstrasser 1996; Martindale et al. 1998). In transgenic mice containing the Spinocerebellar Ataxia-1 gene for Ataxin-1, it has been demonstrated that transport of the mutated protein to the nucleus, but not necessarily aggregation, is important for toxicity (Klement 1998). Similarly, transfection experiments have shown that blocking the nuclear localization of huntingtin blocks cell toxicity in cultured cells (Saudou 1988). Nuclear translocation of this cytoplasmic protein may be mediated by the presence of a nuclear localization sequence (NLS) in huntingtin (Bessert 1995).

A number of other studies have suggested that the inclusions themselves are not the pathologic agent in CAG disorders. The formation of nuclear inclusions in rat striatal neurons expressing the NH2 fragment of mutant huntingtin did not correlate with the extent of apoptotic cell death (Sapp 1999). Transient transfection of full-length mutant huntingtin into mouse clonal striatal cells has been shown to result in nuclear and cytoplasmic inclusions and apoptotic cell death. However the formation of the nuclear inclusions did not correlate with the extent of apoptotic cell death (SAPP et al. 1999). In such transfected cultured cells the addition of the caspase inhibitory Z-DEVD-FMK resulted in a reduction of inclusion formation, but had no effect on apoptotic cell death (Kim 1999). In other experiments cultured cells transfected with mutant huntingtin were cotransfected with mutant genes that block ubiquination, the process that marks proteins for degradation in the proteosome. This led to a marked decrease in the abundance of inclusions but worsened rather than ameliorated cell death (Sandou 1988). Treatment of the transfected cells with BDNF and ciliary neurotrophic factor (CNTF) increased the numbers of intranuclear inclusions but decreased cell death. Saudou et al. concluded from these experiments that the formation of intranuclear inclusions is dependent upon ubiquination but that nuclear localization of the soluble protein, and not the inclusions, is required for toxicity. Indeed, the ubiquinated inclusions could be protective if they sequester the mutated huntingtin protein and block its pathologic interactions with essential nuclear proteins.

III. Reversibility of Molecular Pathology

A major question raised by the recent advances in the understanding of the molecular basis of HD is whether the pathology might be reversible. Importantly, recent transgenic mice experiments using a tetracycline-responsive promoter for the transgene demonstrated that the aggregates of huntingtin can be degraded if production of mutated protein is halted. In these experiments, feeding doxycline to the transgenic mice turned off expression of the mutant huntingtin gene. It was then observed that brain aggregates disappeared from the nucleus and cytoplasm, gliosis partially resolved, striatal shrinkage halted, and mice experienced some ameliortion of abnormal motor signs. This is of course an intriguing finding, suggesting that the pathology and disability in humans with HD may be stopped, or even partially improved, by therapies that decrease expression of the mutant protein (Yamamoto 2000).

IV. Apoptosis

Apoptotic form of cell death is considered likely to contribute to the neurodegeneration in HD . The mere presence of the polyglutamine expansion into a nonrelated hypoxanthine phosphoribosyltransferase gene in a transgenic mouse model has been shown to be sufficient for the development of a neurological phenotype and the presence of intranuclear inclusions (Ordway

1997). In vitro experiments have shown that cells transfected with mutated HD gene constructs were more vulnerable to staurosporine, an apoptosis-inducing agent (Cooper 1998). Striatal cells have been reported to be more vulnerable to apoptotic cell death after transfection with mutant huntingtin than with hippocampal cells (Sandou et al. 1998). Remarkably similar to the findings in human HD, the met-enkephlin containing striatal cells were more susceptible to apoptotic cell death caused by the mutant huntingtin than neurons that did not stain for met-enkephalin. The degree of cell death in striatal cells transfected with mutant huntingtin was dependent upon the length of the polyglutamine stretch. Huntingtin-mediated cell death was blocked by coexpression with the antiapoptotic protein BclX$_L$ and treatment of the cells with BDNF or CNTF (Sandou et al. 1998).

The apoptotic process may also effect the biology of huntingtin protein. Huntingtin is cleaved by specific enzymes in a tissue specific manner (Mende-Mueller 2001). The pattern of normal huntingtin proteolysis was seen to differ in cortex, sriatum, and cerebellum in human brain. 45 and 43-kDa N-terminal fragments were prominent in striatum from brains from individuals dying with Huntington's disease but were absent in control human brains. Ubiquination occurred on specific fragments but not on full length huntingtin. Huntingtin, as well as other mutated proteins in CAG repeat disorders, contains sites of cleavage by caspases which may play a role in the molecular pathophysiology of the disease. (Wellington 1988). Caspase cleavage products of huntingtin have been found in human brain tissue (Kim 2001). Elevation of mature IL-1 β levels is a marker of caspase-1 activation and IL-1 β has been measured to be over twofold higher in R6/2 transgenic HD mice (Ona 1999; Goldberg et al. 1996). Caspase activation leads to reduced levels of the normal huntingtin protein, as well as reduced levels of BDNF, which may be upregulated by normal huntingtin (Zuccato 2001). Most importantly therapeutically, experiments have demonstrated that inhibiting caspase cleavage of huntingtin ameliorates cellular toxicity and reduces aggregate formation (Wellington 2000).

D. Genetic Animal Models of HD

A major breakthrough in HD research has been the development of genetic animal models. The available models currently include: "knock-out," transgenic and "knock-in" mice models (Chesselet 2001) as well as a drosophila model with progressive loss of the light-gathering rhabdomeres in the fly's compound eye (Steffan et al. 2001). Several different strains of genetically altered mice have been developed; some contain the full-length transcript, others contain truncated forms of the huntingtin gene (Mangiarini et al. 1996; Reddy et al. 1998; Schilling et al. 1999). In one transgenic model, affected mice containing the expanded CAG repeat in a truncated form show progressive weight loss despite increased caloric intake. Soon thereafter, they develop a movement disorder and seizures (MANGIARINI et al. 1996) or behavioral abnormalities, including loss of coordination, tremors, hypokinesia, and an abnormal gait before dying prematurely (SCHILLING et al. 1999). The mice have also been found to have diabetes (HURLBERT et al. 1999). Neuronal cell loss occurs at the end-stage of illness (MANGIARINI et al. 1996). Neurodegeneration with gliosis occurs in the striatum, cortex, thalamus, and hippocampus (REDDY et al. 1998). Other transgenic mice have been developed, including some that express 48 or 89 CAG repeats in a full-length HD gene driven by a CMV promoter. These mice show hind limb clasping at 2 months of age, hyperactivity at 4 months of age, and then a period of hypokinetic activity, at which time striatal degeneration is evident (REDDY et al. 1998). A YAC mouse model with full-length mutant huntingtin shows hyperkinesias at 7 months and selective neuronal degeneration in the striatum at 12 months. "Knock-in" models in which the endogenous mouse huntingtin gene is targeted for insertion of a CAG repeat expansion or chimeric exon-1 have been developed with 48-150 CAG repeats (WHEELER et al. 2000). These models, which are genetically most similar to the human disease, in general have the least severe phenotype. They demonstrate nuclear inclusions only at a late age and only one knock-in model has been shown to result in neuronal loss. However, there is decreased enkephlin mRNA in the striatum (MENALLED et al. 2000), increased sensitivity to NMDA (LEVINE et al. 1999), clear behavioral abnormalities (SHELBOURNE et al. 1999; LIN et al. 2001), and wide variation in the CAG repeat length in striatal neurons (KENNEDY et al. 2000). Genetic mouse models of Huntington's disease offer the opportunity to test pharmacological agents for their ability to alter the disease phenotype. As discussed below, slowing of disease in mouse models of HD has so far been reported using treatments that improve metabolism or inhibit caspase activity.

Several of these transgenic mice lines develop cytoplasmic and intranuclear inclusions prior to the onset of symptoms (DAVIES et al. 1997; BECHER et al. 1998). Similar inclusions have been found in human HD, but never normal brain (ROIZIN et al. 1976; DIFIGLIA et al. 1997; Ross 1997) and in the other CAG repeat diseases (PAULSON and FISCHBECK 1996; BECHER et al. 1998). The presence of nuclear inclusions in all the CAG repeat disorders raised the question of a common mechanism of cell death.

In persons with HD, inclusions appear to relatively spare NADPH-diaphorase, acetylcholinesterase, and calretinin interneurons, which are similarly spared in HD. Nuclear inclusions are primarily composed of cleaved NH2 terminal products (COOPER et al. 1998), are ubiquinated, and are more prevalent in the cortex than striatum (DIFIGLIA et al. 1995; TROTTIER et al. 1995a; BECHER et al. 1998; SAPP et al. 1999). Biochemical studies have demonstrated that the formation of amyloid-like aggregation of huntingtin is dependent on the degree of the polyglutamine expansion (SCHERZINGER et al. 1997).

Ultimately, these genetically engineered models should provide us the opportunity of understanding the processes that cause neuronal dysfunction and, subsequently, cell death in HD.

E. Potential Therapeutic Targets

The precise cascade of events leading to apoptosis in HD is an area of active investigation, but it is apparent that a complex chain of events transpire. Each step in the chain presents a potential target for therapeutic intervention (Fig. 3). The genetic mouse models offer an unprecedented opportunity to test multiple different strategies and agents. Specific molecular targets, specific

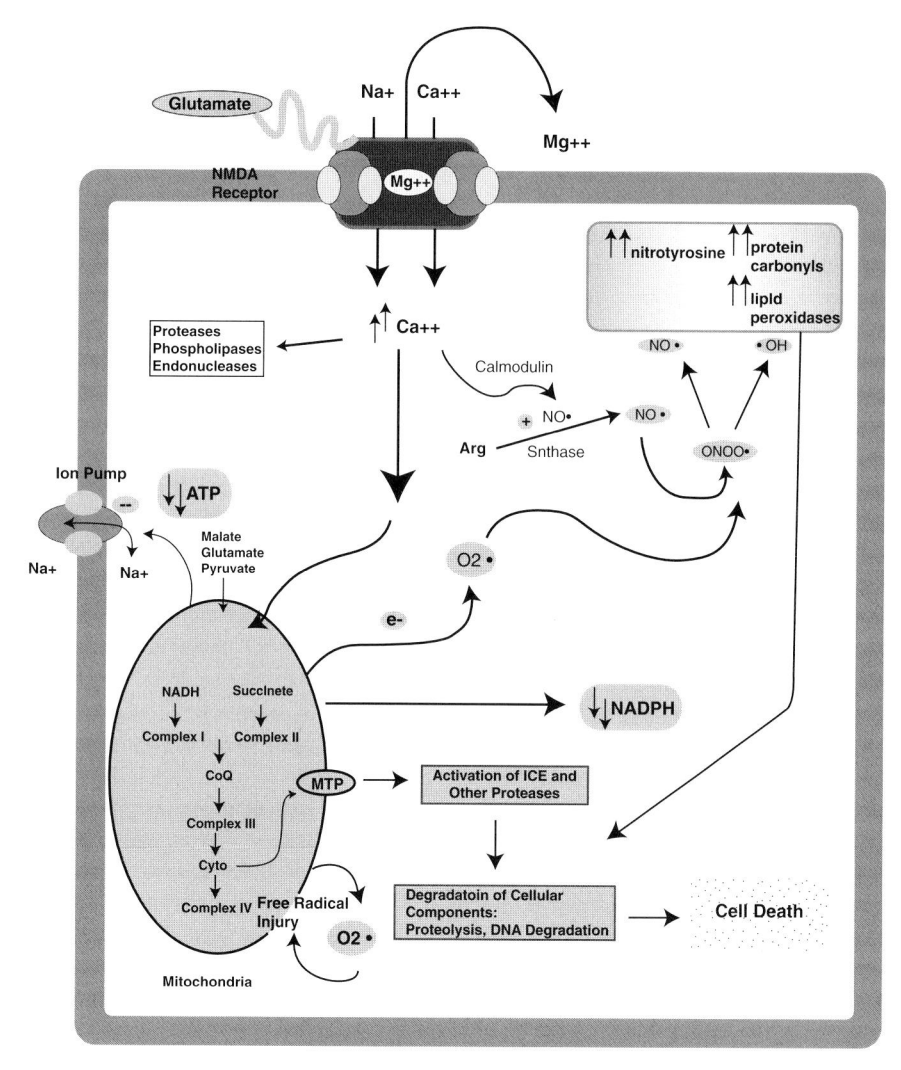

Fig. 3. A complex cascade of events is believed to take place. Excessive glutamate may lead to removal of the voltage-dependent Mg^{2+} block of the NMDA-linked calcium channel, thereby resulting in excessive intracellular levels of intracellular Ca^{2+}. This excessive Ca^{2+} can turn on a cascade of events, including free radical formation, activation of proteases, and the irreversible opening of the mitochondrial transition pore, eventually leading to cell damage and cell death

protein–protein interactions, BDNF upregulation, huntingtin altered gene transcription, decreased expression of the mutant transcript, etc. can form the basis of high throughput molecular screening for interesting compounds. Defined stages of deterioration which occur over weeks or months in the mice models of HD can also be used to test multiple compounds rapidly. Speed becomes a problem in bringing an agent to clinical trials due to the slow progression of the illness in human Huntington's disease. A robust clinical marker that changes over months to a year and that is tied to the neurodegeneration would help to determine whether a promising compound from animal studies shows effectiveness in the human condition.

I. Glutamate-Mediated Excitotoxicity in HD

Overactivity of excitatory amino acids has been suspected to play a pivotal role in the neuronal degeneration of HD based on observations that an HD-like, cell specific cell death occurs in animals given striatal injections of agents which activate glutamate receptors (COYLE and SCHWARCZ 1976). This mechanism is likely to be employed by the striatum, which receives extensive excitatory glutamatergic projections from the neocortex (GRAYBIEL 1990) and the parafasicular/centromedian complex of the thalamus (LAPPER and BOLAM 1992), and where a number of different subtypes of glutamate receptors are present. In two genetic mouse models, the transgenic R6/2 and the "knock-in" CAG94, neurons from striatal and cortical slices displayed enhanced sensitivity to NMDA receptor activation (LEVINE et al. 1999) Neurons from HD mice demonstrated increased cell swelling in response to NMDA exposure as compared to controls. This effect was specific for NMDA as there was no difference in the degree of cell swelling seen after exposure of the slices to the non-NMDA agonists, α-amino-3-hydroxy-5-methyl-4-isoxazole propionic acid (AMPA) or kainate. Resting membrane potential was also more depolarized in the HD mice. Somewhat surprisingly, HANSSON and colleagues reported the transgenic R6/1 mice were remarkably resistant to quinolinic acid-induced striatal toxicity (HANSSON et al. 1999). Quinolinic acid is a mixed glutamate agonist that when injected into mice or rat produces pathologic changes similar to those seen in human HD brain. No explanation for the reduced sensitivity to quinolinic acid was uncovered, NMDA expression was normal in the transgenic HD mice as was superoxide dismutase levels, Bcl-2, and citrate synthase as a marker of mitochondrial density. The authors speculate that upregulation of antiapoptotic defenses might account for the finding. Protection of this type is seen in animal models in which there has been a prior decortication.

The actions of glutamate are mediated via either ionotropic or metabotropic receptors (mGluRs). Ionotropic receptors are composed of multiple subunits, which form multimeric ligand-gated ion channels. Iontropic receptors (iGluR) can be subdivided by their response to specific agonists and the type of cellular response they elicit. *N*-methyl-D-aspartate (NMDA) receptors, α-amino-3-hydroxy-5-methyl-4-isoxazole-propionic acid (AMPA) receptors, and kainate receptors (KA) (BOULTER et al. 1992; MAIONE et al. 1995; KITA 1996) are ionotropic receptors that directly mediate transmembrane ion fluxes. In contrast, metabotropic receptors are made up of a single seven-transmembrane-domain protein, which is coupled to G-proteins to intracellular second messenger systems such as the inositol phosphate cascade and regulation of cyclic AMP (NAKANISHI 1992). These are classified according to sequence homology, pharmacology, and proposed signal transduction mechanisms (NAKANISHI 1992). Group 1 mGluRs, mGluR 1 and mGluR5, increase phosphatidylinositol turnover; group 2 mGluRs, mGluR2 and mGluR3, and group 3, mGluR4, -6, and -7 receptors both decrease forskolin-stimulated cyclic AMP formation (NAKANISHI 1992; TANABE et al. 1992). Both types of receptors are involved in the regulation of glutamate. Neostriatal spiny neurons mainly express mGluR5 and mGluR3.

II. NMDA Receptor

As is true of other ligand-gated ion channels, the NMDA receptor is composed of multiple protein subunits (NAKANISHI 1992), which form multimeric ligand-gated ion channels. Several families of subunits have been identified in the rat. At least 15 subunits of ionotroptic glutamate receptors have been identified: NMDAR-1, NMDAR-2A, and NMDAR-2D belong to the NMDA receptor, subtype GluR1and GluR4 belong to the AMPA receptor subtype, and GluR5–7, KA1, and KA2 belong to the KA receptor subtype (GASIC and HOLLANN 1992; SOMMER and SEEBURG 1992). NMDAR1 subunits are encoded by a single gene containing 22 exons, three of which may be alternately spliced to give rise to a total of 8 isoforms that have distinct structural and pharmacological properties (SUGIHARA et al. 1992). NMDAR2 subunits are encoded by 4 distinct genes; each appears to be present as only a single isoform. In situ hybridization and immunocytochemical studies have suggested that the subunit composition of ionotropic glutamate receptors is differentially expressed by different groups of striatal neurons (LANDWEHRMEYER et al. 1995; CHEN et al. 1996). An alternative hypothesis is that a selective striatal mitochondrial defect may link the expanded CAG repeat and abnormal energy metabolism (SCHAPIRA 1997).

NMDA receptors appear to be involved in a variety of functions within the basal ganglia. Ligand binding studies have demonstrated that NMDA binding sites are very abundant in the neostriatum and have suggested the presence of more than one pharmacological subtype of receptor (MONAGHAN 1991); moreover, the subunit composition of ionotropic glutamate receptors may result in different physiological and pharmacological properties (MONYER et al. 1994; GOTZ et al. 1997). The medium spiny neurons most vulnerable to neurodegeneration contain predominantly NMDAR1 and NMDAR-2B, which have been implicated as mediators of excitotoxic cell death in a number of pathological processes (CALABRESI et al. 1998; STANDAERT et al. 1999). In contrast, the relatively preserved aspiny interneurons are rich in GluR-1 AMPA inotropic glutamate receptor subtypes.

The putative mechanisms leading to excitotoxicity involve the inappropriate actions of excessive intracellular calcium. The NMDA receptor forms a channel permeable to both Na+-K+ and Ca^{2+} in a voltage-dependent manner (ASCHER and NOWAK 1988) and which is gated by voltage-sensitive magnesium block (MAYER and WESTBROOK 1987). In primary murine striatal cultures, NMDA-receptor agonists cause an increase in the intracellular accumulation of calcium, which can initiate a number of intracellular processes, including protein phosphorylation and activation of the phosphoinositide system. Excessive activation of NMDA receptors may lead to the massive influx of calcium; this could lead to unchecked activation of calcium-dependent enzymes, including constitutive neuronal nitric oxide synthase (NOS), which produces nitric oxide and free radicals, and the protease calpain (SIMAN et al. 1989; DAWSON et al. 1991) and could eventually result in excitotoxic cell death (CHOI and ROTHMAN 1990; COYLE and PUTTARCKEN 1993; DAWSON et al. 1993). A selective loss of NMDA receptors in the postmortem brains of patients who die in the early stages of HD suggests that neurons with high densities of NMDA receptors are particularly vulnerable to premature death (YOUNG et al. 1988), with a disproportionate depletion of NMDA receptors in HD putamen as compared to quisqualate, muscarinic, GABAergic, and benzodiazepine receptors.

Excessive glutamatergic input may also result from impairment or dysfunction of glutamate transporters, which play an essential role in terminating the excitatory glutamatergic signal at postsynaptic receptors and in protecting neurons from excitotoxic effects. Such dysfunction has been reported in motor neuron disease (COYLE and PUTTARCKEN 1993). Decreases in NR1 and NR2 mRNAs with corresponding increased number of GLT mRNA expressing astrocytes in postmortem brains of HD patients in early stages of disease have been reported (ARZBERGER et al. 1997); the upregulation of glutamate transporters may represent an early compensatory mechanism to protect neostriatal neurons from glutamate excitotoxicity.

Abnormalities in other modulatory neurotransmitter systems are also believed to play a crucial role in the pathogenesis of disease. Striatal medium spiny neurons contain high densities of mGlu5 (group I) mGluR receptors, which are positively coupled to phosphoinositide turnover and mGluR3 (group II) mGluR receptors, which are negatively coupled to adenylate cyclase. Group I mGluR subtypes have been proposed to be involved in neurotoxic cell death whereas activation of group II has been found to be neuroprotective. mGluR2 mRNA is expressed in cortical neurons projecting to the striatum and in striatal interneurons, but not in striatal projections neurons (TESTA et al. 1994). These mGluR2 receptors are normally responsible for regulating glutamate release from presynaptic corticostriatal terminals (CALABRESI et al. 1996); therefore, it is hypothesized that the loss of these presynaptic receptors might result in unregulated release of synaptic glutamate (CHA et al. 1998).

Activation of metabotropic glutamate receptors in striatal medium spiny neurons has been shown to enhance NMDA-induced membrane responses (PISANI et al. 1997). Metabotropic receptors mediate excitatory synaptic transmission, and also modulate the activity of voltage-operated Ca^{2+} channels, thereby affecting some of the intracellular events that lead to excitotoxicity (CONN and PATEL 1995). Differential expression of mGluRs in different components of the basal ganglia circuitry have been hypothesized to subserve different functions. For example, immunohistochemical analysis of rat striatal tissue has demonstrated the presence of mGluR2/3 on the terminals of corticostriatal afferents, where they might regulate glutamate release (TESTA et al. 1998). NMDA receptors are involved in the regulation of γ-butyric acid (GABA), acetylcholine, neuropeptide and glutamate release in the neostriatum.

Dopaminergic transmission also appears to influence the selective vulnerability of neurons in the striatum. Striatal medium spiny neurons which make up the direct pathway contain D1 receptors; those in the indirect pathway contain predominantly D2 receptors. D1 and especially D2 receptors are markedly reduced in presymptomatic gene-positive individuals, suggesting that loss of dopaminergic innervation contributes to the pathophysiology of disease (WEEKS et al. 1996).

III. Selective Neuronal Loss

The mechanisms underlying the cell type-specific vulnerability of striatal neurons is a topic of much debate. It has been proposed that the variable rates of cell death and the selective vulnerability may reflect the neuronal distribution of glutamate receptor subtypes and/or subunits differentially expressed by striatal neurons (CHEN et al. 1996; CALABRESI et al. 1998). Electrophysiological studies of striatal neurons have shown that both medium spiny projecting neurons and large cholinergic interneurons respond to cortical stimulation with a depolarizing synaptic potential mediated by iGlu-Rs, but that they express differential sensitivity and vulnerability to various glutamatergic agonists: spiny neurons are more sensitive to kainate, AMPA, and NMDA (CALABRESI et al. 1998).

Additional information regarding the various neurotransmitter abnormalities present and their role in excitotoxicity is being provided by transgenic mouse models. Transgenic mice, R6/2 mice, which express only the N-terminal huntingtin gene fragment of exon 1 and an expanded CAG repeat length of between 115 to 156, develop the progressive neurological phenotype, characterized by choreiform movements, involuntary stereotypic movements, tremor, and epileptic seizures (MANGIARINI et al. 1996). Certain mGluRs, such as mGluR1, mGluR2, and mGluR3 are reduced, but mGluR5 levels are unaffected; similarly, kainate receptors, dopamine, and muscarinic cholinergic receptors are reduced, but there are normal concentrations of NMDA and GABA receptors (CHA et al. 1998). mGluR2 receptors are normally responsible for regulating the release of glutamate from presynaptic corticostriatal terminals (CALABRESI et al. 1996); therefore, their loss might result in unregulated glutamate release (CHA et al. 1998). It is of interest that D1 and D2-like dopamine receptor binding was dramatically reduced in animals prior to the onset of clinical symptoms; decreased D2 receptors concentrations have been reported in presymptomatic gene-positive individuals using [^{11}C] Raclopride-positron emission tomography (ANTONINI et al. 1998).

IV. Selective Vulnerability: Toxins

Numerous glutamate analogues have been used to develop animal models for HD in an effort to elucidate the precise role glutamate-mediated excitotoxicity has in HD. Intrastriatal injections of kainate or quisqualate, both non-NMDA receptor agonists, causes loss of both spiny and NADPH aspiny neurons, but spares cells whose axons terminate in the striatum. Intrastriatal infusion of quinolinic acid, an endogenous metabolite of tryptophan, causes the selective loss of spiny interneurons, with sparing of the NADPH diaphorase aspiny neurons and a reduction in the area of the matrix compartment of the striatum compared with the patches (FERRANTE et al. 1993), the same pattern seen in HD. Moreover, intrastriatal administration of quinolinic acid also reproduces the symptoms of HD (JENKINS et al. 1996; BEAL et al. 1986; TURSKI and TURSKI 1993; BROUILLET et al. 1995; PALFI et al. 1996; SCHULZ et al. 1996; BEAL et al. 1986; ELLISON et al. 1987; QINET et al. 1992, 1996). The concentration of calbindin, a calcium-binding protein, is also increased in degenerating striatal spiny neurons in HD (FERRANTE et al. 1991) and intrastriatal administration of quinolinic acid in rats results in an increase in calbindin immunoreactivity in the soma, proximal and distal dendritic branches and spines of surviving spiny neurons (HUANG et al. 1995).

Ibotenic acid, another glutamate analogue, results in a marked astroglial reaction strikingly similar to the gliosis seen in HD (ISACSON et al. 1987). However, local injection of S-4-carboxy-3-hydroxyphenylglycine, an agonist of group II and antagonist of group I mGluR, protects striatal neurons against quinolinic-acid-induced lesions (NICOLETTI et al. 1996). Both the symptoms and lesions may also be partially blocked or reduced by competitive and non-competitive NMDA antagonists (BEAL et al. 1993; GREENE and GREENAMYRE 1995), deafferentation of cortical glutamatergic inputs (BEAL et al. 1991), free radical scavengers (SCHULZ et al. 1995; FALLON et al. 1997), or treatments aimed at improving cellular energy stores.

F. Potential Therapeutics Agents

Given the prominent role of glutamate and activation of NMDA receptors in excitotoxic neuronal death in HD, therapeutics targeted at either reducing glutamate levels or preventing activation of NMDA channels are promising agents in preventing excitotoxic cell death. More promising are drugs with selective actions on specific subtypes of NMDA receptors. Alternative approaches are to alter either the upstream mechanisms that trigger NMDA activation or the downstream effector mechanisms responsible for excitotoxicity.

I. Glutamate Release Inhibitors

Riluzole (2-amino-6-trifluoromethoxy benzothiazole) is an antiglutamatergic agent which may be useful in HD. Its precise mechanisms of action are unknown, but it may act by: (1) presynaptic inhibition of glutamate release (CHERAMY et al. 1992; MARTIN et al. 1993), (2) blockade of voltage-gated sodium channels (STUTZMANN and DOBLE 1994; STEFANI et al. 1997), or (3) an indirect neurotrophic effect (ESTEVEZ et al. 1995; MARY et al. 1995). Riluzole is neuroprotective against excitotoxic amino acid toxicity (MARY et al. 1995) and 1-methyl-4-phenyl-1,2,3,6-tetrahydropyridine (MPTP)-induced neurotoxicity (STUTZMANN and DOBLE 1994) in vivo. Riluzole also improved the motor abnormalities associated with striatal lesions produced by chronic 3-nitropropionic acid (3-NP, a mitochondrial toxin) administration in a baboon model (PALFI et al. 1996). A preliminary study done in humans also demonstrated improvement of chorea, but no effect on lactate concentrations, a marker of impaired energy metabolism (JENKINS et al. 1993; ROSAS et al. 1999). Inhibition of these receptors, however, might compromise fast excitatory synaptic transmission and complicate chronic therapy.

Lamotrigine is an antiglutamatergic agent which works presumably to prevent glutamate release (LEACH et al. 1993). It has been shown to exert neuroprotective effects against kainic acid-induced striatal lesions in vivo (McGEER and ZHU 1990). A pilot clinical trial using lamotrigine in HD patients did not show evidence of slowing disease progression (HAYDEN 1997).

II. NMDA Channel Blockers

Remacemide hydrochloride is a noncompetitive NMDA-receptor antagonist which blocks the ion-channel site of the NMDA-receptor complex (MUIR and PALMER 1991). A tolerability study of oral remacemide in HD suggested a trend for improvement of chorea and was well tolerated by subjects (KIEBURTZ et al. 1996). Oral remacemide was studied in a National Institute of Neurologic Disease and Stroke (NINDS) clinical trial called CARE HD. Over the 30 months of the study no effect on slowing of disease was noted at the doses used. There was some decrease in chorea scores in the remacemide treated group (CARE HD INVESTIGATORS 2001).

MK801 is another NMDA receptor antagonist; administration of MK801 significantly attenuated the lesions produced by intrastriatal injection of malonate (BROUILLET and BEAL 1993). It has been shown to significantly reduce lesions produced by intrastriatal injection of quinolinic acid in rats (MARY et al. 1995). It is not being developed for human study. A number of NMDA receptor antagonists have been developed, and some have been tested in trials in stroke patients. However, all require intravenous administration.

III. Metabotropic Receptor Modulators

Local injection of an agonist of group II and antagonist of group I mGlu receptors, 4C3HPG, protects striatal neurons against degeneration induced by quinolinic acid (Nicoletti et al. 1996). Group II mGluR agonist may be expected to decrease corticostriatal glutamate release. Therefore, drugs targeted at metabotropic glutamate receptors may deserve further testing as neuroprotective agents in HD.

IV. Free Radical Inhibitors

Increases in cellular calcium have been associated with activation of nitric oxide synthase (NOS). Nitric oxide is a free radical which may react with superoxide to form peroxynitrite (Beckman and Crow 1993). Peroxynitrite appears to be a critical mediator of cell death; it can mediate one or two electron oxidations and result in the oxidation of proteins, lipids, or DNA. Inhibiting the formation of peroxynitrite would be expected to be neuroprotective.

V. Nitric Oxide Synthase Inhibitors

Pretreatment of cultured cortical neurons with quisqualate, which preferentially kills NOS neurons, can block glutamate excitotoxicity in cultured striatal neurons (Dawson et al. 1993). Cultured neurons from NOS knock-out mice appear to be resistant to NMDA neurotoxicity (Dawson et al. 1996). Furthermore, mice deficient in NOS are resistant to NMDA striatal lesions (Ayata et al. 1997). The activity of aconitase, an enzyme of the Krebs acid cycle which may be particularly sensitive to ONOO-, is markedly decreased in HD striatum and cortex as well as in the brain of the R6/2 transgenic HD mouse model (Tabrizi et al. 1999).

Nitric oxide inhibitors have been shown to block excitotoxicity. Administration of selective nitric oxide synthase inhibitor 7-nitroindazole blocks MPTP toxicity in mice and baboons (Schulz et al. 1995; Hantraye et al. 1996). S-methylthoicitrulline, another selective neuronal nitric oxide synthase inhibitor, also can attenuate both malonate and MPTP-induced toxicity (Matthews et al. 1997).

Nicotinamide is a precursor of nicotinamide-adenine dinucleotide, which is an electron carrier and substrate for complex I of the electron transport chain as well as several dehydrogenases in the tricarboxylic acid cycle. It inhibits poly (ADP-ribose) synthetase, which may mediate nitric oxide neurotoxicity (Zhang et al. 1994).

Therapy with CoQ and nicotinamide protects against 1-methyl-4-phenyl-1,2,3,6-tetrahydropyridine (MPTP) toxicity; MPTP is a neurotoxin which inhibits NADH-coenzyme Q reductase (complex I) or the electron transport chain and which produces a parkinsonian syndrome in both man and experimental animals presumably by inducing striatal dopamine depletion (SCHULZ et al. 1995). Nicotinamide and CoQ were found to decrease brain lactate levels in some HD patients as measured by magnetic resonance spectroscopy (MRS) (W. KOROSHETZ, unpublished observations) (KOROSHETZ et al. 1997).

G. Energy Metabolism and Its Relationship to Neurodegeneration

Mitochondria are specialized organelles which convert energy into forms that can be used to drive cellular processes. Three distinct complexes pump protons out of the mitochondrial matrix across the mitochondrial membrane using energy provided by nicotinamide adenine dinucleotide (NADH), ubiquinone, and cytochrome C to produce an electrochemical gradient which drives the production of ATP. This process is called oxidative phosphorylation. The interruption of oxidative phosphorylation results in decreased levels of ATP. The only metabolic abnormality that has been proven in genetic mice models of HD is diabetes (HURLBERT 1999). R6/2 mice were found to have elevated glucose. Weight loss is a common early abnormality noted in the HD transgenic mice and may occur when caloric intake is still normal; feeding is impaired in the later stages of the mouse illness.

Impairments in mitochondrial energy metabolism/oxidative phosphorylation lead to enhanced vulnerability to excitotoxic injury and to the production of free radical species. Free radicals are especially injurious molecules which cause lipid, protein, and DNA peroxidation, so-called oxidative stress (COYLE and PUTTARCKEN 1993; BEAL et al. 1993b; BROUILLET et al. 1994; BEAL 1995; JENKINS et al. 1996) of cortical biopsies form juvenile, and adult onset HD cases (GOEBEL et al. 1978). There is some evidence linking oxidative stress and mitochondrial dysfunction with the pathologic process of HD. Mitochondrial abnormalities in HD were first identified in ultrastructural studies of cortical biopsies from juvenile and adult onset HD cases (GOEBEL 1978). Mitochondrial dysfunction and free radical damage have been reported to occur in the R6/2 transgenic mouse model of Huntington's disease (TABRIZI et al. 2000). As in HD brain, the R6/2 mouse brain had significant decrease in the levels of aconitase and mitochondrial complex IV in striatum and decreased complex IV in cortex. Increased immunostaining for inducible nitric oxide synthase and nitrotyrosine was also seen in transgenic mouse brain.

Glucose metabolism has been shown to be markedly reduced in the basal ganglia and cerebral cortex of symptomatic HD patients and presymptomatic gene-positive individuals (KUHL et al. 1984; MAZZIOTTA et al. 1987; KUWERT et al. 1990). MR spectroscopic studies have shown increased cortical and striatal lactate, a marker of impaired energy metabolism in HD patients (JENKINS et al. 1993) (Fig. 4). The relative increase in lactate concentration was furthermore shown to correlate with the duration of disease, implying that normal energy metabolism is progressively impaired.

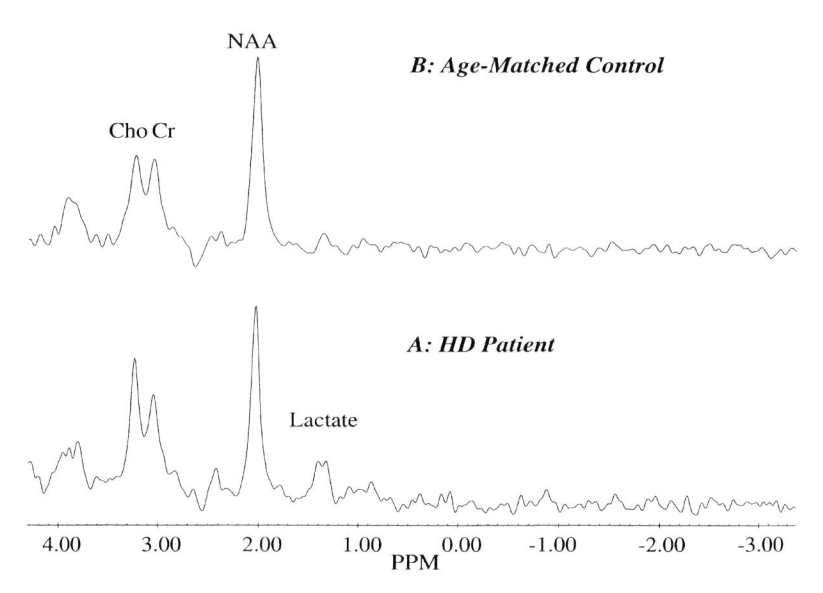

Fig. 4. Proton spectra from the basal ganglia of an HD patient and an age-matched normal control subject. Of note, lactate levels are markedly elevated, potentially an indicator of impaired metabolism. Choline levels are also elevated, as is seen in gliosis

Abnormal mitochondrial function has been reported in human HD and in animal models of the disease. Biochemical studies of postmortem brain tissue have demonstrated reduced pyruvate dehydrogenase levels in basal ganglia (SORBI et al. 1983), decreased complex II/III activity in striatum (GU et al. 1996; BROWNE 1997; TABRIZI 1999), and decreased levels of aconitase activity (key enzyme in Krebs acid cycle) in striatum and cortex (SORBI et al. 1983). Reduction of oxygen utilization and of the activity of succinate dehydrogenase (complex II), cytochrome bc1 (complex III), cytochrome c oxidase (complex IV), and pyruvate dehydrogenase have also been described in HD (BRENNAN et al. 1985; MANN et al. 1990; BEAL 1995; GU et al. 1996). One study has demonstrated that the mitochondrial abnormality in HD may be confined to the striatum (BROWNE et al.). Defects in complex II/II parallel the severity of neuronal loss in HD brains. The study also found severe deficiencies of aconitase activity, suggesting that free radical production is increased. Abnormally low levels of complex IV activity has been reported in the R6/2 transgenic mouse model of HD (TABRIZI et al. 1999).

Increased apoptosis in lymphocytes from patients with HD has been reported to occur associated with CAG length-dependent mitochondrial depolarization (SAWA et al. 1999).

These data may implicate a significant role for abnormal energy metabolism in the pathogenesis of HD. A possible link between defective energy metabolism and NMDA receptor-mediated excitotoxic cell death has been proposed by several investigators. Compounds, such as 1-methy-4-phenylpryidinium, which impair the function of the electron transport chain can produce excitotoxic damage (STOREY et al. 1992). Intrastriatal administration of iodoacetate, an inhibitor of glyceraldehyde 3-phosphate dehydrogenase, has also been shown to produce striatal lesions that can be attenuated by decortication. Decortication removes the corticostriate afferent glutaminergic input to the striatum, and is protective in many excitotoxic striatal models (MATTHEWS et al. 1997). Intrastriatal injections of malonate, a reversible inhibitor of succinate dehydrogenase, or 3-nitropropionic acid, an irreversible inhibitor, also result in the regional and selective neuronal loss seen in HD. 3-NP lesion formation in rats can be blocked by glutamate release inhibitors and by glutamate receptor antagonists, suggesting that 3-NP toxicity is mediated by secondary excitotoxic mechanisms (SCHULZ et al. 1996). Remarkably even systemic administration of 3-NP causes striatal specific neurodegeneral in animals and human (BROUILLET et al. 1995) Intrastriatal injections of quinolinic acid have recently been shown to produce mitochondrial dysfunction, as determined by a marked decline in mitochondrial oxygen consumption and decreases in ATP, NAD, aspartate, and glutamate concentration (BORDELON et al. 1997).

The actual mechanisms linking bioenergetic defects, oxidative damage, and neurodegeneration in HD remain to be elucidated, but one hypothesis is that bioenergetic defects could lead to neuronal death via "secondary excitotoxicity" (ALBIN and GREENAMYRE 1992). Inhibition of oxidative phosphorylation or Na-K+ ATPase can produce partial membrane depolarization, thereby removing voltage-dependent magnesium blockage of the NMDA-linked calcium channel, which may result in excitotoxic cell death in the presence of normal concentrations of glutamate (NOVELLI et al. 1988). The concomitant increase in intracellular Ca^{2+} may trigger free radical production. This hypothesis is supported by findings that normal levels of excitatory amino acids become toxic in the presence of inhibitors of oxidative phosphorylation or of the sodium-potassium pump (NOVELLI et al. 1988). The basal gawglia are especially vulnerable to cellular hypoxic as occurs in carbon monoxide poisoning, pure hypoxic injury, and mitochon drial diseases.

I. Mitochondrial Permeability Transition Pore

ATP is essential to maintain voltage gradients across neuronal membranes, including Na+/K+-ATPase pumps which restore the resting membrane potential after depolarization, and ATPases which regulate intracellular levels of Ca^{2+}. Impaired Na+/K+-ATPase pump activity resulting from reduced or impaired ATP production will inhibit membrane depolarization, which may result in prolonged or inappropriate opening of voltage-dependent ion channels, and if prolonged sufficiently, may remove the voltage-dependent Mg^{2+} NMDA receptor blockade. The ensuing excessive influx of calcium ions could set in motion a cascade of neurotoxic events.

Calcium influx into mitochondria is tightly regulated. Entry into the mitochondrial matrix occurs presumably via a channel transported in the mitochondrial membrane. The matrix contains several Ca^{2+} sensitive dehydrogenases that are members of the TCA cycle; changes in cytoplasmic Ca^{2+} concentrations produce long-acting activation of these enzymes. Calcium presumably leaves mitochondria by one of two routes. The first is an $Na+/Ca^{2+}$ exchange system in the inner membrane; the second is via a channel in the inner mitochondrial membrane called the mitochondrial permeability transition pore (MTP). This is a voltage-gated, cation-permeable channel, whose opening is favored by depolarization, elevated intramatrix calcium, and oxidizing agents; its closing is favored by low matrix pH and adenine nucleotides. MTP opening leads to the transient collapse of the mitochondrial proton gradient and outward movement of calcium.

One possible link between excitotoxicity and the mitochondrial permeability transition pore is that activation of glutamate receptors could trigger the opening of calcium permeable channels and calcium entry; mitochondria would pump out protons in an effort to keep pace, burning ATP and generating increased amounts of reactive oxygen species cultured. If mitochondria were unable to keep up with energetic demands, MTP opening would become irreversible, resulting in pathologically increased concentrations of intracellular Ca++ and unchecked activation of calcium dependent enzymes, including constitutive neuronal nitric oxide synthase (NOS), which produces NO and the protease calpain (SIMAN et al. 1989; DAWSON et al. 1991). NO may react with the superoxide anion to form peroxynitrite, both of which may damage cellular elements, including proteases, lipases, and endonucleases, ultimately leading to cell death. It follows then that impairments in energy metabolism and thus the generation of ATP would predispose cells to apoptotic cell death.

There are other links between mitochondria and apoptosis. Members of the bcl-2 family may exhibit proapoptotic or antiapoptotic activity. The mitochondrial membrane has been shown to be the site of interactions between CED4, an antiapoptotic member, and activator caspases (CHINNALYAN et al. ••; WU et al. 1997). Release of cytochrome c from mitochondria is a potent activator of caspases leading to cell death (GARLAND and RUDIN 1998). The mitochondrial-caspase interactions may serve to sequester activated caspases by the mitochondria, thereby preventing apoptosis (WU et al. 1997).

The evidence to date suggests that apoptosis may be an important mode of cell death in HD. HD brains show increased levels of DNA strand breaks as measured by terminal transferase-mediated deoxyuridine triphosphate nick end labeling (TUNEL) assays (DRAGUNOW et al. 1995; THOMAS et al. 1995).

H. Neuroprotective Strategies

The ability to study cell death in genetic models of HD brings a powerful new tool. Prior to the genetic advances, interesting therapeutic strategies were tested on neurotoxin models of striatal injury. It is now possible to study the molecular pathway of cell death that is caused by introduction of the mutated gene. The more widespread availability of transgenic mouse models of HD affords the opportunity to screen a wide range of compounds for their ability to prevent disease progression. Therapies effective in preventing cell death, inclusions, and neuroreceptor alterations in cell systems, fly and mouse models of HD will offer tremendous promise for clinical trials in patients. A number of agents, either available or in development, target those cell processes thought possibly important in causing the pathology of HD.

I. Antioxidants

Alpha-tocopherol, or vitamin E, blocks lipid peroxidation by trapping peroxyl radicals (HALLIWELL and GUTTERIDGE 1985). It has been reported to reduce oxygen radical damage to cell membranes, but demonstrated no effect on the neurological or neuropsychiatric symptoms of HD patients treated with 3000 IU daily over the course of 1 year (PEYSER et al. 1995). It is possible that the duration of treatment was insufficient to adequately determine efficacy, especially in light of the slow progression of the disease. Additionally, limited CNS penetration may hinder the effectiveness of vitamin E.

OPC-14117, $C_{26}H_{35}N_3O_3$ is a potent lipophilic free-radical scavenger and antioxidant which has been shown to: (1) block iron-induced lipid peroxidation in rat brains, (2) prevent release of hydrogen peroxide free radicals, and (3) be neuro-protective in an MPTP mouse model of idiopathic Parkinson's disease. (MPTP is converted to an active metabolite MPP+, by monoamine oxidase B. MPP+ is selectively taken up by the dopamine transporter and accumulates within mitochondria, where it disrupts oxidative phosphorylation (BEAL 1998). OPC-14117 was shown to be generally safe and well-tolerated in subjects with HD, but without evidence of clinical efficacy after 12 weeks of treatment (GROUP 1998).

II. Free Radical Spin Traps

Spin-trapping agents readily react with oxygen-free radicals to form more stable products (OLIVER et al. 1990). They may prevent oxygen-radical-mediated neuronal damage.

Alpha-phenyl-*t*-butyl-nitrone (PBN) and 5,5-dimethyl-1-pyrroline-*N*-oxide (DMPO) are free radical spin trap agents which can capture toxic-free radicals. PBN can reduce the volume of striatal lesions in rats induced by glutamate receptor agonists and mitochondrial toxins which appear to interfere with the downstream consequences of mitochondrial inhibition (SCHULZ et al. 1995; NAKAO et al. 1996). DMPO has been shown to protect against striatal lesions produced by systemic administration of 3NP (REF).

III. Energy Boosters

If oxidative stress which results from mitochondrial dysfunction is important in the pathogenesis of HD, agents that enhance mitochondrial function may be neuroprotective. Mitochondrial dysfunction and free radical damage have been reported to occur in the R6/2 transgenic mouse model of Huntington's disease (TABRIZI et al. 2000). Few clinical data are available on the efficacy of this strategy in preventing neurodegeneration; a number of agents have been used in animal models of HD with therapeutic promise.

Coenzyme Q10 is an essential component of the electron transport chain; it shuttles electrons from complex I and complex II to complex III (CRANE and NAVAS 1997). It is also an effective antioxidant (FRE et al. 1990). Treatment with CoQ increases the activity of the mitochondrial electron transport chain in vitro and in vivo (OHARA et al. 1981). It protects against glutamate toxicity in cultured cerebellar neurons (FAVIT et al. 1992) and has been shown to attenuate striatal lesions produced by malonate or aminooxyacetic acid (BEAL et al. 1994; BROUILLET et al. 1994). Idebenone, a benzoquinone derivative of CoQ which has better CNS penetration, can act as a free radical scavenger (MIYAMOTO and COYLE 1990).

CoQ has been used to treat patients with mitochondrial cytopathies with reduction of lactate and pyruvate levels, and improved platelet mitochondrial function (BRESOLIN et al. 1988). Idebenone administration resulted in symptomatic improvement, indicating that this drug may prove to be a potential treatment of disorders associated with mitochondrial dysfunction (IHARA et al. 1989). A pilot study in subjects with Huntington's disease demonstrated that administration of CoQ over 1–2 months significantly reduced cerebral lactate concentrations, which returned back to baseline when the drug was discontinued (KOROSHETZ et al. 1997). In CARE HD, patients were randomized to receive 600mg of Coenzyme Q10, placebo, remacemide or remacemide plus Coenzyme Q10 daily over a 30-month period. No change in progression was seen at 6 months to 1 year but thereafter a 15%–20% slowing of disease progression was seen on three HD-functional scales; the total functional capacity scale, the independence scale, and the functional checklist scale. The p values for this effect were 0.15, 0.06, and 0.05, respectively. Occipital brain lactate was measured in a subset of patients using MRS and was elevated above normal in the cohort and decreased in those patients randomized to Coenzyme Q10 alone. Coenzyme Q10 has been shown to prolong survival in transgenic HD mice (••. BEAL, personal communication).

Idebenone protects neurons against excitotoxicity induced by intrastriatal injection of kainic and quisqualate, but not quinolinic acid (MIYAMOTO and COYLE 1990). These agents are generally well-tolerated and have low intrinsic toxicity; as such, they are promising neuroprotective agents which may prove to be effective in slowing disease progression in HD.

Agents which improve energy metabolism, such as creatine or cyclocreatine may compensate for an energetic defect and reduce vulnerability to excitotoxic-mediated cell damage. Creatine and cyclocreatine are both substrates for creatine kinase and may increase phosphocreatine or phosphocyclocreatine levels and ATP generation (HOLTZMAN et al. 1997). Phosphocreatine shuttles energy from mitochondria to the cytoplasm, where it regenerates ATP. It may then be used to subserve a variety of functions, including maintenance of the membrane potential by the Na+/K+-ATPase pump and for Ca^{2+} buffering by Ca^{2+}-ATPase (HEMMER and WALLIMANN 1993).

Both creatine and cyclocreatine are phosphorylated by mitochondrial creatine kinase. Cyclocreatine in the diet resulted in an accumulation of cyclocreatine in the brain, with sparing of ATP levels produced in a mouse model of ischemia (WOZNICKI and WALKER 1998). In another study, creatine supplementation increased levels of creatine phosphate and prevented anoxic damage to hippocampal slices (CARTER et al. 1995). Both creatine and cyclocreatine have been shown to offer protective effects against striatal toxicity associated with the mitochondrial toxins malonate and 3NP in rats. Creatine also protects against 3-NP-induced ATP depletion in rats and prevents the rise in markers of oxidative stress associated with 3-NP treatment (MATTHEWS et al. 1998). R6/2 transgenic HD mice were found to have prolonged survival, decreased rate of brain atrophy, body weight loss, and striatal neuronal loss when supplemented with creatine in their diet. Performance on a motor task was also seen in the creatine treated mice. MRS identified increased brain creatine and increased *N*-acetlyaspartate, a neuronal marker, in the creatine treated animals (FERRANTE et al. 2000). A separate group has also reported that creatine supplementation improved survival and delayed motor symptoms in another transgenic (N171-82Q) mouse model (ANDERASSEN et al. 2001). A trial of creatine in persons with HD is currently underway.

Dichloroacetate has also been reported to slow down disease progression in the R6/2 and N171-82Q transgenic HD mice models (ANDERASSEN et al. 2001). Dichloroacetate stimulated activity of pyruvate dehydrogenase complex, deficits in which can produce mitochondrial cytopathy and which has been reported to be reduced in HD (SORBI et al. 1983).

IV. Targets Against the Mitochondrial Permeability Transition Pore

Drugs that bind to the translocator of the MTP complex, such as atractyloside and bongkrekic acid, are powerful regulators of MTP function. Opening of the mitochondrial permeability transition pore is inhibited by low levels of cyclosporin A (NOVGORODOV et al. 1990).

V. Neurotrophic Factors

Although the role of trophic factors in the pathogenesis of HD is unclear, their potential role as restorative or neuroprotective agents remains an active and exciting area of investigation. This is more appealing given recent evidence that overexpression of wild-type huntingtin in mice leads to increased production of BDNF and that mice with the full length mutant HD gene have deficient production of BDNF (ZUCCATO et al. 2001). Withdrawal of growth factors is a potent stimulus of apoptotic cell death, for which there is some evidence in HD (see above). Under a variety of different scenarios, the administration of trophic factors may protect vulnerable striatal neurons. Growth factors are known to inhibit neuronal death due to excitotoxic insults (MATTSON et al. 1993). Grafting of trophic-factor-secreting cells has proven to be of benefit in several experimental models of HD.

Ciliary neurotrophic factor is a member of the helical cytokine superfamily, which can prevent striatal neuronal loss and protect against cognitive and behavioral deficits in a rodent model of HD (ANDERSON et al. 1996). Intrastriatal implantation of fibroblasts genetically engineered to produce human CNTF in a nonhuman primate model of HD demonstrated a neuroprotective effect on striatal neurons, including GABAergic, cholinergic, and diaphorase-positive neurons. Additionally, it prevented retrograde atrophy of layer V cerebral neurons and exerted a protective effect on the globus pallidal and nigral neurons, and striatal output neurons, suggesting protection not only of vulnerable striatal neurons, but also of neurons involved in cortical-striatal circuits (EMERICH et al. 1997).

NT-4/5 is a potent trophic agent for striatal GABAergic, calbindin, and calretinin neurons and protects neurons against glutamate toxicity in vitro (ARDELT et al. 1994; CHENG and MATTSON 1994; WIDMER and HEFTI 1994). It has been shown to partially protect against the degeneration of a subset of GABAergic neurons in an animal model of HD (ALEXI et al. 1997).

Basic fibroblast growth factor, bFGF, is protective against Ca^{2+} toxicity, NO toxicity, and free radical production (MATTSON et al. 1993). Getting FGF to the brain in patients with an intact blood-brain barrier, however, would likely require intracisternal administration.

Alteration in BDNF production has been proposed as a potential molecular mechanism linked to the Huntington's disease mutation (see Sect. C.). BDNF is made in the cortex and released by cortical axons into the striatum. Pretreatment of cultured rat cortical neurons with BDNF pretreatment has been shown to exert a partially protective effect against glutamate-induced neurotoxicity (SHIMOHAMA et al. 1993). BDNF elevates glutathione disulfide reductase and protects mesencephalic cultures from 6-hydroxydopamine-induced toxicity; it also increases the activity of superoxide dismutase and glutathione disulfide reductase in hippocampal neurons (MATTSON et al. 1995). BDNF was studied in a clinical trial in amyotrophic lateral sclerosis. Because of inability to access brain by other routes, it is administered via a subcutaneous-placed device which pumps drug into the CSF. Given the evidence for decreased BDNF production due to the HD mutation, along with its powerful trophic role in the striatum, a trial of BDNF in HD seems warranted.

Nerve growth factor has been shown to be protective against glutamate-induced neurotoxicity in cultured cortical neurons (SHIMOHAMA et al. 1993); it has also been shown to elevate glutathione levels in PC12 cells (JACKSON et al.

1994). In a human neuroblastoma cell line, it upregulates catalase and protects from direct addition of H_2O_2 (JACKSON et al. 1994). Grafts of nerve-growth-factor-secreting fibroblasts prevent degeneration of striatal neurons destined to die from excitotoxic insult or mitochondrial dysfunction (SCHUMACHER et al. 1991; FRIM et al. 1993). Intrastriatal implants of immortalized central nervous system progenitor cells genetically engineered to secrete NGF can also prevent neuronal degeneration in a rodent model of HD (MARTINEZ-SERRANO and BJORKLUND 1996). Implants of epidermal growth factor (EGF)-responsive stem cells, which upon differentiation into astrocytes secrete NGF, prevent neuronal degeneration in a rodent model of HD (KORDOWER et al. 1997).

I. Future Potential Therapeutics

I. Caspase Inhibitors

Caspases are a family of cysteine proteases which mediate proteolysis of apoptotic proteins, leading to the irreversible commitment of cells to undergo apoptotic death (MIURA et al. 1993; ALNEMRI et al. 1996). Interleukin-converting enzyme (ICE) and its homologues belong to this family. ICE normally catalyzes the proteolysis of inactive precursors of interleukin-1 and it is homologous to the product of the Ced-3 gene of the nematode *Caenorhabditis elegans* (MIURA et al. 1993), which is essential for programmed cell death in this organism. Overexpression of ICE or one of its homologues in mammalian cells results in apoptosis; therefore, they have been suggested to play a role in neurodegeneration. However, there is no effect on apoptosis in ICE-knockout mice (LI et al. 1995), suggesting that there are other crucial components in the protease cascade that leads to apoptosis.

The active site is conserved throughout the family of ICE homologues (FERNANDES-ALNEMRI et al. 1994); each is expressed as an active proenzyme awaiting a specific heretofore unknown signal for activation. These enzymes remain a prime target for agents that suppress or inactivate them, thereby inhibiting apoptosis.

Several classes of reversible and irreversible inactivators have been designed against ICE and its homologues. Irreversible inactivators include halomethyl ketones, which alkylate the active site cysteine residues, but may be toxic due to nonselective reactions with serine- and thiol-containing enzymes (SHAW 1990); diazomethyl ketones, which are more selective for cysteine rather than serine proteases (SHAW 1990); and acyloxymethyl ketones, which are more selective for ICE versus other cysteine proteases. Reversible inhibitors include YVAD-H (Ac-Tyr-Val-Ala-Asp-aldehyde), which is one of the most selective ICE inhibitors developed at the present time (FERNANDES-ALNEMRI et al. 1995). Cotransfection of the potent ICE inhibitor crmA (KOMIYAMA et al. 1994) with ICE or Ced-3 inhibits apoptosis (MIURA et al. 1993). In vivo studies have been limited to animal models of sepsis, inflammation and arthritis, with only limited success thus far. The majority of these compounds are too unselective or metabolically unstable for human use.

Caspase 3 appears to be directly responsible for the proteolytic cleavage of key nuclear proteins, including those involved in genome surveillance and DNA repair (WELLINGTON et al. 1997), as well as those which could induce survival pathways, such as the extracellular signal-regulated kinase and phosphatidylinositol-3 kinase/Akt pathways (WIDMANN et al. 1998). Caspase cleavage appears to be required for cytotoxicity in Kennedy's disease, another CAG triplet repeat disorder (ELLERBY et al. 1999). Huntingtin is a substrate for caspase-3; both normal and mutant huntingtins are cleaved near the NH2-terminus (GOLDBERG et al. 1996; WELLINGTON et al. 1998). Cells transfected with huntingtin truncated at the predicted caspase-3 cleavage site develop perinuclear aggregates and are more susceptible to apoptosis (MARTINDALE et al. 1998). This suggests a role of caspase cleavage of huntingtin in the formation of aggregates. Conversely, caspase inhibitors have been shown to increase cell survival in striatal cells transiently transfected with mutant huntingtin cDNAs (SAPP et al. 1999); however, this protective effect occurred without affecting inclusion formation. Caspase inhibition by intraventricular injection of zVAD-fmk or coexpression of a dominant-negative caspase-1 mutant, has been shown to prolong survival in mouse model of HD (ONA et al. 1999). Minocycline is a tetracycline derivative with ability to inhibit upregulation of caspase-1, caspase 3 and inducible nitric oxide. Minocycline treatment of R6/2 transgenic HD mice has been reported to delay disease progression in the mice (CHEN et al. 2000) A pilot clinical trial of minocycline in individuals with HD is currently underway.

Exciting recent work with a new transgenic mouse, a crossbreed between R6/2 mice and transgenic mice expressing a dominant negative mutant of caspase-1 in the brain, which can act as an inhibitor of caspase-1 pathways, showed delay of motor weakness, nuclear inclusion formation, neurotransmitter receptor alterations, and mortality (ONA et al. 1999). Additionally, markedly elevated levels of IL-1β were found in moderate grade human HD brains; however, the precise association between IL-1 and neuronal cell death in HD remains to be elucidated.

II. Bcl-2

Proteins of the bcl-2 family are potent inhibitors of both apoptotic and necrotic cell death. Bcl-2 is a protooncogene initially characterized by its ability to prevent both apoptosis. It is widely expressed in the nervous system and localized to the outer mitochondrial membrane, endoplasmic reticulum and nuclear membrane. Expression of bcl-2 has been shown to inhibit the opening of the MTP under stressful conditions. This results in the stabilization of the mitochondrial membrane potential, Ca^{2+} homeostasis, and reactive oxygen species production and inhibits the release of cytochrome c (REED 1997). Bcl-2 has been shown to prevent death induced by oxidative damage induced by depletion of glutathione (KANE et al. 1993) and can prevent against H_2O_2-induced damage in a dose-dependent manner (HOCKENBERY et al. 1993).

III. Gene Transcription and Histone Deacetylation Inhibitors

Recent evidence discussed above suggests that huntingtin with an expanded repeat can be cleaved and fragments enter the nucleus where they cause disordered gene transcription. It has been proposed that the ability of the mutant huntingtin to interact with the acetyltransferase sites of transcription factors, including CBP, leads to a decrease histone acetylation as the underlying mechanisms of huntingtin's nuclear effect (STEFFAN et al. 2001). Treatments which enhance histone acetylation, such as n butyric acid or related compounds have been suggested as promising compounds in therapeutic trials (SANTINI et al. 2001).

IV. Neural Transplantation

Transplantation has been used successfully and effectively in the treatment of Parkinson's disease, another neurodegenerative disease. Fetal striatal transplantation may be as safe, feasible, and potentially effective in the treatment of HD. Intrastriatal transplantation of cross-species fetal striatal cells into the lesioned striatum of primates reduced abnormal movements (HANTRAYE et al. 1992). Whether transplantation stimulates a beneficial reaction in the striatum, such as growth factor synthesis, or leads to functionally appropriate synaptic connections is not known.

A recent report of transplantation of striatal allografts in the primate neostriatum, demonstrated recovery in a test of skilled motor performance (KENDALL et al. 1998). Graft survival in a pilot study of fetal striatal tissue into HD subjects demonstrated survival of the graft, without adverse effects (KOPYOV et al. 1998). Whether or not transplantation demonstrates therapeutic efficacy by either slowing disease progression or reversing the motor or cognitive symptoms in humans undergoing transplantation remains to be determined. A small trial of transplantation of porcine fetal cells into the striatum in HD was conducted without apparent benefit.

J. Challenges for Clinical Research

In spite of the many advances in our understanding of the basic pathophysiologic mechanisms of HD and the development of animal models, numerous challenges are still faced by the clinical scientist. Issues regarding accurate assessment of the precise time of onset, which, if any, biological markers are available to monitor progression of disease, and the precise correlation between disease onset, disease progression, and modifying factors, are only a few of the challenges. These are important insofar as determination of an appropriate endpoint for clinical trials is vital to assessing efficacy of future therapies.

At present, we have validated functional scales, such as independence scores and total functional capacity, which provide information on an individual's ability to work, handle finances, and care for activities of daily living, or on the level of independence in daily activities. Both of these are measures are contained within the Unified Huntington's Disease Rating Scale (UHDRS), a rating scale designed to quantitate neurologic, psychiatric, cognitive, and functional deficits in individuals with HD (HUNTINGTON STUDY GROUP 1996). Thus far, the functional, chorea, and dystonia scores have been the most widely used in several clinical trials of symptomatic individuals. On the functional scale used in the UHDRS, patient groups generally decline by just less than 1 point per year (FEIGIN et al. 1995). Improved outcome on the functional scale was the primary endpoint in the NINDS-supported CARE HD study. This study is powered at the 80% confidence level to detect a 40% decrease decline as measured by the total functional capacity (TFC) score. The CoQ effect was only 15–20% slowing of disease progression and thus did not achieve statistical significance on the primary endpoint. The power calculations based on expected rate of decline on the TFC called for randomizing 340 patients and following them for 30 months. A variety of secondary efficacy measures unique to HD are also included in the CARE HD study. In CARE HD a 2 × 2 randomization is performed to test two drugs and their potential interaction. One-quarter of the patients receive placebo, one-quarter Coenzyme Q10, one-quarter remacemide. and one-quarter receive both Coenzyme Q10 and remacemide. Such a design may detect beneficial effect of either drug, or beneficial effect of a drug interaction, but it suffers statistically if there is a negative drug interaction. As CAG number may have some influence on disease progression, the analysis of the endpoint in a clinical trial may benefit from incorporation of this information on the randomized patients.

Various neuroimaging techniques look promising in affording direct brain measures as targets for new therapies. PET studies using labeled glucose have demonstrated hypometabolism in the striatum. More recent studies using radionucleated transmitter receptor agonist binding have demonstrated abnormal reductions of D1 and D2 receptors in the striatum of HD subjects; [^{11}C] raclopride binding has been shown to be reduced in presymptomatic gene-positive individuals (ANTONINI et al. 1998). Magnetic resonance spectroscopy (MRS) has been used to demonstrate evidence of an energetic defect in HD as quantified by abnormally elevated lactate levels, a marker of impaired energy metabolism (JENKINS, KOROSHETZ et al. 1993). MRS has also been used to screen for therapeutic interventions which may improve energy dynamics as manifested by partial normalization of brain lactate.

As Huntington's disease is characterized by major brain atrophy, methods to accurately measure brain volumes in HD patients over time may provide attractive clinical endpoints (Fig. 5). Striatal volumes have been studied using

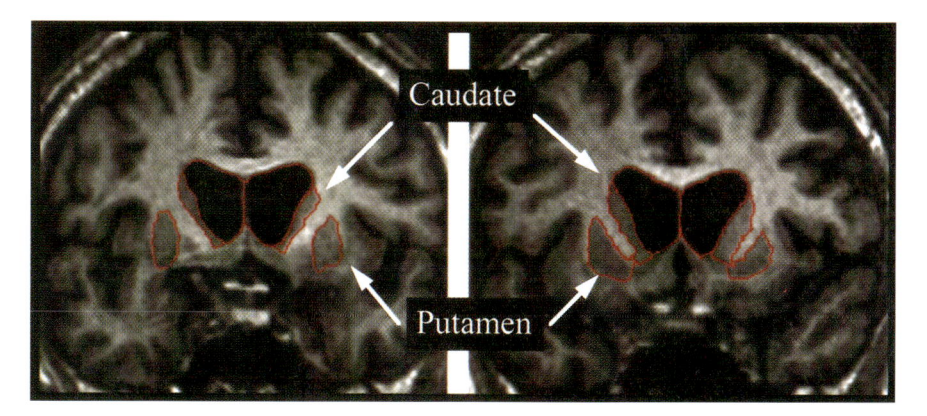

Fig. 5. Morphometry of the basal ganglia of an HD patient

MR imaging; reduced basal ganglia volumes have been reported even in presymptomatic HD subjects. A longitudinal MR study demonstrated progressive basal ganglia atrophy in HD (AYLWARD et al. 1997), amounting to a tissue loss of $0.37\,cm^3$ in the caudate in 21 months. A therapeutic strategy found to retard this progressive brain volume loss in HD patients, even in a small sample size, would be extremely interesting as the focus of a clinical efficacy trial. Newer MR imaging techniques at higher magnetic field strength hold the promise of even greater structural and neurochemical resolution. Imaging endpoints are especially attractive for phase II screening in clinical trials of compounds with potential for biologic efficacy. Our personal preference is to combine MRS measurements of key neurochemicals with fine morphometric measurements. Chemical shift MRS combined with accurate volume measurements should enable monitoring of neurodegeneration as defined by decline in total caudate levels of N-acetylaspartate (a neuronal marker) and increase in total caudate choline (a marker of gliosis).

A small change in the rate of progression of HD may seem insignificant to those already disabled. Even the subtle cognitive, psychiatric, or behavioral symptoms that may be present up to several years before motor symptoms develop and the definitive diagnosis is made can be disabling. The value of showing any ability to slow progression of HD comes if treatment applied to presymptomatic persons substantially delays disease onset. The finding of families in which HD does not manifest until the patient is elderly is evidence for some genetic effect that makes the HD mutation relatively benign. The ability to perform accurate genetic testing for the illness enables identification of those individuals who might benefit from early initiation of life-long therapy. This should also influence the economic decisions of pharmaceutical firms to develop neuroprotective drugs for HD. There are currently 25 000 persons clinically affected by HD in the U.S.; however, another 75 000 asymptomatic individuals are thought to have inherited the HD mutation (CONNEALLY 1984).

Therefore, the market for a drug taken by individuals with the HD mutation over their entire lifetime is considerable.

Delay of onset age has been proposed as an endpoint for clinical trials. Based on CAG repeat number and parent's onset age and sex, algorithms now exist which crudely predict age of onset. A therapeutic strategy could be tested for its ability to delay predicted onset age. Once a therapy has been proven beneficial in symptomatic patients it may be unrealistic to expect presymptomatic gene positive individuals to refrain from taking the active drug. Therefore, randomization between placebo and drug may be difficult. This strategy may be more valuable in testing the value of adding second-level therapy or testing one therapy against another. It has so far seemed more appropriate to test drugs in symptomatic individuals as this does not require the emotional distress of presymptomatic testing, and efficacy trials can occur over shorter periods of time than that needed for trials of delay in onset. Two studies, PREDICT and PHAROS, are now being carried out to better understand the course of neurologic, cognitive, and neuroimaging changes in patients who are asymptomatic but gene positive.

K. Conclusions

The mechanisms underlying cell death in HD are still being elucidated. Although much progress has been made since the discovery of the genetic defect and the development of transgenic mice, much remains to be done. There is substantial evidence that excitotoxicity and apoptosis may play a major role in neuronal death in HD. Elucidation of the precise mechanisms underlying the apparent selective neurodegeneration, new information from various animal models and transgenic animal studies, and improvements in the methodology for studying neuroprotection in patients will undoubtedly demonstrate the important role for neuroprotective strategies in the treatment of HD. Clinical trial design in HD benefits from the ability to genetically diagnose the illness and stratify patients according to CAG repeat number, as well as the availability of validated clinical scales and newer imaging techniques.

References

Albin RL, Greenamyre JT (1992) Alternative excitotoxic hypotheses. Neurology 42(4):733–738

Albin RL, Young AB et al. (1989) The functional anatomy of basal ganglia disorders. Trends Neurosci 12:366–375

Alexi T, Venero JL et al. (1997) Protective effects of neurotrophin-4/5 and transforming growth factor-alpha on striatal neuronal phenotypic degeneration after excitotoxic lesioning with quinolinic acid. Neurosci 78:73–86

Alnemri ES, Livingston DJ et al. (1996) Cell 87:171

Anderassen OA et al. (2001) Creatine increases survival and delays motor symptoms in a transgenic animal model of Huntington's disease. Neurobiol Dis 8:479–491

Anderson KD, Panayotatos N et al. (1996) Ciliary neurotrophic factor protects striatal output neurons in an animal model of Huntington's disease. Proc Natl Acad Sci 93:7346–7351

Antonini A, Leenders KL et al. (1998) [11 C] Raclopride-PET studies of the Huntington's disease rate of progression: relevance of the trinucleotide repeat length. Ann Neurol 43:254–255

Ardelt AA, Flaris NA et al. (1994) Neurotrophin-4 selectively promotes survival of striatal neurons in organotypic slice culture. Brain Res 647:330–344

Aronin N, Chase K et al. (1995) CAG expansion affects the expression of mutant huntingtin in the Huntington's disease brain. Neuron 15:1193–1201

Arzberger T, Krampftl K et al. (1997) Changes of NMDA receptor subunit (NR1, NR2B) and glutamate transporter (GLT1) mRNA expression in Huntington's disease–an in situ hybridization study. J Neuropathol Exp Neurol 56:440–454

Ascher P, Nowak L (1988) J Physiol 399:227–245

Ayata C, Ayata G et al. (1997) Mechanisms of reduced striatal NMDA excitotoxicity in type I nitric oxide synthase knock-out mice. J Neurosci 15:8419–8429

Aylward E, Li Q et al. (1997) Longitudinal change in basal ganglia volume in patients with Huntington's disease. Neurology 48:394–399

Beal MF (1995) Aging, energy and oxidative stress in neurodegenerative diseases. Ann Neurol 38(3):357–366

Beal MF (1998) Excitotoxicity and nitric oxide in Parkinson's disease pathogenesis. Ann Neurol 44 (Suppl 1):S110–S114

Beal MF, Brouillet E et al. (1993) Age-dependent striatal excitotoxic lesions produced by the endogenous mitochondrial inhibitor malonate. J Neurochem 61:1147–1150

Beal MF, Henshaw R et al. (1994) Coenzyme Q10 and nictinamide block striatal lesions produced by the mitochondrial toxin malonate. Ann Neurol 1994

Beal MF, Hyman BT et al. (1993b) Do defects in mitochondrial energy metabolism underlie the pathology of neurodegenerative diseases? Trends in Neurosciences 16(4):125–131

Beal MF, Kowall NW et al. (1986) Replication of the neurochemical characteristics of Huntington's disease by quinolinicacid. Nature 321(6066):168–171

Beal MF, Swartz KJ et al. (1991) Aminooxyacetic acid results in excitotoxic lesions by a novel indirect mechanism. J Neurochem 57(3):1068–1073

Becher MW, Kotzuk JA et al. (1998) Intranuclear neuronal inclusions in Huntington's disease and dentatorubral and pallidulysian atrophy: Correlation between the density of inclusions and IT15 CAG triplet repeat length. Neurobiol Dis 4:387–397

Beckman JS, Crow JP (1993) Pathological implications of nitric oxide superoxide and peroxynitrite formation. Biochem Soc Trans 3:330–334

Bessert D, Gutridge K (1995) The identification of a functional nuclear localization signal in the Huntington Disease protein. Bran Res Mol Brain Res 33:165–172

Bordelon YM, Chesselet MF et al. (1997) Energetic Dysfunction in Quinolinic Acid-Lesioned Rat Striatum. Journal of Neurochemistry 69:1629–1639

Boulter J, Bettler B et al. (1992) Molecular biology of the glutamate receptors. Clinical Neuropharmacology 15:60A–61A

Brennan W, Bird E et al. (1985) Regional mitochondrial respiratory activity in Huntington's disease brain. J Neurochem 44:1948–1950

Boutell JM, Thomas JP et al. (1999) Aberrant interactions of transcriptional repressor proteins with the Huntington's disease gene product, huntingtin. Human Molecular Genetics 8(9):1647–1655

Bresolin N, Bet L et al. (1988) Clinical and biochemical correlations in mitochondrial myopathies treated with coenzyme Q10. Neurology 38:892–899

Brouillet E, Hantraye P et al. (1995) Chronic mitochondrial energy impairment produces selective striatal degeneration and abnormal choreiform movements in primates. Proc Natl Acad Sci USA 92(15):7105–7109

Brouillet E, Henshaw DR et al. (1994) Aminooxyacetic acid striatal lesions attenuated by 1,3-butanediol and coenzymeQ 10. Neurosci Lett 177:58

Brouillet E, Hyman BT et al. (1994) Systemic or local administration of azide produces striatal lesions by an energy impairment-induced excitotoxic mechanism. Exp Neurol 129(2):175–182

Browne S, Bowling A et al. (1997) Oxidative damage and metabolic dysfunction in Huntington's disease: Selective vulnerability of the basal ganglia. Ann Neurol 41: 646–653

Burke JR, Enghild JJ et al. (1996) Huntingtin and DRPLA proteins selectively interact with the enzyme GADPH. Nature Med 2:347–350

Calabresi P, Diego C et al. (1998) Striatal Spiny Neurons and Cholinergic Interneurons Express Differential Ionotropic Glutamatergic Responses and Vulnerability: Implications for Ischemia and Huntington's Disease. Ann Neurol 43:586–597

Calabresi P, Pisani A et al. (1996) The corticostriatal projection: from synaptic plasticity to dysfunction of the basal ganglia. Trends in Neuroscience 19(1):19–24

Carter AJ, Muller RE et al. (1995) Preincubation with creatine enhances levels of creatine phosphate and prevents anoxic damage in rat hippocampal slices. J Neurochem 64:2691–2699

Cha JH, Kosinski CM et al. (1998) Altered brain neurotransmitter receptors in transgenic mice expressing a portion of an abnormal human huntingtin disease gene. Proceedings of the National Academy of Sciences of the United States of America 95(11):6480–6485

Chen Q, Veenman CL et al. (1996) Cellular Expression of ionotropic glutamate receptor subunits on specific striatal neuron types and its implication for striatal vulnerability in glutamate receptor-mediated excitotoxicity. Neuroscience 73:715–731

Chen M, Ona VO et al. (2000) Minocyline inhibits caspase-1 and caspase-3 expression and delays mortality in a transgenic mouse model of Huntington disease. Nature Medicine 6:797–801

Cheng B, Mattson MP (1994) NT-3 and BDNF protect CNS neurons against metabolic/excitotoxic insults. Brain Res 640:56–67

Cheramy AL, Barbeito L et al. (1992) Riluzole inhibits the release of glutamate in the caudate nucleus of the cat in vivo. Neurosci Lett 147(2):209–212

Chesselet FM (2001) Mouse models of Huntington's Disease. Trends in Pharmacological Sciences 23:32–39

Chinnalyan AM, O'Rouke K et al. (1997) Interaction of CED-4 with CED-3 and CED-9: A molecular framework for cell death. Science 275:1122–1126

Choi DW, Rothman SM (1990) The role of glutamate neurotoxicity in hypoxic-ischemic neuronal death. Annual Review of Neuroscience 13:171–182

Conn PJ, Patel J (1995) The Metabotropic Glutamate Receptors: Regulation of neuronal circuits and behavior by metabotrophic glutamate receptors. Humana Press 195–229

Conneally P (1984) Huntington's Disease: genetics and epidemiology. Am J Hum Genet 36:506–526

Cooper AJL, Sheu K-FR et al. (1997) Transglutaminase-catalyzed inactivation of glyceraldehype-3-phosphate dehydrogenase and a-ketoglutarate dehydrogenase complex by polyglutamine domains of pathological length. Proc Natl Acad Sci USA 94:12604–12609

Cooper JK, Schilling G et al. (1998) Truncated N-terminal fragments of huntingtin and expanded glutamine repeats form nuclear and cytoplasmic aggregates in cell culture. Human Molecular Genetics 7(5):783–790

Coyle JT, Puttarcken P (1993) Oxidative stress, glutamate and neurodegenerative disorders. Science 262(5134):689–695

Coyle JT, Schwarcz R (1976) Lesion of striatal neurones with kainic acid provides a model for Huntington's chorea. Nature 263:244–246

Crane FL, Navas P (1997) The diversity of coenzyme Q function. Mol Aspects Med 18 (suppl):S1

Cudkowicz M, Kowall NW (1990) Degeneration of pyramidal projection neurons in Huntington's disease cortex. Ann Neurol 27:200–204

Davies SW, Turmaine M et al. (1997) Formation of neuronal Intranuclear Inclusions underlies the neurological dysfunction in mice transgenic for the HD mutation. Cell 90:537–548

Dawson VL, Dawson TM et al. (1993) Mechanisms of nitric oxide-mediated neurotoxicity in primary brain cultures. Journal of Neuroscience 16:2397–2661

Dawson VL, Dawson TM et al. (1991) Nitric oxide mediates glutamate neurotoxicity in primary cortical neurons. Proc Natl Acad Sci USA 88:6368–6371

Dawson VL, Kizushi VM et al. (1996) Resistance to neurotoxicity in cortical cultures from neuronal nitric oxide synthase-deficient mice. J Neurosci 13:2651–2661

DiFiglia M, Sapp E et al. (1995) Huntingtin is a cytoplasmic protein associated with vesicles in human and rat brain neurons. Neuron 14:1075–1081

DiFiglia M, Sapp E et al. (1997) Aggregation of huntingtin in neuronal intranuclear inclusions an dystrophic neurites in brain. Science 277:1990–1993

Dragatsis I, Efstratiadis A et al. (1998) Mouse mutant embryos lacking huntingtin are rescued from lethality by wild-type extraembryonic tissues. Development 125: 1529–1539

Dragunow M, Fauli RLM et al. (1995) In situ evidence for DNA fragmentation in Huntington's disease striatum and Alzheimer's disease temporal lobes. Clin Neurosci Neuropath 6:1053–1057

Duyao MP, Auerbach AB et al. (1995) Inactivation of the mouse Huntington's disease gene homolog Hdh. Science 269(5222):407–410

Dure LS, Young AB et al. (1991) Excitatory amino acid binding sites in the caudate nucleus and frontal cortex of Huntington's disease. Ann Neurol 30:785–793

Ellerby LM, Hackam AS et al. (1999) Kennedy's Disease: Caspase cleavage of the androgen receptor is a crucial even in cytotoxicity. J Neurochem 72:185–195

Ellison DW, Beal MF et al. (1987) Amino acid neurotransmitter abnormalities in Huntington's disease and in the quinolinic acid model of Huntington's disease. Brain 110(Pt 6):1657–1673

Emerich DF et al. (1996) Implants of encapsulated human CNTF-producing fibroblasts prevent behavioral deficits and striatal degeneration in a rodent model of Huntington's disease. J Neurosci 16:5168–5181

Emerich DF, Winn SR et al. (1997) Protective effect of encapsulated cells producing neurotrophic factor CNTF in a monkey model of Huntington's disease. Nature 386:395–399

Estevez AG, Stutzmann JM et al. (1995) Protective effect of riluzole on excitatory amino acid-mediated neurotoxicity in motoneuron-enriched cultures. European Journal of Pharmacology 280(1):47–53

Fallon J, Matthews RT et al. (1997) MPP+ produces progressive neuronal degeneration which is mediated by oxidative stress. Experimental Neurology 144(1):193–198

Favit A, Nicoletti F et al. (1992) Ubiquinone protects cultured neurons against spontaneous and excitotoxin-induced degeneration. J Cereb Blood Flow Metab 12: 638–645

Feigin A, Kieburtz K et al. (1995) Functional decline in Huntington's Disease. Movement Disorders 10(2):211–214

Ferrante RJ, Andreassen OA (2000) Neuroprotective effects of creatine in a transgenic mouse model of Huntington's disease. J of Neuroscience 20:4389–4397

Fernandes-Alnemri I, Litwack G et al. (1994) CPP32, a novel human apoptotic protein with homology to Caenorhabditis elegans cell death protein Ced-3 and mammalian interleukin-1 converting enzyme. J Biol Chem 269:30761–10764

Fernandes-Alnemri T, Takahashi A et al. (1995) Mch3, a novel apoptotic cysteine protease highly related to CPP32. Cancer Res 55:6045–6052

Ferrante R, Kowall N et al. (1993) Excitotoxic lesions in primates as a model for Huntington's disease: histopathologic and neurochemical characterization. Exp Neurol 119:46–71

Ferrante RJ, Beal MF et al. (1987) Sparing of acetylcholinesterase-containing striatal neurons in Huntington's disease. Brain Res 411:162–166

Ferrante RJ, Kowall NW et al. (1985) Selective sparing of a class of striatal neurons in Huntington's disease. Science 230:561–563

Ferrante RJ, Kowall NW et al. (1991) Proliferative and degenerative changes in striatal spiny neurons in Huntington's disease: a combined study using the section Golgi method and calbindin D28K immunocytochemistry. Journal of Neuroscience 11:3877–3887

Ferrer I, Marin GE et al. (2000) Brain-derived neurotrophic factor in Huntington disease. Brain Research 866(1–2):257–61, 2000 Jun 2

Fre B, Kim MC et al. (1990) Ubiquinol-10 is an effective lipid-soluble antioxidant at physiological concentrations. Proc Natl Acad Sci USA 87:4879–4883

Frim DM, Uhler TA et al. (1993) Striatal degeneration induced by mitochondrial blockade is prevented by biologically delivered NGF. J Neurosci Res 35:452–458

Garland JM, Rudin C (1998) Cytochrome c induces caspase-dependent apoptosis in intact hematopoeitic cells and overrides apoptosis suppression mediated by bcl-2, growth factor signaling, MAP-kinas-kinase, and malignant change. Blood 92:1236–1246

Gasic GP, Hollann M (1992) Molecular neurobiology of glutamate receptors. A Rev Physiol 54:507–536

Goebel HH, Heipertz R et al. (1978) Juvenile Huntington chorea: clinical, ultrastructural, and biochemical studies. Neurology 28:23–31

Goldberg YP, Nicholson DW et al. (1996) Cleavage of huntingtin by apopain, a proapoptotic cystenin protease, is modulated by the ployglutamine tract. Nature Genet 13:442–449

Gotz T, Kraushaar et al. (1997) Functional properties of AMPA and NMDA receptors expressed in identified types of basal ganglia neurons. J Neurosci 17:204–215

Gourfinkel-An I, Cancel G et al. (1997) Differential distribution of the normal and mutated forms of huntingtin in the human brain. Ann Neurol 42:712–719

Graveland GA, Williams RS et al. (1985) A Golgi study of the human neostriatum: Neurons and afferent fibers. J Comp Neurol 234:317–323

Graybiel AM (1990) Neurotransmitters and neuromodulators in the basal ganglia. Trends in Neuroscience 13:133–254

Graybiel AM, Ragsdale CW (1983) Biochemical anatomy of the striatum. Chemical Neuroanatomy. P C Emerson. New York, Raven 427–504

Greenamyre JT, Shoulson I (1994) Huntington's Disease. Neurodegeneration. Caine. Philadelphia, WB Saunders:685–704

Greene JG, Greenamyre JT (1995) Characterization of the excitotoxic potential of the reversible succinate dehydrogenase inhibitor malonate. J Neurochem 64(1): 430–436

Group HS (1998) Safety and tolerability of the free-radical scavenger OPC-14117 in Huntington's disease. Neurology 50:1366–1373

Group TH s. DCR (1993) A novel gene containing a trinucleotide repeat that is expanded and unstable on Huntington's disease chromosomes. Cell 72:971–983

Gu M, Gash M et al. (1996) Mitochondrial defect in Huntington's disease caudate nucleus. Ann Neurol 39:385–389

Gutekunst C-A, Levey AI et al. (1995) Identification and localization of huntingtin in brain and lymphoblastoid cell lines with antifusion protein antibodies. Proc Natl Acad Sci USA 92:8710–8714

Halliwell B, Gutteridge JMC (1985) Oxygen radicals and the nervous system. Trends Neurosci 8:22

Hantraye P, Brouillet E et al. (1996) Inhibition of neuronal nitric oxide synthase prevents MPTP-induced parkinsonism in baboons.

Hansson O, Petersen A (1999) Transgenic mice expressing a Huntington's disease mutation are resistant to quinolinc acid-induced striatal excitotoxicity. PNAS USA 96:8727–8732

Hantraye P, Riche D et al. (1992) Intrastriatal transplantation of cross-species fetal striatal cells reduces abnormal movements in a primate model of Huntington's disease. Proc Natl Acad Sci USA 89:4187–4191

Hedreen J, Folstein S (1995) Early loss of neostriatal striosome neurons in Huntington's disease. J Neuropath Exp Neruol 54:105–120

Hemmer W, Wallimann T (1993) Functional aspects of creatine kinase in brain. Dev Neurosci 15:249–260

Hochstrasser M (1996) Ubiquitin-dependent protein degradation. Ann Rev Genet 30: 405–439

Hockenbery DM, Oltvai ZN et al. (1993) Bcl-2 functions in an antioxidant pathway to prevent apoptosis. Cell 75:241–251

Holtzman D, Meyers R et al. (1997) In vivo Brain Phosphocreatine and ATP Regulation in Mice Fed a Creatine Analog. Am J Physiol 272:C1567–1577

Huang C, Faber P (1998) Somatic Cell. Mol Genet 24:217–233

Huang Q, Zhou D et al. (1995) Quinolinic acid-induced increases in calbindin D28k immunoreactivity in rat striatal neurons in vivo and in vitro mimic the pattern seen in Huntington's disease. Neuroscience 65:397–407

Huntington's disease collaborative research group (1993) The nove gene containing a trinucleotide repeat that is expanded and unstable on Huntington's disease chromosomes. Cell 72:971–983

Huntington Study Group (1996) Unified Huntington's disease rating scale: reliability and consistency. Movement Disorders 11(1):1–4

Hurlbert MS et al. (1999) Mice transgenic for an expanded CAG repeat in Huntington's disease gene develop diabetes. Diabetes 8:649–651

Ihara Y, Namba R et al. (1989) Mitochondrial encephlomyopathy (MELAS):pathological study and successful therapy with coenzyme Q10 and idebenone. J Neurol Sci 90:263–271

Isacson O, Fischer W et al. (1987) Neuroscience 20:1043–1046

Jackson RA, Sampath D et al. (1994) Effects of nerve growth factor on catalase and glutathione peroxidase in a hydrogen peroxide-resistant pheochromoytoma subclone. Brain Res 634:69–76

Jenkins BG, Brouillet E et al. (1996) Non-Invasive Neurochemical Analysis of Focal Excitotoxic Lesions in Models of Neurodegenerative Illness Using Spectroscopic Imaging. Journal of Cerebral Blood Flow and Metabolism 16:450–461

Jenkins BG, Koroshetz WJ et al. (1993) Evidence for impairment of energy metabolism in vivo in Huntington's disease using localized 1H NMR spectroscopy. Neurology 43(12):2689–2695

Kahlem P, Terre C et al. (1996) Peptides containing glutamine repeats as substrates for transglutaminase-catalyzed cross-linking: relevance to diseases of the nervous system. Proc Natl Acad Sci 93:14580–14585

Kane DJ, Sarafian TA et al. (1993) Bcl-2 inhibition of neural death: decreased generation of reactive oxygen species. Science 262:1274–1277

Kazantsev A, Preisinger E et al. (1999) Insoluble detergent-resistant aggregates form between pathological and nonpathological lengths of polyglutamine in mammalian cells. Proceedings of the National Academy of Sciences of the United States of America. 96(20):11404–11409

Kegel KB, M Kim M et al. (2000) Huntingtin expression stimulates endosomal-lysosomal activity, endosome tubulation, and autophagy. J Neuroscience 20:7268

Kendall AL, Rayment FD et al. (1998) Functional integration of striatal allografts in a primate model of Huntington's disease. Nature Medicine 4:727–729

Kennedy L, Shelbourne P (2000) Dramatic mutation instability in HD mouse striatum: does polyglutamine load contribute to cell-specific vulnerability in Huntington's disease?" Hum Mol Genet. 9:2539–2544

Kieburtz K, Feigin A et al. (1996) A controlled trial of remacemide hydrochloride in Huntington's Disease. Movement Disorders 11(3):273–277

Kim M, LaForet G et al. (1999) Mutant huntingtin expression in clonal striatal cells: dissociation of inclusion formation and neuronal survival by caspase inhibition. J Neurosci

Kim YJ, Yi Y et al. (2001) Caspase 3-cleaved N-terminal fragments of wild-type and mutant huntingtin are present in normal and Huntington's disease brains, associate with membranes, and undergo calpain-dependent proteolysis. Proc Natl Acad of Sci 98(22):12784–12789

Kita H (1996) Glutamatergic and GABAergic postsynaptic responses of striatal spiny neurons to intrastriatal and cortical stimulation recorded in slice preparations. Neuroscience 70:925–240

Klement I, Skinner P (1998) Ataxin-1 nuclear localization and aggregation: Role in polyglutamine induced disease in SCA1 transgenic mice. Cell 95:41–53

Komiyama TK, Ray CA et al. (1994) Inhibition of interleukin-1B converting enzyme by the cowpox virus serpin CrmA J Biol Chem 269:19331–19337

Kopyov OV, Jacques S et al. (1998) Safety of intrastriatal neurotransplantation for Huntington's disease patients. Experimental Neurology 149:97–108

Kordower JH, Chen E-Y et al. (1997) Grafts of EGF-responsive neural stem cells derived from GFAP-hNGF transgenic mice: trophic and tropic effects in a rodent model of Huntington's disease. The Journal of Comparative Neurology 387:96–113

Koroshetz W, Jenkins B et al. (1997) Energy metabolism defects in Huntington's disease and effects of coenzyme Q10. Ann Neurol 41:160–165

Kremer B, Goldberg P et al. (1994) A worldwide study of the Huntington's disease mutation. The sensitivity and specificity of measuring CAG repeats. N Engl J Med 44:1533–1536

Kuhl D, Metter E et al. (1984) Patterns of cerebral glucose utilization in Parkinson's disease and Huntington's disease. Ann Neurol 15(S):S119–S125

Kuwert T, Lange H et al. (1990) Cortical and Subcortical glucose consumption measured by PET in patients with Huntington's Disease. Brain 113:1405–1423

Landwehrmeyer G, Standaert D et al. (1995) NMDA receptor subunits mRNA expression by projection neurons and interneurons in rat striatum. J Neurosci 15:297–307

Lapper SR, Bolam JP (1992) Input from the frontal cortex and the parafascicular nucleus to cholinergic interneurons in the dorsal striatum of the rat. Neuroscience 51:533–545

Leach MJ, Swan JH et al. (1993) BW619C89, a glutamate release inhibitor, protects against focal cerebral ischemic damage. Stroke 1993:1063–1067

Lecerf JM, Shirley TL (2001) Human single-chain Fv intrabodies counteract in situ huntingtin aggregation in cellular models of Huntington's disease. Proc Natl Acad Sci USA 98:4764–4769

Levine MS, Klapstein GJ et al. (1999) Enhanced sensitivity to N-methyl-D-aspartate receptor activation in transgenic and knockin mouse models of Huntinton's Disease. J of Neurosci Res 58:515–532

Li P, Allen H et al. (1995) Mice deficient in IL-1b-convertin enzyme are defective in production of mature IL-1b and resistant to endotoxic shock. Cell 80:401–411

Li X-J, Li S-H et al. (1995) A huntingtin-associated protein enriched in brain with implications for pathology. Nature 378:398–402

Maione S, Biggs CS et al. (1995) Alpha-aminoa-3-hydroxy-5-methyl-4-isoxazolepropionate receptors modulate dopamine release in rat hippocampus and striatum. Neurosci Lett 193:181–184

Lin H et al. (2001) Neurological abnormalities in a knock-in mouse model for Huntington's diseases. Hum Mol Genet 10:137–144

Macdonald ME, Barnes G (1993) Huntington's disease. J Med Genetics. 30:982–986

Mangiarini L, Sathasivam K et al. (1996) Exon 1 of the Huntington's disease gene containing a highly expanded CAG repeat is sufficient to cause a progressive neurological phenotype in transgenic mice. Cell 87:493–506

Mann V, Cooper J et al. (1990) Mitochondrial function and parental sex effect in Huntington's disease. Lancet 336:749

Martin D, Thompson MA et al. (1993) The neuroprotective agent riluzole inhibits release of glutamate and aspartate from slices of hippocampal area CA1. Eur J Pharmacol 250(3):473–476

Martindale D, Hackam A et al. (1998)

Martinez-Serrano A, Bjorklund A (1996) Protection of the neostriatum against excitotoxic damage by neurotrophin-producing, genetically modified neural stem cells. J Neurosci 16:4604–4616

Mary V, Wahl F et al. (1995) Effect of riluzole on quinolinate-induced neuronal damage in rats: comparison with blockers of glutamatergic neurotransmission. Neurosci Lett 201(1):92–96

Matthews R, Ferrante R et al. (1997) Iodoacetate produces striatal excitotoxic lesions. J Neurochem 69:285–289

Matthews R, Yang L et al. (1998) Neuroprotective effects of creatine and cyclocreatine in animal models of Huntington's disease. J Neurosci 18:156–163

Matthews RT, Yang L et al. (1997) Methylthiocitrulline, a neuronal nitric oxide synthase inhibitor, protects against malonate and MPTP neurotoxicity. Experimental Neurology 143:282–286

Mattson MP, Cheng B et al. (1993) Growth factor-mediated protection from excitotoxicity and disturbances in calcium and free radical metabolism. Semin Neurosci 5:295–307

Mattson MP, Lovell MA et al. (1995) Neurotrophic factors attenuate glutamate-induced accumulation of peroxides, elevation of intracellular Ca++ concentration, and neurotoxicity and increase antioxidant enzyme activities in hippocampal neurons. J Neurochem 65:1740–1751

Mayer ML, Westbrook GL (1987) Prog Neurobiol 28:197–276

Mazziotta JC, Phelps MF et al. (1987) Reduced cerebral glucose metabolism in asymptomatic patients at risk for Huntington's disease. N Engl J Med 316:357–362

McGeer EG, Zhu SG (1990) Lamotrigine protects against kainate but not ibotenate lesions in rat striatum. Neurosci Lett 112:348–351

Menalled L (2000) Decrease in striatal enkephalin mRNA in mouse models of Huntington's disease. Exp Neurol 162:328–342

Mende-Mueller LM, Toneff T et al. (2001) Tissue specific proteolysis of Huntingtin (htt) in human brain: Evidence of enhanced levels of N- and C-terminal htt fragment in Huntington's disease striatum. J Neurosci 21:1830–1837

Miura M, Zhu H et al. (1993) Induction of apoptosis in fibroblasts by IL-1B-converting enzyme, a mammalian homolog of C elegans cell death gene ced-3. Cell 75:653–660

Miyamoto M, Coyle J (1990) Idebenone attenuates neuronal degeneration induced by intrastriatal injection of excitotoxins. Exp Neurol 108:38

Miyamoto M, Coyle JT (1990) Idebenone attenuates neuronal degeneration induced by intrastriatal injection of excitotoxins. Exp Neurol 108:38–45

Monaghan D (1991) Differential stimulation of [3H]MK-801 binding to subpopulations of NMDA receptors: Their classes, pharmacology, and distinct properties in the function of the central nervous system. Ann Rev Pharmacol Toxicol 29:365–402

Monyer H, Burnashev N et al. (1994) Developmental and regional expression in the rat brain and functional properties of four NMDA receptors. Neuron 12:529–540

Muir RT, Palmer GL (1991) Remacemide. Epilepsy Res Suppl 3:147–152

Myers RH, MacDonald ME et al. (1993) De novo expansion of a $(CAG)_N$ repeat in spovadio Huntington's disease. Nature Genetics 5:168–173

Nakanishi S (1992) Molecular diversity of glutamate receptors and implications for brain function. Science 89:8552–8556

Nakao N, Grasbon-Frodl EM et al. (1996) Antioxidant Treatment Protects Striatal Neurons Against Excitotoxic Insults. Neuroscience 73(1):185–200

Nicoletti F, Bruno V et al. (1996) Metabotropic glutamate receptors: A new target for the therapy of neurodegenerative disorders? Trends in Neuroscience 19:267–271

Novelli A, Reilly JA et al. (1988) Glutamate becomes neurotoxic via the N-methyl-D-aspartate receptor when intracellular levels are reduced. Brain Res 451(1–2):205–212

Novgorodov SA, Gudz TI et al. (1990) FEBS Lett 270:108–110

Nucifora F, Sasaki M (2001) Interference by huntingtin and atrophin 1 with CBP-mediated transcription leading to cellular toxicity. Science 291:2423–2428

Ohara H, Kanaide H et al. (1981) A protective effect of coenzyme Q10 on ischemia and reperfusion of the isolated perfused heart. J Mol Cell Cardiol 13:65

Oliver CN, Starke-Reed PE et al. (1990) Oxidative damage to brain proteins, loss of glutamine synthetase activity, and production of free radicals during ischemia/reperfusion-induced injury to gerbil brain. Proc Natl Acad Sci 87:5144–5147

Ona VO, Li M et al. (1999) Inhibition of caspase-1 slows disease progression in a mouse model of Huntington's disease. Nature 399(6733):263–267

Ordway JM, Tallaksen-Green S et al. (1997) Ectopically expressed CAG repeats cause intranuclear inclusions and a progressive late onset neurological phenotype in the mouse. Cell 91:753–763

Palfi S, Ferrante RJ et al. (1996) Chronic 3-nitropropionic acid treatment in baboons replicates the cognitive and motor deficits of Huntington's disease. J Neurosci 16(9):3019–3025

Paulson HL, Fischbeck KH (1996) Trinucleotide repeats in neurogenetic disorders. Annu Rev Neurosci 19:79–107

Penney JB, Young AB (1982) Neurology 32:1391–1395

Perutz M, Johnson T et al. (1994) Glutamine repeats as polar zippers: Their possible role in inherited neurodegenerative diseases. Proc Natl Acad Sci USA 91: 5355–5358

Peyser CE, Folstein M et al. (1995) Trial of d-alpha-tocopherol in Huntington's disease. American Journal of Psychiatry 152:1771–1775

Pisani A, Calabresi P et al. (1997) Enhancement of NMDA responses by group I metabotropic glutamate receptor activation in striatal neurons. Br J Pharmacol 120:1007–1014

Higher sedentary energy expenditure in patients with Huntington's disease (2000) Pratley RE, AD Salbe et al. Annals of Neurology 47(1):64–70

Pulsinelli WA (1985) Selective neuronal vulnerability; morphological and molecular characteristics. Prog Brain Res 63:9–37

Qin Z, JJ S et al. (1992) Effects of quinolinic acid on messenger RNAs encoding somatostatin and glutamic acid decarboxylases in the striatum of adult rats. Exp Neurolo 115:200–211

Qin ZJ, Wang Y et al. (1996) Stimulation of N-methyl-D-aspartate receptors induces apoptosis in rat brain. Brain Res 725:166–176

Reed JC (1997) Double identity for proteins of the Bcl-2 family. Nature. 387:773–776

Reddy PH, Williams M et al. (1998) Behavioural abnormalities and selective neuronal loss in HD transgenic mice expressing mutated full-length HD cDNA. Nat Genet 20:198–202

Reddy RS, Housman DE (1997) The complex pathology of trinucleotide repeats. Curr Opin Cell Biol 9:364–372

Reiner A, Albin R et al. (1988) Differential loss of striatal projection neurons in Huntington's disease. Proc Natl Acad Sci USA 85:5733–5737

Richfield EK, Maguire-Zeiss KA et al. (1995) Reduced expression of preproenkephalin in striatal neurons from Huntington's disease patients. Annals of Neurology 37:335–343

Roizin L, Kaufman N et al. (1976) Neuropathologic observations in Huntington's chorea. Progress in Neuropathology III H Zimmerman. New York, Grune and Stratton:447–488

Rosas HD, Koroshetz WJ et al. (1999) Riluzole therapy in Huntington's disease (HD) Movement Disorders

Ross CA (1997) Intranuclear Neuronal Inclusions: A common pathogenic mechanism for glutamine-repeat neurodegenerative diseases? Neuron 19:1147–1150

Ross CA (1997) Intranuclear neuronal inclusions: a common pathogenic mechanism for glutamine-repeat neurodegenerative diseases? Neuron 19:1147–1150

Rubinsztein D, Leggo J et al. (1996) Phenotypic characterization of individuals with 30–40 CAG repeats in the Huntington's disease (HD) gene reveals HD cases with

36 repeats and apparently normal elderly individuals with 36–39 repeats. Am J Hum Genet 59:16–22

Santini VA, Gozzini A et al. (2001) Searching for the magic bullet against cancer: the butyrate saga. Leukemia & Lymphoma 42(3):275–289

Sawa A, Wiegand GW et al. (1999) Increase apophosis of Huntington's disease lymphoblasts associated with repeat length-dependent mitochondrial depolaries Nature Medicine 5:1194–1198

Saudou FS, Finkbeiner S (1988) Huntingtin acts in the nucleus to induce apoptosis but death does not correlate with the formation of intranuclear inclusions. Cell 95: 55–66

Sapp E, Penney J et al. (1999) Axonal transport of N-terminal huntingtin suggests early pathology of corticostriatal projections in Huntington's disease. J Neuropath Expt Neurology

Sathasivam K, Amaechi I et al. (1997) Identification of an HD patient with a (CAG) 180 repeat expansion and the propagation of highly expanded CAG repeats in lambda phage. Hum Genet 99:692–695

Saudou F, Finkbeiner S et al. (1998) Huntingtin acts in the nucleus to induce apoptosis but death does not correlate with the formation of intranuclear inclusions. Cell 95:55–66

Schapira AHV (1997) Mitochondrial function in Huntington's disease: clues for pathogenesis: prospects for treatment. Ann Neurol 41:141–142

Scherzinger E, Lurz R et al. (1997) Huntingtin-encoded polyglutamine expansions form amyloid-like protein aggregates in vitro and in vivo. Cell 90:549–558

Schilling G, Becher MW et al. (1999) Intranuclear inclusions and neuritic aggregates in transgenic mice expressing a mutant N-terminal fragment of huntingtin. Hum Mol Genet 8:397–407

Schulz J, Matthews R et al. (1996) Neuroprotective strategies for treatment of lesions produced by mitochondrial toxins: implications for neurodegenerative diseases. Neuroscience 71(4):1043–1048

Schulz JB, Henshaw DR et al. (1995) Coenzyme Q10 and nicotinamide and a free radical spin trap protect against MPTP neurotoxicity. Experimental Neurology 132(2):279–283

Schulz JB, Huang PL et al. (1996) Striatal malonate lesions are attenuated in neuronal nitric oxide knockout mice. J Neurochem 67(1):430–433

Schulz JB, Matthews RT et al. (1995) Inhibition of neuronal oxide synthase by 7-nitroindazole protects against MPTP-induced neurotoxicity in mice. J Neurochem 64:936–939

Schumacher JM, Short MB et al. (1991) Intracerebral implantation of nerve growth factor-producing fibroblasts protects striatum against neurotoxic levels of excitatory amino acids. Neuroscience 45:561–570

Shaw E (1990) Cysteinyl proteinases and their selective inactivation. Adv Enz 63: 271–347

Shelbourne PF, Killeen N et al. (1999) A Huntington's disease CAG expansion at the murine Hdh locus is unstable and associated with behavioural abnormalities in mice. Human Molecular Genetics 8(5):763–774

Shimohama S, Ogawa N et al. (1993) Protective effect of nerve growth factor against glutamate-induced neurotoxicity in cultured cortical neurons. Brain Res 632: 296–302

Shimohama S, Tamura Y et al. (1993) Brain-derived neurotrophic factor pretreatment exerts a partially protective effect against glutamate-induced neurotoxicity in cultured rat cortical neurons. Neurosci Lett 164:55–58

Siman R, Noszek JC et al. (1989) Calpain I activation is specifically related to excitatory amino acid induction of hippocampal damage. J Neurosci 9:1579–1590

Sommer B, Seeburg PH (1992) Glutamate receptor channels: novel properties and new clones. Trends Pharmac Sci 13:291–296

Sorbi S, Bird E et al. (1983) Decreased pyruvate dehydrogenase complex activity in Huntington and Alzheimer brain. Ann Neurol 13:72–78

Sotrel A, Paskevich P et al. (1991) Morphometric analysis of the prefrontal cortex in Huntington's Disease. Neurology 41:1117–1123

Standaert DG, Friberg IK et al. (1999) Expression of NMDA glutamate receptor subunit mRNAs in neurochemically identified projection and interneurons in the striatum of the rat. Brain Res Mol Brain Res 64:11–23

Stefani A, Spadoni F et al. (1997) Differential inhibition by riluzole, lamotrigine, and phenytoin of sodium and calcium currents in cortical neurons: Implications for neuroprotective strategies. Experimental Neurology 147(1):115–122

Steffan J, Kazantsev A et al. (2000) The Huntington's disease protein interacts with p53 and CREB-binding protein and represses transcription. PNAS 97:6763–6768

Steffan J, Bodai L et al. (2001)Histone deacetylase inhibitors arrest polyglutamine-dependent neurodegenerationin Drosophila. Nature 413:739–743

Storey E, Hyman BT et al. (1992) MPP+ produces excitotoxic lesions in rat striatum due to impairment of oxidative metabolism. Journal of Neurochemistry 58: 1975–1978

Stutzmann JM, Doble A (1994) Blockade of glutamatergic transmission and neuroprotection: The strange case of riluzole. Neurodegenerative Diseases. G J a. J M Stutzmann. London, Academic Press:205–214

Sugihara HK, Moriyoshi K et al. (1992) Structures and properties of seven isoforms of the NMDA receptor generated by alternative splicing. Biochem. Biophys Re Commun 185:826–832

Tabrizi SJ, Cleeter MWJ et al. (1999) Biochemical abnormalities and excitotoxicity in Huntington's disease brain. Ann Neurol 45:25–32

Tabrizi SJ, Workman J et al. (2000) Mitochondrial dysfunction and free radical damage in the Huntington R6/2 transgenic mouse. Ann Neurol 47:80–86

Tanabe Y, Masu T et al. (1992) A family of metabotropic glutamate receptors. Neuron 8:169–179

Testa CM, Friberg IK et al. (1998) Immunohistochemical localization of metabotropic glutamate receptors mGluR1a and mGluR2/3 in the rat basal ganglia. J Comp Neurol 390:5–19

Testa CM, Standaert DG et al. (1994) Metabotropic Glutamate Receptor mRNA Expression in the Basal Ganglia of the Rat. J Neurosci 14:3005–3018

Thomas LB, Gates DJ et al. (1995) DNA End Labeling (TUNEL) in Huntington's disease and other neuropathological conditions. Exp Neurol 133:265–272

Trottier Y, Devys D et al. (1995a) Cellular localization of the Huntington's disease protein and discrimination of the normal and mutated forms. Nat Genet 10: 104–110

Turski L, Turski WA (1993) Towards an understanding of the role of glutamate in neurodegenerative disorders, energy metabolism and neuropathology. Experientia 49(12):1064–1072

Vonsattel J, Myers R et al. (1985) Neuropathological classification of Huntington's disease. J Neuropathol Exp Neurol 44:559–577

Weeks RA, Harding AE et al. (1996) Striatal D1 and D2 dopamine receptor loss in asymptomatic mutation carriers of Huntington's disease. Ann Neurol 40:49–54

Wellington CL, Brinkman RR et al. (1997) Toward understanding the molecular pathology of Huntington's disease. Brain Pathology 7:979–1002

Wellington CL, Ellerby LM et al. (1998) Caspase cleavage of gene products associated with triplet expansion disorders generates truncated fragments containing the polyglutamine tract. J Biol Chem 273:9158–9167

Wellington CL, Singaraja R et al. (2000) Inhibitin caspase cleavage of huntingtin reduces toxicity and aggregate formation in neuronal and non-neuronal cells. J Biol Chem 275:19831–19838

White JK, Auerback W et al. (1997) Huntingtin is required for neurogenesis and is not impaired by the Huntington's disease CAG expansion. Nat Genet 17:404–410

Widmann C, Gibson S et al. (1998) Caspase-dependent cleavage of signaling proteins during apoptosis. Journal of Biological Chemistry 273:7141–7147

Wheeler VCx, Auerbach S et al. (1999) Length-dependent gametic CAG repeat instability in the Huntington's disease knock-in mouse. Human Molecular Genetics 8(1):115–122

Wheeler VC, White JK (2000) Long glutamine tracts cause nuclear localization of a novel form of huntingtin in medium spiny striatal neurons in HDhQ92 and HdhQ111 knock-in mice. Hum Mol Genet 9:503–513

Widmer HR, Hefti F (1994) Neurotrophin-4/5 promotes survival and differentiation of rat striatal neurons developing in culture. Eur J Neurosci 6:1669–1679

Woznicki DT, Walker JB (1998) Utilization of the synthetic phosphagen cyclocreatine phosphate by a simple brain model during stimulation by neuroexcitatory amino acids. J Neurochem 50:1640–1647

Wu D, Wallen HD et al. (1997) Interaction and regulation of subcellular localization of CED-4 by CED-9. Science 275:112601129

Yamamoto A, Lucas J (2000) Reversal of neuropathology and motor dysfunction in a conditional model of Huntington's disease. Cell 101:57–66

Young AB, Greenamyre JT et al. (1988) NMDA receptor losses in putamen from patients with Huntington's disease. Science 241(4868):981–983

Zhang J, Dawson VL et al. (1994) Nitric oxide activation of poly (ADP-ribose) synthetase in neurotoxicity. Science 263:687

Zuccato C, Ciammola A et al. (2001) Loss of Huntingtin-mediated BDNF gene transcription in Huntington's Disease. Science 293:493–498

Subject Index

A1 adenosine receptor   180, 184
A2a adenosine receptor   184
A2A receptor antagonists   195
A3 adenosine receptor   184
N-acetylaspartate   473
acetylcholine   312
N-acetylcysteine   430
acidemia   348
actin   126
acute stroke services   377
adenine nucleotide transporter (ANT)   339, 344
adeno-associated virus   340
adenosine   7, 17, 187, 248
adenosine kinase   183
– inhibitor   192, 198
adenosine receptors   178
adenosine transporter   182
adenosine triphosphate (ATP)   99, 183
adenosine, extracellular   185
adenovirus   340
adhesion molecules   65
agatoxin   229
aged primate   311
aging   100
agitation   375
agonist, partial   188
AIDS   165
allergic encephalomyelitis   166
allopurinol   261
Alzheimer's disease   4, 80, 96, 266, 293, 308
aminoguanidine   20, 160
AMPA   428
– receptors   375
amplicon   340
amyloid   409
– plaques   266
amyotrophic lateral sclerosis   4, 12, 16, 96, 104, 266, 317, 347
amyotrophic lateral sclerosis, familial (FALS)   423
aneurysm   362
– clipping   8

antiarrhythmics   218
α (alpha)-1-antichymotrypsin   417
anticoagulation, oral   363
anticonvulsants   218
antigen   65
antigen presenting cells (APC)   65
antiglutamate   437
antinociceptive   161
antioxidant   19, 128, 471
antisense oligonucleotides   427
antithrombotic agents   364
α (alpha)-1-antitrypsin   68
apolipoprotein E   417
apoptosis   6, 21, 37, 69, 115, 245, 349, 425, 456
apoptosis proteins (IAP), inhibitors   50
apoptosis-inducing factor   339
apoptosome   338
aptiganel   366
arachidonic acid   18, 66, 70, 247
L-arginine   264
L-arginine   157
ARNT see HIF-1beta
ascorbyl radical   255
aspartate   246, 391
aspartyl proteinase   416
asphyxia, perinatal   363
aspirin   68, 80, 378
astrocytes   6, 70, 130
– migration   290
– Type I   6
astrogliosis   73, 286
ataxia   13
ATP see adenosine triphosphate
atrial fibrillation   362
AtT20 cells   414
autoimmune reaction   426
autologous   303
axonal shearing injury   395

bacterial artificial chromosome (BAC)   341
barbiturates   388

basal forebrain cholinergic (BFC) neurons 309
basal ganglia paths, direct and indirect 450
basophils 281
Bax 42, 344, 436
Bcl-2 42, 132, 436
BDNF 313, 455, 474
benzodiazepine 8
bicuculline 198
bioenergetics 469
blood-brain barrier 65, 302, 342
– dysfunction 291
– penetration 370
bradycardia 10
bradykinesia 447
bradykinin antagonists 388
brain derived neurotrophic factor (BDNF) 430
brain injury (brain trauma) 13, 16, 20, 291, 362
brain ischemia 4, 361
brain penetration 391
brain tumors 291, 362
branched chain amino acids 431
BW1003C87 221
BW619C89 221

Ca^{2+} chelators 5
CAD see caspase-activated DNAase
caffeine 177
CAG repeat 447
– length 448
calbindin 118, 464
calbindin D-28K 428
calcium 103, 335
– cycling 98, 115
– intracellular 462
calcium blockers, neuroprotection 231
calcium channel
– antagonists 388
– auxiliary subunits 215
– blockers 217
– inactivation 216
– modulation by neurotransmitters 211
– structure 210
– subtypes 214
calcium chelator 371
calmodulin 103, 116, 123
caloric restriction 265
calpain 51
calreticulin 118
cannabinoids 388
CAPON 164
carbamazepine 218
carbohydrate 105

cardiac arrest 4, 8, 363
cardiac ischemia 193
cardiotrophin-1 437
cardiovascular effects 188
CARE HD 478
caspase 116, 132, 339
– activation 342
– inhibitor 437, 475
caspase-3 44, 476
caspase-8 44
caspase-9 44
caspase-activated DNAase (CAD) 50
– inhibitor (ICAD) 50
catalase 246, 256, 260
catechol-O-methyltransferase 249
cathepsin 68
caudate atrophy 449
cell death 342
– programmed (PCD) 37, 245
cell selectivity 463
cerebellum 101
cerebral vessel rupture 361
cerebrovasoconstriction 158
cerestat 369, 390
CHA 196
chemokine synthesis 294
β (beta)-chemokines 281
chemotaxins 73
chemotaxis 66
chloramines 67
cholesterol 143
– elevated 363
– lowering agents 392
choline transporter 103
cholinergic system 309
– basal forebrain 418
chorea 447
choroid plexus 70
cinnarizine 224
citicoline 367
clinical trial design 389
clomethiazole 366, 370
CNQX 346
CNS1237 224
CNTF 315, 318, 474
coagulation disorders 362
Cochrane Collaboration 380
coenzyme Q10 99, 472, 478
cognitive dysfunction 402
collagenase 68
complement 66
concatamer 341
confusion 375
congo red 410
conopeptides 228
– binding 217

Subject Index 495

- hypotension 232
- neuroprotection 232
cortical culture 346
corticostreoids 388
corticotropin-releasing factor (CRF) 70
C-reactive protein (CRP) 80
creatine 99, 472
creatine kinase 101
CREB binding protein (CBP) 454
cromakalim 9
cyclic AMP (cAMP) 125, 184
Cyclin dependent kinase 5 (Cdk5) 52
cyclocreatine 101
cycloheximide 337
cyclooxigenase (COX) 19, 66, 263
- inhibitor 252
N6-cyclopropyladenosine (CPA) 196
cycloserine 14
cyclosporin A (CyA) 44
cyclothiazide 7
cysteine 7
cytochalasin D 141
cytochrome 468
cytochrome C 44, 99, 133, 339, 344
cytokine 66, 116, 342
- chemoattractant 281
cytomegalovirus 284
cytotoxicity 341

D2 dopamine receptor 194
deacetylation 455
deacetylators 477
deferoxamine 264
deltibant 402
dementia 165
deoxycoformycin 191, 198
depolarization 117
L-deprenyl 267
dextromethorphan 368, 431
dextrorphan 368
Disability Rating Scale (DRS) 397
diazepam 138, 366
dichloroacetate 473
diet 143
2,5-dihydrobenzoic acid 250
dihydroepiandrostenedione (DHEA) 415
dihydrolipoic acid 259
dihydropyridine 11
1,25-dihydroxyvitamin D3 438
dilazep 198
5,5-dimethyl-1-pyrroline N-oxide 251
dimethylsulfoxide 259
dimethylthiourea (DMTU) 258
diphenyleiodinium 162
dipyridamole 191

disability 387
dizolcipine 368
DNA fragmentation 37, 339
DNA laddering 41
domain swapping experiments 284
domoic acid 3
L-DOPA 195
dopamine 314, 463
dopaminergic neurons 313
drug marker 480
Duffy blood group antigen (Duffy Ag) 284
dysfunction, mitochondrial 95

EAAT2 427
ebselen 367
eclampsia 15
edema 343, 388
efficacy trial 392
eicosanoids 66
elastase 68
electron paramagnetic resonance spectrometry 250
eliprodil 366
S-emopamil 226
- neuroprotection 232
endonuclease 18, 54
endoplasmic reticulum 116, 412
endosome 412
enlimomab 372
eosinophils 281
ependyma 70
epidural hematoma 395
epilepsy 197
17 β (beta)-estradiol 415
estradiol 418
estrogen 128, 409
- receptor 415
excitatory amino acids (EAA) 391
excitotoxicity 3, 38, 121, 315, 365, 423, 459
excitotoxin 188

factor XIa inhibitor (XIaI) 411
FAD missense 411
FAD mutant 413
FALS see amyotrophic lateral sclerosis, familial 423
felbamate 222
fibroblasts 303, 319
fibronectin 75
fimbria-fornix-transection 310
FK 506 163
flunarizine 222, 367
- neuroprotection 233
fly model 457

focal brain ischemia 288
fosphenytoin 220, 367, 371, 390
– neuroprotection 231
frataxin 95
free radical 20, 95, 466, 471
– cascade 255
– damage 424, 429
– generation 251
– scavengers 250, 434
Friedreich's ataxia 95
Functional Independence Measure 401
fusin 281

GABA 182, 194
– receptors 8
gabapentin 432
β (beta)-galactosidase 341, 349
γ (gamma)-aminobutyric acid (GABA) 118
gangliosides 74
gavestinel 366, 369
GDNF 313, 318, 321
gelatinase 68
gelsolin 126
gene disruption 335
gene expression 335
gene therapy 301
gene transcription 454
Glasgow Coma Scale 394
Glasgow Outcome Scale 394
gliomas, malignant 323
Glu11-secretase 411
glucocorticoid 70, 131
glucose deprivation 13
glucose metabolism 467
glucose transporter 118
glutamate 3, 74, 102, 115, 196, 246, 335, 391, 426, 459
– dehydrogenase 431
– excitotoxicity 434
– release 8
– transporter 427
glutamatergic input 451
glutathione 128, 259
– peroxidase 260
glycine 14
glycogen synthase kinase-3 (GSK3) 48
Golgi apparatus 412
gp41 166
GP668 193
GP683 192, 199
G-protein 285
graft 307
growth factor 253
GYKI-52466 433
gyrate atrophy 105

H2 antagonists 392
Haber-Weiss reaction 68
hallucinations 13, 375
head injury 388
heat shock proteins 139
hereditary cerebral hemmorrhage with amyloidosis, Dutch type (HCHWAD) 413
herpes virus 340
γ (gamma)-herpes virus 284
heterologous 303
hexokinase 102
HIF-1α (alpha) 348
HIF-1β (beta) (ARNT) 348
hippocampus 17, 101, 131, 342
HIV 165, 293
huntingtin 452
– inclusions 456
– wild type 453
Huntington's disease 4, 96, 268
hydrogen peroxide 126
4-hydroxyalkenal 250
4-hydroxynonenal (HNE) 130
hyperglycemia 70
hypertension 363
– drugs 392
– systemic 158
hyperthermia 70
hypoglycemia 4, 348
hypotension 10, 13
hypothalamus 70
hypothermia 7, 188, 402
hypoxanthine 248
hypoxia 4, 13, 38, 118, 181, 186, 348
hypoxia responsive elements (HRE) 348
hypoxia-mimetic agents 349

ibotenic acid 464
ibuprofen 263
ICAD see inhibitor of caspase-activated DNAase
ifenprodil 15
IGF-1 433
immediate early gene 341
immune response 65
immunglobulins 426
inclusions 458
indomethacin 263
infarction 388
inflammation 65, 79, 286
inheritance 448
injury 65
inosine 248
inositol triphosphate 118
integrin 74, 126, 372

Subject Index

interleukin (IL)-1 69
interleukin (IL)-1β (beta) 136, 343, 476
– converting enzyme (ICE) 69, 343, 436, 475
interleukin (IL)-8 291
– receptor a (IL8RA) 281
interleukin-6 72
intracellular adhesion molecule 74, 372
intracranial mass lesion 395
intracranial pressure 396
ion channels, auxiliary subunits 215
ion-motive ATPase 125
ischemia 38, 185, 287, 335
– cerebral 16
– global 186
– spinal cord 16
ischemic brain injury 294
ischemic injury 72
ischemic preconditioning 79, 187
ischemic threshold 351
ischemic tolerance 347
isradipine 225

JUN N-terminal kinase 47

kainate 4, 7, 137
– receptors 366
kainic acid 198
Kaposi's sarcoma associated herpes virus (KSHV) 284
ketamine 369
α (alpha)-ketoglutarate dehydrogenase 100
kinase 18, 116
Kunitz-type protease inhibitor 410

lactacystin 47
lactate 468
lamotrigine 218, 220, 432
lazeroids 437
lentivirus 340
leptomeninges 70
leukarrest 367
leukemia inhibitory factor (LIF) 321
leukocyte 69
– adhesion molecules 74
– infiltration 343
– movement 283
– recriutment 290
– traffic 285
leukotrienes 66
licostinel 368
lidocaine 218
– structure 219
lidoflazine 223
lifarizine 223

life span 254
lipase 18
lipoxygenase inhibitor 252
lomerizine 223
lubeluzole 224, 367, 371
LY300164 431
lymphadenopathy 285
lymphocyte surveillance 65
lymphocytes 6, 281
lymphotactin 285
lysosomal granules 68

macrophage 76
– response 286
magnesium 369
major histocompatibility 65
malate 102
malonate 99
malondialdehyde 250
mannitol 258
matrix assiciated complex 339
maturation phenomenon 336
MELAS see mitochondrial encephalopathy lactic acidosis and strokes
memantine 14
membrane lipid peroxidation (MLP) 127
meningitis, bacterial 291
meningoencephalitis 291
mepacrine 263
metalloproteinase 71, 416
methionyl peptides 66
1-methyl-4-phenyl-1,2,3,6-tetra-hydropyridine (MPTP) 160
methyl-4-phenyl-1,2,3,6-tetra-hydropyridine (MPTP) 165
N-methyl-4-phenyl-1,2,3,6-tetra-hydropyridine (MPTP) 267
N-methyl-4-phenylpyridinium 267
methyl-arginine 158
methylprednisolone 387, 403
mexilitine 219
mGluR 185, 462
microglia 67, 76
microgliosis 286
microtubule 142
middle cerebral artery occlusion (MCAO) 186, 336
mitochondria 44, 66, 121, 337, 467
– dysfunction 95
mitochondria permeability transition (MPT) 44, 97
mitochondrial calcium 470
mitochondrial encephalopathy lactic acidosis and strokes (MELAS) 106

MK 801   346
MnSOD   165
monoamine oxidase   249
– type B   430
monocytes   281, 289
morphometry   479
motor neuron disease   423
mouse model   458
MPTP   195, 470
mRNA   348
multiple sclerosis   80, 166
muscimol   8
muscle atrophy   423
myelin basic protein   286
myocardial infarction   80, 362

NADP oxidase   249
NADPH diaphorase   156
NAIP see neuronal apoptosis inhibitory protein
nalmefene   366, 370
naloxone   403
National Acute Spinal Cord Injury Study (NASCIS)   403
NBMPR   198
necrosis   37
nerve growth factor (NGF)   309
neural progenitor cells   304
neuro-AIDS   293
Neurobehavioural Rating Scale   397
neurodegeneration   95, 293
neurofibrillary tangles   293
neurofilaments   424
neurointensive care   391
neuronal apoptosis inhibitory protein (NAIP)   52
neuronal circuit activity   7
neuronal death   3
– delayed   336
neuronal necrosis   335
neuroprotection   193, 373, 387
neuropsychometric testing   388, 402
neurotoxicity   417
neurotransmitter release   101
– calcium dependent   229
– calcium independent   230
neurotransmitter transporters   230
neurotrophic factor   131, 308
neurotrophin   415
neurotrophin-3 (NT-3)   320
neutrophil gelatinase-associated lipocalin (NGAL)   78
neutrophil proteinase   478
neutrophil recruitment   70
neutrophils   281
NF-κ (kappa)B   45, 138

NGF   315
nicorandil   9
nicotinamide   99
nicotinamide adenine dinucleotide (NADH)   6, 100
nicotinamide adenine dinucleotide phosphate (NADPH)   155
nimodipine   225, 362, 367, 387, 390, 432
– 209, 231
nitrendipine   209
nitric oxide (NO)   5, 95, 123, 249, 263, 466
nitric oxide synthase   19, 263
nitro L-arginine   158
nitro L-arginine methyl ester   158
7-nitro-indazole (7-NI)   158, 165
3-nitropropionic (3-NP) acid   100
nitroprusside   264
NMDA   4, 117, 183, 428
– antagonists   388
– blockers   368
– receptor   164
– receptor channel   156
– pathology   460
N-methyl-D-aspartate see NMDA
nonsteroidal anti-inflammatory drugs (NSAID)   68, 392
NOS isoforms   155
NOS, endothelial (eNOS)   156
NOS, inducible (iNOS)   157
NOS, neuronal (nNOS)   156
3-NP   465
NT-3   433
nuclear factor-κ (kappa)B   73
5′-nucleotidase   180
nucleus basalis   309

okadaic acid   414
oligodendrocytes   7, 375
oncosis   37
oxidative stress   104, 115, 250, 423
oxygen radicals   74
oxypurinol   261

p53   349
paralysis   423
paraproteinemia   426
Parkinson's disease   4, 96, 156, 165, 189, 194, 199
parvalbumin   428
patient management   396
PEG-catalase   260
PEG-SOD   260
pentoxyfylline   81
penumbra   193

Subject Index

periischemic period 351
peroxynitrite 51, 100, 123, 165, 249
phagocytosis 65
pharmacokinetics 390
phencyclidine (PCP) 368
α (alpha)-phenyl-tert-butylnitrone (PBN) 254
phenytoin 10, 218
phorbol ester 414
phosphatase 18
phospholipase 247
phospholipase A2 (PLA2) 66
phospholipase A2, C 262
phospholipase C 116
phosphorylation 116
– oxidative 338
pilocarpine 197
plasma membrane 117
plasmid-based vector 340
plasticity 418
platelet activating factor (PAF) 67
platelet/endothelial cell adhesion molecule (PECAM) 175
pmn mice 433
poly(ADP-ribose)polymerase-1 (PARP-1) 5
polyglutamine 452
polymorphonuclear cells (PMN) 67
polyribosome 347
porin 102
positron emission tomography (PET) 376
potentiation, long-term 157
preconditioning 347
preconditioning response 253
preeclampsia 15
prehospital stroke care 377
presenilin 411
presenilin-1 121
presymptomatics 480
prodeath gene expression 348
promoter 306
propentofylline 190
prostacyclins 67
prostaglandin 19, 66
prosynap 367
protease 18
– cellular 335
proteasome 47
protein aggregates 437
protein inhibitor of NOS (PIN) 164
protein kinase C 415
protein phosphatase 414
protein phosphorylation 414
protein synthesis 336

protein translation 336
protein-protein interaction 453
proteoglycans 283
proteolysis 141
protumor necrosis factor convertase 416
putrescine 13
pyridyl-N-oxide-tert-butylnitrone (POBN) 251

quinolinic acid 464

reactive oxygen species (ROS) 51
receptor, muscarinic 121
recombinant human CTNF (rhCTNF) 430
regeneration, axonal 319
regulated cleavage of APP 415
remacemide 106
reperfusion 289
repinotan 366
respiratory failure 423
respiratory multiprotein complex 339
retrovirus 304, 340
riluzole 10, 104, 218, 221, 431, 465
– neuroprotection 231
RNA processing 425
RNA virus 305
ROS see reactive oxygen species
R-PIA 198
ryanodine 118

Schaeffer-collateral input 180
Schwann cells 303, 321
α (alpha)-secretase 411
β (beta)-secretase 411
γ (gamma)-secretase 411
sedation 375
sedatives 390
seizures 3, 13, 185
L-selectin 74
selegiline 430
selfotel 366, 369, 390
senile plaques 293
sensor, ischemic 338
SERCA 118
d-serine 14
short term memory 254
sickle cell anemia 362
side effects 310
sleep 183
slow-wave sleep 70
SMA see spinal muscular atrophy
smooth muscle cells 289
SOD-1 transgene 261

sodium channels
- activation   216
- auxiliary subunits   215
- blockers   218
- inactivation   216
- structure   210
- subtypes   212
soluflazine   191
spermidine   13
spermine   13
spheroids   424
spinal cord   319
- injury   41, 387
spinal muscular atrophy (SMA)   52, 427
spiny neurons   451
splenomegaly   285
SR57746A   436
striatal anatomy   450
stroke   3, 72, 199, 289
- acute ischemic   66
- hemorrhagic   361
- ischemic   11, 361
- lacunar   362, 379
- model   375
- prevalence   363
- prevention   363
study design   394
subarachnoid hemorrhage   11, 20, 387
substantia nigra   267
substrate activation   416
sulfate proteoglycan   417
superoxide   126
superoxide dismutase (SOD)   67, 246, 256, 260, 347, 423
suppression subtractive hybridization (SSH)   289
surrogate markers   479

T cells   75
therapeutic trials, ALS   429
N-tert-butyl-α (alpha) (2-sulfophenyl)-nitrone (S-PBN)   257
tetracycline   69
tetrahydrobiopterin   162
tetrodotoxin   10
- binding   217
- neuroprotection   231
- structure   219
TGF α (alpha)   416
theophylline   177, 186
therapeutic window   389
thrombolytics   77, 364
thromboxanes   67
thymidine kinase   323
tirilazad   367, 372, 390, 395, 403

tissue plasminogen activator (t-PA)   361, 364, 389
T-lymphocytes   289
α (alpha)-tocopherol (vitamin E)   259, 431, 434
topiramate   218
TRAF2   47
transcription   135
transforming growth factor (TGF)-β (beta)   69
transgene   304
transgene expression   286
transgenic animals   261
trans-Golgi network   412
transient ischemic attack (TIA)   362
transplantation   302, 477
trauma   13
traumatic brain injury (TBI)   256, 389
Traumatic Coma Data Bank   394
1-(2-trifluoromethylphenyl) imidazole (TRIM)   161
tromethamine   388
tumor necrosis factor (TNF)   69
tumor necrosis factor (TNF)-alpha   71, 136
TUNEL   41
tyrosine   103

ubiquination   457
ubiquinone   99
uric acid   128

valvular heart disease   362
vascular endothelium   68
vascular occlusion   37
vascular permeability   66
vasodilation   66, 336
vasopressors   390
verapamil   226
- binding   218
very late activation (VLA) antigens   75
vigabatrin   9
virion   340
vitamin E see tocopherol
voltage dependent anoin channel (VDAC)   339

xanthine   248
xanthine dehydrogenase   19
xanthine oxidase   246, 261
- inhibitor   252

ziconotide   224
- neuroprotection   209
- structure   228
zonisamide   221